AF517367

HERBACEOUS PLANTS AS NATURAL PROTECTIVE FOOD

DR. M.K. RANA

Professor

Department of Vegetable Science

CCS Haryana Agricultural University

Hisar-125 004, Haryana, India

Published by:

Scientific Publishers (India)
5 A, New Pali Road, P.O. Box 91
JODHPUR – 342 001 (India)

Scientific Publishers (India)
4806/24, Ansari Road, Daryaganj
NEW DELHI – 110 002 (India)

E-mail: info@scientificpub.com
Website: www.scientificpub.com

Print: 2021

ISBN: 978-81-7233-900-5

Typeset by Rajesh Parihar
Printed in India

Preface

A diet that harmonizes with natural cycle enhances one's flow of energy and harmonizes the functional activities of all body organs, however, a diet, which is not in harmony with the movements of nature, drains one's energy and brings diseases. Hence, eating healthy food is fundamental to good health and vitality.

For the centuries, fruits and vegetables have been recognized as a good source of minerals and vitamins to which the body needs to perform a variety of functions suitably and keep healthy as they have the ability to prevent vitamins deficiencies in human body. Vitamin A helps to strengthen immune system, B vitamins help to process energy from food, vitamin C helps to keep body cells and tissues healthy, and vitamin D helps to maintain healthy teeth and bones. The steamed carrots and broccoli maintain a higher proportion of vitamins than boiled or fried vegetables. Fruits and vegetables eaten along with skin also contain high amount of fibre, which helps to maintain a healthy gut and digestive system. The vegetables containing starch, also known as carbohydrates, provide most of the energy. The body converts such vegetables into glucose, which is either used as energy immediately or stored for later use. Vegetables also contain iron, which is needed to make red blood cells to carry oxygen around the body, hence, eat five kinds of vegetable and two kinds of fruit every day for good health. The National Institutes of Health suggests a minimum of five servings per day of fruits and vegetables combined. To increase consumption of fruits and vegetables, people need motivation. Appropriate communication efforts can often provide motivation and sometimes improve a person's ability to increase consumption of fruits and vegetables.

Fruits and vegetables are part of a well-balanced and healthy eating plan. There are many different ways of losing or maintaining a healthy weight. To lose weight, it is not necessary to eat less food though it is necessary to consume low calorie versions of some of favourite dishes by substituting low calorie fruits and vegetables in place of high calorie food ingredients. It is true that fruits and vegetables are low in calorie. Water and fibre in fruits and vegetables add volume to the dishes, so same amount of food with low calorie can be eaten by consuming fruits and vegetables. Since most of the fruits and vegetables are naturally low in fat and calorie, *per se*, eating more fruits and vegetables helps in controlling weight. Consuming more fruits and vegetables along with whole grains, nuts and beans is safe and healthy since they contain antioxidants or phytochemicals, which help in protecting the body against diabetes, heart disease, cancers and other chronic diseases. Research of the past 20 years has shown that eating fruits and vegetables not only prevents malnutrition but also helps in maintaining optimum health through a host of chemical components that are still being identified, tested and measured. It has been noticed that a high intake of fruits and vegetables could lower the blood pressure in individuals with either high or normal blood pressure.

Scientific research suggests that the role of fruits and vegetables in preventing heart disease is a protective one. Risk reduction has been estimated as high as 20-40% among individuals who consume substantial amount of fruits and vegetables. People who are diagnosed with coronary heart disease are able to reduce blockage modestly through exercise and an extremely low fat, vegan-like diet rich in fruits and vegetables. A diet containing substantial and varied amounts of fruits and vegetables may prevent 20% or more of all cases of cancer. The strongest evidence relates to stomach, lung, mouth, pharynx, esophagus, colon and rectum cancer. Studies show that the patients who take dietary supplements in place of fruits and vegetables die earlier due to a higher mortality rate among the supplement users since dietary supplements do not have the same positive effects as eating real fruits and vegetables.

Delayed development of cataracts is another beneficial effect of fruits and vegetables. A five folds reduction in cataract risk is found in individuals who consume a minimum of one and a half serving of fruits and vegetables each day. Carotenoid-rich fruits and vegetables containing zeaxanthin and lutein have been proved the most beneficial but not all carotenoids offer equal protection. Diets that are high in fibre may be able to help in the management of diabetes and may offer the best protection against diverticulitis. Soluble fibre delays glucose absorption from the small intestine and *per se* may help in preventing spike in blood glucose levels that follow a meal, however, the long-term effect may be insignificant due to many other factors that affect blood glucose. Asthmatics if consume fruits and vegetables more than once a day may have better lung function since higher intake of fruits and vegetables increases ventilation function of the lungs.

The life style diseases, especially cardiovascular diseases, cancer, obesity and type 2 diabetes mellitus, kill more people every year than any other cause of death. The reasons behind these overwhelming diseases are poor diet, physical inactivity, stress and use of tobacco and alcohol. However, fruit and vegetables are an important component of a healthy diet and if consumed daily in sufficient amount, may help in preventing these major diseases. World Health Organization estimated that low fruit and vegetable intake causes about 31% of ischemic heart disease and 11% of stroke worldwide. Overall, it is estimated that up to 2.7 million lives may potentially be saved every year if fruits and vegetables consumption is sufficiently increased. Recommendations in this direction tend to complement and reinforce other valid messages based on the long-known health benefits of consuming vegetables and fruits as dietary sources of fibre, vegetable proteins and protective micronutrients. The FAO/WHO nutrition experts recommended the intake of a minimum of 300 g of vegetables (excluding potatoes and other starchy tubers) and 100 g of fruits per day for the prevention of chronic diseases, such as heart disease, cancer, diabetes and obesity as well as for the prevention and alleviation of several micronutrients deficiencies, especially in less developed countries.

Considering the importance of fruits and vegetables in human diet in view, the author felt obligatory to prepare a book, which may sequentially help the teachers, scientists, graduate and postgraduate students, and of course, the common man to earn adequate knowledge about nutritional and therapeutic significance of fruits and vegetables in maintaining good health. The objective of preparing this book is to make the populace aware about health benefits of fruits and vegetables. This book containing very concise and precise information has been written in a very simple language, which can be explicable even to undergraduate students and common man. The information given in this book are truly based on scientific records of scientists working on particular aspects. In addition, this book may be user-friendly to others who have the concern to expand knowledge concerning human health through fruits and vegetable. Earning scientific knowledge will undoubtedly be rewarding to its users and finally to the nation.

Place: Hisar **–Author**

Contents

■■■

1 Living with Five Elements (Panchatatvas)

Satyendra Kumar Misra[1] and Pallavi Bhatnagar[2]

[1] Lucknow University, Lucknow, Uttar Pradesh
[2] Lucknow, Uttar Pradesh, E-mail: <dr.pallavib@gmail.com>

Consciousness is the source of all manifestation and experience though it is beyond description and can be only felt. The twin forces of *Prakarti* and *Purusha* emanate from this consciousness only. It is the interaction of *Prakriti* and *Purusha*, which creates and individualized awareness.

The same reality, which is consciousness, seems to become the fivefold elements and the body because of a slight movement of thought. In the same way, the same consciousness becomes worms and other creatures, metals and minerals, earth and what is on it, water and other elements, thus, the whole world is nothing but the movement of energy in consciousness, which appears as the five folds elements. Thus, the five states of matter or the five elements, which also form the base of *Ayurveda*, may be construed as nothing more than movement in consciousness, which generates cosmic *prana* or the life force energy.

What seems rather amazing is that most of the people are not cognizant of this life force energy or essence of life that is *prana*. It is only during yoga, meditation, or some relaxation intervention when the healer or trainer instructs that people willfully, consciously and mindfully attend to their breathing.

Volumes of evidence exists in both scriptures and modern medicine that when with practice to movement of *prana* can be arrested, it promotes longevity. The Sages and Hermits who have practiced this were like us except they were more aware of the powers of this life force energy or *prana*. What is more significant is that when one is able to detach or vanish thoughts or *prana* even temporarily, it reduces a state of relaxation, an internal peace and a state of well-being. The five material states or elements originate from *prana*.

Everything in the universe is composed of the five elements, *i.e.*, air, fire, water, earth and ether. They represent the five states or qualities of energy or matter and exist in all things, in an infinite variety of proportions. This variety of proportion and attributes allows for the enormous diversity of life. These elements are always in motion and are constantly changing and interacting with each other.

Origin of Five Elements

In the beginning, the world existed in a state of unified consciousness, wherein the subtle vibrations of the cosmos created the sound *aum*. From the vibration, the *Ether*, which is the original element, appeared. When set in motion, it condenses to become *air*, which manifests the idea of motion inherent in the idea of space. *Air* through constant movement and density creates friction that becomes *fire*, which manifests the idea of illumination inherent in the idea of movement. *Fire* when heated condenses to become *water*, which manifests the idea of life inherent in the idea of illumination. *Water* when it cools and coagulates, becomes *earth*, which manifests the idea of form inherent in the idea of life.

तस्माद्वा एतस्मादात्मन आकाशः सम्भूतः। आकाशाद्वायुः।
वायोरग्निः। अग्नेरापः। अद्भ्यः पृथिवी।
पृथिव्या ओषधयः। ओषधीभ्योऽन्नम्। अन्नाद्रेतः।
रेतसः पुरूषः। स वा एष पुरूषोऽन्नरसमयः।

The ethereal space was created by the Supreme Soul, afterwards he created air, after air the fire, after fire, the water, after water, the earth, from earth, the herbs and plants, after herbs *etc.*, the food grains (foodstuffs), from foodstuffs, virile semen was created and from virile semen, a man was born, thus, man is made of the essential substance of foodstuffs (Chapter-I, Taittriya Upnishad, Brammanand Valli).

Hence, all the five elements are different densifications of one great element, *i.e.*, Ether, and all the elements are derivations from the same basic etheric substance. One-tenth of ether becomes air, one-tenth of air becomes fire, one-tenth of fire becomes water and one-tenth of water becomes earth. In this way, the earth contains all five elements. Water contains four elements (all but earth), fire contains three (including air and ether) and air contains two (with ether).

The Sanskrit word *Ayur* means duration of life. The verbal root of *Veda* means knowledge. Thus, the word *Ayurveda* refers to science of the duration of life, better known as health science. *Ayurveda* literally means *science of life, longevity and self healing*, which is a traditional system of natural healing. *Ayurveda* is a part of *Vedic* science.

The three modes of nature (*Satva*, *Rajas* and *Tamas*) combined with the five gross elements to form subtle energy principle called *Tridosha* in *Ayurveda*. It is the five elements, which are responsible for the particular nature (*prakarti*) of *Ayurveda* believes that everything in this universe is made up of combinations of the five elements (*tatvas*). All organic living bodies including those in the vegetable kingdom such as herbs and grains and those in the animal kingdom including man have these five elements in them. Five elements theory explains the similarity between humans and natural world around them. In human body, space (*akash*) is reflected in cavities like mouth, lungs, throat, stomach, while air (*vayu*) is mirrored in movement of heart, stomach, lungs, joints, intestines, mouth, *etc.* The element of fire (*agni*) is seen in all metabolic activity, *i.e.*, the eyes, intelligence and body temperature. Water (*Jala*) is mirrored in all plasma, mucous and saliva, and finally, earth (*prithvi*) is evident in all solid structures. What is significant to understand is that the complimentary relation between earth and human nature is at base of the fact that vegetation can repair and restructure the human system.

All five elements in body correspond to five major organs kidney, liver, heart, stomach and lungs. In order to create balance in the body one needs to increase willingly and consciously the intake of foodstuffs, which strengthen the element in which we are weak. At this point, it becomes relevant to understand each element and its dietetic management.

Table 1: Characteristics of elements at a glance

Qualities	Space (*Akash*)	Air (*Vayu*)	Fire (*Agni*)	Water (*Jala*)	Earth (*Prithvi*)
Qualities	Soft, light, subtle and abundant	Weightless cool dry and subtle	Hot, sharp, dry weightless and rough	Oily, moist, sticky and soft	Heavy, solid stable and slow
Action	Provides space and openness	Motion and evaporation	Radiation of heat and light	Lubrication and cohesion	Resistance and density
Facilitate	Sound and nonresistance	Touch and vibration	Form, colour and temperature	Fluidity and taste	Fragrance and shape
Substance	That is light and etheral	That is dry and airy and creates gas	That is spicy	That is liquid or watery	That is solid or heavy
Example	Hollow and light, popcorn and wafers	Toast, cookies cabbage and beans	Chillies, ginger, pepper, clove and cumin	Drinks, soups and cucumber	Fried foodstuffs, cheese cake and
Intake	Increases softness and lightness of body.	Increases coolness, dryness movement and circulation.	Increases digestion, metabolism, glow and colour of skin	Increases smoothness, softness, coolness and flow of fluids.	Increases heaviness stability, obesity and solidity in body

Contemplation of Space (*Akasha*)

Space is an energy field with certain characteristics. It is all pervasive in directions, known, unknown and yet to be known, including time. Space has no boundaries: it is infinite (in place and time). There is the vastness of space. On the other, each grain of sand and each blade of grass is filled with space. No phenomenon of the galaxy, the planets, human, or plants takes place in space, validating that creation is chaotic. In fact, it is organized and cyclic.

The *organs* that relate to space are lungs and large intestines and both organs constantly move. They provide exchange between the outer and inner world. Both take in matter (be it air or prepared foodstuffs), refine it and eliminate that which is not needed and cannot be used. The *imbalances* are evident in diseases of skin, lung, bowel and psychological problems like depression and emotional holding. *Vata* show lack and imbalance in either air or space. It is the most subtle *dosha* and is normally involved in most physical maladies. The *fluid* that relates to this element is mucus. The *color* linked to this element is white or brown. White is really a combination of all colors, which might be the reason why there is some link to space.

The *taste* associated here is pungent, referring to strong flavours such as cheese, curry, peppery dishes, and in general, spicy food items. People particularly liking or disliking such foodstuffs, have most likely an imbalance in this energy. These foodstuffs are said to open the pores and improve the circulation of saliva and mucus, a desire of the body to help itself in the removal and elimination process. On the other hand, one should not eat too much pungent foodstuffs, for the muscles may become knotted and the fingers and toenails may wither and decay with the excess intake.

The *grain* indicative of the space element is rice, the *fruit* chestnuts. The *meat* is that of the horse. The *vegetable* is the onion. These are foodstuffs, which are said to have a specially balancing effect. Rice is a grain that easily absorbs moisture and also when eaten with the husks (brown rice), carries a lot of fibers encouraging the movement of bowls, helping absorption, condensing and eliminating. Chestnuts are water carrying dry fruits and as such lubricate. Onions are rich in water content and its sharp gases penetrate the nostrils and bronchioles, stimulating their activity, as we have all experienced watering eyes and dripping noses while peeling onion. They clean the air passages. An old house remedy is to make onion tea when children have bronchitis, or even to burn onionskins in the room, helping bronchioles and the general respiratory tract.

Dietetic Management for Space Personality

Space: Associated foodstuffs include rice, raw onions, garlic, radish, turnip, cinnamon, mint, rosemary, cloves, mustard greens *etc.* These foodstuffs have a dispersing effect and promote energy circulation. They most benefit sluggish, damp, lethargic and cold people.

Balancing Foodstuffs: These foodstuffs include black mushroom, white mushrooms, apricot, grapes, pear, groundnut, oat, rice, onion, garlic, vitamin A, C, D and E, biotin, magnesium, water, honey, *etc.*

Stimulating Foodstuffs: These foodstuffs include coriander, leek, bitter apricot, orange peel (dried), rice, garlic, ginger, mustard green, sugar, sugarcane, *etc.*

Stifling (Relaxing) Foodstuffs: These foodstuffs include cabbage, cucumber, eggplant, lettuce, mushrooms, soybean sprouts, soybeans, lily flowers, mustard green, potato, winter melon, apple, banana, strawberry, orange, sunflower, Eucalyptus, tulip, nut, mint, sesame, *etc.* Other foodstuffs include navy beans, almonds, broccoli, celery, mustard, green onion and pear.

Contemplation of the Element of Air

According to *Vedic* thought, the gross state of energy called air is the element of creation and dissolution, having an elusive quality in it. Space being the most subtle, air is already so thin that most of the time we are not aware of it. *Prana* is the cosmic principle of movement and energy and is the most subtle element of existence. *Prana* empowers all other forces in the cosmos. *Prana* is prime moving force in the body and universe. *Prana* is not breath rather it is the power that enables us to breathe. *Prana* is also known as *Vayu* and *Vata*. It is the power, which enables us to think, feel and hear. *Prana* is also the source of three *Ayurvedic* humors as *Vata*, *Pita* and *Kapha*.

In *Ayurvedic* medicines, air is linked with stimulating movement. It is associated with *Vata*, which shows lack and imbalances in either air or space. It is the most subtle *dosha* and is normally involved in most physical maladies, because being closer to the source of energy, it is easier to imbalance, and *Vata* is a constitution with a proneness to certain diseases.

The directions associated with air are east and upwards. The tissues or parts of the body associated with this energy are the tendons, ligaments and muscles. The *sense organ* connected to air is the eye. The fluid connected is tears.

The *taste* associated with air is sour. The origin here might be found in decomposing plant matter, which goes sour or even starts to ferment. Fresh fermented leaves or grass are strongly acid, thus, wood soil is often acid and acid loving plants such as rhododendrons, azaleas, *etc.* thrive on it. The *grain* associated with air is wheat, *fruit* is the peach, *meat* is the chicken and *vegetable* is the mallow.

Dietetic Management for Air Personality

Air: Associated foodstuffs include chicken, liver, wheat, greens, citrus fruits, plum, pineapple, vinegar and yogurt. These foodstuffs can obstruct movement and function as astringents. They most benefit changeable, erratic and scattered people.

Balancing Foodstuffs: These foodstuffs include white mushrooms, sour plum and vitamin A and B complex, folic acid, sulphur, sweet potato, cinnamon cherry, cabbage, wild apple, wheat, *etc.*

Stimulating Foodstuffs: These foodstuffs include leek, squash (butternut and spaghetti), raspberries, litchi fruit, pineapple, plum, garlic, marshmallow roots, orange, *etc.*

Stifling (Relaxing) Foodstuffs: These foodstuffs include beets, grape root, mint, wheat, barley, *etc.*

Other Foodstuffs: These foodstuffs include oats, wheat, rye, green lentil, green gram and lima beans, Brazil and cashew nuts, green bell pepper, green peas, lettuce, string beans, zucchini, avocado, grape, lemon and lime.

Contemplation of Fire

अग्निर्मूर्धाचक्षुषी चन्द्रसूर्यौ दिशः श्रोत्रे वाग् विवृताश्च वेदाः ।
वायुः प्राणो हृदयं विश्वमस्यपद्भ्यां पृथिवी ह्येष सर्वभूतान्तरात्मा ॥4 ॥

Fire is as if the head or mouth of God. The directions of space are as if His ears, the famous *vedas are* as if His tongue, the air (atmosphere) is as if His life organ (or lungs), the whole universe is as if His heart and the earth is as if His feet. He is the soul of all souls, *i.e.*, pervades all creatures.

There is hardly a culture, which does not have a myth about fire and its source, *i.e.*, the sun. For mankind, fire has always had two strong connotations, *i.e.*, (i) being a great helper and (ii) a powerful destroyer.

Fire: It transforms darkness into light, and on the same note, it transforms light into matter, *i.e.*, plants, trees, animals and people. It changes clay into ceramics, carbohydrates into diamonds, foodstuffs into the muscles, flesh and energy of our bodies. In *Ayurveda*, it is said that the source of fire in the body are the heart and the small intestines. Besides, two other organs are associated with fire, *i.e.*, (i) Triple warmer (supplementary fire or the immune system) and (ii) the Pericardium.

Tissue or body-parts: associated with fire are blood vessels. *Sense organ* is the tongue: *Sense* related to fire is speech. Effects and modes of unclear speech, too fast or too slow speech, a lack of willingness or the constant need to express oneself verbally, indicate imbalances.

Fluid: The fluid connected to fire is sweat. Fluid connected to fire is sweat. The taste associated with fire is bitter. Fire needs stimulus, coffee, tea, chocolate and all grilled foodstuffs, which have a bitter flavour as alcohol (beer, gin, *etc.*). Extreme likes or dislikes for this taste point to an imbalance.

Expression: Associated with fire is laughter, which shows as a subtle undertone of everyday speech, a constant giggle, or simply humor. *Sound* of laughter is like embers of the fire, and sometimes, like fireworks. The *attitude* associated with the fire energy is one of the receiving.

Dietetic Management for Fire Personality

Neutral fire foodstuffs include lamb, asparagus, alfalfa, citrus peel, peppers, corn, coffee, green beans, peas, potato and rice. These foodstuffs can reduce heat and dry fluids. They most benefit slow, overweighed and aggressive people.

Cooling Fire Foodstuffs: These foodstuffs include alfalfa sprouts, bamboo shoots, wild cucumber, cabbage, turnip, radish, eggplant, lettuce, lotus stems and roots, button mushrooms, sea weed, tomato, figs, grapefruit, lemon, vitamin B, C, G and K, red chilli and all foodstuffs items pungent in taste and red in colour are maize, magnesium, potassium, manganese, onion, milk, red beans, mint, musk melon, water melon, green bean, black cumin and all raw, sprouted and germinated foodstuffs.

Warming Fire Foodstuffs: These foodstuffs include sweet rice, round pepper, garlic, onion, Indian ginseng (*ashwagandha*), camphor root, all Indian spices and hot beverages like tea, coffee and red wine. The other foodstuffs include amaranth, sunflower, cherry, pepper, sunflower and pistachio.

Fire people (*Pitta*) are moderate in weight with strong bone structure, technical, intelligent, energetic and hardworking with leadership qualities. They are also prone to allergies, skin diseases, headaches acidity, *etc.* Fire people like cold foodstuffs like salads, ice creams, fruits, juices and liquids. They eat heavy meals only two to three times a day because fire digests foodstuffs.

Contemplation of Water

Water represents depths of the earth where rest and cleansing takes place. Water is not only peaceful, soft, gentle, receptive and nurturing but it also has great strength, and at times, it turns into violent expression, *viz.* the excessive monsoon, rain floods, *etc.* The anatomical organs that are attributed to the water element are kidney and the bladder. In *Ayurveda*, these organs and their network constitute the *Kapha dosha.* Water provides cohesion for other elements (the body consists of 78% water). The basis of blood, lymph, endocrine fluids, urine, cerebral spinal fluid, sweat, saliva, tears, sexual secretions and breast milk are all forms of water. Imbalance in any of these leads to disruption in our finely tuned energy system and could results in any manner of symptoms, such as creaking dry joints, dryness, thirst, *etc.* On a mental and emotional level, if there is no fluidity in our thought processes and emotions, unexpressed emotions accumulates leading to emotional problems. Water and earth constitute the sub-dimension for the manifestation of taste, which is object of the gustatory sense organ (Chakra Samhita, 1992).

Smell associated with water element is putrid. When water stands for a long time and matter decomposes in it, it has a special stink. The taste related to water is salty. It is the function of kidneys to maintain the salt level in the body. Most vegetables contain natural salts. Often we consume more salt unknowingly than required by the body. Excessive salt consumption can show up as darkness under the eyes, which denotes salt of the kidneys.

Grains associated with water energy are the beans and peas, *fruit* is the dates, meat is the pork and vegetable is the leek

Dietetic Management of Water Personality

Kidney's positive energy is the gentleness, while negative energy is the fear. The colour is blue or black, the season is winter, the taste is salty and the opening is the ears. The associated hollow organ is the bladder, which has energy of gentleness, while the negative energy is fear. Kidneys serve to clear the water from our body through filtration. The associated hollow organ bladder has four foodstuffs associated with it in particular alfalfa sprouts, winter melon (wax or winter gourd) and watercress (cooling) and alfalfa seeds (neutral).

Water: The associated foodstuffs include fish, shellfish, pork, beans, seaweed, soy sauce, walnuts and black sesame seeds. The *grains* associated with the water energy are beans and peas. *Fruit* is the dates, *meat* is the pork and *vegetable* is the leek.

These foodstuffs have a softening effect and promote moisture and calming of the body. They most benefit thin, dry and nervous people.

Balancing Foodstuffs: These foodstuffs includes vitamin A, B-complex, C, D and E, calcium, chlorophyll, zinc, iodine, cabbage, green beans, lotus seeds and potato.

Stimulating Foodstuffs: These foodstuffs include garlic, marshal root, Indian plum and warming foodstuffs are raspberries, grapes, plum, dandelion leaves, soya sauce, salt, *etc.*

Stifling (Relaxing) Foodstuffs: These foodstuffs include asparagus, cabbage, celery, soybean sprouts, soybeans, spinach, mulberry and millet.

Other foodstuffs: These foodstuffs include buckwheat, black bean, kidney bean, black sesame seeds, walnuts, mushrooms, seaweeds, water chestnut (also known as the *immortal foodstuffs* since it has the same texture cooked or raw), pomegranate, and water melon. Water energy is also associated with bones. Eating well and exercising right can help people to be stronger. Without water, there is no taste. Infect water and the earth elements provide the basis of most of the foodstuffs people eat. Together these five elements create the six tastes.

Contemplation of 'Earth'

Every civilization has generated its own myths about the, seemingly immovable base of stability that is earth which in Indian context is referred as mother (*Prithvi maata*), abundantly supporting life. The earth locates us in space and gives grounding for all forms of life. She holds us, she provides the life force energy in terms of foodstuffs we need to sustain ourselves, she is also the reservoir of all forms of life and even after death she accepts us when we give ourselves back to her. Earth itself teaches us to look into the centre of things and deep into the essence of ourselves. It teaches us to look beyond the appearance and discover our original nature as sacred beings. The physical organs associated with Earth are the stomach and spleen.

Direction associated with Earth is the centre. *Climate* related to Earth is humidity or dampness. Dampness stems from accumulation of moisture, it arises from a build up of fluids. It makes the person feel full and heavy. On the one hand, it causes stagnation and affects the circulation, and on the other, it gives a sense of contentment.

Excess humidity makes the atmosphere heavy. Within the body, it leads to accumulation of phlegm, abundant mucus, water retention especially in the abdomen, joints and extremities. The result is a heavy head, a dull feeling and general lethargy. Usually, dampness is accompanied by the heat, cold, or wind. *Tissues* associated with Earth are the flesh, *e.g.*, muscles, lips, collagen and fat. Flesh signifies and describes body mass. *Sense organ* connected to the Earth is the mouth and the sense of taste. The mouth is like a gate, where we take in and reject physical emotional energy, where we receive foodstuffs, drink and even air. These are life-sustaining energies needed for any organism. *Smell* connected to Earth is a sweet fragrance. *Taste* related to Earth is also sweet.

Grain associated with earth is millet, fruit is the apricot and *meat* is the beef. *Expression* associated with the earth is singing. *Sound* attributed to Earth is belching, which is directly linked to the digestive process or rather the ability of stomach to pass things. An *attitude* associated with the Earth Element is *catching foodstuffs in hand*, which means an attitude where the person catches a bite here or there or nibbles. Having meal is often considered wastage of time and often it gets into a habit of nibbling. Quality ascribed to Earth is what we normally deem *intellectual* or rather what we associate with intellectualism, which also accounts for accumulation of undigested ideas. This indicates an imbalance in the earth element.

Activity associated with the Earth is absorbing. The earth absorbs rain in order to gather humidity needed for growth but she also absorbs matter, like wastes decomposes it and assimilates, thus, becoming the biggest recycling plant. The stomach too has a similar function, when it absorbs minerals, vitamins and nutrients, the mind absorbs knowledge and the soul absorbs pure energy. An inability to absorb whether it is knowledge, nutrients, or experience reflects an imbalance in Earth. The traits that characterize the earth are empathy and compassion. Earth people hold on to what they have, be it ideas or opinions. Stubbornness is a sign of holding on beyond reasonable time or circumstances.

Dietetic Management for Earth Personality

Earth Associated foodstuffs: These foodstuffs include beef, barley, sugar, cooked onion, water melon, apple, cherry, dates, grapes, peaches, carrots, cabbage, potato, squash, almond, coconut, alfalfa seeds, lotus seeds, corn, black mushrooms, peas, potato and rice. These foodstuffs slow down acute symptoms and neutralize toxins. They most benefit people with a dry, nervous and weak or aggressive temperament.

Balancing Foodstuffs: These foodstuffs include alfalfa sprouts, bamboo shoots, cauliflower, carrots, cabbage, celery, cucumber, eggplant, green bean, lettuce, lotus stems, mushrooms, radish, tomato, musk melon, grapefruit, lemon, seaweed, soybean sprout, strawberry, mandarin, orange, vitamin A, B, C, E and K, hydrochloric acid, iron element, cardamom, cabbage, rice, pea, grapes, ground nut, golden seal, squash, iodine, sodium, chlorine, *etc.*

Stimulating Foodstuffs: These foodstuffs include coriander, squash, grapes, orange peel (dried), sweet rice, garlic, date, black pepper, camphor and ginger.

Stifling (Relaxing) Foodstuffs: These foodstuffs include salt, salad, brinjal, cumin, tea, wheat, orange, soybean, pine nuts, pumpkin, spinach, papaya, pineapple, *etc.*

Water and earth people (*Kapha*) generally are overweight with strong bone structure. They tend to be very creative as well as diplomatic and business minded. Subsequently, they tend to be tactful and judicious in money matters. In terms of health issues, they are more prone to muscular pains, circulatory disorders, *etc.* With reference to foodstuffs preferences, they like cakes, sweets, grains and vegetables particularly beans. They tend to eat a lot several times a day because their body seems to be making almost a constant demand for more foodstuffs. Water and earth elements are the basis for taste. Ninety percent of our foodstuffs have the taste predominant in water and earth elements.

Atreya in his book practical *Ayurveda* gives an insight into the understanding of 5 states of material existence. The three humors have been derived from the 5 states of material existence. The Chinese system of medicine before 300 B.C. using the concept of *Yin* and *Yang* along with acupuncture adopted the 5 elements theory for understanding *Ayurveda* and certain branches of Chinese medicine. It is still being followed. It is essential at this point to give 5 states of material existence and their relation with the 3 humors.

Table 2: The five states of material existence and the humors

Five elements	Manifestation	Nature	Action
Ether	Space	Ethric	Field
Air	Motion	Gaseous	Propulsion
Fire	Energy	Heat	Conversion
Water	Cohesion	Fluid	Liquidity
Earth	Mass	Solid	Density

The five elements theory explains that the given constituent, *i.e.*, energy, is the cause of the 5 principles of creation. Energy interacting with the *field* creates movement or *propulsion*, propulsion creates friction or heat, which causes the *conversion* of energy. Conversion causes condensation or *liquidity* as liquidity settles *mass* is formed.

Vata is ether + air + *prana*, *pitta* is fire + water + *prana* and *kapha* is water + earth + *prana*. Svoboda (1989) provides another orientation to address the three humors as biological principles, thus, water and earth do not naturally stay together. Some energy is needed to keep them together. *Kapha* is that force. Water and higher does not mix, as one is likely to extinguish the other unless something like acid (both fire and water) keeps them together. *Pitta* is that force. Air needs space to move. Too little or too much space creates problems, dispersion, or constriction. *Vata* is that force which balances the two. Thus, *vata*, *pitta* and *kapha* are the *pranic* forces that keep the 5 states of matter in check and balance.

Prana is the source of the five states of matter, which combine to make the six recognized tastes given in Table 3.

Table 3: Qualities of the six tastes

Taste	Elements	Potency attribute	Long term effects
Sweet	Earth + water	Cold, damp and heavy	Sweet
Sour	Earth + fire	Hot, damp and heavy	Sour
Salty	Water + fire	Hot, moist and heavy	Sweet
Pungent	Fire + air	Hot, dry and light	Pungent
Bitter	Air + ether	Cold, dry and light	Pungent
Astringent	Air + earth	Cold, dry and light	Pungent

Potency and attributes have three possibilities, *i.e.*, heat, hot and cold, which actually denote the digestion pace and action. How hot the foodstuffs are varies according to the foodstuffs. Similarly, some foodstuffs after eating give a sensation of heaviness, while some others give a feeling of lightness. In accordance, the three humours are manifested *viz*. *Vata*, which increases the pungent, bitter and astringent tastes and decreases sweet, sour and salt taste. On similar notes, *Pitta* increases sour, salt and pungent tastes and lowers sweet, bitter and astringent tastes. Finally, *Kapha* tends to increase sweet, sour and salt and reduces pungent, bitter and astringent.

This discussion would be incomplete if the values and physical effects of the six tastes are not addressed in the framework of *Ayurveda*. The values are stated in Table 4.

Table 4: The six tastes and their value

Taste	Nutritive value	Medical Value	In excess causes	Organ affected in excess
Sweet	Great	Little	Toxins	Pancreas and spleen
Sour	Moderate	Little	Acidity	Liver
Salt	Some	Little	Looseness	Kidneys
Pungent	Some	Good	Burning	Lungs and small intestine
Bitter	Least	Great	Cold	Heart
Astringent	Small	Great	Contract	Colon

Physical Effects

Sweet: Sweet is the primary taste for strength, energy and growth of the tissues. Besides, it also brings with it contentment and satisfaction and is responsible for increasing the reproductive fluids and cooling the body. Sugar is commonly found in a number of foodstuffs. However, overuse of sweet taste can lead to accumulation of toxins, digestion problems, lethargy, obesity, *etc.*

Sour: This serves as a stimulant to the body specifically the digestion. Not only it helps in strengthening the sense organs and building all tissues except tissues pertaining to reproduction but also it helps in dispelling gas. The overuse may cause burning, itching, acidity and even premature ageing.

Salty: It is a stimulant, sedative as well as a laxative. It helps in retaining water in the system and can nullify the other tastes. Excess use causes inflammation, skin diseases, premature ageing and water retention.

Pungent: This is the strongest stimulant helping the digestion and metabolism. This also helps in removing the toxins from the body, *viz.* dispelling the gas and inducing sweating and subsequently cleaning the body. Too of pungent taste not only causes pain and burning sensations but also leads to fever, thirst and drying of reproductive fluids.

Bitter: Largely bitter taste is found in spices and herbs. This is the best blood purifier, and interestingly, it promotes the sensitivity for all other tastes. It also facilitates in reducing burning in the body. However, excess of this causes coldness in the body, colic pain, stiffness, nervousness and reduction of reproductive fluids.

Stringent: The best quality of astringent is that it constricts everything in the body. It even works as a medicine and helps in stopping the bleeding and discharge. Besides healing the mucous membrane and the skin, it is found in foodstuffs and herbs, however, its overuse can cause dryness of the body, constipation, cramps, thirst, nervous disorders and decreased reproductive fluids.

At this point, the first author would like to share some of his experiences of treatment, as he is a practicing naturopath.

Five Elemental Profiles of Four Patients

Case - 1

Name: Sri Yogesh Purwar

Age: 48 years

Address: Sarain, Meer Khan Chowk, Lucknow

Chief Complain: Non-bleeding piles and varicose veins in legs

Management of Five Elements

Outgrowth of the patient indicates that he has Earth element in excess. He needs to reduce the foodstuffs rich in earth element and increase foodstuffs of air element.

Prescription: The prescribed diet included a daily intake of litchi fruit, pineapple, plum, orange, wheat, barley, *etc.* He was also advised to take foodstuffs like oats, wheat, green gram, cashew nuts, green peas, lettuce, lemon and lime. Further, he was asked to reduce the intake of foodstuffs like date, black pepper, camphor, ginger, salt, salad, brinjal, soybean, pumpkin, spinach, papaya, pineapple, *etc.* in daily intake.

Follow Up After One Month: There was a reduction in size and pain of the piles. After three months, there was also a reduction in varicose veins, and the patient reported a lot of relief in pain. After six months, the pain was gone and the patient reported a complete cure of piles as well as large reduction in varicose veins.

Case - 2

Name: Mrs. Neelam Diwarer

Age: 32 years

Address: Ashiyana, Raebareli Road, Lucknow

Chief Complain: Bronchial asthma, along with abdominal disorders like constipation, gastritis, *etc.*

Management of Five Elements

Bronchial asthma reflects disharmony of the lungs. Lung is associated with space element which indicates that the patient's space element seems to be efficient. She needs to increase the intake of foodstuffs rich in space element and reduce fire element rich foodstuffs in daily intake.

Prescription: The prescribed diet included an increase in the intake of foodstuffs like mushrooms, tea, coffee, coriander, orange peel (dried), garlic, ginger, mustard green, sugarcane, raw onions, garlic, radish, turnip, cinnamon, mint, cloves, oat, honey, grapes, pear and vitamin A, C, D and E. She was further advised to reduce the intake of foodstuffs like lettuce, citrus peel, peppers, corn, green beans, peas, potato, rice and sweet. These foodstuffs can reduce heat and increase dryness.

Follow Up After One Month: Reduction in suffocation and abdominal discomfort. After three months, the result of one month was maintained, and after six months, there was a lot of reduction in bronchial asthma and complete removal of abdominal discomfort.

Case - 3

Name: Sri Anil Kumar Singh

Age: 23 years

Address: 4, Neel Kamal market, Malihabad, Lucknow

Chief Complain: Indigestion and impaired immunity

Management of Five Elements

Indigestion and impaired immunity of the patient indicate that his heat and air element are in deficient condition.

Prescription: The prescribed diet comprised of foodstuffs like green leafy vegetables, seasonal fruits, grapes, orange, sweet rice, date, black pepper, camphor, ginger, salt, salad, brinjal, cumin, wheat, soybean, pine nuts, pumpkin, spinach, papaya, pineapple, oats, green gram and cashew nuts, green peas, lemon, lime, *etc.* Intake of foodstuffs rich in water content and spongy in nature was reduced.

Follow up: An improvement in condition within two months was apparent.

Case - 4

Name: Mrs. Shanti Devi

Age: 39 years

Address: Ashutosh Nagar, Krishna Nagar, Lucknow

Chief Complain: Migraine, calf muscle cramps, impaired eye sight, fatty liver and menstrual disharmony

Management of Five Elements

All these symptoms are associated with liver, which is related with wind.

Prescription: The prescribed diet included foodstuffs like orange peel (dried), rice, garlic, ginger, mustard green, sugar, sugarcane, raw onions, radish, turnip, cinnamon, groundnut, *etc.* Besides, citrus fruits, plum, vinegar, salad, brinjal, cumin, soybean, pumpkin, *etc.* were reduced in routine diets.

Follow up: Reduction in symptoms was seen. Within six month, there was a complete removal of migraine, calf muscle pain and periodic disharmony. There was little improvement in eyesight and fatty liver.

Ayurveda consistently takes the holistic view of life and reiterates that *Prana* is responsible for all life on the planet. It is the balance of *Prana* and the five elements, which are detrimental for health.

पृथिव्यप्तेजोऽनिलखे समुत्थिते
पश्चात्मके योग्गुणे पवृते!
न तस्य रोगो न जरा न मृंत्युः
प्राप्तस्य योगाग्निमयं शरीरम् ॥12 ॥

A yogi does not get any disease, weaknesses of old age, or pain of death when he has controlled his body constituted by five substantial elements, *viz.* earth, water, heat, air and ether and the same (his body) has acquired radiance of yoga, and the merits of yoga have manifested in him, *i.e.*, the perception of yoga has appeared to the yogi (Shvetashvtar Upnishad).

2 Concentrated Oxygen

Rajesh Kumar

DG' Technical Cell
Council of Scientific and Industrial Research (CSIR)
Department of Scientific and Industrial Research (DSIR)
Anusandhan Bhawan, 2 Rafi Marg, New Delhi- 110 001
Email: <rajesh@csir.res.in>; <rajeshkr38@nic.in>

The environment scientist suggests for proper environmental management to avoid the quarter of all preventable illnesses, which are directly caused by environmental factors. Human being needs air to breathe, water to drink and food to eat but these resources are decreasing due to consumption and increase in population.

The idea of an oxygen bar is not as new as you would perhaps think and it was originally pioneered in Japan towards the end of 1990s. Its popularity soon spread to the United States where they were increasingly found in spas, executive lounges and other recreational businesses. They have even started to appear inside some regular bars. Impressively, in some countries, you can buy recreational oxygen in cans.

The invention extended by creativity and innovation in science spackle some of the miracle and extracted benefit enrich in daily life covering every field, from hygiene food content to tourism at moon and space. The innovative design of household gadget made the common man's life easier. The extended benefits of science create some challenges for scientist, and most concern proven devil is carbon enrich polluted environment due to the industrialization and consumption of fossils fuel. The reducing oxygen content in the atmospheric air, especially in metro cities, is of great concern, and deficiency of oxygen is creating the medical health problems among the living particularly young children. The article explains some benefit of oxygen on the health of common man and the scope of oxygen bar in India prospective *vis-à-vis* challenges.

Almost the entire countries wants a cure for cancer, and it so happens that such a cure naturally exists. Oxygen therapy as the source of cure, there has been widespread opposition and doubt. Once the world comes to accept this cure, humanity will enter a new consciousness of peace and prosperity. However, the oxygen concentrator is an accepted means of delivery of long-term domiciliary oxygen treatment. Conditions of use, however, need to be carefully defined.

The studies conducted in United Kingdom by using fourteen concentrators for one year by patients with hypoxemic chronic obstructive airways disease, and mechanical reliability, patient compliance with a regimen of 15 hours use a day, smoking habits and variation in arterial gas tensions studied. Though many patients failed to achieve either the desired daily use or the recommended arterial oxygen tension, problems were generally minor and could probably be overcome by careful supervision and planning. Overall, the concentrator appeared to be the most economical means of providing oxygen treatment at home and was much preferred by patients who had previously used oxygen cylinders. The available sources of concentrated oxygen are:

- Cylinders
- Tanks
- Concentrators

An oxygen concentrator, also called an oxygen generator, is a device used to provide oxygen to a person at substantially higher concentrations than those of ambient air, used as an alternative to tanks of compressed oxygen. Oxygen concentrators are also used to provide an economical source of oxygen to the patient/industries.

Portable oxygen concentrators are much easier to set-up than oxygen cylinders. You simply plug the unit into a standard electrical outlet and set the flow meter to your physicians prescribed flow rate. You can easily move from room to room with lightweight portable oxygen concentrators.

Oxygen bar Prospective Scope

The oxygen bars offer sniffers an increased percentage (up to 95%) of oxygen compared to the normal atmospheric content of 21%, as the percentage in Metro cities is lower because of severe pollution. It should be noted that when inhaling normal air (21% oxygen), a healthy human body will, on an average, exhale 14% oxygen, absorbing only 5% of the inhaled oxygen.

Hospitals throughout the developing world have very limited access to oxygen, and when oxygen is available, the equipment required to deliver it is often lacking. WHO/UNICEF has published technical guidelines for oxygen therapy in the management of many diseases in low-income countries, covering the indications for use, sources and equipment for the administration of oxygen.

Health workers did not know when or how to administer oxygen to patient. Most of the concentrators that were available on the pediatric wards were more than 10 years old and did not meet WHO/United

Nations Childrens Fund (UNICEF) specifications. It was decided by World Health Organization to supply oxygen concentrators in pediatric wards of all district hospitals and to develop a package of information and tools to cover all stages from procurement, training, installation and maintenance of oxygen concentrators.

Woman inhales concentrated oxygen inside the first oxygen bar in Paris

The oxygen bar concept started in Japan in early 1990, and now, spread over United States, Europe and other countries. The Times of India reported on 19th February 2008, about the opening of oxygen bar in Paris as a late start and writes but Paris first oxygen bar opened only last week. Flagged as *a revolutionary anti-fatigue, anti-stress and anti-depression concept*, the city's maiden oxygen bar is housed in an up-market beauty institute. Indian market has also entered into the business and couple of oxygen bar opened in Delhi, Hyderabad and other cities. The oxygen bar is now concept proven business model concept for health and hygiene spread all over the world and waiting for penetration in Indian market. The globally available oxygen bar is beautifully designed having lots of fun to use. The possible installation destinations for oxygen bars are salons, health clubs and resorts, restaurants, coffee houses, bars and airports. In some countries, caterers are providing oxygen bars for farmhouse for recreation and entertainment. Yoga studios, chiropractors, acupuncturists, medical spas and massage therapists now see how adding an oxygen bar can benefit their clients.

Chemistry of Human body needs Concentrated Oxygen

There are several forms of oxygen, however, the most common and stable form of oxygen is "O_2" or diatomic oxygen. Oxygen is a vital element to good health and it is our primary source of energy. Ninety per cent of our energy comes from oxygen and the other 10% comes from food and water. With oxygen-enriched air (OEA), the body will metabolize food more efficiently, generate more energy and create a healthier body. The conversion process of food into energy is known as oxidation and what is oxidized (or burned) for the body's energy are carbohydrates, sugars, fats and proteins. However, oxygen is the *fuel* required to oxidize all or any of these foodstuffs.

The human body is scientifically works as an *oxygen factory* and the main raw material of the factory is oxygen. Oxygen acts as a fuel, and on consumption of it, cells in our body use to function. Oxygen is the *food* for our cells and body requires enough oxygen around our cells, organs and tissues. The highlighted benefit from using concentrated oxygen, to various function of human body like memory cell improvement and retention, better immune system, sleep disorders digestion problems, better sex, depression headaches and hangovers, better circulation act as cure for cancer and other disease and burning for calories/fat in human body

One of the matchless ironies of the human body is its requirement for the highly reactive oxygen molecule, which has been clearly implicated in many diseases and the aging rocesses. Oxidants produced by metabolic processes damage cells by starting chemical chain reactions including oxidation of DNA and proteins as well as lipid peroxidation. Damage to DNA can cause mutations and lead to cancer if not reversed by DNA repair mechanisms. Damage to proteins causes enzyme inhibition, denaturation and protein degradation. Lipid peroxidation can cause cell lysis as well as creating mutagenic and carcinogenic byproducts. The human body contains antioxidants and enzymes that together work to prevent oxidative damage to cellular components. Largely antioxidants either prevent these reactive oxygen species from being formed or remove them before they cause damage. There are many theories currently that tout the superior nature of diverse antioxidant combinations. One such theory has been given by Dr. Lester Packer of University of California at Berkley. Dr. Packer puts forth the hypothesis that there is a superlative combination of five antioxidants, which have the ability to *recharge* one another both in the blood plasma and intracellular. This would result in a greater quality of antioxidant protection for an extended time. The current study evaluates Dr. Packer's theory of antioxidant combination from his book *The Antioxidant Miracle*. The decay rate of antioxidants vitamin E, vitamin C, lipoic acid, glutathione and coenzyme Q10 alone and in combination were determined using the oxygen radical absorbance capacity (ORAC) assay. Majority of the antioxidants retained activity for longer periods when tested alone, rather than in combination as Dr. Packer's theory would suggest. The assay was also preformed (using the same antioxidants and combinations) on oxidatively damaged Raji cancer cells. Cell viability and uptake of antioxidants into the cytoplasm were monitored. Finally, a variety of multivitamins was subjected to the ORAC assay and their antioxidant capacity compared to that of the *Packer Combination*. The results suggest that multivitamins are superior antioxidants than the packer ratio listed in the book *The Antioxidant Miracle*.

Oxygen therapy has an incredible healing effect on a wide range of diseases as well as in its ability to rejuvenate the aging body. Oxygen therapy is essential for many functions in our body, especially concerning oxygenating the cells in our body for the health of nervous system and brain. Many cities with high pollution levels have shown an increased cancer rate. The cancer rate in places especially mountain areas where negative ions are emitted was only about 10 to 0% of that in America and Europe, and it was practically non-existent in rural areas of the Himalayas.

Madison Cavanaugh was one of the prominent researchers to investigate that cells are highly oxygenated and the white blood cells destroyed up to three times more microbes than before, after the application of this treatment. Oxygen therapy was found to be beneficial in a wide range of diseases by some health professionals. They found an excellent preventative effect on cancer and cured precancerous conditions such as hyperkeratosis, leukoplasia and chronic mastitis. They are surprised by many of these patients experiencing bursts of energy and euphoria. Mountain regions that are rich with negative ions had less cancer and sickness than those who live in urban cities, according to studies. Experts discovered also that if the treatment is applied. They also found that oxygen therapy was effective with bronchitis, asthma, emphysema and pneumonia. Indian scientists demonstrated how the incidences of tumors of the breast, could be reduced from 100% to between 46 and 57% by the application of oxygen therapy, vitamin C, vitamin A, or selenium. Tumor incidences were further reduced to between 25.9 and 31.8%, in combination of two, and finally, tumor incidence was reduced to only 12%, when all four nutrients were given.

Scientific Principle

Oxygen evolution is the process of generating molecular oxygen through chemical reaction by the process of the photolysis of water during oxygenic photosynthesis, electrolysis of water into oxygen and hydrogen and electro catalytic oxygen evolution from oxides and oxoacids.

X-ray crystallography studies have recently provided detailed models of structure of the oxygen-evolving complex and its manganese cluster. Based on structural and spectroscopic experiments, oxygen evolution involves a core three + one cluster of three manganese ions and one calcium ion with additional manganese, which are oxidized *via* intermediate states called *S-states*.

Oxygen is produced industrially by fractional distillation of liquefied air, chemical means using zeolite, concentration of oxygen in air, electrolysis of water and other means.

The difficulty of obtaining oxygen for the treatment of diving accident victims in remote areas has been a long-standing problem for the NOAA Diving Program. The normal method for transporting oxygen is in metal high-pressure cylinders. These cannot be carried on airliners and contain insufficient oxygen for long-term on-site treatment of diving accident victims or for evacuation to a treatment facility. Chemical production of oxygen by *oxygen candles* likewise produces insufficient oxygen. Atmospheric oxygen concentrators offer the potential of providing a long duration supply of oxygen in remote areas or aboard ships. These devices concentrate the oxygen in air and provide a constant supply of low-pressure oxygen. They have been successfully used as an oxygen source for the production of diver's oxygen enriched breathing gas. The most appropriate type of concentrator for the purposes addressed here are pressure swing adsorption (PSA) units. These devices selectively remove nitrogen from air and produce a breathing gas mixture containing up to 95% oxygen.

They are currently used as a replacement for high-pressure oxygen cylinders to provide oxygen enriched mixtures in the homes of patients who require additional oxygen. These electrically powered devices provide a continuous slow flow of 95% oxygen. The flow rates and pressure output of current units are insufficient for use with the demand type, or free flow oxygen masks currently used to provide pure oxygen to accident victims. However, calculations show that when used with oxygen conserving delivery systems, they will provide sufficient flows of oxygen for the on-site administration of up to 90% oxygen and for the long duration evacuation of diving accident victims, or other patients in need of augmented oxygen. Recently developed PSA units are small, lightweight and can operate on batteries or 12 volts DC electricity, thus, making them practical for use in the field. When used with the oxygen conserving delivery systems identified below, they provide viable methods of providing oxygen for on-site administration and evacuation. The purpose of this study was to evaluate currently available oxygen concentrators and oxygen delivery systems and to provide recommendations to NOAA on the most efficient and cost effective combinations of the systems.

The simplest oxygen concentrator is capable of continuous delivery of oxygen and has internal functions based around two cylinders, filled with a zeolite material, which selectively adsorbs nitrogen in the air. In each cycle, air is flowed through one cylinder at a pressure of air flows through one cylinder at a pressure of around 20 psig, where psig indicates the gauge pressure (138 kPa, or 1.36 atmosphere), where the nitrogen molecules are captured by the zeolite, while the other cylinder is vented off to ambient atmospheric pressure allowing the captured nitrogen to dissipate. Typical units have cycles of around 20 seconds and allow for a continuous supply of oxygen at a flow rate of up to approximately 5 liters per minute (LPM) at concentrations anywhere from 50 to 95%. This process is called pressure swing adsorption (PSA).

Another process has been developed by air products, which they call vacuum swing adsorption (VSA). It uses a single low-pressure blower and a valve, which reverses the flow through the blower so that the regeneration phase occurs under a vacuum. Generators using this process are being marketed to the aquaculture industry enabling the output of oxygen continues uninterrupted. While the pressure in the second canister is at 20 PSI (pounds per square inch), the pressure in the first canister is reduced to zero. This allows nitrogen to be released from the zeolite and returned into the atmosphere. The zeolite is then regenerated and ready for the next cycle. By alternating pressure in the two canisters so that first one and then the other is at 20 PSI, a constant supply of oxygen is produced, while the zeolite is continually being regenerated. Individual units have an output of up to 5 liters per minute with an oxygen concentration of up to 95%.

Two oxygen delivery systems were evaluated. The Hi-OX mask is a partial rebreathing open circuit system, and it was selected because of its high efficiency relative to similar simple mask systems. The design of this mask results in the inspiration of high oxygen supply gas during initial portion of the inspiration cycle from a flexible reservoir, and when that is depleted, the latter portion of the inspired gas is made up of surrounding air. The high oxygen mixture, thus, enters the alveoli first, while the air fills the respiratory dead space. This results in higher alveolar oxygen concentrations than it would result if air were mixed with the inspired gas throughout the respiratory cycle.

The closed circuit diving oxygen rebreather, minolung was selected because of its gastight design. Since the gas mixture used contained a minimum of 5% nitrogen rather than pure oxygen, it had to be tested in a semi-closed circuit mode.

The fundamental difference between the above delivery systems is that the resulting alveolar oxygen concentrations at a particular flow/oxygen concentration setting are dependant on the respiratory minute volume, tidal volume and breathing pattern for the HIOX mask, while for the rebreather, alveolar oxygen concentrations depend primarily on oxygen consumption of the individual and are less dependant on breathing pattern and respiratory minute volume.

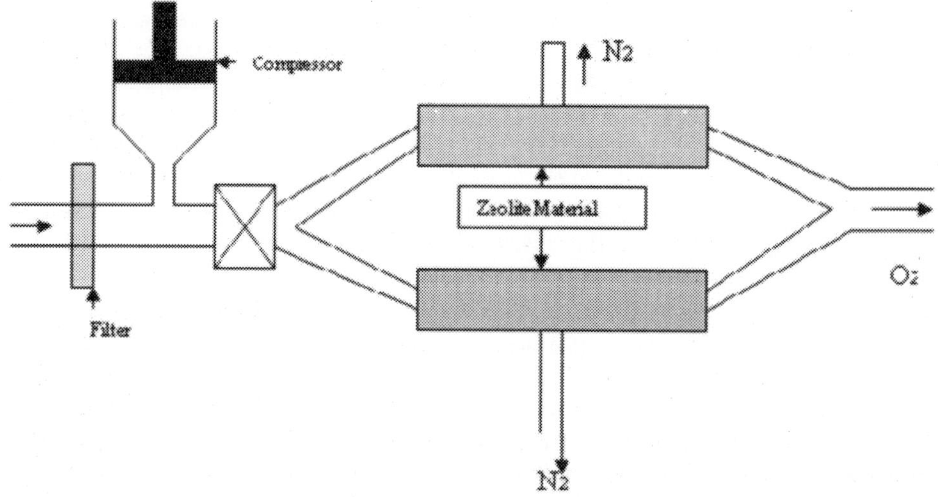

Oxygen Concentrator

Instrumentation for Oxygen Generation, Storage and Administration of Oxygen

The instrumentation for the design of oxygen concentrator with following objectives:

- To identify the sensor for the oxygen concentrator
- Design and development of oxygen concentrator for biomedical applications
- Characterization of oxygen concentrator on different parameter like temperature, humidity, pressure, flow of gases in the chamber and size and material of sieve for biomedical application.
- Generate objective evidence of the effect of concentrated oxygen produced by oxygen concentrator

The recommended instruments for oxygen concentrator are:

1. Oxygen [O_2] Sensor for measuring the percentage of oxygen in the gas

Single Point Continuous Monitor provides field proven technology for the quick detection of low level gases, the next generation of interface free and low maintenance colourimetric gas detection.

2. A pressure regulator

It is used to control the high pressure of oxygen delivered from a cylinder to a low pressure controllable by the flow meter. The Pressure Sensor Selection Checklist lists the specifications you must consider when choosing a pressure sensor. The most important specifications are electrical output, accuracy, operating environment and mechanical coupling.

The engine test stands used with advanced data-acquisition systems require a different type of pressure sensor than those used in the past. Engine hydraulics and pneumatics operate faster and often with higher pressure and temperature. Accordingly, the pressure sensors have to be more reliable and rugged in order to reduce downtime.

In choosing the correct pressure sensor for your application consider several specifications and decide which type of pressure sensor will give you the best results in your application.

Gage pressure sensors: With gage sensors, pressure readings are referenced to the atmosphere, *i.e.*, zero output is at atmospheric pressure. You use this type of sensor when you need to measure both vacuum (negative output) and pressure (positive output).

Vacuum pressure sensors: A vacuum sensor's output is zero at atmospheric pressure, like the gage sensor but the output increases as the vacuum increases. You calibrate vacuum sensors so their output becomes more positive as the pressure becomes more negative.

Differential Pressure Sensors: This type of sensor has two pressure ports as shown in Figure 1 and senses the difference in pressure between the two ports. You can use differential pressure sensors to measure the pressure of liquids or gasses.

Absolute Pressure Sensors: The reference for this type of sensor is full vacuum, *i.e.*, the output is zero at full vacuum. Note that there is no polarity change when the input pressure changes from vacuum to pressure above atmosphere.

Barometric pressure sensors: Barometric pressure sensors are absolute pressure sensors with a limited range. Usually, the output of these sensors is expressed as *inches of Mercury (Hg)*, and the output ranges are 16-32 inches HgA or 26-32 inches Hg with zero output at low number.

3. A flow meter

It is used to control and indicate the flow of oxygen. Typical flow range is 0-15 lpm.

4. Oxygen Temperature sensor

The sensor needs to monitor temperature during the flow in the process of concentration and at the out let available for the use.

5. Humidity Sensor for Air/oxygen

Rapid advancements in semiconductor technology, such as thin film deposition, ion sputtering and ceramic/silicon coatings, have made possible highly accurate humidity sensors with resistance to chemicals and physical contaminants at economical prices. No single sensor, however, can satisfy every application. Resistive, capacitive and thermal conductivity sensing technologies each offer distinct advantages. Resistive sensors are interchangeable, usable for remote locations and cost effective. Capacitive sensors provide wide relative humidity range and condensation tolerance and if laser trimmed are interchangeable. Thermal conductivity sensors perform well in corrosive environments and at high temperature. For most applications, therefore, the environmental conditions dictate the sensor choice.

The most important specifications to keep in mind when selecting a humidity sensor are:

- Accuracy
- Repeatability
- Interchangeability
- Long-term stability
- Ability to recover from condensation
- Resistance to chemical and physical contaminants
- Size
- Packaging
- Cost effectiveness

Additional significant long-term factors are the costs associated with sensor replacement, field and in-house calibrations, and the complexity and reliability of the signal conditioning and data acquisition (DA) circuitry. For all these considerations to make sense, the prospective user needs an understanding of the most widely used types of humidity sensors and the general trend of their expected performance.

Capacitive Humidity Sensors: They consist of a substrate on which a thin film of polymer or metal oxide is deposited between two conductive electrodes. The sensing surface is coated with a porous metal electrode to protect it from contamination and exposure to condensation.

Resistive Humidity Sensors: Resistive humidity sensors measure the change in electrical impedance of a hygroscopic medium such as a conductive polymer, salt, or treated substrate Resistive sensors usually consist of noble metal electrodes deposited on a substrate by either photo-resist techniques or wire-wound electrodes on a plastic or glass cylinder. The substrate is coated with a salt or conductive polymer. When it is dissolved or suspended in a liquid binder, it functions as a vehicle to coat the sensor evenly.

Thermal Conductivity Humidity Sensors: These sensors measure the absolute humidity by quantifying the difference between the thermal conductivity of dry air and that of air containing water vapour. When air or gas is dry, it has a greater capacity to *sink* heat, as in the example of a desert climate. A desert can extremely be hot in the day, but at night, the temperature rapidly drops due to the dry atmospheric conditions. By comparison, humid climates do not cool down so rapidly at night because heat is retained by water vapours in the atmosphere.

Challenges for Oxygen Bar in Context to Indian Market

As oxygen bars entry begin to India, health authorities are beginning to question the merits or otherwise of sniffing oxygen. Some of the scientist claims for using concentrated oxygen is not practically proven, and study is continued but some claims that oxygen therapy is not only safe though it enhances health and well-being, including strengthening the immune system and enhancing concentration. It has been alleged to alleviate hangovers and helps with migraines but no formal studies have yet confirmed any of these claims. Individual flavoured scents (aromas) add to the experience.

There is no scientific basis for the health claims made for the oxygen bar. The medical professional research work shows that the concentrated oxygen is harmful for individuals with respiratory diseases such as asthma and emphysema. The colouring and flavouring are yet to receive the approval from legal authorities. In addition, concentrated oxygen is a flame accelerant, which should be kept away from cigarettes and other sources of ignition.

In Frenchay Hospital, Bristol, United Kingdom, J.P. Dilworth and others investigated the impact of installation of oxygen concentrator on lifestyle of 30 patients in two health districts using a questionnaire and found marked improvements in general well-being (83% of respondents), breathing (82%), mobility (62%) and sleep pattern (52%). The long-term nature of the aims of treatment was understood by 83% of the respondents and the mean time period the patients used the concentrator was satisfactory. However, 34% of respondents had a concentrator with only one outlet and 70% had the concentrator situated in a commonly used room with the possibility of problems with noise, while 31% of the respondents were still smoking. The recommendations given to the patients for the sitting of concentrator and the number of outlets should be improved. However, the oxygen concentrator was found generally well tolerated, and this refutes criticism that patients may find it restricting.

However, skepticism regarding the effectiveness of oxygen shoots runs high in medical circles. The atmospheric air contains 21% of oxygen, which is enough for a healthy human body, which absorbs only 5% of the inhaled oxygen. Thus, it is unclear whether increasing the oxygen concentration during inhalation will result in any more oxygen being absorbed than is normally absorbed.

■■■

3 Wonders of Minerals

M.K. Rana

Department of Vegetable Science
CCS Haryana Agricultural University Hisar- 125 004, Haryana, India
E-mail: <mkrlotus@gmail.com>

All foods ingested have a potential of energy since they are broken down in the body and converted into energy. The foods rich in carbohydrates and fats make significant contribution as a source of calorie, thus, are known as energy foods; the foods containing protein are considered as body building foods, and the foods that are powerhouse of minerals and vitamins, are known as protective foods. In foods, fat provides a convenient and concentrated source of calories. In plants the main carbohydrates of nutritional importance are starch and sugars largely sucrose, glucose and fructose. Carbohydrates that contain C, H and O in proportion of $C(H_2O)$ provide energy and may also be converted into fat and stored temporarily in liver and muscles in the form of glycogen. Much of the sugar absorbed by the body is converted into fat to which the muscles use as fuel for mechanical energy. The nervous system also uses considerable amount of energy to maintain itself in working order since approximately 20% of the basal metabolism takes place in the brain. The nervous system appears to derive energy entirely from carbohydrates, as it seems to have no significant amount of stored energy although it entirely depends upon carbohydrates furnished by blood stream. Water is not at all a nutrient but it is essentially required to maintain the osmotic balance in the cells and to excrete the metabolic waste products from the body.

The daily calorie requirement by a person is termed as Adult Consumption Unit (ACU), which is expressed as calorie or kilocalorie. The consumption of energy by an individual depends on sex, age and nature of work but by and large the daily calorie requirement of an adult man and woman engaged in sedentary work is about 2400 and 1920 kcal per day, respectively. The individuals taking inadequate calorie are known as undernourished. Marasmus, nutrition related disease, is developed due to partial starvation from chronic calorie deficiency, and the symptoms are atrophy (wasting away from want of nutrition) of limb muscles and shrunken body (a shriveled old man face). The subjects, who are lacking in minimum possible daily intake of proteins including essential amino acids, vitamins and mineral elements are designated as malnourished, and over-nourished, the subjects who consume too much food or a single nutrient that leads to specific toxicity. Undernourished subjects are likely to be malnourished too. Undernourishment and malnourishment can cause increased susceptibility to infection and diseases and the stunting of growth in children not only physical but mental as well since the brain and neurological development is adversely affected. Protein malnutrition and vitamins deficiency are also known to make victims of several parasitic infections.

The recommended dietary allowance (RDA) should always be met by a variety of foods as most foods contain more than one nutrient but no single food item supplies all the nutrients in the amounts that are essentially needed by our body. The greater the variety of foods included in diet, the less likely one can develop a deficiency or an excess of any single nutrient. Variety of food also reduces the excessive amounts of toxic substances that occur naturally in foods and additives or contaminants, which may be present in any single foodstuff. A simple approach to adequate nutrition is to consume a variety of food. The total foodstuffs can be divided into four basic food categories, *i.e.*, (i) the foods from milk group supply calcium, protein, and riboflavin, (ii) foods of meat group are the key source of protein, fat, iron and other minerals as well as vitamins, (iii) fruits and vegetables are rich in vitamin C and precursor of vitamin A and (iv) the cereals provide carbohydrates, several B vitamins, and iron. The food items, like tea and certain spices can provide vitamins and antioxidants, which are essential for good health. The detailed descriptions of all the mineral elements and vitamins required by the human body for various functions have been given below along with their various sources of availability.

Minerals

The significance of minerals and vitamins in human nutrition is now well known all over the world. The human body requires at least 45-50 chemical substances, which can be divided into six main groups: carbohydrates, fats, proteins, vitamins, inorganic mineral elements and water. Calcium, phosphorus, iron, sodium, potassium, chlorine, magnesium and sulfur, know as macro-elements, are required in large amounts, and iodine, copper, zinc, fluorine, selenium, chromium and molybdenum, known as microelements, are required in trace amount. At least ten mineral elements, *i.e.*, calcium, iron, phosphorus, sodium, potassium, magnesium, sulfur, chlorine, iodine and fluorine are most essentially needed for normal growth, reproduction and maintenance of good health, and also useful in protecting against some degenerative diseases. Out of these, calcium, iron and phosphorus are required in incredibly large quantities. The rationales of points, which will justify the above-mentioned statement, are given below:

Calcium

Calcium, phosphate and magnesium are the major mineral components of bone. The formation and resorption of bones depend upon a balance between intake and excretion of calcium. Positive calcium indicates bone formation and a negative calcium balance resorption. Food intake plays an important role on bone formation. Sufficient dietary calcium, phosphate, and magnesium supply has a positive effect on the maintenance of mineral balance and bone mass. A high calcium intake, *i.e.*, > 1 g/day, may reduce the rate of bone resorption. A highly calcium available calcium source may be of benefit to aging people, who ingest relatively low amounts of food and have decreased calcium absorption. Bone formation may be optimally supported by combined supplement of calcium, phosphorus and vitamin D. Bone loss may occur despite adequate mineral supply, *e.g.*, as a consequence of a lack of physical activity combined with low estrogen levels. Internal calcium concentration regulates a wide variety of processes. Glycogen breakdown, muscle contraction, and secretion of small molecules like insulin or histamine are the calcium-regulated processes. Initiation of blood clotting process is also regulated by calcium. It also helps in the conversion of vitamin D to its active form by the kidney. The active form of vitamin D not only increases intestinal absorption but also promotes bone resorption directly.

Calcium in body is needed for healthy bones and for resistance to infections. In its absence, children suffer from rickets, pigeon chest, irritability, blood clotting and retarded growth, and teeth become bad due to its deficiency. It is known as anti-rickets mineral element. Osteoporosis in adults is also associated with a negative calcium balance. An improved intestinal calcium absorption alone results in a decreased risk for osteoporosis. Calcium supplements reduce the risk of hip fracture considerably, decrease parathyroid hormone secretion, and increase mineral density as it is believed that increased parathyroid hormone secretion is due to vitamin D deficiency and low calcium intake. Six months supplementation with calcium (1 g/day) and vitamin D (ergocalciferol) reduce the possibility of secondary hyper parathyroidism. Calcium supplements with vitamin D_3 (cholecaciferol) decrease the frequency of fractures in elderly. An adult requires 500-600 mg for normal functioning of heart and nervous system. Amaranth, beet leaf, fenugreek leaves, Bengal gram leaves, radish leaves, mint, drumstick leaves, coriander leaves, portulaca, Swiss chard, onion-greens, parsley, beans, Cole crops, carrot, lettuce, onion, spinach, peas and tomatoes are the chief sources of calcium supply.

Phosphorus

Phosphorus together with calcium is the most abundant mineral in the bone tissues, thus, essential for the cell multiplication of both bones and soft active tissues of the body. Phosphate is actively involved in the regulation of bone formation. It plays an important role in the oxidation of carbohydrates, which liberate energy. Higher concentration of phosphate inhibits bone resorption and enhances bone mineralization. However, too much phosphate in the diet has a negative impact on calcium homeostasis and bone metabolism. Phosphate has a negative interaction with calcium absorption. Increased dietary phosphate increases fecal calcium loss but decreases urinary calcium excretion. Dietary phosphate level does not so significantly affect the calcium balance. High phosphate and low calcium intake result in an increase in parathyroid hormone and a decrease in cholecalciferol (vitamin D_3) synthesis in kidney.

Per day per head phosphorus demand of an adult is about 1430-1520 mg. A deficiency of phosphorus is rarely seen because almost all foods contain phosphorus. Cereals rich in phytic acid reduce the absorption of calcium as the human intestinal tract lacks phytase enzyme. Phytic acid converts the calcium in insoluble form and interferes with calcium absorption. Like calcium phosphorus absorption is regulated by active form of vitamin D. Excessive use of antacids that bind phosphate also causes serious deficiency. High dietary phosphate intakes have been found inversely correlated with bone density. The absorption of inorganic phosphate is more when calcium and phosphorus are taken nearly in equal amounts. Milk is the best source of phosphorus as it contains calcium and phosphorus in equal amounts. The principal sources of phosphorus are protein rich foods and cereal grains. It can also be obtained from potato, carrot, pea, lima bean, drumstick pods and leaves, tomato, cucumber, spinach, lettuce, cauliflower, broccoli, parsnip, globe artichoke and mushroom consumption.

Potassium

About 5% of the total mineral content of the human body is potassium. It is present inside the cells, thus, helps in maintaining the osmotic balance. It is readily absorbed by small intestine and excreted from the body through urine. In combination with sodium and chlorine, potassium is involved in the maintenance of normal water, osmotic and acid-base balance. Its higher level in body can cause the myocardium to arrest in diastole, whereas, a lower level in plasma can provoke systolic arrest. Its deficiency results in a slow growth rate, thinning of hair, necrosis of myocardium and death. The serious deficiency of this element is associated with gastrointestinal tract diseases involving loss of secretions by vomiting or diarrhea. Loss may also take place from the body through urinary excretion. Low concentration of potassium ions in extra-celluar spaces causes muscular weakness, increases nervous irritability, mental disorientation and cardiac irregularities. Sodium is antagonist of potassium, higher concentration of sodium produces same effects as the potassium deficiency. The daily requirement of normal person is about 3-5 g/day.

Sodium

A healthy adult contains about 100 g of sodium, which constitutes 2% of the total mineral content of the body. Sodium is principally absorbed by small intestine and enters the blood stream. About 90% of the total ingested sodium is excreted in the urine. Sodium is also lost from the body through perspiration. In combination with potassium, sodium regulates the water balance and contributes to the osmotic pressure. In association with chloride and bicarbonate, it regulates acid-base balance. It is essential for absorption of glucose and in the transportation of other nutrients across the membranes.

Dehydration due to heavy sweetening, severe diarrhea, or vomiting causes severe depletion of sodium and chloride. The symptoms of its deficiency are weakness, fatigue, nausea, low blood pressure, rapid pulse rate and lack of appetite. A thirst develops that cannot be quenched by drinking water alone but can be quenched by adding salt in water. Foods of plant and animal origin both have enough amount of sodium but animal foods are rich in sodium as compared to plant-originated foods. The principal source of sodium is common salt used in cooking food. The body requirement of sodium chloride is about 10-15 g/day.

Chloride

In human body it is found closely associated with sodium and present to the extent of 0.15% of the body weight. It is essential to maintain the normal acidity required in the stomach for starting the digestion of ingested proteins. Chloride is also important in maintaining water, osmotic and acid-base balance in the human body. Chloride deficiency is generally associated with alterations in sodium metabolism. Its deficiency retards the growth, makes the memory weak and develops psychomotor disturbances. The deficiency of chloride can be cured by simple administration of chloride as it is taken largely as sodium chloride/common salt. Its level may drop in the body if intake of salt is restricted. In such conditions the kidney becomes efficient to reabsorb chloride excreted through urine.

Sulfur

Sulfur is present in the body to the extent of 0.25% of body weight. Every cell of the body contains sulfur. It occurs in sulfur-containing amino acids and in almost all proteins. It is important in enzymes activity and in the transfer of energy. It has several metabolic functions in human body. A proteineous diet can provide the required amount of sulfur depending on the quality of protein.

Magnesium

Magnesium plays critical role in maintaining normal cellular function because it helps in maintaining structural integrity of the mitochondrial membrane. It is an essential part of all the soft tissues and bones and acts as a co-factor for several enzymatic reactions. Magnesium is important for blood and skeleton system and plays a significant role in postmenopausal osteoporosis since it is involved in the regulation of calcium metabolism, in the synthesis of vitamin D and in maintaining bone integrity. Magnesium deficiency in the body decreases osteoblastic and osteoclastic activity and causes a cessation in bone growth, leading to osteopenia and bone fragility. Inadequate supply enhances the progression of osteoporosis. Magnesium supplementation increases bone mineral density.

Magnesium balance in the body depends upon intestinal absorption and urinary secretion. When magnesium intake is low the kidney absorbs almost all the magnesium. Losses of magnesium through urine increase with the use of diuretic and alcohol. Its deficiency may occur in infants with protein energy malnutrition. Intake of low magnesium water may increase the risk of magnesium deficiency. Gastrointestinal disorders also impair magnesium absorption. Prolonged diarrhoea may lead to severe magnesium deficiency and hypomagnesaemia. Alcoholism increases the hypomagnesaemia. Increased level of intestinal fat in combination with magnesium forms insoluble fats and reduces the intestinal absorption of magnesium. Prolonged starvation also causes serious deficiency of magnesium. Hypomagnesaemia develops in 7% of the patients suffering from severe diabetic ketoacidosis. In about 30% of cases acute pancreatitis may result in moderate to severe hypomagnesaemia.

All unprocessed foods contain magnesium since it is present in chlorophyll. Green vegetables are the principal source of magnesium. The highest concentration of magnesium is found in whole seeds like nuts, legumes, and cereal grains. Magnesium is present just below the peel, thus, more than 80% is lost by removal of peel. Meat and dairy products are poor in magnesium. An average weight individual should consume 350 mg/day.

Essential Micro-Elements

Iron

Iron is essential metal in hemoglobin of the red blood cells and also important in the energy transfer system in respiration. It is a best carrier of oxygen in the body. Its deficiency causes anemia, spoon shaped nails, frequent exhausting (fatigue) and pale skin, eyes and mouth/face. Several studies have shown that iron deficiency is more prevalent in vegetarians than non-vegetarians of childbearing age, despite the enhanced absorption conferred by high intakes of vitamin C. Children are more prone to iron deficiency, thus, the food containing haem iron should be incorporated into the diet of infants by 6-8 months of age. Volunteers with borderline anemia as a measured by blood hemoglobin are less able to concentrate than volunteers with higher hemoglobin. Decreased ability to concentrate may be an early indicator that an individual's iron levels are declining. These deficiencies adversely affect the physical and mental development, thus, mental performance tests may be simple way to detect shortage in dietary iron well before other signs of deficiency might show up in blood or urine. A low attention span is corresponded with a subsequent decline in iron levels in the body. Iron deficiency and anemia caused by iron deficiency are still relatively common in pre-menopausal women. Green leaves can easily cure nutritional anemia. An individual requires 10 mg/day, and most of the iron can be had from green leafy vegetables. It can also be obtained from amaranth, Bengal gram leaves, spinach, lettuce, mustard greens, giant spine gourd, pigweed, drumstick leaves, cabbage, peas, beans, tomatoes and water melon.

Copper

Copper enzymes are involved in a variety of metabolic reactions and synthesis of essential compounds. It plays role in transporting iron to sites where hemoglobin synthesis can occur. Copper is a necessary mineral element for tyrosinase activity; its lacking in body impairs iron metabolism, causing anemia and bone loss. The major loss of copper is through stool and trace amount is lost through urine, nails, hairs, sweat and skin. Mild copper deficiency may be involved in the etiology and progression of osteoporosis, and high copper levels have been noticed to induce osteoporotic changes. Copper deficiency aggravates bone loss and even supplementation of this element does not reverse it. Copper deficiency has been observed in prematurely borne infants with signs of diffuse osteoporosis but gets disappeared after supplementation of this mineral element. Supplementation dietary intake with high amount of ascorbic acid may have a negative impact on the copper status. Amino acids and other organic acids occurring in foods stimulate copper absorption. Dietary protein generally enhances the availability of copper ions, whereas, dietary fructose strongly depresses the availability of copper. Copper deficiency has been detected in patients who used zinc as supplement for long time. Zinc produces antagonistic effect on copper availability. Copper deficiency may also occur in infants with prolonged diarrhea, mal-absorption disorders, who have undergone bypass surgery of intestine and who survive on parental nutrition for a long time.

Zinc

Scientists have known for decades that zinc is essential for normal bone metabolism and for maintaining memory and good health. Low zinc leads to faulty memory. After only weeks on a low-zinc regimen the ability to recall the words gets slower down. The men who slowed the most in recalling the memory also have the greatest decrease in blood levels of zinc. Cognitive changes are evident before the biochemical changes occur. It is said to influence every general function of the body, as it is a significant part of the mechanism that turns genes on and off. Genes that are turned on prompt the body to make proteins, enzymes, or hormones; genes that are turned off do not. In healthy individuals, at any given time, some genes are turned on and some are turned off. Without zinc, some genes would not turn on and off properly. Zinc is also a very important component of at least eight-enzymes needed for healthy skin and proper functioning of immune, nervous and digestive systems.

To get zinc from foods into the cells that need it, the body relies on proteins known as zinc transporters. These specialized proteins also take excess zinc out of cells to prevent toxic buildup. Zinc is absorbed into the blood from food in small intestine, and zinc circulating in the blood needs to be transported into cells. The studies indicate that men with prostate cancer have dramatically lower levels of zinc in their cancerous cells than their cancer-free counterparts. Indeed, healthy individuals have more zinc in their prostate than anywhere else in the body; results suggest that zinc supplements may protect men against prostate cancer.

World over more than half of the population is taking diet deficient in zinc, thus, mild deficiency of this element is found existing even in healthy infants, preschool children, school children, pregnant and lactating women, and elderly. In regions where grains and beans make up a significant component of the daily diets, zinc deficiency is a special concern, since zinc in grains and beans is less available to the body than the zinc in beef, pork, and in red-meat chicken. Universally, iron supplementation is common for pregnant women for fetal growth and also for milk production during breast-feeding, however, these iron supplements taken with a meal inhibit zinc uptake during these crucial times. Studies show that the 100 mg iron supplements interferes with the uptake of zinc but 50 mg iron supplements do not. Phytate is another essential body nutrient, but this element in grains such as corn and rice is thought to interfere with the uptake and use of zinc. Foods that are good sources of zinc are beans, whole grains, shellfish, red meat and dark-meat poultry.

Cobalt

An adult human body contains 1.1 mg of cobalt. In human body liver, kidney and bones relatively have higher content of cobalt than other tissues of the body. Liver contains nearby 0.034 μg/g wet weight. No plant contains vitamins B_{12}. It is synthesized by the bacteria in digestive tract and for its synthesis bacteria requires cobalt. The absorption of this nutrient in colon where the bacteria form vitamin B_{12} is found minimum. Humans meet their cobalt requirement largely from food of animal source. Seafood is considered the richest source of cobalt that contains 1.56 μg/g wet weight. The other sources of this nutrient are cabbage, lettuce, spinach, beet leaf and watercress. Cereals and dairy products are poor sources of this nutrient. An average human consumes about 0.3 mg/day. The body may not accumulate this nutrient. Although it is well absorbed in the intestine but most of it (0.26 mg/day) is lost from the body through urine.

Molybdenum

Human body contains a low concentration of molybdenum. It occurs in all tissues and fluids of the body. Most of the molybdenum is present in liver, kidney, adrenal gland and in bones. The body of a normal adult human about a mg. Body absorbs inorganic molybdenum very rapidly from the diet. It remains present in blood and urine as the molybdate ion. It is readily excreted in the urine *via* kidney. No molybdenum deficiency has been observed in humans under natural conditions. A safe and adequate intake of molybdenum per day is 75-250 μg, although the usual daily intake of this nutrient is about 50 to 350 μg. Its intake more than 500 mg/kg of food/water may produce toxicity and may cause diarrhea, anemia associated with failure of red blood cells to mature and high blood uric acid. The deficiency of this element can be observed in patients with an inborn error and patients on total parental nutrition. The inborn err on in clouds congenital abnormalities, feeding difficulties and failure to thrive. The good dietary sources of molybdenum are dairy products, dry seeds of legumes, organ meat (kidney, liver), grain products, and some green leafy vegetables, whereas, fruits, berries, and moist root or stem vegetables are poor sources of molybdenum.

Selenium

Selenium is an essential trace element that helps to keep the immune system strong, and a potent antioxidant that protects the cells from oxidation byproducts known as super oxides. This element is needed for proper growth and reproduction, and found crucial to keep the thyroid glands active and function properly. It is necessary for proper vitamin E function required glutathione peroxidase activity. Selenium deficiency during pregnancy causes cystic fibrosis in infants. Recent studies suggest that selenium helps in fighting cardiovascular disease in those individuals whose arteries do not properly expand and contract since arteries are the stretchy vessels that carry oxygenated blood away from the heart to rest of the body. That expansion and contraction, or vascular responsiveness, is vital for maintaining a healthful flow of blood. It is thought that a selenium-containing enzyme may be involved in signaling the artery to expand. Selenium increased the antibody response to vaccinations and improved the growth of lymphocytes, a type of white blood cells. A study indicates that a very high level of selenium result in lower sperm motility- the ability of sperm to move or swim.

Selenium enriched broccoli may be the best source of an anticancer agent. Garlic, broccoli and Brussels sprouts store selenium in a form of selenium methyl selenocysteine (SeMSC) that appears to be most active against cancer. The body simply snips the end off this amino acid to produce the anticancer agent methyl selenol. Selenium salts (selenate and selenite) can prevent the steps that lead to cancer. The vegetables high in anti-oxidants and contain other substances shown to be active against cancer. The higher dose is representative of selenium level that reduces cancer risk in humans. High selenium broccoli always resulted in fewer pre-cancerous lesions than selenate did about one-third fewer at the 1 mg/kg dose, and the number of lesions decreased as the dose increased. This mineral reduces the incidence of cancer when taken in higher doses than most diets supply. However, the selenium salts in some supplements can be toxic to those individuals who ingested too much selenium. On the contrary, it is harder to get too much selenium through foods. Seafood, meats and grains from regions with selenium-rich soils are good sources of this mineral. Other sources include dairy products and vegetables. Different foods package selenium in different biochemical forms and the body uses these forms differently. Tomorrow's breads, pastas, and other flour-based foods could be specially enriched with selenium.

Manganese

On an average, a normal adult has about 12-20 mg of manganese. Its distribution is normal through out the tissues and body fluids. Liver and pituitary glands contain its highest concentration (2.5-3.3 μg/g) and muscles lowest (0.04-0.18 μg/g). Its absorption from the gastrointestinal tract is poor (about 3%) even in healthy individuals. Iron, calcium, fiber, phytates and sugar produce adverse effects on its absorption. Body regulates its accurate concentration by promoting the excretion of excess metal from the body through urine. Manganese is essential for the activation of glycosyl transferases that catalyze the synthesis glycosaminoglycans and glycoproteins. It acts as cofactor for a number of enzymes. Its concentration in tears is 30 times higher than that present in plasma. This, a mood alteration is controlled by manganese concentration. It is essential for the normal skeletal and connective tissue development. The deficiency of this nutrient causes weight loss, transient dermatitis, vomiting, slight reddening of hair and slow growth of hair. Supplementation of manganese gives very good response of these deficiency symptoms. The typical signs of its deficiency include impaired growth, and reproductive performance, skeletal abnormalities and defects in lipid and carbohydrate metabolism. A deficiency of manganese, during growth phase leads to shortened and thickened limbs, curvature of spines and swollen, enlarged joints. Women with osteoporosis tend to have low blood manganese concentrations. Its deficiency may alter the carbohydrate metabolism and impair the insulin synthesis. Abnormal lipid metabolism is also characteristic of manganese deficiency. Manganese is essentially required for the activity of enzymes necessary for the maintenance of skin integrity. Its deficiency leads to dermatitis, decline in total and high-density lipoprotein and serum cholesterol, and an increase in serum calcium, phosphorus, and alkaline phosphates. During deficiency the manganese stored in bones is mobilized. This dissolution of bone to release this element would also release calcium and phosphorus in blood. An adult requires 2-5 mg manganese per day, whereas, children require 0.3-3 mg per day. Higher doses of manganese (mg/g of food) produce toxic signs. Manganese interferes with iron absorption. On the other hand, iron deficiency in human body results in increased manganese absorption and can make an individual vulnerable to manganese toxicity. Manganese can be made available from grains, cereals, fruits and vegetables. Tea is the major source of manganese, although dairy products, meat, poultry and fish contain very little amount of this element.

Iodine

A normal adult body contains 20-50 mg, and the daily body requirement for iodine is 0.1 to 0.2 mg/day. The concentration of this element in thyroid gland is more than 100 times that in the muscles and 10000 times that in the blood. It is necessary for the activity of thyroid glands that synthesize thyroxin hormone. Fetal survival and development are sensitive to deficiency of iodine. Brain development in fetus is also affected when the level of iodine intake by the mother is less than 25% of normal. Levels less than 50% of normal are associated with goiter.

In food iodine is present in the form of iodide and rapidly absorbed by the stomach and upper part of the small intestine. About 33% of the absorbed iodide is taken up by the thyroid gland and the rest 67% of the absorbed iodide is excreted by the kidney in the urine. The thyroid hormones stimulate cell oxidation and play a pivotal role in the regulation of basal metabolic rate of the adult and the growth and development of children. When iodine deficiency takes place in the body, the thyroid gland attempts to compensate this deficiency by increasing its secretion as a result the thyroid gland starts enlarging to the point where a large nodule is visible in the neck, this condition is known as goiter.

It can be prevented by an intake of iodide salt in the foodstuffs. The symptoms of its deficiency in infants are large birth size, sub-optimal body temperature, an enlarged tongue, lethargy, thick and dry skin and mental retardation. Despite of adequate intake the deficiency of iodine may occur. Some kinds of foodstuff contain those substances that inhibit the transport and utilization of this element by thyroid gland. Such substances are known as goitrogens. The plants of cruciferae family are rich in glucosinolates. During digestion with the help of myrosinase enzyme, these glucosinalates are degraded to isothiocyanate, which is real goitrogen. An adult body requires 150 μg iodine per day. During pregnancy and location period a woman requires an additional 25 and 50 μg, respectively. Intake of iodine more than 50 mg/day disrupts all thyroid functions. It's excess an also cause goiter. Vegetables are generally poor in iodine content. Seafood is rich source of iodine, whereas, dairy products, meat and eggs are source of iodine depending on animal feed composition. Vegetables such as tomato, sweet pepper, carrot, onion, garlic, beet, *etc.* contain an appreciable amount of iodine, and the vegetables grown in iodine rich soils contain more iodine.

Chromium

Chromium is present in the whole body. An adult male body contains 6-10 mg. In infants the chromium level is about 10 times less than the adults. A well-known biochemical role of this mineral nutrient in human body is to potentiate the actions of insulin. Much lower concentration of insulin is required in the presence of biologically active forms of chromium, which are known as nicotinic acid and usually amino acids and glutathione. It also plays role in nucleic acid metabolism and increases liver DNA concentrations. It is also useful in the metabolism of carbohydrates, lipids and proteins, and acts along with insulin in glucose metabolism. The deficiency of this mineral is common in humans. Chromium supplementation brings improvement diabetic patients. Supplementation of 50 μg of chromium in the form of chromium chloride for 12 weeks reduces the cholesterol level in the body and appears to improve glucose utilization and to decrease exogenous requirements of insulin in some diabetics. The American Association has recommended supplementation of total parenteral nutrition with 10-15 μg of chromium per day for stable adults. Subjects with low blood sugars also improve with supplemental chromium. In addition to improvements in blood sugar, hypoglycemic symptoms (drowsiness, blurred vision, shakiness and excessive sweetening), insulin binding to red blood cells and insulin receptor number also get improved following supplementation with 4 g of chromium chloride. The daily requirement of an adult is 0.05-0.12 mg.

Millions of people are believed to have type II diabetes, the kind of diabetes that usually develops during the middle age, and about 90 to 95% of diabetes cases are noticed unaware of it. In such subjects insulin is no longer as efficient at getting circulating glucose into body cells and converted to energy. As the hormone loses its efficiency, the pancreas starts secreting more to metabolize the same sugar load. Eventually, even the extra insulin is not enough, and blood glucose levels start to rise. Finally, the pancreas produces less and less insulin, and blood glucose goes out of control. A person is considered diabetic by National Diabetes Data Group guidelines if the subject has a circulating glucose level above 200 mg/dL within 2 hours after drinking a cocktail high in glucose. Refined sugars are very low in chromium, and they stimulate chromium losses.

Besides keeping body weight down and avoiding too much dietary sugars, an adequate amount of chromium, an essential nutrient, can improve glucose tolerance in majority of cases with elevated blood sugar, provided they still secrete insulin. Chromium is found effective only when the pancreas is still secreting some insulin. It does not seem to improve the efficiency of therapeutic insulin. A growing number of physicians are becoming aware of chromium's ability to keep blood sugar low and recommending that their patients consume more. Though chromium supplementation also increases blood cholesterol and triglyceride levels as it takes part in fat as well as in sugar metabolism, thus, it also increases blood sugar in patients with low levels, or hypoglycemia. Ingestion of more sugar causes more loss or use of chromium. It is also found that tiring exercise can drain the body of stored chromium and can cause physical trauma.

In affluent societies, consumption of more and more processed foods and the recent switch over to nonmetal cookware reduce the amount of chromium in diets. The National Research Council's estimated safe and adequate intake is 50 to 200 μg per day. A normal diet contains about 15 μg of chromium per 1000 calories. Based on this ratio, one should take ≥ 3000 calories a day to get 50 μg of chromium. Although it's very difficult to get enough chromium through the diet but if a good diet is taken, not much chromium supplement is needed. Chromium functions as a nutrient, not as a therapeutic, however, the people who already have plenty of chromium in their body may not get benefits of taking more of a nutrient to insulin levels. It is involved in glucose and energy transformations, thus, an energy boost is noticed when chromium tablets are taken by those whose glucose levels are normal to start with. If inadequate chromium does not cause the problem, then chromium supplement is not going to help at all. Before the availability of insulin, for diabetic patients physicians were use to prescribe brewer's yeast, which is high in chromium. The best sources of chromium are spices like black pepper, eggs, meat products, cheeses, whole grains and unrefined brown sugar. Vegetables are the poor sources of chromium but most of the leafy vegetables particularly crucifers and root crops may provide a trace amount of this mineral element.

Fluoride

All human organs and tissues contain a trace amount of fluoride. An adult human body contains less than 1 g of fluorine, about 99% of which is present in bones and teeth. Fluoride is available more in water than foodstuffs. The soluble fluoride is absorbed in the range of 30-60%. Most ingested fluoride is readily taken up by the bones and teeth and rest 50% is excreted in the urine.

It has positive effects on skeletal and dental health. An appropriate level of it prevents dental caries. Fluoride is required for the transformation of osteocalcium phosphate apatite. Higher fluoride content in body increases the crystallinity of apatite. Fluroroapatite in tooth enamel is less soluble in acid and more resistant to cariogenic action of acids in the mouth. The fluorine content in outer layer of teeth enamel is more than 20 times that found in inner layers. Fluoride produces protective effect on teeth by inhibiting the synthesis of acid by bacterial enzymes in the plaque on the teeth of persons receiving fluoridated water. Intake of fluoride in appropriate amount during young age provides protection against osteoporosis in old age.

It helps in the retention of calcium accompanied by reduction in bone demineralization. It promotes positive calcium balances and causes remineralization of the skeleton. Its deficiency in body results in increased incidence of osteoporosis in old age and dental caries. The recommended dietary allowance for fluride is 1.5-4.0 mg/day for an adult person. Chronic ingestion of excessive amounts of fluoride is dental fluorosis and osteosclerosis. Fluoride is found present in minute amounts in all kind of foodstuffs and water. Vegetables, fruits cereals and meat contain about 0.2-1.5 ppm, seafood 5-15 ppm and tea 75-100 ppm.

It prevents dental decay but its excess causes loss of luster and roughness of teeth and bone disorders.

Ultra-trace Minerals

Silicon

All body cells contain silicon in the range of 3-60 μg/g wet weight. The loss of silicon through urine has been noted in the range of 12-16 mg/day. The excretion of silica increases with the rise in dietary intakes and kidney is the main excretory organ. It appears to be well absorbed. Its concentration remains constant in the blood even over a wide range of intake because it is excreted easily in both stool and urine. Silicon is essential for growth, calcification and in mucopolysaccharide formation. Silicon plays a regulatory role in the formation of connective tissue matrix and subsequent calcification. In whole body cells the calcium to silicon ratio is found 5: 1. The deficiency of this nutrient has not ever been noticed in humans. Silicon does not appear to be toxic in the levels usually present in the diet but silicon toxicity can be as silicosis (a lung condition) due to inhalation of airborne silicon oxide dust.

Nickel

The total amount of nickel in an adult body has been estimated about 10 mg and found distributed equally in the whole body. No specific concentration accumulates in a particular tissue. Most of the ingested nickel occurs unabsorbed and is excreted in the feces. The kidney is the primary route of excretion for absorbed nickel, which does not involve directly in biological functions of human body. However, it functions either as a cofactor or structural component in specific metalloenzymes. Nickel was identified in the structure of urease. It remains associated with DNA and RNA and helps in the maintenance of their structure. It plays a role as activator of arginase and DNAase. During pregnancy in later stage the serum level of this nutrient is absorbed significantly less, however, after delivery a greater than 15 folds increase occurs within 5 minutes and returns to normal level within an hour. The transient hyper nickelemia may be important to separate the placenta. If nickel deficiency is severe, growth and haematopoiesis are depressed. Iron utilization gets impaired. Nickel has been described as nutritionally essential but oral intake of nickel has been found to react with skin in some nickel sensitive individuals. Foods that are rich source of nickel include nuts and leguminous seeds. It occurs mainly in foods of plant origin. Cereals and vegetables are the good source. The canning of some fruits and vegetables increases their nickel content.

Vanadium

Vanadium is widely distributed in human tissues. Adult human tissues contain about 25 mg of vanadium. It is present in fatty tissues, blood serum, and bones and teeth. Vanadium may play role in the regulation of Na^+, K^+, ATPase, adenyl cyclase and protein kinase, and may affect iodine metabolism and thyroid function. Vanadium *via* respiratory system show signs of toxicity that includes depressed growth, diarrhoea and decreased appetite. The physiological function of vanadium is its apparent anticarcinogenic action. The most unabsorbed vanadium is excreted *via* feces. Urine is the major excretory route for absorbed vanadium and major part of the vanadium is retained by bones. Food rich in vanadium are mushrooms, dill seed, black pepper, spinach and parsley but fresh fruits and vegetables contain little amount of this nutrient.

Boron

In human body about 48 mg boron is present. Earlier it was considered essential for plants. However, later findings indicate that boron is nutritional importance to human. Its deprivation affects the functions or composition of several body systems including skeleton and brain, and increased the urinary excretion. The element affects mineral metabolism *via* a regulatory role involving a hormone at cell membrane level. Supplementation with 3 mg boron in post-menopausal women leads to a marked increase in their serum levels of 17-β-estradiol and testosterone, and a decrease in their urinary excretion of calcium and magnesium. The increase in steroid levels in serum is of great importance because 17-β-estradiol is the most biologically active form of human estrogen, and testeoporosis is its precursor. Regulation of estrogen is the best mean to slow down the loss of calcium from bone that generally occurs after menopause. Boron has been found very important nutritional factor that slows down osteoporosis by maintaining a relatively high serum estrogen level. Boron in food in the form of sodium borate or boric acid is readily absorbed and excreted largely in the urine. The deficiency of this element has not yet been reported in humans but its intake/supplementation shows positive effects on mineral metabolism. The daily requirement of this element in human body is at least 1 mg/day. Its concentration above required level might cause mild toxicity, which includes nausea, vomiting, diarrhoea, dermatitis, and lethargy.

There are some evidences that boron is also important in human nutrition, especially under conditions of nutritional or metabolic stress, that it contributes to maintaining cell membrane function and stability and that it is involved in enzymatic reactions. Boron works closely with calcium, magnesium, and vitamin D in the preservation of bone mass and the prevention bone demineralization and stimulate the immune system and inflammatory and hormonal responses. Therefore, it may have significant roles in the prevention and treatment of osteoporosis and osteoarthritis. It positively affects calcium and magnesium metabolism and may be needed for energy substrate use.

Tin

Every tissue of the body except brain contains tin and an adult human contains about 14 mg tin. The organic forms of tin compounds are soluble in lipids and found in tissue lipids. In biological system it can take participation in oxidation-reduction reactions. Tin may be an essential trace element in human nutrition, because it is poorly absorbed and poorly retained by the body tissues. Canned vegetables, fruit juices and fish can provide as much as 35 mg tin per day, whereas, a diet composed of fresh meat, cereals and vegetables can supply 1 mg tin per day. The richest sources of dietary tin are foods that have been contaminated by packaging and processing utensils containing tin. Nuts are relatively rich in tin. A variety of nuts contained from 0.3 to 3.5 μg/g dry weight.

■ ■ ■

4 Wonders of Vitamins

M.K. Rana

Department of Vegetable Science
CCS Haryana Agricultural University, Hisar-125 004, Haryana, India
E-mail: <mkrlotus@gmail.com>

Vitamins are potent organic compounds, which are required by the human body only in trace amounts for normal metabolism, and all these are different in their chemical structure and physiological or biochemical actions. Vitamins never supply energy to the body although they stimulate a variety of biochemical and physiological processes that are essential for survival, however, the human body cannot synthesize these vitamins it self, thus, they must be supplied to the body as nutrient precursors for their synthesis. Including a variety of foodstuffs in the diet plays an important role in providing essential vitamins that are inevitable for preserving the body immune function and protecting the body from the prevalence of infectious diseases and cancer. Supplementing all the vitamins, the life can be prolonged.

Based on their solubility in different mediums the vitamins can be divided into two major groups. The vitamins A, D, E and K are fat-soluble, and vitamins of B group and vitamin C are water-soluble. In foods, the fat-soluble vitamins are found associated with lipids and are generally better absorbed in the presence of dietary fats. Hence, the conditions that are unfavorable for the absorption of fat also impair the absorption of fat-soluble vitamins. The body stores the unutilized amount of fat-soluble vitamins and utilizes them as per its requirement, thus, the intake of these vitamins every day is not required. On the other hand, body does not store water-soluble vitamins in significant amount. These vitamins unlike fat-soluble vitamins are not found associated with dietary lipids and their absorption is not impaired by the conditions that are unfavorable for the absorption of fats. Since these water-soluble vitamins are not stored in the body appreciably, the unconsumed quantity is usually excreted from the body through urine so these vitamins should be supplemented in daily diet to avoid their deficiency.

The fat-soluble vitamins like vitamin A and D sometimes may be toxic but vitamin E is substantially less toxic than vitamin A and D. Since body stores the surplus vitamin A the intake of it higher than recommended dietary allowance could destroy cell membranes and produce toxic signs like drowsiness, headache, vomiting, appetite loss, hair loss, scaly skin eruption, double vision and weakness, thus, the excessive use of fat-soluble vitamins should be avoided without medical prescription. The water-soluble vitamins can also be toxic and can produce adverse effects but comparatively lesser than the fat-soluble vitamins as their higher levels in the body can interfere with the action of other nutrients. The level of toxicity of any vitamin depends on the supplement dose and body reaction of an individual.

In recent years, a common practice is to enrich the food products with certain vitamins so now a day the foods rich in all the vitamins essentially required by the body are available in the market. The objective of supplementing these vitamins from external sources is to make-up the losses that usually occur while processing the food and to enrich the foodstuffs that are as usual low in vitamins. The break fast cereals are often supplemented with B-complex vitamins. In advanced countries, milk is generally supplemented with vitamins. The therapeutic significance of all the vitamins on the whole required by the human body for various functions have been described below along with their various sources of availability.

Vitamin A

The most commonly known vitamin A is retinol, which is also known as vitamin A alcohol and vitamin A_1, and the other forms of vitamin A are retinal (retinaldehyde) and retonic acid- a metabolite of retinal. Retonic acid supports growth of animals that are as usual deficient in vitamin A, however, it does not support vision, or fertility since it is not stored in liver. At ordinary cooking temperature, it remains stable but it becomes sensitive to oxidation after dissolution in dilute solution in the presence of oxygen and light, especially at high temperature. On oxidation, the retinol is converted to retinal and further oxidation results in the formation of retonic acid. However, the presence of vitamin E in the medium inhibits its oxidation. Dehydroretinol (vitamin A_2) that occurs in fresh water fishes and amphibians can be converted to retinal in *in-vivo*. The animals have the mechanism to convert the precursors into the active form of vitamin A so the foods of animal origin (milk butter, cheese, egg yolk, liver, and marine fishes) contain vitamin A in achievable form. Foods of plant origin are rich in carotenoids, which are known (as pro-vitamin A) as precursors that are on ingestion converted in the body by an enzyme-catalyzed reaction to active form of vitamin A.

Carotenes, the members of a family of fat-soluble yellow, orange and red pigments, usually found in fruits and vegetables may be essential for good health. The body can convert the carotenes present in the form of α-, β- and γ-carotenes and cryptoxanthin (pro-vitamins) into vitamin A, which is required for proper growth and reproduction and also for prevention from certain infectious and non-infectious diseases, however, supplementation of β-carotene in diets of subjects who are deficient in vitamin A may not be sufficient to prevent the blindness and death caused due to lack of vitamin A. This vitamin is essential for vision in dim light. It develops resistance against infection, increases longevity and decreases senility. It provides protection against cold and strengthens eyesight. Epithelial cells secrete mucous (a viscous fluid) that keeps the lining of gut, respiratory tract, nails, hair and eyes moist. Vitamin A abets the growth of the epithelial cells, present not only in the outer skin but also in other outer tissues. It is also essential for the development of teeth and enamel forming epithelial cells. The light receptors of eyes contain rod and cone cells of the retina. Their outer segments have light sensitive pigments, which require vitamin A for their synthesis and normal functioning. The pigmented epithelium cells that store vitamin A surround the outer segments of rod and cone cells. The deficiency of retinal produces deleterious effect on structures of epithelial cells that are found in the interior linings of all the body-opening parts. These cells also form the outer protective layer of the skin. Most of these cells secrete mucus but the deficiency of this vitamin reduces the mucus secretion.

Oxidation leads to the release of free radicals that have been found associated with increased incidence of cancer, heart stroke, arteriosclerosis and cataracts. The concentration of compounds in blood known as carbonyls remains nearly 50% higher during low-carotene phases, and an indicator of oxidative damage called thiobarbituric acid reactive substances is increased in plasma about four times during depletion. Some of the carotenes act as antioxidants but not all carotenes can provide protection against oxidative damage. Epidemiological studies demonstrate that β-carotene has numerous functions that are independent of its antioxidant role and beyond serving as vitamin A precursors. Evidences show that other carotene oils occurring in fruits and vegetables may have anticarcinogenic properties, and a possible mechanism for potential anticarcinogenic effects of β-carotene is its ability to scavenge reactive oxygen species that cause oxidative DNA damage. The intake of carotenoids is associated with a diminished risk of various cancers. High dietary intake of lutein and zeaxanthin has been found associated with decreased risk of lung cancer and of β-cryptoxanthin with decreased incidence of cervical cancer. Some evidences further support the significance of carotenes as antioxidants that may reduce the risk of cancer, heart stroke, arteriosclerosis and cataracts. It has also been reported to influence the body immune function favorably.

Night blindness (inability to see in twilight, eyes sensitivity to bright due to softening of cornea) is the first sign of vitamin A deficiency. Xerophthalmia, which is characterized by drying of eye, is an eye disease due to deficiency of vitamin A. In its absence, the lachrymal cells in eyes stop secreting tears. If this condition is not cured timely, may cause infection and ulceration. Its deficiency results in failure of bone growth, formation of stones in kidney and gall bladder, pimples on face and abnormally malformed dry and rough skin due to development of epithelial cells and retardation of children growth. Its deficiency increases susceptibility to bacterial infections in respiratory and digestive systems, thus, this is known as anti-infectious vitamin. Its deficiency also enhances susceptibility to carcinogenesis. The loss of appetite, which is attributed to a change in the taste buds, is due to deficiency of vitamin A. Because of too low carotenes and other carotenoids, abnormal menstrual cycles may develop in most of the women. Perhaps a higher amount of β-carotene is required to prevent menstrual cycle abnormalities, or other compounds in addition to carotenoids found in highly pigmented fruits and vegetables might be needed. The deficiency of vitamin A decreases the synthesis of estrogen hormone in female and also interferes with placental development. The linkage between vitamin A and thyroid hormones is well documented. Low-carotene regimens may stimulate the increases in a key thyroid hormone, thyroxine.

The human body daily requires 5000 IU vitamin A (0.6 μg β-carotene = 1 IU vitamin A) that can be supplied through meals from sources such as beef, chicken, turkey, pork, fish, liver oils, *etc.* Eggs, cheese, butter, ghee, curd and whole milk are rich in vitamin A so people eating any of these foods need to be sure that they are getting an adequate supply of this vitamin from other sources. These findings are vital for people cutting back on the amount of meat and dairy products they eat. Dozens of fruits and vegetables having bright yellow, orange and dark-green color provide lot of carotenes, however, the individuals differ in their ability to absorb β-carotene and convert it to vitamin A. Several fruits such as dark yellow-fleshed mango and papaya, apricot, cantaloupe, winter melon and yellow-fleshed musk melon are also good sources of β-carotene. The vegetables that are good source of carotenes include deep-orange carrots, yellow-fleshed sweet potatoes, amaranth, spinach, beet leaf greens, lettuce, collards, parsley, mustard greens, fenugreek greens, drumstick leaves, portulaca, coriander greens, green onion, green pepper, green capsicum, kale, cabbage, broccoli, tomato, peas, pumpkin, winter squash, turnip, arum, yam and tapioca. Leafy vegetables dried in sun lose their carotenes but cooking does not destroy them. Fresh form of vegetables supplies more carotene than the staled ones.

B-Group Vitamins

Homocysteine, a toxic amino acid known as bad boy, is rivaling the status of high blood cholesterol as a major contributor to cardiovascular heart disease and to other maladies as well. The research findings reveal that most cases of mildly elevated homocysteine in an elderly population are related to low-intake of vitamin B_2 (riboflavin) and vitamin B_6 (pyridoxine), to high-intake of alcohol and caffeine, and to chain-smoking and hypertension. The research findings also demonstrate that the body specially requires folate, vitamin B_6, and vitamin B_{12} (cyanocobalamine) to convert homocysteine to other amino acids that are nontoxic to the lining of blood vessels. Blood folate levels appear to have the special influence on the levels of homocysteine in human body. The elevated homocysteine, a key player in the loss of cognitive function (memory loss), has also been correlated with Alzheimer's disease, especially in elderly. Homocysteine itself may be very toxic to nerve cells. The B-group vitamins are involved in the synthesis of chemicals crucial to brain function. The survey reports exhibit that the high folate level in blood seems to be protected from memory loss even if the homocysteine level of the subject is high, hence, a subject in old age should keep his folate level up.

The men and women both who have a blood homocysteine level >12 micromoles per liter are found liable more than twice to have an experience of heart attack, however, the blood homocysteine level has not been related to heart attack in women who have not reached menopause, whereas, the relationship is established strong in men of the similar age-group. The findings support that women may be protected from heart attack by their high estrogen status as estrogen helps to keep the blood homocysteine levels low. Pre-menopausal women, those who are pregnant or use oral contraceptives have homocysteine level in their blood lower than men and postmenopausal women except those who take estrogen replacement. According to research reports, higher estrogen status is associated with a low mean serum homocysteine level, independent of nutritional status or muscle mass, which is a possible contributor.

Vitamin B_1 (thiamine)

Thiamine, an essential vitamin for growth and reproduction, was first isolated from rice bran and was considered vital for life and known as anti-beriberi principle. It acts as a coenzyme in the metabolism of carbohydrates, thus, its lacking prevents normal metabolism of pyruvic acid in respiratory cycle. For energy metabolism, all body cells need vitamin B_1 in the form of coenzyme. The secretion of HCl in the stomach and the digestive enzymes in gastrointestinal tract depends on energy supply as in absence of energy the secretary glands do not function in the digestion of food, the neuronal activity gets impaired, and central nervous system does not work properly. Depending on the extent of deficiency, the synthesis of lipids is declined and causes increased nerve irritation and pain. It is necessary for good appetite, normal digestion, lactation, and normal functioning of nervous tissues. The brain and nerve cells derive energy mainly from carbohydrates, the first food components to be affected by its deficiency.

The classic example of its deficiency in human is beriberi disease (dropsy). Acute continuous deficiency may lead to paralysis. Heart muscle weakness due to deficiency of thiamine causes cardiac failure. Its deficiency also results in loss of appetite, weight loss, and fall in body temperature. Many other problems, like gastrointestinal (anorexia and indigestion) and neurological disorders, may develop depending on severity of deficiency and stress factors. The depletion of thiamine in body also develops numbness of feet, heaviness of legs, fatigue and anesthesia due to accumulation of edema fluid in the legs, face and trunk. The common signs of thiamine deficiency are mental confusion, ataxia, anorexia, muscular weakness, peripheral paralysis and edema (wet beriberi). The development of edema is favored by carbohydrate intake and increased physical activity. The deficiency symptoms appear more severe and suddenly in infants than in adults. In Asia where polished rice is a staple food, a large population suffers with its deficiency, which is noticed most frequently among alcoholics and in individuals taking diet consists of white rice, white flour, or diet with low thiamine.

All foods of animal and plant origin contain thiamine in more or less amounts but the important source of thiamine are the seeds of plants. In cereals the thiamine is present mainly in germs and outer coat of seeds that are the major source of this vitamin because people eat cereals so often in big quantity, however, this vitamin often gets lost during milling and refining. The thiamine rich foods are sunflower seeds, wheat germs, legumes and water melon. Whole and enriched grains, asparagus, green beans, peanuts, and mushrooms are also good sources of it. All green vegetables (amaranth, beet leaf, lettuce, cabbage), roots (carrot, radish and turnip), green pepper, garlic, leek, pea, giant taro, globe artichoke, ripe tomato, musk melon, fruits and dairy products (except butter) have significant amounts of thiamine but cannot be said rich sources. Most vegetables per 100 g fresh-weight contain 0.05-0.06 mg thiamine. The best way to obtain good amount of thiamine is to eat a variety of foods. Some fishes contain thiaminease enzyme, which destroys thiamine, however, cooking the fish properly can destroy this enzyme. It can also be supplemented in diet with pharmaceutical sources. Being a completely soluble in water, the body may not store this vitamin, thus, question does not arise to be toxic. Even excess use of this vitamin may not be toxic but few cases of sensitization have been reported.

Vitamin B_2 (riboflavin)

Riboflavin, an orange-yellow color substance, presents in foodstuffs as flavin mononucleotide (FMN), flavin denucleotide (FAD) and free riboflavin, and all these forms of riboflavin can meet the body requirement. Lacto-flavin, a yellow-green florescent pigment in milk-whey, was recognized as milk form of riboflavin. It is noticed stable to heat in acidic and neutral medium but not in alkaline medium in solution form, it starts decomposing in the presence of light. It is necessary for cells to synthesize energy that helps in cellular metabolism. It is considered a part of enzyme system in the conversion of foods to chemical energy. Riboflavin is essential for growth and participates in multiple functions including formation of red blood cells in bone marrow, metabolism of fatty acids, synthesis of glycogen and the effectiveness of thyroid hormone in regulating enzyme activity. It is present in the eye retinal pigment where it helps in the adaptation of light. The vitamin B_2 appears to be non-toxic and no case of toxicity from excess intake of riboflavin has been reported. The excess amount of this vitamin is excreted primarily as free riboflavin in the urine to which it imparts a yellow fluorescent color.

The main effect of riboflavin deficiency is on lipid metabolism. Its deficiency causes soreness and burning of lips, mouth and tongue, pellagra (cracking of skin), alopecia (falling of hair) and loss of appetite and body weight. Some other signs are swollen red tongue, fissuring at mouth and lips corners, seborrhatic dermatitis, especially at nasal-labial folds, eyelids, and ears, and skin abnormalities around the valve and anus and at the free border of prepuce. Riboflavin deficiency may cause anemia because of secondary iron deficiency since its deficiency reduces the absorption of iron. The anemia associated with riboflavin deficiency is a result of impaired vitamin B_6 metabolism due to deficiency of riboflavin. The depletion of vitamin B_2 decreases oxidation of dietary vitamin B_6 to pyridoxal. A riboflavin deficient subject is relatively resistant and keeps a lower burden than adequately nourished individuals.

Riboflavin is obtained commercially from a culture medium of several molds, which produce this vitamin in plentiful amount. It can be made available in diet from plants and animal sources but the most important food sources of this vitamin are milk, curd, and cheese, therefore, one-fourth riboflavin of our total body requirements comes from milk and milk products and rest of the riboflavin comes from the diet containing white bread, eggs, meats, mushroom and green leafy vegetables. Fruits and vegetables are normally considered the moderately good sources of this vitamins but are not consumed in plenty amounts to meet the entire body requirements. Asparagus, beet leaf, spinach, lettuce, celery, chicory, endive, parsley, portulaca, mint, pigweed, watercress, radish leaves, beetroot leaves, cowpea leaves, hot pepper, sweet pepper, broccoli, cauliflower, kohlrabi, Brussels sprouts, giant spine gourd, garlic, winged bean and okra are considered the good sources of it, however, leafy vegetables are considered high in riboflavin and other vegetables are low in riboflavin. Exposure to ultra-violet light causes breaking of riboflavin rapidly, thus, the foodstuffs rich in riboflavin should be protected from direct exposure to ultra-violet radiation so should be packaged properly to avoid sun exposure.

Vitamin B_3 (niacin or nicotinic acid)

Niacin plays an important role in many synthetic reactions since it is part of coenzyme system in cellular metabolism. Every cell in the body contains niacin. The deficiency of this vitamin in human body causes pellagra disease (skin roughness or eruption), disorders of gastrointestinal tract and central nervous system, and abnormality in tongue and cellular respiration. The initial symptoms of its deficiency are the burning pain in the oral cavity and tongue as a result it gets red and swollen. Pigmented and scaly skin lesions are observed on body parts that are exposed to sun. The other symptoms of its deficiency are anorexia, nausea, vomiting, abdominal pain, constipation, insomnia, irritability, vertigo, headache and hallucination.

The daily body requirement of niacin is 19 mg, and about half of it is met from the conversion of dietary tryptophan to niacin. It is well established that tryptophan is the precursor of niacin. Most of the niacin is obtained from the diet rich in cereals, however, corn based diet is poor in niacin as the protein (zein) present in corn is very low in tryptophan, which is a major source of this vitamin. The break fast cereals made of grains have been fortified with niacin. As much as 90% of this vitamin is present in bran, thus, whole grains contain more nicotine acid than the flour. Excessive intake of dietary protein supplies the niacin. Animal protein is especially rich in tryptophan. Eggs and milk also contain abundant of tryptophan, hence, indirectly contribute a good amount of niacin. Being a heat stable vitamin a little amount is lost during cooking. The rich sources of this vitamin are the wheat bran, mushroom, peanuts, and asparagus. Coffee and tea supply little amount of this vitamin. Normally vegetables and fruits are the poor sources. The vegetables containing niacin in trace amount are beet leaf, amaranth, spinach, fenugreek, lettuce, bitter gourd, giant spine gourd, pointed gourd, bottle gourd, pumpkin, green pepper, radish, carrot, pea, cowpea, hyacinth bean, okra and sweet potato.

Vitamin B_4 (biotin)

Biotin is a sulfur-containing vitamin, which is essential for human body. The deficiency of this vitamin is not seen so frequently in humans since the body requires it in minute quantity and the dietary food supplies it readily and sufficiently. Biotin deficiency in humans is characterized by anorexia, glossitis, mental depression, nausea, loss of appetite, vomiting, alopecia, dry scaly dermatitis, and an increase in serum cholesterol and bile pigments. Long-term total parenteral nutrition causes hair loss. Biotin deficiency can be developed in humans by taking low biotin diet containing large amounts of un-boiled raw egg white continuously for ten weeks. The deficiency may also develop due to a low content of biotin in milk or defective digestion and persistent diarrhea. Urinary excretion of biotin generally found in pregnant women, and some alcoholics. The elderly also develop deficiency. In such individuals, the symptoms are seen as alopecia, irritability, waxy pallor, lethargy and organic acidosis.

The major sources of biotin are the various dietary foods and synthesized by the gastrointestinal tract microflora. Food contains biotin in two forms, *i.e.*, free form and bound to protein. The dietary proteins containing bound biotin produce biocytin. The biotinidase enzyme presents in pancreatic juice and intestinal mucosa cleaves releases biotin from biocytin. The intestine absorbs only free biotin. In general, most of the meats, grains, fruits, and vegetables are poor in biotin, however, a trace amount of this vitamin can be met from meats, soybean, rice bran, peanuts, vegetables such as cauliflowers, mushrooms, carrots, tomato, spinach, beans, peas and fruits like orange and apple. It can also be met from milk, cheese and eggs.

Vitamin B_5 (pantothenic acid)

The word pantothenic has been derived from the Greek word *Pantos* means everywhere. Pantothenic acid is considered an essential B-group vitamin for humans. Being essential component of coenzyme-A, it plays a basic role in the metabolism of carbohydrates, fats and proteins. It is also required for the synthesis of cholesterol, steroid hormones and protoporphyrin (component of hemoglobin). Human body needs pantothenic acid but its deficiency does not occur in humans due to adequate dietary intake. Intestinal microflora also synthesizes and contributes to the human needs but do not contribute the required adequate amount of pantothenic acid.

Severely malnourished individuals show signs of deficiency diseases. The symptoms include numbness, tingling in toes, burning pain in feet and neurological and mental disturbances. The relief in burning foot syndrome can only be achieved by a therapy with pantothenic acid but not by thiamine and niacin. The other sings of deficiency, like mental depression, vomiting, burning cramps, abdominal distress, paresthesia of hands and feet and sleep disturbances develop in different individuals. Its deficiency lowers down the blood serum cholesterol and alters the response to corticotrophin. The deficient individuals have a functional impairment of gastrointestinal, reproductive, hormonal, and neuromuscular systems. Dietary deficiency of this vitamin results in a variety of abnormalities, like retarded growth, infertility, abortion, frequent neonatal deaths, hair abnormalities, skin pigmentation and sudden death. Pantothenic acid depletion in human body also results in muscle weakness, fatigue, insomnia, tenderness in heals, ulceration, increased susceptibility to upper respiratory tract infections, increased insulin sensitivity, gastrointestinal complaints including severe pain with depressed gastric acid secretion in response to insulin and gastrin.

Pantothenic acid is universally present in most natural foodstuffs. Yeast, mushroom, peanuts and eggs are the richest sources of pantothenic acid. Meat and milk are also the important sources of this vitamin because they are consumed in large amounts. Cereals have substantial amount of this vitamin, however, about 50% of the total amount gets lost during milling. Vegetables like broccoli and sweet potatoes contain moderate amounts of this vitamin.

Vitamin B_6 (pyridoxine)

Vitamin B_6 is naturally present in three forms, *i.e.*, pyridoxine, pyridoxal and pyridoxamine, which are equal in biological activity. The synthetic commercial form of this vitamin is pyridoxine hydrochloride, which is sensitive to light. This vitamin is needed for normal metabolism of tryptophane to nicotinic acid. It is also considered involved in food metabolism, in functioning of nervous tissues and also in anaemia. Its deficiency causes ulceration at the corners of mouth and over the tongue. Deficiency for a long period is required to observe its effects but the symptoms can be produced by the use of structural antagonists. Skin lesions around the eyes, mouth, and nose can be seen. Depression and confusion may occur in individuals taking diet deficient in pyridoxine. In infants, inadequate growth, nervousness, irritability, and seizures have been noticed due to deficiency of pyridoxine.

Pyridoxine is widely present in foodstuffs and synthesized in large amounts by the intestinal microflora, thus, the deficiency of this vitamin is rarely seen. Pyridoxine is stored in muscle tissues of animals so meat of animals is the best source of this vitamin. In meat, it is present in readily available form than that in foods of plant origin. The whole grain cereals including bran are also good source of pyridoxine, however, it gets lost during milling and coking of grains. Nuts, peanuts, banana, soybean and leafy vegetables are the good sources, whereas, other fruits and vegetables are moderate in this vitamin.

Vitamin B_9 (folic acid)

In metabolism of amino acid and synthesis of nucleic acid, folate acts as coenzyme that transports single carbon fragment from one compound to another. Its deficiency leads to impaired cell division. Although the folate forms have variability in stability under various conditions but heat, oxidation and ultra violet light may make it inactive. This vitamin is very important for women, requiring 1 mg per day.

The deficiency of this vitamin is associated with low intakes, inadequate absorption, alcoholism, increased body requirement during pregnancy and especially associated with vitamin B_{12} deficiency, excessive excretion and longstanding diarrhea. Chronic folate deficiency develops large immature cells in entire gastrointestinal tract and contributes to reduced absorption capacity of tract and longstanding diarrhea. Its deficiency also disrupts the synthesis of white blood cells. The deficiency to be noticeable takes time as the body keeps the folate stored. Folate deficiency can easily be dig-nosed by examining the red blood cells. The major symptoms of folate deficiency are the inflammation of tongue and mouth, diarrhea, poor growth, mental confusion, depression, abnormal pigmentation and problems in liver function. The deficiency of folate is also characterized by glossitis, macrocytic anemia and gastrointestinal disturbances involving impaired digestion and absorption. Pregnancy increases the folate deficiency due to marginal intakes of this vitamin. Those who take vegetable and fruit juice or ingest fresh uncooked fruits and vegetables regularly maintain adequate folate status, however, who take diet consists of fine particles foods cooked for a longtime and people who fail to consume fruits and vegetables in sufficient amount because of either poverty or physical problem suffer with the deficiency of folate. Individuals taking drugs and those who smoke have increasing folate needs.

The foods of plant origin such as orange juice, green leafy vegetables (spinach, beet leaf, lettuce, cabbage), asparagus, broccoli, cauliflower, cowpea, French bean, wheat germ and seeds of legumes and sunflower, grain products fortified with this vitamin including bread flour, cornflakes, pasta, and rice, dried beans and peas, and most berries are rich in folate. The other less folate rich sources are baked products and eggs. Cooking and processing of food can destroy folate from 50-90%. Heating, oxidation, and exposure of food to UV-light destroy the folate being extremely susceptible. It is useful to eat fresh fruits and lightly cooked vegetables to obtain more folate. The cooking must be done in a minimum water and minimum time to retain more folate. Steam cooking is the best mean to retain the nutrients. Addition of ascorbic acid in food during cooking protects folate from oxidative destruction. Processed and highly refined food may not be good sources of folate unless supplemented. Ingestion of banana, lima bean and eggs increases the uptake of folate up to 70%, and supplements of folate 15 mg/day for one month may produce deleterious effects like sleeplessness, irritability and gastrointestinal symptoms.

Vitamin B_{12} (cyanocobalamine)

Vitamin B_{12} (cyanocobalamine) is imperative in the metabolism of folate, participates in several body reactions, and helps in the prevention of anaemia, thus, the deficiency of this vitamin also develops the deficiency of folate, and deficiency of any of the vitamins shows the same symptoms. Irreversible nerve destruction may develop before the anemia. Lacks of this vitamin in the body affect the blood forming tissues of the bone marrow and gastrointestinal tract. The symptoms of pernicious anemia caused by the deficiency of vitamin B_{12} are weakness, a red and painful tongue, weight loss, indigestion, anorexia, diarrhea and parenthesia in the extreme conditions. Mental slowness, memory loss, cognitive decline, walking difficulties, balance disturbances to nerve damage and mild dementia are also the common symptoms of vitamin B_{12} deficiency. Infants that are breastfed by vegetarian mother can develop vitamin B_{12} deficiency and long-term neurological problems like retarded brain growth, poor intellectual development and degeneration of the spinal cord. These problems may arise during pregnancy if the mother is suffering from the deficiency of vitamin B_{12}.

One reason of its deficiency is that the vitamin B_{12} in natural food sources such as meat, eggs and dairy products is tightly bound to the proteins and released by cooking, acidification in stomach and by proteolysis so unless the subject has enough stomach acid to break those bonds may not absorb adequate amount of vitamin B_{12}. The sources of vitamin B_{12} that are not bound to protein such as supplements and fortified cereals appear to be protective. The food in stomach liberates vitamin B_{12} by digestive action of gastric juice and that is bound with protein. This vitamin B_{12} complex form reaches the small intestine where pancreatic proteases release this vitamin from protein. In large intestine microflora synthesize vitamin B_{12} but the synthesis takes place far beyond from the site of absorption, thus, this vitamin produced by the intestinal microflora is not utilized. The absorption of vitamin B_{12} can be disrupted by the use of antacids that reduce acid production. Aging contributes to the gradual lessening of vitamin B_{12} absorption. In body, liver can store 5-12 mg of vitamin B_{12}, which remains stable in the body; only little amount of it gets lost, thus, injection of this vitamin once in a month is enough to prevent its deficiency.

Inadequate dietary intake may cause deficiency of vitamin B_{12}. Vegetarians living in highly hygienic conditions also suffer with the deficiency of this vitamin, whereas, the major cause of its deficiency is the stomach, intestine, and panaceas disorders that result in defective absorption of this vitamin. Dietary deficiency of vitamin B_{12} is seen rarely, however, 95% deficiency of this vitamin is due to defects in absorption, which is reduced due to lack of intrinsic factor (trypsin), intestinal infections and by gastrectomy. Atrophic gastritis particularly in old age results in B_{12} deficiency. The people who take supplements with multivitamins or regularly eat vitamin B_{12} fortified foods such as breakfast cereals more than four times per week are significantly less likely to have vitamin B_{12} deficiency.

Higher plant and animals are not capable to synthesize vitamin B_{12}, only the intestinal microflora are capable to synthesize vitamin B_{12}, thus, microbiological process of food is essential to synthesize this vitamin. Smokes may also cause a deficiency of vitamin B_{12} due to its conversion to non-absorbing form. Foods from animal sources such as fish, milk, and eggs are the rich sources of this vitamin but the foods of plant origin do not contain this vitamin, thus, vegetarians may have the deficiency of this vitamin. Water or food contaminated with soil bacteria can provide this vitamin.

Choline

Choline and folate, the B vitamins, work in body in a balanced way. Choline helps the body to absorb and use fats, including those that become part of membranes, which keep body cells intact. Choline is essential for the synthesis of acetylcholine, which is a neurotransmitter needed for keeping the memory sharp and muscle control. Besides, the deficiency of choline can result in fat accumulation in liver and subsequent liver damage. Choline and folate hold methyl group, which the body uses to generate genetic material (DNA). The body also uses methyl groups from folate and choline to convert the homocysteine, which at high amounts is harming amino acid, into nontoxic amino acid- methionine. This choline and folate based conversion of homocysteine prevents the body from accumulating a high level of homocysteine, which is associated with increased risk of cardiovascular heart disease (CHD). Choline, folate and methionine are the body's primary sources of methyl groups. If the supply of folate is low the body may compensate this vitamin by taking methyl groups from choline, and under such conditions, the body might not be able to synthesize enough choline, as a result, the body will need more choline to meet this increased demand. High-protein foods such as meats and dairy products are rich in choline. Nuts and liver both contain choline in sufficient amount. Probably the best way to get enough choline is to consume a varied balanced diet that includes variety of animal foods that provide choline.

Vitamin C (Ascorbic Acid)

Vitamin C is not synthesized by human body as the other animals do. It is necessary for good health. Because of its water soluble nature body requires this vitamin in large quantity (70 mg/day). Its deficiency results in scurvy disease in children and adults, tooth decay (gum's bleeding), unhealthy gums, poor wound healing, poor formation of connective tissues, loss of energy, fatigue and anemia, enlargement and damage to heart muscles and increased disease susceptibility (infection as common cold). Green vegetables, like drumstick leaves and pods, coriander leaves, parsley, fenugreek greens, green pepper, radish leaves, beet leaf, lettuce, amaranth, peas, cabbage, broccoli, kale and other green vegetables contains appreciable quantity of vitamin C. Raw or boiled vegetables are better sources of vitamin C than roasted and fried ones because cooking destroys a part of this vitamin. Tomatoes that can be consumed daily in sufficient quantities, furnish sufficient amount of this vitamin to meet the entire body requirement successfully. Synthetic ascorbic acid can also be added as a food supplement.

Vitamin D

Calciferol is a generic name for closely related compounds that prevent rickets (softening of bones) in children, thus, it is known as anti-rickets vitamin. The bones become thin, soft, brittle and misshapen because of failure of mineralization or inadequate deposition of calcium and phosphorus in newly formed bones. Bowing of legs due to weight may occur when a child begins to walk. Since vitamin D helps in the metabolism of calcium and phosphorus salts, a good supply of this vitamin is essential for proper bone formation and their strength. A progressive demineralization leads to osteomalacia in women of childbearing age who become calcium deficient due to repeated pregnancies. The signs of osteomalacia developed due to deficiency of vitamin D are bone pain, skeletal deformities and muscle weakness. Its lacking is associated with inadequate exposure to ultraviolet light. People living in temperate regions are found at risk of rickets and osteomalacia. This problem is most often seen in low-income group people with poor dietary habits, with minimal exposure to sunlight, strict vegetarians and kidney and liver problems, with fat mal-absorption and with druggists. Increased exposure to sunlight and dietary supplements of vitamin D both are preventive and curative. Being essential for calcium metabolism it also plays an important role in preventing hypocalcaemia or milk fever in high yielding dairy animals, predominantly when a sizable quantity of calcium lost in the process of colostrums (mothers' first milk) production is not replaced by external sources. Severe deficiency causes them to lose nerve and muscle function, including the ability to stand, and eventually they lapse into a coma if not treated timely.

Calcitriol together with parathyroid hormone and calcitonin is useful in maintaining calcium and phosphorus hemostasis. During low levels of calcium in blood, calcitriol mediates important biological actions in intestine, bones and kidney. It increases the intestinal absorption of calcium and phosphorus. In bones, calcitriol along with parathyroid hormone stimulates the release of calcium into the blood possibly by increasing the production of enzyme that accelerates break down of bone matrix. Calcitriol in association with parathyroid hormone increases bone resorption and inhibits urinary calcium excretion. No loss of bone calcium takes place when vitamin D in the body and dietary calcium are enough, however, parathyroid hormone, and calcitriol result in demineralization of bones to maintain normal plasma calcium level, especially when the calcium in body and diet is low. Vitamin D is a stable compound and is dissolved in a wide range of organic solvents but not in water.

Vitamin D_3, the form found in humans and animals, may be toxic only if given too much but high levels of vitamin D_2, the less toxic form may prevent milk fever and the bovine illness. Body exposure to UV-light converts the 7-dehydro cholesterol mostly present in skin to vitamin D_3 (cholecaliciferol) and the ergo-sterol mainly found in plants to vitamin D_2 (ergocalciferol), which is generally used as a dietary supplement, thus, exposure of skin to sunlight (ultraviolet light) prevents rickets. The pigment present in skin provides protection against over production of this vitamin in dark skinned persons in the topical regions. The metabolite 1,24-dihydroxyvitamin D_2, which is produced as a result of vitamin D_2 consumption, enhances calcium absorption and bone calcium resorption. For vitamin D to help build up strong and healthy bones and teeth, it must be activated usually into 1,25-dihydroxyvitamin D_2 form. Both 1,24-dihydroxyvitamin D_2 and 1,25-dihydroxyvitamin D_2 forms work in the intestine and kidney to raise calcium in blood. If this vitamin D activation system breaks down, the metabolic diseases such as milk fever in dairy animals and osteoporosis in women may develop. In early studies, 1,24-dihydroxyvitamin D_2, despite showing promise as a less toxic form of vitamin D for replenishing lost calcium in domestic animals, proved problematic as it creates some unwanted side effects.

Various studies have suggested that vitamin D can treat or prevent cancer but its toxicity does not permit its use. High doses of this vitamin result in irreversible calcification of heart, lungs, kidney and other soft tissues, thus, only short exposure is required for the production of sufficient vitamin D. Excessive intake (5 times the RDA) over a prolonged period leads to toxic effects such as anorexia, nausea, vomiting, weakness, nervousness and impaired renal function, as well as calcification of blood vessels and kidneys. If animals receive too much vitamin D, their blood calcium levels can rise to 40-50% above normal, and if this elevated blood calcium continues for several days the animals may become sick and eventually die, however, most of the mammals can bear an increase in blood calcium of 20-30% for 3-5 days without any harm.

Cereals, fruits and vegetables usually contain no vitamin D, thus, normal diet does not have excessive amount of vitamin D. It is principally obtained from animal sources. In small amounts, it is found only in few foodstuffs, *e.g.*, diary products (cream, butter), yolk, fatty fish and its oil and meat, especially in liver. In many countries, margarines and milk products are fortified with vitamin D_2 or D_3. Fish liver oils and salt-water fish are the rich source vitamin D.

Vitamin E

Aging is an inescapable physiological process and often inevitably increases vulnerability to diseases in elderly. Medical science has not so far invented a fountain of youth. However, dietary supplementation with antioxidants, especially vitamin E, can decelerate the deteriorating body immune functions and lessen the incidence of illness in elderly since antioxidants in human body inhibit chemical reactions such as the breakdown of certain fats. The immune system corresponding to aging process gets weaken and slower to react. The most important factor liable for degenerating immunity in old age is the nutritional deficiency. Diet and nutrition thus play a significant role in maintaining immune function and protecting against infectious diseases. Providing all the essential vitamins preserves the immune function. Clinical studies demonstrated that the subjects who took supplemented diet with vitamin E showed significant improvement in immune cell function known to decline with ageing.

Human studies have shown that exercise and cigarette smoking trigger the production of free radicals. Vitamin E being fat-soluble and chain-breaking antioxidant provides protection against exercise-induced oxidative damage in both young and elderly. During exercise, the blood delivers extra oxygen to the muscles but some of that oxygen escapes the controlled metabolic process as single oxygen ions, known as free radicals. These negatively charged ions behave as pickpockets, stealing positively charged hydrogen ions from fatty acids in cell membranes and damage the membranes that are the gatekeepers of muscle cells- breaching their integrity. All body cell membranes contain this vitamin in sufficient amount. Being fat-soluble vitamin E is oxidized readily and serves as a lipid antioxidant and reacts with free radicals to convert them to a less reactive and non-toxic form and protects cellular and sub-cellular membranes from degeneration caused by them. The study suggests that older people are better protected. An individual ingesting large amount of unsaturated fats should consume supplemental vitamin E to secure protection against formation of free radicals. Supplemental vitamin E can slow down such reactions and aging.

Vitamin E is known as anti-sterility vitamin since its deficiency in body influences first the reproductive system, thus, its supplementation is considered most essential for normal reproduction. It plays an important role in cancer prevention. An adequate intake of vitamin E apart from scavenging of free radicals provides protection against free radical damage associated with diseases, like ischemic heart disease, muscular dystrophy, sterility, menstrual disorders, pregnancy toxemias, spontaneous abortion, arthritis and cataract. However, supplementing this vitamin may lessen the incidence of these disorders to some extent. Tocopherol protects vitamin A, carotene and vitamin C in foods from oxidative demolition. Vitamin E is fairly stable to heat and acid and unstable to alkali, UV-light and oxygen.

Chain-smokers, patients undergoing radiation therapy after surgery and the subjects suffering from infectious diseases, metabolic disorders and surviving on restricted diet are supposed to be liable to the deficiency of vitamin E. The signs of its deficiency in body can be seen in reproductive, nervous, muscular, glandular, vascular and hemopoietic systems. Its deficiency leads to the production of immotile sperm and may cause fetus abortion. Muscle paralysis can also be seen under very severe deficiency. The disorders associated with fat mal-absorption have been linked with low plasma tocopherol levels. Muscle weakness appears to be due to deficiency of vitamin E. Impaired vision is associated with chronic advanced deficiency. Prematurely born infants with low vitamin E levels have impaired fat absorption, thus, they need an oral supplement of α-tocopherol. Low level of α-tocopherol in plasma may develop hemolytic anemia in infants and decreases red blood cell survival time in adults.

The Recommended Dietary Allowance of α-tocopherol up to 600 mg/day is appeared to be safe for most people. However, its requirement depends upon severity of exercise and mental stress. The patients suffering from vitamin K deficiency are advised not to take vitamin E as supplement since it can arise the complicacy of blood clotting. It is also advocated not to combine it with iron supplements as it destroys the vitamin E. Although it is always better to include a variety of foodstuffs in diet, predominantly that are originated from plants rich in oil. Almond and wheat germ oil are very high, and fish and coconut oils are low in vitamin E content. Cereals and eggs are also high in tocopherol. The other sources that provide vitamin E are the oils of plant origin such as dry nuts, safflower, olive, peanut, corn, soybean and cotton. It is also found in green leafy vegetables like cabbage and lettuce.

Vitamin K

The vitamin K is available in two forms: phylloquinone (vitamin K_1) isolated from alfalfa (lucerne) is the only form that occurs in green plants and the bacterial form the menaquinoness, which is known as vitamin K_2 and a third purely synthetic water-soluble form with vitamin K activity is menadione. Phylloquinone is considered nontoxic, while menadione is potentially toxic and must be used cautiously. This vitamin is known for its critical role in blood coagulation (clotting) and considered very important for maintaining bones integrity as it activates at least three proteins involved in bones health. It is useful in the transportation of calcium and deposition of calcium in bone matrix. The soft tissues that transport calcium ha.. either γ-carboxyglutamic acid containing proteins or vitamin K dependent carboxylating system, and vitamin K works as an essential cofactor for carboxylase enzyme, which converts specific glutamic acid residues of precursor protein to γ-carboxyglutamic acid residues in new protein.

The Recommended Dietary Allowance for adult females is 65 μg/day and for adult males 80 μg/day but it is barely possible that the individuals in age range of 18-44 years may get enough vitamin K today since the current analytical methods show that this vitamin is not abundant in the diets. The intake of this vitamin nationwide is very low. Elderly can get much benefit from increasing intake of vitamin K, thus, the people over 65 years age consume more phylloquinone than those of 20 to 40 years age group, and as a result they tend to have higher blood levels of phylloquinone, which is the most common form of this vitamin. Another form of vitamin K produced during the hydrogenation of oils is dihydro-phylloquinone. As much as 30% of the total vitamin K intake may come in the form of dihydro-phylloquinone but it is less biologically active than phylloquinone. In fact, it is half as active with a clot-forming protein, and is completely inactive with a bone forming protein so hydrogenated oils should not be considered an important source of vitamin K.

The deficiency of vitamin K prolongs the blood coagulation time and increases hemorrhage incidence, however, its deficiency is not common in healthy individuals because the intestinal microflora can synthesize a considerable part of daily body requirement of vitamin K, thus, its deficiency may not be due to primary factors but may be due to secondary factors, like impaired absorption of fat, liver infection, lack of pancreas activity, celiac disease and regional enteritis. Anticoagulant drugs such as coumarin work as anti-metabolites by inhibiting conversion of vitamin K to its active form. Broad-spectrum antibiotics that sterilize the intestine can lead to its deficiency. In infants, the deficiency signs persist unless a bacterial flora is developed in the intestine. Newly born infants have a relatively sterile intestinal tract and do not contain microflora to synthesize vitamin K. Premature infants are seen more susceptible to fully mature infants.

Some oils, especially soybean oil and dark green fresh vegetables such as kale, spinach, lettuce, broccoli and Brussels sprouts, provide the icon share of this vitamin to diet, whereas, peas, cauliflower, carrot and cereals are the poor sources of vitamin K. Overall no differences are noticed between vitamin K from oils and from broccoli but it is good to take from green leafy vegetables as they contain so many other nutrients too. One serving of spinach or two servings of broccoli provide four to five times of RDA of phylloquinone but nutritionists have assumed that people absorb more from oil or oil based supplements than from vegetables. Foods of animal sources are poor for this vitamin unless they have undergone bacterial decomposition. *Escherichia coli*, the normal flora in the human digestive tract synthesizes this vitamin in significant amount and meets out the body's requirement for this vitamin that is why a healthy individual never seems to be dependent on supplemental vitamin K in the diet.

■■■

5 Vitamin E: An Elixir for Human Life

M.K. Rana

Department of Vegetable Science
CCS Haryana Agricultural University, Hisar- 125 004, Haryana, India
E-mail: <mkrlotus@gmail.com>

A dietary factor vitamin E was isolated first time in early twenties from wheat germ oil and chemically named tocopherol, which was derived from Greek word *tokos* means childbirth and *pherin* means to bear. The last word *ol* indicates its relation with alcohol. Vitamin E has been proved reasonably stable to heat and acid, however, unstable to alkali, ultraviolet light and oxygen. It preserves the oxidizable vitamins and unsaturated fatty acids easily. Being antioxidant tocopherol protects α- and β-carotenes (precursors of vitamin A) and vitamin C in foods from oxidative destruction.

This is a tale of mice and man and a tale of help for the elderly. No science has yet discovered a fountain of youth, but one aspect of aging that is often accepted as inevitable, increased vulnerability to disease, may not be so inevitable after all, according to the findings of an animal study that focused on elderly rodents. Certain dietary antioxidants, particularly vitamin E, can stimulate some of the flagging immune responses in older men and women as well as in aged mice. Antioxidants inhibit chemical reactions such as the breakdown of certain fats.

The immune system appears to parallel the aging process. It gets weaker and slower to react. Nutritional problems may contribute to declining immunity in old age, and appropriate dietary intervention or supplementation may improve immune function and reduce the burden of illness in the elderly. The immune systems of 2 years old mice, old age for mice, responded more like those of young mice when the animals were fed vitamin E at 18 times the recommended dietary allowance for 6 weeks. Another antioxidant, glutathione, also stimulated immunity in old mice but neither compound completely restored it to the level of young animals.

In a study of 32 men and women, about two-third of the volunteers who got vitamin E supplements had significant improvement in the immune cell function known to decline with aging. However, there was no improvement in antibody production. Half the volunteers, who were al lover 60 years, took 800 international units of the vitamin E daily for a month, while the other half got a placebo. At 53 and 67 times, the Recommended Dietary Allowance for men and women, respectively, the 800 IU dose was not out of line with what some people are taking as supplements. Although vitamin E is substantially less toxic than other fat-soluble vitamins such as A and D but excessive use should be avoided without medical supervision.

The 16 men and women taking supplements averaged three times more vitamin E in their plasma after the 30 days trial than before. Diet and nutrition play an important role in maintaining immune function and protecting against infectious diseases and cancer, *e.g.*, restricting animal calorie intake while providing all the essential vitamins and minerals prolongs their life and preserves their immune function. Just how diets or nutrients affect the immune system, biochemically speaking, is still very sketchy or unknown.

Its deficiency affects the reproductive mechanisms thus recognized very essential for normal reproduction and that is why it is called anti-sterility vitamin. Although it is called the fertility vitamin but at present there is not enough evidence to indicate that vitamin E could be put to use either as a single formula of treatment or as a complementary therapy in infertility patients.

Functions of Vitamin E

Human studies have shown that exercise induces the release of oxygen free radicals, and during exercise, the blood transfers extra oxygen to the muscles in the form of O_2 (super oxide), but some of that oxygen escapes the controlled metabolic process as single oxygen ions, which are known as free radicals. Vitamin E provides protection against exercise induced oxidative damage in both youths and old people. These negatively charged ions behave like hijacker, and steal positively charged hydrogen ions from the fatty acids present in cell membranes. This breaks the integrity of membranes, which function as the gatekeepers of muscle cells; by donating the hydrogen ions these free radicals are looking for, vitamin E protects cell membranes.

Free radicals are unstable toxic chemicals, which are released during the normal processes of the body and on exposure of an individual to polluted environments particularly to cigarette smoke. The free radicals when present in excessive numbers influence the normal cells and predispose them to the development of early heart attack, cataract and some kinds of cancer. These toxic chemicals also cause inflammation, arthritis, brain hemorrhage, circulatory disorders, damages to skin and premature aging. Fortunately, our body has the natural defense system to minimize the generation of free radicals or limit the damage caused by free radicals. If the body defense system is sluggish, supplements like vitamin E, which belongs to a class of substances called antioxidants (which include β-carotene, vitamin C and selenium), may be very effective in neutralizing the harmful effects of free radicals.

Free radicals are the principal factors, which are found closely associated with the aging process. Influencing the factors responsible for the impairment of cells can slow the rate of ageing down. Potent evidences are available, which show the key role of vitamin E in providing protection against development of cardiovascular diseases. Although mechanism is not known but research-reports suggest that vitamin E most probably works by preventing the accumulation of low-density lipoprotein (bad cholesterol) and its absorption by blood vessels. Alpha-tocopherol is fat-soluble and chain-breaking antioxidant, which protects the lipoprotein from per-oxidation damage. The other antioxidants that react with free radicals and provide protection against their harmful effects by inhibiting tissue-damaging oxidations are β-carotene, vitamin C and selenium.

Being fat-soluble, all body cell membranes contain this vitamin, which is readily oxidized and served as a lipid antioxidant. Vitamin E reacts with free radicals to convert them to a less reactive and non-toxic form and protects cellular and sub-cellular membranes from degeneration caused by them. An individual who ingests large amount of unsaturated fats should consume supplemental vitamin E to gain protection against the formation of free radicals as its supplementation can slow down such reactions and aging process. Vitamin E is considered as a panacea for delaying the onset of host of age related degenerative diseases. It is believed that vitamin E plays a significant role in the prevention of cancer because of having free radical neutralizing properties.

The effects of vitamin E in youths and old age people have been noticed different due to different levels of antioxidant enzymes as these are reduced with age that is why supplementation of diet with vitamin E is more beneficial for older people than the younger ones. Although oxidation is diminished in people of both groups but the older people are defended better. The supplemented individuals in both age groups showed a significantly smaller increase in urine by products that indicate fatty acid oxidation by the 12th day of exercise. However, an intermediate by product of fatty acid oxidation tends to be lower in the individuals consumed diet supplemented with vitamin E.

Dietary antioxidants inhibit chemical reactions like break down of certain fats. Research reports show that exercise can defer the damage and repair the muscle fibers in individuals that take vitamin E as supplement regularly for two weeks. The immune system appears to parallel the aging process; it gets weaker and slower to react. Malnutrition problems may contribute to declining immunity in old age. Certain dietary antioxidants, particularly vitamin E, can energize some of the flagging immune responses in old age individuals. Another antioxidant, glutathione, also stimulates immunity in old age, and befitting dietary supplementation may improve body immune system and reduce the scope of illness in the old age people.

Vitamin E as diet supplements produces notable progress in the immune cell function, which is known to decline with ageing. Nutrients rich diet plays an important role in maintaining immune function and providing protection against cancer and infectious diseases. Restricting animal's calorie intake while a subject is furnishing all the essential minerals and vitamins prolongs the life and preserves the immune function. How nutrients rich diet affects the immune system. It is still unknown. In human studies, vitamin E increases the levels of intereukin-2- a substance that augments the growth of certain infection-fighting white blood corpuscles. It is well-established fact that the white blood corpuscles work less efficiently with age and these white blood corpuscles respond better to vitamin E.

Deficiency Disorders

The signs of vitamin E deficiency in human body can be seen in reproductive, nervous, muscular, glandular, vascular and hemopoietic systems. The deficiency of this vitamin in the body leads to the production of immotile sperm and may cause fetus abortion. Impaired vision is associated with chronic advanced deficiency. Muscle becomes weak due to vitamin E deficiency and muscle paralysis can be seen when the body is sufficiently deficient in this vitamin.

An individual suffering from disorders associated with impaired fat absorption has been noticed to have low level of plasma tocopherol, which develops hemolytic anemia in infants and decreases survival time of red blood cell in adults. Infants, particularly born at pre-mature stage with low vitamin E level have fat mal-absorption; thus, they need an oral supplement of water-soluble α-tocopherol.

Apart from scavenging of free radicals vitamin E prevents ischemic heart disease, muscular dystrophy, menstrual disorders, toxemia of pregnancy, spontaneous abortion and sterility. An adequate intake of vitamin E provides protection against diseases like arthritis, ageing, cataract and cancer, which are normally associated with free radicals damage. It is particularly useful in the treatment of nocturnal cramps prevail in old age when the calf muscle goes into severe spasms.

Vitamin E supplements for at least two years may reduce the risk of heart diseases up to 40% in healthy men and women, and its supplementation has been found associated with reduced incidence of heart attacks by 75% when taken daily by patients suffering from heart diseases. It is believed that free radicals produced by atmospheric pollution and sunlight can harm the skin rigorously, that is why, all skin creams available now a days in the market contain vitamin E, which negates the harmful effect of free radicals on skin.

Doses and Sources

It is very difficult to detect the cases of vitamin E deficiency. Those liable to be deficient in vitamin E are smokers, patients under going radiation therapy or convalescing after surgery and persons suffering from infections or metabolic disorders as well as the persons survive on a restricted diet. It has been estimated that 20% of the United States population consumes vitamin E constantly; it is perhaps the fact that vitamin E has no significant side effects in normal individuals so it is considered a relatively safe drug. No doubt, vitamin E is positively less toxic than other fat-soluble vitamins such as 'A' and 'D' but even than excessive use should always be avoided without medical prescription. Stress of any kind and exercise affect the amount of vitamin E a person needs, but for most people the doses of alpha-tocopherol up to 600 mg/day appear to be safe.

Vitamin E is strictly prohibited for patients suffering from vitamin K deficiency as this can complicate the blood-clotting defect, which is directly associated with vitamin K deficiency. Research evidences further show that iron destroys the vitamin E, therefore, iron supplements should be always taken at a different time of day than vitamin E supplements. It may also change the tissue lipids and alter body steroid hormones so it is always better to eat a variety of food from all the basic food groups to meet out ones entire requirement of vitamin E.

Vitamin E is primarily found in foods of plant origin and vegetable oils. Walnut oil and wheat germ oil are considered higher in tocopherol content, whereas, coconut oil and fish oils have low concentration of vitamin E. Grains and other seeds contain vitamin E in their germs. The richest sources are the fats of vegetable origin, principally in the seeds and oils of safflower, olive, corn, soybean, dry nuts, peanut, coconut and cotton. Whole grain cereals (fortified breakfast cereals) and margarine butter are a good source of vitamin E.

■■■

6 Gender and Health Development: Cultural Critique

Madhu Nagla

Department of Sociology
Maharishi Daynand University, Rohtak-124001, Haryana, India
E-mail: <bnagla@yahoo.com>

Impressive achievements and intolerable shortcomings characterize India's health programme over the past 50 years. On one hand, there has been a decline in mortality rates. Life expectancy and infant survival conditions have improved over a period. It is very high when we compare the infant mortality rate with other countries. The World Bank Report notes that public health financing in India is characterized by an emphasis on hospital rather than primary care, urban rather than rural population, medical offices rather than paramedics, services that have larger private rather than social returns and family planning and child health exclusion of wider aspects of female health. In fact, gender differences in health and mortality have not only perpetuated but they also have been accentuated. Therefore, gender and health in terms of cultural explanation from the development perspective will be discussed in this present chapter.

Health is in general capability enhancing. Good health is sought not only for pleasure, nor for reducing pain but also for expanding a person's significantly important capabilities and freedoms. Role is given to health and longevity in the concept of human development. Life expectancy at birth is one of the three principal components of the so-called Human Development Index. Indeed longevity data figure prominently in the development literature not only as one of the indicators of medical achievement, but also as one of the widely used social measures of progress in general. Similarly good health should be seen as a basic human right.

The promotion of human rights proceeds on the understanding that there is much to be promoted, including the expansion of feasibility of the recognized rights. In fact, that can be an important part of the motivation for focusing as the right to health. So what can recognize health as a human right actually does? In fact, it can do a lot. A wide range of policies and actions has power to influence on our health and all of them demand attention in the causes of promoting this critically important human right as a premise that development as freedom.

It is widely recognized that the determinants of health are social and economic rather than purely medical. The poor health of people from the lower castes, their social exclusion and the steep social gradient are due to unequal distribution of power, income, goods and services. The restricted access of those from the lower castes to clean water, sanitation, nutrition, housing, education, health and employment due to a toxic combination of poor social policies, programmes, unfair economic arrangement and bad politicians.

Many health intervention programmes are unable to work properly due to inequality in social structure, which is based on socio-cultural issues, unequal access to health care in lower strata of the socially excluded communities. The systemic discrimination of lower castes based on cultural tradition and religion needs to be tackled if interventions have to work.

Tradition and culture maintain their stronghold on inequality; poverty and social exclusion have a multiplicative effect on the social determination of health and those at higher risk for diseases also having a higher probability of being excluded from health care services. There is need to increase resource allocation for the social-cultural determinants of health and to reinforce the government's primary responsibility in providing for health needs.

Gender equity and social and political inclusion of the poor and lower castes in policy and decision-making are required. There are many health interventions, which are related to women but many of these programmes are not doing well because the potential beneficiaries are unable to take opportunity to access it due to a number of socio-economic factors including gender discrimination. If gender relation is improved through empowering women, it is expected that there will be demand for these services by women.

Despite several programmes are in operation, there is a discrepancy between the actual health status and the desired health status of the society. This discrepancy can largely be set right through medical interventions and from healthcare services but these are influenced by various factors like the degree of awareness about illness, the educational level and the accessibility, availability and affordability of healthcare services, provision of drinking water and a host of other factors like culture, caste, choice and preferences.

It is to be noted that modern health interventions derived from supply side conditions have to recognize that such interventions were not made in a social and historical vacuum, so to say, health seeking behaviour in, let us say, a high mortality low income, traditional community rests on two sets of factors:

- Customary caring and coping strategies at the community, familial and individual levels
- Popular medical knowledge, which has evolved over centuries, is useless without indigenous innovations.

However, both these factors in turn crucially depend on the important point that the customary health seeking behaviour is embedded in definite socio-cultural values, beliefs and systems of meanings attached to death and disease, pregnancy and childcare.

The demand for modern health care would then depend not only on accessibility or on availability in physical terms but more importantly in social and behavioural terms. Thus, in the case of modern health service, social and behavioural factors can dampen demand independent of physical accessibility. Therefore, there can be a significant gap between perceived health needs and effective utilization due to non-price and non-economic factors. Difficulties of physical access and cost of availing health care would only further reduce demand by affecting ability to use.

It is influenced by various factors like the degree of awareness about illness. The educational level and the accessibility, availability and affordability of health care services, drinking water facility and other factors like caste, sex and religions, choice and preference pattern. Therefore, there is a need of policies to reduce discrimination against the affected groups, and in turn, it will increase the access of healthcare for social development of the society. Indeed, in dealing with policy making, it is important to go beyond health care facilities into general economic and social arrangements (change in the present structure), which can have very important effect on health.

Of course, health is influenced by healthcare and the use of medicine, but not just by them. It also depends on general economic prosperity and social cohesion as well as on general economic prosperity and social inequalities. Studies on social determinants of health have sharply brought out that there are seriously negative effects of inequality of status and decisional power on health. Those at the bottom of the hierarchy seem to suffer disproportionately from bad health and premature mortality. Other ways are also there in which inequality of power can lead to the loss of human priorities in public and private arrangement for health care. Gender inequality is an important factor, which can contribute negatively for access to health care of significant section of population.

Actions for enhancing health activities and expanding the corresponding human capabilities come in many different forms and shapes.

There are important roles of many *non-medical* arrangements, including legal framework (such as patent law), economic facilities (including the generation, distribution and use of income) and even social hierarchy since stratification and the loss of freedom of those at the bottom can make them more prone to avoidable illness and premature death.

Health is influenced not just by health care and the use of medicine but also for example by general economic prosperity and social cohesion as well as by economic and social inequalities. It is also important to recognize that even though economic prosperity helps in the achievement of good health, economic growth alone may not be able to override other problems in generating good health in society.

There is a link between a healthy society and human rights. Health fitness of this world has many different causes. A wide range of policies and actions have strong influence on human health, and this have to be comprehensively taken onto account and closely scrutinized in understanding the ways and means of advancing good health.

The current way of interventions will take many years in reducing the inequality of social and health outcome for majority of the people. There is need to change social structure. The small achievement in different sphere in social and healthcare cannot overcome the large contradiction in Indian society. Unless there is fundamental shift in politics, power, or changes in social values, it is distant to achieve a large outcome.

Understanding the role of institutions and culture and gender in facilitating the access to healthcare would go a long way in this regard. Favourable gender relation will lead to empowerment of women. This will change the power relation to decision making in favour of neglected section. Ultimately, empowerment of women will increase their opportunities in getting access to many developmental benefits including health care facilities. This in turn will lead to change in the pattern of development and lead to increase in the quality of life and freedom.

Health care not merely the distribution, not merely the transaction between doctor and patient but the medical technology itself (which is based on certain assumption about the nature of disease possess, the causation and cure of disease, the relations of individuals to their own bodies and to social processes) embodies the social relations created by capitalistic society.

The social relations are contained in the patriarchy, and therefore, seeking health care is far from a matter of purely academic interest. Understanding these social relations is the key to understand how these relations seeking permission, ignoring illnesses and avoiding taking medicine, no rest, having low nutrition diet. Thus, the cultural critique has major political implications for health policy.

Culture

They are living entities, which are continuously being renewed and reshaped. As with culture more generally, gender definitions change over time.

Cultures and traditions are simply the forms of social interaction accepted by particular communities at particular times, and according to their world views and historical experiences, such that there are several alternatives and systems of values selected for their usefulness. In other words, contrary to popular notion, culture is not past values and practices that have fossilized at a particular point in history, never to change and recounted only orally in the present values and practices that remain in the past and are no longer practices should not be considered (for all practical purposes) part of an active culture and tradition.

Societies and cultures are not static. They change over time in accordance with the interpretive values, beliefs and practices of the group, whose members define and live by the ideals of those practices and values. Cultures are living entities that are continually being renewed and reshaped. Cultural change occurs as communities and households respond to social and economic shifts associated with globalization, new technologies, environmental pressures, armed conflict, development projects, *etc.* Cultural values are continually being reinterpreted in response to new needs and conditions. Some values are reaffirmed in this process, while others are challenged as no longer appropriate.

Very often culture is blamed for certain ill-understood health practices, when in fact the culprit is the lofty interpretation of culture for the maintenance of imperial dominance and hegemony. This has been particularly true when westernized professionals have addressed issues of health beliefs and practices in our society.

Culture is a system of interrelated values active enough to influence and condition perception, judgment, communication and behaviour in a given society. Culture is as much a structure as it is economic and political, and it is rooted in institutions such as families and schools as well as in communication industries. Culture must be understood within the context of its transformative possibilities rather than according to the unilineal measure of the West that has come to be known as development. Harry S. Truman introduced the phrase *underdeveloped countries* into popular discourse. The hidden agenda for development has been nothing less than the westernization of the world. This process has entered a new discourse with the valorization of globalization. Globalization, as a concept produced within the context of global change, has tended to focus only on economic production, capital restructuring and their attendant political implications, with little regard for negative cultural changes in global societies, within health and human development constructs like its predecessor, development promises a cultural world, where an individual can be at home anywhere.

Relationship between Gender and Culture

Gender identities and gender relations are critical aspects of culture because they shape daily life. Changes in gender relations are often highly contested, in part because they have immediate implications for everyone, women and men. This immediacy also means that gender roles and particularly women's role as wives and mothers can be potent symbols of cultural change or cultural continuity.

The status of women is a complex issue. It is not amenable to any simplistic explanation of social reality. Literature on the status of women has been varied, and all of them have been addressing the key issues affecting the women in various areas of development. However, the areas of development are manifold, such as education, employment, political, social, legal, health, *etc.* The constitution of India provided equality to women. The introduction of adult franchise along with the removal of all discrimination on sex ground, provided towards the complete emancipation of women but the inequalities inherent in traditional structures have a significant bearing on women in different spheres of life.

The International Conference on Population and Development in Cairo in 1994 placed Reproductive Health at the Centre Stage of demographic objective instead of fertility reduction. The Cairo Conference recognized women's rights as individual reproductive rights. Gender equity and equality was accepted as a goal for sustained and substantial development. Gender equity and equality, infant, child and maternal mortality reduction and the provision of universal access to reproductive health services, including family planning services are some of the areas for achieving the gender equity, which are necessary to reduce the threat to women's reproductive health.

Relationship between men and women are shaped by culture. Gender identities and gender relations are critical aspects of culture because they shape the way daily life is lived in the family but also in wider community and the work place. A gender like race or ethnicity functions as an organizing principle for the society because of cultural meanings given to being male or female. This is evident in the division of labour according to gender. In most societies, there are clear patterns of women's work and men's work both in household and in the wider community, and cultural explanations differ among societies and change over time. While the specific nature of gender relations varies among societies, the general pattern is that women have less personal autonomy, fewer resources at their disposal and limited influence over the decision making process that shape their societies and their own lives. This pattern of disparity based on gender is both a human rights and a development issue.

The argument that women are a culturally constructed gender category rather than simply a biological sex led to a much broader feminist challenge to existing knowledge of the relations between men and women. In 1982, an attempt was made to avoid biological determinism while still explaining women's subordination as a universal phenomenon, and it was asked what it was that would lead every culture to place a lower value on women than on men, and that every society made a distinction between nature and culture. Woman's social ties to children through pregnancy and breast-feeding ensured that women appeared to be closer to nature than men were. Simon de Beauvoir is one of the chief protagonists of cultural dualism. It is based on the perception that the secondary status of women is universalistic. Man controls nature through culture so he is superior to nature and women. Elaborating the perspective of cultural dualism, what is certain is that today it is very difficult for women to manifest at the same time their autonomous individuality and their feminine destiny. The western woman seeks liberation as an individual, while the Asian woman seeks greater satisfaction by the improvement of family's economic situation and status within the family (Ibid). The primary role of woman is always manifested in their family. The society at large views their economic role is secondary so that they are pushed out of the market. Though theory of cultural dualism lacks coherence it provides a useful analysis by supporting a sophisticated view of the internal dynamics of culture as it relates to the position of women.

Gender and Patriarchy

The word patriarchy literally means the rule of a father or a *patriarch*, and originally, it was used to describe a specific type of *male-dominated family*. This refers to the power relationships by which men dominate women and to characterize a system whereby women are kept subordinate in number of ways. The conception of patriarchy as universal is logically based on the conception of woman as a universal category. Whether or not women are explicitly conceived of as a class whose interests are opposed to those of men as a class (a position taken, *e.g.*, men are seen as having and defending power over women.

Thus, patriarchy has been used by feminist scholars as a theoretical construct to explain the structure underlying the subordination of women and as one hindering the development of women. The structure of the patriarchal system has been conceived at psychological, cultural and material levels. The centrality of the family is a primary unit as an entity in which the productivity activities are undertaken. This unit in which the production and reproduction of social life is carried on has definitely a major role in the development of women. Women's experience within the household varies with class, ethnic group, caste, culture and religion, thus, it is necessary to explore the differing principles and notions of family and kinship. It is the notions of family and kinship that shape the woman's position in the society. The process of socialization starts from this basic social unit, *i.e.*, the family so here start the process of subordination.

Besides the above theoretical explanations of women's subordination at global level, and in the Indian society, the roles women play in the society and the images of them have developed are not simply from exigencies of biology and social situations but are rather deeply rooted in the myths, legends, religion and culture. Therefore, it is here one observes that the underlying causes that keep woman in subservience as *second sex* devoid of any identity. This tradition has gained much significance and with the passage of time came to be accepted as unwritten law of the land. Moreover, the demographic indicators like mortality rate, infant mortality rate, life expectancy and reproductive diseases will be able to spell out the prevailing inequality among males and females.

A theoretical approach to the problems and prospects of women's development is extremely essential to have a thorough understanding of the subject. Laws by themselves cannot bring social change. We need to understand the social values, social structure and role of the productive process. One assumption to all the perspectives is that the position of women has not registered significant improvement despite the efforts at modernization. Many feminist scholars have used patriarchy as a theoretical construct to explain sexual inequality and as one hindering to the development of women.

Gender and Health Development: Critical Perspective

Health is an indicator of quality of a society. It reveals how a society deals with gender in various experiences as birth and death, pair, disability, suffering and aging. Viewed through the prism of health care system, India is developing fast in terms of its health indicators related to political economy of the country. However, the political economic critique concentrates its fine on the equitable distribution of health services. After 1990 or in the phase of liberalization, the unprecedented entry of the private players has led to mushrooming of corporate health services in urban and metros. The poor are worst off, and of course, they simply cannot afford good care (or in some cases, any care at all) but finances are not the only barriers to care. Poor geographic distribution of doctors and hospitals, *e.g.*, the lack of services in rural areas, lack of doctors in primary health centers and sub-centers, all act deny many poor people's access to acceptable health care. On the other hand, the uncoordinated mélange of private practitioners, private hospitals, corporate hospitals, poor privately owned drug and medical supply companies are not in business for poor people's health. As long as this situation exists, it remains impossible to plan and regulate the health system in the interests of equity and service.

The political economic critique of course acknowledges that there are problems with health care other than distribution. Services are all too often of low technical quality. Doctors and political workers may discriminate among patients because of their gender, class, or caste. Services are often bureaucratically organized and unnecessarily fragmented and so on but the cultural critique disputes the value of the ideology itself, which prevails at the individual, family and at the societal level. The patriarchal values, which are deeply rooted in the family and in turn in the society, do not allow the seeking of health care the women in the Indian society.

Feminists point out that the treatment of women is often patriarchal to the point of medical irrationality, *e.g.*, standard obstetrical procedure features the women unconscious and/or strapped down to a table while the lordly obstetrician extracts the body with forceps. There is considerable evidence that anesthesia, the supine position and forceps are unhealthy for mother and baby. The effect of this kind of treatment is to promote feelings of positively and low self-worth on the part of women, and hence, to reinforce oppressed roles as wives and workers.

Demographic Indicators, Health and Culture

Although there has been overall improvement in health condition of the both males and females but women still get differentiated from men when it comes to providing medical care to them, which implies comparative low status of women in our society. According to UNICEF, 12 million girls are born in India every year, out of which 25% do not survive more than 15 years. The fact is that as the mortality rate of 0 to 5 years age group is about 20 times greater than any other 5 years age group. Female mortality rate up to the age of 40 years is higher than the rate for mortality. Though an increase in the life expectancy for women has been recorded in absolute numbers in comparison to the life expectancy of males, it is still lower. Strikingly, maternal mortality rates are particularly high.

In the world, the sex-ratio at birth tends to be favourable to females even though the males outnumber the females at younger ages due to slightly more males being born than females (the sex-ratio at birth varies from 102 to 107 boys per 100 girls for biological reasons) but this advantage of males gets neutralized by about the age of 20 due to higher mortality among boys than girls. At national level, the *Child Sex Ratio* in population aged 0-6 years decline from 962 in 1981 to 945 in 1991. In 2001 census, the Child Sex-Ratio had declined further by 18 points from 945 in 1991 to 927 in 2001.

At birth, the chances of survival in general for a male child are less in comparison to a female child, but after the neo-natal stage, social factors such as care, immunization, nutrition of the mother and child and sanitation come into force and govern the survival prospects of an infant. On the other hand, adverse sex ratio is accorded because of high risk of female deaths from infancy to the end of reproductive period. Traditionally, inculcated attitude of self-effacement among women and infant mortality rate is higher for females than males. The same negligence is evident in exceedingly high maternal mortality rates mainly since health care workers are fewer in attendance than half the births.

Gender inequality also persists among the elderly women. There have been improvements in women's life expectancy relative to men's life. In the beginning of 1950s, the life expectancy at birth both men and women was around 32 years and by the mid 1980s, it had increased to 55 years with women gaining in the late adult years. Thus, by the age of 60, a woman could expect to live an additional 16.4 years, while a man could expect to live an additional 14.6 years. Because of higher life expectancy and other cultural reasons, more women tend to be widowed than the men are, *e.g.*, in 1988, nearly two-thirds of the Indian elderly women were widowed compared to about a fifth of the men. This happens due to several reasons in general, women tend to be younger than their husbands are, and cultural and social norms inhibit women from remarriage. Second, the problem of elderly women in the female population is currently greater than the corresponding proportion of elderly man. In addition, the rate at which the female population is ageing is faster than that of male.

Ageing is often perceived as deterioration in health. Most research and services tend to concentrated on women of reproductive age ignoring those beyond these years. While the process of growing old may bring some improvements and benefits to a woman in the post-middle age period, these may diminish as she advances into old age, or when she is widowed. A woman often derives power from authority from her association with the male head of family. If this association suffers erosion, her access to resources for care and sustenance is reduced, making her vulnerable. The risk increases for women who have no assets for survival such as education, possessions, or social status. This vulnerability may be compounded by failing health and disability. Thus, an indigent and physically frail, elderly widow is the most defenseless in the Indian context. It is important to note that the elderly women suffer not only from ailments specific to ageing but also from ill health accumulated over the life cycle, which may manifest in old age in an aggravated form. Several health conditions affecting elderly women such as changes in the skeletal, cardiovascular, nervous, skin, genitourinary and gastro-intestinal systems are caused by declines in ovarian hormonal levels triggered by menopause. Due to the absence of ovarian steroids, chronic diseases such as osteoporosis, coronary heart disease and cerebrovascular develop. Gynecological cancers prolapse and fistulae, which are the result of child bearing, are some conditions afflicting the elderly women. Elderly women may find the nature of their illness difficult to disclose to younger members, especially if there are gynecological problems. This is one reason why conditions such as prolapse tend to be neglected.

Nutritional Pattern and Culture

Poor nutrition is not only food and poverty problem but also socio-cultural problem of women in the Indian society. It cannot be denied that poverty is a major cause of malnutrition and undernourishment of women. Moreover, it is more unfavorable socio-cultural value operating against women in distribution of food.

Optimum standard of nutrition is determined based on body weight, sex, climate and nature of the work performed. More than half of the women in India do not get the recommended dietary intake of nutrients. Generally, the poor nutritional status of Indian girls and women is part of a vicious cycle that has particularly devastating consequences for pregnant and lactating women and their infants. Malnourished women are more likely to give birth to low-birth weight babies, and if the underweight baby is a female who survives, she in turn is likely to continue to be undernourished throughout her childhood, adolescence and adult life. This lack of nourishment has detrimental effects on her reproductive and lactating capacities.

All nutritional programmes are directed towards the needs of pregnant and lactating mothers. However, in spite of these programmes, there is nutritional deficiency in women starting from infancy to pregnancy, and then, again after crossing the childbearing age. It cannot be denied that poverty is a major cause of malnutrition and undernourishment but its more unfavourable impact on women's nutritional standards shows the connivance of poverty and socio-cultural values operative among them. Poor nutrition of girls, especially in childhood and adolescence, has serious consequences. Anemia is a major health problem among Indian women. Studies carried out by the Indian Council of Medical Research found that over 65% of girls ages from 1 to 14 surveyed in the cities of Hyderabad, New Delhi and Calcutta were anemic. Anemia is particularly widespread among women during pregnancy, when iron requirements increases nearly five folds. A variety of studies has shown that a high percentage of pregnant women in India are anemic, particularly in the last trimester. One study conducted in the rural districts of Gujarat and Maharashtra found that about 90% of the pregnant women suffered from anemia. Other studies have found rates of anemia between 50 and 90% among pregnant women. A recent population-based survey of pregnant women showed some degree of anemia and that 56% had severe anemia. An analysis of the data from the referral hospital for the area revealed that severe anemia contributed, directly or indirectly, to 35% of all in-hospital maternal deaths.

Poor nutrition becomes evident among females during infancy, persists through childhood and tends to increase with age. In a study of two Delhi slum populations, only 50-65% of female infants below the age of 1 year received adequate nourishment or was only mildly malnourished. For female children ages between 5 and 9, the percentage fell to about 30-35%. It is probable that the result of this anthropometric study reflect not only differential nutrition but also differential morbidity. The NNMB surveys in 1980 documented low intakes of vitamin A and iron among girl children and adolescents. Vitamin A deficiency, which has been firmly linked to high mortality and morbidity in children, is likely to be an underlying cause of high levels of respiratory and genitourinary tract infection in women.

Among many Indian families, the largest share of food is usually given to the bread earner, the next to boys, to the old or sick and last to the young girls and women in the household. Girls are often neglected in matters of feeding and health care. Discrimination in feeding may begin soon after birth, as girls are breast fed less frequently as for shorter duration than boys are. In recent years, increasing attention has been given to differentials in the allocation of food within households. However, rates of malnutrition among children and women are high. Ethnographic literature suggests that in northern India, the females are not fed as well as the males are. Studies in Uttar Pradesh, Andhra Pradesh and Tamil Nadu indicate that female children are discriminated against when it comes to the allocation of food within households. Marked differences exist between what is fed to boys and girls. The discrepancy increases with age. According to figures obtained from the National Nutrition Monitoring Bureau in the age group of 10-15 years, the boys are fed 31 g of pulses a day but girls get only 25 g a day. Boys get more of leafy vegetables, more of fruits and more of calcium. As children grow up and take up a sedentary or active job, the gap in the degrees of nutrition widens even more. For instance, a woman in a sedentary job gets 403 g of cereals a day. A man in a similar job gets 475 g.

The malnutrition prevalent in a significant proportion of adult Indian women can be attributed primarily to inadequate food intake. The households that have theoretically enough food but the way it is distributed may leave women inadequately nourished. Typically, adult men and male children are fed first. Women eat only after the men have finished, and a young wife must allow her mother-in-law to eat fist. Whatever is left is divided among the young mother and her female children. This disparity in the distribution of food may be as much a matter of poor communication as of deliberate practice since men are normally unaware of how much women eat. Whatever the reason and given the nutritional demands of child bearing and lactation, the lack of nourishment puts women at particular risk during their childbearing years. Indian women may be malnourished because of the poor nutritive quality of food.

Unequal Access to health Care and Culture

Medical and health services too are not evenly available to men and women. For every three men who avail facilities of medical institution, only one woman does so. This is not because of greater *healthiness* in women but due to low importance accorded to women's ailments women are not only neglected by others but they also neglect themselves. Fewer women declare themselves ill as compared to men.

A young paediatrisian in Jammu hospital came across a patient, a girl about 5 years, who had lost a lot of blood in an accident. The girl had to be given blood immediately. The doctor approached the girl's father and asked him to donate a unit of blood as his blood group matched with the blood group of her daughter. The father was reluctant. She is to go away after her marriage so why should I harm my health for her, 'the father said. Even after a lot of persuasion, he did not relent and walked away. The doctor was stunned. The girl, it was learnt, died a few days later.

Access to health care is determined by *need, permission, ability* and *availability*. Use of health system is determined by perceived need and often *need* is limited to the child bearing years. Even in the childbearing year's, the use tend to be limited, *e.g.*, the NFHS reported that of the births which occurred in the previous 4 years, 62% of mothers had received anti-natal care, while 26% had delivered in medical institutions and 34% were delivered by trained medical persons. The reason for such low levels of use is that often women perceive the services to be unnecessary. Social and economic norms determine *permission* and *ability*. Often women need permission to seek health care. When either there is dependence, physical, or financial on caregivers, the access is determined by the willingness and ability of the caregivers to provide treatment, *e.g.*, women who needed treatment for visual disability if they had no sons or could not find alternative escorts. Availability depends upon the geographical proximity ones purchasing capacity.

There is considerable evidence that male children receive more and better health care at an earlier age than the female children receive. At the All India Institute of Medical Sciences, of all the surgeries, which are conducted on children and infants, 70% are on boys and only 30% on girls. From 1986 to 1999, the number of neo-natal surgeries carried out in intensive care was 2,089. Of these, 1,441 were boys and 648 were girls. On an average, in these 15 years, around 69% of the surgeries were conducted on boys. In 1998 as compared to 117 boys operated in intensive care, only 57 girls were operated. In 1999 of the total 152 surgeries in this category, only 47 were girls. On an average, the same number of boys and girls fall ill and develop complications post-birth, but as compared to boys, less than half the numbers of girls are brought to hospitals. It may be because when a child has to undergo surgery, the entire family gets involved. Such sufferers are usually major and involve expenditure. Though the treatment may be free, there are several overhead costs in surgery. This requires longer hospitalization and for a family from outside Delhi, this involves the cost of living in the city.

One study in the Ludhiana district of Punjab found that lower caste female children under the age of 3 received less medical care than other children and that the medical care provided tended to be of poorer quality. Two decades later, Dasgupta found that the situation had not changed. A hospital study in Ludhiana concluded that gender bias prevented 3 out of 4 girls who were ill enough to require hospitalization from receiving care they needed. In rural Karnataka, twice as many boys as girls were brought to the primary health centre, and in a study of one rural area in western Uttar Pradesh, over a week period roughly 3 times as many as boys and girls were brought to the primary health centre. Recent state-level reports for Uttar Pradesh confirm these earlier findings. Seventy-three percent of male children compared with 62% of female children were taken to a health provider or health facility in the event of an acute respiratory infection. In 1992-93 for India, overall more boys than girls were vaccinated and treated for acute respiratory infections and fever.

Like female children, adult women are at disadvantaged with respect to health care. Household surveys typically report more incidence of illness for females than for males. This finding is especially striking because female morbidity is likely to be under reported for several reasons, *i.e.*, women respondents may be reluctant to reveal details of their illnesses to interviewers, they may purposely downplay any health problems to avoid disrupting domestic duties, or consistent with their socialization, they may not wish to imply that they are asking for medical care. Community-based studies make it clear. However, that one reason for the relatively high morbidity rates among women is that women receive less medical treatment than the men receive. Hospital and health centre records provide supporting evidence. Fewer women than men are hospitalized or visit clinics. Community-based studies also reveal that a high proportion of women receive no treatment at all for their illnesses and that among those who do self-care, home remedies and the traditional medical care are the most common methods used and men are more likely to receive modern medical treatment including high-quality institutional care.

In addition, our health system is institution based rather than domiciliary. This generates inhibitions among the rural population about alien environment of the hospital. Moreover, women are burdened with greater disabilities than the men are. They have paucity of time, monetary resources and childcare facilities. Mobility of women is far more restricted than that of men, thus, they are deprived of utilization of facilities rendered by the health system.

Gender, Culture and Development: Value loaded/Free

Development is about change. Development initiatives by government, NGOs, or development agencies, are investments in promoting social and economic change. Some development initiatives aim to change values and practice that shape social relations- consider, *e.g.*, the investment made in family planning implies about family structures. Development models also incorporate cultural values-consider, for example, the concern with the transition to market economies and the support for private property as cultural values. Other types of initiatives less obviously concerned with cultures nevertheless have impacts on the social relations that characterize a culture, *e.g.*, consider the possible impacts of an improved road network linking rural and urban areas. Many villagers could benefit from better access to health services for health care.

Decisions made in planning an initiative shape the type of impact that it will have on culture. Even if gender equality is not explicitly considered, decisions made in the planning process will have an effect on gender equality.

An infrastructure that restructures the health delivery system could include consideration of gender aspects of reproductive right, affordability and availability of health care, or not, in which case, it misses the opportunity to include maternal and child healthy specific programme.

These examples show that the decision taken in planning is not neutral with respect to gender equality, even where gender issue is not included.

Conclusion

Futuristic approach to health development is a challenge to move beyond the decolonization of health. Such a challenge requires that the field of health draw from the theories and practices of cultural studies and other areas of humanities rather than rely solely on the medical model in health problems of women in India.

Those who develop health programmes must examine carefully the differences as well as the similarities in cultural perceptions to understand the health beliefs and practices more fully and to address them appropriately within them particular contexts. Culture develops within the process of a people wrestling with the natural and social environment, and thus, embodies the moral, aesthetic and ethical values that are manifested in the people's consciousness.

Eurocentric approaches to global health processes should be deconstructed and reconstructed to reflect varied cultural expressions. Attempts to affirm cultural differences in communities often manifest themselves in calls for programmes and curriculum integration, although programme and curriculum reformation/transformation is a more progressive approach to centralizing diversity in learning process.

■■■

7 Creating Positive Mental Health with Food

Pallavi Bhatnagar

Lucknow University, Lucknow, Uttar Pradesh, India

E-mail: <dr.pallavib@gmail.com>

अन्नाद्वै प्रजा: प्रजायन्ते। या: काश च पृथिवी ङ्क श्रिता:
अथो अन्नेनैव जीवन्ति। प्रथैनदपि यन्त्यन्तत:। अन्न
ङ्क हि भूतानां ज्येष्ठम्। तस्मात्सवौ षधमुच्यते। सर्व वै ते
ऽन्नमाप्नुवन्ति। ये अन्नं ब्रह्योपासते। अन्न ङ्क हि भूतानां
ज्येष्ठम्। तस्मात्सवौ षधमुच्यते। अन्नाद् भूतानि जायन्ते।
जातान्यन्नेन वर्द्धन्ते। प्रद्यते प्रति च भूतानि। तस्मादन्नं
तदुच्यत इति ॥1 ॥

All species of beings resting on this earth are begotten from food. Food protects the lives of all beings. All beings finally return to food, verily those who worship the eternal as a food (giver of food and life) attain the mastery of food. Food is the greatest of all elements (things), and for this reason, it is called a medicine-like. All being are born and develop from food. Food is called *anna*, which is eaten by all beings (Chapter-II Taittirya Upnishad)

Food is not only the best medicine but also the best long-term treatment available in any system of medicine all over the continent. It would not be an over statement that wrong diet is at the base of 40% of diseases.

Past 5 to 6 decades have brought about a complete change in how we approach food from the point of production to preparation and consumption. Taking a spectrum from affluence on one extreme to below poverty line hard way of living, on the other hand, the attitude for food seems to have different shades of health imbibed in it. On one extreme, we see a mindless eating to either combat boredom or dealing with loneliness, stress, tension, or even for addressing arguments and anger where one is often unaware of the actual need for food. Accessibility of food seems to match with the purchasing power for this end of the continuum. On the other extreme, the need to sustain against all odds is reflected in malnutrition and number of diseases attached to it, helplessness, anger, pain of deprivation and at denial of even two square meals, often resulting in disease, poor quality of life, depression and at times volatile expression of crime and violence for food.

Nutrition and health are not synonymous but without good nutrition, health cannot be at its best. The Nutrition Society defines public health nutrition as the science and art of preventing disease, prolonging life and promoting health through the medicine of sustainable and equitable improvement in the food and nutrition system.

The need of the hour is to reinforce the public health education, particularly salience of healthy food habits and life style for physical, mental and emotional well being by the community health workers. There is voluminous evidence that people living in difficult circumstances below poverty line, have unhealthy diets more because of there poor socio-economic context then irresponsible choices. Eating the *right* balance of food and setting the *right* balance between energy input from eating and drinking and energy expenditure through activity and exercise are of critical importance.

Good nutrition in *Caraka Samhita* is eating the right kind of food with the right beverage, eating the right amount of food at the right time with the right company, in the right place and right frame of mind.

Volumes of literature exists on the relation between food and physical health, however, the connection between mental health and food is less talked about and needs greater attention of the professionals dealing with food and mental health. More so, when health literacy and health improvement initiatives are emerging as significant aspects of health promotion policies and mental health are being integrated with physical health. Good mental health has been defined as the emotional and spiritual resilience which enable enjoyment of life and the ability to survive pain, disappointment and sadness and as a positive sense of well being and an underlying belief in our own and others dignity and worth.

Food and Mental Health Link

Just as brain influence each cell of the body, research validates that our diets affect how our brain works. Though there is no *magic bullet* or a single nutrient, which holds the key to good mental health summary of major research findings on combination of nutrients most commonly detrimental for positive well being and good mental health are being given below:

Polyunsaturated Fatty Acids: Particularly the omega-3 types are found in oily fish and some plants.

Minerals: The minerals such as zinc is found present in whole grain meat and milk, magnesium in green leafy vegetables, nuts and whole grains and iron in red meat, green leafy vegetables, eggs and some fruits.

Vitamins: The vitamins such as folate is found in green leafy vegetables, a range of vitamin B in whole products, yeast and dairy products and antioxidant vitamins such as C and E in a wide range of fruits and vegetable.

People eating diets, which lack one or more of this combination of polyunsaturated fat, minerals and vitamins, seem to be at higher risk of developing condition like attention deficit and hyperactivity disorder (ADHD), *i.e.*, a range of depressive conditions and schizophrenia dementia including Alzheimer's diseases. Though there is not much evidence that diet management alone can cure these conditions, however, there is evidence that the combination of polyunsaturated fats, minerals and vitamins may help to reduce the side effects of some medication, improve the efficacy of medication and alleviate some symptoms too. Food can have an immediate and lasting effect upon mental health and well-being. Certain foods have been found to help people manage their mood and feelings better.

Food is one of the most salient physical supplies and love is an equally significant emotional supplies. There is an intimate relation between the two, the base for which is created in infancy itself when the mother feeds the child, it is not just the supply of nutrition but also the caressing, cuddling and affectionate talking and loving handling of the child while feeding that tends to get associated with food.

Few people are aware of this unspoken bond. The object relations theories, *i.e.*, an extension of Freud's classical psychoanalytical theory clearly and explicitly reiterate that the mother is the first love object for the child and the satisfaction of the relationship needs would be detrimental for her/his later mental health.

During years of counseling and facilitating people, I have come across some facts pertaining to food intake, which validates this theory. A number of patients particularly adolescents, when asked about the food intake have confessed that they do not eat food when handling pain, anger and frustration. Keeping his identify confidential, I would share a conversation with a 17 years old boy.

Counselor: I understand from what you have told me that you get angry often.

Client: Yes, it is right; I am a very angry person. I get angry every now and then.

Counselor: What do you do when you get angry?

Client: Well, I just do not eat. I just sit quiet or lie down and I do not eat food the whole day.

Counselor: You do not eat the whole day. How often does it happen? Let us say once in a week.

Client: No, It happens four days in a week. Actually, I do not eat anything whatever mummy prepares. At times, I cook maggi or pasta.

Somewhere, the client (boy) had been feeling rejected by his mother for a long time. Now he was refusing the food giving a message of rejection of mother. When parents having a turbulent relationship with their children are asked about the food habits and intake by the child, they say that either the child *refuses to eat/eat with us* or more often the complain is that child is fussy and would eat junk food or anything outside but does not eat food cooked at home. The child's denial for food at home is an issue in family therapy for parents particularly the mother to work upon in terms of the emotional supplies. In most mental and emotional disorders, the sudden change in food habits, either excess or withdrawal needs to be taken seriously. Often it serves as one of the symptoms.

Feeling Low: Something Lacking in Your Diet

Often people are unaware of what is it, which is lacking in their diet and could be attributed to some extent to their mood state. A brief account of some emotional states of every day life and the contributing role of food element, which is lacking is given below in Table 1.

Table 1: Deficiency/near absence of food substance and resultant negative mood state

Deficiency/near absence of food substance with example	The Resultant State
Vitamin B_3 (whole grain and vegetables)	Depression
Vitamin B_6 (whole grain and bananas)	
Vitamin C (vegetables and fresh fruits)	
Folic acid (leafy green vegetables)	
Magnesium (vegetables, nuts and seeds)	
Omega-3 fatly acids (nuts, mustard oil, fish oil and flax seed oil)	
Selenium (wheat grain, fish garlic, sunflower and whole grain)	

Deficiency/near absence of food substance with example	The Resultant State
Tryptophan (fresh fruits, lean meat and nuts)	
Zinc (seafood, nuts, fish and green leafy vegetables)	
Vitamin B_6 (whole grain and bananas)	Irritability
Magnesium (vegetable, nuts and seeds)	
Selenium (wheat grain, liver, fish, garlic, sunflower seeds and whole grains)	
Zink (seafood, nuts, seeds and fish)	Lethargy
Magnesium (vegetable, nuts and seeds)	Insomnia
Omega-3 fatty acids (nuts, mustard oil, fish oil and flax seed oil)	Poor memory
Vitamins B_5 (whole grain and vegetables)	
Vitamins B_6 (whole grain and bananas)	
Vitamins B_{12} (meat, fish, dairy products, boiled and poached eggs)	

Processed food: Watch out ....

Studies contend that higher levels of behavioural and emotional problems were found to be associated with the western style of eating, *viz.* red meat, confectionary, soft drinks, white bread and unrefined cereals. One in five children was expected to develop some form of mental health problem by the time they reach adulthood, while 50% of the adult's mental health problems developed during adolescence. However, in a study on 1600 West Australian 14 years old, such problems were less in adolescents who ate more of fruits and vegetables. A current trend since 80s is an increase in the consumption of soft drinks and processed food, and thereby increasing energy level and paradoxical increase in sedentary behaviour *viz.* watching television and using computer. The total energy supplied by food and beverages has increased, as the food has become more processed and more energy dense. Another research (Arnold 2010 in British Journal of psychiatry) on a sample of 3486 patients on whole and processed food preference came up with result that those in the highest 20% of processed foodstuffs, like fried food and sweetened deserts and high fat dairy products, intake showed a 58% increases in risk of depression, while those on the whole food diet had a lower depression risk. Thus, processed food has a number of potential health hazards if consumed on a regular basis. Associated with processed food is another fact.

Food is cold and cool/warm and hot. Excessive intake of uncooked and cold food will impair yang-qi in spleen and stomach and create dampness, causing abdominal pain and diarrhea. Excess intake of hot food adds up heat in the stomach and the intestine, leading to thirst, abdominal fullness, distension and pain, constipation, hemorrhoids, *etc.*

Eighty per cent of the world's population is consuming caffeine daily largely. It is used in coffee, tea, chocolate, cocoa and cola drinks with the effect of arousal and mood elevation. Coffee has achieved symbolic status as recreational and exotic beverages. Canned food is equally detrimental, even poisonous to our system. They do not allow our body to absorb the nutritional value within the food. These are spiritually empty food. Since these foods are altered by the process in which they are prepared, they are termed as *empty foods*. Fast food is another example of empty food. The processing, canning and freezing of this food removes most of the nutrients from the food. It is like eating a cardboard. It is important to remember that fast food may put you on the fast track but it could be a track, which is spiritually empty. It is advised that instead of refined white flour one should go for whole grain. White flour, *viz.* white bread and products made from refined white flour, is yet another class of empty food.

The body gets into a state of imbalance when it is fed too many processed foods and additives. White flour and sugar contribute to ill health, as the excessive amount of meats, dairy products and processed foods do. All of them have a tendency to burden the body with toxins.

Do you know that your depression could be due to some food allergies like citrus fruits, coffee, dairy products, nuts, soybean, or even tea?

Cigarettes are substitute for many things. They can be a smoke screen to keep people away, a substitute for companionship, a way to control feelings, a coping for negative and depressed feelings, or even a method to control weight. No matter whatever the reason is, once a person starts smoking, it becomes an addiction soon. My concern here before you is that taking another cigarette, just ask yourself whether you really need it. Take ten to fifteen deep breaths and confront the real reason. Remember if you feel good about yourself and care for yourself, you will work on it.

Remember, stress can be responsible for letting nutrients into the bloodstream before complete digestion. This causes and aggravates the allergy. Hence, just take time and reflect. When did you last have the allergy? Check it out were you stressed out at that point of time.

The healthy option to processed food or other junk food is wholesome food. Foods that maintain the balance of the seven tissues and the three humors are wholesome, and those that imbalance the humors and tissues are unwholesome. Agentives further gives the twenty best food articles for balancing the humors and twenty worst foods for unbalancing the humors.

Make sure that you take whole grains, such as rice, barley and whole meal bread in at least one meal, green vegetables, fruits, nuts seeds, monounsaturated fats such as olive oil, omega-3 polyunsaturated fatty acids found in oily fish, soybean, walnut, *etc.* and minimize trans fatty acids such as airy products, cakes, biscuits, *etc.* and saturated fats found mostly in animal products.

According to *Caraka Samhita*, the wholesome food causes the growth of living beings and unwholesome food the growth of disease. According to *Ayurveda* system of medicine, balance diet for pertains to balance in the three humors of *vata*, *pitta* and *kapha* and again which is the digestive ability. Food intake needs to be in terms of the constitution of the person. Another way of looking at it is that if food grows, eat it, if it does not grow, do not eat it, says Hay. Fruits, nuts, vegetables, grain all grow. Refreshing colas and canned food articles do not grow so should not be taken. It is well understood that food items, which grow, have a nourishing quality in them. On the other hand, the processed food items may be good and tempting to look at but they do not sustain, add anything positive, or nourish the body and mind. Hence, remember that we are what we think and what we eat.

Food for Brain

The brain eats more than any organs but few people realize that brain also happens to be the busiest organ in the body and it is working even while we are sleeping. Hence, it requires a vast amount of nutrients to function optimally. The brain takes up about 3% of an adult's body weight yet it consumes 25-30% of its fuel in the form of blood sugar or glucose. Interestingly unlike liver, there is no storehouse in the brain for storing the nutrients, which speaks for the constant demand of nutrients. A lack of nutrients to the brain means an immediate cutback on production of the chemical that supports the mind and keeps it peaceful.

According to Markus for the optimal functioning of brain, it requires plant and animal foods. Plant foods are fruits, nuts, berries, rice, oats, beans, sprouts, vegetables, seeds, leaves, whole wheat, millet, peas, potatoes, sea vegetables and legumes, and on the other hand, the animal food comprises of eggs, chicken, beef, lamb, cheese, fish, turkey, pork, milk and sour cream.

Plant foods are a rich source of proteins. Hence, it is essential to discuss briefly how proteins are important to the brain. The word protein is derived from the Greek word protein, which means of first importance. Proteins contain 20 or more amino acids, which are not only the basic units of growth but are also essential for the functioning of brain. With the consumption of proteins, more nor-epinephrine and dopamine are available, both of which increase the alertness. On the other hand, serotonin is the feel good neurotransmitter. Amino acids from proteins make up the enzymes that regulate neurotransmitters. Enzymes help in neuronal-receiving, processing and interpreting vital information. It has been empirically established now that fruits and vegetables not only reduce the chances of heart attacks and strokes but also contribute towards longevity. Foods rich in protein are construed as food for thinking so how about checking up your protein intake.

Fruits, vegetables, nuts, seeds and food grains contain the vitamins to which the brain needs, hence, it is relevant at this juncture to throw some light on the relation between vitamins and the brain. The word vita originates from the French word *Vital*, which could be taken up as life. Thus, vitamins promote life in the brain. Vitamins are co-enzymes, as they are essential for the neurotransmitters. Some of the major vitamins important for the functioning of brain are thiamine (vitamin B_1), which breaks carbohydrates from food, converting them into glucose, which provides fuel to the brain for functioning. Pyridoxine (vitamin B_6) helps in the synthesis of all monoamines, which are a class of neurotransmitters that include serotonin, dopamine and norepinephrine. These are the natural mood enhancing neurotransmitters. Vitamins without minerals are of no use. In fact, vitamins have a controlling relation with minerals. Brain function depends upon minerals, which activate neural communication. Besides, they also regulate brain metabolism and protect the brain from toxic metal contamination. According to Marcus, minerals are the spark plug to the brain. Do you know that zinc is known as the intelligence mineral since it is involved in many functions?

To mention a few, it is required for mental development, maintaining blood sugar control, crucial to mood, memory and behaviour. Besides, it is also needed for the development of healthy reproductive organs, particularly the prostate gland and for protein synthesis. Humans cannot function without zinc. Zinc deficiency has been recognized as a factor in various disorders including growth retardation, slow wound healing, psoriasis, delayed sexual maturation, poor immune functioning, irritability, poor stress control, attention deficit disorder, hyperactivity, *etc.* Zinc deficiency often also leads to an increase in the level of copper in the blood. Research evidences clearly show an association of elevated copper with hyperactivity, depression and even schizophrenia. Food sources for zinc are ginger, roots, potatoes, oysters, round steak, lamb, peas, *etc.*

Chromium is yet another mineral very crucial for the brain. Chromium works with insulin in the metabolism of sugar, thus, ensuring the efficiency of insulin. Remember, if you are eating too much of the empty calorie sugary foods, it would deplete chromium level in the body. Looking for the chromium rich food, well some of them are broccoli, ham, turkey, grapefruit, grape juice, *etc.*

Magnesium is yet another mineral needed for the optimum functioning of brain. Magnesium not only keeps the brain cells alive but also minimizes the negative effects of reduced blood flow and is responsible for increasing the fluidity of cell membranes, thus, improving the capacity of neurons to receive nutrients. The role of magnesium in converting blood sugar into energy and maintaining nerve and muscle functioning makes it a most vital mineral. Magnesium is used in more than 300 enzymes, many of which are brain enzymes. Magnesium also is said to produce a calming effect on the brain. Most people are not aware that low levels of magnesium have been found to have correlation with aggressive behaviour, alcoholism, anxiety, attention deficit disorder, over excite ability, PMS, dementia, depression, fatigue, autism, insomnia, learning disabilities and even schizophrenia. To mention a few, they are raw nuts and seeds, whole grains, dark green leafy vegetables, bananas, dried apricots, dry mustard curry powder and cocoa.

When we are talking about food for the brain, I am sure you must be thinking about the role of fat. Like many other organs, brain is largely made up of fats. However, the issue is what kind of fats is healthy for the brain. There are two major types of essential fat, *i.e.*, omega-3, which are oils from flax seeds and fish, and omega-6, which are oils from plant sources such as natural vegetable oils and evening primrose oils. As with carbohydrates, fats are the sources of energy. In fact, fats are the most concentrated sources of food energy. Each gram of fat provides nine calories is good enough of evidence. Vitamin A, D, E and K are fat-soluble vitamins. We all need good amount of fat in our diets and on the bodies for not only production of hormones and protection of the internal organs, insulating them from cold but also to maintain mental stability and concentration and avoid carbohydrates cravings for keeping the bowels regular. Women may also experience fertility problems without enough body fat.

The brain needs certain amount of saturated fat, polyunsaturated fat, cholesterol. However, hydrogenated fats and partially hydrogenated fats synthesized from natural fats need to be avoided.

Failure to eat enough omega-3 fat is scientifically linked to an array of modern mental disorders and problems like depression, poor memory, low intelligence, learning disabilities, dyslexia, attention deficit disorder, schizophrenia, senility, Alzheimer's disease, degenerative neurological diseases, multiple sclerosis, alcoholism, poor vision, irritability, hostility, inattention, lack of concentration, aggression, violence and suicide.

Serotonin-Stress-Food Cycle

Although there are several neurotransmitters, which play an important role in elevation or deflation of mood but serotonin is one of the most significant neurotransmitter. Serotonin is responsible for a feeling of calm and relaxed state of well being. Serotonin is formed is the brain from the amino acid tryptophan after we eat carbohydrates. The foodstuffs, which lead to greatest serotonin production, are the starches and sugars so need to be eaten in moderate amount.

Some people crave starch-laden food items like bread and potatoes and sugary foodstuffs such as cakes, sweets, *etc.* Eating such food articles gives a sense of feel good. This comfort food craving is an eating pattern often driven by high level of chronic stress.

Let us understand this, when there is chronic stress, it leads to the production of excess stress hormones in the body, particularly cortisol. When cortisol is produced, it inhibits serotonin production. People tend to feel anxious and depressed when serotonin level drops and sleep is disrupted. Subsequently, cortisol levels are increased and serotonin levels are depleted. This becomes a vicious cycle associated with insulin resistance and weight gain. That explains why many people under chronic stress experience intense craving for starch and sugar. Resisting this carving is difficult since it is so strongly physiologically driven. Stress management is the only answer to this. I would say that good nutrition, sufficient sleep and regular exercise help one manage the stress effectively and the stress driven carving for such food.

Most people feel relaxed and lazy after a big feast. The question that arises is- if you have eaten well, why you feel tired. Changes in brain biochemistry may be the reason. The food we eat supplies the precursor molecules for manufacturing neurotransmitters that influence our emotions and mood. Some researchers also contend that by selecting the right kind of food, we can encourage relaxation and alertness.

Big meals, especially those with a lot of fat, take a long time to digest, and with a full stomach, we feel like relaxing rather than working. On the other hand, smaller meals, which are low in fat, take less time and energy to digest and leave us feeling more energetic and alert. Meals, which comprise predominantly of carbohydrates, encourage production of the neurotransmitter *Serotonin* that makes us feel drowsy and relaxed. High carbohydrates meals are a prescription for relation and may be the reason behind overeating by some, which makes them feel good. A small high-carbohydrates snack before bedtime can help in sleeping. Many people find that a meal or snack with carbohydrates but little protein, especially in the middle of the day leaves them drained out and tired. Meals, which include a small serving of protein food with or without carbohydrates, encourage alertness by favouring production of neurotransmitters, such as *dopamine* and nor-epinephrine. A small lunch that includes protein foods is best for working people who need to stay alert in the post-lunch period.

Food to Build Your Stress Resistance

Once you have your blood sugar and neurotransmitter levels under control, you need to focus on your food intake and here are some suggestions, which would help you, make positive choices about your food.

Eat More Fruits, Vegetables and Grains

Intake of fruits, vegetables and grains not only provide energy, vitamins, minerals and fiber but also help in regulating digestive functions of the body. Some vegetables such as soybeans and dried beans are good source of protein especially for vegetarians. These foods are also referred as nutrients dense, which means they supply lot of nutritive value per calorie. Choice of salad, steamed or raw vegetables and root vegetables such as potatoes or carrots may not only add nutrition to your diet but you will also feel more robust from within to fight out stress.

Drink Plenty of Fluids

Often tiredness results from a state of dehydration or inadequate fluids in the system. Dehydration causes fatigue and irritability. A note of caution is that caffeine and alcohol should substitute water and other fluids. The intake of caffeine and alcohol not only increase stress but they also dehydrate you, and thus, increase your fluid needs. Remember that the urine is pale in adequate hydrated state and dark coloured urine is a sign of dehydration.

Monitor Your Intake of Fat and Sugar

Too much dietary-fat is said to contribute to artery disease, stroke obesity and even cancer. The dietary replacement of facts is fibers and starch diet. High fat diet also leads to obesity and associated risk factors such as Type II diabetes and hypertension. As fat is calorie dense, it offers little nutrition but lot of calories. Such individuals should avoid the intake of fried food, baked food and snacks. Instead of potato chips, roasted nuts, cakes, cookies, whole milk cream, ice cream and butter, one should take non-fat yogurt and cottage cheese and should take care of everyday sugar intake. Most of the nutrients dense food items have sugar and most of the time people are not even aware of how much sugar they are consuming.

Focus on Fitness not Fatness

This is for the people who are weight conscious and go for dieting. I would share with you at this point a concept of weight cycling, which means frequent dieting interspersed with periods of normal or greater than normal caloric intake. One should avoid weight cycling as it has a number of harmful side effects like hypertension and coronary heart disease. Instead focusing on long-term life style improvement, increasing physical activity, managing stress and developing healthful eating habits would improve your stress resistance. Our attitude defines life, which is best for those who want to live it, however, life is difficult for those who want to analyze it and life is worst for those who want to criticize it. Our attitude defines life. Enjoy your life, laugh so hard that even sorrow smiles at you, live life so well that even death loves to see you alive, fight so hard that even fate accepts its defeat.

Reduce the Intake of Empty Foods

Be careful and consciously decrease the intake of sweets, empty calorie food and junk food, which contains very few vitamins and minerals. Your diet each day should include the followings:

Whole grains such as barley and whole meal bread, vegetables, including green leafy vegetables approximately 4 cups, fruits about 4 cups, nuts and seeds about 1 cup. It is always good to maximize the consumption of monounsaturated fats such as olive oil, omega-3 polyunsaturated fatty acids (PUFAs) found in oily fish, soy, walnut, *etc.* and minimize the intake of trans fatty acids, *viz.* lamb, beef, dairy products and margarine (cakes and biscuits) and omega-6 polyunsaturated fatty acids (PUFAs) in animal products and vegetable oils.

At this point, I would like to explain the difference between omega-6 and omega-3. The fatty acid omega-6 originates inflammation, constricts blood vessels, encourages blood platelets to stick together, forms rigid cell membranes and even counteracts free radicals, which destroy cells. Omega-3 reduces inflammation, dilates blood vessels, restrains blood platelets from sticking together and also forms cell membrane flexible. If you do not eat fish, take flaxseed oil (having more omega-3). However, if there is a problem and you cannot even have that, then use olive oil, which would give you some amount of omega-3.

Looking for the Happy Food

The biochemical base tryptophan is the precursor for serotonin so food-containing tryptophan is another way of being happy. Hormones including happiness hormones come from amino acids. Some of the amino acids are produced in the body, while some have to be supplemented in terms of food.

Key Chemicals

Dopamine: It is the neurotransmitter that makes one seeks pleasure. It not only produces enthusiasm, exuberance and pleasure but also creates motivation and anticipation. In other words, it is driving force within. When a person has a pleasant experience, dopamine reinforces it and wants to have the same experience again. Now, you know why it is also associated with addiction. Dopamine is an amino acid and found in milk and wheat germ.

Noradrenalin or Nor-epinephrine: It is synthesized from tyrosine. Although it is typically associated with stress, yet in right circumstances, it elevates mood.

Endorphins: These are the body's natural painkillers or opioid named after *endogenous*, which are made up of body and *morphine*. Hence, like morphine's endo-morphine tends to create euphoria. The endorphins can be released by hot spices.

Phenyl ethylamine (PEA): This is the amphetamine like substance, which produces a feeling like *walking on air*. It is found in chocolates and some soft drinks and also produced during vigorous exercise.

Serotonin: It is a neurotransmitter, which makes one feels peaceful, unaggressive, monogamous and contented. However, too much of it can make one lethargic and sleepy.

Testosterone: It is the male sex hormone, though it is also present in women at lower levels. It increases sex drive but too much of it causes aggression and irritability. Intake of meat increases testosterone.

The Happy Food: Some Facts

Caffeine: It is psychoactive drug, which can boost mood and increase mental energy. The dose and the individual make up are also important. Caffeine resembles the structure of a brain chemical, which is referred to as adenosine. The role of adenosine is to dampen activity, however, caffeine can displace it from the nerve cell receptor sites, and subsequently, the mind is more alert. The optimum does is around 100 to 200 mg in the morning and same again in the afternoon but hold on large does of caffeine triggers anxiety. In susceptible individuals, even one or two cups can cause panic attacks. Also remember that coffee is addictive.

Hot Peppers: Hot peppers give boost to happiness. Actually, it contains capsaicin- a chemical responsible for burning sensation but capsaicin is also a painkiller. When one eats hot peppers, the burning sensation sends messages to the brain, which responds by releasing endorphins. Thus, with each additional pepper one gets an additional sense of feel good. Further, hot pepper also clears the sinuses, acts as a decongestant and keeps the blood mobile.

Chocolate: The mood-enhancing chemical is cocoa, hence, the type of chocolate, which has 60% cocoa, which works as a mood elevator. Generally, cocoa has chemicals like caffeine, theobromine- a mild stimulant, which creates a sense of well being, phenyl ethylamine (PEA) - an amphetamine like compound, which also gives a *feel good* feeling, anandamide boosting chemicals, which increases a sense of pleasure and procyanidius, which relaxes muscles.

Complex Carbohydrates, Dried beans, Vegetables, Cereals, Bread and Pasta: which can be clubbed into the category of complex carbohydrates, which also work as mood elevators. The fuel of brain is glucose. However, eating sugar is bad method of providing energy. Brain needs a constant and steady supply, which comes best from complex carbohydrates. They break down slowly and release energy at a constant level. It may be said that complex carbohydrates increase the production of serotonin. Hence, they are *happiness foods*. Hello! Just a note of caution does not upload you with a diet of complex carbohydrates because carbohydrates do not work the same way for everybody. Some become more lethargic and dull after eating the complex carbohydrates food. So, be sensitive and monitor what food makes you feel good.

Folic Acid: It is a kind of vitamin, which gives a sense of well being. It is found in green leafy vegetables so increase intake of vegetables like spinach.

Oats: You will be surprised what all oats can do. Since oats contain tryptophan, which is a precursor for serotonin as well as vitamin B, calcium and magnesium, it will not only give sense of well being but also lower the undesirable cholesterol (LDL) by about 20% and increase the good cholesterol (HDL) by about 15%. More than that, oats also have anti-cancer properties.

Garlic: Garlic is a pack of magical chemicals for health both physical and mental. Garlic not only boosts mood but also works well in reducing anxiety, irritability and fatigue. What more, it improves the immune system, fights cancer, acts as a blood thinner and it is an expectorant, decongestant and a powerful antibiotic, hence, include one or two raw cloves of garlic in your diet at a regular level. Do not worry about the smell, as it will go with regular use.

Besides, researches also reveal that alcohol, caffeine, white rice, fried foods, sugar and animal fat can make you feel more energetic initially but are only good for a quick burst of energy. This is especially true for caffeine and sugar. After a while, you feel more tired than before taking them. Alcohol is a depressant and when taken in excess tends to make a person low more so watch out and be more mindful of your choices.

Associated with these stress driven cravings a need to *feel good just for this moment* is the concepts like emotional eating and compulsive overeating eat. Hence, an attempt is being made to address them briefly.

Emotional Eating

When one eats in large amount in response to one's feelings, especially when one is not hungry it is referred to as *emotional eating*. In other words, instead of body, the emotions dictate when and how much to eat, while some emotional eaters binge when they are sad or confused. Some others resort to eating to avoid thinking about some problems or taking any positive action to solve them. Often, the emotional eaters eat without realizing that they are eating. Five common cues of emotional eating are:

- **Anger:** When angry with self/others/situation, the person stifles feelings using food rather than confronting and releasing them. It seems easier to smother a problem than to deal with.
- **Hopelessness:** When there are thoughts like those that nothing is ever going to change/get better for me, thus, why should I care about my health or weight? Besides, eating makes me feel better.
- **Lack of Control:** When one feels as if she/he is losing control of herself/himself with thoughts like "I have nothing in my life that I can control, everyone controls me and I cannot do anything on my own except eating". I can eat whatever I want or whenever I want it.
- **Feeling unappreciated:** When one works for significant appreciation from others especially after some achievement and the effort go unnoticed, one may be tempted to go for binge eating.
- **Boredom:** When one feels there is nothing to do, no place to go to, even feels lonely and unable to find any concrete work, one tends to indulge in eating as time pass.

Emotional eating tends to multiply one's problems. Instead of working on the problem, the person goes on a binge eating, creating the problems of weight gain, guilt about eating, and subsequently, health problem managing emotional eating and moving out of the vicious cycle would entail not just identification of one's triggers but also being more aware of one's body and hunger signals. Besides, keeping a check on stocking food and making mindful efforts not to skip meals as well as generating options for food items would help. The final and most important one is not to give up the effort of moving out of it. Keeping a food diary may be a good option to become more mindful by recording one's feelings before after eating various foods.

Compulsive Overeating

Compulsive overeating is also referred to as food addiction and is characterized by an obsession or compulsion for food. An individual suffering from compulsive overeating disorder engages in frequent episodes of uncontrolled eating/binge eating during which she/he may feel out of control, eating beyond comfort. Such form of binging is often followed by feelings of guilt and depression. Compulsive over eaters do not attempt to compensate for their binging with purging behaviours such as fasting, using laxatives, or vomiting unlike bulimia nervosa, which is another eating disorder.

Compulsive eaters will typically eat even when they are also not hungry. Their obsession is evident in the excessive amount of time and thoughts devoted to food and they secretly plan or fantasize about eating alone. Compulsive overeating often leads to weight gain. However, it does not mean that anyone who is obese is also a compulsive overeater. At times compulsive, overeater may also engage in grazing behaviour that is a behaviour in which one returns to pick at food through the day, which results in a large overall number of calories consumed even if the quantities eaten at any one times may be very small.

If left untreated, compulsive overeating may lead to serious medical conditions including high cholesterol, diabetes, coronary heart disease, hypertension, major depression disorder, arthritis, kidney problem and even stroke.

Some people have a craving for fatty foods often referred to as *fat gram* eating excess amount of fat which has its consequences like blocked arteries and weight gain to mention a few. What seems unfortunate is that most of us in our childhood were brought upon a high fat diet, and therefore, it becomes a challenge to start on simple food.

The conditioning is so strong between food and high fat diet that even the semantic of health, healthy person and mentally healthy person in Indian context generate the concept of one who eats well (*Jo khoob khata peeta ho doodh dahi ghee ki acchi khurak khata ho mota taja ho vohi sehatmand hai*).

In two of the mental health disorders associated with eating, *i.e.*, anorexia nervosa and bulimia nervosa, the most significant factor is self-devaluation and self-rejection for what one is manifested in binge eating and purging. The rejection for self tends to be translated on the physical self. The unstated message is that if only I was thin enough, pcoplc would lovc mc or I would bc smarter prettier. I am not good enough as I am.

I would like to state here that they both need counseling for self-acceptance and self-affirmation. When we learn to love ourselves, we automatically tend to make healthy choices including that of the food. We need to make this the turning point in our life, before deciding about your food choice, no matter what it is, just ask yourself do I make a difference. Do I value and love myself? Believe me most often, you will pause to answer, and that pause if mindfully taken would make the nontoxic healthy choice.

Remember the more positive and accepting you are about yourself, the more healthy food choices you would make. We tend to gain weight because our bodies are toxic and we have stuffed them with the wrong kind of foodstuffs for a long time. In fact, I read it somewhere that strict diets are a form of self-hatred. Remember, when there is true love for what one is, there would be no need for dieting, the change would be a spontaneous and natural processes.

Nutritious Eating is the Major Principle of Happy Eating

If you want to be happy and healthy, be aware of yourself and your choices, be mindful of your choice of what you eat, be conscious of your eating and pay attention to what your body says when you eat various food items.

When you clean a cooking pan, the first rub of the scrub brings out two realizations that the residuals of cooking are badly stuck up to the pan, and secondly, some of the food stuck to the pan comes out, looking dirty and messy. The first desire is to leave it, and soak it with warm water and soap. However, if you keep scrubbing the pan, it comes out as good as new in a short while. Similar is the case when we want to say good-bye to the thoughts/mindset and unhealthy/toxic food. The first two days, there may be withdrawal symptoms, a craving for the familiar comfort zone of the same mindset and toxins create stress with a desire to go back to the same old system. However, if you continue to persist on change of the mindset and healthy eating, in no time, you will be a happy and healthy soul thanking yourself to make the choice to change and sustain the initial change hangover. So know what is it that you wan and know about what food you want.

प्रन्नं न निन्द्यात्। तत् व्रतम्। प्राणो वा प्रन्नम्।
शरीरमन्नादम्। प्राणे शरीरं प्रतिष्ठितं शरीरे प्राणः प्रतिष्ठितः।
तदेतदन्नमन्ने प्रतिष्ठितम्। स य एतदन्नमन्ने
प्रतिष्ठितं वेद प्रतितिष्ठति। प्रन्नवानन्नादो भवति। महान्
भवति प्रजया पशुभिर्ब्रह्मवर्चसेन। महान् कीर्त्या।

Food must not be censured, for that is the vow of the saint. *Prana* (life) depends upon food, which is eaten by the body. The body rests on *prana* and *prana* rests in the body. From this, one can deduce that food rests (establishes) upon food. One who knows food, thus, he becomes strong in body and *prana*, and great in food, progeny, cartel wealth, radiance of devotion of God and glory (Chapter VII Taittriyiya Upnishad).

To conclude, I would say that community based food initiatives are a method of transferring skills and direct health promotion efforts to lower socio-economic population. There is a strong need to address nutritional inequalities at the level of empirical research as well as community preventive intervention level. Programmes highlighting healthy food, positive and healthy life style and relationships between food and heath mental physical and emotional need to be the inputs for such public health community mobilizing programmes.

By proper intake of vitamins and nutrients, the fraction of one's life during which one is happy becomes greater "said Linus Pauling."

■ ■ ■

8 Dimensions of Health

Gaytri Tiwari

Department of Human Development and Family Studies, College of Home Science, Maharana Pratap University of Agriculture and Technology, Udaipur, Rajasthan, India E-mail: <tiwarigaytri@gmail.com>

Health is the biggest gift of god. In true sense, those are the richest people who are healthy. Here, health is usually misunderstood by only being physically healthy. This is very wrong that complete health is being free from physical, mental, social and spiritual ailment, thus, we can say that health comprises of five aspects, *i.e.*, physical, mental, economical, social and moral. According to WHO, health is a state of complete physical and social wellbeing and not merely an absence of disease or infirmity.

Physical Dimension

It implies the notion of perfect functioning of the body, good complexion, clean skin, bright eyes, firm flesh, not too obese, a sweet breath, good appetite, sound sleep, regular activities of bowl and bladder and coordinated bodily movements. A person should be free from every type of pain and diseases. One should be well balanced in weight, height and physique and should not be dependent on any body for everyday jobs. For this, one needs proper and balanced diet and physical exercise, which need not be very expensive.

P-physical exercise: Our bodies need regular exercise. Physical exercise tones the muscles and relaxes the body. However, strenuous exercises are harming. One should not waste time in watching television. Physical exercise is important for maintaining physical fitness and can contribute positively to maintaining a healthy weight, building and maintaining healthy bone density, muscle strength and joint mobility, promoting physiological well-being, reducing surgical risk and strengthening the immune system.

Exercise also reduces levels of cortical, a stress hormone that builds fat in the abdominal region, making weight loss difficult. Cortisol causes many health problems, both physical and mental. Frequent and regular aerobic exercise has been shown to help prevent or treat serious and life-threatening chronic conditions such as high blood pressure, obesity, heart disease, type 2diabetes, insomnia and depression. Endurance exercise before meals lowers blood glucose more than the same exercise after meals. According to the World Health Organization, lack of physical activity contributes to approximately 17% of heart disease and diabetes, 12% of falls in the elderly and 10% of breast and colon cancer. Both aerobic and anaerobic exercise also work to increase the mechanical efficiency of heart by increasing cardiac volume (aerobic exercise), or myocardial thickness (strength training). Such changes are generally beneficial and healthy if they occur in response to exercise.

H-have self-confidence: While approaching people, do so with self-confidence, not with a sense of fear in whatever we do. There are bound to be setbacks but these should not deter us. It is unlikely that everyone will respond or reject our approach but think of the majority that will respond positively and go ahead with confidence. A negative feeling will not only dampen our spirits but will also reduce our chances of success.

S-sleep well: Make sure to have adequate rest and sleep to wake up fresh in the morning. Insufficient rest and sleep cannot ensure a good performance. Many of us want to sleep as little as possible or feel like we have to. There are so many things that seem more interesting or important than getting a few more hours of sleep but just as exercise and nutrition are essential for optimal health and happiness, so is the sleep. The quality of sleep directly affects the quality of waking life, including mental sharpness, productivity, emotional balance, creativity, physical vitality and even weight. No other activity delivers so many benefits with so little effort.

Y-yoga: Try yoga since these *Asanas* take care of every part of your body and can cure many ailments like sinus, blood pressure and diabetes. Beside, it is neither strenuous nor does it have any side effects.

Yoga is a way of life, an art of righteous living or an integrated system for the benefit of body, mind and inner spirit. This art originated was perfected and practiced in India thousands of years ago. The references to yoga are available in *Upanishads* and *Puranas* composed by Indian Aryans in the later Vedic and post-Vedic period.

Aim of Yoga is the attainment of physical, mental and spiritual health. *Patanjali* has recommended eight stages of Yoga discipline given as under:

- *Yamas* (abstentions or restrains)
- *Niyamas* (observances)- austerities, purity, contentment, study and surrender of the ego
- *Asanas*- Physical postures or exercises
- *Pranayama*- Control of vital energy (breathing control)
- *Partyahara*- Withdrawal of the senses
- *Dharana*- Concentration of the mind (contemplation)
- *Dhyana*- Meditation
- *Samadhi*- Attainment of the super conscious state

I-involve yourself: Cultivate satisfying relationships, social interaction and support, has been found helpful in reducing stress, help cognitive functioning and prevent depression.

C-challenging the mind: Learning new skills and regular mental activity promote healthy mental functioning. Regular mental workouts of only 10-20 minutes daily can improve one's cognitive function and brain processing speed, and one can expect to see results far quicker with brain training than from physical workouts. The exercises provide you with immediate feedback with respect to your performance (speed, accuracy, consistency and perceptual threshold) as brain processing speed increases, the exercises automatically become more challenging.

A-awareness: Change is inevitable. One should know how to go along with the changes to lead a fruitful life. It is important to be alert and conscious or else it looks like one is merely existing rather than living. Awareness is the state or ability to perceive, to feel, or to be conscious of events, objects, or sensory patterns. In this level of consciousness, sense data can be confirmed by an observer without necessarily implying understanding. More broadly, it is the state or quality of being aware of something.

L-live every moment: Enjoy life. Many a times, people miss out small pleasures of life because they take their life too seriously. Have some fun and feel the joy of life. Involve in any such activities that being loved doing. It could be singing, dancing, painting, cooking, or anything. Take some time out to enjoy them at least once in a week. That will lead to make the surroundings more favourable and conducive.

Mental and Emotional dimension

Although mental health and emotional health are listed in the category of *physical wellness* dimension but they really are a dimension in their own right. Mental health can be seen as *knowing* or *cognition*, while emotional health relates to *feeling*. It is not mere absence of mental illness. Mentally healthy person is free from internal conflicts, well adjusted, *i.e.*, able to get along well with others and he accepts criticism and does not upset easily, searches for identity, a firm sense of self esteem, knows himself (his needs, problems and goals), good self control and faces problems and tries to solve them. A person who is physically healthy even remains mentally healthy and energetic. It can be said that a healthy mind lives in healthy body. For being mentally healthy, it is necessary that one should not abuse anyone by negative thoughts, verbally or through actions. Because every action has back reactions, so by harming others somewhere guilt is felt unconsciously. One should always try to follow the inner voice. It is even said that actions are the mirror image of our inner thoughts. If our thoughts are pure definitely will our actions be. Our thoughts should always be like a flowing river, not like a stale water pit.

M-mend relationships unconditionally: It is indeed very easy to make enemies. Others hurt us or we hurt others unintentionally, the result is the same, *i.e.*, strained or broken relationships. A breakdown in relationships leads to a breakdown in our communication with others. Deep in hearts, one does not like such a situation but hesitate to make a start and wait for the person. Making a start to mend relationships requires a lot of courage as well as humility but when, decided to take the first step and make good relationship with others, helping to become free of hatred, bitterness and jealousy. Moreover, God does give us the courage to be humble. As the bible says, "do unto others what you like them to do to you". This is the finest statement on interpersonal relationships. It will bring peace and joy, which is difficult to explain but is best experienced.

E-express feelings: There is a tendency to rebel against others or to retreat by swallowing the hurt. Both are ineffective. Why not express the other person exactly how you feel about his/her behavior towards you? The chances are that the other person will at least start thinking about it and change will come gradually. One should express feelings productively, confront someone respectfully when bothered by others behaviour, express difficult feelings without attacking the self-esteem of the person, clarify for you and the other person precisely what you feel, prevent feelings from building up and festering into bigger problem, communicate difficult feelings in a manner that minimizes the other person's need to become defensive and increases the likelihood that the person will listen.

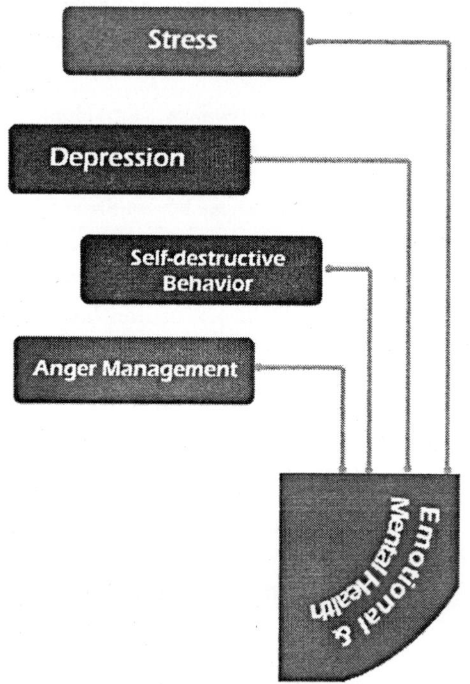

N-negotiate: Negotiate means a compromise. It further strengthens invitation to an understanding. It is better to meet halfway than not to meet at all. Start listening to yourself and become the best friend and supporter. No one is going to work on happiness, so find the power and motivation stored up inside self and use it to direct into the path of true happiness and satisfaction. Do anything to set mind, and once having stopped and gotten in touch self, one will learn and realize just how wonderful and capable one really is, and how always have been. After realizing these things, one will begin loving others, which will not only lead to achieving the things that make happiness but will guide into a world of many new dreams come true.

T-think well of yourself: Every one has born to be a winner. No human being can be tempered as a useless person. Each one has some very special God given talent, which makes a unique person. Hence, one has a right and a reason to feel good about one-self. Remember, if one does not think well of oneself, nobody else can think. Therefore, rejoice and be glad for the talents and specialty.

A-appreciate others: Everybody young and old alike, desires and deserves appreciation because it motivates everyone to do better. Unfortunately, appreciation is lacking in many, and as a result, people are not motivated to put in their best efforts. When appreciation is shown in a person, encouragement is sure for others to participate in activities with renewed enthusiasm. Learning to appreciate others will help us to live a better life. We never loose anything in appreciating others. Instead, we become a better person. Appreciating others will help us to enhance our skill, and it will boost the doers to contribute more for the good of the people.

Every time, someone does something good for others or us and one needs our genuine appreciation that sprouts in the land of jealous free zones. Learning to appreciate others' ideas, sincerity, contribution, good deeds, *etc.* enhances one's life. It is much better to say a word of appreciation while people are alive than writing a beautiful biography or bringing a bouquet on their death ceremony. The dead men cannot see or smell beautiful flowers or read biography, which is written in praise of them after their death. When we appreciate and support the great deeds of others, we feel good and boost the others to work harder. Appreciating others and talking good of others after they died is one of the weakest attitudes of mankind in this world.

Many a times, we do not appreciate others due to some of the reasons, *i.e.*, jealousy, ignorance attitude and callous attitude. When we open up our heart to appreciate others, it will boost our present work and in their works. Those who appreciate and support others learn from others and improve in their works but those who are so callous about other good deeds fail to learn and improve their work. Appreciation should be from our heart.

L-laugh: Nothing is more embarrassing to a critic than a subtle laugh. Almost everyone who criticizes would like other to be shameful, sheepish and serious. However, avoid laughing loudly as if it were a great joke or even being sarcastic since it will only aggravate the situation. A sense of humor helps a great deal. The ability to seek and find humor and laughter in any given situation is a gift but one that takes work. Let us face it that life is funny with all of its quirks and contradictions but it may not always seem so funny when encountered a tough situation. However, the ability to gain a new perspective on a bad situation through humor is a real testament to ones character. Look for the humor in everything and use it to gain new perspectives, to reduce stress and to make the others smile.

Economical and Vocational dimension

Vocational health is one's attitude about work and career. This dimension involves preparing for and participating in work that provides personal satisfaction and life enrichment. This includes continued development of occupational skills, finding balance between work and leisure activities, participating in paid and volunteer activities and finding challenging work. It is a new dimension of life. It refers to mental and physical adaptation to work (human goals, capacities and limitations).

E-economic protection: Being economically healthy does not mean staying rich and prosperous but it means that a person or family is able to meet its basic-necessities of living like food, shelter and clothes. If these requirements are met, surely many other troubles will be solved or get ridden off so less or more one need to have some sort of earning.

C-calculative in expenses: Always keep a record of every penny spent. Keeping a record will help to check how much money is spent and how much is overspent. If one knows, about the expenditure of money spent on the weekend is the sum of money one earn working the weekdays, one is less likely to spend more. When a budget is made, it helps in saving a lot of money for and also helps to keep over expenditure in check. Why even governments have yearly budget plans so make it a habit to make a budget for every month and even stick to it and cut unnecessary expenses. If one wants to save money, one needs to cut on unnecessary expense. Keeping a record of what one spend will help cross out the places one is likely to spend more money. In addition, when one knows about the repercussions of extravagancy, one will be calculative in spending next time. It also helps in deciding the priorities on what is more important and what is unnecessary. If spending your weekend eating, drinking and partying, cut on it if possible. If too much money is spent buying clothes, make an objective of spending less on clothes in coming months.

O-organize priorities: Traditionally, decisions are frequently arrived at by providing an answer to the question "What are our tops or 'A' priorities?" This can lead to several forms of confusion and poor judgment. The more options one has, the greater the chances are of making poor priority decisions. It simplifies the selection of true top priorities by filtering out all of the lesser important issues.

N-notice opportunities: In present scenario, there are many opportunities scattered around, one need to analyze the self choices, priorities and requirements and grab the economic opportunities by doing SWOT (strengths, weaknesses, opportunities and threats) analysis of oneself.

O-optimize resources: Try to use resources to fullest be it human or non-human. The idea of optimize resource techniques is developed to plummet the amount individual, firm and institutions spends on items that add no significant value to its living standard, long-term portfolios and asset-basins, greater control over outgoings and incomings, both in personal and business perspective.

M-money management: is the process of managing money. It includes investment, budgeting, banking and taxes. It is also called investment management. Money management is a strategic technique employed at making money yield the highest of interest-yielding value for any amount of it spent. Spending money to provide answers to all cravings (regardless of whether they are justifiable or not to be included in budget basket) are a natural human phenomenon.

I-initiate planning: One needs to coordinate planning process with outside people or organizations.

Having others involved may also apply if there are nearby organizations with which facilities or community resources are shared, if there are other organizations to which one's business has financial or contractual obligations, *e.g.*, loans, leases, *etc.* and if one will need access to local police, fire and rescue and insurance agencies.

Not all of these external organizations need to be deeply involved in planning process. However, involve them as early on as possible, as they may be able to supply planning resources or relevant knowledge to contingency plan. At the very least, they need to know about the contingency plan so that their own plans do not clash.

External organizations and individuals will probably have different root contexts. While they are important to planning process, their input will be more useful in Context Analysis phase, rather than Scenario Building process.

C-compromises in luxuries: Stop buying stuff. One can live without another CD or latest model DVD player. Before buying anything, remember to ask three questions:

- Is it needed?
- Is it needed now?
- Can get it cheaper somewhere else?

Remember that human beings have survived several thousand years without the latest edition of Grand Theft Auto- One will not drop down dead if one attempts to do the same. Besides, if electricity bill is not paid, what is one going to play it on anyway? Cut down on entertainment budget by entertaining at home instead of going out. Live below your means, live a comfortable life, not a wasteful one, do not spend to impress others and do not live life trying to fool you into thinking wealth is measured in material objects. Manage money wisely since money does not manage you and always live well below your means.

Social Dimension

It implies harmony and integration within the individual, between each individual and other member of the society and between individuals and the world in which they live. Usually, it is seen that a person is said to be socially happy if one is well to do but a true socially healthy person is one who tries to give back society as much as he has taken from it. People usually feel their duty is done towards society by donating some money. A socially healthy person is that who can evaluate himself and his life in true sense. Society is basically a cluster of families so if the duties are met honestly in families itself than a healthy society can be automatically formed.

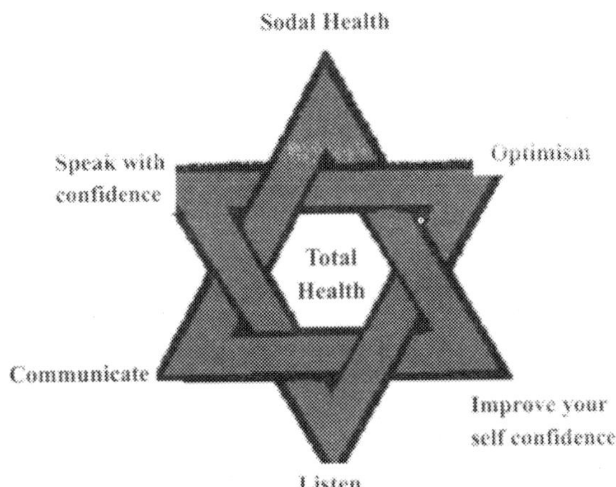

S-speak with confidence: If one is sure about the facts, avoid pushing them down the other persons throat yet be firm in tone of voice and serious about the opinion. The 3 seconds rule: while speaking with someone, take 3 seconds before responding. This may seem awkward at first but it gives a chance to do things a lot before saying something regret regretful. In addition to letting think about what one is going to say, it also allows room for them to continue talking. If one feels uncomfortable in the silence, you have gained the upper hand. Have a plan, go into the conversation with a plan, think about the responses likely to get and play it out like a chess game. Where each move has an effect on others and *vice versa*, be comfortable and use your plan to leave room for awkwardness on their part. If others feel like they are fumbling for words and trying to gain confidence in them, then you are in control. Be comfortable and relaxed enough so that you can say what you want to without feeling stupid. Do not talk so much. When nothing much is said, people have to try to assume what one is thinking. Getting inside others head is much harder than knowing what other is feeling or thinking. It is better to keep mouth closed and let people think you are a fool than to open it and remove all doubt.

Think about what to say. One is bound to run into snags but if one can recover well, it will look planned so take a second and think before start talking. Diction is very important. Enunciate and speak well. Make sure to pronounce well. Try some tongue twisters to improve diction. Use of proper English makes you sound, smarter, more confident and feel better. Colloquialisms are great for friendly conversation but not for confident speaking. Body language also plays a significant role in speaking publicly. Do not smile so much since people take smiling as a sign of warmth but also a sign of emotional involvement. The more distant one is, the harder it is for someone to disregard. Always look into the eyes of other person while talking. This shows confidence and makes the other person know one is business. Do not look down since people believe that if one is looking down has something to hide, they might also think that one is lying. Fight the urge to look at feet, look into the other person's eyes instead and show less emotion. The less emotion shown the more people are likely to listen really. Now, this is not to say that one should not be passionate. Just do not get crazy excited. Instead, use passion to fuel speech. Do not blink so much. Next time while watching a movie during a serious scene, watch the actor's eyes. If they are good, notice they barely blink, the more people can one's eyes the more sincere they believe one to be. In general, make sure that body language and speech patterns match. If one is exciting and the other dull, this will confuse listeners and make them feel unsure about what saying. Practice voice, speaking in front of a mirror and practice with people one is comfortable with those. The more one does it, the easier it will come.

O-optimism: Remember that no human being is a total failure to be considered useless and defeated. Everybody is gifted in some or other ways. All that one need to do is rediscover true self to unable to decide, career path. In order to build a successful career one needs not only be a doctor, engineer, lawyer, charted accountant, or MBA. One can be successful even as a carpenter, electrician, truck driver, nurse, or teacher.

C-communicate: Keep talking and share your feelings and opinions with elders, even if not agreeing with theirs. Remember, everyone has a right to give own opinions, hence, stand by them, but avoid arguments to avoid a fight. While communicating, do express freely and find that many of the minor issues automatically sort themselves out.

P-prove your self-confidence: Consider yourself a success in life. Put in the required effort in whatever is done. Achieve something small each day, read and discuss matters of current interest with friends and family members rather than stay glued all the while with television. Take initiatives wherever and whenever possible. One will discover that slowly and surely self-confidence has improved.

A-avoid arguments: In an argument, no body likes to give up because there is a feeling of defeat. Argument usually ends up in violence, sometimes verbal, if not physical. In convincing others, one becomes like salesperson selling an idea if not a product. Therefore, an argument only weakens the position. If views differ, first agree with the others right to express views, and then, express own. Follow the technique "you have a point in what you say (pause a little) but my view is...." if others still do not see the opinions objectively, remember that it is better to lose an argument than win an argument and lose a friend. Respect elders, respect minors and respect everyone. There are no boundaries or classes that define a group of people that deserve to be respected. Treat everyone with the same level of respect you would give to your grandfather and the same level of patience you would have with your baby brother.

L-listen: Since having one mouth and two ears, all are required to listen to others twice the amount than to speak but often its found people acts as if they have two mouths and one ear. Listening with attention diffuses the heat and just melts like ice, even the hottest headed person. Listen more and talk less since the more one listens and the less one talks, the more one will learn and the less one will miss.

Moral and Spiritual dimension

It refers to that part of individual, which reaches out and strives for meaning and purpose in life. Moral health depends on our life values. These values should not be imposed externally one should decide them internally and follow them. Our life is very busy knowingly or unknowingly everybody is following the blind race resulting in high blood pressure, cardiac arrests and depressions different sorts of tensions. One should know and believe that God has planned fortune of every one. One will be more moral and positive in efforts when one will start believing in destiny.

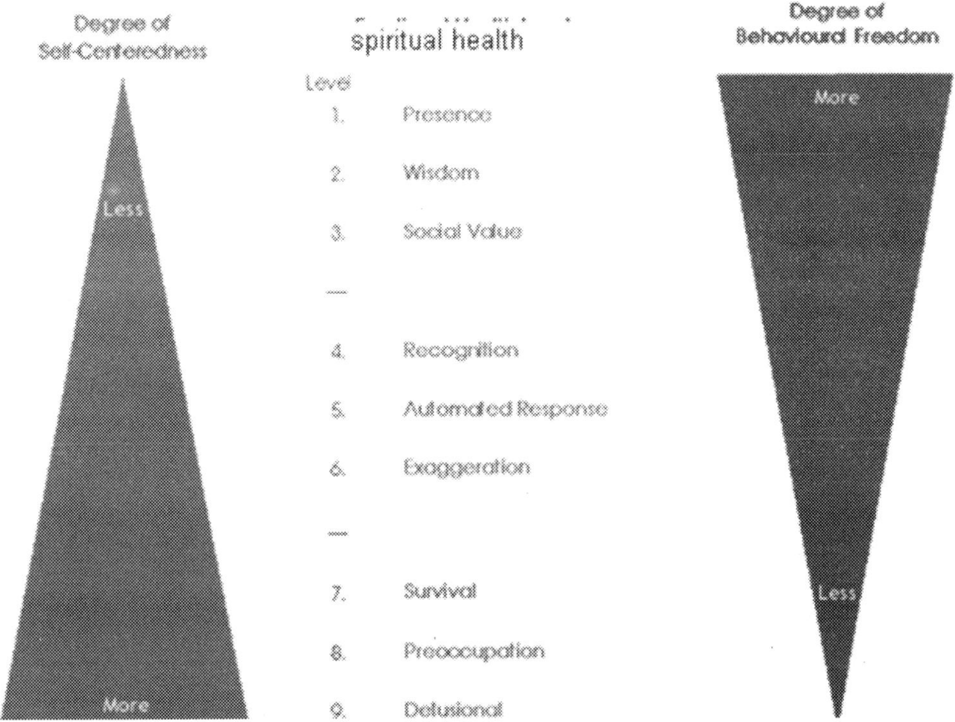

M-meditation: Whether one is looking for relief from an existing health condition or practicing preventative maintenance, the benefits of meditation can enhance not only spiritual health but also the physical and emotional/mental health as well. Additionally, since it costs nothing and can be performed in the comfort of home, it is one of the most affordable, convenient, holistic and natural self-care strategies to be found. Meditating has been found to be beneficial for improving a variety of conditions including high blood pressure, stress, heart disease, insomnia, depression, anxiety, chronic pain management, and drug and alcohol addiction. Another excellent and a great deal during meditation, is a high level of creativity. Some best thoughts, ideas, inspirations, words, *etc.* come during in its most basic form. Meditation can be described as the practice of calming, quieting, or clearing mind to perform inner contemplation or concentrated attention on the mind. The goals are many, but for purposes, one simply want to achieve a more relaxed, less stressful frame of mind, inner peace, balance and perhaps achieve a higher state of consciousness and awareness, all of which will enhance spiritual, psychological and physical health.

O-owe to nature: The stressors of life can really take its toll physically, spiritually and emotionally, and it is essential to renew from time to time. This is especially true for anyone living with a chronic health condition. Communing with nature on a regular basis is a supreme method of reducing stress, nourishing and renewing ourselves, feeding spirituality and optimizing mental, physical and spiritual health.

Universe, in its magnificent beauty, offers endless opportunities for enriching our life. It costs nothing and it surrounds continually no matter what. Get in touch and intimate with nature. This can be something as simple and basic as watching the sunset or sunrise, observing cloud formations, or gazing at the moon and stars. Let the warm summer breeze titillate skin and nurture senses, or hypnotize with the soft patter of rain, or the fall of beautiful snowflakes. Immerse in the healing rays of beauty, which radiate from the colours of fall leaves and delight in the beautiful crunching of them beneath feet. Hug a tree, lie on the ground, or touch a rock. Feel energy of the Universe around and let it rejuvenate and energize deep in core. Take a daily walk to get exercise and connect with nature simultaneously, therefore, doubling the physical and emotional benefits. Savor the natural joys of simple pleasures. The best things in life are to be free. They come in the form of simple pleasures and they appear right in front at various locations and arbitrary times. Beautifully governed by Mother Nature and situational circumstance and captured by mindful awareness. It is all about taking a moment to notice the orange and pink sunset reflecting off the pond water as one holds hands with someone very much loved. Noticing these moments and taking part in them regularly will bring unpredictable bursts of happiness into life.

R-remain thankful: Cultivating a thankful attitude of gratitude is a popular slogan in the self-help field. When faced with great suffering and pain, it may seem a little trite on the surface but look a little deeper since it has some value. Even if not living with a chronic health condition, just the stress and hustle and bustle of every day life can drain resources and cloud perspective. Having gratitude for the small blessings in life will help to keep focused on what is important and keep from drowning in the negativity of illness. It is a valuable tool in the quest to live life to the fullest. Try to appreciate what you do have and what you are able to accomplish. This does not mean that minimizes, repress or deny suffering or loss but rather look for the blessings that are present in life in addition to or in spite of the pain and difficulties.

A-add humor and Inspiration: When faced with a lot of pain, suffering and challenges on a daily basis, it can be easy to lose sense of humor and hard to find inspiration though this is understandable. Finding a way to keep them present in life is not only a powerful coping method but also essential food for the body, mind and spirit. In midst of our most difficult times, a little laughter can literally be a lifesaver. Some of the most valuable ways in which humor helps to cope is that it improves the mood, decreases feelings of powerlessness, makes life lighter and more manageable and helps to see the things in a new light. This stimulates feelings of hope, and thus, feeling of being more capable of handling the problems in our life occurs. It replaces emotional distress, because it is impossible to be angry, depressed, or anxious in the process of laughter and improves overall quality of life. Humor and inspiration temporarily draws our attention away from our pain and suffering and gives a much-needed break. Though it may be brief, it shortly brings joy into our life.

Not only it is good for our mental health but humour also boosts our immune system, reduces stress and provides pain relief. Laughter decreases our stress hormones, increases antibodies that fight infection and releases endorphins, which are the bodies natural painkillers.

L-law of Attraction: Set intentions, visualize, affirm, be grateful and believe to have the power to achieve it, however, take additional steps to address anything that may actually sabotage or hinder progress. Love, accepting and patient if outside forces interfere with desired goals. Allow for and accept the possibility that it may not be the cure for absolutely everything. Remain open to all the good things since the Universe can possibly bestow, while at the same time, remaining aware that there are other factors to take into consideration. Never become boxed in by one particular belief system.

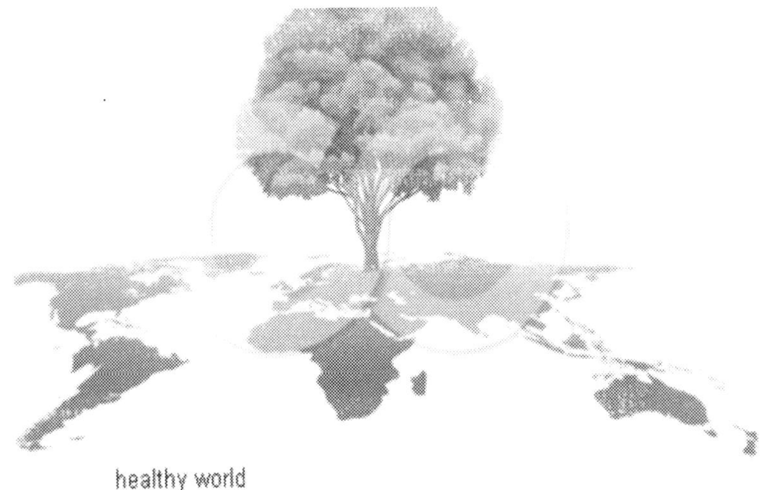

healthy world

Hence, it can be said that if a person follows three H, Head, Heart and Hand on right time and right way than positive health is certain to be achieve.

■■■

9 Nutrigenomics: Gene-Based Personalized Nutrition

Anjana Munshi[1] and Y.R. Ahuja[2]

[1]Department of Molecular Biology, Institute of Genetics and Hospital for Genetic Diseases, Osmania University, Begumpet, Hyderabad-500016, India
E-mail: <anjanadurani@yahoo.co.in>
[2]Department of Molecular Medicine, Vasavi Hospital and Research Centre, Hyderabad, Andhra Pradesh, India

With the success of Human Genome Project and the powerful tools of Molecular biology, the sequencing of human genome has laid the foundation for one of the most significant scientific contribution to mankind. Although all the human individuals are genetically similar, yet each retains a unique genetic identity. We have entered the era of genetic nutrition. Propelled by the recent unraveling of human genome, nutritional sciences are discovering the application of the so-called *omics* sciences. This has the potential of identifying and validating targets to improve personalized nutritional health, and thus, serves to define the added value for next generation foods and the crops.

Nutrition research has primarily focused on nutrition deficiencies to evaluate the relation between nutrition and health. In 1996, Ghai and co-workers filed a seminal patent, which highlighted the potential for the development of foods or supplements that could alter the expression of genes associated with human diseases. They demonstrated that certain flavonoids in citrus peel enhanced the expression of a gene involved in the human body's natural defense against cancer. This was the beginning of study of the relationship of food as nutrition with our genes. The publication of the human *blue print* and the advent of genomics have created unprecedented opportunities for increasing our understanding of how the gene and protein expression is modulated by nutrients and ultimately influencing cellular and organism's metabolism.

The new era of molecular nutrition unfolds in dichotomous direction- *Nutrigenetics* and *Nutrigenomics*. Nutrigenetics refers to the study of how genetic variation manifesting as single nucleotide polymorphisms, copy number polymorphisms and epigenetic phenomenon affect susceptibility to nutrient intake. On the other hand, nutrigenomics, a combination of molecular nutrition and genomics, focuses on the genome-wide influence of nutrition to unravel mechanisms underlying the physiological and molecular effects of nutrition.

Nutrigenomics has, however, also embraced the study of how nutrients influence the consequence of gene expression, *i.e.*, synthesis of mRNA (transcriptomics), protein synthesis (proteomics) and metabolite production (metabolomics). This integrated approach is called *Systems Biology Approach*.

The basic idea is that there are genes that affect the risk of getting illnesses like heart disease, cancer, osteoporosis and diabetes, and the impact of those genes can be modified by what you eat. Everybody carries one version or another of each of those genes. So why not find out what gene versions you have and base dietary advice on that. People already make dietary choices based on their genetics such as switching to soymilk for lactose intolerance, using cholesterol-lowering spreads and by dietary phenylalanine restriction, *etc.* As the science of nutrigenomics is evolving, it will reveal not only which foods can have direct impact on our genes but also how and why. The ultimate goal is able to prevent and potentially even treat diseases through targeted nutrition. This is not very different from the opinion of Hippocrates- Father of medicine (460-360 BC) who said that leave your drugs in the chemist's pot if you can heal the patient with food.

Nutrition and Human Genetic Variation

Even prior to the sequencing of human genome nutritional requirements were recognized as complex traits subject to modification by genetic background and nutrient-nutrient interaction. Nutrigenomics must address how individual genomes respond to complex nutrients and chemical mixtures that comprise food. It is understood that certain genotypes are more severely affected by specific dietary factors than others. One thing is clear that no genotype is free from the deleterious effects in inadequate diet.

Genes express themselves through proteins. Enzymes are special proteins designed to get things started. Our genome instructs ribosomes to produce many enzymes that destroy toxins. Some food articles such as cauliflower, broccoli and Brussels sprout contain chemicals that actually prod our genes to direct biosynthesis of these enzymes. The response to this prodding may be hindered by the genetic make-up, *e.g.*, faba bean contain a toxin usually degraded by a chemical produced in the body. Without this chemical, the toxin leads to destruction of red blood cells (RBC). Eating faba bean results in loss of red blood cells and serious health problems in individuals whose genes do not direct the production of enough of these chemicals. In some individuals, genes give unclear instructions for making an enzyme that metabolizes the amino acid, phenylalanine. As a result, this amino acid builds up, thereby causing brain damage. A diet restricting this amino acid will help stop the damage if detected in early infancy.

Transfer of nutrients from gut to cells requires carrier and receptor proteins. Some individuals have genes that direct overproduction of iron carrying proteins. The resulting iron overload is extremely toxic and may lead to death. Similarly, GLUT4 is a protein that transports glucose across skeletal muscles and fat cell membranes. Individuals with type-2 diabetes lack the ability to grasp the message instructing the ribosomes in skeletal muscles and fat cells to produce GLUT4. As a result, glucose stays in the blood causing a number of problems.

Currently, several nutrition related issues benefit from genetic research. A number of conditions like Phenylketonuria (PKU), caffeine intake and bone loss in post-menopausal women, folic acid and heart disease, obesity, anorexia nervosa, vitamin C supplementation to reduce cancer risk and low fat diet for high blood cholesterol levels, show the influence of genetic variation on nutrition advice.

It is clear now that there is heterogeneity in response to dietary regiments. Nutritional genomic research will determine the genetic basis for that heterogeneity, clarify the molecular basis by which specific genotypes respond to dietary components and other environmental signals, generate analytical tools for determining the genotype of a person and provide targeted therapeutic interventions for disease amelioration and prevention.

Biomarkers are the genetic variants that predict the risk of disease and improve diagnosis and risk assessment. For the testing of efficacy of bioactive functional food ingredients, an urgent need exists for biomarkers. Genetic polymorphism may be partially responsible for variations in individual's response to bioactive food components. Single nucleotide polymorphisms (SNPs) are becoming increasingly recognized to have an important influence on disease risk, *e.g.*, inherited polymorphism in BRCA1 and breast cancer susceptibility. Some common single nucleotide polymorphisms in genes involved in nutrient metabolism, metabolic activation, and/or detoxification could establish the magnitude or there is a positive or negative response to food components. Women, who consume less fruits and vegetables than required, were reported to be at a greater risk of developing breast cancer because of a polymorphism, which causes a change from valine to alanine (Val to Ala) in 9 position in the signal sequence for the enzyme manganese dependent superoxide dismutase. Consumption of well-done meat was correlated with increased breast and colon cancer susceptibility among individuals with a rapid/intermediate N-acetyl transferase genotype but not among individuals with slow acetylator genotype.

Several epidemiological studies have shown correlations between mildly elevated homocysteine levels and cardiovascular disease risk. Methylene tetrahydrofolate reductase (MTHFR) catalyzes a reaction, which produces 5-methylenetetrahydro-folate, a cofactor donating a methyl group to a reaction converting homocysteine to methionine. The single nucleotide polymorphisms (C677T and A1298C) are associated with reduction in MTHFR activity. This might lead to increase in plasma concentrations of homocysteine and thereby to venous thrombo-embolic disease, ischemic arterial disease and neural tube defects. Treatment with folic acid supplementation helps to overcome the effects of these polymorphisms in MTHFR gene.

Cardiovascular diseases (CVDs) are a group of multi-factorial conditions associated with atherosclerosis, hypertension and thrombosis. Of these, atherosclerosis is a key element in the pathogenesis of cardiovascular diseases. Risk of atherosclerosis can be detected at a preclinical stage by quantitative assessment of blood lipid levels. Genetic polymorphism in key genes that have a bearing on blood lipid levels has been addressed in several recent reviews. Nutritional counseling to reduce the dietary fat would be beneficial to overcome the negative health effects of certain polymorphisms.

Similarly, a number of polymorphic genes have been implicated to cancer development, *e.g.*, variants of melanocortin 1 receptor gene (MCIR) have been found to be associated with several types of skin and prostate cancers. There are claims of dietary supplements that may protect against diseases like cancer. New foods are being developed as functional foods, and introduction of the concept of *Kinetics of biomarkers* would be an important step in the validation of such claims.

Obesity has become a major public health problem. Mutations in genes, like those of leptin and leptin receptor, have emerged as leading candidates towards predicting obesity. In addition, mutations in melanocortin 4 receptor and melanocortin 5 receptor genes and in the non-coding regions of the gene for neuropeptide (NPY5R) receptor Y have also been shown to be strongly correlated with the risk of obesity. Once the mutations are detected in the family, the physician might be in a position to offer diet restriction/intervention at an early stage of life.

Evidence is accumulating that variability is induced by single nucleotide polymorphisms in the genes associated with the inflammatory process as well. This variability influences the responsiveness of individuals to changes in nutrient intake, *e.g.*, single nucleotide polymorphisms may be responsible for modulating the ability of fish oil to reduce the production of pro-inflammatory cytokine tumour necrosis factor (TNF).

To understand our personal uniqueness with regard to nurtigenomics, single nucleotide polymorphisms have opened the door to manage and optimize our health through tailored nutrition. Of course, this is a lofty and distant goal but the fact is that we have already started the journey towards it.

Some nutrients are toxic at elevated intake levels, and therefore, a tolerable upper intake level (UL) may be established for an individual nutrient. Genetic variation has been shown to influence upper intake level also. A single nucleotide polymorphism in the hereditary haemochromatosis linked gene (HFE) increases the risk for hemocheromatosis (an iron storage disease) which can result in up to 50 folds increases in storage deposits.

A database of functionally important single nucleotide polymorphisms would make it possible for a nutrition specialist to use single nucleotide polymorphism profiling to make practical recommendation as part of clinical practice.

Effects of Nutrients on Gene Transcription

The nutrients interact with genes at various levels to affect the transcription:

A. Direct interaction: Nutrients sometimes after interacting with a receptor behave as transcription factors that can bind to DNA and induce gene expression. There are different examples of nutrients acting as transcription factors and modifying gene expression. Retinoid derivatives of vitamin A interact with retinoic acid receptor proteins and these complexes either activate or repress transcription when they bind to motifs, *e.g.*, retinoic acid response elements, in promoter region of the gene.

B. Epigenetic interactions: In addition to single nucleotide polymorphism, copy number polymorphisms (CNPs) and epigenetic phenomenon too have shown to affect the susceptibility to diet. Epigenetic phenomenon like DNA methylation, histone acetylation and RNA interference add to the complexity of gene regulation. Methylation is the best-characterized epigenetic process. Around 3-5% of the cytosine residues are methylated in mammalian cells. These methylated cytosine residues appear as 5-methyl cytosine (5MeC) and 70-80% of these residues are clustered in CpG islands. If these CpG islands are present in the promoter of the genes, this suppresses their expression (gene silencing). The presence of 5MeC may interfere with the binding of transcription factors or other proteins, and hence, block the transcription. It has been reported that increased methylation in the CpG island in the promoter region of estrogen receptor (ESR I) gene occurred during ageing and tumour development and was associated with the loss of expression of the ESRI protein.

Further more, DNA methylation protects the genome from selfish DNA elements like transposons, endogenous retroviral long terminal repeats and other dispersed repeat sequences. On the other hand, the loss of methylation from these elements leads to genomic instability and activation of tumour inducing genes. There are a number of reports suggesting intake of several nutrients like folate, selenium, polyphenolics and food contaminants like arsenic can affect DNA methylation pattern. However, one has to go a long way to understand the molecular mechanisms by which dietary components affect DNA methylation patterns.

Histones are a dynamic component of chromatin. Individual histone molecules get modified as a result of modifications occurring in their individual amino acids especially at their tails. These include adding acetyl groups (CH_3COO) to lysines, phosphate groups to serines and methyl groups to lysines. These specific modifications can either stimulate or inhibit gene expression by allowing or denying access to proteins involved in transcription. Diet-derived factors can also influence the histone modifications, *e.g.*, addition of 4-Phenyl butyrate to the diet of *Drosophilia* leads to increased acetylation of histones H3 and

H4, and thereby, alters pattern of gene expression. Similarly, diallyl sulphide found in garlic leads to increase in histone acetylation, which in turn re-expresses the previously epigenetically silenced genes and also alters the cell growth kinetics.

C. Genetic variants: Common genetic variants like single nucleotide polymorphisms also alter the expression of genes. It has already been discussed.

Systems Biology Approach

The inadequate success of single biomarkers in predicting chronic diseases attests to the need for other approaches like transcriptomics, proteomics and metabolomics or in other words thinking beyond the interaction of genes and nutrition. Moving from single biomarkers to other approaches is necessary because lowering the unilateral risk of one disease in an individual simply increases the risk of another disease in the same individual. Single nutrients may have multiple and unknown biochemical targets and physiological actions, which may not be easily addressed with chemical biomarkers.

An important challenge in the development of functional foods for the prevention of complex (multi-factorial) diseases is to obtain a better and improved overview (holistic) picture of the early processes. Here, the concept of *systems biology* steps in. It is related to the integration of all the information at different levels of genomic expression, *i.e.*, mRNA, proteins and metabolite, which depict the effects of bioactive food components on systems biology.

Currently, transcriptomics is relatively a mature technology in comparison to proteomics and metabolomics. We can get an overview of expression of almost all the genes in a single micro-array experiment but it has not been possible to measure the whole proteome or metabolome to date. The expression of genes can be evaluated at an unprecedented scale, using cDNA micro-arrays. These techniques are based on quantitative assay of the relative concentrations of specific mRNAs in tissue samples, which in turn reflects the level of gene regulation. This information can be used to examine the factors (including nutrients) that influence the key gene expression patterns.

Bioactive ingredients directly from food pass through our cells into the nucleus where they interact with DNA to affect transcription of RNA and ultimately translation of proteins. Although the mechanisms by which dietary components influence gene expression are poorly understood but it appears that transcription-factors (TFs) are the main targets or nutrient sensors. In this context, the nuclear receptor super-family of transcription factors with 48 members in humans is gaining much attention. Many nutrients and their metabolites bind to numerous receptors in this super-family, *e.g.*, peroxisome proliferator activator receptor-α (PPAR-α) binds fatty acids or liver X receptor-α binds cholesterol metabolites. PPAR-α is present in liver and there are 3000 to 4000 target genes where this transcription factor is involved. During ligand binding, the nuclear receptors undergo a conformational change. These nuclear receptors go and bind to the specific nucleotide sequence in the promoter region. The conformational change (brought about by the nutrients) in nuclear receptors leads to a change in the level of DNA transcription of specific genes. The transcription is promoted or inhibited as required for the appropriate response to the environment. For example, during fasting, the free fatty acids are released from the adipose tissue. These fatty acids then travel to liver and undergo partial or complete oxidation. In addition to this, these fatty acids also bind to PPAR-α, which in turn increases the expression of a number of genes by binding to specific sequences in the promoter regions. Fasting null mice (lacking functional PPAR-α) suffer form a number of defects like hypoketonemia, hypothermia, elevated levels of plasma free fatty acids and hypoglycemia.

Proteomics refers to all the studies related to structure of proteins, their expression levels, biochemical activity, protein-protein interaction and cellular localization. The protein content of an organism varies from cell-to-cell and it depends on the type of function carried out by a particular cell. Proteins can be analyzed before and after diet intervention therapies.

Plasma proteomics would be of great importance to nutrigenomic research. Progress in this area is likely to generate a wealth of information about important proteins like cytokines or hormone levels form a small plasma sample. Major plasma proteins need to be separated for further analysis. Determining nutrition status, by proteome-derived biomarkers, may be possible before too long.

In fact, a gene product often requires post-translational modifications after it is produced for functioning properly. The control levels in the genomic pathways are truly multilevel. Brudnak has called nutritional supplements capable of altering the integrity and expression or fidelity of genes, collectively as genomoceuticals, for their ability to act on the genome in a manner related to pharmaceuticals acting on biochemical pathways.

Relatively few studies relating diet to proteomics are available. However, recent research work has shown the potential of proteomics to reveal unknown associations between particular protein and metabolic disorders in contrast. There is a global surge in optimism for a role of metabolomics in human nutrition. Metabolomics is emerging as an exciting post-genomics science with applications that span the scope of biotechnology and medicine. It makes nutrition an invaluable tool for determining the distribution of metabolite concentrations in humans, the relationship of these metabolite concentrations to disease and the extent to which nutrition can modulate metabolite concentrations. So why not change the single biomarker concept into one in which the entire range of plasma metabolites (plasma metaboloma) is measured simultaneously. This is going to be bolder step and the databases of metabolome information will be used to provide a significant reflection of the entire biochemical status of an individual at a given point of time.

For broad screening of metabolites, the mass spectroscopy and nuclear magnetic resonance spectroscopy seem to be appropriate and robust technologies. The great expectation of metabolomics is that it might offer the prospects of the ultimate biomarker of overall dietary pattern. Preliminary results show that dietary patterns could have a powerful effect on the metabolic profiles. Therefore, the emerging field of nutrigenomics requires a multidisciplinary and international collaborative approach. A global initiative on nutritional metabolomics has been already initiated by the NUGO, the European Nutrigenomics Society (www.nugo.org).

Nutritionally Improved Agricultural Crops

The first generation of biotechnological crops or the so-called GM crops were produced to provide growers with complimentary and sometimes alternative crop management solutions to herbicides, pesticides, *etc.* Selected genes from other plants or non-plant sources are transferred to the desired plants. The new or altered protein expression resulting from these modifications confers on the plant a desired physiological trait. Second generation modifications are being aimed at identification of plant genes of nutritional importance, and then, transferring them to desired crops. Already, some foods have been modified using biotechnology to provide better nutrition. Golden rice is one example, with its high carotene levels that will help in preventing blindness in children. From a consumer perspective, the focus on improved nutrition is of great interest.

Micronutrients malnutrition that exists in various population groups throughout the world could be reduced by a more nutritious food supply, especially if efforts are focused on staple crops such as rice, maize, wheat, beans and cassava. Identifying the genes needed to increase the levels of essential micronutrients in staple food crops is an immediate goal that would have significant impact on human nutrition worldwide. For essential minerals and vitamins that are limiting in the world diets, improvement strategies should be pursued keeping in mind not only improving the levels but also the upper safe limits for each nutrient.

The nutritional genomics approach has been applied to vitamin E biosynthetic pathway in plants. The researchers identified the first step of pathway from *Arabidopsis* with fungal and human orthologs as database queries. This *Arabidopsis* sequence provided a genomic stepping-stone to identify an ortholog in 10 genes operon in photosynthetic bacterium, *Synechocystis PCC6803.* Gene disruption experiments showed that this operon also encoded the gene for final step in vitamin E synthesis, γ-tocopherol methyltransferase (γ-TMT). The *Synechocystis* γ-TMT gene allowed isolation of an ortholog from the *Arabidopsis* database whose overexpression increased vitamin E levels nearly nine folds in *Arabidopsis.* Therefore, by using database, DNA homologies and insilico-computer searches it is possible to rapidly move experimentally between organisms while remaining focused on the single pathway or enzymatic reaction of interest of the target organisms.

Continued efforts in gene discovery and understanding functional genomics, secondary metabolism and nutritional physiology must be pursued, as these efforts will provide the necessary molecular insights to help us formulate sensible high-impact improvement approaches. The use of model system to assess the efficacy of various transgenic strategies will remain an important proof of concept tool but the researchers must continue to look towards the transformation of agronomically important crops as soon as an effective strategy is identified.

Ethical Issues

As nutrigenomics is involved in public health and food, there are some ethical concerns. Firstly, people need to be educated on nutrigenomics. As there are functional foods that are claimed to have health promoting/enhancing effects on the individuals, acting at the DNA level, it is not known whether they are suitable for all, as each person differs genetically.

Secondly, there needs to be a basic understanding of not only which foods have direct impact on the genes but also as to how and why? Some companies like Sciona Inc. form Boulder, Colorado (USA) have already started giving personalized nutrition advice based on his/her genetic profile, since 2001. Some scientists are skeptical about this, as there is a lot of research to be carried out. At present, genetic science is still too immature to promise such sweeping advice. Hence, there should be serious concerns about quality control in this area.

Thirdly, the role of professional nutritionists needs to be carefully thought in terms of genetic testing and conveying information. There has to be guidelines established while undertaking population screening since sometimes the disorders may not be as clear-cut as they are in PKU. In addition to this, the information from genetic tests may become increasingly attractive to outside parties who stand to gain from it. There is a concern that employers or insurers could use genetic information to the unfair disadvantage of some people. Hence, the question arises where to store genetic information relevant to individuals and who should have access to it. How would the nutritionists get access to this information? Would there be access to large population database, or the individuals carry a smart card containing their genetic profile? Large population databases need controls relating to who shall have access and on what terms.

Fourthly, it would not be long before each individual gets his/her own personalized nutrition, *i.e.*, what a person should eat, which would promote health and what food products the individual has to avoid. Can this be looked upon as invasion of freedom of food choice?

Fifthly, even though the analysis of nutrient-gene interaction is carried out in cell cultures and other animal models, it has to be clinically tested on human subjects. With an ethnically diverse population, the testing has to be carried out on ethnic and culturally diverse groups. Moreover, the question of testing on human subjects is sensitive with the risk of some researchers/institutes using humans as laboratory animals and also exploiting the socio-economically poor.

The availability of DNA is limited to blood for nutrition research. One cannot solely rely on the DNA available from this since gene expression in each tissue is different. The only way to get DNA from liver, kidney and gut is by carrying out biopsies. This method of collecting samples is invasive and is a bottleneck in our pursuits to taking a reasonable decision on personalized nutrition.

Conclusion

The integration of genomics into nutritional sciences would spawn a revolutionary way of viewing the food just not for sustenance but as a pharmaceutical capable of reversing disease and stalling the rigors of ageing. The nutrigenomics approach resembles pharmacogenomics, which looks at the relationship between single nucleotide polymorphisms in gene and response of patients to drugs to personalize medicine. The two are closely linked and without nutrigenomics, pharmacogenomics data cannot be interpreted correctly because diet may affect the expression of genes involved in drug metabolism. Pharmaceutical companies should include nutrigenomics in the design of new drugs because *what you eat* affects a drug's efficacy. The combined knowledge from these two *omics* could pave the way for most successful dietary recommendations.

The real-challenge for the nutrigenomics research is to target the genes involved in the major human diseases like cancer, heart disease, arthritis and obesity. In fact, many of the genes involved in these polygenic chronic diseases are yet to be identified. The bioactive ingredients of food may turn on some genes and simultaneously turn off other set of genes involved in causation of a disease. The success of this science will require meaningful biomarkers and completion of clinical trials that monitor biomarkers. The broader and much more exiting aspect of this technology is the generation of metabolic content of a cell or organism at a given point of time. The knowledge of metabolomics will ultimately guide us on wholesale renovations of agricultural products of nutritional importance.

There has to be global sharing of knowledge for a successful progress, as expertise from various fields is required. With the growing information about nutrigenomics generated by micro-array technology, there is a need for databases to store information securely. This database (http://a-yo5.ch.a.u-tokyo.ac.jp/index.phtml) is built on an open source database system and is freely accessible.

■ ■ ■

10 Functional Foods (Nutraceuticals)

M.K. Rana

Department of Vegetable Science
CCS Haryana Agricultural University, Hisar- 125 004, Haryana, India
E-mail: <mkrlotus@gmail.com>

Fruits and vegetables, which are most fundamental food for good human health, are the potential source of non-nutrients bioactive substances. In recent years, these bioactive substances have attracted significant attention of the populace because of reported health benefits not only as antioxidants but also as anticarcinogenic and antiarteriosclerotic agents. It is believed that flavonoids, which are secondary plant metabolites, are mainly responsible for these actions. Foods contain more than 50 flavonoids and all these flavonoids have been categorized into six main classes, *i.e.*, flavones (apigenin, luteolin), flavanones (hesperetin and naringenin), isoflavones, flavonols (myricetin and quercetin) and anthocyanidins (cyanidin and delphinidin) based on the structure and conformation of heterocyclic oxygen ring (C-ring) of the basic molecule. Most of the food flavonoids have glucose or some other sugar attached.

This awareness of people about the importance of food components in maintaining good health has activated the scientists to highlight the role of various substances, like soya protein, β-carotene, flavonoids, kaempferol, curcumin and coenzyme Q in human health. Right now, a big push is there in both science and society to produce foods that promote better health and prevent diseases. Various parts of fruits, vegetables, spices, and herbs have been used in *Ayurvedic* system of medicine since ancient time to cure human and animal ailments. The fundamental actions of these medicines were earlier not so transparent. However, the principal constituents in *Ayurvedic* medicine, which performed significant role in eradication of ailments, were the non-nutrients bioactive substances. Thus, the thinking that natural food components could have a positive impact on human health is not neoteric. Although these components form the basis of *Ayurveda* but the earliest well acclaimed function of food was only the sustenance of man's life, which was later followed by the enjoyment of organoleptic properties of food as a subsidiary function. In last decades, the alliance between man and food has been localized on the ability of foods to change physiology and biochemistry and thereby conferring protection against human diseases including aging. The active principles in foods and mechanism of their action against human diseases have been identified and explained well at this time.

Functional Food

The term functional food is used to represent a food offering health benefits beyond that of elementary nutrition. In appearance, functional foods are just similar to customary foods and are consumed as segment of a regular diet. In functional foods the specific chemical components, which impart beneficial impact on health, are given the term non-nutrients bioactive substances. These chemicals are essentially of plant source, thus, they are referred to as phytochemicals. These phytochemicals are physiologically very active, hence, termed as bioactives. The key classes of functional food ingredients covered are prebiotics, probiotics, vitamins, phytoestrogens, soluble fibres (gum, depolarized guar gum, cornhusk and bran of other cereal crop), polyunsaturated fatty acids, calcium, and other minerals, proteins, peptides and amino acids and phospholipids. The products that contained these advantageous ingredients are largely placed in the category of functional foods, which have also been given different names such as nutraceuticals, designer foods, smart foods, health foods, medicinal foods, therapeutic foods, or pharma foods. Recently, food components that possess biological characteristics, such as anticarcinogenicity, antimutagenicity, anticholesterol activity, antioxidative activity, and antiaging activity, have received much attention from food and nutrition scientists. Scientific knowledge of the potential health benefits of traditional and novel ingredients has been pivotal in the evolution of functional foods.

The most remarkable trend in food trade is to make the food healthier. Consumers are now more conscious of their well-being and becoming increasingly aware of the relationship between diet and health, and the food manufacturers have provided products to fulfill this need. Now a number of pharmaceutical and nutraceutical companies that are making extended efforts in the direction of developing new functional foods are in fray. Johnson and Johnson, the pharmaceutical and dietetic company has become the leader in nutraceuticals market. The food and drink companies (Kellogg, Nestle, Novartis, Tropicana and Unilever) seem to be moving in making and supplying the nutraceuticals to the people through markets. Other large firms, such as DuPont, also seem ready to stake their claim of high-profit margin bioactive foods. Hindustan Lever's is publicizing the cardiovascular benefits of drinking tea and recently has made patronage efforts to raise the prominence of an ordinary place product to that of health promoting food. The coffee board has also recently rendered acuminate attention to coffee products, which are rich source of antioxidants. The incorporation of these bioactive substances in foods makes the taste no doubt terrible; however, these nutraceutical firms are going to take some expertise in developing technology for suppressing this terrible taste and making the taste of nutraceuticals more delicious. In last couple of years the pharmaceuticals manufacturers have introduced some kind of nutraceuticals and functional foods in the form of pills and capsules in Indian markets.

Balanced Diet

A properly balanced diet contains four food groups, *i.e.*, meat, dairy products, grains and fruits and vegetables, however, the representation of all foods of these four groups must be in a set proportion in the diet to make it balanced. Science has discovered a host of these substances in food that may contribute to human health in one or another way and the balance of these four food groups has tipped decidedly in favor of fruits and vegetables, the main carriers of the beneficial new substances. Epidemiological studies have shown that populations whose diets include plenty of fruits and vegetables rich in flavonoids have lower rates of cancer, heart disease and other ailments of aging; that's why fruits and vegetables are high in the list of foods recommended by the dieticians. One big reason for this upward push in servings of fruits and vegetables is the increasing recognition that non-nutrient bioactives within them play a significant role in long-term maintenance of good health. However, the more concentrated form of these bioactives makes the diet less ideal.

High doses of β-carotene have been found to increase the risk of lung cancer among smokers, hence, it is a useful caution against moving too quickly toward high doses. The diet can be made more balanced and ideal if these substances are consumed with traditional foods of the region. It would be practically impossible to duplicate the β-carotene doses by eating raw carrots but if these substances are extracted and concentrated, then things get out of balance and may cause danger. The biological activities of these substances work in very complex forms. When the balance between these substances is lost, it is difficult to predict what the effect might be.

A broad-based nutritional strategy must include education about balanced diet, fortification, and supplements of nutrients and bioactives in food products, being brought every day new in the market. The bioactive substances are still relatively new in the world of food industry. However, considerable progress has been made in our knowledge about the significance of macro- and micronutrients in human diet though not of bioactives. These bioactives are the minor health-promoting components in food.

Although nearly all the macro- and micronutrients that are essential in human diet have been known for many years but the significance of phytoceuticals in human diet was earlier not known. Now, it is becoming increasingly recognized that there is a wide range of phytochemicals in our diets that are bioactive. These chemicals that confined to fruits and vegetables have no traditional use in human life, however, are almost definitely responsible for the health promoting effects, which can be seen from the consumption of diet rich in fruits and vegetables.

The diet plays significant role in the uptake of specific nutrients and phytochemicals. Those that are lipophilic are absorbed much more readily from a lipid rich diet. Many steps in the food production chain of fruit and vegetable products can have large influence on the final intake of health protective phytochemicals. Processing of tomatoes in oil dramatically improves the absorption of lycopene compared with the consumption of fresh tomatoes. Raw carrots, which have high levels of pro-vitamin 'A' carotenoids, are poorer sources of β-carotene than gently cooked carrot. Healthiness of consumed products is determined by the level of bioactive components in the final product and by their bioavailability from the final product. The bioavailability of certain trace elements is increased on cooking or processing of foodstuffs.

Unfortunately, the stability and doom of phytochemicals, such as flavonoids or glucosinolates in the production chain of fruits and vegetables, have not been investigated to the same extent. The lack of knowledge on the effects of processing and storage has implications on the reliability of intake data used in epidemiological studies. Due to lack of this quantitative information, the optimum intake levels for majority of phytochemicals are still not known. The benefits of phytochemicals to the developing world are potentially great, and majority of the world's population is dependent on plants as virtually their main source of food. Consequently, many people are affected by severe nutritional deficiencies because of either the lack of food or its poor nutritional quality.

Free Radicals (Oxidants/Super Oxides)

Oxygen, which is essential for the sustenance of life, may also cause aging and sickness. Free radicals are the byproducts of oxygen burning in body during normal metabolism of food or produced when the body is exposed to cigarette smoke, radiation, chemical toxins, pollutants, or any kind of stress. The oxidants that travel through the cells and kill them directly contribute aging and cause many degenerative diseases of aging, such as wrinkles, dry skin, loss of elasticity, and discoloration of skin. These free radicals occurring in environment around the man can trigger chain reactions, which cause oxidative damage to sensitive biological structures such as DNA or cell membranes and can substantially demolish collagen and elastin support system, devastate indispensable enzymes, cause widespread tissue deterioration and abate immune functions. A current theory holds that arthritis, cancer, heart disease, jet lag, multiple sclerosis, senility, swollen extremities, varicose veins, and autoimmune diseases of aging result from cumulative damage to cells by free radicals.

A dynamic equilibrium between the factors that promote oxidation and the antioxidant defense mechanism that inhibits this process exists in human body under normal conditions. An imbalance originating from an increased rate of oxidation or from a reduction in bodily status of antioxidants can develop an inability to attenuate the reactive oxygen species and secondary free radicals, which have the ability to react with several cellular components and genetic material. The free radicals contribute progressive aging and are said to be the major source of chronic degenerative diseases such as Alzheimer's disease, atherosclerosis, cancer, cataract, coronary heart disease, stroke, muscular degeneration, Parkinson's disease, phlebitis, rheumatoid arthritis and others. A wide range of plant foods contain many naturally occurring antioxidants and the consumption of plant foods that are rich in antioxidants can extensively develop natural defense mechanism in human body against these diseases.

Antioxidants

Antioxidants are the substances that neutralize the actions of free radicals or act as super oxides scavenger. The term scavenger is the ability of a substance to provide electrons so that the free radicals can attach themselves to the antioxidant electrons and may become harmless. They considerably lessen the adverse effects of reactive oxygen species (free radicals), reactive nitrogen species, or both on normal physiological functions and make the bodily defense mechanism strong against oxidative damage. These antioxidants form the largest single group of naturally available protective agents. Based on their functional competence to act as an antioxidant although they have been put in one group but they are chemically different in their structure. In foodstuffs generally occurring potent antioxidants are flavonoids, isoflavones, carotenoids, zeaxanthin, ascorbic acid, tocopherol, lutein, lycopene, anthocyanin, polyphenols, quinones, coumarins, terpenoids, glucosinolates and selenium. In literature, it has been demonstrated that flavonoids present in fruits and vegetables are very strong antioxidants and have free radical scavenging ability, and are much more effective than vitamin C and E at protecting cells from free radical damage. Various flavonoids have been shown to prevent oxidation, stimulate the immune system and also reduce an allergic response, prevent formation of carcinogens, impede cancer cell growth and protect against bacteria and viruses.

Fortunately, India has diversified climate in different parts, which offer us a variety of fruits and vegetables that are powerhouse of antioxidants. The detection of phytochemicals present in fruits and vegetables and their inherent potent antioxidant potential in scavenging free radicals produced in human body have fascinated immense attention of the scientists working in this field. However, a common man is yet not acquainted with the usefulness of these substances in his body.

Bioactive Substances

A number of phytochemicals, which act as powerful antioxidants that keep the body cells and organs safe from damage caused by free radicals, neutralizing their deleterious effects, are present in fruits, vegetables and other plant species. They are the biologically active substances that bestow color, odor, aroma, and flavor to the plants and their parts and provide them protection against several diseases too. Fruits and vegetables from the very old time are considered as protective food because they supply sufficient quantity of minerals and vitamins, which perform not only basic nutritional role in human body but also as a powerful medicine and help to reduce the risk of several chronic diseases. They also supply some non-nutrient bioactive substances, which play vital role in human body to develop immunity and resistance against several diseases and aid specific bodily functions too and demonstrate physiological and biochemical benefits, thus, fruits and vegetables have been given the status of functional foods by the nutritionists/dieticians. Inclusion of fruits and vegetables in diet can reduce the risk of cancer and reduce total low-density lipoprotein cholesterol, which reduces the risk of coronary heart disease.

A comprehensive list of some of the important fruits and vegetables along with bioactive substances, which play a significant role in maintaining good health, is given below:

Table 1: Bioactive substances present in some foodstuffs

Commodity	Bioactive substance acts as antioxidant	Commodity	Bioactive substance acts as antioxidant
Aonla	Ascorbic acid	Beans	Genistein
	Polyphenols		Isoflavonoid
Apple	Flavonoids		Phytosterols
	Polyphenols		Saponins
Bael	Ascorbic acid	Broccoli	Glucosinolate
	Riboflavin		Selenium
Citrus	Ascorbic acid		Sulforaphane
	Bioflavonoids	Cantaloupe	α-carotene
	Chalcone	Carrot	α-carotene
Crane berry	Flavonoids		Lycopene
	Phenols	Chilli	Capsanthin
Grape	Flavonoids	Cruciferous vegetables	Glucosinolates
	Phenols		Isothiocyanate
Grapes (blue)	Anthocyanins		Sulforaphane
Olive	Polyphenols	Dark green vegetables	α-carotene
Palm oil	Toctrienols		Lutein
Straw berry	Flavonoids	Ginger	Gingerol
	Phenols	Onion	Quercetin
Rice bran oil	γ-oryzanol	Pumpkin	α-carotene
Coffee	Phenol ester	Radish	Isothiocyanate
Tea (green)	Catechins	Soybean	Daidzein
	L-theanine (amino acid)		Genistein
Tea	Catechins		Isoflavones
	Flavonoids	Sweet corn	Zeaxanthin
	Polyphenols	Tomato	Lycopene
Meat/fish	Co-enzyme Q	Turmeric	Curcumin

Foods and drugs are considered to be similar enough because both are consumed into the body and both require a high degree of quality control to ensure purity. Medical research has increasingly emphasized the pivotal role of diet rich in bioactive substances in preventing common killers such as heart diseases and cancer. It is reasonably clear now that many substances in foods, such as antioxidants, flavonoids and estrogen simulating, obviously have beneficial health effects beyond mere sustenance.

Functions of Phytochemicals

Antioxidant and anticarcinogenic activity

Soya bean contains substances such as daidzein, genistein and isoflavones, which are responsible for multifarious health benefits and are known to have weak estrogenic activity, thus, prevent cancer in organs having relation with hormone. The individuals having high levels of isoflavonoids in urine and plasma are less predisposed to the risk of cancer. The ingestion of these isoflavonoids attenuates the vasomotor symptoms in women having menopausal stage and reduces the oxidation of low-density lipoprotein. New research from Japan suggests that the postmenopausal women who consume a diet rich in soya protein have stronger bones. The new studies also indicate that plant estrogens found in soya can help to abate the symptoms of menopause, estrogen levels drop augmenting the risk of bone thinning and other symptoms including hot flushes, depression and aching joints. However, isoflavones-compounds found in soya with a similar chemical structure to the female estrogen hormone- estradiol are believed to duplicate the effects of natural estrogen and lessen some associated symptoms. The researchers are trying to extract soya proteins that may eventually prevent diabetes. Now several nutraceuticals have come in the market in which either isoflavones or soya proteins have been incorporated. The addition of phytosterol esters in fortified foods is a prominent example of prosperous beginning of a functional food.

Plant sterols and stanols are group of chemicals that are universally present in membranes of plant cells, and hence, are ubiquitously consumed as component of plant originated foods. These chemicals somewhat arrest the absorption of cholesterol when present concurrently in small intestine. Margarine containing phytosterol derivatives, which remarkably attenuates the levels of serum cholesterol, is being sold commercially in United States and other European countries. The esters of plant sterol and plant stanol may lessen the jeopardy of coronary heart disease by lowering down the levels of cholesterol in blood. L-theanine, which is available in green tea, is a primary amino acid and promotes cells tranquility without any weariness. Bioflavonoids are said to be associated with vitamin C and increase the effectiveness of vitamin C and are recognized as potent antioxidants, which are supplied by fruits and vegetables in sufficient quantity. Quercetin universally present in red onion has anticarcinogenic activity, and inhibits the growth of cancerous cells too. Saponins, which are usually present in soybean, hold pledge because of their significant role in lowering down the serum cholesterol. Fibers are the indigestible carbohydrates and oligosaccharides together with soluble fibers and have been recognized to have multifarious commending significance in many human ailments such as appendicitis, coronary heart disease, gallstone, colon cancer, diabetes and hyper-cholesterolemia. They can be brought expansively in practice by designing them in such a way that they may favor the activity of useful bacteria and inhibit the actions of pernicious bacteria present in the colon.

Use of colors in foods is of growing importance, and functional colors are adding a new dimension to their use. More recently, the food research has revealed that health benefits may be realized from using plant-derived colors. Carotenoids such as β-carotene, which is responsible for yellow and orange color, can be transformed to vitamin A in the body. Flavonoids from grape skins and other fruits are believed to have antioxidant activity, as does the yellow pigment curcumin, while chlorophyll, which provides green color to plants, is also believed to have healing properties.

Fruits and vegetables besides β-carotene contain hundreds of other pigments as well as non-pigment compounds that may play a role in preventing oxidative damage. These phytochemicals appear to maintain health by boosting the immune system, reducing inflammation and allergies, detoxifying contaminants and pollutants, and/or activating enzymes that block unbridled cell division.

The chemical form of phytochemicals is of profound importance while considering the biological relevance of specific chemicals and their levels in the diet. Some phenols might be better antioxidants than others might. The compounds that are easily absorbed in the human intestine, are not quickly degraded in the tissues, and are able to reach the target sites easily. Flavonoids that are not absorbed by the intestine undergo degradation by gut microorganisms and may only play a limited role in preventing oxidative damage in the colon.

Epidemiology studies on phytochemicals (bioactive compounds), *i.e.*, carotenoids, flavonoids, folates, and glucosinolates present in fruits, and vegetables show an increasing number of evidences for protective effects in human body. It is thought that they play an important role in the prevention of various diseases, most importantly aging diseases like cancer and coronary heart diseases. The orange pigment- β-carotene has been linked with low rates of certain cancers with a high intake of fruits and vegetables containing lots of β-carotene.

The diets of most people supply enough amounts of two more carotenoids, *i.e.*, lutein and lycopene than they do β-carotene. Lycopene possesses exceptionally high antioxidant capacity compared to other carotenoids and may be useful in preventing cancer; it counteracts the assaults of free radicals, and spares the crucial biological chemicals from damage too. In addition to lycopene, the carotenoids, lutein, and zeaxanthin are added in functional foods, based on their competency to guard against vascular degeneration, which is considered the foremost cause of adult blindness, and there are some evidences that carotenoids and other metabolites stimulate the enzymes involved in detoxifying contaminants or blocking the unrestrained increase in cancer cells.

Lutein may help to prohibit the initial cause of blindness in the elderly macular degeneration, which results from long-term deterioration of center of retina, known as macula. People with the highest intake of fruits and vegetables especially those rich in lutein, had half the risk of macular degeneration as those with the lowest intake. Lutein along with several other carotenes gives summer squash, apricots, peaches, and oranges their yellow light orange color. But its richest sources are kale, spinach and collard greens, in which the yellow pigments, are masked by the more abundant greens, chlorophyll pigments. Lycopene gives tomatoes their bright red color. High intakes of these bioactives may protect the humans against prostrate gland cancer.

Catechins, which are structurally flavonols, are water-soluble colorless compounds and impart bitterness and astringency to the food. Almost all the characteristics of food including its color, aroma, and taste are directly or indirectly associated with catechins. These substances also limit the consumption of α-tocopherol and allow it to act as a scavenger within cell membranes whilst catechins scavenge aqueous peroxyl-radicals near the membrane surface. Catechins in green and black tea are well known for their role in lowering down the cholesterol and sugar in blood and reducing the incidence of cancer. Numerous recent research reports have emphasized the effects of catechins on coronary heart diseases and inhibition of carcinogenesis. These inhibitions were observed at all three levels of cancer advancement namely, induction, progression and transmutation. The nutritional antioxidant properties of catechins in foodstuffs can improve the marketing potential of cereals, cakes and biscuits as well as traditional health food products and dietary supplements.

The main flavonols present in foods are quercetin, kaempferol, and myricetin. Flavonols are predominantly present as glycosides rather than as their non-glycosylated form (aglycones). Quercetin, kaempferol, and myricetin were found to be able to inhibit carcinogen-induced tumors in rats and mice. The former two flavonols are effective in protecting cells from both super-oxide and peroxide. Quercetin is by far the most important flavonol followed by kaempferol.

Antimicrobial and deodorizing activity

Among all microbes related diseases dental caries is the most common. The polyphenols present in fruits and vegetables can prevent teeth from decaying by inhibiting biological activities of the carcinogenic *Streptococci*. It is believed that these substances reduce the concentrations of acids on the tooth enamel for the whole time that the starch is trapped in the teeth. Polyphenols have a stronger deodorizing activity than sodium copper chlorophyllin and so should improve breath freshness.

There has been a concern about the inhibition of iron absorption through polyphenols forming insoluble complexes in the gastrointestinal tract, thus, reducing iron bioavailability when consuming food rich in polyphenols. They have been found to be strongly potent to inhibit the absorption of non-haem iron to a significant extent but have no inhibitory effects if the haemoglobin is taken in cooked form. According to several reports published in medicine literature, the supplementation of ascorbic acid or drinking of milk may counteract the inhibitory effect of tea (rich in polyphenols) on iron absorption.

The antibacterial and deodorizing effect of catechins slows tooth decay (anti caries) and improves breath freshness so providing natural added value for tooth pastes, mouthwashes, chewing gums and breath fresheners, with the potential for other body, skin and hair product applications. Many shampoos, perfumes, moisturizing creams and sunscreen lotions contain catechins naturally extracted from vegetative parts of the plants having more quantity of these substances. These are believed to have a soothing effect on the skin as well as acting as antioxidants to protect the skin from free radicals. Catechins could be used to protect the foods against any kind of bacterial infections.

Glucosinolates as Anticarcinogenic

Glucosinolates, ubiquitously present in abundant quantity in plants of cruciferae family, are thioglucosides and the potent precursors of bioactive compounds such as isothiocyanates and indoles. However, the formation of these compounds depends on the enzymatic degradation by the endogenous enzyme myrosinase- becomes active during chewing of the plant tissues (rupturing of plant cells) or by bacterial enzymes in the human gut. In cabbage, the main glucosinolates are sinigrin, progoitrin, and glucobrassicin, contributing to a total glucosinolates content of between 360 and 2754 mg/kg fresh weight. These three glucosinolates also predominate in a variety of other brassica vegetables including Brussels sprouts, broccoli and cauliflower. Researchers have become increasingly concerned about glucosinolates in cruciferous crops and their effects on human body. The American Cancer Society and the National Research Council have recommended increasing consumption of cabbage as a means of decreasing risk of certain types of cancer at several sites in humans. There are increasing evidences in the literature that the consumption of cabbage and other cruciferous vegetables is associated with reduced incidence of some forms of cancer due to glucosinolates-derived bioactive substances but there is no evidence that the glucosinolates present in brassica vegetables are detrimental to human health.

Conclusion

Foods can offer benefits that reach beyond elementary nutrition to functional nutrition for disease evasion and health improvement. Today, functional foods are getting tangible attention and representing one of the rapidly growing segments of the world food industry. One of the captivating ideas in human nutrition is that selected foods and their constituents can improve physical or mental performance or mitigate disease risk. Thus, beyond meeting nutrition needs, diet may regulate various body functions and may play useful roles in the headway of some diseases. Epidemiological studies have led to recommendations that people should consume at least two servings of fruits and three servings of vegetables daily.

■ ■ ■

11 Geriatric Nutrition

Rajni Modgil

Food Science and Nutrition, College of Home Science,
CSK Himachal Pradesh Krishi Vishvavidhyalya, Palampur, Himachal Pradesh,
India, E-mail: <rajni_modgil1@yahoo.com>

Both the number and the proportion of older persons defined as aged 60 and above are growing in virtually all countries, and worldwide trends are likely to continue unabated. In 2002, there were an estimated 605 million older persons in the world, nearly 400 million of whom were living in low-income countries. Greece and Italy had the highest proportion of older persons (both 24% in 2000). By 2025, the number of older persons worldwide is expected to reach more than 1.2 billion, with about 840 million of these in low-income countries. World population of people above 60 years of age is estimated to be 96 million in 2010. The Life span has increased by 60% in 60 years. India has today second largest population of Senior Citizens. There is sharp increase in population of Young Old 60 years to 69 years and Old- old 80+. With increasing life span, the need is to provide good nutrition to elderly so that they can lead a healthy life.

Geriatric Nutrition

Geriatric nutrition means nutrition for elderly. In other words, we can say that it deals with nutritional requirements of aged persons. Aging is a continuous biological irreversible process, which results in reduced capacity for various activists. With increase in age, the body has reduced ability to repair tissues and form new tissues and cells. More cells are destroyed as compared to rate of building of new cells. The process of aging begins with cessation of growth and development. The process of aging brings about physiological, psychological and immunological changes that influence the nutritional status. Chronic diseases such as cardiovascular diseases, diabetes, high blood pressure, *etc* are more prevalent in elderly. These diseases might require nutritional modifications. Disability poverty social and personal problems can also affect nutritional status.

Changes during old age

Various changes, which can be genetic, physical, or biological factors such as exposure to radiation, sunlight, smoking, pollution, *etc.*, take place during aging. Taste and smell are essential to proper nutrition. In elderly, the peripheral sensory receptors decline, causing the appetite to wane. Taste and aroma are inextricably intertwined in determining the palatability and acceptance of food, *e.g.*, during an upper respiratory infection, the olfactory receptors are blocked. Food becomes tasteless, it loses both flavour and aroma and the appetite declines. As a result of aging, the taste buds on tongue and the olfactory receptors in roof of the nasal cavity regress. In addition, the gustatory and olfactory nuclei in the brain decline, causing a reduction in appetite and diet.

Physiological changes: Physiological changes occur slowly over time in all body systems. The *Basic Rule* is that by the age of 70, the organ systems and organ functions diminish by 50%. Clinical observations support this *basic rule* in that pulmonary function, cardiac output, liver function and kidney function all diminish by the age of 70 to about 50% of that observed at the age of 25. Secondary, aging increases the rate of diminishing function observed in the *basic rule*. Aging is a *normal* process as disease and disability are disruptions.

Aging produces a change in the ability to participate in physical activity, which is accelerated beginning at the age of approximately 50. This is due to multiple factors which results in an increased difficulty for delivering oxygen to the tissues of body. Decreased chest wall elasticity results in the moving of adequate air more difficult, thus, more energy should be expended to obtain additional oxygen required for exercise. There is decreased ventilation to perfusion ratio. The amount of available blood output from the heart to supply the skeletal muscles is decreased as well. There is an actual decreased in skeletal muscle mass.

Change in body mass: With age, the lean body mass is lost. Reduced muscle mass includes skeletal muscle, smooth muscle and muscle that affects vital organ function, with loss of cardiac muscle perhaps the most important. Cardiac capacity can be reduced and cardiac function impaired by chronic diseases such as atherosclerosis, hypertension, or diabetes. Changes also occur in the kidneys, lungs, liver and in our ability to generate new protein tissue. In addition, aging can slow the response of immune system in making antibodies.

The most significant result of the loss of lean body mass might be due to the decrease in basal energy metabolism. Metabolic rate declines proportionately with the decline in total protein tissue. Loss of lean body mass also means reduced body water- 72% of the total body water is in lean muscle tissue.

Total body fat increases with age, as fat tends to concentrate in the trunk and as fat deposits around the vital organs. However, in more advanced years, the body weight declines.

Changes in bone density: Bone density is lost. After menopause, the women tend to lose bone mass at an accelerated rate, which results in osteoporosis. Severe osteoporosis is debilitating and serious. Fractures and their associated illness and mortality are certainly a concern. Vertebral compression fractures can also change the chest configuration. This, in turn, can affect breathing, intestinal distension and internal organ displacement.

Sensory changes: These include a decline in sight and peripheral vision, hearing, smell and taste. The losses are neither total nor rapid but they do affect nutritional intake and health status.

- Loss of visual acuteness may lead to less activity or a fear of cooking, especially using a stove. Inability to read food prices, nutrition labels, or recipes may affect grocery shopping, food preparation and eating. This could have an adverse effect on nutritional status.
- Loss of hearing may lead to less eating out or not asking questions of the waiter or store clerk. Changes in smell and taste are more obvious. If food does not taste, appetizing, or smell appealing, the food intake is affected. If there is a cut back on salt, sugar, or fat, the desire to eat is reduced.

Factors Influencing Nutrition and Food Intake during Old Age

Basic metabolic rate (BMR): This is the amount of energy required to carry on the involuntary work of the body and to maintain the body temperature. About one-third of this energy is used to maintain the functional activities of various organs such as the heart, kidneys and lungs. The rate of basal metabolism is affected by size, shape and weight of an individual, body composition and age. The basal metabolism is high during rapid growth, *i.e.*, childhood and adolescence until adult age is reached. After the age of 25 years, there is a gradual decline in metabolic rate with each decade.

Digestion: Digestion in later years is affected in a number of ways. A reduction in the tones of the muscles of stomach, small intestine and colon leads to less mobility (movement) resulting in constipation and abdominal distention, *i.e.*, feeling of gas. Abdominal distention from certain foodstuffs is greater, *e.g.*, whole legumes, beans, *channa*, *rajmah* lobia and vegetables like cabbage, cauliflower, leafy vegetables, *etc.*

Fat tolerance: Fats are poorly tolerated by the elderly. Hence, the intake of fats should be reduced, *e.g.*, pure ghee, butter, cream and deep fried food articles like *puri*, *samosa*, *pakora* and nuts and dried fruits.

Hormonal Imbalance: Hormonal imbalances occur more frequently during later years in both men and women. There is a reduction of thyroid activity, which leads to reduced energy requirement. Therefore, requirement of calories is less. Hence, amount of food intake (quantity) should be reduced.

Elevated blood urea: Many older persons, especially men, have tendency for elevated blood urea, nitrogen and blood uric acid. Hence, it is advisable to cut down on meat, meat products, shellfish, whole legumes, whole grain cereals, oatmeal, dried peas, beans, coffee, cocoa, alcohol, and other foods rich in proteins

Role of Nutrition in Aging

Nutrition has been recognized as an important factor in influencing the functional outcome of ageing. To overcome the problems of aging, nutritious food can be one of solutions. A nutritious food is one that contains all the essential nutrients such as carbohydrates, fats, proteins, minerals and vitamins. A well balanced diet should contain all these in correct proportions and adequate amounts. Carbohydrates, fats and proteins provide energy required for various activities. Minerals and vitamins do not supply energy but play an important role in the regulation of several essential metabolic processes in the body.

Restoration and maintenance of optimal nutritional status in long-term care, setting will help in enhancing the quality of life for geriatric patients. The elderly are at increased risk for poor nutrition due to age-related physiological and psychosocial changes, as well as the added toll of chronic disease. All of these may have a negative impact on adequate food intake. Nutrition can be one of the factors responsible for all the changes. However, the slowing of normal action of the digestive tract plus general changes has the most direct effect on nutrition. Digestive secretions are diminished markedly, although enzymes remain adequate. Adequate dietary fiber, as opposed to increased use of laxatives, will maintain regular bowel function and not interfere with the digestion and absorption of nutrients as occurs with laxative use or abuse.

Many other factors can influence the diet of an older adult, *e.g.*, eating alone has shown to influence nutrients intake. Research shows that adults who live or eat alone have been found to eat less and to be at greater nutrition risk than those live and dine with others. The process of ageing also affects other nutrients needs, *e.g.*, the requirements for some nutrients may reduce and the requirements for other essential nutrients may in fact rise in later life.

Improper or inadequate intake of nutrients can lead to under nutrition because the energy requirements of a person decrease with increase in age. This is because of a lowered basal metabolic rate and lessened physical activity. There is an 8% reduction per decade from the age of 55 to 75 years. Proteins provide the essential amino acids and the amount of protein in the diet is an important measure of the adequacy and quality of the diet.

Nutritional principles apply to all age groups but are most important during old age. The process of aging brings about physiological, psychological and immunological changes that influence the nutritional status. Heredity and good nutrition slow down the process of aging. Individual enjoys physical and mental vigour in old age also. Goal of nutritional care should be to help the aged live a healthy, purposeful and independent life.

Nutritional Requirements of Elderly

These requirements are based on physiological changes that take place during old age. After the age of 40 years, the basal metabolic rate decreases due to the decrease in muscle mass and other metabolically active tissue mass. There is decrease in the production of endocrine glands, and the physical activity reduces with age. The calorie intake should be adjusted to maintain a constant body weight. In case of obese who need to loose weight because of some illness such as diabetes, arthritis of knee joints *etc.*, and the calorie intake should be reduced to create a moderate negative calorie balance, which can bring down the body weight to normal limits gradually but steadily. The nutritional requirements of elderly are discussed below:

Energy requirement: The body needs energy for maintaining body temperature and metabolic activity and for supporting physical work and growth. Change in muscle mass affects the energy requirements during old age.

Daily Allowances of Calories

Gender	Age	kcal
Men	55-75	2208
Women	55-75	1748
Men	75+	2160
Women	75+	1710

Muscles are metabolically highly active. Basal metabolic rate, which accounts for major portion of daily energy expenditure, is directly related to body muscle mass. With age the muscle mass decreases as a result energy requirements decreases as compared. Basal metabolic rate influences the calorie requirement of a person. Between the age of 35 and 55 years, the calorie requirement reduces 5% for each decade. From the age of 55 to 75 years, it is decreased by 8% for each decade and from 75 years and beyond there is a further decrease of 10%.

Protein requirement: Protein, which forms the important constituents of muscles and other tissues and of vital fluids like the blood, is the chief substance in body cells. Protein supplies the building material for the body and makes well the wear and tear of tissues. Several substances concerned with vital life processes, *e.g.*, enzymes, which help in the digestion of food and antibodies, which are body defenses against infection, are mainly protein in nature. Proteins are, therefore, one of the most important nutrients in a person's diet.

Proteins are essential structural components of all the cells and are needed to build and repair tissues, which are constantly undergoing breakdown and changes. They are also required for maintaining output of essential body secretions such as enzymes and hormones. Protein needs are affected by several factors like energy intake, nutritional and physiological state of individual. The protein content of adult body diminishes with increasing age. Studies suggest that total body protein declines with age, a change that accelerates after the age of 65 years. This accelerated decline in total body protein contributes to the age-associated loss of skeletal muscle, a process called sarcopenia. Because sarcopenia exposes the elderly to an increased risk for disability and injury, strategies to counteract a loss of skeletal muscle using nutritional interventions and exercise have been investigated. Changes in muscle mass are related to whole-body protein turnover and the protein synthesis rate.

Due to decreased appetite and poor digestive capacity, old people are likely to consume less protein but the fact remains that despite a decrease in calorie requirement, the need of protein remains the same, hence, the food should be a little richer in protein than the other normal food.

Deficiency of protein is common in the elderly and is one of the contributing factors to edema, anemia and lowered resistance to infections. Wound healing also depends on the adequate proteins in body. In elderly, up to 12-14% of the total calories should be from proteins, but due to decreased appetite and poor digestion, the elderly tend to consume less protein.

Most food items contain proteins but in widely varying amounts. Animal food articles such as meat, fish and eggs are rich in proteins. Milk is also a good source of protein. Among the food of vegetable origin, pulses and nuts are the richest source of protein. The common cereals such as rice and wheat are relatively poor source of protein. Rice contains less protein than wheat does but the protein of rice is of better quality. In general, the normal protein requirement is 1 g/kg body weight.

Carbohydrates requirement: Carbohydrates are a class of substances, which include glucose, cane sugar, milk sugar and starch. Grain foods are largely composed of starch and foodstuffs like cane sugar and glucose are pure carbohydrates, which form the main source of energy to the body. Carbohydrates form the bulk of an Indian diet since the foodstuffs are cheaper.

Besides starch and other digestible carbohydrates, many foodstuffs contain cellulose and hemicelluloses, which are also carbohydrates. These types of carbohydrate are also called fibre or roughage. They are not digested in the human digestive system and are passed out as such. Though they do not contribute to the nutritive value of diet, the presence of roughage in the diet is necessary to the mechanics of digestion and elimination of wastes. The contraction of muscular walls of digestive organs is stimulated by fibers, thus, counteracting the tendency of constipation. Vegetables like okra (ladyfinger) and green leafy vegetables like beet leaf, spinach, amaranth and fruits are rich in fibre, whereas, foods like root vegetables and potato are poor sources. It is recommended that once a day, a green leafy vegetable should be included in the diet and one fruit must be taken daily. This will give enough fibre and help in preventing constipation.

The carbohydrates content in the diets of elderly persons has to be reduced due to reduced physical activity. About 60 to 70% calories should come from carbohydrates. Elderly person should be encouraged to eat whole grain cereals instead of refined flour.

Fats requirement: Like protein, fat is a necessary ingredient in the diet and it is of value to the body in a number of ways. Animal fats such as butter and ghee contain vitamin A. Fat is a concentrated source of energy. One gram of fat provides 9 calories, whereas, 1 g of carbohydrate or protein provides only 4 calories. Some fats, especially vegetable oils, provide essential fatty acids, *i.e.*, linolic, linolenic and archidonic acids to the body. Like vitamins, the essential fatty acids in the diet play important role in general metabolic reactions, and a deficiency of these acids in the diet leads to a skin condition known as

Phrynoderma. Part of fat intake should be in the form of vegetable oils, which are rich in essential fatty acids, and part of it should be in the form of animal fats, which are rich in fat-soluble vitamins A, D, E and K. Fat is a concentrated source of energy. It adds palatability to food. The diet should contain 30-40 g of fat and half of it should be in the form of vegetable oil, which is rich in essential fatty acids. A diet with high content of saturated fatty acids (ghee, butter, coconut oil and unrefined oils) tends to increase the level of cholesterol in the blood. The total amount of fat from direct or indirect sources should not exceed 20-30 g daily.

Minerals requirement: The minerals are also essential for maintaining the elderly person's body. The minerals essentially required by the elderly have been described below:

***Calcium*:** The need for calcium increases during old age especially for women after menopause. There is increased demineralization and osteoporosis of bones in the later years of life. This can give rise to spontaneous fractures and increased incidence of tooth decay. Calcium is very essential for an average elderly person. As people become older, the bones become dematerialized, which is known as osteoporosis so calcium intake should be not less than 400 mg per day. Foods rich in calcium are *Ragi*. The daily requirement of calcium in a woman around the age of 50 is about 1000 mg/day. Food alone may not be capable of meeting the entire requirement. In addition, calcium is available in limited number of food articles. Hence, calcium supplements might be required. Good food sources of calcium are milk and milk products like cheese and curd. Green leafy vegetables like fenugreek, beet leaf, spinach, amaranth, drumstick leaves and sea food are also a good source of calcium.

***Iron*:** Iron deficiency leads to anemia so the diet of the elderly should contain sufficient amount of iron. Green vegetables are rich in iron, cheaper than other vegetables and available easily most of the time. Elders should include some variety of greens in their diet every day. Other foodstuffs rich in iron are wheat flour, *Ragi*, jaggery, dates and liver. The daily iron requirement and recommended daily allowance is about 30 mg/day. If anemia is there then about 10 mg/day extra iron can be given. The good sources of iron are liver, green leafy vegetables like amaranth, beet leaf, spinach, fenugreek, whole wheat bread, iron rich dry fruits and fruits such as apple and gooseberry.

***Sodium*:** There is no evidence that aggressive sodium restriction will prevent or delay the onset of cardiovascular disease. Moderate amount of salt in diet helps to improve the palatability and acceptability of the diet. Total salt free diet should not be followed, except under the advice of a physician for the treatment of a specific disease. Salt is naturally present in almost all fruits and vegetables. Food items that have a very high concentration of sodium such as pickles, chutneys and *papads* should be avoided as a part of daily diet.

***Zinc*:** Some features of old age, such as delayed wound healing, decreased taste and anorexia, are also findings associated with zinc deficiency. However, the normal diet suffices for the body's target for zinc, and very rarely, there may be the need to take a zinc supplement.

***Phosphorus*:** Phosphorus is a mineral colleague of calcium within the body. A proper balance of phosphorus must be present in the bloodstream for good teeth and bones or nourish the brain and nerves.

Phosphorus is a soft, non-metallic element found in the body only in combination with other substances. As a result, it is neither luminous nor inflammable. In human body, phosphorus is a part of the energy transporting systems in cells and is also a component of phospholipids.

An adult human body contains about 400-700 g of phosphorus as phosphates. At least two-thirds of this amount is found in chemical combination with calcium in the bones and teeth and the rest in other tissues.

Phosphorus is essential for the proper utilization of not only calcium but also other minerals like iron, magnesium, potassium and sodium. Phosphorus is indispensable for all active tissues. In combination with calcium, it feeds the nerves, aids the growth of hair and helps in counteracting the fatigue. This mineral is important for the regular functioning of heart and for normal functioning of kidneys.

Phosphorus is essential in the chemical processed involved in the utilization of carbohydrates and fats. It liberates their energy at the rate demanded by the body. It also helps in regulating the acid-alkaline balance of blood, which is vital for the maintenance of health and prevention of disease.

Phosphorus is also necessary for the formation of phospholipids, *i.e.*, lecithin and cephalin, which are integral parts of a cell structure and also act as intermediates in fat transport and metabolism. This mineral is also an essential constituent of nucleic acid and nucleo-proteins of body cells.

Recommended Daily Allowance of Phosphorus

Gender	Recommended Daily Allowance (mg)
Men	1200
Women	1200
Children	800

The most important food sources of phosphorus are whole grain cereals, milk and fish. Vegetables such as carrots, and leafy vegetables and fruits like black currants, raspberries, raisins and apricots are good the sources. The other sources of this mineral are soybeans, lentils and other pulses and legumes.

Vitamins for growing older: Poor appetite, impaired absorption, constipation and general malaise may be caused in part by inadequate intake or improper absorption of vitamins. While the body's demand for vitamins does not increase in old age, it is particularly important that the factors interfering with utilization should be corrected or that the intake should be sufficient to compensate for the poor utilization.

Vitamins are involved in the utilization of major nutrients like protein, carbohydrate and fat in the body. They cannot be made by the human body in adequate quantities, thus, they have to be present in the food that we eat. Vitamin A, D, E and K are fat soluble that are stored in the body and vitamin B complex (thiamin, riboflavin, niacin, pyridoxine, biotin, folic acid and cynocobalamine) and vitamin C are water-soluble that are not stored in the body and must be replaced each day. In the elderly when exposure to light is poor, *e.g.*, in those who are house bound or institutionalized, vitamin D status is likely to be poor. Milk should be included in diet of such people.

Vitamin D is as important to the elderly as it is to children. It plays a vital role in the body's ability to absorb calcium, which is required to keep the bones healthy. With little or no calcium, bones can lose their mass, become brittle and bring about osteoporosis. Very few elderly people consume the proper amount of dairy products to ensure that their body gets ample amounts of vitamin D.

Another serious condition confronting seniors is the oxidation damage caused by free radicals. These are unstable molecules, which when left on their own, run rampant throughout the body, destroying healthy cells. Oxidation is the same process that causes rust to develop on cars.

Antioxidants work to neutralize free radicals- a process that inhibits the amount of damage they cause. The best sources of antioxidants are fruits and vegetables. They also contain beta-carotenes and a number of other important vitamins such as vitamin A, C and E, yet few seniors consume the required amounts.

As healthy cells become damaged, seniors increase their risks of developing heart disease, arthritis, certain forms of cancers, cataracts, Parkinson's disease and many other disorders. Antioxidants also keep the immune system functioning properly. As healthy skin cells break down, skin loses its elasticity and begins to sag, wrinkle and look older.

The body needs vitamin K to keep the body's blood clotting mechanism functioning properly, yet once again, few seniors get enough. The absence of vitamin K is believed to be linked with the onset of Alzheimer's disease.

These vitamins are easily destroyed or washed out during food storage and preparation. The B-complex group vitamins are found in a variety of foodstuffs such as cereal grains, meat, poultry, eggs, fish, milk, legumes and fresh vegetables. Citrus fruits are good sources of vitamin C. The use of mega doses of vitamin C is not recommended. As an anti-oxidant, vitamin C helps to mop up cell-damaging free radicals. In other words, vitamin C is believed to slow down premature ageing and help in keeping the disease at bay so while everyone benefits from enough vitamin C, one might be particularly fond of it if he/she wants to stay younger for longer.

Vitamin C may also help the body to absorb iron. So to get the most iron out of your meal if you're having an iron-rich product, such as a piece of steak, try drinking a glass of orange juice before your meal, or tucking into some tomatoes or peppers as a side dish. Citrus fruit, strawberries, blueberries, green vegetables, peppers, tomatoes and potatoes are all great sources of vitamin C and they are all delicious too.

Role of Free Radicals in elderly

Free radicals have an unnatural molecular structure, which is caused by unnatural forces such as pollution, cigarette smoke, radiation, fried foods, cured meats, stress (mental, emotional and physical), pesticides and other toxic chemicals. Cigarette smoke, including passive smoke, puts billions of free radicals into our blood stream. Unnatural environment and lifestyle have created an over abundance of free radicals, which play a role in every degenerative disease known to man.

Free radicals have been linked to diseases such as Alzheimer's, Parkinson's, AIDS, cancer, premature aging, collagen deterioration, varicose veins, arthritis, asthma, cataracts, retinitis, angina, rheumatism, cataracts, stress, hemorrhoids, heart disease, stroke, senility, swollen extremities, kidney and liver disorders.

Free radicals alter the molecular structure of our body and destroy our cells but we can fight them even in old age. At molecular level, there is a constant battle going on in our body between antioxidant nutrients and free radicals.

A free radical is a molecule or molecular fragment with the spin of one electron, which is not paired with a companion electron. This is a very hazardous, unnatural and unstable state since electrons normally come in pairs.

This odd, unpaired electron in a free radical causes it to collide with other molecules, hence, it can steal an electron from them, which changes the structure of these other molecules and causes them to become free radicals. This can create a self-perpetuating chain reaction in which the structure of millions of molecules are altered in a matter of nanoseconds (a nanosecond is a billionth of a second), reeking havoc with our DNA, protein molecules, enzymes and cells.

Human immune system has a means of defending against free radicals but it is very easy for human body to find itself outnumbered, this is why, man, particularly in old age, needs a generous supply of antioxidant nutrients. Human immune system uses antioxidants to stabilize and eliminate free radicals. Antioxidants are able to give free radicals- an electron, which becomes a companion to their unpaired electron, thus, eliminating threat of that free radical.

The elderly needs to take supplements of antioxidant nutrients. The most commonly known antioxidants are vitamins A, C, E and beta-carotene, which is converted to vitamin A in the body. Other nutrients, such as the minerals, *i.e.*, copper, selenium, zinc and manganese, and certain amino acids, are considered antioxidant nutrients since they invigorate the body's own natural antioxidants. A vitamin C antioxidant blend is an ideal way to accomplish this, as it supplies vitamin C, which bolsters the immune system as well as antioxidants that the strengthened immune system can utilize.

Bioflavonoids are another significant category of antioxidant. Bioflavonoids can offer a double benefit to our immune system because in addition to eliminating free radicals, they also enhance our assimilation of vitamin C.

Water: Water is very essential as it stimulates peristalsis, and thus, aids in combating constipation. The kidneys also need fluid to dispose waste more effectively. Some elderly individuals have a fading sense of thirst and go for longer periods without water. Others avoid drinking enough water for the fear of incontinence. This may lead to dehydration, which in turn can give rise to confusion, headache and instability. The elderly should be advised to consume fluids at regular intervals even if they are not thirsty. Water can be consumed as such or in the form of buttermilk, soups and juices. The daily requirement of water is 1.5-2.0 liters per day.

Dietary fibers

Dietary fiber, also known as roughage or bulk, includes all parts of plant foods that your body cannot digest or absorb. Unlike other food components such as fats, proteins or carbohydrates, which your body breaks down and absorbs, fiber is not digested by your body. Therefore, it passes relatively intact through your stomach, small intestine, colon and out of your body. It might seem like fiber does not do much but it has several important roles in maintaining health.

Fibers are commonly classified into two categories, *i.e.*, (i) insoluble fibers- those that do not dissolve in water and (ii) soluble fibers- those that dissolve in water.

- **Insoluble fiber:** This type of fiber promotes the movement of material through your digestive system and increases stool bulk, so it can be of benefit to those who struggle with constipation or irregular stools. Whole-wheat flour, wheat bran, nuts and many vegetables are good sources of insoluble fiber.
- **Soluble fiber:** This type of fiber dissolves in water to form a gel-like material. It can help in lowering the blood cholesterol and glucose levels. Soluble fiber is found in oats, peas, beans, apples, citrus fruits, carrots, barley, *etc.*

The amount of each type of fiber varies in different plant foods. To receive the greatest health benefit, eat a wide variety of high-fiber foodstuffs.

Benefits of a high-fiber diet for elderly

A high-fiber diet has many benefits, which include:

- **Normalizes bowel movements:** The dietary fiber increases the weight and size of your stool and softens it. A bulky stool is easier to pass, decreasing your chance of constipation. If you have loose, watery stools, fiber may also help to solidify the stool because it absorbs water and adds bulk to stool. For some, fiber may provide relief from irritable bowel syndrome.
- **Helps maintain bowel integrity and health:** A high-fiber diet may lower your risk of developing hemorrhoids and small pouches in your colon (diverticular disease). Some fiber is fermented in the colon. Researchers are looking at how this may play a role in preventing diseases of the colon.
- **Lowers blood cholesterol levels:** Soluble fiber found in beans, oats, flaxseed and oat bran may help in lowering the total blood cholesterol levels by lowering low-density lipoprotein (bad cholesterol) levels. Epidemiologic studies have shown that increased fiber in the diet can reduce blood pressure and inflammation, which is also protective to heart health.
- **Helps control blood sugar levels:** Fiber, particularly soluble fiber, can slow the absorption of sugar, which for people with diabetes can help in reducing sugar levels in blood. A diet that includes insoluble fiber has been associated with a reduced risk of developing type 2 diabetes.
- **Aids in weight loss:** High-fiber foods generally require more chewing time, which gives your body time to register when you are no longer hungry, so you are less likely to overeat. A high-fiber diet also tends to make a meal feel larger and linger longer, so you stay full for a greater amount of time and high-fiber diets tend to be less *energy dense*, which means that they have fewer calories for the same volume of food.
- **Uncertain effect on colorectal cancer:** Evidence that the dietary fiber reduces colorectal cancer is mixed- some studies show benefit, some show nothing and some suggest increased risk. If you are concerned about preventing colorectal cancer, adopt or stick with a colon-cancer screening regimen. Regular testing and removal of colon polyps can prevent colon cancer.

Recommended fiber intake should be 38 g for men and 21 g for women above the age of 50 years. It is suggested that one should eat whole grains and whole-grain products, fruits, vegetables, beans, peas and other legumes, nuts and seeds for increasing fiber content in the diet.

Refined or processed foodstuffs such as canned fruits and vegetables and pulp-free juice, white bread and pasta are lower in fiber content. The grain-refining process removes the outer coat (bran) from the grain, which lowers its fiber content. Similarly, removing the skin from fruits and vegetables decreases their fiber content. These should be avoided.

Whole food articles rather than fiber supplements are generally better since fiber supplements do not provide minerals, vitamins and other beneficial nutrients that the high-fiber food articles do. However, some people may still need a fiber supplement if dietary changes are not sufficient, or if they have certain medical conditions such as constipation, diarrhea, or irritable bowel syndrome. Always check with your doctor if you feel you need to take fiber supplements.

How to Increase Fiber Content in Diet

- **Start the day with a high-fiber breakfast cereal:** Opt for cereals with *bran* or *fiber* in the name, or add a few tablespoons of unprocessed wheat bran to your favorite cereal.
- **Switch to whole grains:** Eat breads (loafs) that are made from whole-wheat flour or another whole grain. Look for a flour brand with dietary fiber. Experiment with brown rice, wild rice, barley, whole-wheat/cereal products, *etc.*

- **Mix it up:** Add pre-cut fresh vegetables to soups.
- **Get a leg up with legumes:** Eat more beans, peas and lentils.
- **Eat fruit at every meal:** Apples, bananas, oranges, pears and berries are good sources of fiber.
- **Make snacks count:** Fresh and dried fruit, raw vegetables and low-fat popcorn and whole-grain crackers are all good choices. An occasional handful of nuts are also a healthy high-fiber snack.

High-fiber foodstuffs are good for health but adding too much fiber too quickly can promote intestinal gas, abdominal bloating and cramping. Fiber content in the diet should be increased gradually over a period of a few weeks. This allows the natural bacteria in the digestive system to adjust to the change. Drinking plenty of water is also good since fiber works best when it absorbs water, making the stool soft and bulky.

The elderly require sufficient fibre or roughage in their diet to avoid constipation, which negatively affects the quality of life of elderly, long-term care resident. In practice, constipation is defined as less than three bowel movements per week. Elderly are at an increased risk of constipation secondary to lower dietary fiber intake, use of constipating medications, immobility and disease condition. Complications of constipation include anorexia, hernia, early satiety, cardiovascular events, psychological effects and gastrointestinal obstruction or inflammation.

Fibre is the indigestible carbohydrate present in the food and has been linked to gastrointestinal disease (constipation, diverticulosis and irritable bowel syndrome), cancer, coronary heart disease and metabolic disorders such as diabetes, hyperlipidemia, and obesity. Rough fibre is not well tolerated by the intestine in old people but the tender fibre of vegetables, fruits and whole-grain cereals will encourage normal bowel movements. The elderly tend to use harmful laxatives and mineral oils. This should be substituted by a fibre-rich diet and adequate fluid intake. Some good sources of dietary fibre are *Ragi*, wheat, Italian millet, horse gram, green leafy vegetables, plantain stem, drumstick, bitter gourd and fruits like dates, figs, guava, wood apple and sweet lime.

Malnutrition and Older Persons

Older persons are particularly vulnerable to malnutrition. Moreover, attempts to provide them with adequate nutrition encounter many practical problems. First, their nutritional requirements are not well defined. Since both lean body mass and basal metabolic rate decline with age, an older person's energy requirement per kilogram of body weight is also reduced.

Many of the diseases suffered by older persons are the result of dietary factors, some of which have been operating since infancy. These factors are then compounded by changes that naturally occur with the ageing process.

Dietary fat seems to be associated with cancer of the colon, pancreas and prostate. Atherogenic risk factors such as increased blood pressure, blood lipids and glucose intolerance, all of which are significantly affected by dietary factors, play a significant role in the development of coronary heart disease.

Degenerative diseases such as cardiovascular and cerebrovascular disease, diabetes, osteoporosis and cancer, which are among the most common diseases affecting older persons, are all diet-affected. Increasingly in the diet/disease debate, the role that micronutrients play in promoting health and preventing non-communicable disease is receiving considerable attention. Micronutrients deficiencies are often common in elderly people due to a number of factors such as their reduced food intake and a lack of variety in the foodstuffs they eat.

Another factor is the price of food articles rich in micronutrients, which further discourages their consumption. Compounding this situation is the fact that the elderly often suffer from decreased immune function, which contributes to this group's increased morbidity and mortality. Other significant age-related changes include the loss of cognitive function and deteriorating vision, all of which hinder good health and dietary habits in old age.

Elevated serum cholesterol, a risk factor for coronary heart disease in both men and women, is common in older people and this relationship persists into very old age. As with younger people, drug therapy should be considered only after serious attempts have been made to modify diet. Intervention trials have shown that reduction of blood pressure by 6 mm Hg reduces the risk of stroke by 40% and of heart attack by 15%, and a 10% reduction in blood cholesterol concentration will reduce the risk of coronary heart disease by 30%.

Dietary changes seem to affect risk-factor levels throughout life and may have an even greater impact in older people. Relatively modest reductions in saturated fat and salt intake, which would reduce blood pressure and cholesterol concentrations, could have a substantial effect on reducing the burden of cardiovascular disease. Increasing consumption of fruits and vegetables by one to two servings daily could cut cardiovascular risk by 30%.

Adequate nutrition and a well balanced diet are of vital importance in old age to prevent and control the common hazards of ageing. Many factors like poor income, decreased mobility, social isolation and depression are known to affect the health and well being of the elderly. Under nutrition, obesity, diabetes, cardiovascular diseases and osteoporosis have been identified as the most important and commonly prevalent nutrition related health problems in old age.

Diet and Feeding Pattern

Since good nutrition is of immense significance during old age, care should be taken that the diets of elderly are nutritionally adequate and well balanced. With the advancement of age, the energy needs are reduced as a result the total quantum of food intake is lowered, while the requirement of most of the other nutrients remains unaltered. Therefore, it becomes all the more important to provide adequate amounts of all the nutrients within the decreased energy levels.

Nutritional Related Problems in Old Age

Many a times, seniors have trouble eating well. Some people may have trouble chewing certain foodstuffs, while others do not feel hungry. Sometimes, you do not want to cook just for yourself, if you are living alone, or the fact that food does not always taste the same anymore keeps you from eating well. These are some of the common problems related with eating right faced by the seniors in United States.

- Improper nutritional habits
- Lack of appetite
- Altered sense of taste and smell
- Bad condition of dentures
- Half or full prostheses
- Higher vitamins and minerals requirement to counter the effects of medication
- Unhealthy digestion

- Poor nutrients uptake
- Frequent suffering from various diseases
- Side-effects of drugs
- Psychological disorders
- Social isolation
- Financial problems

Factors Influencing Balanced Meals for Elderly

- Intake of energy rich foodstuffs like sweets, fried, or high fat food items, cereals and starches needs to be reduced, while liberal amounts of milk and milk products, fresh fruits, vegetables particularly green leafy vegetables should be included to meet the vitamins and minerals needs.
- An adequate intake of calcium, in particular, should be ensured to compensate for its losses due to gradual demineralization of bones associated with ageing.
- Since sufficient exposure to sunlight is essential to meet the body's needs for vitamin D, in case of elderly individuals confined to bed, supplements of this vitamin need to be provided.
- Food rich in fats, especially saturated fats, should be avoided and instead oils containing high levels of polyunsaturated fatty acids such as sunflower oils, fax seed oil soybean oil, safflower oil, *etc.* should be used to prevent and control the condition of hypertension and other cardio-vascular diseases.
- Intake of simple sugars should be reduced as these provide only empty calories.
- Dietary fibre has a beneficial effect in various conditions associated with ageing such as constipation, diabetes and cardiovascular diseases, adequate amount of food rich in dietary fibre should be included in the diet.
- With the advancement of age, the capacity to digest and tolerate large meals often decreases. Therefore, the quantity of food given at a time needs to be decreased. If required number of meals can be increased as per the individual's tolerance.
- Due to the loss of teeth particularly if dentures are not used, modifications in consistency needs to be done. The diet should be soft, well cooked and should include foodstuffs that need little or no mastication such as milk and milk products, soft cooked eggs, tender meats, gruels, soft cooked vegetables, grated salads, fruit juices, soft fruits like banana or stewed fruits.
- The food for the elderly should be colourful, attractive and tasty and should be served in pleasant surroundings so as to arouse their appetite and interest in the food.
- Apart from the diet, the elderly are advised to continue with their physical activity and light exercises to keep fit and prevent the occurrence of diseases like arthritis, gout and obesity. Regular or periodic health checkups and weight monitoring also helps in physical fitness and early detection of complications, if any.

■■■

12 Pre- and Probiotics

M.K. Rana[1] and Kamla Malik[2]

[1]Department of Vegetable Science; E-mail: <mkrlotus@gmail.com>
[2]Department of Microbiology, CCS Haryana Agricultural University, Hisar, Haryana, India; E-mail: <kamlamalik@rediffmail.com>

The major causes of death for general population are the infectious diseases, including food- and water-borne infectious diseases but in affluent societies, the key causes of death and disabilities (cardiovascular heart disease, diabetes, obesity, constipation and certain cancers, especially colorectal cancer) are related to diet and lifestyle, thus, there is a substantial increasing public interest in food and food components to promote health and lower the risk of those non-infectious death causing diseases that are largely associated with diet and human lifestyle, and now with the augmenting cost of health care, the people are looking for alternate approach to the prevention and treatment of ailments. It is well establish fact all over the world that holistic approach to health prevention has always been better than cure. The human intestinal tract is the home of about 100 trillion bacteria. This staggering number includes many beneficial microbes that help in keeping the body healthy by fending off pathogenic ones that otherwise might colonize in intestinal tract. These beneficial microbes attempt to prevent the food borne pathogens from establishing *beachheads* by adherence in human guts, where they thrive, proliferate and inflict harm. If the beneficial microorganisms are overwhelmed, the pathogenic ones win battles by causing infections and diseases. Fortunately, the beneficial microbes have strategies *weapons* to rout immobilize and defeat the enemy. These weapons are called *probiotics*. These microbes promote physiologic balance, which is essential for good health in humans. More than 400 species of bacteria live in the gastrointestinal tract. The large intestine or colon provides a cozy home for 100 million to 100 billion probiotic bacteria. These bacteria comprise about 3.5 pounds of the average adult human's body weight. Ideally, the microflora colony in the digestive tract should be composed of 85% friendly bacteria to 15% harmful bacteria.

In advanced countries, the people have developed their interest in the potentials of probiotics as therapies and preventive agents since probiotics could be of great use in such conditions through either direct competition with pathogens or the synthesis of specific antibacterial or antiviral compounds. Oral ingestion of probiotics has been documented in several reports to reduce the duration of infectious diseases and the severity of dental caries.

The potentials of probiotics to provide such health benefits can be coupled with the intake of particular live organisms in the form of fermented food articles, like curd and yogurt, and the health promoting features of these exogenous live organisms, mostly species of lactic acid bacteria (LAB), are just contrasted with the putrefactive actions of intestinal flora. Intake of fermented-milk products enriched with specific strains of *Lactobacillus acidophilus* or *Bifidobacterium bifidum* by healthy people leads to enhance phagocytosis of *Escherichia coli*. Probiotics in gastrointestinal tract keep the pathogenic agents under control.

Probiotics are supposed to assist in lowering of plasma cholesterol and reduction of cardiovascular heart disease risk. The mechanism for such health benefits is still not so clear but it is believed that only a relatively small proportion of ingested organisms survives passage through the stomach and small intestine and colonizes the large intestine, *e.g.* bifido bacteria are recognized to have very important healthy effect such as cancer prevention, better food digestibility and protection against pathogens. To target such microorganism to the body, many steps are involved (Figure 1) in probiotic processing in gastrointestinal tract under optimum physiological conditions.

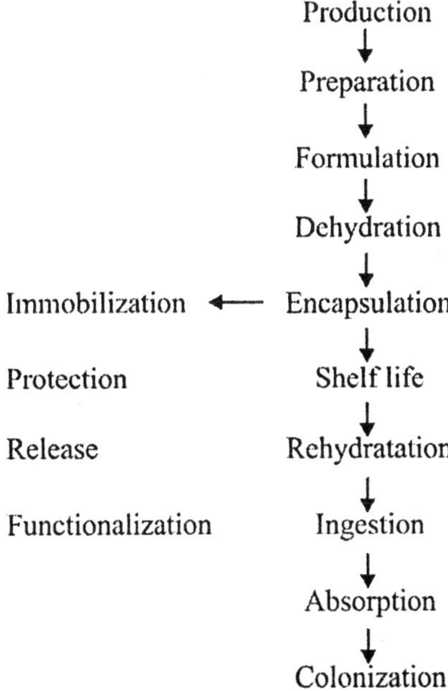

Figure 1: Probiotic processing in gastrointestinal tract

The consumption of probiotics if discontinues, the organism is washed out of the gastrointestinal tract due to a capacity of the host's microflora to remain stable in the face of exogenous organisms, therefore, to exert sustained effect, incessant ingestion of probiotics appears to be obligatory. Enhanced colonization seems to be a viable option, which means promoting the survival of ingested probiotics to ensure that greater numbers of live organisms reach the large bowel and/or enhancing their residence time in that viscous environment. The caecum and colon are the major sites of bacterial colonization in the human gastrointestinal tract. Relatively few bacterial species and numbers are found in the stomach or small intestine. An ingested probiotic must survive passage through these internal organs and take up active residence in the large bowel. This process can be facilitated by prebiotics, which when combined with probiotics, are synbiotics.

Historical Concepts of Probiotics

As soon as the man is born, he becomes habitat of intense microflora, and later the gastrointestinal tract, the mucous membranes and the skin become the reservoir of microflora, however, all these microorganisms in healthy individuals are inoffensive and sometimes even very useful, thus, probiotic bacteria are generally said to be of human origin and considered non-pathogenic. The origin of cultured dairy products is thought to be as old as the sunrise of civilization and is found mentioned in the sacred books of Hindus and the Bible. Numerous traditional soured milks or cultured dairy products such as *kefir,koumiss*, *leben*, yogurt and curd were developed since the dates back and therapeutically used before the existence of bacteria was documented. By the 1950s, curd or yogurt was often used to re-establish the balance of gut flora generally upset by regular use of antibiotics and to alleviate or prevent conditions such as constipation, diarrhoea, dyspepsia, cystitis, chronic ulcerative colitis, mucous colitis and dermatitis.

In the beginning of 20th century, the positive role played by certain bacteria was first introduced by Russian scientist and Nobel laureate Eli Metchnikoff and presumed to be beneficial to human health. He wrote a book entitled *The Prolongation of Life,* published in 1907, the author hypothesized that the bacteria (*Lactobacillus bulgaricus* and *Streptococcus thermophilus*) involved in the fermentation of yogurt suppress the putrefactive fermentation of intestinal flora, and *per se* ingestion of yogurts played a significant role in maintaining good health. Undeniably, the author of this book attributed the prolonged life of Bulgarians to their consumption of yogurt full of *Lactobacillus* species.

Bifidobacteria were first isolated from a breast-fed infant by Henry Tissier who also worked at the Pasteur Institute. The isolated bacterium named *bacillus bifidus communis* was later renamed to the genus *Bifidobacterium* (Figure 2a), which played a important role in maintaining good health. *Bifidobacteria* in breast-fed infants also played a key role in maintaining good health and degraded the growth of pathogenic bacteria in the gut. In 1935, certain strains of *Lactobacillus acidophilus* were found to be very active when implanted in the human digestive tract. The fermented milk products also containing *L. acidophilus* (Figure 2b) bacteria have therapeutic effects, especially a settling effect on digestion but their colonization and growth in human gut are essential for effectiveness.

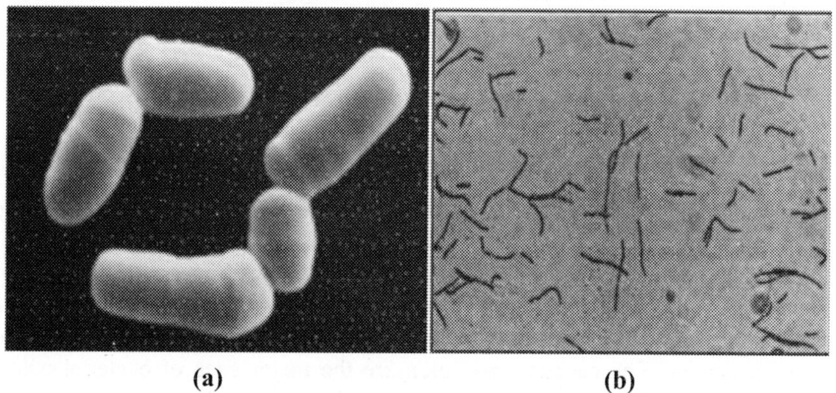

(a) (b)

Figures 2a: Scanning electron micrograph (SEM) of *Bifidobacterium bifidum* and 2b: *Lactobacillus acidophilus* under the light microscope

A successful *Lactobacillus* therapy depends on definite principles, *i.e.*, (i) it is necessary to use only that organism (*L. acidophilus*), which is found to be a natural inhabitant of the intestine, and (ii) the organism should be non-pathogenic and capable of establishing itself in the gut and should retain viability during processing and transit through the gastrointestinal tract since a large number of viable cells (10^7-10^9) are the prerequisite for swift establishment of this useful flora, and a further important factor for successful therapy is the intake of a high milk-based diet.

Composition of probiotics

Probiotics can be bacteria, moulds, or yeast but mostly probitics are the bacteria. Among bacteria, the species of *Lactobacillus* and *Bifidobacterium* are the most common type of bacteria used as probiotics. A probiotic may be made out of a single bacterial strain or it may be a consortium as well. Probiotics can be in the form of powder, liquid, gel, paste, granules, or available in the form of capsules, sachets, *etc.* Across the globe, about 20 probiotic strains, singly or in combination of lactic acid bacteria and a probiotic yeast culture are used in dietary supplements and functional foods or in mainstream food products, which help in maintaining health. The live microbiotic strains currently being used predominantly as probiotics in food ingredients are mentioned in Table 1.

Table 1: Microbiotic strains currently being used as probiotics in food ingredients

Lactobacilli	Bifidobacteria	Non-lactic acid bacteria and others
Lactobacillus acidophilus DDS-1, SBT-2062, NCFM and LA-5	*Bifidobacterium adolescentis*	*Bifidobacterium cereus*
Lactobacillus casei DN114001, CRL431, F19	*Bifidobacterium animalis*	*Clostridium butyricum*
Lactobacillus crispatus	*Bifidobacterium bifidum*	*Escherichia coli* M-17, *Escherichia coli Nissle* 1917
Lactobacillus johnsonii La1	*Bifidobacterium breve*	*Saccharomyces boulardii*
Lactobacillus paracasei F19 and St11	*Bifidobacterium infantis* UCC 35624	*Saccharomyces cerevisiae*
Lactobacillus plantarum 299V	*Bifidobacterium lactis* BB-12	*Propionibacterium freudenreichii*
Lactobacillus reuteri ATTC55730 and SD 2112	*Bifidobacterium longum* BB536	*Enterococcus faecium* and *E. faecalis*
Lactobacillus rhamnosus GG, ATCC 53013 and LB21	*Bifidobacterium align*	*Bacillus subtilis* and *Bacillus vietnam*
Lactobacillus salvarius UCC 118	–	*Streptococcus salivarius* and *S. thermophilus*
Lactobacillu delbrueckii subspecies *bulgaricus* 2038	–	*Streptococcus cremoris*, *S. alivarius* and *S. intermedius*
Lactobacillus brevis and *Lactobacillus sporogene*	–	*Pediococcus acidilactici*
Lactococcus lactis L1A	–	*Leuconostoc* spp.
Lactobacillus cellobiosus and *Lactobacillus curvatus*	–	*Aspergillus cerevisiae*, *A. niger* and *A. oryzue*

Lactobacilli	Bifidobacteria	Non-lactic acid bacteria and others
Lactobacillus fermentum and *Lactobacillus gasseri*	–	*Clostridium Pintolopesii*
Lactobacillus heloticus and *Lactobacillus furciminus*	–	–

Definition of Probiotics

The term probiotic has been derived from the Greek word meaning *'for life'*, and over the time, scientists of various working group evolved the definition of probiotics. In 1960s, reaction against the side effects of antibiotics led to augmented interest of scientists in the use of potential live bacteria as food supplements. In 1965, Lilly and Stillwell first time employed the word probiotic to describe the substances secreted by one organism to stimulate the growth of another, and in 1989, Fuller defined the probiotic as a live microbial food supplement, which beneficially influences the host by improving its intestinal microbial balance. Huis in't Veld and Havenaar defined probiotics as a mono- or cocktail of cultures of live microbiota, which when given orally to man affects beneficially by escalating the effectiveness of native microflora, and this definition developed the concept of probiotics in a number of ways. It did not restrict probiotic actions to only the gut microflora but included the possibility of application to microbial communities at other sites, like respiratory tract, digestive tract, urogenital tract and skin. A short time ago, an European Commission supported group of European scientists suggested that probiotics for use in human diet are best defined as live microbial food ingredients that are beneficial to human health.

The term *probiotics* by the commonly accepted definition is applied to live nonpathogenic microbial food supplements that favourably influence the intestinal microbial balance and contribute to the health of the host. An optimum gut-microflora balance is considered one in which beneficial bacteria such as the Gram-positive *Lactobacilli* and *Bifidobacteria* predominate over potentially injurious bacteria. The good bacteria are known as probiotics for reasons that they help to support life, they have a job to perform and without them life would be impossible.

The Food and Agriculture Organization and the World Health Organization (FAO/WHO) define the probiotics as live microorganisms (specific bacteria, fungi and yeasts), which, when administered in adequate amounts, confer a health benefit on the host.

The probiotics that are usually given orally to help the host to fight against illness and ailments are very beneficial for human, however, for these beneficial effects, the probiotic bacteria must have the ability to colonize (a biofilm) in human gut, adhere to intestinal cells, resist to acidity and bile toxicity, antagonize towards pathogens, yield antibiotic substances, offer proven health-benefits and a history of safe use in human beings. Probiotic bacteria that satisfy all these criteria are numerous, *i.e.*, *Lactobacilli*, *Bifidobacterium*, *L. bulgaricus* and *Streptococcus thermophilus*. Some of these organisms also known as *friendly-bacteria* or *designer bugs* are commercially available in the form of pure and mixed cultures. These good bacteria may also support dental health as they contain calcium and other teeth healthy nutrients. However, they also have properties that improve the health of the mouth. *Bifidobacterium* species are also likely helpful in the mouth. It reduces gingival and periodontal inflammation. Hence, probiotic offers a wide range of health benefits to help for fight to viruses. Supplementation with *L. rhamnosus* GG may also help to prevent dental caries in children, possible by reducing carcinogenic bacteria in the mouth.

Pathogens such as *Salmonella* and *Campylobacter* are the main causes of food borne illness from meat consumption. Since the nonpathogenic bacteria get to the intestinal sites first and able to colonize the gut, they diminish the opportunity for pathogenic bacteria to establish. One the most commonly known probiotic, *Lactobacillus acidophilus*, is naturally occurring bacterium in food articles such as curd, grains and meat products. *Lactobacilli* play an important role in the control of undesirable microbiota in the gastrointestinal and urogenital tracts. It guards the health of host digestive tract.

Besides indigenous *Lactobacilli*, which naturally reside in the human gastrointestinal tract, distinct *Lactobacillus* strains from fermented food products have shown beneficial effects on gut health. These probiotic bacteria are live microbial food supplements with health benefits to the host by increasing the population of intestinal microorganisms. Thus, probiotics are a viable microbial dietary supplement that provides beneficial effects in the intestinal tract. They can be used as complementary and alternative medicine (CAM). Probiotics are available in food articles and dietary supplements such as capsules, tablets and powders, and in some other forms as well. The examples of foodstuffs containing probiotics are yogurt, fermented and unfermented milk, *miso*, *tempeh* and some juices and soy beverages. In probiotic foods and supplements, the bacteria may have been present originally or added during preparation.

Definitions of Prebiotics

Prebiotics are the non-digestible part of the foodstuffs that the probiotics need to stimulate the metabolism of food articles. They feed the beneficial bacteria and modify the composition of intestinal microflora; as a result, the probiotics can predominate. The latest concept of prebiotics envisages the use of non-digestible carbohydrates, predominantly oligosaccharides and growth factors, which are specific to a limited number of bacterial species from the colonic microflora. They are designed to favour the proliferation and activity of useful bacteria and to inhibit the activity of harmful colonic bacteria. Prebiotics are the complex carbohydrates, *i.e.*, fructo-oligosaccharides (FOS) or oligosaccharides such as inulin, which is found in onions, asparagus and banana, short chain sugars and lactose derivatives such as lactulose and lactitol, which are being extensively used. They have provided a positive effect on human health since inulin is a fiber, which helps in maintaining normal bowel function, decreases constipation, lowers cholesterol and triglycerides and helps in normalizing the blood sugar levels. Inulin has the potential to have a beneficial effect on cancer prevention as well. *Bifidobacteria* digest inulin to short chain fatty acids like propionic acid and butyric acid. Butyric acid has been used for protective agent for colon tissues from damage and colon cancer and ulcerative colitis in the large intestine. They pass undigested food from lower intestine to the colon where beneficial bacteria consume them, releasing vitamins and other nutrients. They may also curb the growth of food pathogenic bacteria such as *Salmonella* and *Campylobacter* by changing the colonic environment.

Synbiotics

Synbiotics are the food articles containing both probiotic bacteria and prebiotic ingredient to provide a diet in which the intestinal growth of probiotic bacteria is enhanced by the presence of the prebiotic, thus, promoting the chance of probiotic bacteria becoming establish in gut and conferring a health benefit.

Prebiotic functions mainly through the selective modification of indigenous intestinal microflora. This selectively modified microbe is directly or indirectly at the origin of prebiotic's effect.

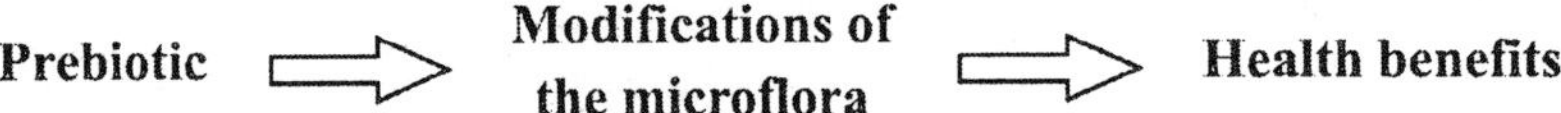

A probiotic mainly provides an action *via* its own qualities (generally genus and species specific). The newly introduced strain of microflora is the direct cause of probiotic's effects.

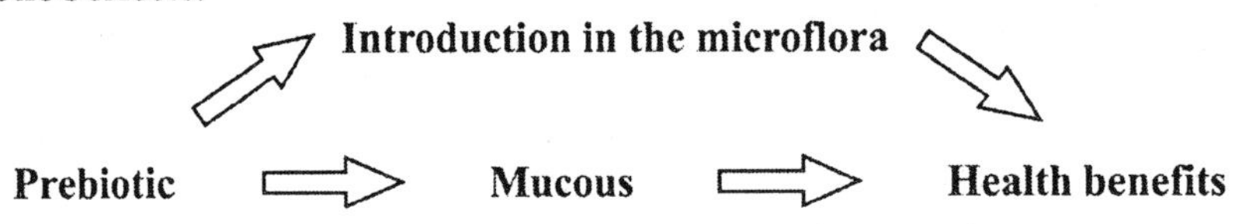

Figure 3: A systematic representation of pre- and probiotics

Prebiotics as Supportive to Probiotics

Prebiotics are growth substrates directed specifically towards potentially beneficial bacteria already resident in the colon. Prebiotics are important because of the followings:

- It allows specific changes, both in the composition and activity of gastrointestinal microbiota, which helps in improving health.
- Currently, prebiotics in use, especially inulin and its derivatives and galacto-oligosaccharides (GOS) are relatively cheap to manufacture or extract from plant sources, and having beneficial effects on gut microbiota.
- They are also valuable functional ingredients in food articles with the potential to give fat-based spreads and dairy products improved organoleptic properties.

Prebiotics are the growth substrates for beneficial colonic bacteria and being non-digestible food ingredients. The prebiotics affect the host beneficially by stimulating the growth and activity of a number of colonic bacteria and *per se* improve the health of the host. A prebiotic acts to stimulate the growth of microflora in the large bowel, however, this concept fails to notice the fact that synbiotics have the health potential, and that increasing the numbers and the activity of ingested organisms within the colon by enhancing their survival as they pass through hostile environment of the upper gastrointestinal tract will have the same net effect as stimulating growth of related organisms in the large bowel. A capsule containing probiotics could be a prebiotic if it increases the population of favourable organisms in the large bowel.

Prebiotics are different from colonic foodstuffs in that the latter serve as general fuel for endogenous colonic microflora, thus, providing the host with energy, metabolic substrates and essential micronutrients. Prebiotics, *i.e.*, oligosaccharides, principally the fructo-oligosaccharides and galacto-oligosaccharides are usually indigestible by human digestive enzymes and have well-documented effects on the large-bowel microflora. Although they are classified as colonic food articles but they also satisfy the criteria for prebiotics, however, other food ingredients and components regarded formerly as colonic foodstuffs, *e.g.*, partially hydrolyzed guar gum appear to promote the populations of probiotics. One dietary component showing great potential as a prebiotic and colonic food is resistant starch and fermentable fiber, which are emerging as a major factor in the bacterial ecology of the gut.

Three prebiotics, oligofructose, galacto-oligosaccharides and lactulose, alter the balance of the large bowel microbiota by increasing *Bifidobacteria* and *Lactobacillus* numbers. These prebiotics are based on inulin and shorter fructo-oligosaccharides. Prebiotics can be supplemented into foodstuffs (often cereals and breads), or may be found naturally within plant foodstuffs, such as artichokes, leeks and onions, are rich sources of the fructo-oligosaccharide type of prebiotic. They are resistant to stomach acids and indigestible by human enzymes but are degraded by specific bacterial species in the colon, including *Bifidobacteria* and *Lactobacilli*. Since prebiotics stimulate the growth and activity of resident intestinal bacteria, they do not suffer from the viability problems that can be associated with probiotics. Some bacteria that feed on fermentable carbohydrates produce many beneficial substances, including short-chain fatty acids (SCFAs) and certain B group vitamins (Figure 4). Additionally, there are some evidences that they may promote further absorption of some minerals that have escaped the small intestine, including calcium and magnesium. The other possible benefits include:

- Lower the cholesterol and triglycerides
- Improve insulin sensitivity and glucose metabolism
- Improve immune system function

Starch, which offers more than 50% of the daily energy intake largely from grains, is a sizeable part of human diets, however, in affluent societies, where interest in prebiotics and probiotics is greatest, the average consumption is much lower, possibly as little as 25% of the daily energy intake. Starch is the only naturally occurring food polysaccharide, which can be digested by the intrinsic digestive enzymes. Earlier it was thought that starch digestion in small intestine was completed since a very little of starch was noticed in faeces, but later, the extensive studies demonstrated that an ample proportion of starch escapes in the large bowel. This escaped starch was named as resistant starch and was defined as the sum of starch and products of starch degradation not absorbed by the small intestine of even healthy individuals. Starch may be digested slowly in the small intestine but if the rate of food passage is sufficiently slow, digestion may complete and no resistant starch, which includes oligosaccharides and other products of small intestinal starch hydrolysis, will be left undigested.

In foodstuffs, the resistant starch exists because of many reasons. Raw starches in unripe banana and other starchy fruits are digested poorly. However, their digestibility can be enhanced by cooking, especially in water, which gelatinizes the raw starch giving greater access to α-amylase enzyme. Disruption of food structure by either milling or chewing also enhances digestion by activating the enzymes involved in digestion of food. However, cooking increases starch digestibility but subsequent cooling of cooked food leads to the formation of crystallites that are usually resistant to digestion. This process is termed retro-gradation and the resistant starch content of foodstuffs appears to increase when they are subject to increasing number of heating and cooling cycles. Chemical structure is an important factor in starch digestibility, especially amylose (amylopectin).

Most food starches largely have about 70% amylopectin, and after digestion, the remainder is amylose. The greater the content of amylose, the more difficult the starch is to gelatinize and the more susceptible to retro-gradation. Ungelatinized high-amylose starches (60-70% of the total starch) are resistant to amylolysis and are used commercially as an ingredient to raise the resistant starch content of processed food. The factors such as chewing may also have impact on the amount of resistant starch in a food, as chewing dictates the particle size of an ingested food, with highly masticated foods having a smaller particle size. Large particles travel more rapidly through the gut than smaller ones, thus, greater mastication would be expected to increase digestibility. Studies have shown that the digestibility of finely ground rice is greater than that of coarse rice. However, not all species of resistant starch have the similar effects, especially as prebiotics on microflora in the large-bowel.

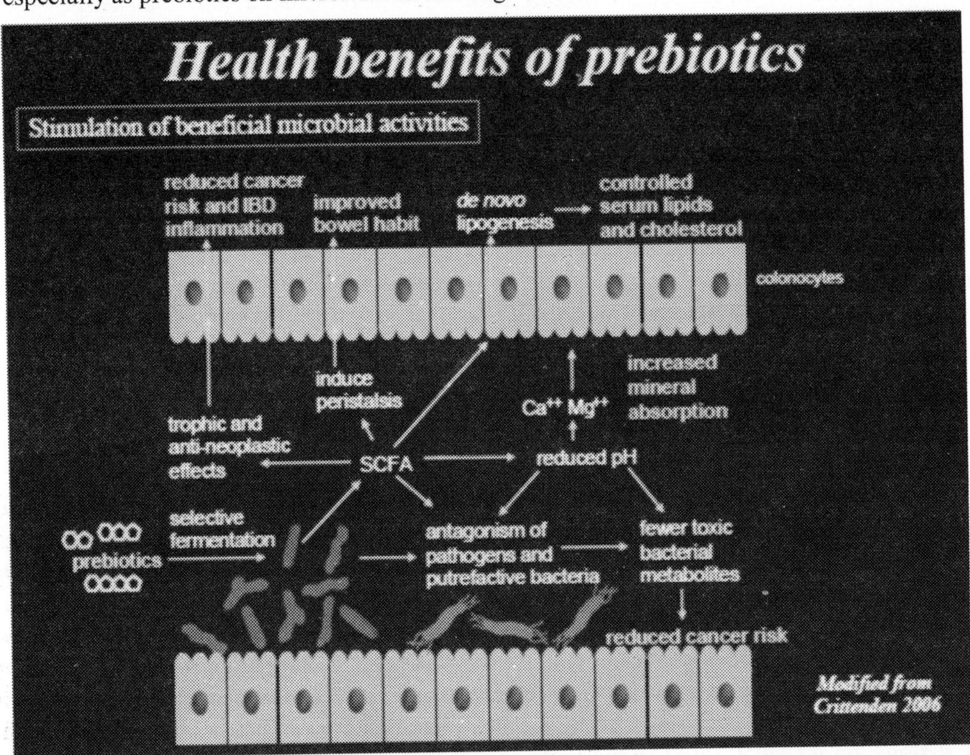

Figure 4: Health benefits of prebiotic

The fermentation of complex carbohydrates by the microflora present in large-bowel is important for good human health. The metabolic products, especially short-chain fatty acids (SCFA) have emerged as important metabolic fuels for colonocytes as well as having specific actions that promote normal colonic function. It is assumed that non-starchy polysaccharides (major components of dietary fiber) are the principal fermentative substrates. The studies show that majority of population is generally consuming <20 g non-starchy polysaccharides per capita per day, which is too below than the normally required quantity (60-80 g substrate per day per capita) to sustain 10^{13}-10^{14} organisms population needed in large bowel of healthy human, and resistant starch is the substance that most likely fills this carbohydrate gap. Indeed, it appears plausible that some populations at low risk of large bowel disease including cancer consume relatively little non-starchy polysaccharides but their diets are high in starch. It could be suggested that eating food items containing resistant starch can lower the frequency of probiotics consumption.

Most probably resistant starch has health-promoting actions on colonic microflora beyond the prebiotic effect. Studies show that ingestion of resistant starch plus the usual hydration therapy reduces fluid loss in children with cholera-induced diarrhoea and also reduces the recovery time to half. Both resistant starch (as green banana) and NSP facilitate recovery in children from infectious diarrhoea. Ingesting cooked rice, which is good source of resistant starch, lowers the occurrence and severity of disease. This benefit is thought to be due to increased fluid absorption through greater short-chain fatty acids production, as these acids stimulate the uptake of water and cations (Na^+, K^+ and Ca^{2+}). Short chain fatty acids also appear to modulate the muscular activity of the large bowel and to promote the flow of blood through the viscera. Both these actions could assist in lowering the severity of diarrhoea. In addition to these well-documented upshots, it is possible that resistant starch could limit the viability of cholera organism in the gut. The bacteria are supposed to adhere to starch granules very much as the *Bifidobacteria* and *per se* removed from the site of infection. Both, the high-amylose starch and the rice product also lower the gut pH, which is a further means of biocontrol for infection causing latent pathogens. In this way, giving such probiotics orally to the patients can speed up the recovery from diarrhoea, and the maximum effectiveness can be achieved, if the prebiotic is used as a mixture of oligosaccharide and resistant starch, as these agents in combination seem to have additive effects.

Inclusion of Prebiotics in Infant Formulas

The foremost idea of adding prebiotics to infant formulas is to modulate the intestinal microflora, and the general aim is to promote long-term effects on the health of infants. The galacto-oligosaccharides obtained from lactose by an enzymatic reaction are incorporated in infant formulas. Human milk oligosaccharides are multifaceted, and the biological effects of prebiotics in infants could be associated with this mixture of different oligosaccharide structures, thus, a mixture of high molecular-weight fructo-oligosaccharides and low molecular-weight galacto-oligosaccharides is added to infant formulas. The infant formula containing metabolites obtained from the fermentation of milk with *Bifidobacterium breve* and *Streptococcus thermophilus* appears to have active components to stimulate the growth of *Bifidobacteria*, which are considered as traditional prebiotics and the key factors that produce beneficial effects on intestinal microflora and infant health. In adults, the ingestion of fructo-oligosaccharides and galacto-oligosaccharides stimulate the population of *Bifidobacteria* and *Lactobacilli* in faecal but do not show similar actions in infants.

In infants receiving prebiotics supplemented formulas, there was a dose-dependent softening of stool, a decrease in stool pH and an increase in defecation frequency, and infant receiving no supplemented formula had the problem of diarrhoea during the course of study. It would be expected that the use of infant formulas supplemented with prebiotics would benefit some aspects of the intestinal immune system. Since prebiotics are involved in the modification of stool consistency and frequency and in decreasing faecal pH, their use has been proposed in the treatment of constipation and bowel discomfort. Fructo-oligosaccharides consumption has been found to be associated with a reduction in febrile events, vomiting, antibiotic use and day-care absenteeism.

Mechanisms of Probiotics to be Defensive

The production of secondary metabolites, *i.e.*, lactate and specific compounds such as bacteriocins, inhibits the growth of disease causing pathogens in large bowel, and this inhibitory effect could be easily enhanced through the immune stimulatory effects of probiotics. The dietary complex carbohydrates help in emptying large bowel where they provide preventive and therapeutic actions and consequently influence the health of large bowel either through physical bulking or through fermentation of resistant starch and oligosaccharides.

The foodstuffs containing high insoluble fiber (cereals bran) are particularly effective in promoting bowel emptying. Human studies suggest that some resistant starches increase butyrate production, whereby raising butyrate levels in large bowel could be of positive health benefit and particularly important in the distal colon, which is the major site for most of the large bowel diseases. Short-chain fatty acids production and availability predominate in the proximal large bowel where fermentation is greatest, reflecting substrate supply. Resistant starch can modify large bowel and faecal short-chain fatty acids favourably. Lactic acid bacteria (*Lactobacilli* and *Bifidobacteria*) are found most abundantly in gastrointestinal tract of adult and milk-fed infants where *Bifidobacteria* fermentation appears to predominate. Yogurt does not contain natural free butyrate at high levels but the currently available probiotics stimulate butyrate production in the human colon.

Mucosal epithelial surfaces such as urogenital tract, respiratory system and gastrointestinal tract are the sites at which the host encounters a numerous species of microorganisms derived from the surrounding atmosphere, and through which, the pathogens invade the host or colonize to initiate infections. It is well known that the gastrointestinal tract harbors a rich microflora, which entertains a symbiotic interaction with its host, though such extensive mucosal surfaces are well protected. In addition to host-derived innate immunity, colonized mucosal organs are also protected by the native microflora, which creates a barrier effect against harming pathogens, and body systems have the different mechanisms such as competition for metabolic substrates, occupation of available niches at mucosal level, and the production of regulatory short-chain fatty acids and bacteriocins. Increasing beneficial components of gastro-intestinal microflora, health benefits can be rendered to host at the non-sterile mucosal surfaces, thus, in diet, live bacteria (probiotics), which interact with and improve the functions of intestinal microflora, should regularly be inoculated, and a bacterial substrate, *i.e.*, oligosaccharides (prebiotics) that promote the growth of beneficial microbiota such as bifidobacteria should also be administered in the diet.

Stomach is such an organ that is almost sterile because of very low pH (1-2). This constitutes a very efficient non-specific mechanism of defense against infections because it constrains the survival of pathogenic bacteria, thus, gastric acidity is the rust barrier effect usually occurs in the gastrointestinal tract, however, probiotic live bacteria that are administered to healthy people also have to face the stress of high acidity, hence, very few probiotic microbiota, *i.e.*, *Lactobacillus reuteri* (various strains), *Lactobacillus gasseri* strain ADH, *Lactobacillus johnsonii* strain La1 (formerly referred as *Lactobacillus acidophilus*), *Lactobacillus caseilrhamnosus* strain GG, *Lactobacillus casei* (various strains), *Lactobacillus plantarum* 299, *Bifidobacterium* spp. and *Saccharomyces boulardii* are actually able to survive in high numbers passage through the stomach and the intestine.

Probiotics protect the host against pathogens by two mechanisms, *i.e.*, (i) the barrier effect or colonization resistance and (ii) modulation of the host's own defense mechanism, thus, in order to antagonize small infection causing agents such as *Salmonella* spp., *Listeria monocytogenes*, *Escherichia coli*, *Shigella flexneri*, *Botrytis*, *Yersinia* spp. and certain viruses such as reovirus and poliovirus, the most beneficial strains of probiotic organisms have to be present in an environment where no truly stable microflora exists. The barrier effect of probiotics in small intestines needs the presence of the protective strains at cellular target of the pathogen. Studies have demonstrated that adhesive strains of lactic acid bacteria produce a better mucosal barrier effect against disease causing pathogens in small bowel than non-adhesive ones. Few strains of *Lactobacilli* are able to secrete bacteriocins (anti-bacterial substances) that are actively against *Salmonella typhimurium.* Probiotics can modulate the host immune system, both at a local mucosal immune level and systemically. In using probiotics to promote immune function, and thereby, afford the host, better protection the foremost aim is to enhance non-specific immunity without triggering a pro-inflammatory response that could potentially be harmful. Human studies show that both *L. johnsonii* La1 and *L. salivarius* UCC 118 can modulate the immune system in healthy humans, boosting innate immune defense mechanisms and these strains do not induce inflammatory responses.

The gastrointestinal tract represents a complex ecosystem in which a delicate balance exists between the intestinal microflora and the host. The microflora is principally comprised of facultative and obligate anaerobes. Approximately 95% of the intestinal bacterial population in humans is comprised of obligate anaerobes and 1 to 10% is comprised of facultative anaerobes. Most of the bacteria are present in the colon where the bacterial concentration ranges between 10^{11} and 10^{12} colony forming units (CFU) per milliliter. The intestinal microflora is important for the maturation of immune system, helping in prevention in the attachment of pathogenic microorganisms and the entry of allergens. They also contribute to the body's requirement of certain vitamins, including biotin, pantothenic acid and vitamin B_{12}. Thus, an organism may be considered as an effective probiotic, if the following criteria are fulfilled:

- It should be isolated from the same species as its intended host, and produce acid, hydrogen peroxide and bacteriocins that are antagonistic to pathogens growth.
- It should be safe, non-invasive, non-carcinogenic and nonpathogenic and co-aggregate to form a normal balance flora.
- It should have large number of viable bacteria and must be able to survive prolonged periods.

Mechanisms of Probiotic Impact on Microbiota

The potential mechanisms whereby probiotic microorganisms might influence the intestinal microbiota showed in Figure 5. Consumption of probiotic cultures may modulate the microbiota or change its metabolic properties by competition for nutritional substrates. This competition is probably not restricted to the intestine, *e.g.*, using two modified *Lactobacillus* strains leads to be changed in faecal concentrations of short-chain fatty acids and marked changes in faecal levels of diverse metabolites including choline, acetate, ethanol, a range of putative N-acetylated metabolites (NAMs), unconjugated bile acids (BAs) and tauro-conjugated bile acids. Similarly, exopolysaccharide (EPS) also produced by probiotics including lactic acid bacteria could act as a growth substrate for microbiota. Different steps involved for indirect mechanisms of probiotics on the intestinal microbiota are as follows:

- Adherence and colonization of the gut
- Suppression of growth or epithelial cells by pathogenic bacteria and production of antimicrobial substances
- Improvement of intestinal barrier function
- Controlled transfer of dietary antigens
- Stimulation of mucosal and systemic host immunity
- Severity of subclinical tissue damage and release rates of host-derived micronutrients that could translate into impacts on the microbiota.

By an unknown mechanism, probiotic bacteria could thus bolster innate immune activity against transient pathogens or non-commensal elements in the microbiota, leading to subtle changes in long-term overall composition. However, more studies are required to substantiate the mechanisms in the probiotic-host interactions and to investigate if they do in fact impact on the microbiota.

Proposed Mechanism

The various proposed mechanisms by which probiotics act are given as under:

- Prevention of gastrointestinal tract colonization by a variety of pathogens is a primary mechanism of beneficial effects mediated by probiotics.

- Probiotic bacteria attach to enterocytes, and thus, inhibit the binding of enteric pathogens to the intestinal mucosa by production of inhibitory substances like bacteriocins, lactic acid and toxic oxygen metabolites.
- Of the toxic oxygen metabolites, the production of hydrogen peroxide is of prime importance as in combination with lactoperoxidase- thiocynate milk system, it exerts a bacteriocidal effect on most mucosa.
- The production of butyric acid by some probiotic bacteria affects the turnover of the enterocytes and neutralizes the activity of dietary carcinogens, such as nitrosamines.

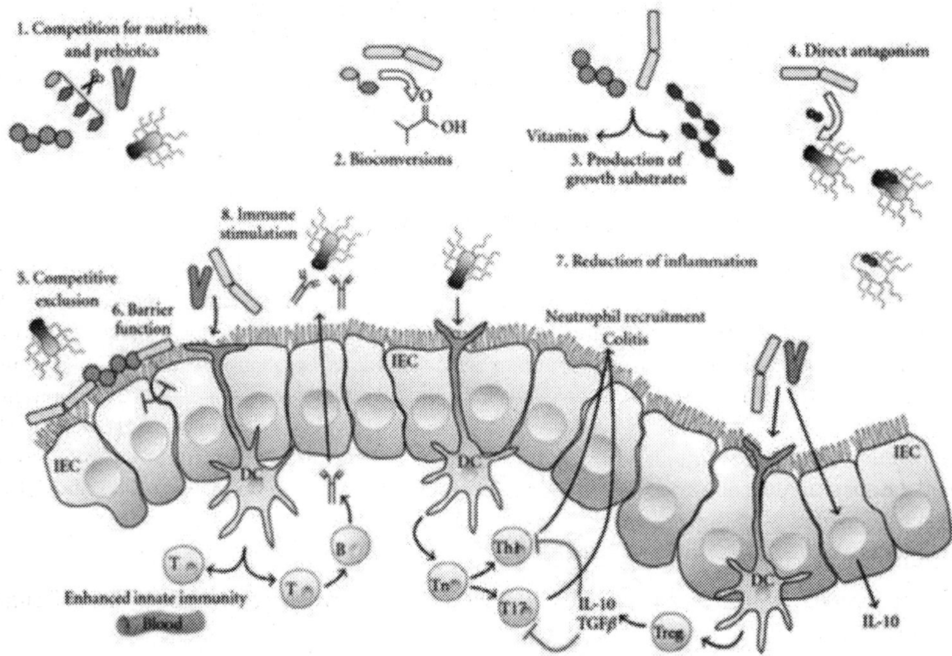

Figure 5: Schematic diagram illustrating potential or known mechanisms whereby probiotic bacteria might influence the microbiota. These mechanisms include: (i) competition for dietary ingredients as growth substrates, (ii) bioconversion of, *e.g.*, sugars into fermentation products with inhibitory properties, (iii) production of growth substrates, *e.g.*, EPS or vitamins for other bacteria, (iv) direct antagonism by bacteriocins, (v) competitive exclusion for binding sites, (vi) improved barrier function, (vii) reduction of inflammation, thus, altering intestinal properties for colonization and persistence within, and (viii) stimulation of innate immune response by unknown mechanisms. IEC= epithelial cells, DC= dendritic cells and T= T-cells.

Therapeutic Competence of Probiotics

Probiotics have the potential to offer major benefits to human health. Increasingly probiotics are being used as prophylactic with promising results among potentially vulnerable groups, and being evaluated as adjuncts in the treatment of many ailments such as inflammatory bowel disease, food allergy, oral rehydration therapy, *Helicobacter pylori* infection, bladder cancer and urogenital infections. The scientifically proven positive health effects of probiotics are energy salvage, modulation of cell growth and differentiation, antagonism against pathogens, innate immunity against infection, immune stimulation, reduction of blood lipids and serum cholesterol, overcoming lactose intolerance, correcting gastrointestinal disorders, inhibiting *Helicobacter pylori* responsible for peptic, gastric and duodenal ulcers, anti-carcinogenicity, diminishing the recurrence of superficial bladder tumors, reduction of harmful intestinal microbial enzyme activity, decreasing faecal mutagenicity, reduction of duration of rotavirus diarrhoea and production of vitamins. They stimulate the antibody and cell-mediated immune responses. In healthy gut, the immune system is able to create a balance between protective mucosal immunity and tolerance to dietary antigens, but in food allergies, this balance is impaired. Probiotic intervention may help to alleviate the symptoms of food allergy by modulating the immune response through beneficial modification of the intestinal flora.

In patients suffering from hyperacidity, the regular use of antacids or H_1 blockers (Cimetidine, Ranitidine, Lansprazole, Omeprazole, Acidoprazole, Rabeprazole, Esomeprazole, *etc.*) that suppress acid secretion leads to reduced gastric acidity (increased stomach pH) and subsequent colonization by potential pathogenic bacteria such as *Helicobacter pylori*, the causative organism in many peptic ulcers and strongly associated with gastric cancer, becomes more prevalent with age and the consequences are generally mal-absorption and higher risk of infectious problems due to Gram negative bacteria. Numerous reports indicate that probiotics may contribute a lot to the control of *H. pylori* infections. The recent epidemiological reports reveal that the consumption of whey-based medium fermented by *Lactobacillus johnsonii* La1 strain is noticed associated with lower incidence of ulcers and bladder cancer to some extent in specific populations. Some strains of lactic acid bacteria have been shown to inhibit *H. pylori*. Results from European cross-sectional population study suggest that high intake of fermented milk is found associated with decreased risk of ulcers, and the habitual intake of *L. casei* reduces the risk of bladder cancer.

In hospitalized patients during treatment, diarrhoea associated with antibiotic is a major complication because of frequent use of antibiotics, which disrupt the steady state of colonic microflora and consequently overgrowth of opportunistic disease causing pathogens in small intestine. Consumption of foodstuffs such as fermented milks containing probiotic strains and prebiotic factors have been advised to improve physical condition of general population. Probiotics are actively useful in avoidance of intestinal infections. Furthermore, it seems increasingly important to find alternatives to standard antibiotic treatments for gastrointestinal infections. The regular consumption of probiotics and prebiotics in a preventive fashion could diminish the requirements for antibiotic treatments in subjects that are at risk of infections. Probiotics could prevent medical complications such as antibiotic-associated diarrhoea or septic conditions. Administration of probiotic bacteria can be beneficial in reducing the incidence and duration of acute diarrhoea, though it is imperative that safe probiotic strains are used, and probiotic species should be adapted to different microenvironments of gastrointestinal tract to accomplish their protective function. The probiotics such as *Saccharomyces boulardii*, *Lactobacillus casei* GG 4 and *Enterococcus faecium* have been suggested for use either to prevent or to treat the diarrhoea associated with antibiotics. A mixture of *Lactobacillus acidophilus* and *Lactobacillus bulgaricus* has showed high efficacy in this respect. *Clostridium difficile* can also be associated with recurrent colitis, and the use of probiotics has been shown to be more beneficial in this situation. The probiotic yeast *S. boulardii* has been proven to have some beneficial effects in diarrhoea associated with human immunodeficiency virus (HIV).

The term inflammatory bowel disease (IBD) is implied for a variety of incurable immune mediated diseases with unknown etiology that results in chronic relapsing inflammation of the gut. The two major clinical forms of IBD that significantly impair the quality of life of patients and substantially increase the risk of intestinal cancer are Crohn's disease and ulcerative colitis. Besides genetic predisposition, environmental factors including the composition and activity of intestinal microbiota also play a role in the development of these diseases. However, probiotics may be useful in controlling IBD by modulating the useful intestinal microbiota. The reduction in inflammation is also translated to a reduction in cancer incidence.

A reduction in *Enterococci* and *Clostridia* having strong association with colon cancer has also been noticed in healthy humans upon consumption of probiotic *L. salivarius* UCC 118, which establishes beneficial modifications to the intestinal microbiota, resulting in reduced inflammation, and therefore, a decrease in cancer incidence. In most cases, the probiotics improve the quality of life, and thus, patients are requested to continue probiotic therapy. It is safe to use the probiotic strain *L. salivarius* UCCl18, which does not exacerbate Crohn's disease. Probiotics may bestow prophylactic benefits against infection, especially in infants born at premature stage and in HIV positive subjects.

Beneficial effects of probiotic strains

Probiotics provide defense against the followings:

- Increased nutritional value (better digestibility, increased absorption of vitamins and minerals)
- Promotion of intestinal lactose digestion
- Positive influence on intestinal and urogenital flora (antibiotics and radiation induced colitis and yeast infections)
- Prevention and reduction of intestinal tract infections (bacteria or virus induced, *Candida enteritis* and *Helicobacter pylori*)
- Regulation of gut motility (constipation, irritable bowel syndrome)
- Decreased incidence and duration of diarrhoea (antibiotic associated, *Clostridium difficile*, travelers and rotaviral)
- Maintenance of mucosal integrity
- Improvement and strengthening of immune system
- Prevention of breast and colon cancer
- Reduction of catabolic products eliminated by kidney and liver
- Prevention of osteoporosis
- Better development (growth)
- Anti-carcinogenic, anti-mutagenic and anti-allergic activities
- Feeling of well-being
- Anti-Candida properties
- Hepatic encephalopathy
- Regulation of inflammatory conditions such as inflammatory bowel disease (Crohn's disease and ulcerative colitis)
- Reductions and elimination of small bowel bacterial overgrowth
- Relieving urinary tract infections
- Positive influence on autistic children
- Provides antagonistic environment for pathogens

- Blocking adhesion sites from pathogens
- Inactivating enterotoxins
- Preventing food allergy and eczema
- Prevent asthma and cardiovascular diseases
- Decelerating ageing
- Decreasing effects of the depression
- Lowering cholesterol and blood pressure

Many scientific studies demonstrated that probiotic bacteria might mediate a variety of health effects through numerous proposed mechanisms (Figure 6).

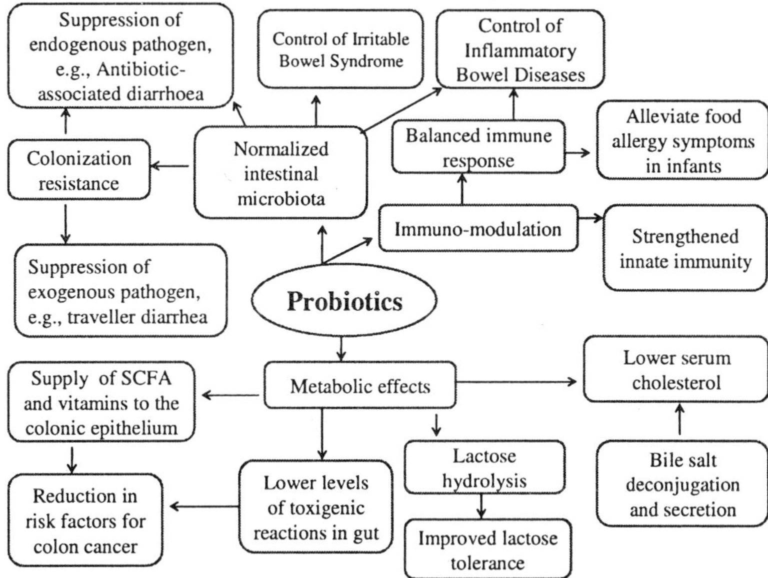

Figure 6: Health benefits stemming from consumption of probiotics

The health benefits also depend upon the strains of probiotic, *e.g.*, some strains of probiotics have ability to inhibit the reproduction of many harmful bacteria. The *Lactobacilli* and *Bifidobacteria* are producing organic compounds -such as lactic acid, hydrogen peroxide, and acetic acid that decrease the acidity of the intestine. Probiotic bacteria also produce some antimicrobial substances called bacteriocins, which act as natural inhibitor to kill undesirable microorganisms. Similarly, some probiotics strains have also ability to participate effectively in digestive process by improving the digestibility of certain components found in diet, by releasing various enzymes into the intestinal lumen and exerting potential synergistic effects on digestion and alleviating symptoms of intestinal malabsorption. For instance, beta-galactosidase (lactase), which is secreted by certain probiotics, aids in the hydrolysis of ingested lactose.

Market Dynamics

The concept of probiotic is currently widespread among the general population and is being exploited commercially on large scale in advanced countries. The modulation of composition of microflora might be more easily obtained by ingesting prebiotics that favour the production of antibacterial metabolites such as short-chain fatty acids or bacteriocins that are implicated in the control of newcomers into a steady existing microflora. There are validations that the currently used probiotic microbes are safe in their use. Commercialization of this probiotic concept has played a key role in growing awareness of the role of gut flora in health and in publicizing information concerning potential benefits of probiotic therapy. Marketing of probiotics, which is introducing a new paradigm for food safety professionals, *i.e.*, positive physiological effects beyond those of good nutrition, has undoubtedly become the fastest growing field for the development of new food products.

Since 1919, Japan has been selling fermented milk drinks, and now the market in Japan is copiously mature with a sale of 110 million liters fermented milk drinks per year. In last few years, the sale of probiotic yogurts containing *Lactobacillus acidophilus* and *Biofidobacterium* cultures has increased in Australia and United States of America too. In Asian and European countries too, the consumers are now buying prebiotic formulated food products to boost the populations of *Biofidobacterium* bacteria in the gut. Presently, the five-bifidobacteria species that have captivated the attention in dairy industry for manufacturing probiotic milk products are *Bifidobacterium adolescentis, B. bifidum, B. breve, B. infants* and *B. longum.*

Safety of Probiotics

The nutraceutical and pharmaceutical industries are widely using the probiotic microorganisms without causing any health problems to the consumers. It has been demonstrated by the working group scientists that probiotic strains in certain cases can significantly ameliorate measurable clinical end-points of disease and prominently can make the life easeful. In most of the cases, probiotics have been shown to be safe and considerably promise, however, some pathologies and side effects have been described for *Saccharomyces boulardii* and *Lactobacilli*. Risk factors have been identified for *Lactobacilli* that are involved in the production of some proteinases and glycosidases, platelet aggregation. One should take care of that the probiotic microorganisms should not boost up gut permeability, which is the prime cause of allergy or inflammation.

As probiotics increase in popularity, it is prudent to consider whether there are any safety concerns associated with the resulting increased exposure to live microbes. It is tempting to presume that the question of safety is only about the probiotics strain being used. In fact, how the strain is used and who is consuming it are the important considerations as well. For example, a safety assessment of a strain used in yogurt at 10^9 CFU per serving would be different for the same strain administered at 10^{12} CFU per *enteric-coated* capsule. A safety assessment, therefore, must consider the nature of the specific microbe being consumed, how it is prepared, how it is administered, what it delivers and the health status of the consumer. People have believed for more than a century that live microorganisms in foodstuffs may be beneficial for human health. More recently, probiotics have appealed to consumers who want nutritional products with added health attributes. The efficacy of probiotics, thus, offers vast opportunities for the development of human health. However, the addition of novel bacteria strains of food articles and therapeutic products presents a challenge in safety assessment for regulatory authorities. The introduction of new probiotic culture demands that it should be at least as safe as its conventional counterparts should. The most commonly used probiotic (*Lactobacillus* and *Bifido*) are generally regarded as safe because of their long history of use as dairy starters. The issue of safety becomes more complicated if one considers organisms such as *Enterococcus* species as probiotics. These bacteria are present in relatively high numbers in the intestine and are often caused diseases. However, *entrococci* have emerged as an important cause of nasocomial infection, and isolates are increasingly vancomycin resistant.

Bacteria that produced lactic acid as a major end-product of metabolism are the most common commercial available probiotics agents. They mainly include the species belonging to genera *Lactobacillus*, *Pediococcus*, *Leuconostoc*, *Entrococcus* and *Bifido*. The probiotic concept dictates that these organisms manifest properties that are advantageous toward human health. It is also vitally important that are completely safe for human consumption. Over many years, a consistently important area of probiotic research has been those studies focusing on the suppression of pathogenic activity of microorganisms in the digestive tract. The various components of human gut microbiota have potentially health promoting groups. From the viewpoint of harmful properties, the intestinal bacteria may be involved in the onset of localized or systematic infections, intestinal putrefaction, toxic formation and production of mutagenic and carcinogenic substance. Alternatively, some intestinal organisms, *e.g.*, *Bifidum* and *Lactobacillus* may confer general health promoting benefits such as vitamin production, stimulation of immune system through non-pathogenic means and inhibition of growth and establishment of harmful microbial species.

Probiotics are generally considered safe. As evidenced by epidemiologic studies, bacteraemia or sepsis from lactobacilli is extremely rare. Numerous probiotics have a long history of safe use and no health concerns have been observed. The safety of probiotics in conditions where the mucosal integrity of gastrointestinal tract is compromised requires more studies before comprehensive safety recommendations can be formulated.

In recognition of the importance of assuring safety, even among a group of bacteria that are generally recognized as safe (GRAS), the working group recommends that probiotic strains should be characterized at a minimum risk with the following tests:

- Determination of antibiotic resistance patterns
- Assessment of certain metabolic activities
- Its side effects on human beings
- Testing for toxin production *etc.*

Future perspectives

There is evidence that specific strains of probiotics are able to influence functioning of immune system. This includes both stimulation and regulation of the immune function depending upon the immunological status of the host. The probiotic intake is also effective in prevention and management of acute gastroenteritis and intestinal inflammatory disorders and pediatrics atopic diseases. Probiotics plays an important role in maintaining human health. Having a better understanding of how they elicit these effects will help researchers better target certain conditions or diseases with specific probiotic strain. Many researchers from the world are trying to better identify the mechanisms by which probiotic strains affect the body. There are efforts to develop probiotic strains that have an anti-inflammatory effect, as well as strains with effect on oral health and allergy prevention. These new strains will be the basis of new generation of probiotics with targeted effect on human health. These microbes seem to offer many health benefits to human beings.

■■■

13 Medicinal Benefits of Probiotics

Charu Sharma[1], Shreesh Ojha[2] and Kamla Malik[3]

[1]BIO5 Institute, University of Arizona, Tucson, Arizona, USA
E-mail: <shree.charu@gmail.com>
[2]College of Medicine, University of Arizona Health Sciences Center, Arizona, USA
[3]Deaprtment of Microbiology, CCS Haryana Agricultural University, Hisar-125 004

Foodstuffs are used as a preventative against diseases and being made an essential part of medicine since ancient time in natural medicine. Being important to human existence and a major determinant of health the food intake and the particular choice is within our control. Eating certain food articles daily and on certain auspicious days as well as drinking beverages is unique in every culture. One eats food to survive, to taste and to be healthy.

The idea of using food as a medicine is quite old and interesting. Along with rapid development in medical science, understanding of human diseases and the development of therapeutic innovations, modern science also emphasizes on the preventive aspects of medicines. In recent years, there is a rapid rise in the use of functional foods for improvement of their general health, well-being and prevention of the incidence of diseases.

In last decades of the twentieth century's, interest in probiotics among the public steadily increased due to (i) the incidence of premature death from diseases and digestive disorders, (ii) growing concerned about the increased and sometimes excessive use of pharmaceutical products like antibiotics often with associated side effects and (iii) the concept of holistic health, in which overall body systems are improved rather than targeting a specific diseases, has grown in popularity. Due to this context, the public has been aware for improving the intestine and immune system by adding *friendly bacteria* or probiotics. Now a day, probiotics are available in a variety of food products and supplements and have wide applications in controlling cholesterol, cancers, allergies, *etc.*

Among these, foods containing beneficial friendly bacteria termed as probiotics, *i.e.*, yogurt, fermented and unfermented dairy products and soy, are on rapid rise. These foodstuffs are believed to be good for health and represent a major segment in food industry across the world. Probiotics are consumed in the form of fermented food products since ancient time almost in every culture to promote health and well-being.

Probiotics are available in foods and dietary supplements or in some other forms. Examples of foods containing probiotics are yogurt, fermented and unfermented milk, *miso*, *tempeh*, tofu, some juices and soy products and beverages. Some probiotic foods date back to ancient times, such as fermented foods and cultured dairy products. Interest in probiotic food and supplements has been growing worldwide when campaign started among every society for promotion of good and healthy food in an approach to increase the life span.

The consumption of probiotic food, supplements and other dairy products are rising immensely and are now used as a preventive medicine. Probiotics are in general consumed for health benefits beyond providing basic nutritional value. They cooperatively maintain a delicate balance between the gastrointestinal tract and immune system. However, in another term, prebiotics are the non-digestible food ingredients that encourage the growth and activity of favourable intestinal bacteria are quickly gaining attention as functional foods.

In beginning of the 20th century, Eli Metchnikoff wrote in his book *Prolongation of Life* that consumption of some friendly microbes present in fermented foods plays a beneficial role in human health and increase life span. However, it is only in the past two decades that the influence of these gut colonizing host friendly bacteria on human physiology has been fully recognized. The term *Probiotics* was coined in 1965 for these host friendly microbes but was popularized in 1989 and defined as live, non-pathogenic microorganism in the form of microbial supplements that benefit the host by improving its intestinal microbial balance.

The definition of probiotics has evolved over the years from *organisms and substances, which contribute to intestinal microbial balance* to a live microbial feed supplement that beneficially affects the host animal by improving its intestinal microbial balance to a viable minor mixed culture of microorganisms. Probiotics when applied to animals or man beneficially affects the host by improving the properties of indigenous microbiota and finally to a microbial preparation that contains live and/or dead cells.

Further, it also includes the microbial metabolites, which are intended to improve the microbial or enzymatic balance at mucosal surfaces or to stimulate immune mechanisms. However, the related term *prebiotics* is defined as non-digestible food ingredients that benefit the host by selectively stimulating the growth and/or activity of one or a limited number of bacteria in the colon, which have the potential to improve host's health. Another term *symbiotic* was also introduced and it is defined as a mixture of probiotics and prebiotics, which benefits the host and improves the health by adding dietary supplements in the gastrointestinal tract.

Generally, pro-, pre- and symbiotics can be found in dietary products like yoghurt, cereals, baby milk products and margarine. The present article emphasizes on the therapeutic benefits of only probiotics in several diseases considering its benefits based available scientific data.

In probiotic foods and supplements, the bacteria may have been present originally or added during preparation. Most probiotic food and supplements have bacteria similar to those naturally found in people's guts, especially in those of breastfed infants (who have natural protection against many diseases). The probiotic bacteria usually come from two groups, *Lactobacillus* or *Bifidobacterium*. Within each group, there are different species (for example, *Lactobacillus acidophilus* and *Bifidobacterium bifidus*) and within each species, there are different strains or subspecies or varieties. Some of the probiotic food and supplements contain fungus such as *Saccharomyces boulardii*, which are different from bacteria. These are popularly called as yeasts and known to be a major component of fermented products.

Gut Microbiology

Bacteria exist in our planet since the late Achaean eon. They colonize amazingly in a wide variety of habitat and survive in extreme environmental stress where the nutrients are extraordinarily scarce. Trillions of bacteria habitat and colonize in the human digestive system. These bacteria are referred to as gut flora.

In a healthy adult body, there is roughly 2 kg of bacteria. The gut flora or the beneficial bacteria take part in digestion and absorption of food, they produce a number of enzymes that break down proteins, carbohydrates, fiber and fats and can actively synthesize a number of nutrients, *i.e.*, vitamin K_2 (menaquinone), B_1 (thiamine), B_2 (riboflavin), B_3 (niacin or nicotinic acid), B_5 (pantothenic acid), B_6 (pyridoxine), B_9 (folic acid), B_{12} (cyanocobalamine) and some amino acids that are essential for human. The human colon is a complex ecosystem, which harbors more than 400 different species of microbes, although on a quantitative basis, about 20 genera probably predominate.

This is likely to be an under estimation of the diversity of gut microflora as not all microorganisms can be cultured. Though microbes that present in other parts of gastrointestinal tract, colon is most heavily colonized with 10^{10}-10^{11} microorganisms per gram of colonic contents and a considerable concentration on the mucosa.

The composition of human gastrointestinal flora is listed in Table 1. Fetus is sterile in uterus but progressively colonizes bacteria soon after birth from the mother's vagina, faeces and also from the environment. Diet exerts a major effect on the composition and activity of gut microflora. Breast feed infants have a natural predominance of *bifidobacteria*, while the formula fed have a more complex and heterogeneous profile similar to the adult indigenous microbiota with *Enterobacteria, Lactobacilli, Bacteroids, Clostridia, Bifidobacteria* and *Streptococci.* After weaning, the microbial status alters and resembles to that of the adult.

Table 1: Composition of human gastro-intestinal flora*

Bacterial flora	Stomach	Jejunum	Ileum	Faeces
Total bacterial count (/ml)	0-10	0-10^5	10^3-10^7	10^{10}-10^{12}
Aerobic or facultative bacteria				
Enterobacteria	0-10^2	0-10^3	10^2-10^6	10^{10}-10^{12}
Streptococci	0-10^3	0-10^4	10^2-10^6	10^3-10^{10}
Staphylococci Lactobacilli	0-10^2	0-10^3	10^2-10^5	10^4-10^7
Fungi	0-10^3	0-10^4	10^2-10^5	10^6-10^{10}
Anaerobic bacteria				
Bacteroides	Rare	0-10^2	10^3-10^7	10^{10}-10^{12}
Bifidobacteria	Rare	0-10^3	10^3-10^5	10^8-10^{12}
Gram-positive cocci	Rare	0-10^3	10^2-10^3	10^8-10^{11}
Clostridium spp.	Rare	Rare	10^2-10^4	10^6-10^{11}
Eubacteria	Rare	Rare	Rare	10^9-10^{12}

Partially adapted from Simon and Gorbach (1984)

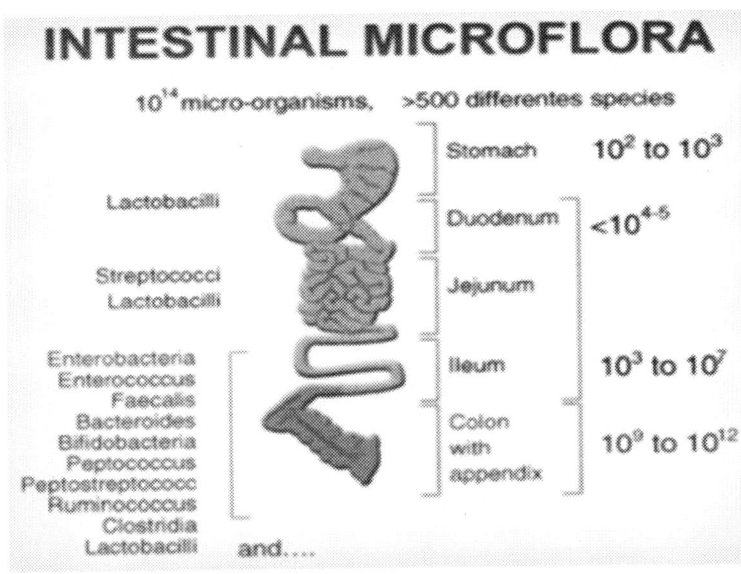

Source: www.knowabouthealth.com

A complex balance exists between bacterial population and the host. Some microbes have detrimental effects on the host, whereas, others exert beneficial effects (probiotics). Hydrogen sulphide (H_2S) produced by sulfate reducing bacteria in human gut is highly toxic and may cause ulcerative colitis. Similarly, the organisms having proteolytic activity, *e.g.*, *Clostridium*,*Peptostreptococcus* and *Bacteroids*, produce ammonia, phenolic compounds and amines. Saccharolytic bacteria, *e.g.*, *Bifidofacterium*, *Eubacterium* and *Lactobacillus*, on the other hand produce short chain fatty acids, *e.g.*, acetate, propionate butylate and lactate, which contribute in energy metabolism of colonic mucosa and colonic cell growth and are also metabolized systemically by host in the liver, muscle and brain. For a healthy gut, the microbiota a large number of saccharolytic and reduced number of proteolytic microbes are preferable.

The fermentative capacity of gut microflora influences digestive function of large bowel, acting on dietary compounds and host secretions allowing recovery of nutrients. The colonic microflora is involved in bowel motility, enterohepatic cycling of primary bile acids and possibly the metabolism of cholesterol (resulting in the production of fecal neutral sterols), maintaining homoeostasis (by preventing invasion of microorganisms to colonize) and preventing many diseases.

The representative species in probiotics generally include *Lactobacillus acidophilus*,*Lactobacillus johnsonii*, *Lactobacillus gasseri*, *Lactobacillus casei*, *Lactobacillus rhamnosus*, *Lactobacillus plantarum*, *Bifidobacterium longum, Bifidobacterium breve, Bifidobacterium bifidum*, *Bifidobacterium infantis*, *Enterococcus faecalis* and *Enterocuccus faecium. Bifidobacterium* species specifically inhabit the intestinal tracts of animals, such as *Bifidobacterium thermophilum* and *Bifidobacterium pseudolongum*, are used in animal probiotics. There are some bacteria, which do not normally inhabit the intestinal tract, may also come under the category of probiotics. Mainly *Lactobacillus bulgaricus*, *Streptococcus thermophilus* and *Leuconostoc* and *Lactococcus* species are used as starters in dairy products. Most often, these bacteria do not colonize the intestinal tract and their effect on intestinal microbial balance is expected to be small.

The role of probiotics bacteria in dairy fermentations is to assist in the preservation of the milk by the generation of lactic acid and possibly antimicrobial compounds and other metabolites that will provide a product with the organoleptic properties desired by the consumer to improve the nutritional value of food, as in, for example, the release of free amino acids or the synthesis of vitamins and the provision of special therapeutic or prophylactic properties as cancer and control of serum cholesterol levels.

Potential benefits may result from growth and action of the bacteria during the manufacturing of cultured foods. A therapeutic benefit also includes prophylaxis against some types of gastrointestinal infection, improved digestion of lactose against lactose mal-digestion of lactose-containing foods. Lactose mal-absorption may compromise their intake of protein and calcium, and these microbes are capable of providing numerous health benefits beyond basic nutritional value.

Desirable Properties of Probiotics

Selected strains of microbes, which already exist in human colon, can be consumed for health benefits. Most probiotic strains have been isolated from human intestinal contents, or from raw milk, dairy products, or other sources. Commercial products containing probiotics bacteria are presented in Table 2. Whatever the source is, these strains should be nontoxic and nonpathogenic with accurate taxonomic identification. The certain required features are:

- Bacteria used in probiotics are desired to express their activities in the host body. Therefore, the first consideration is the bacteria that normally inhabit the intestinal tract, and ingestion of these bacteria may affect the intestinal microbial balance.
- The probiotics should be viable and active in the intestinal lumen.
- It should be resistant to hydrochloric acid secreted by stomach, pancreatic enzymes and bile to ensure passage through the intestinal tract and survival.
- An ideal probiotic strain should adhere to human intestinal cells to protect the cells of intestinal epithelium against invasion by pathogens.
- It should be able to colonize in large intestine of human and produce antimicrobial substances, antagonize both carcinogenic and pathogenic bacteria.
- Upon consumption, probiotic strain should be beneficial for human health and host friendly.
- It should be safe in clinical use in large amounts and should not pose any health risk to host environment.
- Other desirable properties of probiotics strains should be of its genetic stability, good growth characteristics and prolonged shelf life during storage and economic production in large scale, desirable organoleptic properties, resistant to industrial processing, *etc.*

Table 2: Commercial strains sold as probiotics*

Strain	Commercial products	Source
Lactobacillus acidophilus NCFM *Bifidobacterium lactis* HN019 (DR10) *Lactobacillus rhamnosus* HN001 (DR20)	Sold as ingredient	Danisco (Madison WI)
Saccharomyces cerevisiae (boulardii)	Florastor	Biocodex (Creswell OR)
Bifidobacterium infantis 35264	Align	Procter and Gamble (Mason OH)
Lactobacillus fermentum VRI003 (PCC)	Sold as ingredient	Probiomics (Eveleigh, Australia)

Strain	**Commercial products**	**Source**
Lactobacillus rhamnosus R0011 *Lactobacillus acidophilus* R0052	Sold as ingredient	Institut Rosell (Montreal, Canada)
Lactobacillus acidophilus LA5 *Lactobacillus paracasei* CRL 431	Sold as ingredient	Chr. Hansen (Milwaukee WI)
Bifidobacterium lactis Bb-12	Sold as ingredient	Chr. Hansen (Milwaukee WI)
Lactobacillus casei Shirota *Bifidobacterium breve* strain Yakult	Yakult	Yakult (Tokyo, Japan)
Lactobacillus casei DN-114 001 (*L. casei* Immunitas) *Bifidobacterium animalis* DN173 010 (Bifidis regularis)	DanActive fermented milk Activia yogurt	Danone (Paris, France) Dannon (Tarrytown, NY)
Lactobacillus reuteri RC-14 *Lactobacillus rhamnosus* GR-1	Femdophilus	Chr. Hansens (Milwaukee WI) Urex Biotech (London, Ontario, Canada) Jarrow Formulas (Los Angeles, CA)
Lactobacillus johnsonii Lj-1 (same as NCC533 and formerly *Lactobacillus acidophilus* La-1)	LC1	Nestlé (Lausanne, Switzerland)
Lactobacillus plantarum 299V	Sold as ingredient; Good Belly juice product	Probi AB (Lund, Sweden); NextFoods (Boulder, Colorado)
Lactobacillus rhamnosus 271	Sold as ingredient	Probi AB (Lund, Sweden)
Lactobacillus reuteri ATCC 55730 (Protectis)	BioGaia Probiotic chewable tablets or drops	Biogaia (Stockholm, Sweden)
Lactobacillus rhamnosus GG ("LGG")	Culturelle Dannon Danimals	Valio Dairy (Helsinki, Finland) The Dannon Company (Tarrytown, NY)
Lactobacillus rhamnosus LB21 *Lactococcus lactis* L1A	Sold as ingredient	Essum AB (Umeå, Sweden)
Lactobacillus salivarius UCC118		University College (Cork, Ireland)
Bifidobacterium longum BB536	Sold as ingredient	Morinaga Milk Industry Co., Ltd. (Zama-City, Japan)
Lactobacillus acidophilus LB	Sold as ingredient	Lacteol Laboratory (Houdan, France)
Lactobacillus paracasei F19	Sold as ingredient	Medipharm (Des Moines, Iowa)

Strain	Commercial products	Source
Lactobacillus paracasei 33 *Lactobacillus rhamnosus* GM-020 *Lactobacillus paracasei* GMNL-33	Sold as Ingredient	GenMont Biotech (Taiwan)
Lactobacillus plantarum OM	Sold as Ingredient	Bio-Energy Systems, Inc. (Kalispell, MT)
Bacillus coagulans BC30	Sustenex, Digestive Advantage and sold as ingredient	Ganeden Biotech Inc., Cleveland, Ohio
Streptococcus oralis KJ3 *Streptococcus uberis* KJ2 *Streptococcus rattus* JH145	ProBiora3 EvoraPlus	Oragenics Inc. (Alachua FL)
Lactobacilli rhamnosus PBO1 *Lactobacilli gasseri* EB01	EcoVag	Bifodan (Denmark), www.ecovag.com

*Adapted from http://www.usprobiotics.org/products.asp#commercial

The strains listed in this table may or may not have been adequately evaluated. The purpose of this table is to give the reader a sense of what is commercially available, not to provide recommendations for probiotic strain use. Efficacy and safety dossiers should be reviewed to determine adequacy of substantiation by those interested in using these strains.

How do probiotics benefit our health?

There are many beneficial effects of probiotics on human health. Many scientific reports suggested that the lack of probiotic bacteria in the gut is the main cause of many diseases. Probiotics are best taken as a supplement. The best probiotic supplement would be a probiotic blend of the dominant species of bacteria (LAB) in capsule or powder form. Probiotics support human health (Figure 1) by:

- Enhancing the immune function
- Digesting food and improving digestive disorders
- Enhancing the body's absorption of food and nutrients
- Repopulating depleted intestinal flora
- Protecting the gut wall against harmful bacteria that can cause *leaky gut* since a leaky gut allows the pathogens to break through the gut wall causing havoc in the body.
- Supporting vitamins, short-chain fatty acids (SCFA) and amino acids production

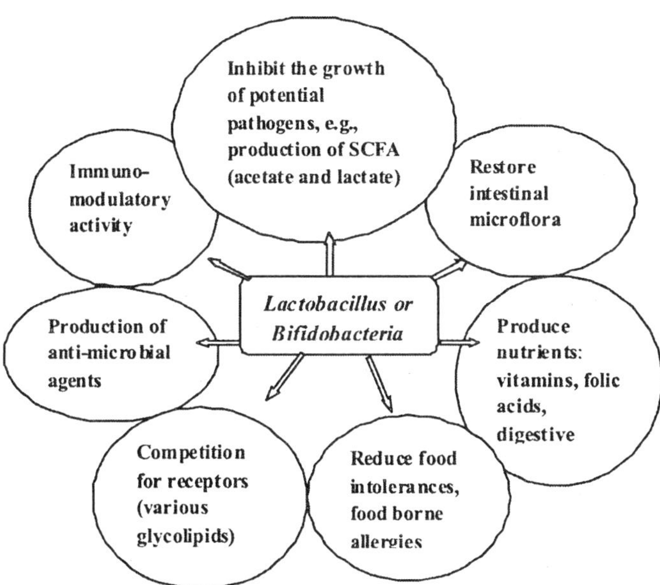

Figure 1: Beneficial effects of *Lactobacillus* and *Bifidobacteria* on human health

Probiotics Uses for Health Purposes

In recent years, people are interested in probiotics for health purposes since the world as well as the human body (in and on the skin, in the gut and in other orifices) is full of microorganisms including bacteria. Friendly bacteria are vital to proper development of immune system, to protection against microorganisms that could cause disease and to the digestion and absorption of food and nutrients. Each person's mix of bacteria varies. Interactions between a person and the microorganisms in his body and among the microorganisms themselves can be crucial to the person's health and well-being. This bacterial *balancing act* can be thrown off in following ways:

- Use of antibiotics kills friendly bacteria in the gut along with unfriendly bacteria. Some people use probiotics to offset side effects from antibiotics like gas, cramping, or diarrhea, and sometimes, use them to ease symptoms of lactose intolerance, a condition in which the gut lacks the enzyme needed to digest significant amounts of major sugar in milk and which also causes gastrointestinal symptoms.
- Unfriendly or bad microorganisms such as disease-causing bacteria yeasts and fungi can also unbalance in gut. Many researchers have studied whether probiotics could halt these unfriendly agents in the first place and/or suppress their growth and activity in conditions like:
  - Infectious diarrhea
  - Irritable bowel syndrome
  - Inflammatory bowel disease, *e.g.*, ulcerative colitis and Crohn's disease
  - Infection with *Helicobacter pylori*, a bacterium causing most ulcers and many types of chronic stomach inflammation
  - Tooth decay and periodontal disease

- Stomach and respiratory infections the children acquiring in day care
- Skin infections and allergies

Probiotic Benefits

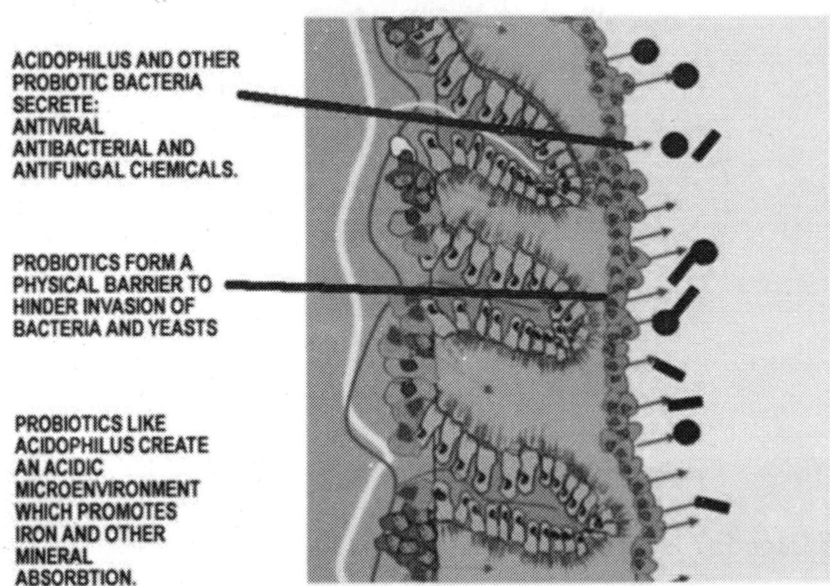

Source: www.greaterimmunity.com/Images/probiotic

Health Benefits and Therapeutic Effects of Probiotics

It is well understood that several bacteria, viruses and protozoa invade gastrointestinal tract and causes infection, which results in serious inflammation. Local inflammation in gut is responsible for various inflammatory diseases such as Crohns disease (CD), ulcerative colitis (UC), food allergies and atopic dermatitis. The exact pathophysiology of these illnesses has not been well understood. Probiotics have been claimed to be effective in treatment of gastrointestinal inflammatory diseases. Specific strains of probiotics are recommended in diverse pathologic situations such as treatment and prevention of acute diarrhea, immune disorders, high serum cholesterol levels and prevention of formation of intestinal tumors. The probiotic bacteria and their beneficial effects are presented in Table 3. The mechanism of action of probiotics is not fully understood.

Table 3: Probiotic bacteria and their beneficial effects*

Strains	Beneficial effects
Lactobacillus acidophilus	LA1 adherence to human intestinal cells
	Balances intestinal microflora
	Enhances immunity
	Decreases significantly diarrhea in patients receiving pelvic irradiation

Strains	Beneficial effects
	Decreases polyps, adenomas and colon cancer in experimental animals Prevents urogenital infection with subsequent exposure to three pathogens *Escherichia coli, Klebsiella pneumoniae* and *Pseudomonas aeruginosa* Lowers serum cholesterol levels Lowers the activity of faecal enzymes Decreases the faecal mutagenicity Removes constipation
Lactobacillus GG	Prevents antibiotic-associated diarrhea Cures rotavirus diarrhea Cures diarrhea caused by *Clostridium difficile* Increases IgA secretion in Crohn's disease Prevents recurrent chronic vaginitis Reduces food allergies and atopic dermatitis Increases interleukin-10
Lactobacillus casei Shirota	Prevents intestinal microbiota disturbance Produces positive effects on bladder cancer Lessens duration of colds and flu's in older people Helps in immune system and dlgestive health
Lactobacillus gasseri	Reduces carcinogenic-associated enzyme Restores normal microbial flora Acts as immuno-stimulatory agent
Bifidobacterium lactis	Helps in immune system and digestive health
Bifidobacterium infantis	Alleviates symptoms of irritable bowel syndrome
Bifidobacterium bifidum	Prevents viral diarrhea Reduces incidence of neonatal necrotizing entero-colitis
Bifidobacterium animalis	Improves gut health and fast digestion
Propionibacterium freudenreichii	Stimulates growth of other *friendly* bacteria
Enterococcus faecium	Decreases the duration of acute diarrhea from gastroenteritis

Strains	Beneficial effects
Lactobacillus strains	Helps in the administration of multiple organisms, predominantly *Lactobacillus* strains Effective in ameliorating pouchitis Improves lactose digestion, decreases diarrhea and symptoms of intolerance in lactose intolerant individuals, children with diarrhoea and in individuals with short-bowel syndrome Useful in microbial interference therapy- the use of nonpathogenic bacteria to eliminate pathogens and as an adjunct to antibiotics Improves mucosal immune function, mucin secretion and prevention of disease
Lactobacillus plantarum	Reduces incidence of diarrhea in daycare centers when administered to only half of the children Especially effective in reducing inflammation in inflammatory bowel, *e.g.*, enterocolitis in rats, small bowel bacterial overgrowth in children, pouchitis Reduces pain and constipation of irritable bowel syndrome Reduces bloating, flatulence and pain in irritable bowel syndrome in controlled trial Produces positive effect on immunity in HIV+ children
Lactobacillus reuteri	Shortens the duration of acute gastroenteritis Shortens acute diarrhea
Saccharomyces boulardii (yeast)	Reduces recurrence of *Clostridium difficile* diarrhea Produces effects on *Clostridium difficile* and *Klebsiella oxytoca* resulted in decreased risk and/or shortened duration of antibiotic-associated diarrhea Shortens the duration of acute gastroenteritis Decreases only functional diarrhea but not any other symptoms of irritable bowel syndrome
Bacteroides species	Useful in chronic colitis, gastritis and arthritis (increased bacterial urease activity in chronic juvenile arthritis)
Lactobacillus salivarius	Suppresses and eradicates *Helicobacter pylori* in tissue cultures and animal models by lactic acid secretion
Lactobacillus rhamnosus	Enhances cellular immunity in healthy adults in controlled trial

*Adapted from Journal of Applied Microbiology 100 (2006), pp. 1171-1185

Benefits of Probiotics against Cancer

Currently, cancer is a very serious global public health problem. The incidences and morbidity as well as mortality from cancer have been on high-rise throughout the world. Although we have now good control over communicable diseases with the rapid advent of antimicrobials but the high rise of cancer and cardiovascular diseases has a huge impact on quality of life. In recent years, a large number of epidemiological studies have suggested the link between diet and disease incidence.

Consumption of natural products, beverages and antioxidant rich diets along with exercise based life style has suggested reducing the incidences of cancer at large extent. There are numerous epidemiological studies supporting the protective role and benefits of probiotics against cancer. *Lactobacilli* have been shown to exert many beneficial effects in the gastrointestinal tract. Recent studies indicated that the consumption of fermented dairy products with lactobacilli might exhibit chemo-preventive and anticancer effects.

These effects are attributed to the inhibition of mutagenic activity, the decrease in several enzymes implicated in the generation of carcinogens, mutagens, or tumor-promoting agents, suppression of tumors and epidemiology correlating dietary regimes and cancer. The cellular components of *Lactobacilli* believed to induce strong adjuvant effects including modulation of cell-mediated immune responses, activation of the reticulo-endothelial system and augmentation of cytokine pathways. It has been believed that probiotics decreases the exposure to chemical carcinogens by several mechanisms given as under:

- Detoxifying ingested carcinogens
- Altering the environment of the intestine
- Decreasing populations or metabolic activities of carcinogen generating bacteria
- Promoting apoptotic pattern of cells
- Augmenting the antioxidant defense
- Producing compounds which inhibit the growth of tumor cells
- Enhancing immunity of the host against cancer cell proliferation

Numerous experimental studies demonstrate that the effect of probiotic consumption on cancer is quite encouraging by reducing the risk and incidences of the cancers of colon, liver and bladder. The chemo-preventive effect of probiotics against cancer incidence and progression is attributed to the binding of mutagens by intestinal bacteria, which may suppress the growth of bacteria that convert procarcinogens into carcinogens, and thereby, reduces the amount of carcinogens in the intestine. Increased activity of antioxidants and adaptogenicity is believed to be the mechanistic explanation for anticancer and chemo-preventive effect.

Benefits of Probiotics in Immunological Diseases

Probiotics have been shown to modulate specific immune responses in gut. They normalize increased intestinal permeability and improve intestinal barrier functions. They also alleviate intestinal inflammatory responses by restoring and maintaining the normal composition of gut microbiota and provide immunomodulatory stimuli for the intestinal immune system. These make it to be useful in impaired mucosal barrier functions and inflammation along with pathological changes in gut microbiota. Probiotic cells exhibit immune reactions, modulate immune system and maintain intestinal homeostasis.

Probiotics could have a beneficial action on intestinal microflora, functioning of intestinal lining and regulation of critical components of immune system such as lymphocytes, antibodies and natural killer cells following probiotic-mediated signalling in the gut.

Benefits of Probiotics in Diarrhea

Probiotic are involved with intestinal health. Diarrhea is an indication that the state of intestinal health has been compromised, whether due to toxin, pathological bacteria, rancidity, medication reaction, or something else.

Antibiotic associated diarrhea (AAD) is a frequent complication, which results from a microbial imbalance (decreased colonization, resistance and fermentation capacity) and toxic effects of antibiotics on the bowel mucosa along with pharmacological influence of antibiotics on gut motility. It has been reported that 20% of the patients receiving antibiotics suffer from antibiotic associated diarrhea in both adults and children. Many laboratory and human studies provide convincing evidence regarding the therapeutic potential and benefits of probiotics in the treatment of antibiotic associated diarrhea including the bacteria *Lactobacillus rhamnosus* GG (LGG), *Lactobacillus acidophilus* combined with *Lactobacillus bulgaricus*, *L. acidophilus* combined with *Bifidobacterium infantis*, *Lactobacillus sporogens*, *Bifidobacterium lactis* combined with *Streptococcus thermophilus*, *Bifidobacterium longum*, *Enterococcus faecium* SF68, *Bacillus clausii*, *Clostridium butyricum* and the yeast *Saccharomyces boulardii*. Probiotics are useful in the treatment of childhood diarrhea, rotavirus caused diarrhea and *Lactobacillus* GG, *Saccharomyces boulardii* and *Bifidobacteria* demonstrated best effects. Several studied proposed that *Lactobacilli* strains have been reduced the duration of rotavirus diarrhea under potential mechanisms of, *viz.* competitive blockage of receptor sites in which *Lactobacilli* bind to receptors, enhanced immune response, signals from lactobacilli that regulates the secretary and motility defences designed to remove perceived noxious substances and that lactobacilli produce substances that inactivate the viral particles.

Probiotics with oral dehydration therapy (ORT) are found successful in the prevention of AAD. In such combination, oral dehydration therapy replenishes the depletion of fluids and electrolytes, and probiotics correct disturbances to the structure and function of the gastrointestinal tract or to the resident microflora. However, the effectiveness of probiotics in treating diarrhea depends on the strains of bacteria, its suitability to an individual and the ability to displace pathogens.

Clostridium difficile associated diarrhoea (CDAD) is a commonly occurring serious disease among older adults, and it usually develops 2 to 3 weeks after cessation of antibiotic therapy with metronidazole or vancomycin. Antibiotics alter the normal fecal flora and this disturbance of gastrointestinal micro-ecology leads to a loss of colonization resistance predisposing the gut to colonization by potentially pathogenic microbes, such as *C. difficile*. It causes a range of illness from mild diarrhea to pseudomembranous colitis and death. Several strains of probiotics produce proteases that directly degrade *C. difficile* toxins or increase the immunity. Most studies confirmed the benefit of probiotics such as *S. boulardii* or different species of *Lactobacillus* combined with high-dose of vancomycin for the prevention of *C. difficile* infection in adults. Total flora replacement (TFR) or fecal bacteriotherapy has been reported as a treatment alternative in severe *C. difficile* infections. It is based on the transfer of fecal flora from a healthy individual, often a close relative, to a severely sick patient. Total flora replacement has been used occasionally in severe *C. difficile* cases but also in severe constipation, irritable bowel syndrome and inflammatory bowel disease. Several studies were performed with the use of probiotics like *L. acidophilus*, *L. plantarum* 299v, *Lactobacillus rhamnosus* GG (LGG) and *B. bifidum* for the prevention of CDAD.

Benefits of Probiotics in Inflammatory Bowel Disease (IBD)

In both, humans and animals, bacteria strongly influence the intestinal inflammation and majority of inflammatory bowel diseases are more severe in the presence of indigenous flora. Probiotics has shown an important role in the interference with this abnormal inflammatory response. Lower number of *Lactobacillus* sp. and *Bifidobacterium* sp. in patients with Crohn's disease has been reported. Probiotic therapy has been reported to reduce inflammation and benefits in colitis and inflammatory bowel diseases.

Benefits of Probiotics in Irritable Bowel Syndrome (IBS)

Irritable bowel syndrome is a chronic painful abdominal condition arising from an altered intestinal function and not accompanied by discernible underlying structural changes in the gut. Irritable bowel syndrome frequently appears after gastro-enteritis or during antibiotic therapy with symptoms such as abdominal pain, varying bowel function (constipation or diarrhea), flatulence and bloating or even dyspepsia. The probable etiological reasons of IBS are motility disturbances, psycho-physiological and colonic mal-fermentation. The latter has been most often linked with an increased numbers of *Klebsiella* spp. and *Enterococci* and decreased numbers of *Lactobacilli* and *Bifidobacteria*. *Lactobacillus plantarum* by releasing nitric oxide affecting gut motility relieves pain and other symptoms of IBS.

Benefits of Probiotics in Allergic Diseases

Microbial exposure in early age appears to be effective against allergy. The influence of gut microflora on the development of immune system suggests a close relationship between allergic sensitization and the development of intestinal microflora in the infancy. Thus, the modification of the gut microflora by probiotics may decrease the occurrence of allergic diseases. The rationale behind probiotic therapy in allergic disorder is to provide a microbial stimulus for immune system of the host by supplementing beneficial live microorganisms that are characteristic of the healthy gut microflora. *Lactobacillus* is found useful in the treatment of allergic inflammation and other allergies. *Lactobacillus* protects infants if supplementation to their mother is prenatally and during breast-feeding. *Lactobacilli* and *bifidobacteria* aid in the management of atopic dermatitis and reduces its incidence in infants at risk.

Benefits of Probiotics in Maldigestion and Lactose Intolerance

The problem of lactose mal-digestion and intolerance is worldwide and is most predominant in young children more frequently with milk rather than yoghurt and curd. In infants, lactose intolerance is very common characterized by diarrhea. Consumption of yogurt has been shown to improve the absorption of lactose in lactase deficient patients and limit the digestive symptoms. Yoghurt and other dairy products contain lactase, which digests lactose in gut lumen and improve lactose intolerance and digestion. In addition, the gastrointestinal transit time of yoghurt in slow as compared to milk. Similarly, *Saccharomyces cerevisae*, a yeast contains sucrase enzymes improve intolerance in infants with sucrase deficiency. However, the probiotic therapy had not prevented cow milk allergy.

The probiotic bacteria present in fermented milk products, such as yogurt, can help in decreasing the symptoms of lactose intolerance. Lactose intolerance is a physiological state in human beings where they lack the ability to produce an enzyme named lactase or β-galactosidase, which is essential to assimilate the disaccharide in milk that needs to be split into glucose and galactose. Individuals lacking lactase will not be able to digest milk and this condition often poses a problem in newly born infants. People with lactose intolerance problem express abdominal discomfort, diarrhoea, cramps, flatulence, nausea, vomiting, *etc.*

Another problem associated with lactose intolerance is calcium deficiency. A person suffering from lactose intolerance will be advised to take non-milk diet. The resident bacteria in the colon ferment the undigested lactose, producing acid and gas, causing symptoms such as abdominal pain, bloating and diarrhoea.

Yogurt contains less lactose than milk and delays gastric emptying, which partly explains why lactose-intolerant individuals tolerate yogurt. However, yogurt tolerance is mainly due to the supply of lactase activity from the lactic acid bacteria present in the yogurt itself. Calcium requirement of the body is met only through milk since milk is the richest source of calcium. Hence, a person consuming non-milk diet will naturally develop calcium deficiency, leading to osteoporosis. Studies confirmed that lactose deficiency leads to calcium malabsorption and thereby to osteoporosis. Calcium malabsorption may be due to deletion of diets with milk to avoid the complications of lactose intolerance. Calcium absorption is better and more in acidic conditions. Hence, if lactose is converted to lactic acid, pH of the gut decreases, *i.e.*, it becomes acidic favouring enhanced absorption of calcium. If probiotics are fed to lactose intolerance patients, the milk lactose is hydrolyzed by probiotic strains, lactose is assimilated and calcium absorption is favoured.

Benefits of Probiotics in Weight Loss

A probiotic study, published in the *Journal of Gastrointestinal Surgery*, revealed that after three months, the patients taking probiotics saw a 47.6% weight loss, while the control group had only a 38.5% weight loss. Another reported benefit of probiotics is that they help in maintaining the levels of vitamin B_{12} in patients after gastric bypass surgery. The study showed that patients taking the supplements had higher vitamin B_{12} levels three months after surgery. This study could be beneficial in patients undergoing gastric bypass since gastric bypass frequently causes deficiency in vitamin B_{12}.

Benefits of Probiotics as Antimutagenic and Anticarcinogenic agent

Probiotics can also be used as antimutagenic agents in several ways. The *Escherichia coli* and *Clostridium perfringens* are enteropathogens and produce enzymes such as β-glucuronidase and nitroreductase. Beta-glucosidase and urease can convert procarcinogens to proximate carcinogens. The colonizing cells of *Bifidobacterium* produce lactic acid, lower the intestinal pH, create a bactericidal environment for enteropathogens and develop a favourable microenvironment, which modulates the bacterial enzymes. Further experiments suggest that purified bifidobacteria has cell wall antitumour activities and induces activation of phagocytes to destroy the growing tumour cells.

Necrotizing Enterocolitis (NEC)

Necrotizing enterocolitis (NEC) is a life-threatening gastrointestinal disease most commonly found in infants with very low birth weight (<1500 g). The main factors of pathogenesis include immaturity of intestinal and immune function and gas-forming bacteria. Many clinical trials on very low birth weight infants suggest that the probiotic mixture (*Lactobacillus acidophilus*, *B infantis*, *Streptococcus thermophilus* and *Bifidobacterium bifidus*) significantly reduced the incidence and severity of NEC without any adverse effect.

Safety and Tolerance of Probiotics

Most often, a majority of probiotics foods and supplements are marketed as functional foods, foodstuffs or as drugs. Therefore, consideration of the safety of probiotics is of utmost importance. The most commonly used organisms in probiotics are the *Lactobacilli*. These are found in large numbers in the gut of healthy animals and are in use since ancient time, therefore, United States Food and Drug Administration has termed them *Generally Regarded as Safe*. The safety of the microbes that have been used traditionally in probiotics has been confirmed through a long period tested safety and efficacy.

The absence of pathogenicity and infectivity is a requisite of probiotic safety. The frequent isolation of lactic acid bacteria from clinical infections in recent years has raised debate over the safety of these bacteria whether the bacteria are actually infective. Another requisite of probiotics is that the probiotic bacteria should not produce harmful substances by metabolic activities.

One test is to determine whether the bacteria convert food components or biological secretions into secondary substances harmful to the host, *e.g.*, some intestinal bacteria act on proteins and their digested products to produce ammonia, indol, phenols, and amines.

Platelet aggregating activity has been considered a required test in the assessment of safety. Aggregation of platelets by bacteria is thought to contribute to the progression of infective endocarditis. The platelet-aggregating activity of strains of *L. rhamnosus* and *L. paracasei* subsp. paracasei isolated from infective endocarditis, laboratory strains of the same species and *Lactobacillus acidophilus*, *Lactobacillus fermentum*, *Lactobacillus oris*, *L. plantarum* and *L. salivarius* have been shown in some isolates.

Recently, a review of human trials shows the beneficial effects with a wide spectrum of safety. On long-term therapy, it is found that intake of 10^6-10^9 CFU (colony forming units) daily of *Bifidobacteria* and *Lactobacillus* is safe. A wide spectrum of safety, tolerability and oral ingestion make it more common to consume for a long time too.

- People use probiotic products to prevent and treat certain illnesses and support general wellness.
- There is limited evidence supporting some uses of probiotics. Knowledge that is much more scientific is needed about probiotics, including about their safety and appropriate use.
- Effects found from one species or strain of probiotics do not necessarily hold true for others, or even for different preparations of the same species or strain

What need to know before consuming Probiotics?

Before taking probiotic food, one must look at colony forming units (CFU). One needs to be ascertained of the strains because kinds of bacterial strain affect the effect of product. After knowing the strain, knowing CFU is really useful. The minimum common dose for anyone is 1 billion CFU, which is the least amount of organisms that gives a measurable effect. The human digestive tract is inhabited by numerous microbes. The balance of this microbial flora greatly influences the intestinal environment. Among numerous intestinal microbes, those that are expected to affect the host beneficially by improving the intestinal bacterial balance, and hence, are selected as probiotics, include species of the genera.

There is no guarantee to maintain the potency of probiotics if they have not been refrigerated. Since probiotics are containing live bacteria, temperature plays an important role in keeping them potent. The probiotic products and supplements need to be properly labeled for their storage and shipping conditions. Some probiotic products are quite shelf stable and need not to be refrigerated but once they are open to use they must be refrigerated. It is said that probiotics are fragile since they are living bacteria. They need to be kept alive, not only in the body but also in foods when they are consumed. This is essential to give consumers health and wellness benefits that probiotics provide. The problem is that it is difficult for probiotics to survive in the conditions of heat, humidity, or exposure to oxygen that are characteristic of processing, food manufacturing as well as transport and storage.

Probiotics should not be consumed by severely immuno-compromised patients, infants with short bowel syndrome, post-cardiac surgery patients, patients with pancreatitis or Type 1 diabetes and patients in the Intensive Care Unit. Probiotics products or supplements should be taken with physician's advice by the patients.

Some Points to Consider About Probiotics Use

- Using a probiotic product, one must consult the health care provider first.
- Effects from one species or strain of probiotics do not necessarily hold true for others, or even for different preparations of the same species or strain.
- If you use a probiotic product and experience an effect that concerns you, contact your health care provider.

Concentration of probiotic required for beneficial health effects

The minimum concentration of probiotic bacteria for effective function is still unknown. Nevertheless, adequate numbers of viable cells, namely the *therapeutic minimum* needs to be consumed regularly for the transfer of *probiotic* effect to consumers. Consumption should be more than 100 g per day of bio-yogurt containing more than 10^6 CFU/ml. Scientists suggested a minimum viable number of 10^6 CFU/ml or per gram but recommended 10^8 CFU/g to compensate for reduction passage through the gut. Yogurt is a functional food with probiotics. Yogurt with probiotics, called bio-yogurt, should contain living bacterial cells. According to regulation, yogurt should contain 2×10^6 living bacteria in 1 ml at the end of the recommended storage period. The daily dose of probiotic microorganisms should reach 1×10^9 cells. The titre of bacteria in fermented drinks reaches 10^8 to 10^9 per ml, which later decreases with storage. Another possibility to use tablets or capsules as additives to foodstuffs that contain lyophilized cultures of bacteria. The intake of functional dairy products is physiologically more acceptable for patients as well as consumers health.

Conclusion

Probiotics are live microbial food products and supplements or components of bacteria, which have been shown to have beneficial effects on human health. Normalization of the properties of unbalanced indigenous microflora of intestinal tract by ingestion of specific strains of healthy microflora forms the rationale of probiotic therapy.

The probiotics industry is flourishing, and interest in establishing scientific credibility has attained importance for many companies and scientists. Fundamental knowledge of intestinal bacteria and their interactions with each other and with the host are a prerequisite for successful probiotic research and development.

It is now clear that microbial probiotics are useful in promoting health and disease prevention. Many studies have reported the beneficial potential of microbial probiotics in health and various diseases but we are still lacking a comprehensive understanding about their modes of action, long-term clinical outcomes, as well as the degree of variation within and among individuals. Therefore, more clinical-trials are needed to evaluate the safety and tolerability of microbial probiotics to make it more common to consume for a long time too.

■■■

14 Take Care of Your Amazing Brain

Umashashi C. Hegde

Deputy Director, Senior Grade (Retired), National Institute for Research in Reproductive Health (ICMR), Parel, Mumbai, Maharashtra
Present Address: Chandrama, Post Kemmannu, Udupi, Karnataka
Email: <hegdeuc@yahoo.com>

Human brain is the most complex and sophisticated system to be found anywhere on planet earth. Its power and versatility are awesome. The brain contributes to our lives, memory, pleasure, attention, anticipation and physical coordination, and it can do them all at the same time. To keep the brain healthy, a good balance of nutrients, *i.e.*, carbohydrates, proteins and fats plus minerals and vitamins is essential since the right balance of all these in the diet can protect our mental faculties.

Brain through Life

Over 90% of people over 65 show little deterioration in mental abilities. Even late in life, nerve cells in the brain seem capable of forming new connections with other neurons. This continuous building of common connections between brain cells is what happens during life long learning, and it is the powerful argument for the importance of education throughout life.

However, certain abilities, such as being able to solve new problems, seem to collectively decline somewhat later in life. These abilities referred to by psychologists as *fluid intelligence* may be affected by a natural process of reduced efficiency in the functioning of our brain cells, as we get older.

The Mind Body Link

Eastern traditions believe that mind and body are intrinsically linked. The brain it seems where our physical and mental selves come together. According to WHO, health is a state of complete physical, mental and social well being and not merely the absence of disease or infirmity we are what we eat. The food we eat has been shown influence greatly on our brain function.

Benefits of Exercise

Physical exercise is good for health. It is well known that it reduces the chances of developing heart disease. What is less known is that exercise can also improve psychological well being by reducing anxiety and depression and boosting self-esteem and the ability to cope up with stress? On physiological level, it may be that increased physical activity causes more blood to circulate around the brain, both supplying essential nutrients and oxygen and removing waste products including carbon dioxide. Research has established that incidents of depression may be due to a low level of neurotransmitter, serotonin. Therefore, instead of taking antidepressant drugs to increase the amount of serotonin available in the brain, regular exercise could provide serotonin boost required. Among the most popular exercises practical today are walking, running, cycling as well as yoga.

Get Adequate or Quality Sleep

Related to stress is the other plague of modern society, lack of deep and restful sleep. With diet and lifestyle, sleep is a fundamental pillar of *Ayurveda*. Optimal physical and mental performance is impossible if not all three of these factors are addressed.

Getting eight hours of sleep is less important than striving for quality sleep, which, according to *Ayurveda*, occurs when the mind is totally disconnected from the senses. Quality sleep recharges and rejuvenates the mind, improving both short- and long-term mental prowess. How much or how little you need is an individual matter, go by what your body and mind tells you.

Water is good for brain. Scientists at King College London found that water is good for brain. They have found that just 90 minutes of steady sweating can shrink the brain by as much as a year of aging. Starved of water, brain is forced to work harder to process the same information. Drinking a glass of water or two the brain returns to normal, say the researchers after scanning the brain of teenagers after an hour and half of cycling. The sweaty teens did just as well when asked to play a computer game that tests the ability to plan and solve the problems but the scan showed they had to use more of their brain to do it.

Cutting down on stimulants such as caffeine, alcohol and nicotine, making work-related materials taboo in the bedroom and winding down the evening with calming activities, instead of stimulating ones, to help prepare the mind for restful sleep are key to getting your sleep.

Food for Brain and Thought

Brain is the greediest organ in the body but its proper nourishment is vital to creative thought, positive mood, memory and good overall health. It is no surprise that what you eat affects how you think, feel, remember and potentially even increase intelligence. If it is possible to eat our way to genius, who would not want to do it? We know that the foodstuffs we eat affect the body but they can have even more influence on how well our brain functions. What we eat can have a powerful affect on energy of our brain, how the mind handles tasks and our general mood. The focus here is on those particular foods with nutrients that enhance neuron firing and cross-linking in the brain. The foodstuffs listed below can help us to concentrate, the serotonin boost required for good memory, tune sensor motor skills, keep us motivated, speed up our reaction time, control stress, and even slow down the aging of brain cells as in dementia and Alzheimer's. Foodstuffs that are canned or preserved with artificial additives, preservatives and artificial colours and flavours are not good. Cooking with fresh organic ingredients can take a little more time and effort but our brain and indeed our entire physiology will thank us for it and repay the effort with accelerated performance.

Make shopping for food fun, visit farmer's markets if there is one in your area for the freshest produce to try exotic new vegetables and grow fruits organically in your own garden.

Food as Fuel for Life

The next time you find yourself at your food outlet, keep in mind the old adage *Garbage IN equals Garbage OUT*. If you want to feel healthy and alive, your kitchen island should harbour only the best wholesome foodstuffs. If we keep our bellies full with food good for our mind and body and our kitchen sinks clean our body will thank us.

Now, that we have discussed the requirements and problems the brain throughout life, we shall see what type of food is best suited for the mental well-being. Start early to pay attention to this important aspect of preserving brain function. Here is a list of different food types that we should incorporate in our diet since they are beneficial for brain function.

Use It or Lose It

No matter, what your age or your occupation is, your brain needs to be constantly challenged to be at its peak in terms of performance. Whether it is doing logic puzzles, memorizing lines from Shakespeare, or learning a new skill, keep your brain busy if you do not want it to rust away like a car in a junkyard.

Carbohydrates

These provide energy, and as the brain uses up a fifth of the body's energy requirements, they are important for mental functioning. People should eat more complex carbohydrates found in foodstuffs such as grains and pulses and minimize the intake of simple carbohydrates such as sugar. Complex carbohydrates take longer to digest, thus, they release their energy gradually, providing ongoing energy and stamina. Simple carbohydrates are digested quickly and they can cause too much sugar to be released into the blood stream. This sudden rise in the blood sugar is countered by the rise in insulin. The brain is alternatively flooded and starved of essential glucose, which may lead to unpleasant and potentially dangerous symptoms, including dizziness, anxiety, headaches, thirst, mental confusion and sleepiness.

With the dark days of winter, about 35 million Americans fall into a depression known as seasonal affective disorder (SAD). It is believed that light deprivation alters brain chemistry in biologically susceptible persons. Such depressed persons often crave sweets and starches. Carbohydrates are self administered to fight off winter blues according to Norman Rosenthal a leading researcher at the National Institute of Mental Health. He explains that normal people after eating carbohydrates, like sugar, are less alert and less energetic, regardless of the seasons but people with SAD get the opposite effect. To them, carbohydrates are antidepressants. Some believe, eating carbohydrates helps overcome this problem by boosting serotonin levels and/or serotonin activity.

Proteins

These are vital for brain function since neurons like all other cells made from them. Good sources of protein include beans, nuts, egg, milk and cheese as well as fish and lean meat. Vegetarians and vegans prove it possible to live healthy without meat and dairy products so long as a varied source of proteins is maintained from pulses and cereals.

Fats

These insulate the nerve cells in the brain and other vital organs. In fact, about one third of the brain is made from fats. The two types of dietary fat, *i.e.*, saturated and unsaturated, are classified according to their chemical structure. Vegetable fats are mostly unsaturated and are particularly important since some of them are used to make essential messenger molecules such as prostaglandins. Animal fats are mainly saturated, and in excess, they can contribute to heart disease and other problems. Though fats are useful source of energy, most experts agree that they should comprise no more than 30% of our calorie intake.

Minerals

These are needed for a healthy brain. Minerals are derived from the earth, and so like, most vitamins can be directly obtained from plants. Calcium, usually associated with healthy bones, is needed by nerves, muscles and blood. Magnesium, potassium and sodium work together to maintain water balance, and nerve and muscle impulses. Since people often consume too much sodium, they may need more potassium to counter balance. This may be obtained from fruits such as bananas, vegetables and whole grains. Zinc, one of the *trace elements* minerals that are essential in very small quantities, is necessary, as it is good for mental functioning. Extra zinc may be needed if people have too much iron in their diet or if zinc levels are depleted by stress, alcohol, or smoking. Other trace elements needed by the brain include manganese, copper and selenium.

Vitamins

While all vitamins are necessary, some are particularly important for mental functioning, especially the B group vitamins. Vitamin B_1 (thiamine) helps to make the neurotransmitter acetylcholine, making it vital for maintaining an efficient nervous system. Prolonged deficiency of vitamin B_1 can produce mental disturbances, and there is some evidence of B_1 deficiency being associated with lower levels of intelligence. Supplements of vitamin B_3 (niacin) can be effective for migraines and headaches, and it has been used to treat schizophrenia. Pantothenic acid (B_5) contributes to acetylcholine production and it is believed to boost memory. It is also required by the adrenal glands, and so is sometimes known as *anti-stress vitamin*. Vitamin B_{12} is needed to make DNA, and it helps to form the myelin sheath that insulates nerves. Vitamin B_{12} deficiency can also cause mental disorders.

Folic acid

Another B group vitamin is not only good for your mood but also good for dementia and schizophrenia. It is found in various foodstuffs, including spinach, orange juice and yeast. Adequate levels of folic acid are essential for brain function, and folate deficiency can lead to neurological disorders such as depression and cognitive impairment. Folate supplementation, either by itself or in conjunction with other B group vitamins, has been shown to be effective in preventing cognitive decline and dementia during aging and enhancing the effects of antidepressants. The results of a recent randomized clinical trial indicate that 200 milligrams of folic acid, *i.e.*, the amount in about three-quarters of a cup of cooked spinach, was given to half the group for a year. The others got a placebo. It was observed that those getting folic acid had a dramatic relief of depression compared to those given placebo. McGill University's Dr. Young finds that folic acid deficiency can contribute to depressed mood and 200 to 500 milligrams of folic acid a day may fight depression in certain susceptible people. That much is easily attainable in food. High folic acid doses can be toxic he cautions.

Now that we have discussed the requirements and problems the brain through life, we shall see what type of food is best suited for the mental well-being. Start early to pay attention to this important aspect of preserving brain function. Here is a list of different food types that we should incorporate in our diet, as they are beneficial for brain function.

Wholegrain

Whole grain is a great brain stimulator because it contains carbohydrate- a mood elevator and high percentage of folic acid. Make sure, you are eating a diet rich in whole grain breads, cereals, barley, popcorn, beaten rice (*poha*), *etc.* because they can boost blood flow to the brain. Every organ in the body is dependent on blood flow, especially the brain. Wholegrain breads and cereals are rich in vitamin B_6, an important brain vitamin. Wheat germ additionally contains memory-improving thiamine.

Nuts

A nut a day keeps the blues away. Recent researches show that certain foods like nuts and seeds can boost your self-esteem and stabilize your mood. Whether you believe it or not, if you consume large quantities of any kind of nut (almonds, hazelnuts, walnuts, *etc.*) and seeds (sunflower seeds, pumpkin seeds, *etc.*) you will experience mood balance and you will start thinking clearly and be happier. Everything from the most common nuts, such as walnuts, hazelnuts, cashews, groundnuts or peanuts and almonds to the more exotic seeds and nuts can clear up that *brain fog*, and enable us too clear and be happier than you ever imagined. These are positive mood enhancers.

Walnuts: The most important nut recognized for its favourable effect on brain is walnut, which looks a lot like the human brain. Literally and figuratively speaking, walnuts are *brain food.* The thin outer green cover, which is taken out before the walnuts are sold, is similar to the scalp. The hard shell of a walnut is like a skull. The thin sheet inside, with its paper-like partitions between the two halves of the walnut, is like the membrane. The shape of the walnut itself represents the human brains two hemispheres. The beneficial effect of walnut on brain function is scientifically well documented.

Can Some Foods Make You Smarter?

Walnuts have often been thought of as a *brain food,* not only because of the wrinkled brain-like appearance of their shells but also because of their high concentration of omega-3 fats. The brain is more than 60% structural fat. For brain cells to function properly, this structural fat needs to be primarily the omega-3 fats found in walnuts, flaxseed and cold-water fish. This is because the membranes of all our cells, including our brain cells or neurons, are primarily composed of fats, and cell membranes are the gatekeepers of the cell. Anything that wants to get into or out of a cell must pass through the cells outer membrane and omega-3 fats, which are especially fluid and flexible, make this process a whole lot easier, thus, maximizing the cell's ability to usher in nutrients while eliminating wastes. This definitely is a good idea, especially when the cell in question is in our brain. Walnuts are made up of 15 to 20% protein and they contain linoleic (omega-6 fatty acids) and alpha-linoleic acids (omega-3 fatty acids), vitamin E and vitamin B_6, making them an excellent source of nourishment for our nervous system.

Walnuts may also help correct the human brain's serotonin levels. Serotonin is an important brain chemical that controls both our moods and appetite. Walnuts may be able to relieve disorders like insomnia, depression, overeating and other compulsive behaviour, commonly treated with antidepressant drugs like Prozac, without the dangerous side effects

Epidemiological studies in various countries including the United States suggest a connection between increased rates of depression and decreased omega-3 consumption, and in children, the relationship between low dietary intake of omega-3 fats and attention deficient hyperactive disease or ADHD has begun to be studied. A recent Purdue University study showed that kids low in omega-3 essential fatty acids are significantly more likely to be hyperactive, have learning disorders and display behavioural problems. In the Purdue study, a greater number of behavioural problems temper tantrums, and sleep problems were reported in subjects with lower total omega-3 fatty acid concentrations. In the study, more learning and health problems were also found in the children who had lower total omega-3 fatty acid concentrations.

Walnut as a source of bio-available melatonin: Effect of sleep is well recognized for good brain function. If you wish a good night's sleep, try sprinkling your dinner's tossed green salad, fruit salad, or steamed vegetables with a handful of walnuts or enjoy a baked apple or poached pear topped with walnuts for dessert.

Melatonin, a hormone produced by the pineal gland, which is involved in inducing and regulating sleep and is also a powerful antioxidant, has been discovered in walnuts in bio-available form, making them the perfect evening food for a natural good night's sleep.

Melatonin has been shown to help in improving sleep for night shift workers and people suffering from jet lag but maintaining healthy levels of this hormone is important for everyone over the age of 40 since the amount of melatonin produced by the human body decreases significantly as we age, and this decrease in antioxidant protection may be related to the development of free radical-related diseases later in life.

In a study published in Nutrition, Russell Reiter and colleagues at the University of Texas have not only quantified the amount of melatonin present in walnuts between 2.5 and 4.5 ng/gram but have demonstrated that eating walnuts triples blood levels of melatonin and also increases antioxidant activity in the bloodstream in animals. The authors theorize that by helping the body resist oxidative stress (free radical damage), walnuts may help in reducing the risk of cancer and delaying or reducing the severity of cardiovascular disease and neurodegenerative disease such as Parkinson's or Alzheimer's disease. Walnuts, best known as a heart-healthy nut, are also a rich source of another highly cardio-protective nutrient, *i.e.*, omega-3-fatty acids, hence, Reiter and his team will next investigate possible synergy between walnuts' omega-3 fats and melatonin. To those who are interested in human health issues, this sounds like a familiar theme in Nature's symphony in which whole, wholesome foods each provide a wealth of nutrients whose harmony promotes our optimal health.

Enjoying good health is not restricted to a single organ of our body or a single food. It depends on the overall well being of our body and mind. Therefore, it may be relevant to include additional health benefits of walnuts in our diet.

Help in preventing gallstones: Twenty years of dietary data collected on over 80,000 women from the Nurses' Health Study show that women who eat at least 1 ounce of nuts, peanuts, or peanut butter each week have a 25% lower risk of developing gallstones. Since 1 ounce is only 28.6 nuts or about 2 tablespoons of peanut butter, preventing gallbladder disease may be as easy as having a handful of walnuts as an afternoon pick me up, or tossing some walnuts on your oatmeal or salad.

Omega-3-rich walnuts protect bone health: Research carried out by Griel and co-investigators and published in Nutrition Journal has demonstrated that alpha linolenic acid, the omega-3 fat found in walnuts, promotes bone health by helping to prevent excessive bone turnover when consumption of foods rich in this omega-3 fat results in a lower ratio of omega-6 to omega-3 fats in the diet.

Other studies have shown that diets rich in omega-3s from fish (DHA and EPA), which also naturally result in a lowered ratio of omega-6 to omega-3 fats, reduce bone loss. Researchers think this is most likely because omega-6 fats are converted into pro-inflammatory prostaglandins, while omega-3 fats are metabolized into anti-inflammatory prostaglandins. Prostaglandins are hormone-like substances made in our bodies from fatty acids.

In this study, 23 participants ate each of three diets for a 6 weeks period with a 3 weeks washout period in between diets. All three diets provided a similar amount of fat, but their ratio of omega-6 to omega-3 fats was quite different.

Diet 1 provided 34% total fat with omega-6 and omega-3 fats in amounts typically seen in the American diet, *i.e.*, 9% polyunsaturated fats (PUFAs) of which 7.7% was omega-6 and only 0.8% omega-3 fats, resulting in a pro-inflammatory ratio of 9.6 : 1.

Diet 2, an omega-6 rich diet, provided 37% total fat containing 16% polyunsaturated fats of which 12% was omega-6 and 3.6% omega-3, a better but still pro-inflammatory ratio of 3.3 : 1.

Diet 3, which provided 38% in total fats, was an omega-3 rich diet, containing 17% polyunsaturated fats of which 10.5% was omega-6 and 6.5% omega-3, resulting in an anti-inflammatory ratio of 1.6:1.

After each diet, subject's blood level of N-telopeptides, a marker of bone breakdown, was measured and was found to be much lower following Diet 3, the omega-3 rich diet than either of the other two.

The level of N-telopeptides seen in subject's blood each diet also correlated with that of a marker of inflammation called tumour necrosis factor-α (TNF-α). Diets 1 and 2- the diets that had a significantly higher ratio of omega-6 to omega-3 fats-also had much higher levels of TNF-α than the Diet 3, which was high in omega-3 fats from walnuts and flaxseed. These studies show that in addition to protecting your brain, omega-3 fatty acids also help to protect your bones by making anti-inflammatory factors. Therefore, omega-3-rich flaxseed and walnuts as well as cold-water fish are important contributors to healthy way of eating.

Protective omega-3 levels greatly improved by eating just 4 walnuts a day: Enjoying just 4 walnuts a day significantly increased blood levels of the health-protective omega-3 essential fatty acids, *i.e.*, α linolenic acid.

Over 2,000 scientific studies have demonstrated the wide range of problems associated with omega-3 deficiencies. The American diet is almost devoid of omega-3s, except for nuts, such as walnuts, seeds and cold-water fish. In fact, researchers believe that about 60% of Americans are deficient in omega-3 fatty acids, and about 20% have so little that test methods cannot even detect any in their blood.

Cashews: While you are in the nut aisle shopping for walnuts be sure to pick up some cashews, almonds, pecans and peanuts too. Each nut can enhance your mental health in its own way. Cashews are high in magnesium, known to open up the blood vessels in your body. More oxygen-rich blood means better brain function.

Almonds: Phenylalanine found in almonds can do wonders for your neurological and mental health. Phenylalanine has the rare ability to cross the blood-brain barrier where it stimulates the brain to generate natural mood-boosting neurotransmitters called dopamine, adrenaline and noradrenaline. Additionally, almonds are high in riboflavin, which is known to boost memory.

Brazil nuts: Most selenium in the diet comes from grains, seafood, cereals and meat but the real selenium booster is Brazil nuts. Eating single nut a day will guarantee, you are never deficient, says Donald J. Lisk, Director, Toxic Chemical Laboratory at the Cornell University. He found Brazil nuts, since they are grown in selenium rich soil, to be extremely high in the mineral, which is about 2500 times of selenium-rich as any other nut. He cautions not to eat too much since intake of high concentrations of selenium can be toxic

Pecans: Pecans and peanuts provide choline, another important nutrient for optimal brain function. Choline aids in both memory and brain development.

Plant Seeds

Sunflower Seeds: Like nuts, many seeds and nuts can boost your mood and brainpower. Sunflower seeds contain tryptophan, an important amino acid that the brain converts to serotonin, which is a natural way to relieve mild depression and insomnia. Additionally, sunflower seeds are high in thiamine, an important B vitamin, which increases memory and cognitive function.

Pumpkin Seeds: Amazingly, the most powerful part of pumpkin lies in its least used part. The seeds of pumpkin are a power food, rich in many nutrients including zinc, vitamin A and E and the precious Omega 3 and Omega 6 fatty acids. The zinc found in pumpkin seeds plays a vital role in enhancing memory and thinking skills.

Fruits

Blueberries: Eating blueberries and a diet rich in deep pigment from fruits and vegetables helps in preserving the brain machinery and boost the potency of neuron signals. Blueberries literally strengthen the brain. They have compounds that turn on key systems in the brain enable other proteins to help with memory or other cognitive skills. In a recent study, subjects who ate one cup of blueberries a day for two weeks showed an increased production of brain cells in the hippocampus (region responsible for memory), and scored significantly higher in classroom tests than those subjects who did not.

Blueberries are also known to protect the brain from oxidative stress and may reduce the effects of age-related conditions such as Alzheimers disease and Dementia. In addition, blueberries also contain ellagic acid, another phytochemical that has been shown to prevent cell damage.

Strawberries: Antioxidant-rich strawberries can prevent age-related neurological declines by improving brain cell abilities to send and receive the *signaling* molecules. The brain uses these signaling molecules to communicate. Remarkably, these same studies showed that the powerful antioxidants in strawberries, spinach and blueberries can improve the ability to communicate even among brain cells already showing signs of age-related damage.

Blackberries: Blackberries contain an amazing class of nutrients called anthocyanins. Our brain is particularly vulnerable to oxidative damage but anthocyanins help in protecting our brain from oxidation stress, which in turn fights against degenerative brain diseases. One study even found anthocyanin-rich supplements to reverse age-related neurological deficits in subjects.

Cherry: Scientists found that the antioxidant compounds, anthocyanins, which give cherries their bright red colour, possess anti-inflammatory properties that could work like pain medications but without the nasty side effects. Non-steroidal anti-inflammatory drugs or NSAIDs can cause adverse effects, some of which can have severe side effects like stomach bleeding and heart attack, especially among the elderly but cherries do not irritate the stomach the way manufactured drugs do and they also contain compounds that keep platelets in the blood from clumping together.

Berries like blueberries, raspberries and strawberries are jam-packed with antioxidants and anti-inflammatory phytochemicals that help to reduce brain inflammation and brain oxidative stress, both of which have been associated with dementia and Alzheimers also confer some inhibitory effect on dementia indirectly.

Berries like cherries, raspberries and cranberries also contain some but lower amounts of this flavonoid.

Apple: There are now even more reasons to take an apple a day. Quercetin, found in abundance in the apples skins, has been found to protect the brain from damages associated with Alzheimers disease and other neurodegenerative disorders in studies conducted by Cornell University. Other studies have also suggested that eating apples may also help to reduce the risk of cancer. Capers, a common ingredient in Mediterranean cuisine, lovage and red onion are also excellent sources of quercetin.

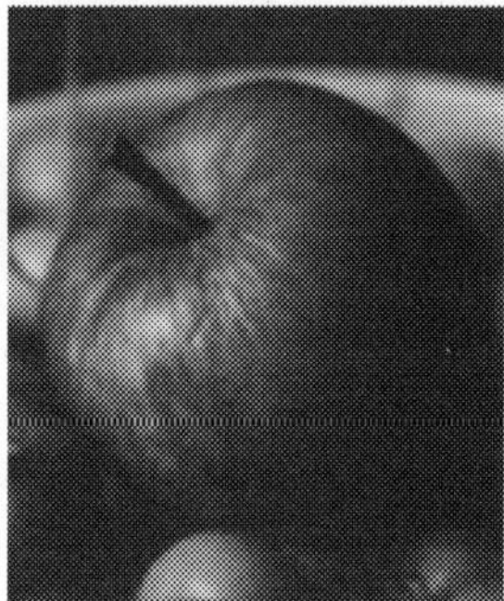

Avocados: For brain health, avocados are nearly as good as blueberries. Avocados contain mono-unsaturated fats, which contribute to healthy blood flow, the main requirement for a healthy brain. To include avocados to your diet, add ¼ to ½ of an avocado to one meal daily as a side dish.

Vegetables

Asparagus: A Korean study published in 2008 found that individuals who were folate deficient were 3.5 times more likely to develop dementia. What the more surprising is that those who were not folate deficient but had low folate measures were also at significantly increased dementia risk. To make sure, you get sufficient folate, make asparagus a frequent addition to your meals since one cup of these green spears will fulfill nearly 66% of your daily folate needs.

Other great sources of folate include citrus fruits, beans (be sure to sprout them to maximize their nutrients and enhance absorption by the body), broccoli, cauliflower, beets, lentils and leafy green vegetables such as spinach and turnip greens.

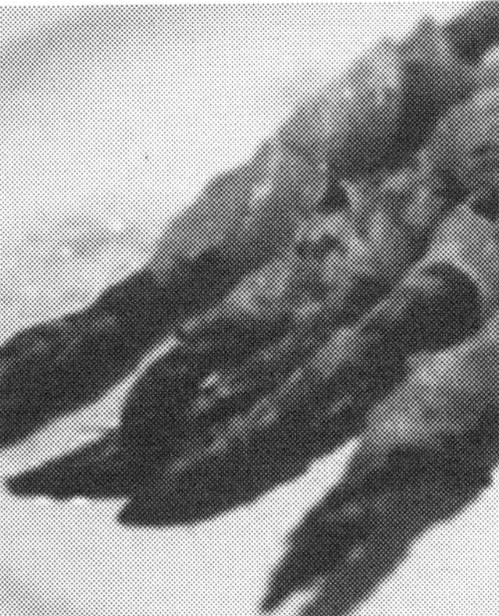

Tomatoes: Lycopene, an amazing antioxidant found in tomatoes, could help in protecting against free-radical damage to cells, which is believed to be a primary factor in cases of dementia and particularly the Alzheimers disease.

Red cabbage: Red cabbage is full of an antioxidant called polyphenol. Polyphenols reduce brain cell damage and are especially helpful in the prevention and treatment of Alzheimers disease.

Broccoli: It is labeled as super food due to its high overall nutrient content. It is a great source of vitamin K, which enhances cognitive function and improves brainpower.

Eggplant: Eggplant skin contains a nutrient called nasunin, which keeps our brain sharp by enhancing communication between our brain cells and messenger molecules. Remembering to use the skin pays tremendous benefits in vastly improved focus.

Spinach: Spinach slows down the effects of age-related declines in brain function and helps in protecting the brain from oxidative stress. Researchers suggest that a diet rich in spinach can significantly improve learning capacity and motor skills.

Spices

Garlic: Garlic is considered a mood elevator. Investigators testing garlic for its positive effects on blood and cholesterol noticed that garlic eaters also experienced a decided lift in mood. This was especially striking to German researchers at the University of Hanover who tested special garlic preparation on people with high cholesterol. Garlic takers experienced measurably, less fatigue, anxiety, sensitivity, agitation and irritability. This mood-lifting bonus of garlic is important and in sharp contrast to the adverse side effects of pharmaceutical preparations. Garlic supplements are the largest selling over counter drug in Germany.

Ginger: It is a close cousin of turmeric with similar anti-inflammatory properties. Though study on its effect on the brain is limited, gingers ability to reduce inflammation may also confer some inhibitory effect on dementia indirectly.

Turmeric: Turmeric is used extensively in Indian cuisine. Several neighboring countries also use it in curries. There are encouraging results to demonstrate that the low incidence of age related decline in mental function in Alzheimers disease could be due to the regular incorporation and intake of turmeric in food. Plaques in the brain are thought to contribute to the degradation of brain cells and lead to Alzheimers disease. A study found that curcumin in turmeric is an effective substance that prevent and removes plaques from the brain. Turmeric, an anti-inflammatory food used in India since ancient times, is commonly used as a spice in curry for the colour and earthy flavour. Countries with populations that eat curry regularly, such as India, have been observed to report lower rates of dementia. Add this inexpensive versatile spice into your meals as often as you can to cool inflammation and ward off dementia. Studies on the effect of turmeric on Alzheimers have b een discussed later in the chapter.

Turmeric powder

Adjunct foodstuffs

Yogurt: Calcium rich foodstuffs such as yogurt, milk and cheese improve nerve function. Yogurt contains an amino acid called tyrosine, which is responsible for producing the neurotransmitters dopamine and noradrenalin. In short, yogurt helps in improving alertness and memory.

Green Tea: Green tea is believed to be a wonderful beverage, and when freshly brewed, it enhances memory and focus and fights mental fatigue. It is rich in a type of flavonoid known as catechins, which can reach as high as 1 gram in a single cup. This helps you to relax mentally, yet also keeps your wits sharpened. Numerous studies have shown the health benefits of green tea including its ability to cut the risk of cancer as well as neurodegenerative diseases. Its ability to keep weight in check also makes green tea a popular choice in many health-promoting diets.

Green tea also helps in maintaining positive mood states and fights against many brain disorders. Polyphenols are powerful antioxidants found in green tea that can boost the availability of important signaling brain substance dopamine in brain circuits. Dopamine is vital in creating positive mood states. Polyphenols also help the brain and body in running smoothly by maintaining a steady supply of bodys primary fuel, *i.e.*, glucose. These powerful polyphenols also help in preventing cancer and heart attacks.

Chocolate: What can be better to end with? It is hard to believe that anything as incredibly delicious as chocolate can actually be equally good for you as well. Dark chocolate has powerful antioxidant properties and contains several natural stimulants, which increase the production of endorphins while enhancing focus and concentration. The stimulants found in dark chocolate also improve mood. It has high content of flavanols that facilitate blood supply to the brain and enhance cognitive skills. Studies carried out by Norwegian researchers found that the flavonoids in cocoa increase blood flow to the brain and may help to protect against conditions with reduced cerebral blood flow like dementia and stroke. To get the most benefits, buy chocolates with low sugar but high cocoa content (70% or more is the best).

However, while cocoa is rich in beneficial compounds, it is also high in saturated fats. Many a times, it is also paired with high-fat ingredients like full cream in desserts. Therefore, it is best to keep dark chocolate to a small once in a week treat.

Milk chocolate jump-starts impulse control and reaction time. It has also been known to improve visual and verbal memory. More is not necessarily better when it comes to chocolate. Unfortunately, this is one super food, which you have to indulge in moderation.

It is not possible to go into a detailed account on the effect of all these foods listed above on the brain. An attempt will be made to describe three foods namely turmeric, walnuts and brahmi on which extensive studies have been carried out and are considered the most important brain foods in this list.

Red Wine: The Norwegian study on dark chocolate also credited modest wine drinking for conferring protective effect on cognitive function and decreasing the risk of Alzheimers disease and dementia. This is due to the high levels of flavonoids and possibly other polyphenols such as resveratrol that are present in red wine.

However, alcohol is a double-edged sword that should be used with great caution. Because, ironically, many studies have also found that excessive alcohol intake can lead to dementia as well as a host of other serious health conditions such as cancer by triggering chronic inflammation. Hence, if you do drink, limit yourself to not more than one glass a day. It is also important to note that in the first few hours after drinking wine, alcohol may actually slow thinking and interfere with the memory temporarily. If you are preparing for an important test or going to sit for an examination, drinking wine is probably not the way to boost your performance.

Another school of thought recommends us to eat organically grown red and black grapes as mother earth has made them and get all the benefits of the constituents present in it. Conversion of grapes to wine is not likely to increase the beneficial effects on the brain.

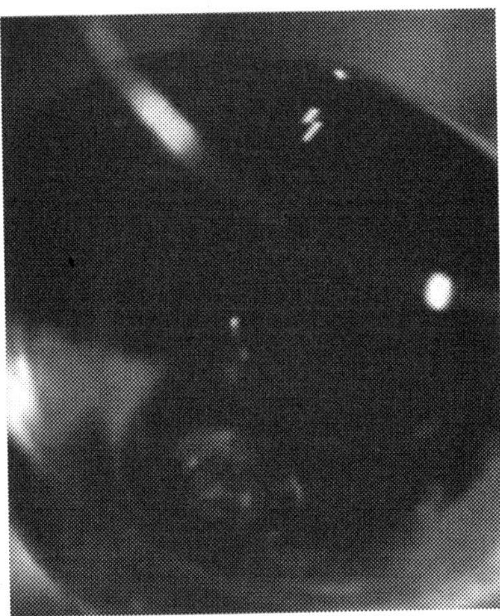

Food from Animal Kingdom

Eggs: Eggs indeed offer a very impressive nutritional profile for their 70 calories. They are a precious source of high-quality proteins and rich in vitamins and minerals but there is more. A substance known as choline, found in eggs, can help boost the memory center in the brain. Researchers have found choline to increase the size of neurons, which helps them fire electrical signals more strongly and rebound faster between firings. Two antioxidants found in egg yolk called lutein and zeaxanthin help in preventing the risk of age-related cataracts and macular degeneration, two of the most prevalent age-related eye conditions.

Seafood: You have probably heard that fish is brain food. The reason is that seafood is extra high in trace mineral selenium. How selenium influences mood is unknown but its antioxidant powers may be involved. In a study, involving elderly people given selenium plus vitamin E or other antioxidants had significant improvement in mood and mental functioning as well as increased blood flow to the brain. In addition, in a study of patients with early Alzheimers disease, antioxidants including selenium boosted mood and mental performance. However, these speculate that selenium may have some unknown neural function.

Clams, Oysters and Mussels

Shellfish can be eaten raw, baked, steamed, fried, or made into chowder. Clams provide the most vitamin B_{12} with 98.9 μg per 100 g serving, accounting for 1648% of the recommended dietary allowance (RDA). Mussels and oysters are also good sources of B_{12} providing 600 and 400% of the RDA, respectively. Researchers in the Oxford Project to Investigate Memory and Aging (OPTIMA) found that elderly adults with low vitamin B_{12} levels had more than four times the usual risk of Alzheimers. It is thought that when vitamin B_{12} levels are low, blood levels of homocysteine will raise, significantly increasing the risk for dementia, heart attack ad other ailments. To get a healthy dose of vitamin B_{12}, look no further than clams (this shellfish packs a whopping 98.9 μg of vitamin B_{12} in just 100 g serving).

Rich sources of vitamin B_{12} are found in seafood and animal sources including fish, shrimps, scallops, liver of most animals and beef. Lower levels of vitamin B_{12} can also be found in seaweeds, yeasts and fermented foods like miso and tempeh.

Wild Salmon: According to a study conducted by Tufts University in Boston, subjects who consumed an average of three servings of oily fish a week had almost 50% lower risk of Alzheimers disease and dementia. These individuals had significantly higher levels of omega-3 fatty acid, *i.e.*, docosahexaenoic acid, in the blood than those who took less fish in their diets. Salmon is an excellent source of omega-3 essential fatty acids as well as other important nutrients such as vitamin D and B_{12} that can help to prevent neurodegenerative disorders but as wild caught salmons, frozen or canned, generally contain more omega-3, less omega-6 and fewer toxins, you are better off buying the wild version whenever possible. Aim to eat at least 2 to 3 servings of oily fish each week to supply your body with inflammation-fighting compounds. If you are concerned about environmental toxins such as PCB in seafood, you can also consider taking whole fish oil supplement derived from wild salmon.

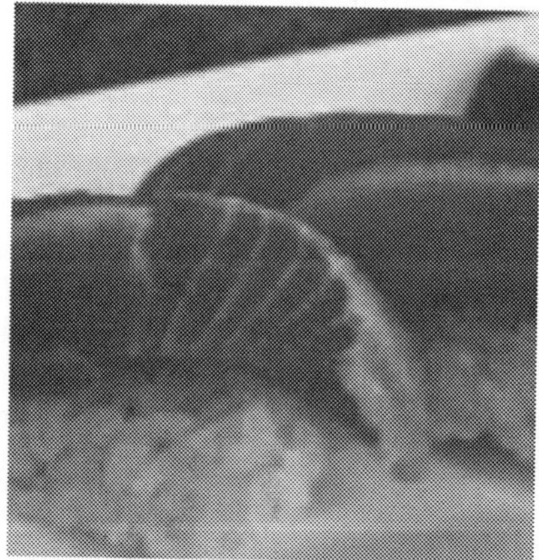

Anchovies and sardines are also very good sources of omega-3 fats. For people who do not eat fish, walnuts, flaxseeds and dark green leafy vegetables are rich in plant-based omega-3 precursor, *i.e.*, α-linolenic acid (ALA) but they lack eicosapentaenoic acid (EPA) and docosahexaenoic acid (DHA), the two primary omega-3s. Though technically ALA is converted by the body to EPA and DHA, in reality, the conversion rate is very low (only about 1 to 2%) and is further reduced if your omega-6 intake is high. Thus, relying solely on walnuts, flaxseeds and dark green leafy vegetables may not provide your body with adequate omega-3 fats. In this case, you can boost your omega-3 intake either by selecting more EPA and DHA-fortified foods, take vegan omega-3 supplements, or both.

Medicinal Plant

These are described below in detail:

Brahmi: Brahmi is well known in our country as a brain food. In addition to commonly used foodstuffs to which most of us incorporate in our regular diet are believed to be good for brain functioning. There is a special herb known as brahmi, used in *Ayurvedic* preparations for its beneficial effects on the brain. Unlike other foodstuffs described so far, this important brain food is recognized in India for the centuries. However, the fresh herb is not generally available in the market like other greens that we use as vegetables. This will be described in detail here. If some garden space is available, it can easily be grown in the garden. If not, it can also be grown in pots and used on a regular basis.

Bacopa monnieri

Brahmi has a pleasant flavour and can easily be used in salads, soups and *chatnis* In South India, *Brahmi chatni* is eaten with rice. Some people eat a few fresh leaves of *Brahmi* first thing in the morning for better memory. A wide variety of *Ayurvedic* preparations is available in the market.

Species and names known as Brahmi: *Bacopa monnieri*, *Centella asiatica* (gotu kola) Babies tear, *Hespestis monniera*, Nirbrahmi, Indian Pennywort, Jalanevari and water hyssop are the names that generally are used for *Brahmi*. In English, it is commonly called as the *Herb of Grace*.

Cultural importance: Brahmi is the name derived from the Lord Brahma. It is said that the use of Brahmi for the memory goes back 3000 years or more in India. Brahmi represents the creative energy of this earth.

Location: Brahmi is generally found in marshy areas near streams and ponds throughout India especially in the Northeastern regions. It is found at elevations from sea level to altitudes of 4,400 feet in Uttar Pradesh, Punjab, Haryana, Bihar, Bengal, Tamil Nadu, Kerala, Karnataka, foot hills of Himachal Pradesh and Uttaranchal. It is also widely grown in the Bandhavgarh National Park in India. According to Dr. John

Anne, the traditionally venerated *brahmi* of *Ayurvedic* medicine is popular as water hyssop in the west. Other popular names used in different parts of the world are thyme-leafed gratiola and moneywort. It is an indigenous Indian herb, which grows in large numbers near the marshy areas. *Brahmi* is a small creeping herb with white or pinkish flowers. Apart from India, it also grows in China, Nepal, Sri Lanka and even as far west as in the Florida state of United States.

Centella asiatica

Cultivation methods: *Brahmi* is propagated by seed, cuttings and also by the root division. Before establishing the plant cutting, the field should be ploughed properly and must be free of weeds. The cuttings should be 4-5 cm long and contain roots and leaves. Hand weeding is required every 5-20 days. It requires moderate to high temperature (33-42°C) to grow and high humidity (65-80%) and prefers wet spots.

Though*brahmi* is considered a bog plant that likes wet spots, it is found on herb farm, will grow in pots, or in the garden in shade. It can be grown in the pots in the house garden and can also grow in both the sun and shade. Fertilize plants regularly with seaweed or other organic fertilizer. Plants can be grown in hanging pots, and as the stems cascade over the sides, they are easily nipped off for use. Plants can be grown in hanging pots and as the stems cascade over the pots. *Brahmi* being shallow rooted requires water from time to time and organic fertilizer regularly for its growth. It gives flowers and fruits in the summer season.

Medicinal uses: In addition to its reputation as a brain food, it is also used for a wide variety of health problems in traditional Indian medicines for over 3000 years. These include treatment of bronchitis, chronic cough, allergies, asthma, hoarseness, arthritis, rheumatism, backache, inflammatory conditions, fluid retention, blood cleanser, high cholesterol levels, constipation, hair loss, boils, ulcers, fevers, digestive problems, epileptic fits, depression including post-natal depression, diarrhea, impotence, premature ejaculation, irritable bowel syndrome, frigidity, mental and physical fatigue and many more. It is also used to treat all sorts of skin problem like eczema, psoriasis, abscess and ulceration. It stimulates the growth of skin, hair and nails. *Brahmi* is known to help cancer patients. It is used for relief from stress and anxiety. According to the *Ayurveda,brahmi* has antioxidant property. It has been reported to reduce oxidation of fats in the blood stream, which is the risk factor for cardiovascular diseases. Most important use of *brahmi* is that it is considered as the main rejuvenating herb for the nerve and brain cells.

Brahmi's medicinal propertiesare derived from the two main alkaloids, *i.e.*, brahmine and herpestine, present in it. Apart from these, there are other chemicals in the *brahmi*, including several acids, saponins and esters, which make it a valuable herb in *Ayurvedic* medicine. In *Ayurvedic* parlance, the chemicals present in the herb are also known as *brahmi*.

Properties

Brahmi is known in *Ayurvedic* medicine for the following curative properties:

- *Brahmi* is a bitter and cooling herb with an astringent taste.
- Children in Indian families have been consecrated with *brahmi* for the centuries to give them an intelligent mind. Recent studies have shown that *brahmi* improves the intellect.
- *Brahmi* is both a nerve as well as a cardio tonic.
- *Brahmi* is a diuretic, and therefore, helps in solving urinary problems.
- *Brahmi* helps in adapting and acclimatizing to a new place.
- *Brahmi* is a nootropic, *i.e.*, a drug that increases cognitive ability.
- *Brahmi* has antioxidant properties due to which it protects the nutrients from getting oxidized (and hence destroyed) in the body.
- *Brahmi* is a blood-purifying agent.
- An ethanolic extract of the herb has given evidence of anti-carcinogenic activity in humans but this is yet to be concretely studied.

Uses of *brahmi*

Brahmi has been used for ages for the treatment of following diseases:

Dandruff: *Brahmi* oil replenishes the hair roots. If *brahmi* oil is applied each night before going to bed, it will cure all dandruff problems and also make the hair stronger.

Hair Loss: *Brahmi* is renowned in *Ayurvedic* medicine for the proper growth of hair. Its antioxidant properties allow the proper nourishment to reach the hair roots and thus promote their growth. *Brahmi* is an essential component of many herbal oils. Using such oils make the head cool as well as allow the hair to grow luxuriantly and with luster. The oil also prevents split ends and discolouration of the hair.

Epilepsy: The entire *brahmi* plant is useful in the treatment of epilepsy. The juice of this plant is extracted and used in the preparation of medicine. In the treatment of epilepsy, *brahmi* is traditionally used in conjunction with other plant products.

Insomnia: *Brahmi* is a cooling agent. When applied on the scalp in the form of oil, *brahmi* relaxes the brain and induces good and restful sleep throughout the night. The powdered mixture of *brahmi* with other plant products is to be consumed thrice a day in the dose of one teaspoon. The powder can also be taken dissolved in a cup of water or milk.

Mental Problems: Perhaps the most important use of *brahmi* is in the treatment of several mental problems in humans. Taken on a regular basis, *brahmi* helps to increase the concentration of humans and improves their memory and retention capacity. It is believed that consecrating children with *brahmi* oils makes them into intelligent individuals. The regular use of *brahmi* oil is effective in treating mental imbalances, emotional disturbances and also in the prevention and cure of geriatric mental problems such as amnesia and Alzheimer's. *Brahmi vati*, an important *Ayurvedic* preparation containing *brahmi* along with *jatam Brahmi*, has been used for the treatment of the following diseases:

Syphilis: *Brahmi* has antibacterial properties. A paste of its leaves is applied on syphilitic sores. This helps the sores to heal quickly. Regular application of the paste of *brahmi* leaves can treat the syphilis completely.

Other uses: *Brahmi* is rich in vitamin C and can be used in the salads, soups and sandwiches. *Brahmi* oil restores and preserves the memory. In India it is given to the infants to boost memory power intelligence.

Special *Ayurvedic* Preparations

Brahmi is the main ingredient in followings *Ayurvedic* preparations:

- ***Brahmi ghrita*:** *Brahmi* is boiled in pure cow's ghee along with some other herbs. This preparation is known as *brahmi ghrita*, which is taken orally one teaspoon twice a day on an empty stomach for people who are suffering from amnesia.
- ***Brahmi taila*:** *Brahmi* oil is used for application on the head. Besides *brahmi*, this oil can also contain amalaki, camphor, *etc.*
- ***Brahmi vati*:** It is prepared by mixing the powder of *brahmi* with *jatamansi* (muskroot) and serpagandha (rauwolfia). This is used as a brain tonic in improving mental capacity, becoming more active and reducing anxiety and stress.

Precautions: *Brahmi* has some drawbacks, for which its use must be restricted. An excess of *brahmi* prevents the oxidation of fats in bloodstream. This makes the fats accumulate in the blood, increasing the risk of cardiovascular disorders.

Holistic Approache

While western medicine excels in treating acute problems, the holistic approach, long favoured by the eastern healing systems aims to prevent problems before they cause illness. Holistic treatment focuses on the whole person, including life style and mental state. To enjoy good health and prevent physical and mental diseases, include yoga, *pranasyama* and meditation in your day to day life.

Brain Food in *Ayurveda*

The recommended *Ayurvedic* diet is vegetarian. This is supported by the recent investigations. The link between animal products and many different diseases is as strongly supported in the scientific literature as the link between cigarette smoking and lung cancer. It was found in a carefully designed study that subjects, who ate meat, including poultry and fish, were found to be twice as likely to develop dementia (loss of intellectual function with aging) than their vegetarian counter parts, *e.g.*, grains, beans and legumes provide nutrients for the brain, hence, be sure to include plenty of cooked leafy greens. Several servings of fruits and vegetables are recommended. Sweet juicy fruits are excellent cleansers and they help in flushing the toxins out of the body. Freshly squeezed fruit juices deliver nutrients quickly to your mind and body.

According to *Ayurvedaic* system of medicine, ghee is considered a good fat. Hence, fat should not be entirely excluded from the diet since fat performs some essential functions in the body, including brain. The Council of *Maharishi Ayurveda* Physicians recommends ghee, which is valued for its ability to transport the benefits of herbs and spices to the brain and to other parts of the body. Use ghee in moderation. Include some dairy products if you are not lactose-intolerant, such as whole organic milk and fresh yogurt. Almonds and walnuts are also excellent brain foodstuffs.

Cut down on *empty* junk food articles and sugared drinks that interfere with the absorption of nutrients by the body. Sweet fruits, raisins and dates are a much better option when you need something sweet.

Spices are crucial additions to a brainpower-enhancing diet. Many spices enhance digestion and facilitate the clearing of toxins from the body. This means that the channels of the body, including those that serve the brain, will be clear. Black pepper in small quantities is particularly recommended for clearing channels of the brain and promoting free flow of oxygen to the brain.

Exposure to heavy metals such as lead and aluminum has been shown to be damaging for the brain. Also avoid alcohol and caffeine. According to *Ayurveda*, good eating habits are as important as what you eat. Sitting down to eat, focusing on the food rather than on work or other activities and savoring the meal are important. Ice-cold drinks disrupt digestion and impair the assimilation of nutrients by the body and thus should be avoided.

Isolating an active substance from a food or a plant and taking it in the form of a pill or a capsule can sometimes do more harm than good, *e.g.*, research has shown that β-carotene, an antioxidant, taken as a supplement, actually increases the risk of some forms of cancer. It is best to take your antioxidants the way Nature intended, in food articles, where they occur in a balanced form.

Use Herbs and Aromas to Enhance Mental Potential

Ayurveda talks about a special class of herbs called *Medhya* herbs- the herbs especially useful for mind. These herbs individually enhance learning, retention and recall and also promote the coordination among the three factors. They provide powerful nourishment for the brain.

Shankhapushpi, Aloe weed, the authentic *Brahmi* or *Herpestis monniera* and Indian Pennywort or *Centella asiatica*, also known as Gotu Kola, are some well-known *Ayurvedic* herbs, which are known to promote health and functioning of mind, *i.e.*, aloe weed enhances memory and has the problem-solving ability even under situations of day-to-day stress. Gotu Kola is a memory-enhancer with *vayasthapana* or age-related decline.

Synergistic formulas that contain not only these herbs but also other supporting and balancing herbs such as *ashwagandha* (winter cherry) to help in boosting resistance against stress are ideal, especially for individuals in high-stress job environments who need to be at the peak of their mental potential on a consistent basis.

Youthful Mind helps in maintaining mental clarity and memory no matter what your age. Some essential oils are excellent aromatherapy for the mind. Try rosemary for mental clarity and alertness or basil oil to help dispel mental fatigue and enhance your mind's innate ability to focus. Peppermint oil and lemon oil are also effective pick-me-ups for the mind.

Manage stress levels

We live in difficult times. Stress and anxiety, whether caused by time pressures or difficult work relationships or worrying about losing a job, can have a significant negative impact on mental performance. A calm mind, serene and centered in its equilibrium, is unquestionably superior in terms of focus, attention span, retention of knowledge and problem-solving ability. It is important to manage stress before it becomes a severe problem since stress has been known to damage or destroy brain cells, and many prescription pills for anxiety can have side effects that can further erode mental performance.

Planning tasks ahead and managing timelines, taking occasional short (even five minutes) breaks during periods of intense mental activity and maintaining a healthy balance between work and relaxation can all help. The transcendental meditation technique, yoga and *pranayama* practiced twice a day for 20 minutes, has been shown to be extremely effective for managing stress. Listening to relaxing music, diffusing aromas that help relax you, such as sandalwood or lavender, or a quiet evening stroll, alone or with a nurturing companion, can all help in balancing the mind and emotions.

The daily *Ayurvedic* massage is excellent for both body and mind. It is a cleansing, toning and rejuvenating experience that can charge you up to meet a full day ahead.

Effects of *Pranayama* on the Brain

As we have mentioned elsewhere in this article, brain is the greediest organ for oxygen consumption or it is the most important brain food. *Pranayama* or controlled breathing has been practiced in our country for the centuries by Yogis, revived and popularized today by Yoga Guru Swami Ramadev. Baba Ramadev illustrates the beneficial effect of *Pranayama* on various types of chronic diseases including hypertension and diabetes in addition to stress release. Increased oxygen intake during *Pranayama* is bound to have positive effect on the brain function. Therefore, it is of interest to see what information is available in this area. Sannyasi Sivagyana (Richard Budden) of Australia has carried out research on the effect of *pranayama* on brain. According to him, these practices contribute initially to changing the physiological state of brain and are said to awaken *prana* or psychic centers, within the human body. This confirms the ability of *pranayama* to influence an individual's brain activity. It has been demonstrated that extensive *pranayama* practices can significantly influence the physical, *pranic*, mental and psychic aspects of human brain.

Swami Niranjanananda Saraswati defines prana or vital energy as the essence of all created, manifest forms whether animate or inanimate, the force, which determines the existence of matter and the elements. Prana nigraha is the manipulation of breath to control prana. When practiced regularly, prana nigraha leads to pranayama or expansion of the vital energy. Without prana, the body and mind are dead.

Modern science states that out of ten areas of brain we are using only one at our present stage of evolution. To use the other 90% involves the distribution of *prana* to awaken these areas. The subconscious mind and its relationship to the conscious mind are dealt with in *pranayama* by the establishment of an interface between the conscious and the subconscious minds in the area of brain called the reticular activating system (RAS), which is trigger for other parts of the brain. Man is able to affect the RAS through the breath only. No other function of the autonomic nervous system can be controlled by conscious human activity. Control of brain through the RAS by means of conscious breathing is a method by which other functions of the body may be controlled, *e.g.*, heart rate, blood pressure, digestion, excretion and absorption. Therefore, control of the subconscious is achieved through conscious activity of *prana nigraha*, and then, *pranayama*.

Four *pranayama* practices are examined for their effects on brain or other parts of the human body. These practices are selected based on their importance in the practice of yoga and their stated influence on physiological and psychic bodies.

Kapalbhati: Van Lysbeth states that kapalbhati influences the circulation of blood within the brain. Kapalbhati changes the volume of brain according to the respiratory rhythm, and therefore, increases the irrigation of brain matter. Normal respiration consists of 12-18 massages per minute, whereas, kapalbhati can involve up to 120 massages per minute, which leads to a significant increase in blood volume throughout, and thereby, improves irrigation of the brain.

The capillaries are opened up, and the brain cells related to the pineal and pituitary glands receive significant stimulation. It is logical to conclude that increased brain irrigation with blood is accompanied by elevated *pranic* levels and ensures even and harmonious distribution of *prana* throughout the body.

Van Lysbeth supports this conclusion as follows: "Together with the acceleration of blood circulation in the whole body, this stimulation of the brain and thereby of the central nervous system produces special *relation* of the body that invigorates and tonifies each cell."

Kapalbhati reduces the ratio of outward breath to the inward breath in kundalini yoga to one quarter. This is known to increase the breath control, stretching it to the limit, and dramatically affects the carbon dioxide, chemical, acid and alkalis in the blood. Carbon dioxide is the trigger for inhalation of breath into the lungs and the body and therefore, the body is very sensitive to carbon dioxide levels.

Kumbhaka: In practice of kumbhaka or breath retention, which may be *antar* (internal) or *bahir* (external) tolerance to starvation of oxygen and buildup of carbon dioxide is achieved. Kumbhaka, practiced over a period, will allow the body to retain carbon dioxide and become accustomed to reduced oxygen levels to achieve hypometabolism, *i.e.*, a slowing down of the metabolic rate. The production rate of carbon dioxide is thereby reduced which causes a subtle effect to take place with conscious control of breathing. This effect influences the brain and body chemistry and reduces the need to breathe when carbon dioxide buildup is experienced.

External *kumbhaka* also affects the body physiologically by causing the mental process to stop because of the vacuum created inside the body. This action is very useful in the practice of *pratyahara*, sense withdrawal and *dharana*, concentration, as a prerequisite to achieve the state of meditation.

Kumbhaka stops vital body rhythms and affects the brain waves. Control of the brain waves is the key to control all brain rhythms. While the effects of *bahir kumbhaka* are many, in broad terms, the body and mind learn to stay calm under stress.

Nadi shodhana or Anulom vilom: *Kumbhaka* is used in the practice of *nadi shodhana* or alternate nostril breathing. *Nadi shodhana* is the *perfect balancing practice*, which stimulates equally the left and right side of the brain and body, the major *nadis* or *pranic* channels, are balanced which, in turn, modifies the human thinking process to balance introversion and extroversion. The ancient yogis have recorded that once these are balanced and purified, the central *nadi* begins to flow, leading to increased awareness and the state of meditation.

Nadi shodhana imposes a rhythm on the brain and the *nadis* over irregular state that normally exists. Modern living has removed the regular rhythms of nature from the human body and *nadi shodhana* assists in bringing the body, *prana* and mental activity into balance. Research has shown that *nadi shodhana* affects the brainwaves by superimposing a regular sine wave over the normal irregular brain activity, imposing a discipline on irregularities of the mental process, and eventually, the autonomous body rhythms.

Kumbhaka in *nadi shodhana* places a momentary block on the body rhythms, changing the usual carbon dioxide/oxygen relationship, and thereby, affecting the whole system.

Ujjayi or the psychic breath, produced by a slight contraction of the throat, has a subtle effect on brain activity *via* four processes:

- *Ujjayi* increases pressure of air in the lungs and expands effective use of the lungs. This ensures transfer of oxygen to each cell within the lungs rather than a significantly smaller percentage used during normal respiration.
- Increased oxygen transfer in the lungs enhances blood flow throughout the body, while the body is in a relaxed state. The effect is similar to that achieved when the body is physically active with the advantage of the whole body being in a relaxed state.
- Conscious awareness is transferred into the unconscious mind, which affects the nervous system governing respiration. A smooth rhythm is exerted on the nervous system, which has a profound effect at psychic level of the mind.
- The contraction of throat caused by *ujjayi* affects the carotid sinuses to regulate blood pressure in the arteries. *Ujjayi* exerts a slight pressure on carotid sinuses, which over time lowers the blood pressure, which leads to reduced tension and slows the thought processes of the mind.

The effect of the practices of *prana nigraha* outlined above has been substantiated in part through work carried out and reported at 5th annual convention of the International Association for Religion and Parapsychology in 1977. The research revealed that Ramananda Yogi, who had practiced *pranayama* for many years, had the ability to control heart muscle itself and was, therefore, able to control his heart function. During *pranayama*, Ramananda Yogi was able to reduce his pulse rate from 100 per minute to 65-80 per minute, although such changes would be dangerous for persons who had not practiced *pranayama*. It was also concluded at the conference through biological tests that Ramananda Yogi was able to control his basal metabolic rate through *pranayama*.

The effects of *pranayama* on brain as detailed by Swami Niranjanananda and the results of clinical trials carried out by the International Association for Research for Religion and Parapsychology, substantiate the profound effects of *pranayama* on physical and mental human body.

Today, Yoga Guru Swami Ramadev advocates practice of *pranayama* for a variety of diseases and for physical and mental well being or self-healing without resorting to allopathic drugs. Millions of his followers have vouched for the benefits of regular practice of *pranayama*.

It is recognized by research that the effects of the practices of *pranayama* on a regular basis support the ancient yogic view that *pranayama* exerts profound effects on the human brain. The limited experience of writer also supports the view that the practices of *pranayama* can have subtle effects on brain, human well-being and influence the individual's level of spirituality.

Age related decline in brain function

Some of us will have to cope with chronic illness at some stage in our lives ranging from high blood pressure, arthritis, hearing loss or physical or mental debility. At a given time, about half the population is suffering from a chronic illness of some kind. As we live longer, more and more of us may have to learn to manage a condition that may not have a cure. Some chronic illnesses have little immediate effect on everyday life, *e.g.*, high blood pressure has few obvious symptoms, but it must be regarded as a serious condition since it can cause heart disease and stroke. By contrast, Alzheimer's disease will affect a person's ability to live an independent life in quite a short time. Therefore, it is desirable to try to prevent or delay the age related decline in physical and mental well being

As far back as 1906, Alois Alzheimer observed plaques and tangles in brain in the first reported case of disease, though he did not know that they were due to amyloid and tau protein. Today, plaques and tangles are well documented as the key features of Alzheimer's disease (AD), and they seem to occur in certain parts of the brain more than they occur in other parts.

The problem is that many people start to accumulate plaques and tangles just as a matter of getting old. The question arises, what else besides aging turns the brain toward Alzheimer's disease in some instances and toward vascular dementia in others. It is possible that in some people *silent* strokes in the brain, even though they are clinically undetectable, predispose a person to a more aggressive form of Alzheimer's disease. These people waste a lot of time trying to find things they have kept somewhere or trace their way back home. They find it extremely stressful to take decisions and are unable to live independently.

Alzheimer's disease and Dementia in India

In India, not much attention is paid to the problems associated with the brain functions generally encountered by the aged. The prevalence of Alzheimer's and dementia is reported to be low in India and other developing countries compared to the developed countries. There are limited studies in this area of human health in India. With increasing medical care, life expectancy, urbanization and nuclear family, the mental well being of Indians is affected to a great extent. In recent years, one can see more elderly, especially women suffering from these devastating diseases. This not only affects the individual but also the family members and the caregivers. We still do not know why some elderly are affected, while others manage well through out their life. The incidence is more prevalent in urban population compared to rural. Is it due to the isolation of elderly in the urban population?

Does the higher incidence of Alzheimer's and dementia in the elderly in developed countries and urban population suggest the type of life style including intake of junk food and processed food or the side effects of prolonged use of certain medications such as anti-hypertensive for decades?

Let us see what information is available in India in this area to understand the gravity of situation. We can then see how one of the most important factors, food that are beneficial for fine-tuning our brain can be chosen. With the ancient wisdom from *Ayurveda* and modern research, we should be able to incorporate brain food into our diet to activate brain function of the young so that they excel in their studies and profession and also avoid or delay of dementia in our elderly populations so that life may be more comfortable for them as they grow old.

Incidence of Alzheimer's Disease in a Rural Community in India

The Indo-US study was carried out in the department of Epidemiology, University of Pittsburg, United States of America in 2001 to determine overall and age-specific incidence rates of Alzheimer's disease in a rural, population-based cohort in Ballabgarh, India and to compare them with those of a reference United States population in the Monongahela Valley of Pennsylvania. A 2 years prospective epidemiologic study of subjects aged ≥ 55 years utilizing repeated cognitive and functional ability screening, followed by standardized clinical evaluation was standardized against the age distribution of the 1990 United Sates Census. The overall incidence rate in those aged ≥ 65 years was 4.7 per 1000 persons per year, substantially lower than the corresponding rate of 17.5 per 1000 persons per year in the Monongahela Valley. These are the first Alzheimer's disease incidence rates to be reported from the Indian subcontinent, and they appear to be the lowest among ever reported. However, the relatively short duration of follow-up, cultural factors and other potential confounders suggest caution in interpreting this finding.

Does use of turmeric in curries explain low occurrence of Alzheimer's disease in India?

It is suggested that the curcumin, a naturally occurring phytochemical in turmeric, the earthy spice that gives curry its intense yellow colour, may prevent, delay, or even reverse the build up of neural plaque, which is implicated in Alzheimer's Disease.

The World Health Organization's regional forum article entitled "Alzheimer's disease: of Emerging Importance" points out a much lower incidence of Alzheimer's disease in India.

Studies in Indian villages where curried food is a staple found a less than 1% incidence of Alzheimer's disease in people over the age of 65 years. Turmeric, one of the ingredients in curry, contains curcumin, a bioactive phytochemical that some researchers believe may, at least in part, explain why Alzheimer's disease is so uncommon in India compared to Western countries.

Alzheimer's affects more than 4 million Americans and many more millions worldwide. It is most prevalent among the elderly. The prevalence of Alzheimer's among adults of age between 70 and 79 in India, however, is 4.4 times less than the rate in the United States.

A joint UCLA-Veteran's Affairs study concludes that in view of curcumin's efficacy and apparent low toxicity, this Indian spice component shows promise for the prevention of Alzheimer's disease.

Curcumin has been used for thousands of years as a safe anti-inflammatory in a variety of ailments as part of Indian traditional medicine, said Gregor Cole, the lead UCLA researcher, as quoted by News-Medical Net. "What we really need, however, are clinical trials to establish safe and effective doses in aging patients," said Cole. As a culinary spice, turmeric not only imparts a rich yellow colour but also a characteristic earthy flavour with dill-like undertones. Often called *poor man's saffron* because of its yellow colouring potency, turmeric is an ingredient in many of the-shelf products, and of course, is one of the components in curry, the spice blend that predominates in the cuisines of India, Pakistan, Nepal, Bangladesh, Kashmir and even Jamaica, and which can be readily found in ethnic restaurants in cities around the world.

With turmeric readily available and the enormous upside health benefit potential, it may well be worthwhile to introduce more turmeric into your diet- be it in mustard, other condiments, curries, or take standardized curcumin extract supplement such as curcumin capsules, pills, or similar supplements. Commercially curcumin preparations are available as Curcumin Capsules, Curcumin Pills, and Curcumin Standardized Extract Supplements.

Alzheimer's disease in Minority Community

In an International study, a group of native Africans (Yoruba tribe) in Ibdan, Nigeria was compared with a group of African Americans in Indianapolis. Epidemiological data of a 5 years study showed a lower prevalence and incidence of Alzheimer's disease in the native tribe living in Africa. This study is not an isolated data and has been supported by other investigators. These studies also found that there was no association with either life expectancy or the presence of Apo-E allele, a risk promoter gene associated to late onset of Alzheimer's disease. If one inherits two copies of the gene, he has 95% chances of developing late onset of Alzheimer's disease by age of 85 years.

If it were neither life expectancy, nor the ApoE-4 allele, it would be of interest to see what is accounting for this difference. Is it diet, environment, or lifestyle? These questions continue to be debated in the literature and in scientific circles.

There seems to be something about living in a developed country. Currently, there is no consensus among researchers about what may account for the higher incidence of Alzheimer's and dementia in ethnic-minorities in the United States. Therefore, we learn from studies comparing similar groups outside the United States. In addition to the Indianapolis-Ibadan study, data are there from an ongoing study tracking dementia prevalence in eastern India. It turns out that the prevalence of Alzheimer's disease in rural India is very similar to the prevalence in Ibadan, Nigeria. Therefore, this is another developing country with low prevalence. In India, it is thought that it may be due to dietary influences since many of the people in India are vegetarian and eat curry frequently. Some studies have suggested that Indian curries have an ingredient called curcumin, an antioxidant, which may be protective against dementia

These emerging data suggest that in developing countries, the rate of dementia is low. There is something about how the disease develops in North America, England, Australia and other developed countries, which is different than how it develops in developing countries. One important issue is that whether this may be in part because there is more stroke-related dementia (also called vascular dementia or multi-infarct dementia) diagnosed in developing countries. Experts believe that this may not be true.

These observations would seem to point to some kind of multi-factorial influence, *i.e.*, combinations of dietary, gene expression, and/or environmental factors. The question is when does the prevalence change? In other words, if you have low prevalence in an African cohort, does it change in first-generation immigrants to developed countries? If Yoruba Nigerians from Ibadan were to immigrate to New York or Indianapolis, does their risk for Alzheimer's disease change in that same generation, or does the risk become increased only in subsequent generations? If it changes in subsequent generations, which would suggest that might be there are lifestyle, dietary, or environmental factors that come into play in the United States to account for the difference. Is it because food is grown, processed and consumed differently, or are there other factors that have not yet been identified by epidemiologists who study disease trends in populations?

Another possible explanation is exercise. In developing countries, people tend to be more physically active, not in the sense of *working out* but because there is typically less access to automobiles and public transportation. It is not yet clear how non-genetic factors affect risk.

Most epidemiologists agree that the two biggest risk factors for Alzheimer's disease are aging and family history associated with genes or is it related to vascular risk factors such as diabetes, high blood pressure, or high cholesterol? What risk can be attributed to each of these factors? Is hypertension more significant than diabetes or is it the co-occurrence of these conditions, which is problematic? Are there common or unique biomarkers to the type of dementia that appears in these populations? These issues have not been sorted out.

While it has not been proven, the clinical sense is that vascular risk factors are probably different in minorities *vs.* whites. Researchers are debating whether cognitive impairment in an elderly person is stroke-related, so-called *pure* Alzheimer's disease, or a mixture of both (called mixed dementia). Many people think that these exist on a continuum or a spectrum. Pure Alzheimer's disease refers to the presence of particular degenerative processes confirmed by histological changes of varying density seen in selected brain regions upon autopsy. The changes include plaques, which are outside of neurons and are primarily related to a protein called beta-amyloid and neurofibrillary tangles, which occur inside neurons and are primarily related to a protein called tau.

Harness the power of antioxidants

The hundred billion cells in your brain continuously use oxygen to stoke the mental fire but because the brain is such a heavy user of oxygen, it is particularly susceptible to oxidative damage. Free radicals, reactive oxygen-based chemicals that thrive on stress, pollution and chemicals in your food can build up, damaging cells and brain function over time. Therefore, antioxidants are crucial for continued effective functioning of the brain. Recent research showed that blueberries are powerful brain food, primarily because of their antioxidant value.

Antioxidant Prevention of Alzheimer's Disease: Lessons from *In Vitro* and Animal Models

In contrast to conventional NSAID/antioxidants such as ibuprofen, omega-3 fatty acid DHA and vitamin E, the unconventional NSAID/antioxidant curcumin was effective in lowering oxidative damage, cognitive deficits, synaptic marker loss and amyloid deposition in cell culture and animal models for Alzheimer's disease.

Extensive research in the last few years has shown that the pathway that activates this transcription factor can be interrupted by phytochemicals derived from spices such as turmeric (curcumin), red pepper (capsaicin), cloves (eugenol), ginger (gingerol), cumin, anise and fennel (anethol), basil and rosemary (ursolic acid), garlic (diallyl sulfide, S-allylmercaptocysteine, ajoene) and pomegranate (ellagic acid).

Beneficial Effects of Omega-3 on Brain

In addition to helping protect us from heart disease and cancer, a balanced diet and regular exercise can also protect the brain and ward off mental disorders.

Food is like a pharmaceutical compound that affects the brain, said Fernando Gómez-Pinilla, a UCLA professor of neurosurgery and physiological science who has spent years studying the effects of food, exercise and sleep on the brain. Diet, exercise and sleep have the potential to alter our brain health and mental function. This raises the exciting possibility that changes in diet are a viable strategy for enhancing cognitive abilities, protecting the brain from damage and counteracting the effects of aging.

Gómez-Pinilla analyzed more than 160 studies about food's affect on the brain. The results of his analysis appear in the journal Nature Reviews Neuroscience. "Omega-3 fatty acids found in salmon, walnuts, kiwi fruit and flax seeds provide many benefits, including improving learning and memory and helping to fight against such mental disorders as depression and mood disorders, schizophrenia and dementia" said Gómez-Pinilla, a member of UCLA's Brain Research Institute and Brain Injury Research Center. Synapses in the brain connect neurons and provide critical functions. Much learning and memory occur at the synapses. Omega-3 fatty acids support synaptic plasticity and seem positively to affect the expression of several molecules related to learning and memories that are found on synapses. Omega-3 fatty acids are essential for normal brain function, he added.

Dietary deficiency of omega-3 fatty acids in humans has been associated with increased risk of several mental disorders, including attention-deficit disorder, dyslexia, dementia, depression, bipolar disorder and schizophrenia. A deficiency of omega-3 fatty acids in rodents results in impaired learning and memory. Children who had increased amounts of omega-3 fatty acids performed better in school, in reading and in spelling and had fewer behavioural problems.

Preliminary results from a study in England show that school performance improved among a group of students receiving omega-3 fatty acids. In an Australian study, 396 children between the age of 6 and 12 who were given a drink with omega-3 fatty acids and other nutrients (iron, zinc, folic acid and vitamins A, B_6, B_{12} and C) showed higher scores on tests measuring verbal intelligence and learning and memory after six months and one year than a control group of students who did not receive the nutritional drink. This study was also conducted with 394 children in Indonesia. The results showed higher test scores for boys and girls in Australia but only for girls in Indonesia.

Getting omega-3 fatty acids from food rather than from capsule supplements can be more beneficial, providing additional nutrients. Scientists are learning which omega-3 fatty acids seem to be especially important. One is docosahexaenoic acid (DHA), which is abundant in salmon. DHA reduces oxidative stress and enhances synaptic plasticity, learning and memory. It is the most abundant omega-3 fatty acid in cell membranes in the brain. The brain and the body do not synthesize DHA though it has to come through our diet. Eating omega-3 rich foodstuffs along with a balanced diet is essential for your brain function. A healthy diet and exercise can also reduce the effect of brain injury and lead to a better recovery.

Effect of Fasting and Calorie Restriction on Brain Function

Recent research also supports the hypothesis that health can be passed down through generations, and a number of innovative studies point to the possibility that the effects of diet on mental health can be transmitted across generations. A long-term study that included more than 100 years of birth, health, death and genealogical records for 300 Swedish families in an isolated village showed that an individual's risk for diabetes and early death increased if his or her paternal grandparents grew up in times of food abundance rather than food shortage. Evidence indicates that what you eat can affect your grandchildren's brain molecules and synapses. Researchers are trying to find the molecular basis to explain this phenomenon. Controlled meal-skipping or intermittent caloric restriction or fasting might provide additional health benefits. Excess calories can reduce the flexibility of synapses and increase the vulnerability of cells to damage by causing the formation of free radicals. Moderate caloric restriction could protect the brain by reducing oxidative damage to cellular proteins, lipids and nucleic acids. The brain is highly susceptible to oxidative damage. Blueberries have been shown to have a strong antioxidant capacity.

Trans- and Saturated-Fats Bad for the Brain

Studies indicate that in contrast to the healthy effects of diets that are rich in omega-3 fatty acids, diets high in trans-fats and saturated fats adversely affect cognition. Junk food and fast food negatively affect the brain's synapses. Brain synapses and several molecules related to learning and memory are adversely affected by unhealthy diets. Emerging research indicates that the effects of diet on brain, combined with the effects of exercise and a good night's sleep, can strengthen synapses and provide other cognitive benefits. In Okinawa, an island in Japan where people frequently eat fish and exercise, the lifespan is one of the worlds longest, and the population has a very low rate of mental disorders.

Effect of Folic Acid on Brain

The results of a recent randomized clinical trial indicate that a 3 years folic acid supplementation can help in reducing the age-related decline in cognitive function.

Effect of Food on Brain-Derived Neurotrophic Factor (BDNF)

In patients with major depression and schizophrenia, the levels of a signaling molecule known as brain-derived neurotrophic factor (BDNF) are reduced. Antidepressants elevate BDNF levels, and most treatments for depression and schizophrenia stimulate BDNF. Here, too omega-3 fatty acids are beneficial, as is the curry spice curcumin, which has been shown to reduce memory deficits in animal models of Alzheimer's disease and brain trauma. Brain-derived neurotrophic factor is most abundant in the hippocampus and the hypothalamus, brain areas associated with cognitive and metabolic regulation.

The high consumption of curcumin in India may contribute to low prevalence of Alzheimer's disease on the subcontinent. In humans, a mutation in a BDNF receptor has been linked to obesity and impairments in learning and memory. Brain-derived neurotrophic factor is reduced in the hippocampus, in various cortical areas and in the serum of patients with schizophrenia. Brain-derived neurotrophic factor levels are reduced in the plasma of patients with major depression.

Smaller food portions with the appropriate nutrients seem to be beneficial for the brain's molecules, such as BDNF. It has been shown that exercise can have an effect on the brain by elevating levels of BDNF. While some people have extremely good genes, most of us are not so lucky and need a balanced diet, regular exercise and a good night's sleep.

Most epidemiologists agree that the two biggest risk factors for Alzheimer's disease are aging and family history associated with genes or is it related to vascular risk factors such as diabetes, high blood pressure, or high cholesterol? What risk can be attributed to each of these factors? Is hypertension more significant than diabetes or it is the co-occurrence of these conditions that is problematic? Are there common or unique biomarkers to the type of dementia that appears in these populations? These issues have not been sorted out.

There is an ongoing multi-center international study being led by Vladimir Hachinski, M.D., London Health Sciences Center in London, Ontario, in collaboration with the National Institute of Neurological Disorders and Stroke. This study developed harmonization criteria for vascular cognitive impairment that may help answer these questions eventually. The researchers are following people diagnosed with vascular cognitive impairment (where brain damage results from problems with the blood vessels in the brain) to determine how many go on to develop *pure* vascular dementia and how that is differentiated from mixed dementia and cerebro-vascular disease.

It is well documented that these kinds of vascular risk factor are more common in African-Americans. Let us see whether it is true with other minorities as well.

A few researchers are beginning to look at other minority groups, *e.g.*, Helena Chui (M.D.) and colleagues from the University of Southern California are following an Asian-American cohort in California along with a group in Hawaii. Her study is called the Honolulu-Asia Aging Study. One of the researchers, Helen Petrovitch (M.D.) is following Japanese-American males. Another group at the University of Pittsburgh, led by Mary Ganguli (M.D.) is tracking dementia incidence in a rural population in eastern India.

While the progress is being made at this point, a lot of us have questions more than answers. We are getting one set of information from the epidemiologists who do population studies and a different set of information from the clinicians who are trying to follow these groups over long periods to understand the differences. Unraveling all of this work is very much in progress.

African-Americans and Hispanics tend to come to the attention of physicians only in the middle or later stages of the illness, when behavioural problems such as agitation, aggression, or wandering start to manifest. That is a much more difficult stage to treat, because the disorder has already progressed. The medications currently we have that are FDA-approved work best in the earlier stages should be started as early as possible.

Barriers to Seek Earlier Diagnosis and Treatment

Whether this phenomenon of seeking treatment later than would be ideal is due to educational, financial, or cultural barriers is being debated. The cultural aspect is interesting, because in many minority settings, the family may be more likely to ignore signs of forgetfulness, thinking that they are just a matter of getting older and not really a problem. Some researchers now identify a condition called Mild Cognitive Impairment (MCI) but many believe MCI is really an early stage of Alzheimer's disease. The first task is to educate all individuals including minorities to recognize MCI. Once that is done, we have to see whether we can treat those affected individuals. Perhaps it is better to take the suggestions made here to keep mental well being of the aged rather than introduce drugs at this point although the medical fraternity believes that the centrally acting cholinesterase inhibitors, which are currently FDA-approved should at least be tried in MCI. The problem is that none of these drugs is curative, and they only modestly affect the progression of symptoms.

If we start treatment in advance, might we be able to prevent the development of Alzheimer's disease? That is the hope but unfortunately a lot of the antioxidants that have been touted as potentially preventative has turned out to be ineffective. The most recent one to fail is ginkgo biloba, based on a large 6 years study investigating its effectiveness in delaying or preventing the onset of Alzheimer's disease.

Some people feel that controlling vascular risk factors may help. If we know somebody has a family history of cognitive problems, we recommend doing everything possible to ensure that they do not develop diabetes, high cholesterol, or hypertension. Alternatively, if they do develop one of those conditions, they should be treated aggressively as soon as they are diagnosed. Of course, there are similar barriers to the early and effective treatment of these conditions, especially in minorities, as there are with early identification and treatment of Alzheimer's disease.

Educational and financial barriers play a major role. A published study from Emory University's nursing group and neurology department compared African-Americans presenting to a public, inner-city hospital, to African-Americans who went to a hospital next to the Emory campus in an upper-middle class area of Atlanta. The researchers found a clear delay in the presentation of possible dementia patients at the inner-city hospital compared to those at the hospital next to Emory. Why might that be? It may be that the African-Americans who go to the hospital near Emory are better educated, and therefore, better informed about early identification of Alzheimer's disease, hence, when they notice problems with memory, they take their loved one in hospital to get the patient evaluated sooner. In contrast, often those persons who receive care at the inner-city hospital may not be as well educated or as well off financially.

In usual setting in which many minority patients receive healthcare, there is likely to be high volume. That is because minorities are more likely to be receiving medical care from primary care physicians, who typically see many people and may have less time to spend with individual patients. Most physicians use something called the Mini-Mental State Exam (MMSE), which is a test that was validated and originated by two geriatric psychologists in Baltimore. Some psychologists who test minority groups, such as Yaakov Stern (Ph.D.) and Jennifer Manley (Ph.D.) at Columbia University, contend that minority patients tend to test lower on the MMSE because the Baltimore study subjects on whom the test was validated were largely college-educated and white.

It is a matter of the families needing to be advocates and seeking out medical care for their loved one early on and for the physicians taking care of elderly patients to look for problems as early as possible.

What physicians and researchers need to do is to make sure that not everybody who is forgetful has Alzheimer's disease. They have to be tested and screened. We have to emphasize that there is good evidence that early treatment delays eventuality of placement of the affected in nursing homes. We need to communicate clearly that this is a poorly understood condition, which seems to occur among older individuals but we are still trying to understand who gets it, in what proportions, how fast it progresses and what we can do about it.

Important Facts about the Brain

Human brain is the most complex system in the universe. Its 100 billion cells and intricate neural fibers give it immense processing power and incredible versatility that far outweighs that of any computer. Here are the facts:

- The brain is the fastest growing organ in the developing fetus. It grows up to a quarter million cells per minute. After birth rapid growth continues. By the time the child is 2 years old, the brain is nearly as big as that of an adult.
- No two brains are identical. Even identical twins have visible differences in brain structure by the time they are born. These are caused by the minute differences in nutrients and sensory stimuli that each fetus experiences.
- During childbirth, the mother's brain is flooded with the hormone oxytocin that helps build mother child bond and dull the memory of pain of childbirth.
- Cuddling the baby or physical affection causes dopamine secretion, which triggers the release of growth hormone. Babies deprived of close physical contact with others may therefore be slow to develop physically and physiologically.
- The brain regulates the appetite. If left itself without any outside pressure, the brain provokes hunger only when glucose levels fall below the required level. In this situation, a person's weight would vary less than 5% during the entire adult life.
- The brain has about 100 billion neurons and each neuron has up a 1000 *docking points* where it can connect with others. If all the connections were made, there would be 100 thousand billion information-exchanging links. In practice, only a tiny fraction of these connections is made.
- The brain is able to alter almost any body function. Heart rate can be slowed, the bowel relaxed, blood vessels opened and closed just through thought.
- A happy brain can help to fight infection. The body's immune system directly responds to changes in the brain. A sad event, such as losing a loved one- can produce measurable depletion of disease fighting white blood cells within four days.
- The brain has powers to mend the damage after stroke. Although the nerve cells affected by stroke usually die but the connection between surviving cells can re-grow, bypassing damaged areas. This way the brain can regain some of the functions lost at the time of stroke.
- The body constantly monitors every thing happening at the body's skin or surface in every muscle and joint. By doing this, the brain keeps track of all activities of the body. This enables it to

maintain balance and carry out automatic as well as deliberate movements. Most of this input is registered unconsciously but it is cut off by brain injury or sensory deprivation the body collapse.

Some interesting and hopeful statistics were offered at the World Alzheimer's Conference 2000 held in Washington, United States of America. It was said that the rate of occurrence of Alzheimer's disease doubles every 5 years for those between 65 and 85 years of age but if its onset were delayed by 5 years, the number of cases worldwide would be halved. Thus, serious attention needs to be paid to the risk factors and preventive measures that may be taken to postpone the onset, if not prevent the appearance of Alzheimer's disease. Epidemiological data already suggests that certain communities in Asia and Africa may have a lower risk for Alzheimer's disease compared to western countries. Researchers in India have suggested a gene and environment interaction in the causation of Alzheimer's disease. Although genes cannot be altered but the environmental risk factors, which interact with the genetic factors, can possibly be modified, and thereby, reducing the risk. This promising concept must be pursued, as it affords a ray of hope for the future.

■ ■ ■

15 Drug Abuses in Medical Nutrition View

Viroj Wiwanitkit

Wiwanitkit House, Bangkhae, Bangkok, Thailand-10160, Visiting professor, Hainan Medical College, China, E-mail: <wviroj@yahoo.com>

Drug is an important basic need for everyone. Drug is the thing for treatment of disease. Since everyone cannot avoid getting sick hence drug is an important thing in everyone life. At present, several kinds of drugs can be seen in daily life. It is no doubt that drug can cause usefulness to human beings, getting rid of disorder. On the other hand, drug can also cause unwanted effect to the human beings.

The use of drug seems to be very important thing in biomedical science. A specific branch of science to study on drug is the pharmacology. It is well known that the use of drug has to be well controlled. To prescribe for a drug, physician has to carefully investigate on the patient and decide that there is indication for using drug. The control of drug distribution has to be done by the pharmacist.

Table 1: Role of medical personnel on drug

Personnel	Roles
Physician	Judge the indication for using of drug, prescribe for drug and administer of drug.
Dentist	Judge the indication for using of drug (some specific drugs), prescribe for drug (some specific drugs) and administer of drug (some specific drug).
Nurse	Giving drug to the patient under physician's order and administer of drug under physician's order (some specific drug).
Pharmacist	Prepare drug, giving drug to the patient according to the prescription, and giving basic drug to the patient without prescription.

However, the problem on the control of drug use by non-expert or layman is a serious concern at present. Many drugs can be available without control, without physician's prescription and without pharmacist's recommendation. The incorrect usages or abuses of drugs seem to be a big problem in public health, which needs proper management.

The Pathway of Drug

The drug that we can see in daily life has its long pathway. Firstly, drug has to be produced. This means that there must be complete chemical materials for the production of drug. Those chemical materials can be derived from several ways including:

- **Natural product:** This is the classical way of getting a new drug. It has to pass several trials and errors. Several natural materials, from both plants and animals, are known for their pharmacological properties.
- **Newly synthesized materials:** This is the new way of getting drug. Many biomedical laboratories perform the study to evaluate the new pharmacological properties of many chemicals. With a complete process, the candidate for new drug development can be derived.

In production of drug, the standard manufacturer is needed. Several standards have to be used. This is the important point for controlling quality of the produced drugs.

After production, drugs have to be distributed for usage. Some kinds of drug are allowable to be sold in pharmacist shops without prescription, while the others are strictly controlled and to be used only under physicians' prescriptions. Keeping on the drug before usage is another point to be considered since there are many factors that can insult the quality of kept drugs including moisture, temperature and sunlight.

After kept for a period of time and it is ordered, drug will be passed to the patient. The patient is the final terminal of drug pathway. Patient has to use drug for the improvement of their health status. Some drugs can be used by the patients without any help of medical personnel, whereas, the others have to be administered by medical personnel.

Table 2: Route of drug administration

Route	Details
Oral	This is the main route, which is used for many drugs in the present day. Eating can be practiced by the patients.
Intravenous	This is another important route. This means injection of drug into the venous vessel. This is aiming at rapid and bolus administration of drug.
Intramuscular	This is another important route. This means injection of drug into the muscle. This is aiming at rapid administration of drug.
Intradermal	This is another important route. This means injection of drug into the superficial skin layer.
Subcutaneous	This is another important route. This means injection of drug into the subcutaneous layer under skin.
Dermal	This means application of drug onto the skin (not injection into the skin layer).
Intrathecal	This is a rare administration route. This means injection of drug into the spinal canal *via* lumbar puncture.
Nasal	This is usually a spray form of drug. The administration is mainly aimed at pharmacological action within respiratory tract.
Ophthalmic	This is usually an ophthalmic solution form of drug. The administration is mainly aimed at pharmacological action at eye.
Otic	This is usually an otic form of drug. The administration is mainly aimed at pharmacological action at ear.

Route	Details
Anal	This is usually an anal form of drug. The administration is mainly *via* suppository of drug into anal canal *via* anus. The local action at anus is the aim of administration.
Vaginal	This is usually a vaginal form of drug. The administration is mainly *via* suppository of drug into vaginal canal *via* vagina. The local action at vagina is the aim of administration.
Transurethral	This is usually a transurethral form of drug. The administration is mainly *via* direct administration into urethral orifice.

What is the correct drug use?

In order to know what the incorrect drug use is, it is required to know what the correct drug use is. The correct drug use means:

- **Correct indication:** Correct indication means use drug when there is a medical indication. This means that it is really needed. The judgment for medical indication can be decided by the physician. In case of basic safe drug, it can be judged by pharmacist. Using drug without indication means incorrect usage.

  The main indications of administration of drug include:

  - Treatment of disease
  - Prevention of disease through chemoprophylaxis (vaccine)

  The main contraindications of administration of drug include:

  - Allergy
  - Hypersensitivity
  - Problems that prevent administration of drug (such as infection at injection site)

- **Correct route:** Correct route means correct administration of drug. There are several routes for administration of drug (Table 1). It is needed to follow the guideline of drug usage for each specific drug. Incorrect route of administration means several things. It means the expected pharmacological effect of drug cannot be derived. The incorrect route of drug administration might also mean danger to the drug user. Some drug can be safe only if given in a specific correct route.
- **Correct time:** Correct time is very important. There are many suggestions on using of drug (Table 3). It is needed that the drug user has to follow the correct time while using the drug. Incorrect time in usage means lower pharmacological effect of drug than expected level. It might also imply the adverse effect of drug such as irritation to stomach and stomachache. The problem on use of drug at correct time is very common in general practice. It is the big issue to be discussed in pharmacology and general medical practice.

Table 3: Time for use of drug

Time	Details
Before meal	This means, the patient has to take this drug about 30 minutes before eating food because of the fact that the absorption of drug can be disturbed by the contents of food.

Time	Details
With meal	This aim of this prescribed time is to allow the drug to be absorbed with food.
After meal	This means, the patient has to take this drug about 15 minutes after eating food because of the fact that the drug is usually irritating to gastrointestinal tract.
At bedtime	This means, the action of the drug is aimed at sleeping period.
With symptom	This means, the drug is used for symptomatic treatment. It can be used as frequently as indicated (not mean every time of symptom). The best example is the acetaminophen (paracetamol) which is usually indicated as *with pain or fever* but its actual limitation of usage is not more frequent than every 4 hours.

- **Correct dosage:** Dosage of drug means the amount of drug that has to be taken to reach the pharmacological aim. The dosage of each drug depends of several conditions including age of patients, severity of disease, kind of disease, *etc*. The correct dosage is important since it is the standard for the use of drug. If no correct drug dosage is used, the problem can be expected. The problems of incorrect drug dosage use include:
  - Cause toxicity (if excessive dosage)
  - Cannot treat the disease (if inadequate dosage)
  - Induce resistance (has to increase dosage if future treatment has to be performed)
- **Correct person:** Correct person is a very basic concern. The common problem is the use of drug in incorrect individual. The identification of the patient before administration is the safety rule of drug use. Everyone who has to use drug has to take a concern on his or her personal drug.

Since there are many concerns on the use of drug in correct way, the use of drug has to be concerned. The recommendation for correctly usage of drug includes:

- Get the drug prescription from the physician
- Take the basic drug from the pharmacist
- Consult the pharmacist before using drug
- Avoid using drug without any medical suggestion
- Read the drug information in drug brochure carefully before use
- Use the drug according to all suggestions
- Use the drug according to the prescribed dosage
- Use the drug at correct time
- Carefully read the name of patient at drug package before use
- Do not use the other's drug

Incorrect Use of Drug

Since there is a correct drug usage, there can be incorrect usage. The incorrect usage means opposite usage to correct usage. This includes:

- Incorrect indication

- Incorrect route
- Incorrect time
- Incorrect dosage
- Incorrect person

The cause of incorrect use of drug can be due to several reasons including:

- **Lack for self-concern on drug use of the patient:** Lack for self-concern on drug use of the patient is the big problem. This can be seen in many countries around the world. The problem is very highly prevalent in the developing countries. Sometimes, some people buy drugs without prescription for some pharmaceutical drug stores that do not have professional concern on their standards. Several problems due to this unwanted practice can be seen in medicine. The important problems include:
  - Drug resistance
  - Resistant pathogen
  - Adverse effect of drug
- **Illegal drug distributor:** Illegal drug distributor is the big problem that can be seen around the world. Several drug distributors can be seen. This can be the selling of un-standardized drugs, selling of faked drugs and selling of illegal drug or narcotics.
- **Iatrogenic problem due to medical personnel:** Iatrogenic problem is another important problem to be concerned.

Although safety is the basic need in medical practice but it is frequently reported on the error. The error in drug prescription is very common and this can cause danger to the patients. The good medical accreditation system is the hope to decrease this problem.

Drug Abuse

An important to be discussed is the *drug abuse*, which means incorrect use of drug. It can be due to many causes as already mentioned. However, it generally refers to the incorrectly use with illegal problem. It usually means the use of narcotic drug. At present, any narcotic drugs are considered illegal and there are many laws for control of narcotics. High punishment can be seen in many countries for the ones who use narcotic drug.

Table 4: Types of narcotic drugs

Types	Examples
Mixed	Marihuana
Stimulant	Amphetamine, metamphetamine and cocaine
Depressant	Heroine and opium

Several problems can be expected due to use of narcotic drugs, which include:

- **Drug toxicity:** This is the effect of toxic chemical composition of narcotic drugs. The problems can be due to the type of narcotic as shown in Table 4.

- **Addiction:** This means the dependence on the narcotic drug. The characteristics of addiction include:
  - Dependence
  - Increase dosage requirement
  - Cause health problem
  - Psychological problem
- **Withdrawal:** This means the symptoms if the narcotic drug use is stopped. This is the main cause that one who use narcotic drug cannot stop using.

Drug and Nutrition

Drug and nutrition has a strong correlation. Since nutrition and drugs are the two mains of basic needs of any human being. The interrelationship between drug and nutrition can be seen.

1. Effect of drug on nutrition

There are several effects of drug on nutrition. Since drugs are biochemical substances, they can disturb the normal physiological function within human body. The possible effects of drug on nutrition include:

- Disturb the absorption of nutrients
- Disturb the nutrients distribution in the body
- Directly destroy the nutrients

2. Effect of nutrition on drug

There are several effects of nutrition on drug. Nutrients can disturb the expected pharmacological function within human body. The possible effects of nutrition on drug include:

- Disturb the absorption of drug
- Disturb the drug distribution in the body
- Directly destroy the drug

Table 5: Examples of interrelationship between drug and nutrient

Nutrients	Drug	Effect
Vitamin C	Acetylcysteine	Precipitation
Vitamin B_{12}	Ometprazole	Disturbance on absorption

To help reader better understand on the scenario, some interesting situations will be further detailed and discussed.

Vitamin B_{12} deficiency due to drug

Vitamin B_{12} deficiency is an important vitamin deficiency. This kind of nutritional disorder can be seen in any countries. The signs and symptoms of this disease include:

- Anemia (megaloblastic type)
- Neurological complication

Vitamin B_{12} deficiency can be due to the result of many possible etiologies including:

- Low intake of vitamin B_{12}
- Prolonged vegetarianism
- Alcoholism
- Some drugs

The case of drug induced vitamin B_{12} deficiency is very interesting. One of the most problematic drugs that can cause vitamin B_{12} deficiency is the omeprazole, a basic drug, which is mainly used for treatment of gastric disorders (including peptic ulcer, gastritis, dyspepsia, *etc.*).

In many countries, this drug can be bought from the drug store as an over and over the counter drug. An important profound adverse effect of omeprazole is the reduction of vitamin B_{12} absorption.

Indeed, it is suggested that prolonged usage of vitamin B_{12} should be avoided. The special precautions should also be set in some specific population. This includes the ones who practice strict vegetarianism. The recommendation for anyone who has to take a long-term usage of vitamin B_{12} for treatment of peptic disease is increased amount of ingestion of meat. Meat is the good source of vitamin B_{12}.

Vitamin C deficiency due to drug

Vitamin C deficiency is another important vitamin deficiency. This kind of nutritional disorder can be seen in any countries with a very high prevalence in tropical developing countries. The signs and symptoms of this vitamin deficiency include:

- Bleeding gums
- Fragile capillary

Vitamin C deficiency can be due to the result of many possible etiologies including:

- Low intake of vitamin C
- No or low intake of fruits and vegetables
- Some drugs

The case of drug induced vitamin C deficiency can be seen and is important concern in medical nutrition. One of the most problematic drugs that can cause vitamin C deficiency is the vaccine. Several vaccines are reported for their possibilities to induce vitamin C deficiency. The condition is so called *Barlow's disease*. The problem is significant in the infantile groups, especially those with maternal smoking or fed on formula-milk.

Vitamin A deficiency due to drug

Vitamin A deficiency is another important vitamin deficiency. This vitamin deficiency is one of the most common nutritional deficiencies among the local people in developing countries. The signs and symptoms of this vitamin deficiency include:

- Night blindness

- Poor vision

Vitamin A deficiency can be due to the result of many possible etiologies including:

- Low intake of vitamin A
- No or low intake of fruits (especially of yellow ones) and vegetables (especially of yellow and green ones)
- Some drugs

The case of drug induced vitamin A deficiency can be seen and is important concern in medical nutrition. One of the most problematic drugs that can cause vitamin A deficiency is the vaccine, similar to the case of drug induced vitamin C deficiency. The most problematic vaccine is the mumps-measles-rubella vaccine (MMR). Similar to vitamin C deficiency, this kind of drug induced vitamin deficiency is common in the infantile groups, especially those with maternal smoking or fed on formula-milk.

Another important drug that can induce vitamin A deficiency is the corticosteroid. Corticosteroid is the drug that is used as anti-inflammatory drug. This drug is indicated in some inflammatory diseases such as autoimmune diseases. The corticosteroid can induce the excessive excretion of zinc and further prevent the normal releasing of vitamin A from liver. This can be final with vitamin A deficiency.

Vitamin B_6 deficiency due to drug

Vitamin B_6 deficiency is an important vitamin deficiency. The signs and symptoms of this disease include:

- Glossitis
- Dermatitis
- Neurological complication including mental retardation
- Seizure (in some cases)

Vitamin B_6 deficiency can be due to the result of many possible etiologies including:

- Low intake of vitamin B_6
- No or low intake fruits and vegetables
- Some drugs

The case of drug induced vitamin B_6 deficiency is very interesting. One of the most problematic drugs that can cause vitamin B_6 deficiency is the antituberculosis drug, *i.e.*, isoniazid- a basic drug, which is mainly used for the treatment of tuberculosis as a main drug in antituberculosis regimen. Due to the remerging of tuberculosis in the era of HIV, the problem of isoniazid induced vitamin B_6 deficiency should be kept in mind.

Another important drug that can induce vitamin B_6 deficiency is the hydralazine- a drug, which is used as anti-hypertensive drug. This drug is indicated in some hypertensive patients with very high blood pressure. The hydralazine can disturb the normal absorption of vitamin B_6 and can result in vitamin B_6 deficiency.

Finally, it should also be noted that the MMR vaccine could also induce vitamin B_6 deficiency. An important component within this vaccine, neomycin can cause blocking of vitamin B_6 *in vivo*.

Vitamin K deficiency due to drug

Vitamin K deficiency is an important vitamin deficiency. The signs and symptoms of this disease include:

- Bleeding
- Bruise

Vitamin K deficiency can be due to the result of many possible etiologies including:

- Low intake of vitamin B_6
- No or low intake fruits and vegetables
- Neonatal deficiency (relating to milk intake)
- Some drugs

The case of drug induced vitamin K deficiency is very interesting. The common problematic drug that can cause vitamin K deficiency is the antibiotic. Since antibiotic can destroy normal flora within human gut that naturally help in the synthesis of vitamin K. Hence, prolonged taking of antibiotic might result in vitamin K deficiency. In some countries, antibiotic can be bought as an over and over the counter drug with a good control. Hence, the problem of vitamin K deficiency can be expected.

Glutathione deficiency due to drug

Glutathione deficiency due to drug is a serious problem to be discussed. The problematic drug is the paracetamol, which is widely used around the world. This drug is the most basic over and over- the counter drug, which is sold in every drug store. At a normal dosage, this drug causes no problem on health status but the over dosage and an incorrect usage can cause the toxicity, especially on the liver. An important process that over dosage of paracetamol can cause the problem to liver is the induction of glutathione deficiency. This severely impairs the liver function.

Effect of Drug Abuse in Medical Nutrition

Drug abuse or narcotic drug use in medical nutrition can be seen. Several problems in health can be seen in drug abuse. This is due to many facts including:

- Direct toxicity from narcotics
- Side effect of narcotics
- Loss of appetite due to addiction

The interrelationship between narcotic drug and nutrition can be seen.

- **Effect of narcotic drug on nutrition:** There are several effects of narcotic drugs on nutrition. Since narcotic drugs are toxic substances, they can disturb the normal physiological function within human body and cause toxicity. The possible effects of drug on nutrition include:
  - Disturb the absorption of nutrients
  - Disturb the nutrients distribution in the body
  - Disturb normal eating function
  - Directly destroy the nutrients
  - Destroy the organ, which is used for the absorption and distribution of nutrients
- **Effect of nutrition on narcotic drug:** There are several effects of nutrition on narcotic drug. The possible effects of nutrition on narcotic drug include:
  - Disturb the absorption of narcotic drug

- Disturb the narcotic drug distribution in the body

Table 6: Examples of interrelationship between drug and nutrient

Nutrients	Narcotic drug	Effect
Vitamin C	Amphetamine	Excretion
Vitamin B_{12}	Alcohol	Disturbance on absorption

To help the reader to understand the scenario better, an interesting situation will be further detailed and discussed.

Vitamin C for the treatment of amphetamine toxicity

Amphetamine toxicity is a common narcotic related disorder. This can be seen in many countries and the problem is very common in Thailand. The strange invasive behaviour can be seen. The patients usually attack, injure, or kill the others. The symptoms of the intoxication include:

- Change of behaviour to fierce
- Palpitation
- Tachycardia
- Profound sweating
- Polyurea

In treatment of amphetamine, the control of aggressive behaviour has to be mainly done. The sedation might be used and the increased excretion of narcotic drug is required. The use of vitamin C is recommended for increasing the excretion of vitamin C since this is due to acidic urine induction.

■ ■ ■

16 Drug Residues in Foodstuffs of Animal Origin

Vijay J. Jadhav

Department of Veterinary Public Health and Epidemiology, CCS Haryana Agricultural University, Hisar, Haryana, India, E-mail: <vetviju@gmail.com>

Foods of animal origin include products obtained from farm animals such as meat, milk, eggs, fish and even honey. When consumers pay for such commodities, they expect that the product that they are getting should be genuinely natural and should not contain unwanted substances in the form of contaminants or even deliberately added chemicals. Therefore, thousands of such chemicals that are used in the society are subject of food safety issue. In case of foods of animal origin, the chemical contaminants of primary concern include residues of veterinary medicinal products, hormone and pesticide residues, nitrates, bacterial toxins, mycotoxins, phytotoxins, algal toxins, marine toxins, heavy metals, dioxins and dioxin like compounds, disinfectants, polycyclic aromatic hydrocarbons (PAHs) and processing contaminants such as acrylamide. Consumers are now aware of significance of most of these contaminants in food. However, they are unaware about likely presence of veterinary drug residues in the foods of animal origin and their adverse health effects.

Veterinary drugs are administered to animals to cure ailments and diseases. Drugs are also incorporated into animal feed to promote growth and to reduce the chances of disease occurrence. Whatever the purpose is, once a drug is administered by any route, it is absorbed into the bloodstream from the site of administration. The drug then is distributed into various body fluids and tissues to attain an effective, yet safe, concentration for a sufficient period at the site of action. Subsequently, the drug is inactivated or eliminated from the body, generally by metabolism and excretion. The effectiveness of these processes with respect to time (pharmacokinetics) varies with the particular drug and species of animal. However, 100% of clearance of administered drug does not occur within a very short span. Hence, the occurrence of its residue in various body compartments and in all excretions and secretions is an unavoidable squeal. It means occurrence of drug residues in meat and milk is always a possibility if these products are obtained from treated animals without allowing sufficient time for its complete clearance.

Most of the drugs used in veterinary profession are also used in human medicines and drugs are not supposed to cause any damage to individual. Therefore, one may raise doubt regarding so much concern for exposure of veterinary drug residues to people through consumption of food articles of animal origin. This can be explained by the factual difference between intentional (medical treatment) and unintentional (drug residues in animal products) nature of exposure to these drugs. Unlike medical use of drugs in humans, which involve voluntary treatment for short periods of time under the supervision of medical professionals, the exposure to the traces of these drugs in the food supply are involuntary and uncontrolled. Drug residues exert low dose long time exposure, and therefore, harmful effects of these residues are not always detected immediately. Due to these reasons, the safety consumers of foods originated from animals treated with veterinary drug is of paramount importance.

Classification of veterinary drugs

In context of food safety, the term veterinary drug is defined as any substance applied or administered to any food producing animal, such as meat or milk producing animals, poultry, fish or bees, whether used for therapeutic (for treatment of disease), prophylactic (for prevention of disease), diagnostic purposes or for modification of physiological functions in food producing animals. Major groups of veterinary drugs approved for use in food producing animals are as follows. The examples of drugs included in such groups are given in Table 1.

Table 1: Examples of veterinary drugs used in food animals

Sr.No.	Group	Drugs
1.	Beta-lactams	Amoxicillin, ampicillin, ceftiofur, cephapirin, cloxaxillin and penicillin G
2.	Tetracyclines	Chlortetracycline, doxycycline, oxytetracycline
3.	Aminoglycosides	Streptomycin, gentamicin and neomysin
4.	Macrolides	Tylosine and erythromycin
5.	Sulfonamides	Sulfadimidine, sulfamerazine, sulfadiazine, sulfapyridine, sulfabromomethazine and sulfaethoxypyridazine
6.	Quinolones	Nalidixic acid
7.	Fluroquinolones	Amifloxacin, benofloxacin, ciprofloxacin, enrofloxacin and norfloxacin
8.	Anthelmintics	Albendazole, febantel, levamisole, pyrantel pamoate, piperazine DEC, closantel and ivermectin,
9.	Steroids	Trenbolone, nortestosterone and estradiol
10.	Non-steroidal antiinflammatory drugs	Aspirin, paracetamol and phenylbutazone
11.	Beta-agonists	Clenbuterol, sulbutamol and mabuterol
12.	Coccidiostats	Nitrofurazone, furazilidone and ronidazole
13.	Organochlorines	Dichlorodiphenyltrichloroethane (DDT)
14.	Organophosphates	Dichlorvos, malathion and parathion
15.	Carbamates/pyrithroids	Carbaryl, cypermethrin and deltamethrin

The types of drug used for the treatment of different diseases are being discussed below:

Antimicrobial drugs: An antimicrobial is a substance that kills or inhibits the growth of microorganisms and can be divided into antibiotics and antimicrobial drugs. Earlier the term antibiotic was used to describe only those formulations derived from living organisms, but now, it is also applied to synthetic antimicrobials, such as sulfonamides. In the livestock production, a wide variety of antibiotics such as beta-lactams, tetracyclines, aminoglycosides, macrolides and polymixins are used.

Antifungal drugs: These are mostly available as drugs used on the surface of the skin to cure superficial fungal infections.

Anthelmintics: Invasion of gastro-intestinal system by internal parasites like nematodes and flukes is very common affection of farm animals. These parasites cause distress and substantial economic loss. To get rid of these parasites, administration of wide variety of anthelmintic drugs to farm animals is a routine management practice.

Ectoparasiticides: Ectoparasiticides are commonly known as pesticides. Some chemicals from organophosphates and synthetic pyrithroide group are used as veterinary drugs in suitable formulations against external parasites like fleas, mites and lice that invade skin of cattle, sheep and buffaloes.

Steroid hormones: Anabolic steroids are the drugs that mimic the effects of the male steroids testosterone and dihydrotestosterone. They increase protein synthesis within cells, which results in the buildup of cellular tissue (anabolism), especially in muscles. Therefore, when used in food animals, they act as growth promoters. The other group known as corticosteroid is drug that mimics the actions of cortical. It is a natural hormone released by the adrenal glands and primarily used in veterinary medicine for its anti-inflammatory effects.

Non-steroidal anti-inflammatory drugs: Non-steroidal anti-inflammatory drugs, usually abbreviated to NSAIDs or NAIDs, are the drugs with analgesic (reducing pain) and antipyretic (fever-reducing) effects and which have, in higher doses, anti-inflammatory effects (reducing inflammation). The term *non-steroidal* is used to distinguish these drugs from steroids, which (among a broad range of other effects) have a similar anti-inflammatory action.

Beta-agonists: Beta-agonists are the synthetically produced compounds, which are used regularly in veterinary medicine to treat pulmonary diseases. In addition, these drugs have the ability to promote live weight gain in food-producing animals. They are also called as repartitioning agents because of their ability to exert desirable effect on carcass composition by increasing the deposition of protein while reducing fat accumulation.

Causes of veterinary drug residues in food

Veterinary drugs when used judiciously in food animals, there is hardly any possibility, or at the most, very low and acceptable concentration of residue to remain in the products derived from the treated animals. In several countries, the food safety authorities have established maximum residue limits (MRLs) to safeguard the health of consumer. It is defined as the maximum concentration of residue resulting from the use of a veterinary drug (expressed in mg/kg or mg/kg on a fresh weight basis) that is acceptable in or on a food. It is based on the type and amount of residue considered to be without toxicological hazard for human health as expressed by the acceptable daily intake for that drug. It also takes into account other relevant public health risks as well as food technological aspects and estimated food intakes. In India, such MRLs are available only for sea foods like shrimps and prawns and not yet established for meat and milk. Internationally, Codex Alimentarius Commission created in 1963 by World Health Organization and Food and Agricultural Organization is responsible for the establishment of MRLs for all types of chemical residue in food commodities. The examples of such MRLs for meat are given in Table 2.

Table 2: MRLs established by Codex Alimentarius for some veterinary drug residues in meat (muscle tissues)

Veterinary drug	Food animal species	MRL (µg/kg)
Benzylpenicillin	Cattle, pig and poultry	50
Ceftiofur	Cattle	1000
	Pig	2000
Chlortetracycline/Oxytetracycline/Tetracycline	Cattle, pig and sheep	200
	Poultry	1200
Gentamysin	Cattle and pig	200
Albendazole	Species not specified	200
Febantel/Fenbendazole/Oxfendazole	Cattle, pig and goat	100
Levamisole	Cattle, pig and sheep	10
Zeranol	Cattle	2
Clenbuterol	Cattle	0.2
Dexamethazone	Cattle and pig	1.0
Cyfluthrin	Cattle	20
Cypermethrin	Cattle and sheep	50

Though the residue in products derived from treated animal is imperative, its concentration above prescribed MRL indicates some sort of human error. Various reasons responsible for unacceptable residue are given below:

The major reason for incidence of above MRL residue is failure to observe withdrawal period. Withdrawal period is the time between the last administration of a drug and the collection of edible tissue or milk from a treated animal that ensures the contents of residues in food comply with the maximum residue limit for this veterinary drug. Violative concentration of drug residues is observed when livestock farmers are not keen to maintain treatment records and release the food products procured from such treated animals during the withdrawal period.

Extra-label use of drugs is also another important reason for the occurrence of unacceptable residues. In India, Food and Drug Administration controls the use of veterinary drugs. The use of drug is approved with certain predetermined conditions based on outcome of research and development activities and clinical trials. These conditions are mentioned on the label of the drug container, *e.g.*, the route of administration, species of the animal, *etc.* When these conditions are violated while using the drug, it is referred as extra-label use. In other words, it means use of a drug in an animal in a manner that is not in accordance with the approved labeling. This includes, but is not limited to, use in species not listed in the labeling, use for indications (disease and other conditions) not listed in the labeling, use at dosage levels, frequencies, or routes of administration other than those stated in the labeling and deviation from labeled withdrawal time based on these different uses. Since withdrawal time is established under circumstances similar to the approved use of drug, extra-label use invariably leads to the occurrence of residue even though withdrawal period is observed.

Animal feedstuffs may get contamination with a variety of compounds, and thereby, serves as an important source of unintended application of veterinary drugs to farm animals. Feedstuff contaminated with feed medicated with coccidiostats can lead to its residue in chicken.

The residues of veterinary drugs can be transferred from mother to its offsprings, *e.g.*, calves may get antibiotic residues of penicillin G or amoxicillin by consuming contaminated milk of its mother. The residues of levamisole were detected in eggs layed by hen feeding on diet containing levamisole residues.

Some banned veterinary drugs are illegally used in farm animals to increase production. Repeated injections of oxitocin injection to increase milk production serves good example of such abuse.

Concerns of Antimicrobials Drug Residues in Foods of Animal Origin

Effects on human gut microbiota

The microbiota in the human gastrointestinal tract presents very diverse and stable ecological community (residential microorganisms) containing more than 400 bacterial species. It provides the natural defence by preventing the colonization and translocation of exogenous potentially pathogenic microbes as well as overgrowth of indigenous opportunistic microorganisms. Administration of antimicrobial agents causes disturbances in these functions by killing the useful microorganisms. The extent of disturbances in the ecological balance between host and microorganisms occur depends on the spectrum of the antimicrobial agent, the dose, pharmacokinetic and pharmacodynamic properties and *in-vivo* inactivation of the agent.

Development of resistance by microorganisms

The important issue regarding impact of veterinary use of drugs on public health is the development of resistance in medically important pathogens because it makes very difficult to control the infection caused by such pathogens. Since the year 1957 when it was demonstrated for the first time, the development of resistance of *Escherichia coli*, an enteric pathogen due to use of tetracycline in animal feeds in swine and poultry, the risk of feed antibiotics on human health became the central point of focus because tetracycline is also used in human medicine for treatment of bacterial diseases. This happens because the use of certain antibiotics in animal feeds leads to elimination of susceptible microbial population and allows proliferating resistant population in the gut of animals. This resistance or susceptibility phenomenon is basically genetic in origin. In a closely associated microbial community, the exchange of genetic material is quite possible and well-known process that takes place through transfer of plasmids. When possible human or animal pathogen comes in contact with resistant residential microorganisms, it acquires resistant genes through plasmids. This makes otherwise susceptible pathogen into resistant one. A close relationship between tetracycline, streptomycin, gentamicin and chloramphenicol residues and the resistance of bacteria isolated from the samples was found, suggesting that the presence of low levels of antimicrobials might exert a positive pressure towards the selection and expression of resistance in bacteria colonizing animal tissues.

Hypersensitivity reactions

Drug hypersensitivity is defined as an immune-mediated response to a drug agent in a sensitized patient, which is a reaction mediated by antibodies known as immunoglobulin E. Hypersensitivity reactions may manifest from life-threatening anaphylactic reactions to milder reactions such as rashes. Drug-induced allergic reactions may occur acutely (within 60 minutes of challenge), sub-acutely (1-24 hours) or as latent responses (1 day to several weeks). Antimicrobial drug residues in animal tissues or milk may cause hypersensitivity reactions in humans. An allergic reaction may be triggered by antimicrobial residues in a previously sensitized individual. Beta-lactams appear to be responsible for most of the reported human allergic reactions to antimicrobials. Aminoglycosides, sulphonamides and tetracyclines may also cause allergic reactions. However, only a few cases of hypersensitivity have been reported as a result of exposure to residues in meat. Anaphylactic reactions to penicillin in pork and beef have been described. Failure to associate minor reactions such as urticaria with possible exposure to allergenic residues may be one reason for the lack of reported cases.

Other harmful effects

Aplastic anemia can occur in susceptible individuals exposed to concentrations of chloramphenicol that might remain as residues in edible tissues of chloramphenicol-treated animals. Aminoglycosides can produce damage in urinary, vestibular and auditory functions.

Some antibiotics are considered resistant to quite harsh food processing conditions like pasteurization, drying, or freezing. The residues remaining in the fluid milk interfere with necessary microbiological reactions that are required in the production of certain milk products. Thus, the major problems of antibiotic residues in milk are partial or complete inhibition of acid production by bacterial starter cultures used in the manufacture of dairy products such as cheese, buttermilk, sour cream, or yogurt. As little as 0.05-0.1 IU/mL penicillin completely inhibited the growth of most cultures studied.

(a) Beta-adrenergic agonists: Since the European ban on anabolic steroids was established in 1986, alternative substances such as beta-agonists, the physiological analog of adrenalin, have been introduced as growth-promoting agents. These substances, in addition to their regular role in veterinary medicine as bronchodilatory and tocolytic agents, increase lean meat to fat ratio (repartitioning) and improve feed conversion efficiency. Illegal use of these pharmacologically active substances has resulted in severe cases of food poisoning with adverse effects such as palpitation, tachycardia, agitation, tremors and headaches. The use of clenbuterol in cattle feed resulting into above-mentioned adverse reaction in the consumer was reported.

(b) Residues of hormones, anabolic steroids and other drugs: There has been increasing concern about the impact of environmental compounds with hormone-like action on human development and reproductive health over the past decades. An alternative but neglected source of hormone action that may be considered in this connection is hormone residues in meat from farm animals treated with sex steroid hormones for growth promotion. Treatment of cattle with naturally occurring or synthetic sex hormones may enhance lean muscle growth and improve feed efficiency and is therefore a very cost effective procedure for cattle producers who have used it for decades in some western countries, including the United States of America and Canada. Steroid hormones are implicated in a range of cancers, including breast and cervical cancers in woman, prostate and testicular neoplasia in men and other diseases such as polycystic ovary syndrome and malformations of the external and internal genitalia.

Studies have shown that 17 beta-estradiol, a growth hormone, is a complete carcinogen. This synthetic form of progesterone shares similar characteristics with a class of other molecules called endocrine disruptors, which imitate other human hormones in the body, and have been linked to diseases such as cancer. Zeranol, another progesterone-imitator, has also been linked to cancer in a recent study done at Ohio State University. A study done by a fresh-water research facility of Environmental Protection Agency (USA) found that female fish exposed to trenbolone in water developed bumps on their heads normally only witnessed in male fish when their natural testosterone level was elevated. There was a general pattern where female fish was slightly masculinized and male fish slightly feminized by trenbolone exposure. The reproductive impacts observed in fish raise concerns that hormones used in beef and consumed by people might be a factor in rising incidence of premature development in girls and lower sperm counts in men.

Anabolic-androgenic steroids are widely misused in human sports for enhanced muscle power and stamina. Although it does not serve the case of food safety issue but the residues of such steroids in meat pose risk of failing a sports drug test by athletes who consume it.

Arsenical compounds are used in swine and poultry to promote growth and to prevent bacterial enteritis. The most commonly used arsenic compound for poultry is roxarsone. Most of the roxarsone is excreted unchanged but some metabolites have been detected in hen urine. Inorganic arsenic is a known carcinogen and may adversely affect the circulatory and nervous systems.

Detection and quantitation of veterinary drug residues

Veterinary drug residues in foods of animal origin are generally present in the proportion of parts per million or even parts per billion. Therefore, detection and quantitation at such low concentration is very challenging task for analysts. Fortunately, there are various methods available, which are classified as qualitative, semi-quantitative and quantitative. These include microbial inhibition assay, microbial receptor assay, enzymatic calorimetric assay and receptor binding assay, chromatographic methods and immunoassay. Typically, the residue analysis involves both screening and confirmatory tests. A screening method is the first-hand analysis of the sample to establish the presence or absence of residues. It is usually low-cost and high-sample throughput method. Confirmatory methods involve determination by molecular spectrometric methods so that unambiguous identification of the analytes, *i.e.*, drug of interest is possible.

Chromatographic methods are commonly used for separating the components of solution. In drug analysis, chromatography was originally used to verify drug levels in formulation broths or biological fluids for clinical applications. The commonly used procedures for the detection of veterinary drug residues include high performance liquid chromatography, gas chromatography, thin layer chromatography and mass spectrometry. Chemical methods usually proceed with a preliminary extraction in order to isolate the drugs of interest from the biological matrix. The main objectives of sample treatment are the removal of macromolecules and other matrix constituents that may either adversely affect the chromatographic system or interfere with the detection and enrichment of the analytes in order to achieve the required low limits of detection. Liquid chromatography (LC) has emerged as the method of choice for determination of antimicrobials, which are rather polar, non-volatile, and sometimes, heat sensitive. The development of coupled liquid chromatography-mass spectrometry (LC-MS) has increased the range of antimicrobials for which assays, based on molecular spectrometry, can be developed. Analytical methods are comprised of analyte clean up and detection. Sample clean-up procedures include preferably chromatography, liquid-liquid extraction, solid phase extraction and matrix solid phase dispersion. The mobile phase passes over the stationary phase at a constant rate. The two phases possess different chemical properties. As the analytes in the mobile phase pass over the stationary phase, those with polarity closer that of the stationary phase are retained selectively for a time on the column. Conversely, the analyte molecules with polarity closer to that of the mobile phase tend to remain in the mobile phase, passing through the column faster. Passing through the instrument monitor sequentially, these *groups* of molecules give rise to peaks on the chromatogram.

Prevention of veterinary drug residues- challenges in Indian conditions

The issue of prevention of veterinary drug residues in foods of animal origin in Indian context is very complex. It is not surprising because the issue of food safety itself has not been given due attention. There are many challenges to deal with these issues. The major challenge is due to the lack of necessary infrastructure. High cost involved in building and maintaining the well-equipped residue analytical laboratories makes it very difficult the already voluminous task of surveillance and monitoring of veterinary drug residues. Indian legislation on food safety has not yet established the MRLs for veterinary drug residues in meat and milk indicating that it has not even taken up the issue. Another challenge is due to lack of traceability of the trade of meat and milk, which is attributed to the marketing of considerable volume of these animal originated products through unorganized channel. Further, Indian consumers are unaware about the hazards of these drug residues, and therefore, cannot build the positive pressure necessary for the functioning of government authorities. Therefore, to prevent the veterinary drug residues in foods of animal origin, we have to overcome these challenges by implementing following steps:

- The first step in residue prevention is to make individual consumers and food establishments aware of the problem through education by veterinarians, organizations and literatures.
- Taking into consideration the high cost involved in establishment of sophisticated laboratories, there is a need for the development of rapid and cheap screening procedures for the analysis of antibiotic residue.
- Indiscriminate use of antibiotics in field veterinary practices should be avoided.
- There is need to strengthen the nation-wide monitoring and periodic surveillance system for veterinary drug residues in edible tissues and milk.

Conclusion

Foods of animal origin constitute an important component of human diet. When properly used, veterinary drugs are safe. However, their indiscriminate and extra-label use can lead to the problem of residues in foods of animal origin. Due to which, the consumers have to face undue health hazards associated with the veterinary drug residues. It is therefore important to monitor consistently the quality of foods of animal origin with respect to chemical quality. Further, the safety of foods of animal origin should be addressed at farm level itself by maintenance of proper health records, disease and treatment history and strictly monitoring the withdrawal time of all the veterinary drugs used in farm animals.

■■■

17 Impact of Molecular Biology on Human Health

Amit Mathews

Canadian Food Inspection Agency, 2301 Midland Avenue, Scarborough, Ontario M1P 4R7 Canada, E-mail: <amitmathews@gmail.com>

For many centuries, the scientists had wanted to find out what living matter was made of. Why were living organisms different from inanimate materials? What distinguished a dog from a fish or a bird from a human? Were there similarities between life forms, which show why they were alive? In this chapter, these issues will be briefly discussed and how they led to the dawn of new science of Molecular Biology. We then discuss the most common and important tools used by Molecular Biologists and explore the tremendous impact this new science has had in every aspect of human life, especially human health.

1. The Information of Life

All living matter is made of tiny compartments called cells and only cells can give rise to cells. Cells are the fundamental units of life. Higher organisms like plants and animals are composed of millions of cells grouped in organs and performing different functions. They vary in size from one or two millionths of a meter (1-2 μm) in diameter in bacteria to around 10 μm for most animal and plant cells, however, some specialized cells are much larger. The egg cells of certain birds may exceed 10 cm in diameter, while the nerve cells connecting a giraffe's brain to its toes are 4-5 m long. Cells are complex microscopic bags of chemicals surrounded by a very thin flexible cell membrane that holds together the contents of cell and controls the passage of nutrients into the cell and toxic wastes out of the cell. A rigid outer wall reinforces most plant and microbial cells. The most important property of all living cells is the ability to produce replicas of themselves from animate materials. This capacity for self-perpetuation can be taken as the definition of life. Living matter can produce copies of itself without any instructions from outside, while the lifeless matter cannot.

1.1 The genetic blueprint

To build complex structures, the cell needs reliable construction material and a plan. The equipment and raw materials are the cell and its contents, and the plan (blueprint) is the genetic information that is the same whether it is in the cells of the brain or the liver. The genetic information determines the distinguishing features of all organisms. It controls growth of the cells and their development into tissues. Genetics, the study of the information content of cells, has had a profound impact on how scientists understand the natural world. Knowing how to interpret the blueprint of life allows scientists to draw up plans for the design of organisms with new functions and properties.

In later part of the 19th century, the chromosome was discovered. Chromosomes, seen under a microscope, are dark-staining strands in nucleus of the cell. Chromosomes occurred in pairs and they were highly organized so that when cells divided, each daughter cell received a complete set of chromosomes. In 1903, William S. Sutton, an American biologist suggested that chromosomes were the physical basis of heredity. By 1900, it was known that the chromosomes contained both protein and a molecule called deoxyribonucleic acid (DNA). Chemists had broken DNA down to its fundamental components and analyzed them. The composition of DNA was simple and appeared to very similar in all organisms. On the other hand, proteins were highly complex molecules made up of 20 different subunits called amino acids. Therefore, the prevailing opinion was that proteins were the genetic material. DNA was regarded merely as molecules, which held proteins in a linear conformation while protein duplication took place. In early 1940s, the American chemist Oswald Avery produced a 99.98% pure preparation of DNA, which he used to transform bacteria. Its transforming activity was destroyed by the enzyme DNAase that destroys DNA but not enzymes (proteins). This finally produced convincing information that DNA, after all was the *transforming principle*, the elusive genetic material.

2. The Tools of the Trade

2.1 DNA- The molecule that codes for life

Since beginning of the 20th century, the chemists had been trying to work out the chemical structure of large molecules found in the nucleus of cell, the *nucleic acids*. They knew that these substances always contained phosphate, molecules of a particular kind of sugar and several different kinds of organic compound called bases. The nucleic acid found in the chromosome was named as deoxyribonucleic acid (DNA). There were four different bases in DNA, each containing carbon and nitrogen atoms arranged in rings. These were Adenine (A), Guanine (G), Cytosine (C) and Thymine (T). However, the chemists did not know how these subunits might be assembled. At first, they thought the bases occurred in equal amounts in the DNA of all organisms, a belief which could be held as evidence against DNA being the genetic material. However, Erwin Chargaff and his colleagues showed in the late 1940s that the base composition of DNA varied considerably from species to species. In humans, there is 1.5 times as much Adenine and Thymine than Cytosine and Guanine. In fungus *Aspergillus niger*, there is approximately equal amount of each of the bases. Thus, there was sufficient variation for DNA to be the genetic material.

While Chargaff was analyzing the composition of DNA, the English chemist Alexander Todd was investigating its structure. He showed that each sugar molecule was connected through phosphate groups to two other sugars. He concluded that DNA was a long linear molecule with a backbone of alternate sugar and phosphate molecules and the repeating units were called nucleotides. In early 1950s, Rosalind Franklin and Maurice Wilkins examined the structure of DNA using X-ray crystallography. Their results allowed James Watson and Francis Crick to postulate a structure for DNA in 1953. The classical and well-known double helical structure of DNA was first described in 1953 for which Watson and Crick won the Nobel Prize. They proposed that the DNA is a helix in which two strands of DNA run parallel but in opposite directions. In center of the molecule, the weak chemical bonds called hydrogen bonds hold the two strands together. Adenine was shown to bond always with Thymine and Cytosine with Guanine. Other combinations are not possible since the base pairs cannot form the appropriate hydrogen bonds. The model they proposed was an unprecedented triumph of Science, one of the greatest discoveries of the 20th century. It drew upon and accounted for all known physical and chemical properties of DNA. It also heralded the dawn of Molecular Biology.

2.2 Restriction Enzymes- Molecular Scissors to cut and paste DNA

Until 1970, there were no convenient ways of fragmenting DNA. The molecule could be broken mechanically or biologically into very small pieces by several enzymes but this could only generate undefined and randomly sized DNA pieces. Such methods were not suitable for the purposes of cloning DNA. All that changed with the discovery of restriction enzymes. A restriction enzyme recognizes a unique sequence of DNA and makes a cut in or around that unique sequence at a unique site. Restriction enzymes thus provided an entirely predictable and reproducible way of cutting DNA. EcoRI is a restriction enzyme obtained from the bacterium *Escherichia coli.* It recognizes the six base GAATTC sequence and breaks the phosphate bond between G and AATTC. Different enzymes recognize different sequences and make different kinds of cuts. While the order of the bases dictates where a restriction enzyme will cut DNA, the length of the recognition sequence dictates how often cuts will be made. EcoRI with its six base recognition sequence cuts every 2000 bases or so. Restriction enzymes cut both strands of the duplex DNA. Where the cuts are on opposite ends, the cleaved (restricted) DNA will have short single stranded ends called *sticky* ends. Sticky ends help to bring pieces of DNA together but they cannot hold them permanently. An enzyme called DNA ligase is needed to join and seal the fragments. It closes gaps in the DNA backbone between single strands of DNA held in place by hydrogen bonding. Thus, DNA can be both cut and rejoined in a controlled fashion. The next problem is to get the DNA into a host organism using specialized carrier vehicles called vectors.

2.3 Vectors- Vehicles for DNA travel

Vectors are the vehicles in which new fragments of DNA are transported into the host cells. The three main types include plasmids, bacteriophages (viruses that infect bacteria) and cosmids. Plasmids are circular extra-chromosomal rings of DNA. Only small inserts of DNA (up to 1000 bases) can be shuttled easily using plasmids. This is usually more than enough for cloning single genes and their regulatory sequences since the average gene is about 1000 base pairs long. The bases up to 24000 (24 kilo bases or kb) can be inserted into bacteriophage vectors. Most of these vectors are derived from lambda virus that infects *Escherichia coli.* To insert any more, essential functions need to be deleted or removed. This is exactly what the scientists do to produce the cosmid. All the lambda genes are removed except the cos genes- those required to pack the DNA into the bacteriophage head. Thus, very large DNA fragments up to 50 kb can be packaged and shuttled using cosmids. The bigger problem is to transfer the cloning vector constructed into an organism in which it can replicate and maintain itself. DNA borne by bacteriophages and cosmids can enter host cells by bacteriophage infection. The transfer of plasmid DNA is much less efficient. Plasmids are made to enter *Escherichia coli* by a technique called transformation, wherein plasmid DNA is made to enter the host cell by treating the cell with a special chemical broth or using high voltage electric pulses. Transformation is very important in genetic engineering because DNA from any source can be introduced into suitable cells. Artificially created sequences of DNA that have never existed in nature can also be transferred.

2.4 Polymerase Chain Reaction - DNA goes viral

Molecular Biology has been revolutionized by a technique that allows selective amplification of DNA sequences. The technique is known as polymerase chain reaction (PCR). The principle behind PCR is very simple in theory. Two primer sequences are required one for each strand to start the reaction. Another critical requirement is an enzyme called Taq DNA polymerase, a heat stable enzyme (purified from a bacterium called *Thermus aquaticus* naturally found in hot springs) that can copy DNA sequences. To design a primer, sequence information of the region to be amplified must be available. The primers are synthesized as oligonucleotides and are added to the reaction in excess. The target DNA is denatured by heating (usually about 95°C), and then, the temperature is brought down close to a temperature where the primers would anneal to each strand but the parent DNA would remain separated. The temperature is again raised to 72°C kick starting Taq DNA polymerase into copying DNA. These steps constitute one-step of the cycle. Each cycle doubles the amount of DNA from the target sequence. The reaction is carried out in automated thermocyclers wherein the cycle is repeated about 30-35 times and under 3 hours thousands to millions of copies of the target DNA sequence is obtained. Developed in 1983 by Kary B. Mullis for which he won the Nobel Prize in chemistry in 1993, PCR has found an application in almost every aspect of biomedical science and is now routinely used in research and development.

2.5 DNA Sequencing- DNA goes digital

Scientists represent the bases of DNA as letters- A, T, C and G, and then, the genes become long words and they can be analyzed efficiently by computers. To read the language of DNA, it is necessary to determine the order of bases in the chain. There is no direct way of doing this. Sequencing relies on converting the order of bases into something that can be measured. Scientists have an accurate way of comparing lengths of DNA fragments. It is a process called electrophoresis, which separates different sized molecules using an electric charge. DNA is a negatively charged molecule and in an electric field, it will migrate towards the positive electrode. Small fragments of DNA travel more quickly than the larger fragments and by comparing the distances that the fragments migrate, they can distinguish molecules of DNA with one, two, three and four bases and so on but how is electrophoresis adapted to yield information on the sequence of DNA molecules rather than just their lengths? The trick is to produce DNA fragments in a sequence specific way. Sequence specific DNA fragments can be produced biochemically using compounds called dideoxynucleotides. These are identical to normal components of DNA, the deoxynucleotides, except that they are missing an oxygen atom. They can *fool* enzymes into incorporating them into growing strands of DNA. However, once in the strand, the absence of oxygen atom prevents DNA from growing any further. By blocking DNA synthesis with different dideoxynucleotides, the sequence specific fragment lengths can be obtained which can be analyzed by electrophoresis. This method was developed by Fred Sanger at Cambridge, United Kingdom. The other common method uses a radioactive phosphate molecule, which is inserted at one end of a singe stranded DNA sequence. Next, four chemicals are used to disrupt the DNA chain at each of the four bases. One chemical breaks the DNA chains before G, and another cleaves them before C and so on. Each of the four chemical treatments produces different sets of fragments. Electrophoresis can then be used to compare fragment lengths, and from the pattern of radioactive bands, the scientists can figure out the sequence of the original DNA. Thus, if a molecule six bases in length is produced by the chemical that cleaves before G, then the seventh base must be G. The same chemical might produce fragments of 10, 11 and 16 bases indicating G at position 11, 12 and 17. When all four patterns are combined, the DNA sequence is revealed. This technique was developed in 1977 by Allan Maxam and Walter Gilbert at Harvard University and is commonly called the Maxam-Gilbert method of DNA sequencing.

2.6 Nucleic Acid Hybridization Blots- DNA goes diagnostic

Nucleic acid hybridization is a process in which complementary single strands of nucleic acids combine to achieve a stable double-stranded nucleic acid molecule. This action has been used to establish a molecular or genetic relatedness among organisms and to characterize their genomes. Furthermore, this technique is one of several diagnostic tools useful for detecting a wide variety of conditions. Since the determination of basic principles of duplex formation and stability in the 1950s, many variations of the hybridization techniques have been developed. Edwin Southern, a British biologist first transferred DNA fragments from an agarose gel, after electrophoretic separation, onto nitrocellulose. The technique is now known as Southern blotting. Shortly afterward, a similar Northern blotting technique in which separated RNA strands are transferred from an agarose gel to a suitable solid support was described by James Alwine, David Kemp and George Stark at Stanford University. A logical extension of these blotting techniques has been the dot or slot blot in which the sample is applied directly to the solid support without prior size separation. Over the years, these techniques have been further developed and modified extensively by many researchers across the world. The application of these methods is as varied as the procedures used, *e.g.*, to determine the changes in nutritional state of an environment, to establish taxa genetically, to distinguish pathogenic from non-pathogenic viruses and to analyze gene structure.

Southern Blotting

The DNA sample, which can include blood, tissue biopsies, buccal scraps, amniotic fluid and cultured cells, is generally digested using a restriction enzyme, and then, subjected to electrophoresis in a horizontal agarose gel. After sufficient time has elapsed to achieve adequate separation of the required fragments, the gel is soaked in an alkali solution to achieve denaturation of the double-stranded nucleic acid, then neutralized and prepared for transfer. Transfer to a nitrocellulose, polyvinylidene (PVDF), or Nylon membrane is achieved by a process of blotting, in which buffer is drawn through the gel and the membrane. The fragments carried with the buffer are retained on surface of the membrane. The retention is made more permanent through a fixation process. The blot can then be used in hybridization with labeled probes to identify the fragments of interest.

The membrane of choice is determined by the sensitivity required and the detection method to be used. The quantity of sample has a significant effect on both of these. The use of nitrocellulose usually results in low backgrounds and it is recommended when the level of target is high. Nitrocellulose membrane can bind 80-125 μg nucleic acid/cm^2, which is significantly less than the binding capacity of 400-600 μg/cm^2 for a nylon membrane. Its ability to bind small molecules (< 400 nucleotides) is also poor, and transfer buffers must contain high salt concentrations to ensure efficient nucleic acid binding. Nylon membranes are available in uncharged and positively charged supported forms. Charged nylon has a higher binding capacity and is particularly useful when working with low-molecular-weight nucleic acid. Binding to a nylon membrane is independent of the ionic strength of the transfer buffer. However, backgrounds can be elevated. Where repeated use of a membrane in hybridization assays is needed, the use of nylon membranes is strongly advised. Nylon is also recommended for use with medium- or low- abundance targets and is, in general, the membrane of choice when working with nucleic acids. PVDF membranes behave similarly to uncharged nylon, but because of its hydrophobic nature, use in nucleic acid blotting is limited.

Fixation bonds the target nucleic acid to the membrane. The principal fixation methods of heat and ultraviolet (UV) light can be used with all types of membrane. Heat fixation is very reproducible but requires a vacuum oven for nitrocellulose. Ultraviolet cross-linking, performed using an ultraviolet cross-linker (constant energy setting), is faster than heat. Alkali provides a third alternative method when charged Nylon membranes are used. Original methods describe the use of a DNA probe radioactively labeled by random priming or nick translation with (P^{32})-dCTP to detect specific nucleic acid fragments immobilized on nitrocellulose membrane. Since then, many different methods for labeling nucleic acid probes ranging from short oligonucleotides to longer DNA or RNA fragments have been developed. Nonradioactive labeling kits and reagents are also available, finding favour in a number of niche areas. The role of hybridization buffer is to provide conditions that promote hybridization between the labeled probe molecules and its complementary sequence immobilized on the membrane, and simultaneously, to limit hybridization between sequences that are not perfectly matched. Southern blotting has been applied to the diagnosis of many human diseases at molecular level caused by point mutations, gene rearrangements, or the amplification of genes or specific sequences within the genome. These methods have in common that restriction-digested genomic DNA is size-separated in agarose gel electrophoresis, transferred onto the membrane and hybridized with gene-specific probes.

2.7 Western Blotting- Proteins catch up with DNA

Protein blotting also known as Western blotting refers to the transfer of electrophoresed proteins from the polyacrylamide gel electrophoresis to an adsorbent membrane that binds the eluted macromolecules (*the blot*). This method was first described in 1979 by Towbin. Electrophoretic transfer uses the driving force of an electric field to elute proteins from gels. This method is fast and efficient, and it maintains the high resolution of protein pattern. Different probes can be used to react with the transferred proteins on the blot: antibody for the identification of corresponding antigen, lectin for the detection of glycoproteins, ligand for the detection of blotted receptor components, *etc.* Western transfer with subsequent immuno-detection has found wide application in the fields of life sciences and biochemistry. The blotted proteins can efficiently be detected and characterized, especially those that are of low abundance.

The blot is also widely used with various techniques of protein identification from which the measurement of protein mass (mass spectrometry) or determination of the protein sequence (*N*-terminal Edman degradation, *C*-terminal sequence, or amino acid analysis). The power of protein blotting lies in its ability to provide simultaneous resolution of multiple immunogenic proteins within a sample. The blot analysis generally requires small amount of reagents, the transferred proteins on membrane can be stored for many weeks before their use, and the same blot can be used for multiple successive analyses.

2.8 cDNA Microarray Technology- DNA lands on a chip

Ribonucleic acid (RNA) stands between DNA and protein in the central dogma of life proposed by Francis Crick. Although the majority of messenger RNA species have no direct role in cellular metabolism but the relative abundance of each gene's transcript can reflect the identity and state of each cell. Thus, accurately measuring the pattern of expression across all genes also referred to as the *transcriptome* is a potential mechanism by which to assay cellular biology. Messenger RNA (mRNA) comprises approximately 1% of total cellular RNA but is the template for protein translation and the major focus of expression analysis. The product resulting from mRNA processing for micro-arrays (henceforth referred to as target) depends on the micro-array platform with the two most common forms being single-stranded complementary DNA (cDNA) following reverse transcription of mRNA and antisense RNA (aRNA) resulting from *in vitro* transcription of a double-stranded cDNA template.

Two key features of target synthesis are critical to the successful micro-array analysis. First, the resulting target must be in a form that can hybridize to its complementary sequence. In general, the target is single stranded and not too heavily laden with large macromolecules to interfere with complementary base pair binding. In addition, the final target is often fragmented to shorter lengths to minimize secondary structure that can significantly affect complementary binding. Second, the target must be detectable when bound to its complementary sequence. Although this is achieved by incorporating radioactive isotopes or fluorescent-dye-labeled nucleotides into the final synthesis of target, a balance must be struck between successful detection and steric hindrance of complementary binding.

Once a labeled target is synthesized from mRNA, micro-arrays are used to detect the relative abundance of each transcript. The platforms upon which DNA arrays have been made range from the earliest having cDNA clones on Nylon filters to silicon wafers that use photolithography to create oligonucleotides of known sequence. The two major platforms currently in use for large-scale expression analysis are spotted micro-arrays and synthesized micro-arrays. Spotted micro-arrays are created by adhering individual species from curated cDNA libraries onto a glass slide. Each spotted micro-array element is a known cDNA assigned to a specific location on a two-dimensional surface. Synthesized oligonucleotide micro-arrays use chemistry to create a grid of unique oligonucleotides (called features) complementary to known genes.

Each platform has advantages and disadvantages but all use the basic chemistry of sequence-specific hybridization. After a labeled target is hybridized to the micro-array, the amount of radioactivity or fluorescence at each location on the micro-array (signal) is determined. Because the intensity of signal at any location is dependent on many variables, micro-array determination of expression is relative not absolute. Although this is obvious with the dual-hybridization approach applied to most spotted arrays, it is equally true for synthesized arrays. Thus, detection of a gene on a micro-array is almost meaningless unless evaluated in the context of a well-designed experiment or previously subjected to rigorous analysis to understand the correlation between detected expression and absolute expression. Scientists have seized on new modality of cDNA micro-arrays to assay and observe cellular biology. Investigators began applying micro-arrays to single-cell organisms and they have advanced up the phylogenetic tree. They began with very simple experimental designs in simple model systems and they have progressed to large multifaceted experiments in complex models. As the technology has progressed, the methods by which to analyze and interpret data generated by micro-arrays have also become much more advanced.

2.9 Site Directed Mutagenesis – making mutants becomes a reality

Site directed mutagenesis-using oligonucleotides was first described in 1978 by Michael Smith, who shared the Nobel Prize in Chemistry in 1993 with Kary B. Mullis, the inventor of PCR. Since late 1980s, the protein molecules were altered by site-directed or site-specific mutagenesis of their genes. In this technique, a mutation is created at a defined site in DNA leading to a change in the amino acid of corresponding protein. This method requires the wild-type sequence to be known. The change itself is made by PCR methods where primers containing the desired mutation are used. In first cycle, there is a priming mismatch for the primers binding the template DNA strand, but after the first cycle, the primer-based strand, containing the mutation, will be at about equal concentration to the original template. After successive cycles, its number will increase exponentially and out numbers the original, un-mutated strand, resulting in a nearly homogeneous solution of mutated amplified fragments. For this PCR, it is necessary to design primers that are suitable for the desired changes, considering also their annealing temperature.

Two techniques are commonly used to introduce specific amino acid replacements into a target gene. The first of these is termed the overlap extension method. In this method, four primers are used in the first polymerase chain reaction (PCR) step with two separate PCRs being performed. The primer pairs for these PCRs are 1 and 3 as well as 2 and 4, respectively, with primers 2 and 3 containing the mutant codon with a mismatched sequence. Two double-stranded DNA products containing the desired mutagenic codon are obtained over several PCR cycles. In second PCR step, these two DNA products are amplified using primers 1 and 4 resulting in the mutated DNA. A useful variant of the overlap extension method is the megaprimer method. In this procedure, two rounds of PCR are performed employing two flanking primers and one internal mutagenic primer that contain the desired base substitutions. A benefit of this method is that mutations can be inserted into the flanking primers so that multiple codons relatively far from each other can be replaced at once. The second method for performing site-directed mutagenesis is referred to as whole plasmid, single-round PCR. In this protocol, two oligonucleotide primers containing the desired mutation(s) are extended with DNA polymerase. In this PCR step, both strands of template are replicated without displacing the primers to obtain a mutated plasmid containing breaks that do not overlap. As the original wild type plasmid originates from *Escherichia coli* and is thus methylated on various A and C residues, it may then be selectively digested using Dpnl methylase endonuclease resulting in a circular, nicked vector containing the mutant gene. When this nicked vector is transformed into competent cells, the nick in the DNA is repaired by the cell machinery to give a mutated, circular plasmid. The advantages of whole plasmid, single-round PCR, are that only one PCR needs to be performed and only two primers are required. The disadvantages of this technique relative to overlap extension are that it does not work well with large plasmids (>10 kB) and typically only two nucleotides can be replaced at a time. Several companies offer kits for performing these methods. After the PCR step and the cloning and transformation, expression and purification of the recombinant protein mutants must be performed for testing and evaluation.

2.10 Human Genome Project- The most ambitious biological project in history

Scientists have obtained the complete sequence of thousands of bacteria and hundreds of higher organisms. The most ambitious plan so far is to sequence the entire DNA in the human cell or the human genome. Each cell in our body contains 23 pairs of chromosomes, together coding for every aspect of our biological existence. In these chromosomes, there are approximately 3 billion base pairs of DNA to be sequenced. Such is the enormity of this task that without computers, it is almost impossible to sequence the human genome. The Human Genome Project (HGP) was an international scientific research project to determine and identify the complete DNA sequence of the human genome including mapping approximately 25000 genes of the human genome. The project began in 1990 and it was initially headed by James Watson at the US National Institutes of Health (NIH). A working draft was released in 2000 and a complete draft in 2003. Celera Corporation headed by Craig Venter conducted a parallel project and first came up with a working draft using advanced techniques collectively called the *shotgun* method. Most of the government-funded sequencing was achieved in research centers and universities from the United States, Canada, United Kingdom, Japan, France, Germany, China, India and New Zealand. The project opened a treasure trove of information about our genome, the impact of which will be felt for decades to come. Without any doubt, this project is one of the greatest achievements of mankind, the biological equivalent of going to the moon.

3. Molecular Biology revolutionizes Agriculture and Human Health

Molecular Biology has revitalized every aspect of Agriculture and Human Health. It has revolutionized many areas and has proved to be seminal in improving every conceivable facet of human life.

3.1. Molecular Biology revitalizes Agriculture

Plants are the primary sources of human and animal food. Fueled by light energy from the sun and assisted by a range of helpful microbes, they convert carbon dioxide and nitrogen gases from the atmosphere, and minerals and water from the Earth into their own cells and tissues. Animals and humans are incapable of achieving this remarkable feat. We cannot obtain the chemicals we need for growth and maintenance simply by breathing air, eating soil, drinking water and lying in the sun. We must consume plants and animals who themselves have consumed plants.

Growing plants is the world's oldest, largest and most important technology. We depend on plants for food, medicines, fuel, paper and furniture and most importantly for oxygen. For thousands of years, humans have selected plants to meet their own needs. Varieties have been produced and selected to be more resistant to summer drought and winter cold enabling farmers to sow crops over wider geographical areas. Plants have been interbred so that desirable characteristics from different varieties are combined in a new strain. Until recently, plant breeders had to depend on natural mechanisms of plant reproduction to obtain a better crop strain. To improve the disease resistance of a high grain yielding commercial strain, pollens from a disease resistant variety of wheat might be used to fertilize a commercial strain. The resulting plants would have a mixed genetic composition and would display mixed characteristics, some good and some bad. The undesirable ones could be gradually removed from the new plants by repeatedly breeding them with the high yielding parent (called backcrossing) to produce eventually a high yielding disease resistant plant with no undesirable-features. Selecting, crossing and backcrossing can take a very long time though, sometimes years. Molecular Biology is changing all that. Molecular manipulation (also called genetic engineering) enabled the process with extreme rapidity and the genes no longer have to mixed and sorted randomly. Furthermore, Molecular Biologists could take the desired genes from any source (even bacterial genes) to improve upon a commercial variety. The method is much quicker and better than the traditional way of breeding new plant varieties.

Although plants are far more complex than microbes but the principles of Molecular Biology are much the same in both. The genes are isolated from the donor organism and are inserted in a carrier molecule of DNA- known as a vector. This vector is then introduced into the plant cell where it gives rise to the protein coded for by the inserted genes. The most commonly used carrier of genes in Plant Molecular Biology is the *Ti* plasmid, a bacterial vector that can function in plant cells. It comes from the remarkable bacterium called *Agrobacterium tumefaciens*, which lives in the soil. This bacterium is a plant parasite that infects cuts in plants and causes a cancerous growth called a crown gall. It does so by invading plant cells and releasing its Ti plasmid. Part of the Ti plasmid integrates itself into the plant DNA, and it is then treated like a normal plant gene. The inserted DNA converts the invaded plant into a factory producing food for *Agrobacterium*. Scientists call this bacterium *nature's own genetic engineer*.

Human genetic engineers have adapted the Ti plasmid to transfer foreign genes into plants. The trick is to remove the genes coding causing the cancer and virulence and replace it with the gene of interest. The genes that code for insertion of the DNA into the plant chromosome are also retained. This re-engineered plasmid is utilized to infect the plant. The new plasmid does not cause crown galls, and hence, molecular biologists cannot easily tell which plants have been infected. They use other techniques to detect the successfully infected plants. The foreign DNA introduced into the plant also contains a gene that codes for antibiotic resistance, most commonly against kanamycin. Only the successful variants would survive in the presence of kanamycin and they are selected. *Agrobacterium tumefaciens* only infects broad-leaved plants like tobacco, petunia and tomato. Wheat, rice and maize crops, most important to agriculture, are not susceptible and so other ways to introduce foreign DNA are used for these crops. Methods like electroporation (direct injection using electric pulses) are generally used. Plant cells are stripped off their protective cell wall, which results in a structure called a protoplast. Short duration electric pulses make tiny holes in the cell membrane through which DNA enters the cell. Though these holes exist only for a fraction of econd (the duration of the pulse), it is enough time for DNA to enter the protoplast. The foreign DNA is then integrated into the plant's genome and is thereby treated like its own DNA. A complete plant can be regenerated from the fragile protoplast.

Molecular Biology has also substantially removed the need for plants to be sprayed with insecticides. The toxic protein from the bacterium *Bacillus thuringiensis* kills larvae of moths and insects but has no effect on animals, plants and humans. In 1987, scientists in Belgium and the United States took the gene coding for this toxic protein and inserted into tobacco plants *via Agrobacterium tumefaciens* Ti plasmid. The plants produced the toxin and killed predatory caterpillars feeding on its leaves while themselves growing normally. Viruses are another source of crop damage and one for which chemical treatments are available. Molecular Biology has again produced a solution. Tobacco mosaic virus (TMV) is a virus infecting and damaging tobacco crops. This virus consists of a spiral of RNA surrounded by a protein coat. If the gene for the coat protein is inserted into the DNA of cells from the tobacco plant, the plants that result are immune to the virus. The same result is seen in tomato.

3.2 Molecular Biology in Diet and Disease

Malnutrition and starvation in overpopulated countries might be overcome in the short term by intensive agriculture. Molecular Biology powered Biotechnology will help to increase agricultural productivity, providing nitrogen-fixing *Rhizobium* bacteria with which farmers can inoculate their crops, improving livestock and producing drought resistant plants. Farmers in developing countries could also grow more-nutritious crops with the help of genetic engineering. People in developed nations are now beginning to alter their diet for health reasons. However, eating habits die hard, so there is also a growing demand for diet products that smell, taste and look just like traditional foodstuffs but have very few calories. Sugars in soft drinks and processed food are increasingly being replaced with sweeteners like aspartame.

Diagnosis of diseases depends on information passing from patient to doctor. To penetrate the biochemical secrets of disease, the scientists have designed three main categories of diagnostic aids. These are enzyme assays, antibody assays and DNA probe assays. In all three, biological polymers (proteins and DNA) are used to provide information on molecular symptoms of the disease.

3.2.1 Enzyme Assays

Enzyme tests usually give information for the detection of small molecules like sugars. Diabetics, for instance, use a simple biochemical test for measuring their glucose levels to help them decide whether to take more insulin. The test enzyme, glucose oxidase, produces peroxide in response to any glucose in blood. A second enzyme converts the peroxide produced by glucose oxidase to water and simultaneously changes tetrametyhl benzidine (TMB), another chemical present to a deeply coloured dye. The amount of dye formed corresponds to the amount of peroxide, and hence, of glucose in the sample.

The enzyme reactions can be performed in liquid solutions, but for convenience, the enzymes are often attached to a solid surface. Solutions of enzymes and other reactants can be dried onto strips of cellulose fiber paper or a swab. When needed, a strip can be dipped into a sample of blood or urine. In this form, the tests last a long time without *going off*, and it can be used by non-medical personnel including patients.

3.2.2 Antibody assays

There are thousands of enzymes but few are used in diagnostic tests. Antibodies are much more widely employed. The exceptional capacity of antibodies for specific binding of biological molecules makes them powerful *homing* devices, however, when antibodies bind to antigens, neither of the components changes in a detectable way. In diagnostic assays, signals such as radioactive or fluorescent compounds need to be attached to antibodies. When radioactive antibodies bind an antigen, the antigen effectively becomes radioactive and easily detectable. However, radioactive labeling requires special safety measures and is not always convenient. Other labels are more frequently used, mostly enzymes. For instance, the peroxidase enzyme can be chemically linked to antibodies to combine biological specificity with signal generation.

Light producing enzymes are routinely used and other labels are often attached indirectly to the antibodies by linker molecules. Biotin, a vitamin and avidin a protein than binds to biotin, provide a kind of universal link. With biotin attached to antibodies, anyone of a number of labels bonded to avidin can be used in the assay.

Pathogenic microorganisms can be detected in an impure clinical sample such as a skin swab, pus or feces sample. If a slide on which the specimen has been smeared is flooded with a solution containing fluorescence labeled antibodies against the pathogen, the organism will show up as bright spots. Many antibody assays start by attaching the antigen to a surface. In Enzyme Linked Immunosorbent Assay (ELISA), antibodies bonded to the inside of test tubes effectively extract the antigen from solution placed in the tubes. In next stage, a second antibody with an enzyme attached binds to the antigen. Third stage of ELISA is the production of colour change reaction by the bound enzyme. When macromolecules like proteins or carbohydrates form part of the cell surface, they can be used to identify the type of cell. The *Salmonella* bacteria, which sometimes cause food poisoning, are distinguished by antibody-based tests that identify their carbohydrate coat, a protein appendage called flagellum and other antigens responsible for virulence. These characteristics are equivalent to eyewitness descriptions of criminals. Usually, no single antigen is definitive. Microorganisms also leave molecular fingerprints- single antigens that identify the organism. Infected cells of AIDS sufferers contain proteins produced by the AIDS virus. Their associated antigens can also identify cancers.

Monoclonal antibodies

Activation and clonal expansion of antigen-reactive B cells to mount an immune response is one of our body's most important defenses against foreign materials. Each clone of B cells is able to secrete its own unique antibody such that an invading pathogen will be countered by millions of antibodies capable of binding to different sites on its surface. Such a polyclonal response is ideal for our body's defense. However, many experimental and clinical situations require access to an unlimited supply of a single antibody with a clearly defined specificity and affinity, *i.e.*, a monoclonal antibody.

In short, monoclonal antibodies are identical antibodies produced by hybridoma cell lines derived from fusion of a B cell with a tumor cell. Kohler and Milstein first described the technique of monoclonal antibody production in 1975, a process that has been put to use in a wide range of laboratory, clinical and industrial applications. The importance of monoclonal antibody production and the theories surrounding it earned them, the 1984 Nobel Prize in Physiology or Medicine. Monoclonal antibodies have since become powerful tools in virtually every field of biological sciences and medicine and are a major driving force in medical diagnostics from over the counter home pregnancy tests to precise laboratory-based immunoassays. More recently, monoclonal antibodies have moved into the medical therapeutic field with products for the treatment of various cancers and autoimmune diseases. Therapeutic monoclonal antibodies now represent one of the fastest growing areas of the pharmaceutical industry, with potential market capital of billions of dollars. To date, the United States Food and Drug Administration (FDA) has approved 19 therapeutic antibodies for clinical use and over 150 antibodies are being tested in clinical trials. The future applications of monoclonal antibodies are limitless.

3.2.3 DNA probes

DNA probes take diagnosis of disease beyond the surface of cells and their products to the genetic material. Proteins and other metabolites are only manifestation of the genetic material and may not always provide the best sort of diagnosis. Analysis of DNA is the most intimate and the most-irrefutable form of diagnosis. DNA probes have been used since 1970 for diagnosing infectious diseases. Though antibody based assays are usually quicker and easier to perform, they are often inappropriate for microorganisms that cannot be grown in the laboratory or that vary in their antigens. DNA probes have successfully diagnosed various diseases glandular fever, hepatitis B, malaria, Legionnaire's disease, bubonic plague, AIDS, gonorrhoea and Chlamydia infections.

3.2.4 Genetic disease

Most of the excitement surrounding DNA probes stems from their use in diagnosing, not infectious disease but an inherited or genetic disease. Due to the defects in single genes, there are many diseases. It is possible to diagnose many of these conditions by using a DNA probe to track down a defective gene. According to the location of the gene and its behaviour, the scientists have grouped the genetic diseases in three classes, which are sex-linked. Most sex-linked diseases are X-linked. In X-linked diseases, males with the defective gene will have the disease, and females with it will be the carriers. Women, who are carriers of X-linked conditions, have a 50% chance of passing the disease to their sons or the carrier state to their daughters. One of the most tragic X-linked conditions is Duchenne muscular dystrophy. In autosomal recessive diseases, there are many more carriers of the disease than the sufferer ones. For instance, in 20 white people, one is a carrier of cystic fibrosis but only one in 1600 gets the disease. This is because the normal gene prevents the effects of the mutant form. It takes two defective copies of the gene, one from each parent before the disease appears. In diagnosing genetic disease, doctors attempt to trace inheritance of a disease gene through a family tree. Finding out which gene (healthy or defective) has been inherited is not simple but there are differences or polymorphisms in DNA sequences littered randomly throughout the chromosome. These polymorphisms arose by single base changes many thousands of years ago, and these are now often spread evenly in the population. Restriction Fragment Length Polymorphisms (RFLPs) are a special group of polymorphisms that occur at cut sites of the restriction enzymes. Restriction enzymes will cut a given piece of DNA into a distinctive set of fragments, which can be separated out electrophoretically to form a pattern of bands like a bar code. Scientists compare the child's DNA with that of its normal and affected relations to work out, which of the cutting sites are associated with normal genes and which with mutant types. DNA from parents, grandparents, aunts, uncles and siblings are all examined to build up a pattern for the disease. If the child's DNA shows the same restriction pattern as affected relatives, the chances are that the child has inherited the disease.

3.2.5 Prenatal diagnosis

Perhaps the most important feature of DNA probe diagnosis is that it can be performed before birth. This is not entirely new. Women over the age of 35 years, who are at relatively high risk of giving birth to children with Down's syndrome, often undergo amniocentesis, a harmless procedure for obtaining cells from the fetus. If chromosome abnormalities are apparent, a sure sign of the syndrome, the woman can then decide whether she wants the pregnancy to continue. This is a difficult decision riddled with ethical questions. Genetic diagnosis can even be performed before a pregnancy is confirmed.

3.2.6 Bioproducts

Traditionally, medicines have been extracted from plants, animals, or microorganisms. Molecular Biology provides remedies that are to be found inside our own bodies. Through genetic engineering, proteins responsible for maintaining health can now be produced in amounts large enough to study or use as drugs. Genetic engineers use recombinant DNA techniques to splice new DNA into a plasmid, and then, insert the plasmid into *E. coli.* The genetically engineered microbes are then grown to a high concentration in a fermentation vessel. Plasmids express their genes and produce proteins. If human genes (foreign DNA) are cloned into plasmids, these re-engineered plasmids express human protein. Each organism processes hundreds of copies of the plasmid expressing the human protein, and the substantial amounts of protein are produced quite rapidly. Only very small amounts of human proteins are usually needed to treat patients. One-tenth of a gram of insulin provides between 1000 and 20000 doses. The same amount of growth hormone represents daily treatment for almost 10000 patients.

The human proteins produced by genetically engineered organisms fall into two main classes. Firstly, there are those like insulin, growth hormone and Factor VIII whose deficiencies cause diabetes, dwarfism and hemophilia. Treatment of these diseases aims to restore the natural levels of the missing hormones in the body. Traditionally, insulin has been obtained from the pancreas of cattle and pigs, growth hormone from the pituitary gland of corpses and Factor VIII from donated human blood. Genetic engineering produces alternative sources that are often cheaper, more convenient and safer. Coronary thrombosis has been treated for many years with bacterial enzymes like streptokinase that destroys the blood clots blocking blood vessels. Genetic engineering has produced tissue plasminogen activator (tPA), a human protein that does a better job and with fewer side effects.

There is a second class of proteins, the biological response modifiers, for which genetic engineering is the only realistic source. The biological response modifiers occur at very low levels in the body but are vital in controlling bodily processes. For instance, α-interferon, present in blood as a defense against viral infection, is in such a low concentration that it takes about 90000 donors to produce one gram. Scientists have cloned the genes coding for proteins such as α-interferon, interleukins (immune stimulants) and endorphins (natural painkillers) in microorganisms. Many types of these proteins have already undergone safety tests in animals and humans.

4. Bioinformatics

Bioinformatics was originally defined as an interdisciplinary field involving biology, computer science, mathematics and statistics to analyze biological sequence data and to predict structure and function of macromolecules. With the dawn of genomics era, bioinformatics plays increasingly significant roles in biological and medical research. Informatics has been traditionally a discipline in which mathematicians, computer scientists and statisticians develop technologies to support information management in fields like health care. Bioinformatics is now involved in these activities by organizing biological data related to genomes with a view to applying this information in agriculture, pharmacology and medical science. The biological information is of two kinds- sequence information and structure/function analysis of the gene products. In the case of human genome, this role of bioinformatics means collecting information about biological role of the ~25000 human genes to uncover which ones have the most significant role in disease. Using modern technology such as gene expression micro-arrays, genetic manipulation of genes in cells and organisms, and rapid structural and functional assessment of proteins, many new data on gene function are being assembled.

The closely related field of computational biology also provides computational support for many of the above purposes but there are differences with the support provided by bioinformatics. On one hand, computational biology is concerned with the development of novel and efficient algorithms that can be proven to work on a difficult problem, such as multiple sequence alignment or genome fragment assembly. On the other hand, bioinformatics focuses more on the development of practical tools for data management and analysis, *e.g.*, display of genome information and sequence analysis, but with less emphasis on efficiency and proven accuracy. Bioinformatics supports a broad spectrum of research that includes determining the biological significance of data provides the expertise to organize it and develops practical computational tools needed to mine the data for new information. When the information is of practical importance, bioinformatics assists with practical applications such as the identification of new protein targets for drug therapy. The field of bioinformatics and computational biology is rapidly evolving and it will play changing roles in the use of genome data.

4.1 Protein sequence databases

The development of protein sequencing methods led to the sequencing representatives of several of the more common protein families, such cytochromes from a variety of organisms. Margaret Dayhoff and her collaborators at the National Biomedical Research Foundation (NBRF), Washington D.C. were the first to assemble databases of these sequences into a protein sequence atlas in the 1960s. Dayhoff and her co-workers organized the proteins into families and super-families based on the degree of sequence similarity. From these sequences, a phylogenetic tree was derived showing graphically which sequences were the most related, and therefore, shared a common branch on the tree. Once the trees were made, they were used to score the amino acid changes that occurred during evolution of the genes for these proteins in various organisms from which they originated. Dayhoff and her colleagues contributed in several ways to modern biological sequence analysis by providing the first protein sequence database as well as tables for performing protein sequence comparisons. Amino acid substitution tables are now routinely used in performing sequence alignments and database similarity searches.

4.2 DNA sequence databases

Collecting DNA sequences into the GenBank database was initiated by the Theoretical Biology and Biophysics Group founded by George I. Bell at Los Alamos National Laboratory in New Mexico, United States in 1974. This group of physicists wanted to provide a theoretical background to experimental work, primarily in immunology. The group's research expanded into other areas of computational biology and bioinformatics and into development of the first versions of GenBank under the leadership of Walter Goad and colleagues between 1982 and 1992. Goad first conceived of the GenBank prototype in 1979. Translated DNA sequences were also included in the Protein Information Resource (PIR) database at the NBRF. The European Molecular Biology Laboratory (EMBL) Data Library was founded in 1980. GenBank is now under the auspices of the National Center for Biotechnology Information (NCBI).

Initially, a sequence entry in these databases included a computer file name and DNA or protein sequence files. The entries were eventually expanded to include much more information about the sequence such as function, mutations, encoded proteins, regulatory sites and references. This annotated information was then placed along with the sequence into a database format, which could be readily searched for many types of information. The number of entries in the nucleic acid sequence databases GenBank, and EMBL has continued to increase exponentially from the daily updates. Annotating all of these new sequences is a time consuming, painstaking, and sometimes, error-prone process. However, the process is becoming more and more automated.

4.3 Sequence retrieval from public databases

An important step in providing sequence database access was the development of Web pages that allow queries to be made of the major sequence databases (GenBank, EMBL, *etc.*). An early example of this technology at NCBI was a menu driven program called GENINFO developed by D. Benson, D. Lipman and colleagues. This program searched rapidly through previously indexed sequence databases for entries that matched a biologist's query. Subsequently, a derivative program called Entrez with a simple window-based interface, and eventually, a Web-based interface was developed at NCBI. The idea behind these programs was to provide an easy to use interface with a flexible search procedure to the sequence databases using keywords searching on standardized entry fields. Sequence entries in the major databases have additional information about the sequence included with the sequence entry, such as accession number, name and alternative names of the sequence, types of regulatory sequences, the source organism, known mutations and references. Entrez accesses this information, thus, allowing rapid searches of entire sequence databases for matches to one or more specified terms. These programs also can locate similar sequences based on previous similarity comparisons. When asked to perform a search for one of more terms, simple pattern search programs will only find exact matches to a query. In contrast, Entrez searches for similar or related terms with great ease and lists the found items in the order of likelihood that they matched in the original query. Entrez originally allowed straightforward access to databases of both DNA and protein sequences in separate or the same databases. Eventually, Entrez provided access to all of Medline, the full bibliographic database of the National Library of Medicine (NLM), Washington D.C. This access is provided without cost to any user- private, government, industry, or research- a decision by the staff of NCBI that has provided a shot in the arm to biological research that cannot be overestimated since NCBI already gets several million independent access events to their system each day.

5. Molecular Biology- blazing into the New Millennium

For a new science, Molecular Biology has already changed human life in more ways than anyone could have possibly imagined. It is also poised to be at cutting edge of all biomedical discoveries of the new millennium and would definitely aid in finding cures for diseases like cancer, AIDS and malaria, which have plagued humans for a long time. It has and will continue to revolutionize every aspect of human existence.

■■■

18 Gene Therapy for Nutritional Disorder

Viroj Wiwanitkit

Wiwanitkit House, Bangkhae, Bangkok, Thailand-10160,
Visiting professor, Hainan Medical College, China, E-mail: <wviroj@yahoo.com>

Nutrients are required by everyone. Nutritional disorder is a common problem in medicine. There are many kinds of nutritional disorders including:

- **Under nutrition:** This is the condition that a person gets lower nutrients than basic requirement. This can result in many clinical symptoms.
- **Over nutrition:** This is the condition that a person gets excessive nutrients. A huge among of nutrients more than basic requirement is derived, and it can cause many clinical symptoms.
- **Contaminated food:** Contamination of food is the big problem. This is relating to the sanitation, and the problem is common in any country. Contaminated food is the big issue in public health.

Several diseases are listed in medical textbook. The important etiological factor for nutritional disorders is usually environmental related factors. However, some of those diseases are classified as genetic diseases. To manage these genetic nutritional diseases seems to be difficult. The primary approach is using nutritional management, which is usually the management at the terminal side, not the original side. A new approach for management is using gene therapy. In this specific chapter, the author specifically focuses on gene therapy for nutritional disorders.

Genetics and Nutrition Disorder

For any disease, the two main basic effectors include environment and genetics. Environment can be classified as any external abuse that affects the body, whereas, genetics can be classified as internal factor derived from ancestor and deeply rooted in body. Normally, these two main factors act on body in a good and appropriate way resulting in a good physiological status. However, in some instances, these two factors might irregularly act on body and can result in abnormality with in body, which can be called as disease.

Genetics have many implications in medicine. Genetics can determine several things in a human life. Genetics disease is an important group of medical disorder with these followings characteristics:

- Naturally occur
- Pass from a generation to the next generation
- Relating to reproductive system
- Nucleic acid within nucleus plays important roles in genetics.

- Abnormality within genetic code is the main cause of disorder.
- Genetic diseases can be identified by present molecular biology approach.
- Hard to treat and manage

Genetic diseases can present as the diseases of any organ systems. For sure, it can present as nutritional disorder. The question is how genetics can affect nutritional status and causes nutritional disorder. Indeed, there can be several ways that genetics can affect the nutritional status. The possible ways will be hereby described.

- Genetics disorder of eating organ
- Genetics disorder of gastrointestinal tract
- Genetics disorder of absorption of nutrients
- Genetics disorder of transportation of nutrients in body
- Genetics disorder of cellular uptake of nutrients
- Genetics disorder of cellular metabolism of nutrients

Any of these mentioned ways can result in nutritional disorder. Hence, it can be said that genetic nutritional disease can be due to diverse underlying patho-physiological process.

In brief, the expression of any genetic nutritional disease is due to this rule:

- The starting point is the defect in genetic content in genomics level.
- The abnormality in genomics will results in abnormal translation. The abnormal protein can be the result.
- With abnormal protein, abnormal biochemical process can be seen. This implies biochemical change.
- Due to biochemical change, physiological changes can be expected.
- Due to physiological change, functional change can be expected.
- In severe case, the anatomical change can be seen.
- Based on several minute anatomical changes, gross anatomical changes can be seen.
- The overt expression can be the result. The phenotype change can be observed.

Nutrigenetics and Nutrigenomics

Nutrigenetics is the new term used for describing the genetics study of nutrition. This specific branch of genetics focuses on several aspects of genetics and nutrition including:

- Interrelationship between genetics and nutrition
- Genetic variation, which affects the change of metabolism of difference nutrition
- Gene and nutritional disorder
- Inheritance of nutritional disorder
- Effect of nutrition on genetics
- Nutritional management for genetic diseases

In addition to the basic nutrigenetics, the newer science namely nutrigenomics has just been proposed. This is a new *omics* science in the post genomics era. In brief, nutrigenomics can be described as an *omics* study on nutrition and genetics. This means the study of nutrition, which affects the express of gene or alters product of gene.

The characteristics of nutrigenomics are:

- A kind of biochemeoinformatics
- Linkage between biomedical science and computational science
- Include two important parts, *i.e.*, database management and data manipulation management

As a kind of bioinformatics, several databases of nutrigenomics are available at present. In addition, several data manipulation tools are also available. The examples of important tools and databases will be described further.

Database: There are several nutrigenomics databases. However, a referencing data based is available on an open-source database system that can be freely accessible *via* the World Wide Web (http://a-yo5.ch.a.u-tokyo.ac.jp/index.phtml). This specific database contains minimum information about a micro-array experiment-compliant micro-array expression.

Data manipulation tool: In biochemeoinformatics, the manipulation of data derived from database is an important step since this mean actual usage of the data for making advantage, clarification of the process and prediction of the process. The basic manipulation tools in biochemioinformatics such as structure prediction, functional prediction and ontology can be applicable for nutrigenomics.

Table 1: Steps of nutrigenomics

Sr.No.	Steps	Importance	Activity
1.	Database management	Collection of data and extraction for further manipulation	Database setting Database storage Database query and searching
2.	Data manipulation tool management	Arrangement and processing of the data for usage	Database backup Clarification on structure and function Prediction on structure and function Grouping

Finally, the newest term *Nutritional genomics* should be mentioned. This specific term describes a specific study, which concerns on how nutrition, consumer behaviour and the subsistence have impact on life in the level of cellular, organelle and population. Hence, nutritional genomics includes contents of both nutrigenetics and nutrigenomics.

Important Genetic Nutritional Disorders

Hemochromatosis

Hemochromatosis is a disease presenting with abnormal accumulation of iron. This is the condition with abnormal iron metabolism. This condition can be either primary (genetic) or secondary. In case of primary hemochromatosis, the genetics effect is confirmed. The patients with hemochromatosis usually suffer from liver disease since excessive iron usually accumulates at liver.

Focusing on important clinical presentations of hemochromatosis, the following symptoms can be seen:

- Chronic hepatitis (increased level of liver enzyme)
- Cardiac arrhythmia
- Symptoms of diabetes mellitus
- Skin colour change
- Frequent arthralgia
- Testicular atophy (hypogonadism)
- Hair loss
- Memory loss
- Vertigo

All of these symptoms are due to the accumulation of iron at the affected organs, *i.e.*, liver, brain, heart, testes, or skin.

Table 2: Difference between primary and secondary hemochromatosis

Aspects	Primary hemochromatosis	Secondary hemochromatosis
Cause	Genetics	Usually due to hypertransfusion (which is a common treatment of the patient with hemoglobinopathy and thalassemia)
Clinical presentation	Onset at before age of 40 years old	Onset at before age of 25 years old
Abnormal HFE	Yes	No

Generally, primary hemochromatosis can be genetically transmitted *via* recessive inherit-ance. This means the offspring of the patient can have a chance to develop disease about 25%.

Focusing on treatment, the standard therapy is administration of iron chelating agent. This can be either oral or intravenous administrations. The patient has to get the drug thorough his/her life. If it is not continuously treated, the disease will progress and the patient can end up with death. Without treatment, the patient usually dies before age of 40 years old due to liver failure.

It is recommended that any family with history of primary hemochromatosis should get complete family investigation for genetics pattern to early diagnose of the disease. The method for diagnosis is the determination for the mutation within a specific gene namely HFE. If mutation is detected, it means that there is a *disease*.

For nutritional management, the patient with hemochromatosis has to strictly control for ingestion of iron. The main foodstuffs to be avoided include animal blood, animal viscera and animal liver.

Wilson disease

Wilson disease is another important genetic nutritional disorder. This is a specific disorder of copper metabolism. This disease is more common than primary hemochromatosis and can be inherited *via* autosome. It can be in either sex and usually within all offsprings in a generation of a family.

The main disorder in Wilson disease is due to lack of transporter of copper in blood circulation. This leads to abnormality in getting rid of copper in blood circulation and can result in accumulation of copper in several tissues. The accumulation of copper in tissues is the main patho-physiological process leading to symptoms of Wilson disease.

Focusing on important clinical presentations of Wilson disease, the following symptoms can be seen:

- Chronic hepatitis (increase level of liver enzyme)
- Severe jaundice
- Memory loss
- Movement disorder
- Sunflower cataract

All of these symptoms are due to the accumulation of copper at affected organs, *i.e.*, liver, brain, or eye.

Similar to primary hemochromatosis, this disorder is also chronic and the patient does not usually know that there is a disease within the body. If no treatment is taken, the symptoms will emerge at the age of about 40 years old. If treatment is not taken, death can be expected within 1-2 years. The patients usually develop liver failure before death.

At present, similar to primary hemochromatosis, there is a drug for treatment and control of this disease. The specific copper chelating agent is available and becomes the standard treatment regimen at present. However, similar to primary hemochromatosis, the ingestion of drug has to be continuous thorough the lifespan.

The complete family investigation for genetic pattern is required if there is family history. The specific test to be done is the determination of blood ceruloplasmin level (ceruloplasmin is the transporter of copper within blood circulation). Another important screening is the eye examination to detect the sunflower cataract. If there is evidence of defect, prompt treatment by chelation therapy is suggested.

Focusing on nutritional management for the patient it is recommended that the patient should avoid eating seafood since there is a high content of copper within seafood.

Glycogen storage disease

Glycogen storage disease is the name of a specific group of genetic disorder that shows the abnormality of carbohydrates metabolism. In this disease, the abnormal accumulation of glycogen, a large molecule carbohydrate can be seen. This is a rare but deadly disease.

Focusing on important clinical presentations of glycogen storage disease, the following symptoms can be seen:

- Chronic hepatitis (increase level of liver enzyme)
- Seizure
- Muscle weakness, loss of muscle tone
- Delayed development
- Mental retardation
- Macroglossia
- Cardiac enlargement

This is a genetic disease with recessive inheritance. Offspring of the patient can get disease in one fourth. The main defect is the defect within in GAA gene.

The diagnosis of this disease is due to the pathological examination on histological specimen derived from muscle or liver biopsy. Microscopically, the accumulation of glycogen within lysozome can be seen. In addition, the blood test for enzyme acid alpha glucosidase in white blood cell can be an alternative less invasive diagnostic technique.

The patient with glycogen storage disease usually presents the clinical abnormality in the pediatric life. The patient has to be urgently treated by enzyme therapy to stop progression of the symptoms. The treatment has to be started as early as possible (at least before 6 months). However, specific enzymatic drug for the treatment of patient is very expensive and not easily available.

Table 3: Kinds of glycogen storage disease

Kinds	Details
Von Gierke's disease	This is so called type 1 glycogen storage disease. It is the most common kind of glycogen storage disease. The defect in enzyme glucose-6-phosphatase can be seen. Lactic acidosis and hyperlipidemia can also be seen in the patients.
Pompe disease	This is so called type 2 glycogen storage disease. The onset of this disease can be either infantile, juvenile, or adult-onset forms. The cardiac involvement is usually seen in adult-onset form.
Forbes-Cori disease	This is so called type 3 glycogen storage disease. The defect is due to the defective enzyme, the glycogen debranching enzyme. Severe hepatomegaly and severe muscle weakness can be seen.
Andersen disease	This is so called type 4 glycogen storage disease. The problem is on the enzyme namely transglucosidase. Amylopectinosis is the predominant finding in this disease.
McArdle disease	This is so called type 5 glycogen storage disease. The problem is on the enzyme namely myophosphorylase. The problem is usually seen in muscle.
Hers diseas	This is so called type 6 glycogen storage disease. The problem is on the enzyme namely phosphorylase. The problem can be generalized in the body.
Tarui disease	This is so called type 7 glycogen storage disease. The problem is on the enzyme namely phosphofructokinase. The problem can be generalized in the body. This is the least common disease in the group of glycogen storage disease.

Lipid storage disease

Lipid storage disease is another important group of storage disease. This disease is very similar to glycogen storage disease but the accumulated substance is lipid, not the glycogen.

Focusing on important clinical presentations of lipid storage disease, the following symptoms can be seen:

- Chronic hepatitis (increase level of liver enzyme)
- Muscle weakness (loss of muscle tone)

- Splenomegaly
- Seizure

However, it should be noted that there are wide spectrum of clinical severity of lipid storage disease. In some mild cases, the patient can be alive without any problem until age of 90 years old.

Similar to glycogen storage disease, lipid storage disease has genetic recessive inheritance. This mode means 25% of the offspring will develop disease. The main defect is the lack of enzyme glucocerebrosidase. The lack of this enzyme results in accumulation of sphingolipid within cell and causes destruction of normal cellular structure and function.

The diagnosis of this disease is microscopic detection of glycogen accumulation within lysozome in histological access. In addition, the blood test for enzyme glucocerebrosidase can be another less invasive diagnostic test.

Some patients with lipid storage disease might present the clinical abnormality in the pediatric life, while the others can have no disease thorough their life span. If disease is detected, the patient should be treated by enzyme therapy to stop progression of the symptoms. The treatment has to be started as early as possible, similar to that of glycogen storage disease. However, specific enzymatic drug for the treatment of patient is not cheap and can be available at only some large medical centers.

Treatment of Genetic Diseases

Basically, the concepts of treatment for any medical disorders in medicine are:

- Getting rid of the cause of disorder if possible
- Symptomatic and supportive treatment for the existed symptoms
- Control for further expansion of pathology

These concepts can be applied for any medical disorders including genetic diseases. However, the cases of genetic diseases have some specific issue to be mentioned:

- Genetic disease is naturally existed. The causative agent in disorder is the genetic content within the patient's body.
- The environment has its effect as a modifier or stimulator to overt expression of the symptoms of genetic disease.
- The patient or the host has congenital defect in genetics and prone to develop the symptoms of disease.

It should be noted that the treatment of genetic diseases based on the basic principles might be difficult due to following reasons:

- The cause of disease, the genetic content within the patient's body, is very difficult to correct or manipulate.
- The symptoms of the disease usually progresses worsen and worsen. Sometimes, the symptomatic and supporting treatment seems to be useless. The palliative treatment can only be given.
- Since it is an internal blooming out due to hard-to-correct genetic underlying content, it is very difficult to control further progression of the disease. It might be said that it is not controllable.

In medicine, there are many approaches for the treatments of genetic diseases. These approaches include:

Drug therapy: The drug therapy is the most widely used treatment approach in medicine. The drug is used for the treatment of disease, killing, or getting rid of causative agent, symptomatic and supportive treatment, controlling of progression of disease. However, in case of genetic disorders, drug plays limited role. It is usually used for symptomatic and supportive treatment as well as control of disease progress. The good example of drug therapy for genetic disorder is the use of chelating agent for Wilson disease and primary hemochromatosis.

Enzyme therapy: Enzyme therapy is a specific treatment, *i.e.*, use of enzyme. It might be classified as a variety of drug therapy. Enzyme therapy is indicated in any genetic disorders that have the underlying defects in biological processes, *i.e.*, defects in enzyme systems. Many symptoms occur due to the lack of specific enzymes. Hence, the effective corrective action for those disorders is the enzymatic replacement. This is the basic concept of enzyme therapy. The good example of enzyme therapy for genetic disorder is the use of specific enzyme replacement in glycogen storage disease and lipid storage disease.

Surgical therapy: Surgical therapy is the correction of disorder based on operative procedure. This is usually indicated in the case of trauma and tumor. For genetic disorder, it is used very limited. It might be used in some cases with gross anatomical defects or handicapped structure. The surgical therapy is usually used for the correction of those problematic structures to recover the normal physiological function as much as possible.

Physical therapy: Physical therapy means rehabilitation. This is the action for correction and prevention for further progression of organ destruction (tertiary prevention). Physical therapy, which requires the process of rehabilitation, can be used in the case with permanent structural defect. For sure, some genetic disorders lead to structural defect or functional loss such as brain involvement in glycogen storage disease and lipid storage disease. Those disorders usually require rehabilitation process by physical therapy for the management of patients. The aim of physical therapy in those patients are (i) recovery of function of the defective organ, if possible, (ii) help the patients to take care of themselves as much as possible and (iii) decrease burden to the patients' family.

Gene therapy: Gene therapy is the specific treatment approach. By term, it is a specific treatment attacking to the gene. The details of this new specific therapy will be further detailed:

Concept of Gene Therapy

Due to the present advent in diagnostic technique, many rare diseases can be effectively diagnosed. However, the treatment for those diseases seems to be a present difficult activity. Gene therapy, a treatment on gene, is the new hope for the management of those rare difficult to treat diseases. Focusing on genetic disease, it has its rooted cause as abnormality within gene. Hence, the best approach is the treatment attacking on gene level. The basic principle is treatment at the starting point of the disorder. If a specific genetic disorder is known, treatment can be specifically given to that specific point. Basically, gene therapy can be done *via* several ways including:

Gene augmentation: Gene augmentation therapy is the gene therapy based on gene augmentation process. In brief, this is the process to increase the number of lacked gene. This is aiming at increase the protein, which is the product of gene, which is usually deficit in defective gene. This process can be said as a simplified adding of producer when there is no or lack of producer.

Gene inhibition: Gene inhibition therapy is the gene therapy based on gene inhibition process. Inhibition means stop of the existed process in the body. This has to be done if there is unwanted abnormal biological process. Those unwanted process can be seen in many diseases such as infection, cancer, or metabolic disease. The main aim of this approach is to decrease or cease those unwanted existed biological process. This means reduction of the unwanted protein product that existed in genetic disorder. As a result of protein production cessation, the recovery of the normal physiological function is expected.

Killing of specific unwanted cells: This is similar to gene inhibition but it is different in process. The aim is to stop the protein production. The approach is by direct killing of the unwanted cells, not modification of the gene process. This is performed by adding some new genes into the body. The added gene has its cytocidal effect and will produce new protein to kill specifically the unwanted cell or induce immune system to kill the unwanted cell or produce new specific enzyme to react with pro-drug leading to active form in body to kill the unwanted cells.

With these three main approaches, the gene therapy can be functional in the treatment of genetic diseases. Classifying by type of manipulated gene, gene therapy can be divided into two groups, *i.e.*, (i) somatic gene therapy and (ii) germ line gene therapy:

Somatic gene therapy: This means the gene therapy attacking on autosome or somatic gene. This gene therapy affects on a single generation, not genetically transmit to the offspring. This is usually co-manipulated with stem cell technology; purification of the stem cell, genetic modification of the cell, or therapeutic design and back insertion of the modified stem cell into the patient's body for the treatment.

Germ-line gene therapy: This means the gene therapy that aims at correction of disorder in both parents and offspring. This is the pre-fertilization correction. However, this is an ethical dilemma issue to be discussed. This seems to be the problem if the parents selectively designed the genetic content of their offspring.

Designing of a gene therapy for a specific genetic disorder can follow these steps:

- The problematic gene that causes the genetic disorder can be recognized. This can be done based on database searching *via* genomics approach. If there is no recorded data in the database, it is required to perform genetic content analysis on the patient.
- After getting data on problematic gene, the design of a specific gene therapy can be done. Design of the therapeutic gene will be then done. This is done according to the mode of therapy, *i.e.*, gene augmentation, gene inhibition, or killing of specific unwanted cells.
- The designed therapeutic gene has to be tested before actual usage. This means testing on in silico-model *via* computational manipulation tools of genomics following by *in vitro* testing and *in vivo* testing in animals and volunteer.
- Since there are many phases of trials for gene therapy, it is hardly to say that there is any specific acceptable gene therapy technique in medicine at present.

Gene Therapy for Genetic Nutritional Disorders

Gene therapy for genetic nutritional disorder can be done following the described gene therapy technique. The approach on gene therapy is very new and reported in some diseases. Focusing on the genetic nutritional disorders, some important publications can be seen in the literature. The data on each specific important genetic nutritional disorder are hereby presented and discussed below:

Hemochromatosis: As already mentioned, HFE is accepted as defective gene in primary hemochromatosis. This gene testing can be used to diagnose hemochromatosis but analyses of liver histology and clinical features are still required to identify patients. The gene therapy based on manipulation of HFE is not in use at present. However, there are many reports on trials based on AAV-mediated technique. This procedure is aiming at producing short antisense RNA that leads to normalization of iron transportation process. Enhancement the iron transport could be seen in hemochromatosis after performing gene therapy using AAV-mediated technique. Nevertheless, there are many publications on the use of gene therapy for the treatment of thalassemia and hemoglobinopathy, which can be the causes of secondary hemochromatosis.

Wilson disease: Several gene mutations can be seen in Wilson diseases. Since the present standard treatment is not successful in treatment and management of patient with Wilson disease, the use of gene therapy becomes the new hope. In anima-model experiment gene therapy for Wilson disease, usage of adenoviral and lentiviral gene transfer could lead to a therapeutically effective result, and it can reverse poor clinical presentations. Hence, it is expected to be the new approach for the treatment of Wilson disease, which is better than using of transplantation. However, both viral transfer approaches are still limited and problematic by a more or less transient transgene expression. Hence, further research to overcome this problem is still required.

Glycogen storage disease: The progress of gene therapy in glycogen storage disease is more than primary hemochromatosis and Wilson disease. This might be due to the fact that, there is limited therapeutic option for this disease at present. Adeno and adenovirus-like virus based therapies are widely evaluated. The neonatal gene therapy of glycogen storage disease using a feline immunodeficiency virus (FIV)-based vector is confirmed for its efficacy in animal model. It is reported that the recovery of blood glucose to normal levels and significant decreased accumulation of liver glycogen associated with the disease can be observed after implementation of this therapeutic approach. In addition to FIV-based approach, another important approach is helper-dependent adenovirus vector-mediated gene therapy. Similar recovery in animal model can be seen in this approach similar to that can be observed in FIV-based approach.

Lipid storage disease: This kind of genetic nutritional disorder has the most progress in gene therapy. The use of adeno and adenovirus-like virus based therapies are confirmed for clinical usefulness in the treatment of lipid storage disease. Late infantile neuronal ceroid lipofuscinosis is the specific disease, which has strong evidence of using gene therapy as an effective approach. A phase I study of late infantile neuronal ceroid lipofuscinosis using an adenoassociated virus serotype 2 (AAV2) vector containing the deficient CLN2 gene (AAV2(CU)hCLN2 was recently published in 2010. The report confirms the usefulness of gene therapy in human cases with lipid storage disease. Indeed, this approach is confirmed for safety in many animal models.

■■■

19 Treatment of Nutritional Disease through Stem Cells

Viroj Wiwanitkit

Wiwanitkit House, Bangkhae, Bangkok, Thailand-10160, Visiting professor, Hainan Medical College, China, E-mail: <wviroj@yahoo.com>

Nutrition implies many things relating to eating. In an easy way, study of eating and nutrients can be a simple explanation for nutrition. Nutrients are one of the four basic needs for everyone (food, garment, home and drug). Nutritional disorder is a group of medical disease, which is very common in the present day. It can be seen in any countries around the world. The abnormality in nutrition can be in either quality or quantity aspects.

Abnormality in quality

Contaminated food: The contamination of food is often seen in setting with poor control. Contaminated food, which needs strict management, is the big issue in public health.

Abnormality in quantity

- **Under-nutrition:** This is the condition under which a lower than adequate level of nutrition is gotten. It can result in many clinical symptoms, and it can be seen in any settings, especially for the developing countries, around the world.
- **Over-nutrition:** This is the condition under which a higher than adequate level of nutrition is gotten. Similar to under-nutrition, it can cause many clinical symptoms. The most common problem in over-nutrition group is the *obesity*, which is an important public health problem in many countries at present.

Several nutritional diseases are listed in medical textbook. To manage the nutritional disorder is the basic requirement in routine medicine. This includes treatment of nutritional disease as well as nutritional arrangement for each patient. To manage these nutritional diseases, nutritional management seems to be the basic thing. However, based on the present advanced biotechnology, several new management approaches are available. A new approach for nutritional management is using stem cell therapy.

Nutritional Disorders

As a kind of medical disorder, nutritional disorder is the disorder with following characteristics:

- Naturally occur or induced by environmental factors.
- Can be genetically or not genetically appeared episode.

- Relating to eating and processing of food within the body
- Can be identified by either naked eyes or medical diagnostic techniques.
- If detected, treatment and management are needed.

Nutritional diseases can present as the diseases of any organ systems. For sure, it can present as single organ or multiple organs disorders. The question is how nutritional disorder can occur. Indeed, a nutritional disorder can appear in several ways. The possible ways are hereby described:

- Disorder of eating organ
- Disorder of gastrointestinal tract
- Disorder of absorption of nutrients
- Disorder of transportation of nutrients in body
- Disorder of cellular uptake of nutrients
- Disorder of cellular metabolism of nutrients

Any of these mentioned ways can result in nutritional disorder. Hence, it can be said that nutritional disease can be the result of any basic defect. In brief, the expression in any nutritional disease is gradually due to the following rules:

- The starting point is the defect within cell.
- The abnormality of a single cell might not be problematic. However, if multiple cells abnormality occurs, the problem can be expected.
- Abnormality in cells means abnormality of functions acted by those cells.
- This can lead to the abnormality of tissue.
- Then it can process to the abnormality of the organ.
- Gross change can be seen in case with organ defect.
- The overt expression can be the result and the phenotype change can be observed.

Nevertheless, the cases of nutritional diseases have some specific issues to be mentioned:

- Nutritional disease can be naturally existed or getting by any external insults. The causative agent in disorder is the nutrient within the patient's body. It is really a biochemical story.
- The environment has its great effect in expression of symptoms caused by nutritional disease.
- Nutritional disease is not an anatomical structure though it is a dynamics process, which can be changed time to time.

Cell Abnormality

For any disease, the defect in cell can be seen. Indeed, cell is the smallest unit of any living organism. The characteristics of cell include:

- Have signs of *living*
- Viable
- Can increase in number, which is the basic concept for reproduction.
- Contain genetic content.
- Have small organelles within the cell.

- Can die.

In any cell of high living organism, there are two important parts:

- **Cytoplasm:** Cytoplasm is the cytosol, which can be seen around central part of the cell, which is known as nucleus. Cytoplasm is the place for generating the protein. Cytoplasm also contains many organelles (organs of the cell).
- **Nucleus:** Nucleus is the central part of cell. It is the main regulator of cell and contains the genetic content. It takes the role in cell division, which is the basic primary process for reproduction.

In every human being, the body consists of millions of cells. Hence, it can be said that human body is the cluster of cells. Hence, if there is any defect in the cell, there can be the problem for the whole body.

Table 1: Cells within human body

Types of cell	Description
Somatic cell	Somatic cell is the general cell of human beings. This kind of cell contains full number of chromosomes. The cell dividing process to multiply the number of cell can be possible and this is the basic phenomenon for growth and development.
Sex cell	Sex cell is the specific cell in reproductive system. There are only two types of sex cell in human body, egg and sperm. Egg is the sex cell, which is generated in ovary of the female and sperm is the sex cell, which is generated in testes of male. The main role of sex cell in human beings is for reproduction. Transmission of genetic content to new offspring is the final aim of sex cell functioning. Unlike somatic cell, sex cell contains only a half number of chromosomes. This is for further fertilization process (combination between egg and sperm) to regain the full number of chromosomes.

Cells have many implications in human body. Cells can determine several things in a human life:

- Construction of tissue and organ
- Growth of the body
- Development
- Reproduction
- Repairing
- Counteract with foreign body

From a cellular level to gross phenotypic appearance, there are several processes including:

- **Intracellular phenomenon:** This includes several processes especially for the protein synthesis and sending out of the synthesized proteins.
- **Intercellular phenomenon:** This includes any interaction between cells, which is the basic fundamental process of biological process in human beings.

Focusing on cell defect, there are two main kinds of defect described below:

- **Anatomical defect:** Cell lost, decrease or increase in cell content, decrease or increase in cell size and transformed cells (such as turn into cancer)
- **Functional defect:** Loss of normal physiological process and loss of normal biochemical reaction

It should be noted that the co-presentation of both defects could be seen in some cases.

Cell and Nutrients

The correlation between cell and nutrition is the present focus in nutritional medicine. Indeed, there are complex interrelationship between cell and nutrient. As a basic rule in medicine, the two factors that affect cells are genetics and environment.

- **Genetics:** Genetics is the inherited content that passes from a generation to the next generation. This is an actual intracellular factor, which is mapped from birth thorough everyone's life.
- **Environment:** Environment is a non-genetic factor or the external insult to human body. Several environments are described as important etiologies of medical disorders. Nutrient is a kind of environmental factor, which is widely mentioned in medical literature.

Nutrient that can be seen in food is the specific biochemical substance. The characteristics of nutrient include:

- Naturally exist
- Nutrient is the content within food
- Useful to body
- Can be detected by medical and scientific methods.
- Nutrient is required by human body for living.

The important groups of nutrients within the food include:

- Carbohydrates
- Fat
- Protein
- Vitamins
- Minerals

These five groups are needed for the maintenance of cellular function and body stabilization. Hence, it can be said that cell needs nutrients for living and avoiding death. If cell run off nutrients and cannot find new nutrients, the cell will die.

In an easy word, the nutrients have the following relations to the cell.

- Nutrients are essential for the cell.
- Nutrients are up taken and used by the cell.
- Nutrients can also be synthesized by some cells such as liver cell.
- Nutrients can be accumulated in some cells such as fat cell.
- Nutritional disorder can also be seen in some kinds of cell disorder.

On the other hand, cell has the followings relations to the nutrients.

- Cell has to use nutrients for normal physiological process.
- Using intracellular process, cell can change the nutrients into a new biochemical substance.
- Cell will die in the absence of nutrients.
- Cell disorder can also result in nutritional disorder.

Table 2: Examples of some interrelationship between nutrients and cells

Nutrients	Cells	Medical disorder
Folic acid	Colon, heart and brain	Colon cancer, heart disease and abnormal brain
Vitamin B_{12}	Brain	Abnormal brain
Niacin	Brain	Memory loss
Vitamin C	Eye	Cataract
Vitamin E	Colon and heart	Colon cancer and heart disease
Ferric	Brain	Abnormal brain
Zinc	Brain	Abnormal brain

Focusing on cell defects in nutritional disorders, the following forms of defect can be expected.

- **Anatomical defect:** This means the loss of normal anatomy of cell. This can be seen in some specific kinds of nutritional disorder. The best example is the decrease in size of red blood cells in case of iron deficiency anemia
- **Functional defect:** This means the loss of normal function of cell. This can be seen in some specific kinds of nutritional disorder. The best example is the insulin resistance of cells at the target organ, which is the big problem in diabetes mellitus.

Important Nutritional Disorders in Medicine

1. Diabetes Mellitus

Diabetes mellitus (DM) is a disease presenting with abnormal homeostasis of blood glucose. This is the condition with abnormal glucose metabolism. This condition can be either primary or secondary. The defect of beta cell within endocrine portion of the pancreas can be seen.

Focusing on important clinical presentations of diabetes mellitus, the following symptoms can be seen:

- Polyurea
- Polydipsia
- Polyphagia

All of these three classical symptoms are due to the high blood glucose level. Also, due to the high blood glucose level, several complications of diabetes mellitus can be expected. Those complications include:

- Eye complication, *i.e.*, blindness (retina disorder)
- Renal complication, *i.e.*, chronic renal failure
- Brain complication, *i.e.*, cerebro-vascular disease
- Heart complication, *i.e.*, cardiovascular disease

Table 3: Difference between primary and secondary diabetes mellitus

Aspects	Primary diabetes mellitus	Secondary diabetes mellitus
Cause	Unknown	Many etiologies such as pancreas trauma and pancreatic stone
Clinical presentation	Onset at teenage (type 1) or at age above 35 years old (type 2)	At an age
Abnormal blood glucose	Yes	Yes

Generally, primary diabetes mellitus is more common than secondary diabetes mellitus. In medicine, the primary diabetes mellitus can also be divided into following main groups:

- **Type 1 Diabetes Mellitus:** This is a juvenile onset type. The symptoms usually develop at teenage. The strong genetic evidence is identified in this type of diabetes mellitus. The patients usually have lack of insulin and the pathology of beta cells at pancreas is predominant. The ketoacidosis is common. This is classified as insulin dependant diabetes mellitus.
- **Type 2 Diabetes Mellitus:** This is a degenerative kind of diabetes mellitus. The classical symptoms are slowly progressive and the patients usually have overt symptoms when they are old. The patients have cellular resistance to insulin. This is called non-insulin dependant diabetes mellitus.
- **Type 1.5 Diabetes Mellitus:** This is the type with both characteristics of type 1 and type 2 diabetes mellitus. Early onset with insulin dependence can be seen.
- **Type 3 Diabetes Mellitus:** This is the diabetes mellitus with type 2 characteristic plus brain manifestations. The Alzheimer with diabetes mellitus is the classical picture of type 3 diabetes mellitus.

For nutritional management, the patient with diabetes mellitus has to strictly control for ingestion of hyperglycemic food. The mains foodstuffs to be avoided include fruits and deserts.

2. Hypercholesterolemia

Hypercholesterolemia is another important common nutritional disorder. This is a specific disorder of lipid metabolism. This disease is as common as diabetes mellitus. The diagnosis of this condition is based on the blood lipid determination. In general, there are two kinds of blood lipid, *i.e.*, cholesterol and triglyceride. The cholesterol can be further divided into two types: low-density lipoprotein (LDL) and high-density lipoprotein (HDL). The problematic conditions are high blood triglyceride level and high LDL cholesterol.

Table 4: Clinical importance of blood lipids

Blood lipids	Clinical importance
Total cholesterol	Total cholesterol has a limitation for clinical implication. It requires the data on LDL and HDL for interpretation.
Low-density lipoprotein (LDL)	LDL is the *bad* cholesterol. It induces the pathology within blood vessels and further causes arterial diseases.

Blood lipids	Clinical importance
High-density lipoprotein (HDL)	HDL is the *good* cholesterol. Exercise can increase HDL level. It is useful for the prevention of arterial disease.
Triglyceride	Similar to LDL, triglycerides are considered problematic. They can cause problem on the vascular lining and directly lead to arterial disease.

Hypercholesterolemia is relating to many other disorders. It is proved that hypercholesterolemia is the cause of many diseases including:

- Cerebrovascular disease
- Cardiovascular disease
- Pancreatitis

How to Treat a Disease with Cell Defect

Basically, the concepts of treatment for any medical disorders in medicine are as follows:

- Getting rid of the cause or etiology
- Symptomatic and supportive treatment for any symptoms
- Control for the progression of disease

These concepts can be applied for any medical disorders including cell defects. In medicine, there are many approaches for the treatments of genetic diseases. These approaches include:

- **Drug therapy:** Drug therapy is the most widely used treatment approach in medicine. The use of drug is applicable in cell defect that has functional defect. Drug cannot recover the loss of anatomical structure.
- **Surgical therapy:** Surgical therapy is the correction of disorder based on operative procedure. This might be useful in case of cell defect with anatomical defect. Surgery is usually used for getting rid of unwanted cell.
- **Physical therapy:** Physical therapy means rehabilitation. This is useful in cell defect with anatomical defect. Despite the loss of anatomy of a part within the body, it is possible to be functional replaced by other parts of the body.
- **Stem cell therapy:** Stem cell therapy is the specific treatment approach. By term, it is a specific treatment using stem cell. The details of this new specific therapy will be detailed further.

What is stem cell?

In any human beings, there are numerous cells within the body. However, there is a specific group of cells known as stem cell. Stem cell by definition is the primitive cell. It is the ancestor of other cells within human body. In embryology, the cell formation starts from fertilization, the combining of two cells, sperm and egg. The embryogenic process has many steps. From one cell at fertilization, continuous multiplication of cell occurs until embryo is formed. At first, the cells of embryo are all similar, which is called primitive stage of the cell, and then, the differentiation of cells into various tissues and formation of organs will occur. With these described process, the complete structure of human body can be resulted at birth.

After birth, most cells are completely developed. Further developments can be seen in only organ systems such as reproductive system and blood systems. Hence, the primitive cell can still be seen in the human body. Based on the present advanced biomedical techniques, the primitive cell in embryo and human body are widely focused. This is due to its property that can induce differentiation and formation of the new tissue and organ.

Stem cell is another name of the primitive cell. As a summary, the characteristics of stem cell include:

- Has no specific characteristic of mature cell in the body.
- Has potential for cell dividing process.
- Can be differentiated further.
- It is the main cell in embryo.
- It can be seen in some specific organs in mature human body.
  - Tooth
  - Placenta
  - Sex cell
  - Skin cell
  - Blood cell

By definition, stem cell can be divided into 2 main groups:

- **Embryonic stem cell:** This is a type of stem cell, which can be seen in embryo phase.
- **Adult stem cell:** This is a type of stem cell, which can be seen in mature human body.

Based on the stem cell activity, stem cell can be divided into three groups:

- **Totipotential stem cell:** Totipotential stem cell is the most primitive stem cell. It has the potential to differentiate further into any tissue. This is the nature of the embryogenic stem cell.
- **Pluripotential stem cell:** Pluripotential stem cell is the stem cells that can differentiate into many cells but in the same tissue or organ system.
- **Unipotential stem cell:** Unipotential stem cell is the least primitive stem cell. It has the highest maturation. It can only specific differentiation tract to a specific kind of cell.

Considering three kinds of stem cell, the totipotential stem cell is the specific group that can be useful in differentiation process.

Concept of Stem Cell Therapy

Stem cell therapy is the specific treatment based on stem cell. It is classified as a type of cell therapy or treatment based on cell. At present, the use of stem cell therapy is on its first pioneer phase and its main usage includes:

- Repairing of cell loss due to trauma or accident
- Repairing of senile cell due to degeneration process
- Treatment of some specific disease with destruction of tissues in organ such as ischemic heart disease and cerebrovascular incidence
- Hypersensitivity

- Abnormal brain development
- Some cancers such as leukemia

For the process of stem cell therapy, the following important steps are included:

- **Acquistration of stem cell:** This is the starting point for therapy. If there is no stem cell, the treatment by stem cell therapy cannot be successful. The acquistration of stem cell is usually from the described source of stem cell. The common source includes placenta, bone marrow and tooth bud.
- **Preparation of stem cell for usage:** After deriving cells from collective site, the important step is to clean and classify the cell. This done is to ensure purification of the stem cells and confirm for the potential of derived cells.
- **Stem cell implementation:** This is the important step by which that the prepared stem cell is transmitted into the patient. The most common technique is the stem cell injection.
- **Following up:** This is the step in which the result of stem cell implementation is followed up. In general, after injection, the stem cell is still stem cell. However, it will autonomically run to the defect site for induction of further differentiation for repairing or treatment. This specific process is known as *homing*.

Stem Cell Therapy for Nutritional Disorders

Stem cell therapy for nutritional disorder can be done by following the described stem cell therapy technique. The approach on stem cell therapy is very new and the trails can be seen in some diseases. Focusing on the nutritional disorders, the data on each specific important genetic nutritional disorder will be hereby presented and discussed.

At present, stem cell treatment is available in many tertiary medical centers and accepted as new hope for treatment of previously untreatable diseases. However, further researches are still required on this new treatment at its starting history.

Several diseases can be treated based on stem cell therapy. The stem cell therapy is also applicable to the nutritional disorders.

Stem Cell Therapy in Diabetes Mellitus

Diabetes mellitus is a disease with many reports of using of stem cell therapy. In the easiest way, the stem cell therapy is used for the treatment of an important diabetes mellitus complication, *i.e.*, the diabetes mellitus foot. The diabetes mellitus foot is the condition of loss of cells and tissue at feet of the diabetes mellitus patients due to poor control of blood glucose.

However, a more direct usage of stem cell therapy on the diabetes mellitus, not its complication, is still available. This is the new concept that is introduced when the concept of islet cell transplantation does not go to its best aim. Stem cell therapy is widely studied in the animal models of diabetes mellitus type 1. The results from the experiments are favorable.

This approach is the new hope to support the present usage of islet cell transplantation in type 1 diabetes mellitus cases with end stage renal disease. With additional use of stem cell, the outcome of islet cell transplantation will be better improved. There are some recent new reports on human trials for this new approach for treatment. The results in human studies are also favorable.

Stem Cell Therapy in Dyslipidemia

Since dyslipidemia is not as serious as diabetes mellitus and the antilipemic drug is still accepted for its pharmacological activity. The usage of stem cell in dyslipidemia is not presently in focus. However, stem cell therapy is used in the treatment of some complications of dyslipidemia such as ischemic heart disease and cardiac infarction.

Stem Cell Therapy for Genetic and Nutritional Disorders

There are many important inherited nutritional disorders including glycogen storage disease and lipid storage disease. Stem cell therapy is a new approach for the treatment of those disorders. There are many ongoing researches on stem cell therapy, and the results are favorable. The lipid storage disease has been tested on usage of hemopoeitic stem cell transplantation for years and the new approach seems to be a hope for better management of the patients.

■■■

20 Skin, Hair and Nail Health

Viroj Wiwanitkit

Wiwanitkit House, Bangkhae, Bangkok, Thailand-10160, Visiting professor, Hainan Medical College, China, E-mail: <wviroj@yahoo.com>

Skin is the organ with the largest area of the human body. It covers all parts of the body. Skin can be accepted as the organ with the most area of human body. Skin seems to be an organ of concern for cosmetic. Skin health is very important. In dermatology, skin is the most talked issue. In the present, many skin clinics can be seen and serve for correction and supplementation of the skin.

In addition to skin, the special concerns are also given to its important appendages including nails and hair. Nail and hair are also focused consideration in cosmetic view. Hairdo and nail decoration are the two main services in beauty shops. Here, the author will specifically discuss on skin, hair and nail health.

First, the author will talk on the skin health in general. Structurally, skin can be divided into several layers as described below:

Keratin Layer or Death Skin Layer

This is the layer of already death skin cells. It lies on the highest layer or most superficial part. It can be rubbed out during bathing as keratin. Indeed, this layer is not alive but has important function to protect the skin from external insults. The examples of external insults that can be protected by keratin layer include:

- Water
- Chemical
- Sunlight
- Germs and pathogens

Keratin is produced daily. The production is due to the upward replacement of new skin cells by the deeper skin cells.

In hygienic concern, the cleaning of the body during bathing must include the rubbing out of the keratin. In cosmetic concern, no rubbing of keratin seems to be dirty and is not acceptable.

Superficial Layer or Cutaneous

Cutaneous is the layer of actual skin cells. This is the most superficial lived cells layer, which has sensation. Any injury to the superficial skin layer can cause pain. However, it should be noted that the injury, which is limited to upper part of this layer, results in no scar. Since cutaneous is superficial, it can be insulted by many external factors. It is needed to protect the cutaneous from external insults. The summary of the functions of cutaneous includes:

- Production of keratin
- Prevention from external insults (as described in the keratin layer)
- Exchange between subcutaneous and deeper layers with external environment

In medicine, it can divide the cutaneous into two sub-layers including:

- **Epidermis:** Epidermis is upper part of the cutaneous. As already noted, if injury occurs to this part without any extension, there will be no scar.
- **Dermis:** Dermis is lower part of the cutaneous. If injury occurs to dermis, there will be scar.

Deep Layer or Subcutaneous

Subcutaneous is the deepest skin layer. It contains several important structures including:

- Several cutaneous glands
- Hair root
- Blood vessels
- Nerve fiber

As already mentioned under the heading skin, there are many important glands that can be divided into many groups. The most common and well-known glands include:

- **Sweat gland:** This is a specific gland producing sweat. Sweat is the body fluid, which is caused by hyper-filtration or ultra-filtration. Sweat gland plays important role in several functions including:
  - Excretion
  - Giving out the heat from our body
- **Sebaceous gland:** This is a specific gland containing a fatty content. It can be seen in as much number as sweat gland.

In addition to skin, hair is another important part of dermatological concern. By definition, hair is a structure with filariform budding from the skin. Hair can be seen all around our body except for the two sites given as under:

- **Palm:** This is because the palm is necessary for working on several activities especially for grasping. Palm has vcry thick skin and contains no hair.
- **Sole:** This is because the sole is necessary for walking. Similar to palm, sole has very thick skin and contains no hair.

Hair buds out from the skin and can reflect the skin health. Many uncountable hairs can be seen around our body. However, the site with the most number of hairs is at our head around our scalp. The other sites with many hairs include:

- **Pubic area:** The pubic area has many hairs. The density of hairs at this area is greater than other areas of the body except for the scalp. The main function of pubic hair is related to the sexual process. It might act as shock absorber during sexual intercourse. Also, it is useful in avoiding infection and rubbing of thing responsible of causing injury. This is because of aeration that can be provided by pubic hair. Finally, it should also be noted that pubic hair is a sex symbol that attracts the opposite sex.

- **Axilla:** The axilla is another area with many hairs. Similar to public area, axilla seems to be an organ linking to the sexual function. Specific axilla odour might be attractive to the others. Everyone has two axillas, *i.e.*, one on each side connecting between arm and body (chest area).
- **Perioral area:** Perioral area has many hairs. It is more dominant in male than female. It is accepted as a masculine sign. Shaving of hair at perioral area is the basic practice of any adult male.
- **Chin:** At chin, there are many hairs. Similar to the perioral area, hairs at this area can rapidly grow and this is the reason that shaving of the beard is required. In male, the hair at chin is more dominant.
- **Sternal area at chest:** The hair at sterna area at chest is thick in male. It is the sign of masculine. Also, it seems to be a sexual attraction to the female.
- **Eyebrow and eyelash:** Eyebrow is another part of the body with thick hair. It acts as a shelf preventing the injury from upward direction to our eye. In addition to eyebrow, eyelash is the specific hair connecting to eyelid. The role of eyelash is to prevent external insults that will enter into the eye.

Eyebrow and eyelash seem to be cosmetic structures. The eyebrow might be shaved out in some culture. The best example is the Buddhist religious related practice to become a monk.

In this specific issue, the author will discuss on skin and hair health. The reasons for the close interrelationship between hair and skin are discussed below:

- It is the connecting structure.
- Hair directly buds out from the skin.
- Skin stabilizes the hair.
- Hair externally protects the skin.
- Circulating and nervous feedings on skin and hair are from the same system.

Special focus on nutrition and skin and hair health will be made in this chapter.

Skin Disorders

Skin disorders are important problems in dermatology. There are several skin disorders in dermatology. The important disorders are described and discussed below:

Melasma: Melasma is the specific change of pigment in our skin. Melasma seems to be an important cosmetic problem. The classical appearances of melasma include:

- Change of skin color
- Irregular progression
- Increased density of pigment
- Notable area of skin
- Extension
- Relating to the pigmentation

There are many etiologies of melasma including:

- **Hormone change**
  - Pregnancy

- Perimenopause
- Use of oral contraceptive drug

- **Exposure to sunlight (ultraviolet)**

The ones who take risks for developing melasma include:

- Pregnant subjects
- Perimenopause subjects
- Workers exposing to sunlight
- Workers in the sea such as fisherman

Acne: Acne seems to be the most well known dermatological problem. The most common case in dermatological clinic is acne. Normally, acne can be seen at face and it becomes of cosmetic concern. However, the acne can be seen at other areas of the body including:

- Back
- Chest
- Buttock

The acne is the result from the obstruction of the skin gland. The classical appearances of acne include:

- Skin lesion
- Can be inflammation.
- Can be infected.

There are many etiologies of acne including:

- **Drug induced**
  - Use of oral contraceptive drug
  - Steroid usage
  - Anticonvulsive drug
  - Lithium (psychotic drug)
  - Antituberculosis drug
- **Hormone change**
  - Pregnancy
  - Puberty

The ones who take risks for developing acne include:

- Pregnant subjects
- Teenagers

Table 1: Important pathogenesis of acne

Pathogenesis	Details
Increase production of sebum by skin gland	Increase androgen production in teenage males Menarche in teenage females

Pathogenesis	Details
	Menstruation
	Increase activity of enzyme 5-alpha-reductase
Abnormal keratin production	Hyperproduction
	Ductal hypercornificatin
Infection	*Propionibacterium acne* infection

The severity of acne is relating to many factors including:

- Genetic
- Emotion
- Stress
- Season
- Food
- Cosmetic
- Drug
- Irritant

Nevus: Nevus is a change of pigment at skin similar to melasma. The classical appearances of nevus include:

- Change of skin color
- Usually congenital
- Increased density of pigment
- Notable spot at skin
- Relating to the pigmentation

Nevus is usually confined in its area. Extension of nevus lesion seems to be serious and has to be worked out to rule out malignancy change. Nevus can be seen around our body. In many classical beliefs, nevus is mentioned as a sign for horoscopic prediction. Based on this fact, the removal of undesired nevus is generally practiced in dermatological clinic.

Dermatitis: Dermatitis is the term used for describing the inflammation of skin. Several causes of dermatitis can be mentioned. Those causes include:

- Irritation
- Allergy
- Autoimmune disease (such as in SLE)
- Infection
  - Chicken pox
  - Herpes simple
  - Herpes zoster
  - Papilloma virus

    - Ring worm
    - Fungal infection
    - Scabies
    - Staphylococcal infection

The classical appearances of dermatitis include:

- Change of skin
- Can be acute or chronic
- Red
- Painful or itching
- Swelling
- Hot (if acute)
- Thickening (if chronic)
- Relating to starching by hands

Scar: Scar is the result from wound. It is due to the healing process. Basically, after the occurrence of wound, the following steps will occur:

- Hemostasis process to stop the bleeding
    - Platelet aggregation
    - Coagulation factor system in blood
    - Vascular constriction
- Repairing
- Remodeling of the skin structure
- If fail to return to previous normal appearance, the scar can be observed.

Keloid: Keloid is a kind of scar. Keloid is also due to the healing process. This is excessive remodeling. The classical appearances of keloid include:

- Scar
- Prominent
- Ugly, not good look
- Palpable
- Consist of connective tissue

The prominent scar is the nature of keloid. It is usually noticeable and becomes of cosmetic concern. Keloid is hard to treat and extensive manipulation on the keloid usually fails and can cause increased keloid.

Wart: Wart is the thickening of skin. It is a small tumor at skin, which is caused by tumerogenic virus, *i.e.*, papilloma virus. Wart has to be treated. The surgical excision, using of cyotoxic drug, electrical cauterization and chemical excision are the widely used method for the treatment of wart in dermatological practice.

Corn: Corn is the thickening of skin. It is not infectious but due to chronic irritation that ends up with marked skin thickening. It might be painful. The excision is the standard treatment.

Skin cancer: Similar to any other part of the human body, cancer of skin can be seen. There are several kinds of skin cancer including:

- Basal cell cancer
- Squamous cell cancer
- Melanoma or pigment cell cancer

The cause of skin cancer is still unclear. However, several factors are mentioned including:

- Ultraviolet exposure
- Chronic infection, especially for herpes virus
- Chemical irritant such as arsenic
- Chewing of betel nut (especially skin cancer at mouth)
- Cellular abnormality within nevus

Pellagra skin disease: This is a specific skin disorder, which is due to the deficiency of vitamin B_3 (niacin). Fruits and vegetables are rich in this vitamin. Lack of this vitamin can be sporadically seen. However, it should be noted that the vitamin B_3 deficiency could be the result of other nutrients disorders including:

- Lysine deficiency
- Tryptophan deficiency
- Hyperleucinemia

The classical symptoms of pellagra include:

- Skin disease, *i.e.*, photodermatitis
- Diarrhea
- Neurological deficit, *i.e.*, dementia
- Death in the worst case

However, there can be other uncommon symptoms of pellagra including:

- Psychological disturbance
- Alopecia
- Insomnia
- Weakness
- Edema
- Glossitis
- Cardiac disorder, cardiomyopathy

At present, the pellagra is still prevalent in Africa, China and Indonesia. The poor socioeconomic status of patients is the main factor, which causes pellagra skin disease among the affected cases.

Hair Disorders

Hair disorders are important problems in dermatology. The specific study on hair is sometimes called trichology. There are several hair disorders in trichology. The important disorders are described and discussed below:

Alopecia: Alopecia is the lost of hair. Generally, everyone loses his/her hair everyday. This is the normal process. However, the excessive lost might be problematic and needs proper management. The causes of alopecia include:

- Genetic (especially in male)
- Drug induced
- Irritation (such as from hair lotion or shampoo)
- Hormone (especially for androgen)
- Psychiatric problem (especially for hair pulling madness or trichotillomania)

The specific form of alopecia is called bald, meaning total lost of head hair. This seems to be common in male and becomes an important cosmetic problem.

For the treatment of alopecia, there are many possible options. However, the most basic approach is the use of specific hair growing drug. The most well known drug is minoxidil.

Lice infestation: Lice are the infection of hair by external parasite, *i.e.*, the pathogenic insect. Normally, lice infestation can be divided into two groups:

- **Head lice infestation:** Head lice infestation is common. It implies to poor hygiene, and it is frequently seen in children from poor socio-economical status background.
- **Pubic lice infestation:** Public lice infestation is considered as sexually transmitted infection. It has to be treated on both the couples.

Fungal infection: Similar to the case of skin, fungal infection of hair can be seen. It is common at head and usually relating to the fungal infection of the scalp (tinea capitis). The treatment of fungal infection includes the use of antifungal drug. To prevent fungal infection, good hair cleaning is recommended.

White hair: Change of hair color seems to be a big concern for many people. Basically, hair is black and will turn white when one becomes senile. The white hair in the young seems to be the defect of pigment in hair. This becomes a cosmetic issue. At present, several hairdo products are available. Hair colour change can be done by oneself at home.

Nail Disorders

Nail is another important structure that bugs from skin. Nail can be seen on some specific sites of the body:

- **Finger:** 1 finger 1 nail (totally 10 nails)
- **Toe:** 1 toe 1 nail (totally 10 nails)

The structure of nail is composed of hard keratin. The nail, similar to hair, has outside growing up. This is continuously progressive. More externalization means less viable. Hence, it is no doubt that we can cut off the tip part of nail without any problem. Nail disorders are important problems in dermatology. There are several nail disorders. The important disorders are described and discussed below:

Fungal infection: Similar to the case of skin and hair, fungal infection of nail can be seen. It is common at fingernail and is called tinea ungium. The treatment of fungal infection includes the use of antifungal drug. However, it is very difficult to cure fungal infection of nail.

Nail deformity: Deformity of nail can be seen in medicine and might be the sign of many medical disorders such as:

- Chronic respiratory tract disorder
- Anemia (the most well-known condition is the koilonychias seen in iron deficiency anemia)

Paronychia: Paronychia is a specific bacterial infection of skin connecting to the nail. It can form the pus and the patient will be painful. Drainage of pus has to be done.

Nail fracture: Nail fracture can be seen. This can lead to the subungal hemorrhage and this can cause pain sensation. The removal of nail or drainage of blood clot under nail is indicated.

Nutrition and Skin Health

Skin is an organ of the body. Hence, it requires nutrients. The ways that nutrients will be useful to the skin can be:

- Direct absorption *via* direct skin contact
- From ingested and digested food
- Biosynthesis within skin (vitamin D)

Basically, the skin consists of cells. Hence, the primary requirement of skin is the protein for the formation of skin structure. In addition, skin cells are living cells. Hence, it requires energy from carbohydrates and lipids. However, there are some interesting specific conditions showing the important role of other nutrients on skin health, which are discussed below:

Selenium deficiency and skin fungal infection

Selenium is another important micronutrient. Selenium can be derived from vegetables like broccoli. Selenium deficiency, which can affect the skin health, can be seen. This can reduce resistance to fungal skin infection. Indeed, selenium sulfide shampoo is the widely neutraceuticals for the treatment of tinea capitis.

Vitamin B_1 deficiency and dry skin: Vitamin B_1 (thiamine) can be seen in fruits and vegetables. The deficiency of this vitamin is very common in developing countries. The most affected population is the childhood and the most common presentation of vitamin B_1 deficiency is the paresthesia. However, the skin manifestation as dry skin can be seen. To treat this vitamin deficiency, the supplementation of vitamin B is recommended and favourable outcome can be derived.

Vitamin B_2 deficiency and angular stomatitis: Vitamin B_2 (riboflavin) is a well-known vitamin that can be seen in fruits and vegetables. This is a micronutrient, meaning we require it at very small amount but cannot lack. The deficiency can result in a condition namely angular stomatitis. This condition presents with the dermatitis at skin at angular of mouth. This can cause painful when open and close mouth.

The problem is common in children who do not eat fruits and vegetables. To treat this vitamin deficiency, the supplementation of vitamin B can be done and the dramatic response to vitamin treatment can be seen.

Vitamin B_3 and skin rash: Vitamin B_3 (niacin) can be detectable in fruits and vegetables. However, high dosage and vitamin B_3 intake can cause red flushing. Skin eruption can be seen in excessive intake.

However, on the other hand, the vitamin B_3 deficiency can also cause the skin disorder. The dermatological condition namely pellagra is the most well known complication of inadequate vitamin B_3 intake. The characteristics of skin lesion in this vitamin deficiency include:

- Thickening
- Painful flushing
- Dark colour
- Scaly

It should be noted that either more or less than adequate level of vitamin B_3 intake could cause skin disorder.

Vitamin B_5 and acne treatment: Vitamin B_5 (pantothenic acid) can be seen in fruits and vegetables. This vitamin is mentioned for its usefulness in the treatment of acne. Oral administration of vitamin B_5 is reported to help in:

- Reducing water loss of skin
- Decreasing skin oil
- Decreasing production of sebum

Vitamin B_9 deficiency and skin disease: Vitamin B_9 (folate or folic acid) can be seen in meat and vegetables. The deficiency of this vitamin is well described for relationship to megaloblastic anemia and neurological complication. However, the skin disorder is also described. The classical skin manifestation includes:

- Hair fall
- Brown skin
- Peeling skin disease

However, although the use of folic acid supplementation is effective for the treatment of anemia but the usefulness in the management of skin disorder is still not proved.

Vitamin C and prevention of skin cancer: Vitamin C can be seen in many citrous fruits such as orange and guava. It is accepted as a potent antioxidative agent and is proved for clinical usefulness in the prevention of senility and cancer. It is the best vitamin for skin care. There are many available topical forms of vitamin C product for promoting good skin health. The advantages of local vitamin C skin application include:

- Whiten skin
- Decrease melasma
- Decrease scar
- Decrease skin cancer

Vitamin E deficiency and dry skin: Vitamin E can be seen in fruits, vegetables and oils. Its deficiency is not so common as compared to other vitamin and can cause infertility. However, the skin manifestation as dry skin can be seen. Easy bruising and damaged hair is also noticed. To treat this vitamin deficiency, the supplementation of vitamin E is recommended.

Nutrition and Hair Health

Similar to skin, hair also requires nutrients. The ways that nutrients will be useful to the skin can be:

- Direct absorption *via* direct hair contact
- From ingested and digested food

Basically, the hair is viable, hence, required nutrients. There are some interesting specific conditions showing the important role of nutrients on hair health. These are discussed below:

Vitamin B_9 deficiency and hair lost: As already discussed, folic acid deficiency can also cause hair lose in addition to other skin disorders.

Vitamin E deficiency and damage hair: As already described, vitamin E deficiency can also cause damage to the hairs, addition to dry skin.

Selenium deficiency and hair fungal infection: Similar to that mentioned in skin disorder, selenium is proved useful for the prevention of fungal infection of hair.

Albumin and hair strengthening: Albumin or protein from egg is widely mentioned for its usefulness to hair. It is believed to promote hair growth. There are many available topical form of albumin product, especially for shampoo, promoting good hair health

Nutrition and Nail Health

Similar to skin and hair, nail also requires nutrients. The ways that nutrients will be useful to the skin can be:

- Direct absorption *via* direct nail contact but this is rather difficult since nail is hardly to be penetrated by chemical.
- From ingested and digested food

Basically, nails are viable and require nutrients for continuous growing out. Hence, they require energy from carbohydrates and lipids for their maintenance generating cells in nail beds. Protein is also highly required for structural generation of the nail. However, there are some interesting specific conditions showing the important role of other nutrients on nail health, which are discussed below:

Vitamin A deficiency and ridged nails: Vitamin A deficiency is a common vitamin deficiency that can be detectable around the world. In addition to its common presentation, night blindness, the nail disorder can also be seen. The common nail manifestations include ridged nails and loss of half moon figure at nails.

Vitamin D deficiency and dry nails: Vitamin D deficiency is a common vitamin deficiency. Vitamin D is necessary for supporting hard structure within the human body. Hence, it also plays important role for nail structuring. Vitamin D deficiency can result in dry nails. Similar observations on dry nails can also be seen in calcium deficiency.

Iron deficiency and koilonychia: Iron deficiency is common in many developing countries. This is usually due to the lack of adequate intake of iron (iron can be found in meat and green leafy vegetables) as well as parasitic infestation (hookworm infestation). Lack of adequate iron can cause deficiency status and the most common clinical manifestation is the small red cell anemia. This can results in many anemia symptoms such as pale, weakness, fatigue, dizziness, *etc.*

However, sever prolonged iron deficiency can also present uncommon symptoms. Abnormality of the nails is an interesting problem to be discussed. The condition is called *koilonychias*. This nail defect can be seen accompanied with or without anemia. However, it should be noted that there are also other causes of koilonychias including:

- Exposure to chemicals
- Some genetic disorders

In case that koilonychias is proved to be related to malnutrition, iron deficiency, the iron supplementation treatment is recommended.

Selenium deficiency and nail fungal infection: Similar to that mentioned in skin and hair disorders, selenium is proved useful for the prevention of fungal infection of nail.

Yellow Nail Syndrome and Diabetes Mellitus

Diabetes mellitus is a common defect of glucose metabolism. It is a well-known disease in nutritional medicine. An interesting observation on nail in the diabetic patients is the nail infection by fungus. However, other rare conditions are also observed. Yellow nail syndrome, a disorder with yellow colouration of the nail, can be seen in diabetes mellitus. It is believed to be related to the lymphatic disorder at nail area. On nail examination, yellow colour and slight brushing at nail base is the common finding. It should also be noted that yellow nail syndrome could also be seen in other disorders such as chronic bronchitis and vitamin E deficiency.

■■■

21 Essential Oils and Aromatherapy

Nidhi Dubey[1] and C.S. Raghav[2]

[1]Indian Pharmacopoeia Commission, Government of India,
Sector-23, Rajnagar, Ghaziabad, Uttar Pradesh-201002
E-mail: <nidhiknj@gmail.com>
[2]National Bureau of Plant Genetic Resources, Pusa Campus, New Delhi-110012

What is Aroma?

Odorous compounds present in food reach the olfactory mucosa receptors in two different ways. The first is by direct sniffing volatile molecules into the nose where they interact with the receptors producing the sensation known as *odour*. The second is where the volatile substances released from the food in the mouth reach the olfactory epithelium by a retronasal route, producing the sensation known as *aroma*. The combined sensations of odour, aroma and taste give the overall *flavour* of the food. The odour and aroma fractions are composed of molecules belonging to different classes of organic compounds. These include phenolics which have a low molecular weight and a relatively hydrophobicity. The molecules are present in low concentrations (1 mg/t to 1 mg/kg) are able to escape from foods and are present in the vapour phase. A natural product aroma consists of several hundred volatile compounds. Aroma generally results from the contribution of a more or les important number of volatile compounds each of them being responsible for the typical contribution. Volatile compounds in fruits and vegetables are also present as non-volatile precursors and are released fro these precursors during maturation, pretreatment, or processing by enzymatic or chemical reactions.

Though the words aroma, odour, smell, scent, fragrance and perfume may be technically different, they are generally synonymously used for each other. Aromatherapy, aromatic-bath, aroma-candles, aromatic-oils, aroma-chemicals, aromatic-rice, aroma of tasty food, *etc.* are some of the examples of multiple use of the word aroma. The word aroma is specifically used for pleasant odour. A smell may be good or bad, or in other words, it may be obnoxious or pleasant. The definition of good or bed smell depends on the choice of a person. Some people like rose fragrance, while others like jasmine the most. The best example is the odour of garlic, those who eat garlic as spice, like garlic aroma in food, while those who do not eat garlic for any reasons they do not like the same garlic smell. Thus, smell or aroma of the garlic may have different impacts on moods of two different persons. The aroma or smell is the subject of chemical composition of a substance and sensory organs in the brain. When a substance possesses the aromatic properties and brain perceives it, we feel the sense of smell. Therefore, smell is a function of olfactory and perception together. The other meaning of aroma also includes the taste. Here, smell and taste have separate meanings, while smell is a function of three things, *i.e.*, an aromatic material, nose and perception, and the taste is a function of four things, *i.e.*, an aromatic material, nose, tongue and perception.

Why some materials possess aroma, while others do not?

The answer lies in chemical nature of the constituents of a material. We all know that all materials of the universe are composed of different kinds of chemical, molecule, or atom. Different chemicals and molecules have different nature and properties, which can be classified in various categories. According to one classification, the material may be either volatile or non-volatile. Though volatility is a relative term and highly affected with temperature and other factors, we will take it here in the broader terms. The volatile chemicals are also called aroma chemicals or the volatile chemicals possess aroma. They evaporate in the atmosphere, while non-volatile materials do not evaporate. Chemically, they contain at least one benzene ring in their structure. Majority of the aromatic materials come under a chemical class known as terpenoids. Therefore, any material containing aroma chemicals will emit smell. Some materials like rose, jasmine, marigold, kerosene, petrol and diesel possess high volatile chemicals, while others like grasses, vegetable, fruits and have low volatile chemicals. Non-volatile materials do not possess aroma. Some of the non-volatile materials include metals like steel, copper, aluminum, brass, stones, glass, *etc.*

Origin and sources of aroma

The origin or sources of an aroma may be either natural or synthetic. Almost all living beings weather plants or animals have aroma volatiles in their bodies. These aroma volatiles may be a part of either their bodies or any of their physiological processes in the form of products or byproducts. Apart from natural origin, an aroma may also be created synthetically using various chemicals and reactions. The aroma synthesized through chemical process may be either similar to an aroma already present in nature (nature identical synthetic) or entirely new (purely synthetic), not so far available in the nature. Therefore, there could be an unlimited types and varieties of the aroma from natural or synthetic sources.

How to create or extract aroma volatiles from plants?

As we know, aroma volatiles (aromatic molecules) are a part of any natural products or may be created by chemical reactions. In nature, aroma volatiles are available in plants and animals in the form of essential oils, extracts, exudates, resins, *etc.* The aroma of synthetic nature may be created in the form of single aroma chemical. When such single aroma chemical is isolated from an essential oil, it is called as *isolate*. Blending of various aroma chemicals extracts and isolated together produces fragrance and flavour which are used to give smell to various household and industrial products we use in our day to day life. Perfumes are the best example of created aroma, and the pure essential oils are examples or naturally available aroma. The aroma has various names and faces based on its form, composition, use, extraction and isolation technique, origin, *etc.* some of the varieties of aroma are described below:

Concrete: When a plant material is subjected to solvent extraction and the solvent is evaporated after extraction, the remaining residue is called as concrete. Solvent extraction is nothing but solubilizing aroma of the plant material in solvent such as petroleum ether or hexane in the same manner as we solublize the oil spot of our clothes with petrol or tea flavour in boiling water. In this case, if we evaporate the petrol or water, the residue will be oil in case of cloth and flavour in case of tea. The concrete contains essential oil and other high volatile aroma-chemicals and waxy and resinous part of the plant material. Concrete is the crudest form of essential oil, which is the ultimate fine form of natural aroma. In practice, concrete is the prepared from certain costly flower such as jasmine. When steam or water distillation is not feasible due to quality or economic reasons, the practice of extracting concrete is adopted. Concrete is used as whole in preparation of fragrance and flavour for giving aroma to various products.

Absolute: When the above concrete is extracted with alcohol and the alcohol is evaporated, the residue is called as absolute. The absolute is nearly pure form of essential oil. The absolute is prepared when the plant material contains water-soluble aroma and the water distillation is not economic due to possible loss of aroma in water during distillation process, or the water or steam distillation may damage the heat sensitive aroma constituents of the plant material. Rose or jasmine flowers are used to obtain absolutes.

Tinctures: These are the alcoholic extracts of natural perfumery or flavouring materials in which the finally divided material is allowed to macerate or steep in 95° alcohol over a period of time ranging from some weeks to several months, the solvent remaining in the finished tincture as a diluent.

Infusions: Infusions are prepared almost in the same way as tinctures, but with the application of more or less heat, the treatment usually involving reflux and a period varying according to the material and temperature. Infusions tend to be less mellow and true to type than tinctures, owing to the heat induced reactions involved, especially with the alcohol. For this and other reasons, the use of infusions is very rare in perfumery practice.

Essential oils: The essential oils are steam volatile odoriferous oils from plants or in few cases from animals, *i.e.*, civet from cat, musk from musk dear, ambergris from whale and castoreum from beaver. The term *essential oil* was used for the first time in the 16th century by Paracelsus von Hohenheim, who named the effective component of a drug, *i.e.*, *Quinta essential.*

In plants, essential oils are present in the oil glands, which are freed from them using various methods by rupturing the oil glands. Most simple and common method of obtaining essential oils from the source material is hydro-distillation, in which, when the essential oil bearing plant parts are boiled in water, the oil glands are ruptured due to high vapour pressure and oil gets free in the boiling water. The vapourized mixture of essential oil and water from the boiling water is condensed in a separator with the help of a condenser. The layer of oil is collected from the separator. The oil can be releasing from the glands by various other means also such as steam distillation, hydro-steam distillation, solvent extraction, compression, super critical fluid extraction (SCFE), *etc.*

The basic concept of obtaining essential oil from the source materials remains the same in all methods, as the processes are comprised of two steps, one releasing the oil from oil glands by rupturing the glands and the other distilling the free oil from the solution. Whatever means is used for releasing (extraction) oil from the glands, the isolation or separation technique essentially need hydro (water) or steam distillation. The selection of extraction method and solvent depends on nature of the plant parts and purpose of use. Hydro or water distillation is a process in which the aromatic plant material is subjected to boiling and the steam generated by doing so is condensed with the help of a condenser. The steam volatile oil components come with the steam and make a separate layer in the distilled water.

Essential oils are present in various parts of plants such as root or rhizome in case of vetiver (*Vetiverza zizanioides*), nagarmotha (*Cyperus scariosus*), jatamansi (*Nardostychus jatamansi*), sungandhabala (*Valeriana officinalis*), buch (*Acorus calamus*), ginger (*Zingiber officinale*), turmeric (*Curcuma longa*), kapoor kachari (*Hedychium spicatum*), galangal or kulanjan (*Alpinia galanga*), *etc.*, stem in case of sandalwood (*Santalum album*), agar wood (*Aquilaria malaccensis*), rose wood (*Dalbergia latifolia*), Cedar wood (*Cedrus atlantica*), *etc.*, bark in case of cinnamon (*Cinnamomum camphora/tamala*), cassia (*Sassafras albidum*), *etc.*, gum resin in case of guggal (*Commiphora wightii*), frankincense (*Boswellia carterii*), balsam tolu (*Toluifera balsamum*), *etc.*, leaves in case of patchouli (*Pogostemon patchouli*), rosemary (*Rosemarinus marianum*), basil (*Ocimum sanctum*), tea tree (*Melaleuca alternifolia*), wintergreen (*Gaultheria procumbens*), pine (*Pinus roxburghii*), bay leaf (*Cinnamomum tamala*), thyme

(*Thymus vulgaris*), eucalyptus (*Eucalyptus globulus*), *etc.*, flowers in case of rose (*Rosa damascena*), jasmine (*Jasminium sambac* or *grandiflorum*), champaca (*Michelea champaca*), tuberose (*Polianthese tuberosa*), marigold (*Tagetus erecta* or *minuta*), chamomile (*Matricaria chamomilla*) or Clary sage (*Salvia officinalis*), neroli (*Citrus aurantium*), lavender (*lavandula angustifolia*), ylang ylang (*Canaga odorata*), orange (*Citrus sinensis*), hyssop (*Hyssopus officinalis*), *etc.*, anthers (parts of flower) in case of kesar (*crocus sativas*), berries in case of allspice (*Pimenta officinalis*), juniper (*Juniperus communis*), fruit peel in case of lemon (*Citrus limonum*), lime (*Citrus aurantifolia*), bergamot (*Citrus bergamia*), grape fruit (*Citrus paradise*), orange, tangerine, *etc.*, seed in case of celery (*Apium graveolens* var. duke), cumin (*Cuminum cyminum*), coriander (*Coriandrum sativum*), anise (*Pimpenella anisum*), nutmeg (*Myristica fragrans*), *etc.*, whole plant in case of palmarosa (*Cymbopogon martini*), geranium (*Pelargonium graveolens*), citronella (*Cymbopogon nardus*), lemongrass (*Cymbopogon flexuosus*), mint (*Mentha arvensis* or *spicata*), *etc.*

Types of essential oil

There are various forms of essential oils such as pure, natural, constituted and synthetic. The pure essential oils are obtained from the same species of a plant by steam or water distillation. Constituted essential oil is a preparation by manipulating the quantities of one or more isolates of the pure essential oil from outer sources. Synthetic essential oils are prepared by blending the natural isolates, extracts and synthetic aroma chemicals in the proportion so as to give a smell resembling to the pure oil of a specific species. Natural oil may be pure oil or constituted oil. When isolates from pure essential oils are used to prepare oil, it may be called as natural because no synthetic aroma chemical has been used in blending but it is not necessary that natural oil will necessarily be pure oil. In market, these terms play very important role in sale purchase. One has to be very cautious while ordering oil for purchase. Mere ordering an essential oil may end up the purchaser in getting synthetic or constituted oil. Therefore, the use of adjectives and specifications is most important. The quality of an essential oil also depends on the geographical origin of oil. The lavender from Spain or France will have different quality and smell from that of Indian lavender produced in Kashmir. Some essential oils from the same plant species may have entirely different chemical compositions and the Basil (*tulsi*) is the best example of this variation. Basil essential oil has various chemo-types such as linalool type, camphor type, euginol type, thymol type, citral type, methyl chevicol type, methyl cinnamte type, *etc.* All chemo-types have different smell from each other.

Essential oils from animal origin

Majority of the essential oils in the nature are derived from plant sources, however, a very limited number of essential oils come from animal sources. The animal sources of essential oils are as under:

Musk: The musk is an animal product. It is a secretion of a pouch like internal organ called preputial gland located on stomach of the male Himalayan musk dear (*Moschus moschiferus*). This species of dear is found in India, Nepal, China and Tibet and other parts of Himalaya. Poachers kill the animal to obtain this organ. However, the organ can be removed even without killing the animal. Since catching the animal live is more difficult to them, they kill the animal and take out the pod. Two to five cm pear shaped hairy musk pod contains inside the granular aromatic material of a colour and texture similar to dried blood called *musk grains*. The gland or pod is covered with white and brownish hair in the centre of which is a small orifice. The active musky material inside the pod comprised around 60-75% of the total weight of the pod. The smell of musk in the dear attracts the female dear in rutting season for mating. The dried musk pod contains characteristic animal type, slightly amine like, sweet and very persistent odour. It acts as a fixative, enhances the diffusive properties of perfumes and imparts added lift. The musk odorant along with all other animal originated odorants have mellowing, sustaining and diffusing action on the perfume blending.

Ambergris: Ambergris is a substance of animal tissue- a pathological calculus rather analogous to gall stones, formed in the stomach or more rarely in the bowel or intestine of the sperm or cachalot whale (*Physeter catodon*) and excreted out in the ocean as a result of intestinal irritation caused by ingestion of cuttlefish beaks. The whale sometimes expels it through vomiting. The ambergris may be found floated on various ocean coasts and islands. It occurs in lumps of varying size, shape and weight, which are of a waxy spongy texture, lighter than water and of various colour shades from chalky white and silver grey to dark brown and black solid mass, which melts in boiling water. The quality of ambergris improves with the ageing. The raw ambergris is extracted with alcohol and the resultant resinoid or absolute is used in the form of tincture in preparation of various perfumery and flavour materials. A tincture of 2-3% concentration is prepared in 95° ~ alcohol by soaking and stirring for three months or more. Its odour is rather subtle, reminiscent of seaweed, wood and moss with a peculiar sweet, yet very dry undertone of unequalled tenacity. A good grade of Ambergris does not contain animalic note.

Civet: It is a glandular secretion, of the consistence of butter or honey, taken from glands in the anal pouch of species of the Civet Cat, male and female, preferably the male. The Abyssinian variety, *Viverra civetta* (African Civet Cat), is the most important supplier of this animal perfume raw material. Smaller quantities of civet are collected from other species of the civet cat in India, Indonesia, Malaya, China, Belgian, Congo, Kenya, Somali and occasionally in Haiti. It is of clear yellowish or brownish colour, of a strong musky odour, offensive when undiluted but agreeable when a small portion is mixed with another substance. It is used as a perfume. It has a skatolic odour due to its contents of methyl indole or skatole. It also contains musk like lactone and civettone. It is used in the forms of tinctures, anhydrols, concretes, absolutes, *etc.* It yields about 50-60% concrete of the crude secretion and the larger part of the concrete is alcohol soluble absolute. Civet Tincture may be produced directly from the crude civet by maceration with 95° ~ ethyl alcohol. It is customary to use either 5 or 10 parts by weight of alcohol to one part by weight of civet. The civet is used in a variety of perfumes. It blends well with the nitromusks, coumarin derivatives, quinoline derivatives, vanillin and related materials *etc.*

Castoreum: Castoreum is a secretion from the male or female beaver (*Castor canadensis* and other species). The ointment like secretion is found in pear-shaped pouches or sacs. Castoreum is a term normally used for the entire gland. It is collected in a gland, which produces an oily substance in which the beavers rub themselves in order to protect their fur from being soaked in water. The contents of the gland turn dark, hard and brittle when the pouches are sun dried or smoked. One gland weighs up to 100 grams. It is much less expensive than civet. Castoreum is used in the form of tincture of 5 or 6% strength in 95°~ alcohol as a resinoid or a kind of absolute. The odour of castoreum is very characteristic, suggesting gloves and Russian leather and to a lesser extent galbanum and ripe figs.

Fragrance and Flavour

We come across and use a variety of products in our day to day life which possess aroma. Food items such as ice creams, cold drinks, alcoholic drinks, sweets, squashes, packed milk, cosmetics and toiletries items such as creams, lotions, talcum powders, lip sticks, soaps, detergents, hair oils, perfumes, shampoos, bath salts and other household items such as floor cleaners, insect repellants and killers, incense sticks, new fabric, leather, plastics and many others, all contain one or the other kind of aroma. Have we ever thought as where from the aroma comes in these items? In addition, how this aroma is prepared? All these varieties of aroma come from fragrance and flavour, which are the blends of various aromatic chemicals, isolates and essential oils. The word fragrance in general is used to indicate the aroma for non-edible products, while the word flavour is used for the aroma for use in edible items but this is the oil concept. Now a day, there is no boundary between fragrance and flavour. The use of both is inter-changeable subject to the condition that the aroma has been prepared with non-toxic and safe chemicals for the health. Rose and khus are the examples of common use as fragrance as well as flavour.

Spray Perfumes

Alcoholic spray perfumes and room fresheners are prepared using alcohol as a solvent to dilute and facilitate the homogenous spray of a perfume in smaller quantities. The common composition of any personal perfume may range from 5 to 30% fragrance and 70 to 95% alcohol. It is a simple mixing of concentrated fragrance and alcohol. In case of flavours, the fragrance may remain the same but the solvent will change to propylene glycol or other neutral solvents.

Attars

Attar is the other category of aroma, which is a floral/other plant material distillate over sandal wood oil under a specific distillation process called Deg Bhapka process- a kind of water distillation. There are various kinds of Attars like rose, kewda, hina, mitti, *etc.* The name of attar is based on the material, which has been distilled over sandal wood oil, and the process remains the same for all, *e.g.*, if there are rose flowers distilled over sandal wood oil, the end product will be rose Attar. Originally, the attars were prepared on the sandal wood oil base, but with increasing cost and difficult availability of sandal wood oil in India, the cheaper synthetic variants of attars were introduced in the market. The synthetic attars are prepared over Di Octyl Pthalate (DOP) or other such base instead of sandal wood oil. Due to sandal-wool oil base, the aroma of pure attars is everlasting and homogenous, while the aroma of synthetic attars fades up fast and ends in a very short period leaving an odourless DOP as useless residue. Interestingly, Kannauj city of India is the originator of attars through out the world.

Floral or Aromatic Waters

Floral waters are prepared by water distillation method wherein certain amount of water-soluble aromatic constituents dissolved in the remaining water. Rose water, sandal wood water and basil water may be the some examples of floral or aromatic water. This water may also be prepared using synthetic aroma chemicals or natural isolates.

What is Aromatherapy

Use of aroma to treat body, mind and spirit for overall well-being is known as Aromatherapy. It is also known as Essential Oil Therapy or Aromacology. Aromatherapy is not only the curative therapy but it also works as preventive therapy and adds value to the good health- one can say an additive therapy. There are two schools of thoughts regarding the use of aroma in aromatherapy. According to one school of thoughts, the use of aroma, only through the olfactory (nose) is treated as aromatherapy, while according to the second school of thoughts, the aroma may be used by any means such as through olfaction, through skin, through oral intake, *etc.*

The other aspect of looking at the meaning of aromatherapy may deal with the forms of aroma in terms of natural or synthetic. In most cases, the natural sources of aroma such as essential oils and their isolates, extracts, concretes, *etc.* are used in aromatherapy. However, in many cases the synthetic sources such as synthetic aroma chemicals may also be used. The reason behind supporting the use of synthetic aroma-chemicals is that many natural identical aroma chemicals have the same properties as the natural isolates. Purely synthetic aroma chemicals may also be used in aromatherapy because the impact of aroma on mood is not purely chemical in nature. Various psychological factors including placebo effect also play an important role in therapeutic effects. The prevailing most accepted and doubtless concept is to use pure essential oils and their products for the best benefit of aromatherapy as compared to using synthetic aroma chemicals.

In case of aromatherapy, the source of aroma could be natural or synthetic materials but essential oils and natural extracts are the most acceptable and widely known material used in aromatherapy. Some experts may differ in their opinion but there are no extensive studies, which indicate that synthetics are not beneficial or they are always harmful.

Fragrance has been used since ancient times to deliver a variety of benefits. More recently, it has been discovered that the combined use of touch-therapy and aromatic oils has important role in maintaining wellness. With the growing mainstream acceptance of holistic health care, herbal remedies and nutraceuticals, aromatherapy has become much more broadly defined to include benefits, which are strictly cosmetic in nature. Today, varieties of products ranging from candles to dishwashing liquids are marketed with *aromatherapy* benefits. Although there is no legal definition of aromatherapy in the United States of America and other many advanced countries, specific claims can affect whether a product is considered a cosmetic or a pharmaceutical. Outside of the United States of America, the term aromatherapy itself has very specific connotations, which can affect the regulatory status of potential product introductions. In recent years, a number of new tools such as psychophysical measures and brain imaging techniques have greatly enhanced our knowledge of how touch and aromas are interpreted by the mind and body.

Throughout the ages, plants have been used for their mysterious healing properties. Many ancient cultures believed that they contain magical remedies to cure the body, enhance the mind and repel bad spirits, but despite the facts that the therapeutic benefits of plants have been known for many centuries, the last two decades have seen a tremendous resurgence of interest in our ability to harness these benefits through essential oils the plants yield. While plant constituents are found in abundance in modern day medicine, there is an overwhelming interest in using plants in their most natural forms. Aromatherapy does just this. The healing and mood altering properties of essence from various parts of the plant can be utilized to improve the mind, body and spirit.

The main branches of aromatherapy include Home Aromatherapy (self treatment and cosmetic use), Clinical Aromatherapy (as part of Pharmacology and Pharmacotherapy) and Aromacology (the psychology of odours and their effects on the mind). The practice of aromatherapy may be employed using various materials such as essential oils, absolutes, phytoncides, hydrosols, infusions, carrier oils, *etc.*

These materials may be applied through olfaction, skin, oral, or injection depending on nature of the materials and need of the treatment. Essential oils are generally used in 2-5% concentration with carrier oils such as sweet almond (*Prunus amygdalus*), apricot kernel (*Prunus armeniaca*), avocado (*Persea americana*), borage (*Borago officinalis*), cocoa butter (*Theobroma cacao*), evening primrose (*Oenothera biennis*), hazelnut (*Corylus americana*), jojoba (*Simmondsia chinensis*), kukui (*Macadamia integrifolia*), olive (*Olea europaea*), peanut (*Arachis hypogaea*), pecan (*Carya pecan*), rose hip (*Rosa canina*), sesame

(*Sesamum indicum*), sunflower (*Helianthus annuus*), *etc.* as direct applications of essential oils may be harmful.

History of Aromatherapy

The oldest references of aromatherapy in the form of use of essential oils and other aromatic materials have been cited in the Hindu mythology where the God and Goddesses used the sandal and basil for their aroma. There are several references in Indian mythology including the use of various aromatic herbs by the God of Drugs-Dhanvantri. Thereafter, ever since the man first began to record history, there are references to the use of precious essential oils to honour, seduce, or delight the rich and powerful. It is thought that their use goes back as far as 7000 BC when Neolithic man combined olive and sesame oil with plant fragrances to produce ointments. The use of oil is documented in some of the earliest Chinese writings, and numerous excavations of artifacts produced by the Greeks, Romans and Persians show that they all embraced the power of mysterious, pungent and mood-enhancing essences.

Papyrus manuscripts record that during the reign of the Egyptian pharaoh Khufu, the great Pyramid builder (2700 BC), the temple-incense perfumes, fragrant herbs and oils were highly valued and used in the healing process. Of the many fragrances used in these times, kyphi was one of the most popular. It was viewed as some kind of *wonder drug* comprised over 16 different essences and was used as a sleeping aid, as a dream enhancer, as a stress relaxant and also as a treatment for such diverse ills as asthma and sorrow. Cleopatra would strew her room knee-deep in rose petals to capture the heart of Mark Anthony, believing that when the petals were crushed underfoot the devastating aphrodisiac properties of the oil contained within the petals would be unleashed.

The Greeks learned a great deal from the Egyptians, they recognized the medicinal and aromatic benefits of plants. Hippocrates commonly called the *father of medicine* practiced fumigations for both aromatic and medicinal benefit. A Greek perfumer by the name of Megallus created a perfume called megaleion. Megaleion included myrrh in a fatty-oil base and served several purposes, *i.e.*, (i) for its aroma, (ii) for its anti-inflammatory properties towards the skin, and (iii) to heal wounds. The Geek priestesses Delphi used burning bay leaves to induce a Trans like state which enables them to foretell the fortune, while the handsome virile athletes at the ancient Olympic Games anointed their bodies with scented oil before competing. Roman honoured the use of fragrant oils and those who could afford to bathe in perfume. During the reign of Julius Caesar, communal bathing was always followed by a massage of sweet smelling oils which were then scraped off the body using a special tool. When guests arrived to see the Emperor Nero, he had them sprayed with particular oils to put them in the mood of his choice. Perfume was sprinkled on the spectators at games, and petals were strewn before men of rank so that they could walk on a bed of sweetly scented flowers.

In China, the herbal tradition has a history stretching back to thousands of years. Chinese nobility and upper echelons of society in the ancient times not only bathed in the oils but also used them to fragrance their clothing, their homes and their temples. Even statues of great God Buddha were fashioned from camphor wood. One of the earliest known documents about the uses of aromatic plants dates from around 2000 BC and is known as the Yellow Emperor's Book of Chinese Medicine. The Japanese set up special schools to teach the art of Kodo (perfumery) which still exist today. The ceremonial art of burning incense were taught as well as complex story dances of incense burning rituals.

Having made the connection between mood enhancing aromas and the power of religion, the ancient Britons, particularly the Druids, favoured herbs that induced mind altering states to enhance their mystical experiences. The druidic priests and priestesses burned their offerings during ceremonial worship of their

Gods, during magical rites and for *miraculous* healing of the sick. Many causes of the sickness were attributed to evil spirits entering an unsuspecting person, and hence, exorcism was the only cure. The smoking out of these evil spirits with powerful aromatic herbs was widely used. Juniper, particularly, was linked with purification and was thought to be at its most effective during the winter solstice rituals when the son died and was then reborn. With the arrival of Normans, the custom of strewing floors with aromatic leaves became common, as it was believed quite rightly that the antibacterial properties would help ward off disease and deter flea's lice. It also reduced the stench of unplumbed housing and ameliorated the fact that bathing was considered unhealthy, if not positively dangerous.

During the 15th and 16th centuries, perfumers widened their art putting plant essence to darker uses. Within the 15th century, more plants were distilled to create essential oils including frankincense, juniper, rose, sage and rosemary. A growth in the amount of books on herbs and their properties also begins later in the century. Paracelsus, an alchemist, medical doctor and radical thinker, is credited with coining the term Essence and his studies radically challenged the nature of alchemy and he focused upon using plants as medicines. During the 16th and 17th centuries, perfume creation was considered a form of art and it was more clearly defined as its own field. By the end of 18th century and the beginning of the 19th century, the use of oils and perfumes as medicinal products had declined, as more had become known about synthetic drugs. Oils and essences were relegated to the minor realms of scented toilets waters, which were sold for the smell alone not for any therapeutic, healing mystic, aphrodisiac, or ritualistic purposes. During the 19th century, perfumery remained in the focus of science and industry as the identity and varied use of individual isolates came into existence.

Rene Gattefosse, often referred to as the father of modern aromatherapy, first coined the term aromatherapy in the 1920s while studying the powers and properties of essential oils. His work was put to the test when one day in his laboratory he accidentally burned his hand. Plunging it into a container of pure lavender to cool down the burn, Gattefosse was to discover that the burn quickly lost its redness and swiftly begin to heal. Other highly respected 20th century aroma therapists include Jean Valnet, Madam Marguerite Maury and Robert B. Tisserand. Today, though we have science to reinforce many of the findings of the ancients, the powers of many of the oils have been scientifically backed up by research, while others only have the attribution of many, many skilled practitioners. From the late 20th century and on into the 21st century, there is a growing resurgence to utilize more natural products including essential oils for therapeutic, cosmetic and aromatic benefit. The use of essential oils never ceased but the scientific revolution minimized the popularity and use of essential oils in one's everyday life. Today's heightened awareness regarding the use of synthetics coupled with the increased availability of aromatherapy information within books and the Internet has refuelled the use of essential oils for therapeutic, cosmetic, fragrant and spiritual use.

How Aromatherapy Works?

The action of aromatherapy lies in the chemistry of aromatherapy materials and their interactions with body system. These aroma chemicals may influence genetics and the environment (G x E) interaction in human beings, which may further affect the chemistry and biology of human body and human behaviour. Various aroma chemical groups found in essential oils are:

Terpenes	:	Anti-viral, antiseptic and bactericidal
Alcohols	:	Antiseptic and anti-viral with uplifting qualities
Aldehydes	:	Sedative and antiseptic

Ketones	:	Ease congestion, aid flow of mucus and can be toxic
Acids	:	Preparation of esters and flavouring
Esters	:	Fungicidal, sedative and pleasant
Ethers	:	Flavouring and stimulating
Phenols	:	Bactericidal, antiseptic, flavouring and stimulating

Aromatherapy treatment not only seems capable of boosting positive and emotional well being but also has a direct physical effect on the body, affecting its organs and tissues, as well the nervous system. The hormone like actions of many oils, *e.g.*, Clary sage and chamomile, may help to regulate the endocrine system, while their antiseptic and antimicrobial properties combined with their ability to stimulate the immune system gives them a broad spectrum of other benefits. Some oils seem to have quite specific effects on the body but most of them have a wide range of overlapping properties.

Brain is the master

When we smell aroma, the aroma molecules reach our brain through the breathing process. Located at the top and on both sides of the inner nasal cavity, approximately at eye level is the olfactory bulb, which is covered with a mucus membrane known as olfactory epithelium. This membrane is lined on both sides with a special tissue, comprising of about 10 million olfactory nerve cells covered with a thin layer of mucous. Each individual nerve cell carries a bundle consisting of six to eight tiny hairs or cilia equipped with receptor cells. The chemical constructions of odour molecules fit exactly into a specific place. This olfactory membrane is the only place in the human body where the central nervous system is exposed and in direct contact of environment. The cells of olfactory membrane are the brain cells. The hairs attached to the nerve cells (up to 80 million of them) are capable of carrying an incomprehensible amount of information, a capability that outperforms every known analytical human function. With every breath, we take the minutest pieces of information from our environment.

The act of hearing and seeing requires the energy stimulus of sound and light. Similarly, the sense of smell requires the mere presence of an odour molecule, which we register in the brain when inhaling and discharge when exhaling. Odor molecules stimulate tiny hair bundles carried by nerve cells in the form of electrical impulses. The minute extensions at the site of nerve bundles are located in the nasal cavity, where they pass through the ethmoid bone behind the septum to the brain. There they come in contact of olfactory bulb, which in turn passes along the stimulus to the relevant location in the brain. Pungent, sharp odours like ammonia, acetic acid and carbon dioxide are carried by one set of nerve cells that react to the stimulation of these molecules. Fragrant substances, however, pass on to the limbic system (also known as old brain or olfactory brain) without being registered by the cerebral cortex. They reach the innermost control centers in our brain, the place where fragrances *touch our heart*. In other words, even before we consciously know we are in contact with an aroma, our subconscious receives and reacts to it. The limbic system transfers energy and inspiration through the sense of smell to the cerebral cortex, which can analyze these stimuli before a reaction takes place.

Odour stimuli on limbic system or olfactory brain release various neurotransmitters such as encephalin endorphins, serotonin and noradrenalin. Encephalin reduces pain, produces pleasant, euphoric sensation and creates a factory of well-being. Endorphins also reduce pain, stimulate sexual feeling and produce a sense of well-being. Serotonin helps in relaxing and calming, and noradrenalin acts as a stimulant that helps in keeping us awake.

Role of Perception and Memory

Within the limbic system resides the regulatory mechanism of our highly explosive inner life, the secret core of our body being there is the seat of our sexuality, impulse of attraction and aversion, our motivation and our moods, our memory and activity as well as our autonomic nervous system. Our exotic life is much influenced by odour. Pheromone, hormone like substances in one's personal odour influence partner choice in human as well as in animals. This is particularly true for choosing a suitable mate in the animal kingdom. A dog can recognize the smell of a female in heat over a distance of more than 3 km. We know that a male butterfly can smell a female as far away as 10 km. In humans, male underarm perspiration, for instance, contains substances very similar to the male sexual hormone testosterone. The female menstrual cycle is also regulated by a specific body odour. Since many essential oils contain pheromones, they respond to our sensuality. Love comes through the nose. Personal attraction and aversion are always influenced by odour. Our body has special glands that produce a very personal perfume that changes according to our model, state of health or illness and sexual inclinations. Every illness has its own smell. We smell differently when we are happy and centered than we do when are unhappy, anxious, aggressive, or under stress.

Illness has its beginning in the brain. It may grow in the soil of an attitude or motivates. These subconscious altitudes are stored in the limbic system, so it is possible to heighten or influence them with specific scents. Fragrances can create well-being or discontent. For instance, if we have distinctly coming to the past, feel sense, or seem unable to experience joy, a new fragrance may help create fresh mood or impulse. It may help us change our altitude or find a more positive reaction to a given situation. A pure essential oil also does not create negative feelings such as aggression or fear rather it has a balancing influence. It is said that synthetic essential oils do not match the impact generated by natural pure essential oils.

Regulation of our autonomic nervous system is not usually subject to our conscious influence. Breathing, digestion and heartbeat are automatic but fragrances can influence these functions as they influence the organs in our body. They may slow down as well as deepen the breathing process and affect our heartbeat. The digestive system is particularly susceptible to smell. The wonderful aromas of the spices we use in cooking stimulate our appetite and produce digestive juices, or they may relax a tense stomach. Vegetative dystonie, an illness with varying symptoms, is caused by a disturbance in the automatic nervous system. There, aromatherapy may be very useful in supporting the hearing process.

Miraculous Healing Properties of Essential Oils

When aromatherapy is used for the treatment or prevention of diseases, a precise knowledge of bioactivity and synergy of the essential oil used, knowledge of doses and duration of application as well as naturally, a medical diagnosis is required. Besides the role of aroma chemicals in healing process, the life force of the plant distilled with the oil also plays an important role by balancing the energies of the body and promoting healing or well-being by purging negative vibrations from the body's energy field.

Essential oils are so potent that many of them can irritate the skin and can cause toxic reactions like liver damage and seizures unless diluted with carrier oil. Some hazardous essential oils include ajowan (*Trachyspermum copticum*), bitter almond (*Prunus duleis*), arnica (*Arnica montana*), sweet birch (*Betula lenta*), calamus (*Acorus calamus*), camphor (*Cinnamomum camphora*), garlic (*Allium sativum*), joborandi (*Pilocarpus jaborandi*), yellow sweet clover (*Melilotus officinalis*), mugwort (*Artemisia vulgaris*), onion (*Allium cepa*), rue (*Ruta graveolens*), sassafras (*Sassafras albiolum*), thuja (*Thuja occidentalis*), wormwood (*Artemisia absinthium*), *etc.*

Depending on method, doses and other conditions, single oil can be used for several purposes, *e.g.*, basil oil may be used for muscular pain, depression and fatigue, jasmine oil for mood uplifting as aphrodisiac and as anti-stress, Eucalyptus for char, sinus, sores and insect bites and as disinfectant, lavender for calming, mood uplifting, skin care and pre-menstrual syndrome (pms), cyperus (*Cyperus scariosus*) for female hormonal and skin care and as anti-stress and anise for breathing and digestion.

There are range of various other essential oils which can be utilized for a single property such as cinnamon, Clary sage, frankincense, jasmine, juniper, patchouli, sandalwood and ylang ylang for anti-stress property, Clary sage, jasmine, patchouli and ylang ylang for aphrodisiac property, all spice, anise, bergamot, juniper, lavender, sweet marjoram, myrrh for depression and basil, bergamot and sandalwood for calming property. Believe it or not but *people* have explored the essential oils holding metaphysical uses such as all spice for prosperity, almond for peace and prosperity, amber for success and good fortune, anise for peaceful sleep and youthfulness, bergamot for money and success, carnatum for healing and strength, cedar for purification, love, money and psychic power, cinnamon for attraction and speed, cyperus for comfort and protection, coconut for purification, peace and intuition, frankincense for purification, luck, spirituality and protection, gardenia for love and peace, heliotrope for healing and wealth, *etc.* There is no end to this list. You name the metaphysical problem and there is oil available for it. This is not the end of miraculous properties of essential oils.

There are claims that these oils take care of emotional well-being like anger (bergamot, jasmine, neroli, orange, patchouli, petitgrain, roman chamomile, rose, vetiver and ylang ylang), anxiety (bergamot, cedarwood, clarysage, frankincense, geranium, lavender, mandarin, neroli, patchouli, roman, chamomile, rose, sandalwood and vetiver), confidence (bay laurel, bergamot, cyperus, grapefruit, jasmine, orange and rosemary), depression (bergamot, Clary sage, frankincense, geranium, grapefruit, jasmine, lavender, lemon, mandarin, neroli, orange, Roman chamomile, rose, sandalwood and ylang ylang), fatigue, exhaustion and burnout (basil, bergamot, ginger, rosemary, patchouli, jasmine, vetiver, *etc.*), fear (cedar wood, neroli, Clary sage, lemon, *etc.*), grief (cypress, neroli, rose and sandalwood), happiness and peace (bergamot, geranium, lemon, neroli, orange, rose *etc.*), insecurity (bergamot, cedarwood, sandalwood, frankincense, *etc.*), irritability (lavender, mandarin, neroli, Roman chamomile, *etc.*), loneliness (bergamot, Clary sage, rose, *etc.*), memory and concentration (basil, cyperus, hyssop, rosemary, *etc.*) and panic (frankincense, lavender, neroli, rose, *etc.*).

Factors Affecting Healing

The healing effect of essential oils and aromatherapy is a combined function of various factors such as body physiology, genetic makeup of a person, state of mind, acceptability of person, *etc.* Aromatherapy combined with yoga and meditation, acupressure and acupuncture may leave miraculous effect on body and mind.

Purity of essential oil is a very important aspect in aromatherapy as adulterants and substitutes may alter the healing effects of oil. Finding pure essential oil in the market is not an easy task. Besides adulterants, oil itself may be of various grades, *e.g.*, khus oil is available distinctly in two grades, *i.e.*, (i) North Indian, which is treated as the best and (ii) the South Indian. Mixing of cheaper South Indian khus oil into costly North Indian khus oil is a common practice in the market. Reconstituted oil is another way creating cheaper substitute of oil. For a common user, it is very difficult to differentiate between natural pure and reconstituted oil, while the difference in rates is almost double. Reconstituted oils are cheaper. Cheaper natural oils such as lemongrass, Palmarosa, citronella, basil, pine and mint, are less likely to be adulterated or reconstituted as compared to costly ones such as rose, jasmine, sandal, khus, *etc.* Sometimes, a part (fraction) of oil is mixed in other oil as adulterant, and in other cases, the matching chemical constituents from other cheaper oils or from synthetic sources may also be added to costly oil.

Since adulteration is a common practice in the market, it is advisable to get your oil tested before purchase through a Government laboratory or alternatively purchase the oils as for as possible from Government agencies or reliable sources only. Central Institute of Medicinal and Aromatic Plants (CIMAP), Lucknow, Indian Institute of Integrative Medicine (IIIM), Jammu), Institute of Himalayan Biosphere Technology (IHBT), Regional Research Laboratory, Jorhat and Bhuvneswar, Herbal Research and Development Institute, Gopeswar, South Indian Mint and Aromatic Products (SIMAP), Thirunelveli, Tamil Nadu, International Centre for Aromatherapy Research and Development, New Delhi and Pondicherry are some important reliable agencies wherefrom, some of the commonly known essential oil may be obtained. Some of these centers also provide testing service for essential oils.

Essential oils may be transformed into water-soluble form for use in cosmetic and skin care products such as bath oils. Just put few drop of these bath oils in the bath tub for rinsing the body. The oils contained in the water will be absorbed in the body giving a glow and smooth feel to the body. These preparations also help in removing body odour besides importing therapeutic effect to the body and mind. Aroma candles and vaporizers may be used for spreading desired aroma in the living or bedrooms in the evening. Attars are non-alcoholic fragrances having long lasting effect. Putting a drop of attar on the cloth under arms will give a pleasant aroma for many days. Khus attar or oil is wonderful for summers. Attars are prepared in rose, khus, jasmine (bela and chameli), mitti (earth), hina (lawsonia) and miscellaneous aromas. Massage is another way to apply essential oil on body. Aromatherapy massage is done using various essential oils blended in carrier oils. The massage with these oils not only imparts beneficial effects through the individual oil properties but also increases blood circulation of body and activates acupressure points. The aromatherapy can very well be applied in acupuncture also enhancing the effects of acupuncture. It has close relations with other traditional therapies and sciences like yoga and meditation.

Whether aromatherapy works or not?

This question has always been a matter of debate among scientists. On one side, there is more number of reports on the positive effects of aromatherapy treatments in several stress related problems. Simultaneously, there are certain reports where aromatherapy treatment has not been found beneficial but these negative reports have failed to check the multifold growth of aromatherapy believers and practitioners throughout the world. The probable reasons for ignoring the negative reports by the majority of people are increasing faith and belief in naturals, increasing scientific evidences of effectiveness of aromatherapy and people's own experience. We all know that stress is the cause of many major diseases. Besides meditation and yoga, there are several other therapies also which combat stress and provide relaxation to body and mind. Aromatherapy plays double role. It works directly as independent therapy in many stress related cases such as depression, anger, fatigue, hypertension, pain, *etc.* and indirectly in resultant diseases such as heart attack, diabetes, cholesterol, insomnia, *etc.* It can also play additive role in so many alternative therapies such as acupressure, acupuncture, yoga and meditation, psychotherapy, physiotherapy, *etc.* Aromatherapy is a delicate and deep science connecting ancient sciences of faith and belief and modern material sciences. It is spreading at a fast rate throughout the world.

Scientific Evidences

There are unlimited traditional claims about the benefits of use of essential oils and aromatherapy in various ailments. Many of them have been validated through the scientific research. Several new finding are also coming up day by day on the efficacy and authenticity of essential oils and other aromatic materials widening and strengthening the basis of aromatherapy. Some of the findings are given below:

Cancer

The diverse therapeutic potential of essential oils has drawn the attention of researchers to test them for anticancer activity, taking advantage of the fact that their mechanism of action is dissimilar to that of the classic cytotoxic chemotherapeutic agents. Research findings indicate that essential oil components, especially monoterpenes, have multiple pharmacological effects on tumour suppressive activity. Monoterpenes have been shown to exert chemopreventive as well as chemotherapeutic activities in mammary tumour models, and thus, may represent a new class of therapeutic agents. Essential oils could be utilized to protect the body organs against carcinogenesis. For instance, nutmeg (*Myristica fragrans*) showed a potent hepatoprotective activity against liver damage caused by certain chemicals. Trans-Citral, which is found in many essential oils like, lemongrass oil, posses inhibitory effect on early phase of the hepatocarcinogenesis in rats.

The other essential oils elicit hepatoprotective activity due to their phenolic and/or monoterpene content, *e.g.*, black cumin (*Nigella sativa*) due to thymoquinone, orange essential oil due to *d*-limonene and sweet fennel (*Foeniculum vulgare*) due to both *d*-limonene and beta-myrcene. Dextro-limonene is also a candidate for the chemoprevention of skin cancer. Garlic essential oil is a rich source of volatile organosulfur components (OSCs). These compounds are recognized as a group of potential cancer hemopreventive agents due to their activity for modulating phase I and II drug detoxifying enzymes. Alpha-bisabolol, a major sesquiterpene alcohol in chamomile (*Matricaria chamomilla*) essential oil and geraniol found in palmarosa oil (*Cymbopogon martini* var. *martini*) and geranium (*Pelargonium graviolens*) could be considered as a potential anticancer agent. Dextro-limonene showed antiangiogenic and proapoptotic effects on human gastric cancer implanted in nude mice, thus, inhibiting tumour growth and metastasis. Dextro-limonene, which occurs naturally in many essential oils, especially citrus fruit peel can also induce the formation of apoptotic bodies on BGC-823 gastric cancer cells. Diallyl trisulfide (DATS), a major constituent of garlic essential oil, showed a high activity for arresting the division of human liver tumour cells (J5) at the G2/M phase of the cell cycle. These results may draw the attention of cancer researchers to further extend their clinical testing of essential oils, and it would be interesting to investigate the antitumour activity of a major individual terpene constituent in comparison with its whole essential oil source.

Cardiovascular Diseases

Atherosclerosis is a process in which deposits of plaque build up in innermost layer of the artery, the intima. Plaque can eventually significantly reduce blood flow, leading to serious health problems. Increased concentrations of oxidatively modified low-density lipoproteins (LDLs) in cholesterol play a substantial role in disease initiation. Atherosclerosis can therefore be slowed down or inhibited by preventing the oxidation of LDLs using a high daily intake of antioxidants. Essential oils and their aroma volatile constituents have shown an antioxidative activity against LDL oxidation. Terpinolene, a monoterpene hydrocarbon, can effectively inhibit the oxidation of both the lipid part and the protein part of LDL. Essential oils rich in phenolic constituents such as eugenol and thymol have the highest antioxidative activity against LDL oxidation. Tea tree essential oil has been recommended as an addition to foods and beverages to protect against LDL oxidation and to reduce plaque formation. Black cumin oil (*Nigella sativa* Linn.) was found to decrease plasma concentrations of cholesterol and triglycerides due to the high content of thymoquinone. Oral administration of dill seed (*Anethum graveolens*) essential oil reduced the triacylglycerides level by almost 42%, although total cholesterol level was not reduced. Alpha-curcumin, the major constituent (approximately 65%) of the essential oil of Japanese turmeric (*Curcuma xanthorrhiza*), exerts triglyceride-lowering activity on serum as well as liver triglycerides. Garlic essential oil significantly lowered the serum cholesterol and triglycerides while raising the level of high-density lipoproteins in both healthy individuals and patients with coronary heart disease. The hypolipidemic action of garlic oil is primarily due to a decrease in hepatic cholesterogenesis. One of the consequences of atherosclerosis is hypertension, and some essential oils exert hypotensive activity when applied *in vivo*. Oral administration of combinations of oregano, cinnamon, cumin and other essential oils decreased the systolic blood pressure in rats and intravenous administration of the essential oil from aerial parts of a hybrid mint, (*Mentha villosa*) induced a significant and dose-dependent hypotension associated with decreasing heart rate. This activity was attributed to the volatile component, piperitenone oxide, which represents 55.4% of the oil. The hypotensive effect induced by the oil is probably due to its direct cardio-depressant action and peripheral vasodilation. Intravenous administration of basil (*Ocimum gratissimum*) essential oil induced an immediate and significant hypotension and bradycardia, while intravenous injection of the monoterpene alcohol terpinen-4-ol decreased mean aortic blood pressure.

Antibacterial

The Ancient Egyptians used aromatic plants in embalming to stop bacterial growth and prevent decay, an effect attributed to a great extent to their essential oils. Strong *in vitro* evidence indicates that essential oils can act as antibacterial agents against a wide spectrum of pathogenic bacterial strains including *Listeria monocytogenes*, *L. innocua*, *Salmonella typhimurium*, *Escherichia coli* O157:H7, *Shigella dysenteria*, *Bacillus cereus*, *Staphylococcus aureus* and *Salmonella typhimurium* and many more. Thyme and oregano essential oils can inhibit some pathogenic bacterial strains such as *Escherichia coli*, *Salmonella enteritidis*, *Salmonella choleraesuis* and *Salmonella typhimurium*, with the inhibition directly correlated to phenolic components, *i.e.*, carvacrol and thymol. The same correlation was also confirmed for oils rich in carvacrol alone. Eugenol and carvacrol showed an inhibitory effect against the growth of four strains of *Escherichia coli* O157:H7 and *Listeria monocytogenes*. The presence of a phenolic hydroxyl group, in carvacrol particularly, is credited with its activity against pathogens such as *Bacillus cereus*. Some essential oils demonstrated antibacterial activity against zoonotic enteropathogens including *Salmonella* spp., *Escherichia coli O157*, *Campylobacter jejunii* and *Clostridium perfringens*. Thus, these oils could possibly be used as an alternative to antibiotics in animal feed.

Essential oils with high concentrations of thymol and carvacrol, *e.g.*, oregano, savoury and thyme, usually inhibit Gram-positive more than Gram-negative pathogenic bacteria. However, the essential oil of *Achillea clavennae* exhibited strong antibacterial activity against the Gram-negative *Haemophilus influenza* and *Pseudomonas aeruginosa* respiratory pathogens, while Gram-positive *Streptococcus pyogenes* was most resistant to the oil. *Ocimum gratissimum* essential oil may find a use as a therapeutic measure against shigellosis. Methicillin-resistant *Staphylococcus aureus* can also be inhibited by the application of peppermint and spearmint essential oils.

Essential oils could be used as antibacterial agents against some respiratory tract pathogens. The oil of *Achillea clavennae* showed its maximum activity against *Klebsiella pneumoniae* and penicillin-susceptible and penicillin-resistant *Streptococcus pneumoniae*. The oil also exhibited strong activity against *Haemophilus influenzae* and *Pseudomonas aeruginosa*. An increased density of *Helicobacter pylori* in the gastric mucosa is associated with severe gastritis and an increased incidence of peptic ulcers. Among the individual constituents of various essential oils, carvacrol, isoeugenol, nerol, citral and sabinene exhibited the strongest anti-*pylori* effects.

Essential oils show bactericidal activity against oral and dental pathogenic micro-organisms and can be incorporated into rinses or mouthwashes for pre-procedural infection control, general improvement of oral health, interdental hygiene and to control oral malodor. *Croton cajucara* Benth essential oil was found to be toxic for some pathogenic bacteria and fungi associated with oral cavity disease and may be useful for controlling the microbial population in patients with fixed orthodontic appliances. A 6 months controlled clinical study demonstrated that a mouth rinse containing essential oils showed a comparable antiplaque and antigingivitis activity to that containing the synthetic antibacterial agent, chlorhexidine. Mouthwashes containing essential oils could also be used as a part of plaque-control routine since they can penetrate the plaque biofilm and kill pathogenic plaque-forming micro-organisms by disrupting their cell walls and inhibiting their enzymatic activity. In addition, essential oils in mouthwashes prevent bacterial aggregation, slow the multiplication and extract bacterial endotoxins.

Antiviral

Herpes simplex virus (type I and II) causes some of the most common viral infections in humans and can be fatal. Synthetic antiviral drugs have been used to treat herpes infections but not all are efficacious in treating genital *Herpes* infections. Plant extracts, especially essential oils, may afford a potential alternative to synthetic antiviral drugs. They have virucidal properties, with the advantage of low toxicity compared with synthetic antiviral drugs. Incorporation of *Artemisia arborescens* essential oil in multilamellar liposomes greatly improved its activity against intracellular *Herpes simplex* virus type-1 (HSV-1). *Melissa officinalis* L. essential oil can inhibit the replication of HSV-2, due to the presence of citral and citronellal and the ability to replicate of HSV-1 can be suppressed by incubation with different essential oils *in vitro*. Of these, lemongrass essential oil possessed the most potent anti-HSV-1 activity and completely inhibited viral replication after incubation for 24 hours, even at a concentration of 0.1%. Peppermint (*Mentha piperita*) essential oil exhibited high levels of virucidal activity against HSV-1, HSV-2 and acyclovir-resistant strain of HSV-1 in viral suspension tests. The antiviral activity of oil was confirmed when the virus was pretreated with the essential oil prior to adsorption. The essential oil of *Lippia junelliana* and *Lippia turbinate* showed a potent inhibition against Junin virus. Australian tea tree essential oil and to a lesser extent, eucalyptus essential oil, demonstrated antiviral activity against HSV-1,2.

Antioxidants

Free radicals and other reactive oxygen species cause aging, arteriosclerosis, cancer, Alzheimer's disease, Parkinson's disease, diabetes, asthma, *etc.* Essential oils as natural sources of phenolic components attract the investigators to evaluate their activity as antioxidants or free radical scavengers. The essential oils of basil, cinnamon, clove, nutmeg, oregano, Clary sage, *Achillea millefolium*, turmeric and thyme have proven radical-scavenging and antioxidant properties. The antioxidant activity of essential oils cannot be attributed only to the presence of phenolic constituents. Monoterpene alcohols, ketones, aldehydes, hydrocarbons and ethers also contribute to the free radical scavenging activity of some essential oils. It is clear that essential oils may be considered as potential natural antioxidants and could perhaps be formulated as a part of daily supplements or additives to prevent oxidative stress that contributes to many degenerative diseases.

Antidiabetic

Diabetes is a disease in which the body does not produce or properly use insulin. Investigators have conducted few studies exploring the potential benefits of essential oils as hypoglycaemic agents. Research indicated that an oral administration of a combination of essential oils including cinnamon, cumin, fennel, oregano, myrtle and besides others was able to enhance insulin sensitivity in type 2 diabetes, in addition to lowering circulating glucose in tolerance testing in rats. The essential oil of *Satureja khuzestanica* resulted in significant decreases in fasting blood glucose level in diabetic rats. More research is needed to confirm the veracity of the hypoglycemic activity of other essential oils and to elucidate their mechanism of action.

Useful Aromatherapy Blends (aromaweb.com)

(A) Blends for emotional well-being

Anger reducing blends

Blend #1: 1 drop rose, 3 drops orange and 1 drop vetiver

Blend #2: 3 drops bergamot, 1 drop ylang ylang and 1 drop Jasmine

Blend #3: 1 drop Roman chamomile, 2 drops bergamot and 2 drops orange

Blend #4: 3 drops orange and 2 drops patchouli

Anxiety relieving blends

Blend #1: 2 drops bergamot, 2 drops Clary sage and 1 drop frankincense

Blend #2: 3 drops sandalwood and 2 drops bergamot

Blend #3: 3 drops lavender and 2 drops Clary sage

Blend #4: 1 drop rose, 1 drop lavender, 2 drops mandarin and 1 drop vetiver

Confidence Enhancing Blends:

Blend #1: 2 drops bay laurel and 3 drops bergamot

Blend #2: 3 drops orange and 2 drops rosemary

Blend #3: 4 drops bergamot and 1 drop jasmine

Blend #4: 3 drops grapefruit and 2 drops cypress

Calming/Relaxing Blends:

Blend #1: 1 ounce carrier oil such as sweet almond, 10 drops Roman chamomile and 5 drops lavender (for foot massage)

Depressing reducing blends

Blend #1: 1 drop rose, 3 drops sandalwood and 1 drop orange

Blend #2: 3 drops bergamot and 2 drops Clary sage

Blend #3: 1 drop lavender, 1 drop ylang ylang and 3 drops grapefruit

Blend #4: 2 drops frankincense, 1 drop lemon and drops jasmine or neroli.

Energizing blends

Blend #1: 2 drops basil, 1 drop cypress and 2 drops grapefruit

Blend #2: 3 drops grapefruit and 2 drops ginger

Blend #3: 2 drops rosemary and 3 drops bergamot

Blend #4: 2 drops peppermint, 1 drop frankincense and 2 drops lemon

Fear fighting blends

Blend #1: 3 drops grapefruit and 2 drops bergamot- suited for times when you also need to be energized

Blend #2: 2 drops Clary sage, 2 drops roman chamomile and 1 drop vetiver- suitable for times when you need to be relaxed and calm

Blend #3: 3 drops sandalwood and 2 drops orange

Blend #4: 2 drops jasmine or 2 drops neroli, 2 drops frankincense and 1 drop Clary sage

Grief easing blends

Blend #1: 2 drops rose and 3 drops sandalwood

Blend #2: 2 drops rose and 3 drops cypress

Blend #3: 1 drop neroli, 1 drop rose and 3 drops sandalwood

Blend #4: 1 drop rose, 1 drop helichrysum, 1 drop cypress and 2 drops frankincense

Happiness and peace enhancing blends

Blend #1: 3 drops bergamot, 1 drop ylang ylang and 1 drop grapefruit

Blend #2: 1 drop geranium, 2 drops frankincense and 2 drops orange

Blend #3: 2 drops sandalwood, 1 drop rose and 2 drops bergamot

Blend #4: 2 drops lemon, orange or bergamot, 2 drops grapefruit and 1 drop ylang ylang, rose or neroli

Insomnia/sleeplessness blend

Blend #1: 10 drops Roman chamomile, 5 drops Clary sage and 5 drops bergamot

Irritability reducing blends

Blend #1: 3 drops mandarin and 2 drops lavender

Blend #2: 2 drops lavender, 1 drop neroli and 2 drops roman chamomile

Blend #3: 1 drop neroli and 4 drops sandalwood

Blend #4: 2 drops mandarin and 3 drops sandalwood

Blend #5: 3 drops roman chamomile and 2 drops mandarin

Loneliness easing blends

Blend #1: 3 drops mandarin and 2 drops lavender

Blend #2: 2 drops lavender, 1 drop neroli and 2 drops roman chamomile

Blend #3: 1 drop neroli and 4 drops sandalwood

Blend #4: 2 drops mandarin and 3 drops sandalwood

Blend #5: 3 drops roman chamomile and 2 drops mandarin

Memory and concentration blends

Blend #1: 3 drops rosemary and 2 drops lemon

Blend #2: 4 drops cypress and 1 drops peppermint

Blend #3: 1 drop basil, 2 drops rosemary and 2 drops cypress

Blend #4: 3 drops lemon and 2 drops hyssop

Blend #5: 2 drops peppermint and 3 drops lemon

Panic and panic attack blends

Blend #1: 2 drops helichrysum and 3 drops frankincense

Blend #2: 1 drop rose and 4 drops lavender

Blend #3: 1 drop neroli and 4 drops lavender

Blend #4: 1 drops rose and 4 drops frankincense

Self Confidence Enhancing Blends

Blend #1: 3 drops bergamot, 1 drop jasmine and 1 drop vetiver

Blend #2: 2 drops cedar wood, 1 drop bergamot and 1 drop frankincense

Blend #3: 4 drops sandalwood and 1 drop jasmine

Blend #4: 2 drops frankincense and 3 drops sandalwood

Stress reducing blends

Blend #1: 3 drops Clary sage, 1 drop lemon and 1 drop lavender

Blend #2: 2 drops roman chamomile, 2 drops lavender and 1 drop vetiver

Blend #3: 3 drops bergamot, 1 drop geranium and 1 drop frankincense

Blend #4: 3 drops grapefruit, 1 drop jasmine and 1 drop ylang ylang

Directions: Select one of the blends shown above, and then, choose which blend you would like to use. Before use, the below mentioned directions should be followed:

Diffuser blend

Multiply your blend by 4 to obtain a total of 20 drops of your chosen blend. Add your oils to a dark coloured glass bottle and mix well by rolling the bottle in between your hands. Add the appropriate number of drops from your created blend to your diffuser by following the manufacturer's instructions.

Bath oils

Multiply your blend by 3 to obtain a total of 15 drops of your chosen blend. Continue by using the 15 drops blend in the bath oil recipe.

Bath salts

Continue by using 5 drops blend in the bath salts recipe.

Massage oil

Multiply your blend by 2 to obtain a total of 10 drops of your chosen blend. Continue by using 10 drops blend in the massage oil recipe.

Air freshener

Multiply your blend by 6 to obtain a total of 30 drops of your chosen blend. Continue by using the 30 drops blend in the air freshener recipe.

(B) Blends for physical well-being

Arthritic joints: 57 grams carrier oil such as sweet almond, 20 drops Roman chamomile and 5 drops black pepper (for massage)

Balm for cuts and scrapes: 85 grams vegetable carrier oil such as sweet almond oil or infused oil such as calendula (infused oil of calendula can supply added therapeutic benefit). 1 ounce grated beeswax, 40 drops lavender oil and 40 drops tea tree oil.

Directions: Place the beeswax in a microwave safe bowl and melt in the microwave using a reduced power setting if you have one. You can also melt the beeswax in a pan on the stove using a low heat setting. Beeswax is hard to remove from pans, so please keep that in mind. In a separate pan, slowly and gently heat your carrier or infused oil. Pour the warm carrier or infused oil into a bowl, add the melted beeswax, and then, stir very well. Add lavender and tea tree essential oils and again stir well. Close the jar and wait until the ointment has cooled before using.

Bruises: 1 ounce carrier oil such as sweet almond oil and 8 drops Helichrysum

Directions: Mix the Helichrysum into the carrier oil well. Apply lightly to bruises 1-2 times a day.

Menstrual cramps: 1 ounce carrier oil such as sweet almond, 15 drops peppermint, 10 drops Cyprus and 5 drops lavender

Directions: Mix oils well in a clean, dark-colour glass container. Gently massage a small amount into the abdominal area.

Stuffy nose: 15 drops Eucalyptus, 13 drops Pine and 2 drops peppermint

Directions: Blend the oils together in a clean dark-colour glass container, preferably one with an orifice reducer (a built- in dropper insert). Apply 2-3 drops to a cotton ball and inhale occasionally to help clear a stuffy nose.

Blends for skincare, beauty and hygiene

Anti-wrinkle treatment blend

Frankincense oil 1 drop, neroli oil 1 drop, lavender oil 2 drops, Jojoba oil 1 teaspoon and avocado oil 1 teaspoon.

Skin drencher blend

Neroli oil 1 drop, lavender oil 1 drop, myrrh 1 drop and cream base (unscented)- 3 table spoons

Acne blend: 1 ounce light cold-pressed vegetable carrier oil (such as apricot kernel), 10 drops lavender essential oil, 7 drops tea tree or New Zealand tea tree (Manuka) essential oil, 2 drops bergamot* essential oil and 1 drops geranium essential oil.

or

Without bergamot: 1 ounce light cold-pressed vegetable carrier oil (such as apricot kernel), 12 drops lavender essential oil, 7 drops tea tree or New Zealand tea tree (Manuka) essential oil and 1 drops geranium essential oil

***Special precaution*:** *Avoid exposure to the sun for at least 12 hours after applying this oil since bergamot is a phototoxic oil. There is a special version of bergamot that you can get called *bergaptene free* that eliminates the phototoxic properties of oil.

Directions: Pour the selected carrier oil into a very clean bottle (amber glass is best), and then, add the essential oils and tightly close the bottle. Roll the bottle to mix the oils gently for a minute or two. Apply a small amount to the face, neck, or back but be certain to avoid the eyes, lips, nostrils and inside the ears. Roll the bottle gently before each use to ensure the essential oils are properly mixed.

Bath oil: 2 ounce carrier oil such as sweet almond, 20 drops lavender or 15-20 drops of your own blend of essential oils (be sure they are EOs that do not provide skin sensitization).

Directions: Blend the oils together and store in an amber or cobalt glass bottle. *Do not use all* 2 ounce *of bath oil in one bath.* After you have drawn your bathwater, add about 7-8 ml of the bath oil blend to your bath water. If you intend to use this bath oil with children, ensure that the oils that you have selected are safe for administration to children and use a smaller amount of the blend in the child's bathwater.

Mix well to ensure that the blend has dispersed well in the tub and hop on in. It is best to add the bath oil just before getting in the tub instead of while the water is running so that the oils do not evaporate before you get into the tub. Using this bath oil blend is safer than adding pure essential oils (EOs) directly to the bath water. This is because the EOs can settle in one spot on your skin and cause irritation.

Bath salts: 3 cups Dead Sea salt, regular sea salt, or Epsom salt, or a blend of two or three of these salts. Sea salts typically come in several grain sizes. Combining multiple grain sizes can make your salts more appealing. However, keep in mind that more coarse grains do take longer to dissolve. Add 15-24 drops of your selected essential oil or essential oil blend. Add 1 tablespoon fractionated coconut oil or other carrier oil for miniaturization (optional).

Directions: The salt mixture is always placed into a bowl. If you have chosen to include the optional vegetable oil to your salt recipe, add it to the plain salts and mix well with a spoon or fork, and then, add the drops of your chosen essential oils. Again, mix very well. Add the mixture to a pretty jar, salt tube, or container that has a tight fitting lid. Salts that are kept in a container that is not airtight will lose their aroma more quickly. After a day, you may wish to mix well again to ensure that the oils are well incorporated.

***To Colour the Salts*:** For the most natural bath salt recipe, leave your bath salts uncoloured. Certain exotic salts including Hawaiian Red Sea salt and Black Sea Pink salt are naturally colourful. Try mixing these salts with Dead Sea or plain sea salt for a speckled effect.

If you would like to add colour to your salts, FD and C liquid dye or mica powder can be added before you add the essential oils. When adding FD and C grade liquid dye, be sure to add only a drop at a time and stir well. When adding mica powder, add only a tiny amount (1/16 to 1/8 a teaspoon is usually sufficient) and stir very well. Using too much dye or mica powder can colour the water and discolour the skin, so be very careful. Leave bath salts at a soft pastel colour. It is also important that you make sure that you are using skin-safe colorants and that the user of your bath salt blend does not have any allergies or sensitizations with the colorant that you have chosen.

***To Use*:** Add ½ to 1 cup of the salts to running bath water. Mix well to ensure that the salt has dispersed well in the tub before entering. To keep the essential oils from evaporating too quickly, you can add the bath salts just before getting in the tub instead of while the water is running. Sitting on undissolved chunky bath salts, however, can be uncomfortable, so make sure the salts have dissolved well before entering.

Dry hand lotion: 8 ounce unscented lotion (many sellers of natural products and aromatherapy products sell unscented lotion, shower gel and shampoo), 10 drops patchouli, 20 drops sandalwood and 5 drops carrot seed.

Directions: The best way to blend the oils with the lotion is to pour the lotion into a bowl, add the oils and then mix very well. Once mixed, add the lotion back to the bottle. Patchouli, sandalwood and carrot seed oils help aid with dry skin. The patchouli and sandalwood oils have a lovely rich scent when blended together. The carrot seed oil is wonderful for the skin but is a little harsh in aroma. Borage and evening primrose are wonderful carrier oils for helping the skin. If you have either of the oils or can afford to get a small quantity, add a little to your lotion to give your skin even for more nourishment. It is always wise to do a skin patch test first, on new blends before using it on your skin to ensure that you do not have sensitivity to the blend.

Facial Toner: 1 ounce vodka, 3 ounce witch hazel, 10 drops grapefruit, 5 drops tea tree and 5 drops Cyprus

Directions: Add all ingredients to a 4 ounce bottle and shake to mix all ingredients. Be careful when adding the essential oils. If you make this recipe first time, add only 10 drops at a time and test on a small area of your skin to ensure no reaction before adding more oils to reach your desired strength. Shake the toner well before each use to disperse the oils.

Hair Conditioner: one-tablespoon Jojoba carrier oil and 3 drops rosemary

Directions: Mix the Jojoba and rosemary in a small bowl (tiny condiment bowls work great for this). Wet your hair with warm water and then apply the conditioner. Let it sit on your hair for 15-30 minutes, and then, wash your hair as normal. Jojoba and rosemary are helpful for dry hair. The rosemary is also said to be helpful in aiding dandruff.

Mouthwash: 4 ounce water, 4 ounce vodka (vodka may be substituted for water or a pure rosemary tea infusion), 4 teaspoons liquid glycerine (available at many drug stores) - optional and simply sweetens the mouthwash, 10-12 drops peppermint or spearmint essential oil, 5 drops myrrh (optional- myrrh can assist in mouth and gum care)

Notes: This recipe makes a rather minty mouthwash but as it contains no added flavourings, it is not as sweet or strong as commercially available mouthwashes.

Directions: Boil the water and vodka then add the glycerine. Remove from the heat and let cool for 10 minutes. Add the essential oils and let cool for 1/2 hour and bottle. I like to bottle the mouthwash in a clear plastic bottle that has a flip top to make it easy to pour the mouthwash into a small Dixie cup for use.

Perfume: 4 ounce water, 4 ounce vodka (vodka may be substituted for water or a pure rosemary tea infusion), 4 teaspoons liquid glycerine (available at many drug stores)- optional and simply sweetens the mouthwash, 10-12 drops peppermint or spearmint essential oil 5 drops myrrh (optional- myrrh can assist in mouth and gum care).

Scented hair: 2 drops of rosemary, lavender, or sandalwood

Directions: Apply 2 drops of your chosen essential oil to the bristles of your hairbrush and brush your hair well. The oils will leave your hair with a wonderful aroma.

Scented shower gel: 4 ounce unscented shower gel (many sellers of natural products and aromatherapy products sell unscented shower gel, lotion and shampoo), 40 drops of lavender or your favourite essential oils (be sure that the oils that you choose to use are gentle to the skin and always do a skin patch test first).

Directions: Add the essential oils to the 4 ounce bottle of unscented shower gel and mix extremely well. You do not want to have pockets in the shower gel where the oils collect as the concentration of essential oils could cause skin irritation and also because your gel will not be fragranced thoroughly. Use as you would any shower gel.

Shampoo: 7 ounce of unscented shampoo (many sellers of natural products and aromatherapy products sell unscented shampoo, shower gel and lotion), 1 tablespoon jojoba (optional but can supply the hair with added nutrition), 20 drops patchouli, 20 drops sandalwood and 5 drops ylang ylang.

Directions: Add the essential oils to the bottle of unscented shampoo (you may wish to use the unscented shampoo once to either make more room in the bottle for the oils or discard a little bit) and mix extremely well. You do not want to have pockets in the shampoo where the oils collect as the concentration of essential oils could cause scalp irritation and also because your shampoo will not be fragranced thoroughly. Use as you would your normal shampoo.

Shoe deodorizer: 4 tablespoons cornstarch, 4 tablespoons baking soda, 20 drops tea tree, 10 drops lemon and 10 drops lavender

Instructions for use: Sprinkle the deodorizer lightly into shoes in the evenings or at times when the shoes will not be worn for a few hours. You will not see a *cure* for smelly shoes the first time you use the deodorizer. The magic occurs after regular uses.

Whipped shea butter: Whipped shea butter is a decadent treat for your skin. It is light, soft and creamy and contains all the nourishing properties of the shea butter and the essential oils that you add to it. The consistency of whipped shea butter is similar to that of whipped cream. It is a bit tricky and time consuming to make but the results are well worth it. This recipe will make approximately four 4 ounce jars of whipped shea butter. This recipe may be doubled.

Blends for environmental care

Essential oils can vary in strength. You may want to try 20 drops at first especially if you live in a household with someone who is sensitive to strong aromas. Shake the bottle prior to each use and let it sit for about a day before making a conclusion that the aroma is too weak. The aroma can change after the air freshener has had time to sit. Mist lightly in the room. *Be especially careful not to allow the air-freshener mist fall onto furniture or into open beverages.*

Air Freshener

Blend #1: 4 ounce clean spray bottle with a fine mist setting (do not use a bottle that previously contained cleaning products or hair products such as hair spray). 30-40 drops of your favourite essential oil or essential oil blend keeping the safety data of the chosen essential oil(s) in mind 43 grams of distilled water and 1.5 ounce of high-proof alcohol (vodka is suitable, isopropyl, or rubbing alcohol is not). Alternatively, you may omit the alcohol and increase the amount of distilled water to 3 ounce. The alcohol helps the aroma linger for a longer duration.

Blend #2: 20 drops lime, 14 drops bergamot, 4 drops ylang ylang and 2 drops rose

Blend #3: 15 drops Clary sage, 9 drops lemon and 6 drops lavender

Blend #4: 20 drops rosemary, 8 drops grapefruit, 4 drops peppermint and 2 drops spearmint (optional)

Blend #5: 15 drops bergamot and 15 drops spearmint

Be sure and take heed in the safety data for the oil(s) you choose to use since an air freshener, by its nature, will come in contact with all those in your household.

Directions: Fill the spray bottle with the choice of either 3 ounce of distilled water and 43 grams of alcohol or 85 grams of distilled water (even though a 4 ounce bottle is suggested, leave about 1 ounce unfilled so you can shake the bottle well between uses) and then, add 30-40 drops of essential oil.

Bed linen spray: 4 ounce clean spray bottle with a fine mist setting (do not use a bottle that previously contained cleaning products or hair products such as hair spray). , 30-40 drops of your favourite calming/relaxing, or sensual essential oil or essential oil blend keeping the safety data of the chosen essential oil(s) in mind. 3 ounce of distilled water and 43 grams of high-proof alcohol (vodka is suitable, rubbing alcohol is not) *or* 85 grams of distilled water.

Example blends of essential oils that may be used:

Calming/relaxing blends: 10 drops lavender, 10 drops Clary sage, 10 drops mandarin or bergamot, 30 drops Roman chamomile, (avoid German chamomile due to its weaker calming properties and its deep blue colour that can discolour your linens), 20 drops Roman chamomile (avoid German chamomile due to its weaker calming properties and its deep blue colour that can discolour your linens), 15 drops bergamot.

Sensual blends

Blend #1: 15 drops sandalwood, 15 drops grapefruit or bergamot, 2 drops rose and 2 drops jasmine.

Blend #2: 10 drops sandalwood, 2 drops jasmine, 2 drops ylang ylang, 3 drops ginger, 10 drops bergamot, 3 drops lime (be sure and take heed in the safety data for the oil(s) you choose to use).

Directions: Fill the spray bottle with the choice of either 3 ounce of distilled water and 43 of alcohol *or* 1.5 ounce of distilled water (even though a 4 ounce bottle is suggested, leave about 1 ounce unfilled so you can shake the bottle well between uses), and then, add 30-40 drops of essential oil.

Carpet deodorizer: 10 drops tea tree, 10 drops lemon, 20 drops lavender and around 3 kg box of baking soda

Directions: Add the baking soda to a large bowl, and then, add the essential oils. Be sure to mix the oils into the baking soda well. Since lemon essential oil is normally yellow, it is essential that it is mixed well into the baking soda so that no undiluted lemon oil stains your carpet. This blend may be adapted to use your favourite essential oils but strive to add essential oils that offer antibacterial and disinfectant properties to give the deodorizer an added benefit (the oils in this blend work to accomplish that). After you have blended the oils into the baking soda, the deodorizer can be stored in a glass jar or airtight container.

Instructions for use: Sprinkle the deodorizer onto your carpet (use of a discarded but clean large spice container that has holes for *shaking* out the spice works well for this). Allow the deodorizer to work its magic for 10-20 minutes, and then, vacuum normally.

Potpourri-drying and scenting

Blend #1: An exotic, citrus blend: 3 drop jasmine, 9 drops sandalwood, 12 drops bergamot, 6 drops grapefruit and 5 drops oak moss

Blend #2: A citrus blend with a lovely floral hint: 12 drops bergamot, 5 drops lemon, 8 drops grapefruit, 4 drops ylang ylang and 6 drops cedar wood

Blend #3: A wonderful Christmas time blend: 8 drops fir needle, 8 drops spruce, 7 drops cedar wood and 7 drops juniper

Blend #4: A warming spicy blend: 11 drops orange, 6 drops ginger, 8 drops cinnamon, 1 drop nutmeg and 4 drops frankincense

Unscented plant materials such as whole flowers, petals, leaves, pinecones and wood chips can usually be purchased from stores and online vendors that sell an array of herbal or craft items. It's even more fun to dry your own flowers, petals, leaves and pine cones to make your own very special potpourri.

Making potpourri of essential oil blend: Choose one of the above blends and add your oils to a dark coloured glass bottle and mix well by rolling the bottle in between your hands. By making a small quantity of your chosen blend, you can later *refresh* your potpourri later by adding more drops of your blend. You of course can also experiment with your own blends of oils. It is suggested that your blends include essential oils or absolutes that are considered *fixatives*. Fixatives are oils or materials that increase the *staying power* of the aroma. Oils that have fixative properties include benzoin, cedar wood, frankincense, myrrh, oak moss absolute, patchouli, Peru balsam, sandalwood and vetiver.

Scenting your potpourri: Mix your desired assortment of flowers, petals, leaves, wood chips, *etc.* Choose a bowl that you would like to display your potpourri in and place your potpourri into this bowl. Choosing glass or glazed ceramic bowls works best as the essential oils can absorb into porous surfaces. Using a pipette, dropper, or the orifice reducer that may be built into your blending bottle, sprinkle 5-8 drops of your blend onto your potpourri. Based on the strength of the chosen blend and the amount of your potpourri materials, you may need to adjust the number of drops that you use. When the aroma weakens, add more drops of the blend to refresh the aroma.

Problems of adulteration in essential oils

Roughly 30% of all oils today are adulterated. 50% of lavender is adulterated. Out of 46 oils tested by Aura Cacia 23 were adulterated. Cheaper botanicals are substituted and lavendin is substituted for lavender. Fragrances are commonly synthetically made in a laboratory and solvents are added to the oils. They are not safe. Solvents are odourless, cheap and available. Solvents are cheap additives. Diethyl phthalate (DEP) is the most common adulterant in both oils and body care products in general. All companies should test for the quality of the product by a laboratory. Diethyl-phthalate is the compound, which is most commonly found in the bodies of women in the childbearing years. Phthalates are irritating to the skin. They most likely are carcinogenic and affect male and female fertility. Nitro benzene smells like almond. A cheap almond smelling commercial skin product most likely contains nitro benzene, which is a known carcinogen. Wintergreen and anise are adulterated nearly 100% of the time now. With high priced oils spiking is commonly done. Spiking is when you take the top note of oil and add it to a solvent or oil, which generally smells really strong.

The other common solvents are propylene glycol (DPG) and diethylene glycol isopropyl mystratlae. If you see the product, it lists fragrance but not an essential oil. Fragrance is a synthetic product made in a laboratory. Besides mixing of synthetic chemicals, fractions of cheaper essential oils are also mixed in various oils.

Storage of essential oils

Essential oils could be stored safely in stainless steel or glass containers. Aluminum containers could be used in the absence of above. However, they should not be stored in mild steal containers. Low temperature storage-conditions are best for long-term storage of oils. In most of the cases, oils can be stored for 3 years. Some oils like, vetiver, patchouli and sandal could be stored for longer period. Due to ester formation, the odour value of such oils is improved.

■■■

22 Castor Oil Good for Human Health

M.K. Rana

Department of Vegetable Science, CCS Haryana Agricultural University, Hisar-125 004, Haryana, India, E-mail: <mkrlotus@gmail.com>

Castor (*Ricinus communis*), also known as caster bean though not a true bean, is a species of flowering plants belonging to the spurge family Euphorbiaceae and sub-family Acalyphoideae, however, it is distantly related to cassava, Chinese tallow tree, croton, crown of thorns, poinsettia, true rubber tree and tung oil tree. It belongs to a monotypic genus *Ricinus* and sub-tribe *Ricininae*. Another plant species, *i.e.*, *Fatsia japonica*, is although similar in appearance but it is known as *false castor oil plant*. *Ricinus* is a classical Latin word used for ticks. The seeds of this plant having markings and a bump at the end resemble certain ticks, hence, Linnaeus used this feature as a basis for the genus name of this plant. The name *communis* is also a Latin word meaning common or general. The common name castor oil probably has been derived from its use as a replacement for castoreum, which is a perfume base made from dried perineal glands of beaver, meaning a *castor* in Latin. It has another common name, *i.e.*, palm of Christ, which has been derived from castor oil's reputed capacity to cure wounds and ailments. Although castor is native to the southeastern Mediterranean Basin, Eastern Africa and India, but today, it is widespread throughout the tropical regions, where it grows as a shrub or a small tree.

Castor plant can vary greatly in growing habit and appearance. It is a fast-growing suckering perennial shrub, which can reach a height of about 12 m. In areas with a suitable climate, castor establishes itself easily as a native plant and can often be found on wasteland. If sown early in greenhouse and kept at 20°C temperature until planted out, the castor plant can reach a height of 2-3 m in a year. In areas prone to frost, it is usually a small tree and grown as an annual plant. It can also grow well outdoors in cooler climate, nevertheless, the plant is intolerant to frost and dies as soon as the temperature becomes below 0°C. Its plants grow at an amazingly fast rate, if they are grown in full sun and provided with ample nutrients and water.

The first leaves produced are round, completely unlike subsequent leaves, which are severely serrated. The glossy star-shaped leaves are 15-45 cm long along with stalk, alternate and palmate with 5-12 deep lobes and coarsely toothed segments, which make the plant attractive in groups. However, in some of the varieties, they are dark reddish purple to bronze at young stage, gradually changing to a dark green, sometimes with reddish tinge as and when they mature. The leaves of some other varieties are green from the start of growing, whereas, in other varieties, a pigment masks the green colour of all the chlorophyll-bearing parts, leaves, stems and young fruits, so that they may remain purple to reddish-brown throughout the plant life. Plants with dark leaves can be found growing next to those with green leaves. The stems along with spherical spiny seed capsules also vary in pigmentation. The fruit capsules of some varieties are more showy than their flowers.

The flowers borne in terminal panicle-like inflorescence of green/red shade flowers without petals are relatively inconspicuous. The male flowers are yellowish-green with prominent creamy stamens and are carried in 15 cm long ovoid spikes. The female flowers born at spike tips have prominent red stigmas. They lack petals and rely on wind for pollination. The flowers are monoecious type, meaning male and female flowers are borne separately on the same plant but on different positions. Flowers appear in terminal clusters, with the male blossoms below and the female blossoms above. Male flowers senesce shortly after shedding their pollen, while the female flowers after pollination develop 1.5-2.5 cm long capsules, which remain covered with soft spines.

The fruit is a spiny greenish to reddish-purple capsule containing large-oval shiny bean-like highly poisonous seeds with inconsistent brownish mottling. Castor seeds have a warty appendage called the caruncle, which promotes dispersal of the seeds by ants. The capsules open at maturity, revealing 3 smooth attractive 1.25 cm long seeds that are a mosaic of muted black, gray, brown, yellow-brown, maroon, or white color. Each seed seems to have its own unique colour pattern. They contain valuable oil along with some extremely toxic compounds. Castor seed is the source of castor oil, which has a variety of uses. The seeds contain 40-60% oil or about 50% of the seeds weight, which is rich in triglycerides, predominantly the ricinolein. The seed also contains ricin, a toxic substance, which is found present throughout the plant in lower concentration.

Castor oil

Castor oil, which is extracted from its seeds, is a very pale yellow liquid, containing mostly ricinoleic acid, with small amount of dihydroxystearic, linoleic, oleic and stearic acids. It is fast drying non-yellowing liquid, which is valued for industrial and medicinal purposes. Most of the castor oil is used as lubricant in fine machinery and auto-engines and also used in candles, crayons, inks, dyes, leather preservatives, linoleum, paints, plastics, polishes and waxes. Though castor oil has a strong and unpleasant taste, it is used in cosmetics, soaps, textiles, massage oils and many other everyday products. Evidence available in literature indicates that the oil was used in medicine over 6,000 years ago. Because of its antiinflammatory and antioxidant, purgative, laxative and general cure properties, it has been used over the ensuing years for its therapeutic and medicinal benefits. However, ingesting oil in large quantity can result in poisoning, and many medical professionals feel that the oil is a dangerous ingredient in a variety of folk remedies. It is believed that most of its benefits are derived from its high concentration of unsaturated fatty acids. Although India is the world's leader in the production of castor oil but the commercial production also occurs in California and the southern United States, Australia, Brazil, Venezuela, Morocco, Taiwan, South Africa, Thailand, Haiti, Belgium, Germany, Holland, Peru, China, Argentina, Mexico, Paraguay, Ecuador and Ethiopia. Worldwide, over 500,000 tonnes of castor oil is produced annually.

Health Benefits of Castor Oil

In early civilizations, this oil was used in rituals of sacrifice to please the almighty God, but later, it was used in medicine and other applications. It has been used both internally and externally for the thousands of years due to its many amazing health benefits. The use of castor bean oil in India has been documented since 2000 BC as a laxative, purgative and cathartic in local, *Unani*, *Ayurvedic* and other *ethnomedical* systems of medicine. Castor seed and its oil, mainly prescribed in local medicine for internal use or use in dressings, have also been used in China for the centuries. Since castor oil can strengthen the immune system, it is considered a great remedy to treat cerebral palsy, hair loss, multiple sclerosis, arthritic and rheumatic pain and also in Parkinson's disease. It is regularly given orally to children for de-worming. It has also been found to help in many everyday problems such as constipation, gastrointestinal problems, sunburn, menstrual disorders, migraines, acne, athlete's foot, ringworm, skin abrasions and inflammation and yeast infections. Castor oil is also commonly used to induce labour pain but the pregnant women should always consult a doctor before using it. The major health benefits of castor oil are discussed below:

Remedy for Arthritis: The *Ayurvedic* system of medicine considers it the king of medicine for curing the arthritic diseases. Its anti-inflammatory properties make it excellent massage oil for relieving arthritic joints, nerve inflammations and sore muscles. A small pad made from a piece of unbleached cotton flannelette is dipped in castor oil and placed on to the affected portion. The pad is covered with a plastic wrap and a hot water bottle or heating pad is placed over the plastic wrapped cotton pad. Covering with plastic wrap will prevent the water bottle or heating pad from getting oily. The cotton pad is left over for 45 minutes to an hour only once in a day.

Remedy for Constipation: Studies have shown that its oil is a very strong laxative, which makes it very effective against constipation. Therefore, a teaspoon of castor oil can be taken only once in the morning. Taking this oil directly is disagreeable. Hence, the oil can be mixed with any fruit juice to mask its bitter and unlikable taste without affecting its laxative effects. However, it should not be taken continuously for more than 3 days. In some of the users it may be intolerant. If any adverse symptom persists for more than 3 days, the physician should be consulted without any delay.

Ringworm Treatment: Ringworm is known to be a stubborn condition to treat but one of the active compounds of castor oil, *i.e.*, undecylenic acid, has been found very effective in treating this fungal infection. Its leaves are sprayed with a small quantity of hot water, and thereafter, the leaves are crushed coarsely on to the fungal infection. The leaves are soaked in pure coconut oil and made them warm to a temperature, which may not burn the skin. Thereafter, the leaves are applied as a poultice over the affected area and left onto the skin for at least one hour or optionally overnight. The procedure is repeated every night before going to bed until the affection is cured. However, if any redness or discomfort around the applying area is noticed, this remedy may be discontinued. To prevent the oil from staining bed sheets, the leaves may be wrapped with unbleached cotton fabric before using them.

Remedy for Skin Abbesses: Castor oil is traditionally used onto the skin to prevent dryness, and thus, it is now used as a base for many cosmetics. The studies have approved the use of castor oil for skin infections and other skin abbesses such as abrasions, acne, age spot, athlete's foot, boils, chronic itching, dry skin, inflamed skin, stretch marks, sunburn and warts. A cotton plug is dipped in castor oil and applied onto the affected area in the morning and at night. However, for larger areas, a large piece of unbleached cotton cloth is soaked in castor oil and wrapped over the affected area overnight. If the area is too small, a drop of castor oil is applied onto the Band-Aid and the infected area is covered with it overnight. For stubborn fungal infections affecting the skin or nails, the affected skin is soaked in Epsom Salt for 10-15 minutes to soften and disinfect the skin before applying castor oil. This can of course help in speeding up the healing process.

Stye (oil gland infections on eyelid) Treatment: The anti-bacterial components of castor oil have been found to be effective against styes. To chuck out this problem, a very small drop of its oil is applied 2 or 3 times a day directly onto the stye.

Wrinkle Treatment: Castor oil is a natural emollient that penetrates the skin and helps in stimulating the production of collagen and elastin, which can soften and hydrate the skin. Therefore, it is a staggering natural treatment for wrinkles since it restores and rejuvenates the skin's natural youthful appearance by making the skin smoother, softer and supple. A small cotton plug is dipped in castor oil and applied onto the wrinkled skin during night before going to bed. Only a small amount of oil is usually used when applied to the skin near eyes.

Other applications: The slow burning oil is used mostly to fuel lamps for lighting. Castor oil is a good motor engine lubricant and has been used in internal combustion engines, including those of airplanes and racing cars. It is advantageous over petroleum based oils, especially at very high or low temperature. It is not mixed with petroleum products but has largely been replaced by synthetic oils, which are more stable and less toxic. Some of the countries are using castor oil for the manufacturing of biodiesel. In rural areas, the seeds are used by the children for slingshot balls to increase their weight, size and hardness. In some of the countries, its attractive seeds are used in making necklaces and bracelets. A processed form of castor oil, *i.e.*, polyglycerol polyricinoleate, is currently being used as a less expensive substitute of cocoa butter in the manufacturing of chocolate bars. It is used as body ointment and for improving growth and texture of hair. It is also used to brighten the whites of eyes.

After oil extraction from the seeds, the remaining cake can be used as organic fertilizer or animal feed. For feeding the animal, it first must be carefully boiled, heated, or treated by other means to inactivate the toxins present in it. The toxins in seeds are water soluble, not the oil soluble. Therefore, the toxins are not released during the pressing process though they remain in the seedcake. Its plant residues are used by the paper industries for making papers and wallboards.

Plant-animal interaction

Castor is the host plant for castor butterfly (*Ariadne merione*), silk moth (*Samia cynthia-ricini*) and the castor semi-looper moth (*Achaea janata*). It is also used as a food plant by the larvae of some other species of Lepidoptera, including *Hypercompe hambletoni* and the nutmeg (*Discestra trifolii*). The tribal of Assam use its leaves to feed and rear silkworm larvae. Castor oil is often used to repel moles and voles from lawns. Alcoholic leaf extract is hepatoprotective. The methanolic leaves extract shows antimicrobial properties. The castor seed pericarp shows the central nervous system effects at very low doses. Antihistamine and anti-inflammatory properties are found in ethanolic extract of root bark. A water extract of root bark shows analgesic activity. An extract from its leaves exhibits acaricidal and insecticidal activities against the adult of *Haemaphysalis bipinosa* Neumann and hematophagous fly *Hippobosca maculata* Leach. The toxin ricin in castor plant provides some degree of natural protection from insect-pests, such as aphids, hence, may be used as a potential insecticide. Castor plant is also the source for undecylenic acid, which is a natural fungicide.

Side Effects of Castor Oil

Castor oil is normally safe if it is used in moderate quantity. However, the people with intestinal blockage, acute inflammatory intestinal disease, appendicitis, or abdominal pain and the pregnant and lactating women should not consume it without consulting the doctor or naturopathic professional before consuming it. It is advisable to take precautions while taking the oil internally and also to follow the directions mentioned on the package. The dosage of oil should not be more than half to one full teaspoon a day since over dosage may result in abdominal pain or cramps, diarrhea, nausea and vomiting.

Toxicity Caused By Caster Seeds

The toxicity of castor seeds is world over well known. The extremely toxic component of castor beans is ricin, which is a deadly poison found in its seeds in abundance. Ricin is a water-soluble protein, which inhibits protein synthesis in animal cells, leading to their death. As little as 0.5 mg of ricin (4-8 seeds) can kill an adult and only one seed can kill a child. If raw castor seeds are ingested, the toxicity symptoms of ricin poisoning, *i.e.*, abdominal pain, bloody diarrhea, a burning sensation in mouth and throat, stomach irritation, vomiting, purging, profuse sweating, low blood pressure, increased heartbeat rate, collapse and convulsions commonly begin to occur within 2-4 hours exposure by ingestion or inhalation, but in some individuals, the symptoms may delay up to 36 hours. Within several days, there is severe dehydration, a drop in blood pressure and a decrease in urine. If it is not treated in time, the death can be expected to occur within 3-5 days. However, in most of the cases, a full recovery can be made.

Although there is no antidote for ricin poisoning, but in most cases on ingestion of broken seeds, drinking a teaspoon of cold-pressed castor oil may prevent any poisoning. The only remedy is to give supportive medical care to minimize the developed symptoms. There are some potential medicinal uses of ricin since it is cytotoxic. Hence, it can be used in bone marrow transplant procedures and as an anti-tumor agent. Poisoning generally occurs when animals or human beings ingest broken seeds. The intact seeds are passed by the animals through the digestive tract without releasing ricin, which is incredibly toxic. The toxicity varies among animal species, *e.g.*, 4 seeds are enough to kill a rabbit, 5 a sheep, 6 an ox or horse, 7 a pig and 11 a dog, however, the ducks are having incredible resistance against castor seeds since around 80 seeds are needed to kill them.

Buying Castor Oil

Castor oil is best purchased from a health food store since the oil that can be found in a supermarket may be meant for industrial use only, not safe for human consumption. Only the cold-processed oil should be purchased and it should be confirmed from a naturopathic doctor before consuming it. While purchasing castor oil from the market, always observe its colour. It should be of very light or pale yellow colour.

■■■

23 Olive Oil: Medicine-cum-Cooking Oil

M.K. Rana

Department of Vegetable Science, CCS Haryana Agricultural University, Hisar, Haryana, India;
E-mail: <mkrlotu@gmail.com>

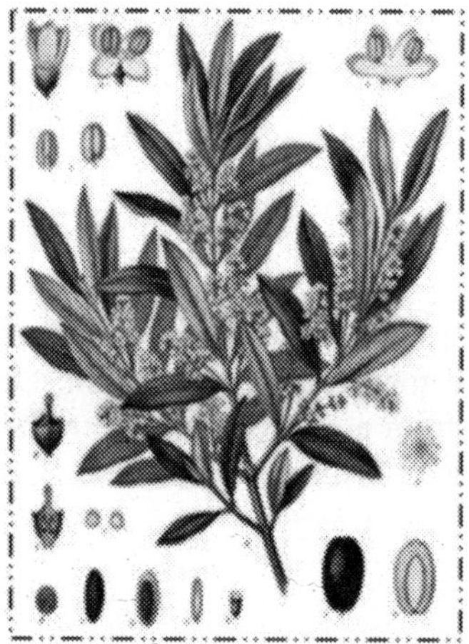

Olive (*Olea europaea* var. oleaster), a species of a small tree, belongs to the kingdom (community) plantae (plant), sub-community spermatophyta (seed bearing plant), division angiosperms (seed enclosed in ovary), class dicotyledoneae (eudicots), order lamiales, family Oleaceae, genus Olea and species europaea and is native to coastal areas of the eastern Mediterranean region, tropical and central Asia and various parts of Africa as well as northern Iran at south end of the Caspian Sea. It has a history almost as long as that of Western civilization. Olive is one of the plants most often cited in western literature. At a site in Spain, carbon dating has shown that olive seed found there to be 8000 years old. Olive has been cultivated independently in two places, *i.e.*, Crete and Syria. It is also commercially grown in Australia, California and South Africa. Its fruits are of major agricultural importance as the source of olive oil in the Mediterranean region. The olive was sacred to Athena and appeared on the Athenian coinage. Lord Monboddo comments on the olive in 1779 as one of the foodstuffs preferred by the ancients and as one of the perfect food articles. Its trees and fruits give its name to the plant family, which also includes species such as lilacs, jasmine, Forsythia and the true ash tree (Fraxinus). The word olive has been derived from the Latin word oliva, which in turn comes from the Greek. The word oil in multiple languages ultimately has been derived from the name of this tree and its fruit.

Olive oil is made from the crushing and then subsequent pressing of olives that are rich in oil, which is reflected in botanical name of the olive tree (*Olea europea*) since the word *oleum* means oil in Latin. Olive oil is available in a variety of grades, which reflect the degree to which it has been processed. Extra virgin olive oil is derived from the first pressing of the olives and has the most delicate flavour and strongest overall health benefits.

Plant Description

The olive is an evergreen tree growing to a height of 8-15 m with a spread of about 9 m. The silvery green leaves are oblong in shape, measuring 4-10 cm long and 1-3 cm wide. The tree can be kept to about 6 m with regular pruning. The trunk is typically gnarled and twisted. The graceful billowing appearance of olive tree can be rather attractive. In all green gardens, its grayish foliage serves as an interesting accent. The attractive and gnarled branching pattern is also quite distinctive. Olives are long-lived with a life expectancy of 500 years. The trees are also tenacious and easily sprouting back even when chopped to the ground.

The olive's feather-shaped leaves grow opposite one another. Their skin is rich in tannin, giving the mature leaf its gray-green appearance. The leaves are replaced every two or three years, leaf-fall usually occurring at the same time and new growth appears in the spring.

The small white feathery flowers with 10 cleft calyx and corolla, 2 stamens and bifid stigma are borne generally on a year old twig, in racemes springing from axils of the leaves. The small fragrant cream colour olive flowers are largely hidden by the evergreen leaves and they grow on a long stem arising from the leaf axils. The olive produces two kinds of flower, *i.e.*, (i) a perfect bisexual flower and (ii) a staminate flower with stamens only. The flowers are largely wind pollinated with most olive varieties being self-pollinating, although fruit set is usually improved by cross-pollination with other varieties. There are self-incompatible varieties that do not set fruit without other varieties nearby, and there are varieties that are incompatible with certain others. Incompatibility can also occur due to environmental factors such as high temperature.

The fruit is a small drupe 1-2.5 cm long, thinner fleshed and smaller in wild plants than in orchard cultivars. Olives are harvested at green to purple stage. Canned black olives may contain chemicals (ferrous sulfate) that turn them black artificially. The olive fruit is a green drupe, becoming generally blackish-purple when fully ripe. A few varieties are green when ripe and some turn a shade of copper brown. The cultivars vary considerably in size, shape, oil-content and flavour. The shape ranges from almost round to oval or elongated with pointed ends. Raw olives contain an alkaloid that makes them bitter and unpalatable. A few varieties are sweet enough to be eaten after sun drying. Thinning of fruits about 2 or 3 per foot of twig will give larger fruit size. This should be done as soon as possible after fruit set. The trees reach bearing age in about 4 years.

Nutritional composition

Olive contains dozens of health-protective nutrients. A 100 g of green olives contain 145 calories, 15.32 g fat, 3.84 g carbohydrates, 3.3 g fibers and 1.03 g protein with no cholesterol. Olive is a remarkable source of antioxidant and anti-inflammatory phytonutrients, *i.e.*, tyrosol and hydroxytyrosol (simple phenols), several terpenes especially oleuropein, erythrodiol, uvaol, oleanolic acid, elenoic acid and ligstroside, flavonoids including apigenin, luteolin, cyanidins and peonidins and hydroxycinnamic acids like caffeic acid, cinnamic acid, ferulic acid and coumaric acid. Their contents depend upon olive variety, stage of maturation and post-harvest treatment. Olives are the chief sources of nutrients including carbohydrates, sugars, fatty acids and amino acids, minerals like sodium, vitamins, soluble and insoluble fibers and a very good source of monounsaturated fat in the form of oleic acid. The greatest exponent of monounsaturated fat is olive oil. It is a natural juice, which preserves the taste, aroma, vitamins and properties of olive fruit. Olive oil is the only vegetable oil that can be consumed. The nutritional composition of olives is given as under in Table 1:

Table 1: Nutritive value of olive per 100 g oil

Constituents	Content	Constituents	Content	Constituents	Content
Water (g)	79.99	**Vitamins**		**Amino acids**	
Energy (kcal)	115	Vitamin A (IU)	1.415	Threonine (g)	0.026
Protein (g)	0.84	Vitamin A, RAE (µg)	1.179	Isoleucine (g)	0.031
Total fat (g)	10.68	β-carotene, (µg)	237	Leucine (g)	0.050
Ash (g)	2.23	β-cryptoxanthin (µg)	9	Lysine (g)	0.032
Carbohydrates (g)	6.26	Thiamin (mg)	0.003	Methionine (g)	0.012
Dietary fiber (g)	3.2	Riboflavin (mg)	0.000	Phenylalanine (g)	0.029
Minerals		Niacin (mg)	1.65	Tyrosine (g)	0.023
Calcium (mg)	88	Pantothenic acid (mg)	1.4	Valine (g)	0.038
Iron (mg)	3.30	Pyridoxine (mg)	0.009	Arginine (g)	0.067
Magnesium (mg)	4	Vitamin C (mg)	0.9	Histidine (g)	0.023
Phosphorus (mg)	3	Alpha-tocopherol (mg)	0.236	Alanine (g)	0.043
Potassium (mg)	8	Phylloquinone (µg)	7.888	Aspartic acid (g)	0.092
Sodium (mg)	872	Lipids	0.086	Glutamic acid (g)	0.093
Zinc (mg)	0.22	Saturated fatty acids (g)	7.770	Glycine (g)	0.049
Copper (mg)	0.251	Monounsaturated (g)	7.888	Proline (g)	0.040
Selenium (µg)	0.9	Polyunsaturated (g)	0.911	Serine (g)	0.031

Among its phytonutrients, no single category of nutrients is more important than its polyphenols since most of the polyphenols have been shown to function both as antioxidants and also as anti-inflammatory nutrients in the body. The polyphenols in olive oil help in explaining the unique health benefits of this culinary oil. While commonly recognized as a high fat containing food (80-85% of the calories in olives come from fat), olives are not always appreciated for the type of fat they contain. Olives are unusual in their fat quality since they provide almost three-fourth of their fat as oleic acid, which is a monounsaturated fatty acid. In addition, they provide a small amount of the essential fatty acid known as linoleic acid and a very small amount of alpha-linolenic acid known as an omega-3 fatty acid.

Health Benefits of Olives

Olives are used as a source of food, medicine and cooking oil. Its wood and twigs are used for fuel and timber. Olive oils obtained from olives are used in cosmetics, soap and in pharmaceuticals. The olive leaf used as a symbol of abundance, glory, peace, power, purity and wisdom. They were also used to crown the victors of friendly games and bloody wars. As emblems of benediction and purification, they were also ritually offered to deities and powerful figures. In the Quran and Bible, olive is praised as a precious fruit. Olive tree and its oil's health benefits have been propounded in Prophetic medicine. The Prophet Mohamed reported that *take oil of olive and massage with it as it is a blessed tree.*

Olives have documented health benefits that extend to most of the body systems such as cardiovascular, respiratory, nervous, musculoskeletal, immune, inflammatory and digestive system. It is believed that many of these diverse systems benefits are actually related to two health-support aspects of olives, *viz.* their unusual antioxidant and anti-inflammatory nutrients. In this section of Health Benefits, the focus will be on antioxidant and anti-inflammatory properties of olives, as well as some anti-cancer benefits.

Antioxidant benefits: Majority of the olive phytonutrients function as an antioxidant and help the body in avoiding redundant problems developed due to oxidative stress. Oxidative stress is a situation in which the cells are inadequately protected from potential oxygen damage, and oxidative stress can often be related to scarce supply of antioxidants. Olives are a good source of the antioxidant vitamin, *i.e.*, vitamin E, and they also contain small amounts of antioxidant minerals like selenium and zinc. However, it is the phytonutrient content of olives, which makes them unique as an antioxidant-rich food. The vitamin E content of olives may increase during early ripening when the total phenolic antioxidants in olives are slightly decreasing. Later on in the maturation process, these trends may be reversed.

The best-studied antioxidant phytonutrient found in olives is oleuropein, which is found solely in olives and has been shown to work as an antioxidant nutrient in a variety of ways. Intake of oleuropein has been shown to decrease oxidation of LDL cholesterol, to scavenge nitric oxide (a reactive oxygen-containing molecule), to lower several markers of oxidative stress and to help in protecting the nerve cells from oxygen-related damage. Olives have the ability to increase glutathione (body's premier antioxidant nutrient) levels in blood. Consumption of olive pulp is found associated with significantly increased glutathione levels in the blood and improved antioxidant capacity.

Anti-inflammatory: There are only few high fat containing foodstuffs, which offer such a diverse range of antioxidant and anti-inflammatory nutrients, some of which are unique to olives themselves. In herbal medicine practices, the olives and their leaves preparations have often been used for the treatment of inflammatory problems, including allergy-related inflammation. Olive extracts have now been shown to function as anti-histamines at a cellular level. The exceptional components in olive extract may help in lessening a cell's histamine response by blocking special histamine receptors, *i.e.*, H1 receptors. Since histamine is a molecule that can get overproduced in allergy-related conditions and can be a key player in the inflammatory process, it is likely that the anti-inflammatory benefits received from olives always involve this anti-histamine pathway. It is also possible that olives may have a special role to play as part of an overall anti-allergenic diet.

Olive extract has been shown to lower down the risk of inflammation by lowering the levels of leukotriene B4, which is a common pro-inflammatory messaging molecule. Oleuropein has also been shown to decrease the activity of inducible nitric oxide synthase, which is an enzyme whose over activity has been linked with inflammation. The anti-inflammatory benefits of olives have been given special attention in the area of cardiovascular health. In heart patients, olive polyphenols have been determined to lower the blood levels of C-reactive protein, which is a widely used blood measurement for assessing the likelihood of inflammation. Olive polyphenols have also been found to reduce activity in a metabolic pathway called the arachidonic acid pathway, which is centre for mobilizing inflammatory processes. Monounsaturated fats, polyphenols and vitamin C in olives because of antioxidant content help in neutralizing the free radicals, reducing the chances of body inflammation. Its anti-inflammatory agents help in reducing the symptoms and treat severe inflammatory conditions such as osteoarthritis and rheumatoid arthritis. Hence, regular intake of 1-2 tablespoons of extra virgin olive oil per day has been shown to be associated with significant anti-inflammatory benefits.

Useful in reducing blood pressure: The clinical studies have shown that the monounsaturated fat found in olives can help in lowering the blood pressure. The oleic acid found in olives once absorbed up into the body and transported to the cells finds its way into cell membranes, changes signaling patterns at a cell membrane level and thereby lowers the blood pressure. These changes at a cell membrane level result in decreased blood pressure.

Cardiovascular benefits: The blood vessels of our body get damaged when free radicals oxidize cholesterol, which sooner or later results in the formation of fats in arteries, leading to a heart attack. However, the presence of antioxidants in olives prevents this cholesterol oxidation and reduces the chance of heart disease. According to dieticians, the inclusion of olives in diet helps in maintaining healthy heart. Olive oil has been shown to lower the risk of lipid peroxidation (oxygen damage to fat) in bloodstream. The fat-containing molecules including low-density lipoprotein in blood need to be protected from oxygen damage since oxygen damage to fat-containing molecules significantly increases the risk of numerous cardiovascular diseases, including atherosclerosis. Protection of the fat-containing molecules in blood from oxygen damage is a major benefit provided by olive oil and its polyphenols. The polyphenols in olive oil provides protection against oxidative damage. Too much clumping of blood cells (platelets) together in blood vessels is sometimes very harming though clumping of blood platelets together in an open wound is desirable to seal off the wound. The polyphenols including hydroxytyrosol, oleuropein and luteolin found in olive oil appear to be helpful especially in keeping the blood platelets in check and avoiding problems of too much platelets aggregation.

The cardiovascular problems including gradual blocking of the arteries and blood vessels (atherosclerosis) are due to oxidative stress, meaning too much damage from the presence of overly reactive super oxides. Another cause of cardiovascular problems is chronic and undesirable inflammation, which may be due to a number of factors, *i.e.*, abnormal metabolism, unbalanced lifestyle, unwanted exposure to ecological contaminants, *etc.* The best ways to avoid oxidative stress and chronic inflammation is to consume a diet, which is rich in antioxidants. There are number of foodstuffs containing a big amount of antioxidants and anti-inflammatory compounds but few food items are especially rich in these compounds as the extra virgin olive oil.

For individuals having heart problems, olive oil is clearly one of the healing fats. Most people do quite well with it since it does not upset the critical omega 6 to omega 3 ratio, and most of the fatty acids in olive oil are actually omega-9 oil, which is monounsaturated. According to clinical studies, olive oil offers protection against heart disease by controlling low-density lipoprotein (bad cholesterol) levels while raising high-density lipoprotein (good cholesterol) levels. The beneficial health effects of olive oil are due to both its high content of monounsaturated fatty acids and its high content of antioxidative substances. No other naturally produced oil has as large an amount of monounsaturated mainly the oleic acid as the olive oil. People who consumed 25 ml virgin olive oil daily for 1 week will have lower oxidation of low-density lipoprotein and higher levels of antioxidant compounds, particularly phenols in the blood but while all types of olive oil are sources of monounsaturated fat, extra virgin olive oil from first pressing of the olives contains higher levels of antioxidants, particularly vitamin E and phenols because of its less processing.

Olive oil is one of the few widely used culinary oils that contain about 75% of its fat in the form of oleic acid (a monounsaturated, omega-9 fatty acid), which is essential for keeping the total cholesterol, LDL cholesterol, and HDL cholesterol in the body in proper balance. The high monounsaturated fat content of olives has been associated with reduced risk of cardiovascular disease. When the diets low in monounsaturated fat are altered to increase the monounsaturated fat content, the researchers experience a decrease in their blood cholesterol, LDL cholesterol and LDL to HDL ratio. All of these changes lower the risk of heart disease.

Anti-cancerous: The antioxidant and anti-inflammatory properties of olives provide protection against cancer since chronic oxidative stress and chronic inflammation are the key factors in the development of cancer. The polyphenols in olives are the natural agents for lowering the risk of certain cancers. Number of cancers gets initiated only when the cells are damaged by oxidative stress and chronic excessive inflammation. Polyphenols in olive oil act as antioxidants and anti-inflammatory molecules and as a result lower the oxidative stress and chronic inflammation risk to cells. Research studies have shown that consuming 1-2 tablespoons of olive oil per day can lower the risk of certain types of cancer, *i.e.*, cancers of the breast, respiratory tract, upper digestive tract, and to a lesser extent, the colorectal cancer.

The triterpene phytonutrients in olives, including erythrodiol, uvaol and oleanolic acid have been shown to help in interrupting the life cycle of breast cancer cells. Spanish researchers suggest that including olive oil in diet may also offer benefits in terms of colon cancer prevention. Their study results showed that rats fed diet supplemented with olive oil had a lower risk of colon cancer than those fed safflower oil-supplemented diets. In fact, the rats that received olive oil had colon cancer rate almost as low as those fed fish oil. Several studies have already linked to a reduction in colon cancer risk. Olives are rich in vitamin E, which helps in preventing cancer and reduces the growth of carcinogenic cells. Hence, olive is referred as one of the most popular vegetables to prevent cancer. It has been seen that olive oil has a preventive result on the formation of skin cancer and tumours too. The scientists from the University of Oxford have concluded olives as one of the effective measures for the prevention of various types of cancer including the colon and the breast cancer.

Digestive health benefits: Regular consumption of olive oils helps in the prevention of various stomach ailments. Being easily digested by human stomach, eating olives helps in easy digestion of food and prevents constipation. It also helps in protecting the stomach from gastritis and ulcers. Olive in the form of olive oil increases the secretion of pancreatic hormones and bile, thus ensuring healthy digestive track. Consumption of olives reduces the formation of gallstone.

Lower rates of digestive tract cancers, especially cancers of the upper digestive tract including the stomach and small intestine, are found in populations that regularly consume olive oil. Many of the anti-cancerous effects in the digestive tract were believed to depend on the polyphenols in olive oil and their antioxidant plus anti-inflammatory properties. In olive oil, numerous polyphenols, *i.e.*, oleuropein, hydroxytyrosol and tyrosol, have been shown to slow down the growth of unwanted bacteria, including bacteria commonly responsible for digestive tract infections, and some of these polyphenols along with other olive oil polyphenols like ligstroside are specifically able to inhibit the growth of *Helicobacter pylori* bacterium. This effect of polyphenols may be especially important since overpopulation of *Helicobacter* bacteria can lead to stomach ulcer and other unwanted digestive problems.

Olive oil is very well tolerated by the stomach. In fact, the protective function of olive oil has a beneficial effect on ulcers and gastritis since olive oil activates the secretion of bile and pancreatic hormones much more naturally than prescribed drugs. Consequently, it lowers the incidence of gallstone formation.

Bone health benefits: Overall bone health is another promising benefit of olive oil. Intake of olive oil keeps the blood calcium levels better. In addition, at least two polyphenols, *i.e.*, tyrosol and hydroxytyrosol, in olive oil have been shown to increase bone formation. A group of researchers has also suggested that olive oil may eventually prove to have special bone benefits for post-menopausal women since they found improved blood markers of overall bone health in female rats who had been fed olive oil after having their ovaries removed.

Hydroxytyrosol in olives is regarded as a potential phytonutrient to prevent bone loss. Several laboratory studies have shown increased deposition of calcium in bones and decreased loss of total bone mass following consumption of this phytonutrient as well as oleuropein, an another key phytonutrient found present in olives. Consumption of a Mediterranean Diet has long been associated with decreased risk of osteoporosis and olives often find themselves on center stage in Mediterranean Diet.

Cognitive benefits: Improved cognitive function, especially among elderly, is a well-known feature of the Mediterranean Diet. A large-scale study on elderly has shown that visual memory and verbal fluency can be improved with regular use of olive oil. Consumption of olive oil helps in offsetting different types of brain-related problems, including unbalanced water content, unbalanced nervous system activity and too easy passage of molecules across the blood brain barrier. Olive oil may provide us with cognitive benefits. The ability to help in protecting our brain during times of imbalance may turn out to be one of the special health benefits offered by this unique culinary oil.

Skin and hair health: Olives being rich in antioxidants and fatty acids hydrate and nourish the skin. Its application protects the skin from harmful effects of ultraviolet rays, and being rich in vitamin E, it provides natural glow to the skin since consumption of vitamin E protects the skin from sun's harmful radiation, forms a shield against premature aging, prevents skin cancer and keeps the healthy skin complexion. The antioxidant properties of olive oil along with vitamin E provide good hair growth and ensure good hair texture.

Cures anaemia: Olives are very high in iron content, thus, they help in the formation of red blood cell, which acts as an oxygen carrier. Iron plays a great role in the secretion of enzymes and in the production of energy. Iron present in the olives also helps in boosting up body immune system.

Useful in reducing cholesterol: Olives are cholesterol-free, hence, consumption of olives ensures healthy heart benefits, *i.e.*, it reduces low-density lipoprotein (bad cholesterol) level, anti-inflammatory, antihypertensive, antithrombotic and vasodilatory effects in humans.

Useful in reducing Alzheimer's: Consuming olives on a regular basis is very useful in reducing the symptoms of disease in Alzheimer's patients.

Anti-asthmatic: Olives contain high amount of vitamin C, which helps in giving relief from the symptoms of asthma.

Culture

Olive requires a long hot growing season especially for proper ripening of fruit. Frost in late spring especially at flowering stage may kill the blossoms. Its green immature fruits are damaged by temperature ≤ -1°C but ripe fruits can withstand somewhat lower temperature. The homegrown olives generally give fruit satisfactorily in the warmer coastal areas. The tree may be grown as an ornamental plant where winter temperature does not drop below -10°C. Hot dry winds may be harmful at the time of anthesis and fruit setting too. Although its tree is quite tolerant to gentle winds but strong winds may uproot the trees. Olives can also be grown in a large container as a bonsai.

The cultivated varieties cannot be propagated by seed since the seed propagated trees revert to the original small-fruited wild variety. Hence, the seedlings are grafted with material from desired cultivars by budding, bark grafting, or top working. Another method of propagation is transplanting suckers that grow at the base of mature trees, and later, these are grafted if the suckers grew from the seedling rootstock. Olive is generally propagated by 30-35 cm long cuttings taken from two years old twig of a mature tree. The cuttings after treating with a rooting hormone are planted in a light rooting medium, which is kept moist. The seedlings grown from such cuttings are further grafted with twig of another cultivar. Over the centuries, a myriad of olive varieties are being propagated, but today, the commercially grown varieties around the world are Manzanillo, Sevillano, Mission, Ascolano, Barouni, Gordal, Mission, Picholine, Rubra and Sevillano.

Olives can be grown well on all types of well-drained soil having a pH of 8.5 but can also withstand mild saline conditions. Applying organic manure with additional supplies of inorganic fertilizers every other year well ahead of flowering is advantageous so that the trees can absorb the nitrogen before fruit set. Though the olive tree can survive even in completely dry periods because of its small leaves, with its protective cuticle and slow transpiration rate, irrigation is essentially required in hot dry summers. A deep irrigation at a month interval of homegrown trees is quite satisfactory.

Proper pruning is important for regulating the production, giving shapes to the tree and also for easier harvest. Careful pruning every year keeps the trees at a desired height and avoids the problem of alternate bearing since olive never bears fruit in the same place twice, and usually bears fruits on the growth of previous year. Pruning of suckers and branches growing below the grafting point is desirable to keep a single trunk. Flowering branches in early summer are also pruned to prevent olives from forming.

Olive fruits, which are to be processed as green olives are picked while they have reached full size but are still green. They can also be picked for processing at full ripe stage. The ripe olives may bruise easily so they should be handled carefully. Development of mould on fruits is also a problem between picking and curing operation. Although there are several classical ways of curing olives but the most commonly adopted method is the lye-cure process in which green or near-ripe olives are soaked in a series of lye solutions for a period to remove the bitter principle and then transferred to water and finally a mild saline solution. The other processing methods are water curing, salt curing and Greek-style curing. Both green and ripe cured olives are the popular snack. From a commercial point of view, the black olives are identical to green olives. The black color is obtained by exposing the olives to air after lye extraction. Home production of olive oil is not at all recommended since the equipments needed are beyond most households.

Though the olive tree has fewer problems than most of the fruit trees, it is attacked by numerous insect-pests and diseases. A number of lepidopterous caterpillars feed on its leaves and flowers but the serious damage is caused by olive fruit fly (*Dacus oleae*) attacking the fruits. A pest, which spreads through olive trees, is the black scale bug. It attaches itself firmly to olive trees and reduces the quality of fruits. The curculio beetle eats edges of the leaves, leaving sawtooth damage. A fungus, *i.e.*, *Cycloconium oleaginum*, can attack the olive trees for several successive seasons, causing great damage. A species of bacterium, *i.e.*, *Pseudomonas savastanoi* pv. oleae, induces tumour growth in the shoots. A bacterial disease known as olive knot is spread by pruning with infected tools during rainy months. Rabbits eat its trees bark and can do substantial damage, especially to young trees. If the bark is removed around the entire circumference of a tree, it is likely to die. There is no effective treatment other than avoiding planting on infested soils and removing damaged trees and branches.

Since olive has fewer natural enemies than other crops, and because the oil in olives retains the odour of chemical treatments, the olive is one of the least sprayed crops.

■■■

24 Safflower Oil an Inexpensive Alternate for Saffron

M.K. Rana

Department of Vegetable Science, CCS Haryana Agricultural University, Hisar, Haryana, India; E-mail: <mkrlotus@gmail.com>

Safflower, also known as carthamine, belongs to the kingdom (community) plantae (plant), sub-community spermatophyta (seed bearing plant), division angiosperms (seed enclosed in ovary), class dicotyledoneae (eudicots), order asterales, family Asteraceae (daisy family), tribe Cynareae, genus *Carthamus* and species *tinctorius*. Safflower, botanically known as *Carthamus tinctorius* L., is also known as *bastard saffron* since its flowers are sporadically used in cooking as an inexpensive alternate of saffron. It is a highly branched thistle-like perennial small tree, but commercially, it is grown as an annual seed crop for the extraction of vegetable oil. The plant, which is related to the sunflower and resembles a thistle, is 30-150 cm tall. Each branch usually bears 1-5 globular flower heads, having yellow, orange or red colours. The flowers eventually drop, exposing the seeds. A single flower head contains about 20-50 seeds.

Safflower is grown as a minor crop with a production of 600,000 tonne. It is cultivated worldwide in more than 60 countries. The most leading countries producing safflower are India, United States of America and Mexico with small scale producing countries Ethiopia, Kazakhstan, China, Argentina and Australia. Lana is a strain of Safflower, which is grown in the southwestern United States, especially in Arizona and New Mexico.

Nutritional Importance

Safflower oil is comprised of linoleic fatty acids 78.71%, palmitic fatty acids 4.85%, oleic fatty acids 12.44%, stearic fatty acids 2.40%, linolenic fatty acids 0.08% and only 8% of saturated fatty acids (Table 1).

Table 1: The comparison of nutritional composition of different types of vegetable oil

Types of vegetable oil	Saturated fatty acids	Mono-unsaturated fatty acids	Polyunsaturated fatty acids			Oleic acid (ω-9)	Smoke Point (°C)
			Total Poly	Linolenic acid (ω-3)	Linoleic acid (ω-6)		
Unhydrogenated							
Canola	7.37	63.28	28.14	10	22	62	204
Coconut	86.50	5.80	1.80	–	2	6	177
Corn	12.95	27.58	54.68	1	58	28	232

Types of vegetable oil	Saturated fatty acids	Mono-unsaturated fatty acids	Polyunsaturated fatty acids			Oleic acid (ω-9)	Smoke Point (°C)
			Total Poly	Linolenic acid (ω-3)	Linoleic acid (ω-6)		
Cottonseed	25.90	17.80	51.90	1	54	19	216
Olive	13.81	72.96	10.52	1	10	71	190
Palm	49.30	37.00	9.30	–	10	40	235
Peanut	16.90	46.20	32.00	–	32	48	225
Safflower	7.54	75.22	12.82	–	78	13	266
Soybean	15.65	22.78	57.74	7	54	24	238
Sunflower (<60% linoleic)	10.10	45.30	40.10	1	68	19	227
Hydrogenated							
Cottonseed	93.60	1.53	0.59	–	.287	–	–
Palm	47.50	40.60	7.50	–		–	–
Soybean	21.10	73.70	0.40	0.10	–	–	–

Values are based on weight percent (%) of total fat

Its oil has the highest content of linoleic acid, which is commercially available in the market. Its oil contains more vitamin E as compared to olive oil, which is comparatively higher in vitamin K. The essential fatty acids omega-3 and omega 6 in safflower oil are similar to sunflower oil in amount. Safflower oil contains less vitamin E than sunflower or any other oil. Safflower and canola oils have a lot of the same cardiovascular disease fighting agents, *i.e.*, monounsaturated fats and omega-3 fatty acids. One of the chief reasons of replacing canola with safflower oil is that the canola cannot be cooked at as high of heat since the smoke point of canola oil is 206°C, while safflower oil is 266°C. Canola oil is much less expensive and is often substituted for safflower to reduce cost. Safflower oil and other vegetable oils have several similar properties. Both can be heated to a high temperature, hence, they are heart-healthy and are often substituted for one another in cooking.

Oil of Safflower

Formerly, it was grown for a number of uses, but now, it is cultivated only for its oil extraction. Safflower both as a plant and seed after pressed into oil has had a rich history in the east and the west civilizations, and it is still very much viable in today's diet and financial system. Its oil is such an item, which bridges the agriculture commercialization from ancient to modern time. For the last 50 years, this plant has been cultivated predominantly for the extraction of oil from its seeds. Its seeds contain two types of oil, *i.e.*, (i) high in monounsaturated fatty acid (oleic acid) and (ii) high in polyunsaturated fatty acid (linoleic acid), which are used for the preparation of different types of oil. The former, which is stable and lower in saturated fatty acids as compared to olive oil, is chiefly used as edible oil and can be stored in a cool, dark and dry place. The monounsaturated oils are often durable in cooking at high temperature. The latter, which is not stable, is used in paints in place of linseed oil, especially in white paints since it does not have the yellow touch, which is found present in linseed oil. The polyunsaturated safflower oil is used as cold oil such as mixed into salad dressings. The polyunsaturated safflower oil must also be kept in a cool and dry place. Polyunsaturated oils should not be used in high-heat cooking since they become rancid, developing chemical compounds which will add an unpleasant taste to the food. Hence, the packages of polyunsaturated oils are labeled what type of safflower oil they contain to avoid confusion.

Uses of Safflower Oil

Safflower oil is used both in personal and commercial applications. In literature, it has been claimed that safflower oil allays the symptoms of arteriosclerosis, type-2 diabetes, high blood pressure and coronary heart disease but these claims have not been proven to date by any of the studies or tests. The health benefits of using safflower oil as claimed by many people are given below:

Cooking food: In cooking, the petals of its plant can be used as a substitute of saffron since they provide flavor and colour similar to saffron, and thus, they have become a popular additive. The safflower seed oil with no special flavour and colour is nutritionally found similar to sunflower oil. Although it is more expensive than other cooking oils but it is useful to use it to high temperature cooking. However, the monounsaturated safflower oil is usually used as cooking oil, which is suitable for high temperature cooking, such as in deep frying and for the manufacturing of margarine. It is a commonly used vegetable oil in yummy delicacies including desserts, breads and salad dressings. It may also be consumed as a nutritional supplement but its most common use is in cooking. For culinary use, safflower oil is favourably comparable with other vegetable oils because of its high smoke point and low saturated fat content. A cup of its oil provides 74.3 mg of vitamin E, which is essential to include in diet. Its oil is totally cholesterol free, which makes it a part of healthy diet. Hence, it can be used as a part of balanced diet to promote the well-being. Safflower concentrate is an ingredient of some types of carbonated soft drink. In birdfeeders, its seeds are normally used as a substitute to sunflower seed but squirrels do not like its taste. The seedcake, which is left over after oil extraction, is used in feed of livestock especially in cows feed.

Industrial use: Earlier, the crop was grown for its flowers and seeds, which were used for colouring and making red (carthamin) and yellow dyes for both food and clothing, especially before the aniline dyes became available. Its dried flowers are used in colouring textiles as a quinone-type natural textile dye. The pigment of flowers responsible for dye is benzoquinone-based carthamin. Although olive green, mustard, khaki, yellow, or even red colours can be obtained from its flowers but the flowers pigment is used mostly for dyeing the cloths of yellow colour in the textiles. All the hydrophilic natural fibers, such as cotton and wool fibers may be dyed with the dye of this plant since it is classified as a direct dye. Polyamide can also be dyed without a mordant agent due to its wool-like chemical structure. Polyester, polyacrylnitryl and other fibers, which are hydrophobic synthetic fibers, can be dyed only with a mordant agent. Its oil is used in the manufacturing of paints and stains and linoleum tile. Safflower oil can be used in commercial production as a drying oil used in printing. Its oil is also appears as a component in a wide variety of commercial products.

Useful in weight loss: Weight gain is often accompanied by the accumulation of abdominal fat, which becomes extremely problematic when that fat surrounds the organs because it develops inflammation, insulin resistance and further fat deposition, and as a result, it increases the risk of heart disease, stroke and certain forms of cancer, hence, appropriate weight loss is essential though it is not so easy. It eliminates the excess fat accumulated in the body, thus, it helps in losing weight. Brown fat, which is a major factor that aids in burning calories, is present in several parts of the body. Therefore, people who are interested in losing abdominal fat should include safflower oil in their regular diet. Taking it oil regularly improves metabolic measures, whereas, the conjugated linoleic acid (CLA) shows no effect on glycemic or lipid control, however, consuming CLA for 16 weeks reduces the total body fat and lowers the women's body mass index (BMI).

Beneficial in diabetes mellitus: Obesity and diabetes have the strong correlation. The cis-linoleic acid present in its oil is converted into gamma linolenic acid (GLA), which triggers the brown fat for the burning of calories. The nutritionists also suggest the obese people suffering from diabetes to use safflower oil. They will not only lose their excess fat but also will have a significant decline in their blood sugar levels. Certain people need a little more of good fat, especially when they are obese women who are already suffering from diabetes mellitus. Using its transgenic plants increases the insulin production in patients suffering from diabetes. It promotes muscle growth and insulin sensitivity in obese postmenopausal women with type-2 diabetes. Higher insulin sensitivity is essential for the transfer of sugar or glucose from blood to the tissues, where it is used for energy as insulin resistance or lowered insulin sensitivity is the hallmark of type-2 diabetes. Safflower oil lowers the fasting blood sugar levels. Blood sugar is considered normal if it is below 110 milligrams per deciliter and this level of sugar can be maintained by supplementing safflower oil for 16 weeks. It also increases adiponectin hormone, which regulates blood sugar, fats and insulin levels. Higher level of adiponectin is also expected to increase the efficiency of dietary fat burning.

Remedy for hair loss: The safflower oil is principally used in commercial production of products like soap and cosmetics. Massaging its oil in hair roots has been shown to improve the hair quality and also giving the hair a lustrous gleam. People with dry and brittle hair as well as colour treated hair are advocated to get a regular safflower oil massage since massaging it into the scalp and on nails improves blood circulation and it promotes growth of healthy and strong hair and nails. The individuals who are having the problem excess hair fall can find effective hair regrowth with the application of safflower oil. Because of its very light texture, it is easily absorbed by the scalp.

Useful in menstrual disorders: Using safflower oil is an effective method of curing premenstrual syndrome. Consuming safflower oil instead of hormonal drugs helps in regularizing menstrual cycles, thus getting rid of the problem of irregular periods, something which is a very common in women these days due to various reasons. Its flowers are included in herbal remedies for menstrual pain and minor physical trauma.

Remedy for skin disorders: Safflower oil is also having many skin benefits. Hence, it is added in body massage oils and lotions since unlike other oils, it is low in comodegenic action and does not clog pores. It is a very good skin moisturizer. Because of its light and non-greasy texture, it is easily absorbed by the skin and makes the skin soft, smooth, healthy, glossy and supple, reduces dryness and roughness and soothes some skin problems, especially during the dry winter months. Therefore, this oil is added to numerous bath, body lotions and cosmetic products such as moisturizers, conditioners and facial cleansers. In skin care products, safflower oil is used as a lubricant with a protective barrier to prevent water loss from the skin. Linoleic acid present in safflower oil provides it a powerful dirt- and oil-removing ability and a softening effect in conditioners. Because of its moisturizing properties, safflower oil is best suited for people with normal to dry skin, thus, people who have dry and sensitive skin should use safflower oil. Appling directly to the skin reduces the visibility of lines and wrinkles, eczema, inflammation and redness of skin. The use of safflower oil is also useful in treating acne. Its application on skin also helps in getting rid of other skin problems like extremely dry flaky skin along with acne and skin scars.

Useful in cardiovascular disease: A high protein level in the blood or body increases inflammation and the risk of cardiovascular disease. Higher unsaturated and lower saturated fats in safflower oil make it a great choice to keep the heart healthy. Hence, it is beneficial to choose food items containing polyunsaturated fatty acids more often than the oils containing polyunsaturated fatty acids since the PUFA oils contain omega-6 fatty acids, which promote harmful cell inflammation. Safflower oil has been scored highest in the category of polyunsaturated oils. It is mild flavored and heat stable, making it ideal for frying. However, monounsaturated fats are usually preferred over polyunsaturated fats but are preferred over saturated fats since monounsaturated fatty acids lower the bad cholesterol (LDL) and triglycerides without affecting the good cholesterol (HDL) levels although maintain it high. Inclusion of polyunsaturated fatty acids in diet lowers the level of high- and low-density lipoprotein. A daily dose of safflower oil (about 1⅔ teaspoons) in the diet is safe to help in reducing the risk of coronary heart disease. Polyunsaturated fatty acids are needed in diet to avoid the occurrence of cardiovascular disease, which is the main killer of man in this country. However, it is important to make sure that consumption of safflower oil is properly balanced with other oils high in omega-3, *e.g.*, hemp oil contains a highly beneficial balance of both omega-3 and omega-6 (linoleic acid), which is converted by the body to arachidonic acid (a 20-carbon fatty acid with 4 double bonds). This arachidonic acid is again converted by the body to a series of eicosanoids, which have direct effect on constricting blood vessels and promoting blood clotting and inflammation.

Other Uses: People with metabolic syndrome generally have excess fat in the abdominal area, high blood pressure, cholesterol problems that foster plaque in arteries, insulin resistance or glucose intolerance and a high level of triglycerides (a form of fat in the blood). Low levels of high-density lipoprotein weaken the immune system, making the body susceptible to infections and at risk for certain cancers. A small change in eating habit to alter the fatty acid content of diet may improve metabolic measures in people already consuming an adequate amount of dietary linoleic acid.

The dried flowers of safflower and its oil are incorporated in traditional Chinese medicine to reduce bruising, to alleviate pain, to increase blood circulation and to energize blood. Safflower oil can be taken as a health supplement since taking daily dosage of 1¾ tablespoons for 16 weeks reduces the abdominal fat and inflammation, lowers bad cholesterol (LDL) and increases good cholesterol (HDL). Its oil contains omega-6 fatty acids, which help in burning the body fat and building prostaglandins, which help in controlling muscle contractions, balancing blood pressure, strengthening cell membranes and proper body functioning and make the body immune system strong, which is essential against certain infections and diseases. Its oil also contains a high amount of vitamin E, which reduces respiratory problems and helps in the circulation of blood, eliminating free radicals from the body and lowering the risk of cancer.

Side Effects and Precautions

Safflower oil is extracted from its seeds by using two methods, *i.e.*, chemical and mechanical. In chemical process, the oil is refined by adding certain chemicals, which actually eliminates the essential nutrients found in the oil and also contaminates it with certain harmful chemicals, which on consumption can cause negative effects. Hence, the consumers should carefully read the label before buying this oil to know the method used for the extraction of oil.

Taking higher doses of its oil everyday may cause a number of side effects, such as allergic reaction, diarrhea, vomiting and stomach cramps, and taking it as a daily supplement may also develop an unpleasant taste and nausea. Hence, in a condition of low blood pressure and cardiovascular disease, a family doctor should be consulted before taking any safflower oil since it can further drop the blood pressure. It is not advisable to take safflower oil during pregnancy as it may induce contractions and labor pain. It may react with some of the medicines taken for any treatment so taking safflower oil during any medical treatment should be avoided. It is strongly advised that always seek advice of a doctor or holistic health professional before beginning any prescribed treatment. Some dermatologists advised the users to be prudent against the use of natural oils, however, the Cosmetic Ingredient Review Expert Panel considers safflower oil to be a safe ingredient in cosmetic products and unlikely to cause skin irritation, yet the consumer should be vigilant while using it.

Storage of Safflower Oil

The safflower oil should be packed in an airtight container and it should be placed at low temperature since doing so will reduce the chance of oxidation and development of rancidity and will keep the oil fresh, hence, it should be bought only in small quantity as per the requirement.

Safflower Cultivation

Safflower is an annual plant like its close relative sunflower. Safflower is cultivated for a number of uses but mainly for the extraction of oil that the seeds contain. It is native to the arid climate with a long dry season and a limited rainy season. Its deep taproot system enables this plant to thrive in such harsh climate since its defensive mechanism under rainy conditions is very poor against fungal diseases, especially after its seedling stage. This greatly restricts its commercial cultivation in areas where the rainfall is more than 250 mm. The plants are also very susceptible to frost injury from stem elongation to maturity.

■■■

25 Cupping Therapy to Cure Physical Problems

Shaista Urooj, MD Medicine

RRIUM, University Campus, Naseem Bagh, Srinagar,
Corresponding address: C/o Dr. Umar Jahangir, Khan Mohalla, Bagaat, Srinagar,
E-mail: <shaista_urooj@dr.com>

Cupping therapy is a treatment modality that utilizes the physical mechanism of negative pressure or vacuum for therapeutic purposes. This technique is implemented through bell shaped cups made of silicon or plastic, hence, the therapy acquired the name cupping. A cup is applied on the skin and negative pressure is created so that the skin and superficial muscle layers are drawn in and held until the desired period.

This negative pressure was initially created by combustion inside the cup. In those times, the cups were made from horns, hallowed bamboos, earthenware like porcelain and ceramic, or glass cups. For the pressure, cups were swapped with alcohol and ignited. These inflamed cups were placed on the body. Alternately, sticks larger than the diameter of the cup were tied with inflammable material and torched. These sticks are kept in the direction away from the body in the cup. Subsequently, the cups are kept on the skin. As the oxygen in the cup is consumed due to combustion, the cup is sealed to the skin. Another method of creating the vacuum is by sucking the air and then sealing the opening with wax. This type of vacuum creation is done in horns. The broad end is placed on the skin, whereas, the cone is punctured and the air is sucked out through the mouth. These days the vacuum is created by a piston into plastic cups. A pull rod siphons the air from the cup through the rubber knob. The vacuum is maintained by a plastic valve bar.

History

Bloodletting for therapeutic purposes has been an old ritual. It has a rich history dating back to Stone Age. Miniature bow and arrow have been evacuated from New Guinea, which was probably used for the same purpose. Cave paintings from South America depict use of bloodletting techniques. Initially, thorns, sharp pointed sticks, bones, flints, shells and even shark teeth were the weapons to prick the skin or vein for bloodletting. Almost all civilizations practiced bloodletting. The ancient civilizations known to practice it are Egyptians, Aztecs, Babylonians, Greeks, Romans and Hindus. The practice of pricking a vein for bloodletting is known as veinesection. Earlier, different veins were attributed to different Gods and so to please that God that specific vein was incised. Latter, the veins were distributed according to the cosmic relation of stars and heavenly bodies. The credit of differentiating them for therapeutic reasons goes to the Greeks.

Another form of bloodletting was through leeches. This too gained a lot of importance in many cultures. During 18th-19th Century, Europe imported 100 million leeches. Leech finds a place in the Pharaoh's tomb. A Solomon Parable is descriptive of leech therapy. Dhanvantri, the Father of Ayurvedic medicine, holds leech in one hand and a pot of nectar in another. Hippocrates, Father of medicine, introduced the medicinal use of leech. These facts depict the importance given to this form of bloodletting. Yet, another form of bloodletting is cupping. Small incisions are made on the skin and either blood is sucked before the cup is applied or the vacuum pulls the blood from the incision. This form of cupping is known as wet cupping. Cupping was practiced by the Nile River Egyptians 1000 years BC. It is also mentioned in the Ebers Papyrus dating back to 1550 BC. Hippocrates in 460 BC postulated the humoural theory and freed medicine from rituals and superstitions giving it the status of science. He stated that the body is made up of and is maintained by four humours, *viz.* phlegm, chole, melanchole and sanguine. Their presence in appropriate proportion is responsible for health. He prescribes herbs and certain techniques for therapeutic purposes in his book *Guide to Clinical Treatment.* Amongst other techniques prescribed is Cupping. The Greek medicine was developed with the influence of Egyptian and African medicine. Galen, a Roman physician, carried the legacy of Hippocrates. He further developed this science and performed cupping extensively. This science traveled from Rome to Italy to Middle East, again back to Europe, and finally, nestled in Indian Subcontinent. The countries under its influence even today are India, Pakistan, Bangladesh and Sri Lanka. In India, it is being practiced as a part of regimental therapy of the Unani system of medicine (Unan means Greece). Here, it is being patronized and researched under the aegis of department of AYUSH, Ministry of Health and Family Welfare, Government of India.

Taoist alchemist and herbalist Ge Hong (281-341 AD) prescribed cupping and wrote at length in his book *Handbook Prescriptions for Emergencies.* Cups back then were made from horns, brass, iron, ceramic, bamboo and glass. Under the Tang Dynasty, another book recommending cupping was written named *Necessities of Frontier Officinal.* In Quig Dynasty, Zhao Xuemin wrote *Supplement to Outline of Materia Medica*, which had an exclusive chapter on *fire jar qi.* Today, Traditional Chinese Medicine (TCM) is the mainstay of health care delivery in China. Cupping in China is used to stimulate acupuncture points and clear *qi* meridians.

Other cultural relevance is from Europe. Cupping was so prevalent in United Kingdom during 18th century that one of the leading journals *The Lancet* was named after it. Lancet is a pointed surgical device used to cut and prick the skin before cupping. The Lancet published 17 papers on cupping in 1826-1915 on various topics like snake bite, *etc.*

In Arabia, cupping is known as Al Hijamah meaning *to reduce in size* or to return the body to its natural state. It is held in great reverence here because it was highly recommended by the Prophet of Islam. Therefore, it is given the status of religious sanctum. High price is paid to get cupping especially wet cupping done by the followers of Islam.

In Poland, it is known as *banki* and singularly as *banka.* In Iran, it is practiced with much reverence and is known as *bod-kesh* interpreted as *pull with air.* In Veitnam, it is known as *guasha* meaning sand scrapping. It has also been a part of Russian, Cyprus and Balkan traditional medicine. Cupping in Malaysia is known as *Bekam.* It is also popular in Eastern European Jewish folk medicine. The Yiddish name for cupping is *bankes.*

Timings for Cupping

Some cultures describe certain times that are best for cupping. The Arabic, Iranian and Malaysian culture and Unani medicine have a prescribed time, which is best suited for cupping. They link the cupping with the lunar calendar. According to them, the best-suited seasons for cupping are the spring and autumn since there is neither too hot nor too cold in these seasons. The best-suited day is the mid lunar day, *i.e.*, the full moon day. The best time to do cupping is the early hours of the day. It is not proper to do cupping after bathing.

Types of Cupping

Cupping is of various types depending on the disease, site and patient. Dry, wet, flash and gliding are some of the different types.

Dry Cupping is done by creating vacuum in cups over intact skin surface for a period of 10-20 minutes. No special care needs to be done after dry cupping. There is a lot of controversy regarding the time for which the vacuum can be maintained. Some say a maximum of 5minutes, whereas, others claim that cupping is not effective unless kept for 30 minutes. Others talk of 10-20 minutes. The time 15 minutes is pretty sufficient if the vacuum created is sufficient. This difference in timing can be due to the difference in the pressure maintained, the thin layer of skin, the room temperature and the patient temperament. Water vapour and sebum are also seen in the interior of the cups after dry cupping.

Wet Cupping is usually preceded by dry cupping and is done by first disinfecting the skin and then by giving incision which are 3-5 mm in depth either with a lancet or a scarifier or a sterile needle. The cup is placed on this surface and vacuum is created. Blood oozes out of these wounds and gets collected inside the cup. This is a very good way of removing toxins from the tissue. The maximum amount of blood that oozes out from these wounds is 10 ml. On removal of the cup, the blood that has coagulated by then is removed and the wounds cleaned with antiseptic. The wounds are finally dressed with a light antiseptic. The patient is advised to maintain hygiene at the site. In some instances blood oozes out of dry cupping from the skin pores.

Gliding Cupping is done over a large surface. Oil is applied on the surface and vacuum is created in the cup. The cups are then pulled along the surface on which gliding cupping is desired. This procedure is more painful than dry or wet cupping. In wet cupping, the pain is only when giving incision. Moving the cup under vacuum is more painful than any other type. The mark of this type of cupping corresponds to the surface over which the gliding was done.

Flash Cupping is done where echymosis of cupping cannot be afforded like face or where the cup cannot stay for long like the soles and palm. The cup size in this type of cupping is usually small. The vacuum is created and the cup is kept for 2-3 minutes and then it is pulled out. This causes redness at the site. The pressure in this type of cupping is very little, so that when the cups are pulled they do not cause much pain. This is done on face, forehead, palms and sole. This type of cupping is relatively new because it could not be done with fire.

Cupping can be done for therapeutic purposes and also for prophylaxis. The cupping for therapeutic purposes is done on the sites indicated for that specific disease. For prophylactic purposes, it is done to eliminate toxins from the blood and to tone up muscles.

Benefits of Cupping

Cupping has the following properties:

- Analgesic
- Anti-inflammatory

- Removes toxins
- Increases blood supply
- Promotes healing
- Removes edema
- Tones muscles

Sites of Cupping

HEAD 1

- **Site:** On forehead lateral to the midline, three fingers above the eyebrows and 3 fingers lateral
- **Benefits:** Headache-migraine, psychosis, facial paralysis, blurred vision, increased ocular pressure and trigeminal neuralgia

HEAD 2

- **Site:** Four fingers above the occipital protuberance
- **Benefits:** Headache, insomnia, seizures and pressure in the eyes

FACE 1

- **Site:** On the chin in the depression below the lower lip
- **Benefits:** Facial paralysis, toothache, ulcer of the mouth, excessive salivation, facial oedema, clears the head and jaws (tanqiyah)

FACE 2

- **Site:** Near the ear in the depression formed when the mouth is opened.
- **Benefits:** Tinnitus, deafness, otitis media, toothache, temporo mandibular stiffness, arthritis and sore eyes

FACE 3

- **Site:** Between the eyebrows
- **Benefits:** Clear the senses, sinusitis, rhinitis, vertigo, headache, dizziness, clears the toxins from the nose and eyes and nasal polyps

FACE 4

- **Site:** On the maxilla near the edge of the nose
- **Benefits:** Sinusitis, blocked nose, nasal polyps, trigeminal neuralgia and facial paralysis

NECK 1

- **Site:** Behind the earlobe in the depression between the mastoid process and mandible
- **Benefits:** Hearing and vision, clears the eye and ear channels and facial paralysis

NECK 2

- **Site:** Behind the ear just below the mastoid process

- **Benefits:** Illnesses of the eyes, night blindness, atrophy of the optic nerve, cataract, tinnitus, vertigo and insomnia

THORAX 1

- **Site:** Manubrium of the sternum
- **Benefits:** Regulates the function of lungs, clears the throat and voice, regulates the thyroid gland, bronchial asthma, bronchitis, pharyngitis, goiter, vocal cord problem and hoarse voice

THORAX 2

- **Site:** In the middle of the body of the sternum
- **Benefits:** Benefits the diaphragm expand, chest pain, bronchial asthma, bronchitis, mastitis, insufficient lactation and wheezing

THORAX 3

- **Site:** Lower end of the body of the sternum
- **Benefits:** Benefits the diaphragm expand, chest pain, bronchial asthma, bronchitis, mastitis, insufficient lactation and wheezing

THORAX 4

- **Site:** Notch below the lateral end of the clavicle
- **Benefits:** Bronchitis, asthma, pneumonia, pulmonary tuberculosis, wheezing, coughing and clears the lymphatic system

THORAX 5

- **Site:** Four finger lateral to the xiphoid process to the left
- **Benefits:** Improve circulation to the heart, chest pain, cardiac spasm and ischaemia

THORAX 6

- **Site:** Four finger lateral to the xiphoid process to the left
- **Benefits:** Hepatitis, jaundice, gallstones, cholecystitis and enlarged liver

ABDOMINAL 1

- **Site:** Two fingers below the lower end of the body of the sternum in the midline
- **Benefits:** Gastritis, peptic ulcer, stomach aches, bloated stomach, hiccups, vomiting and indigestion

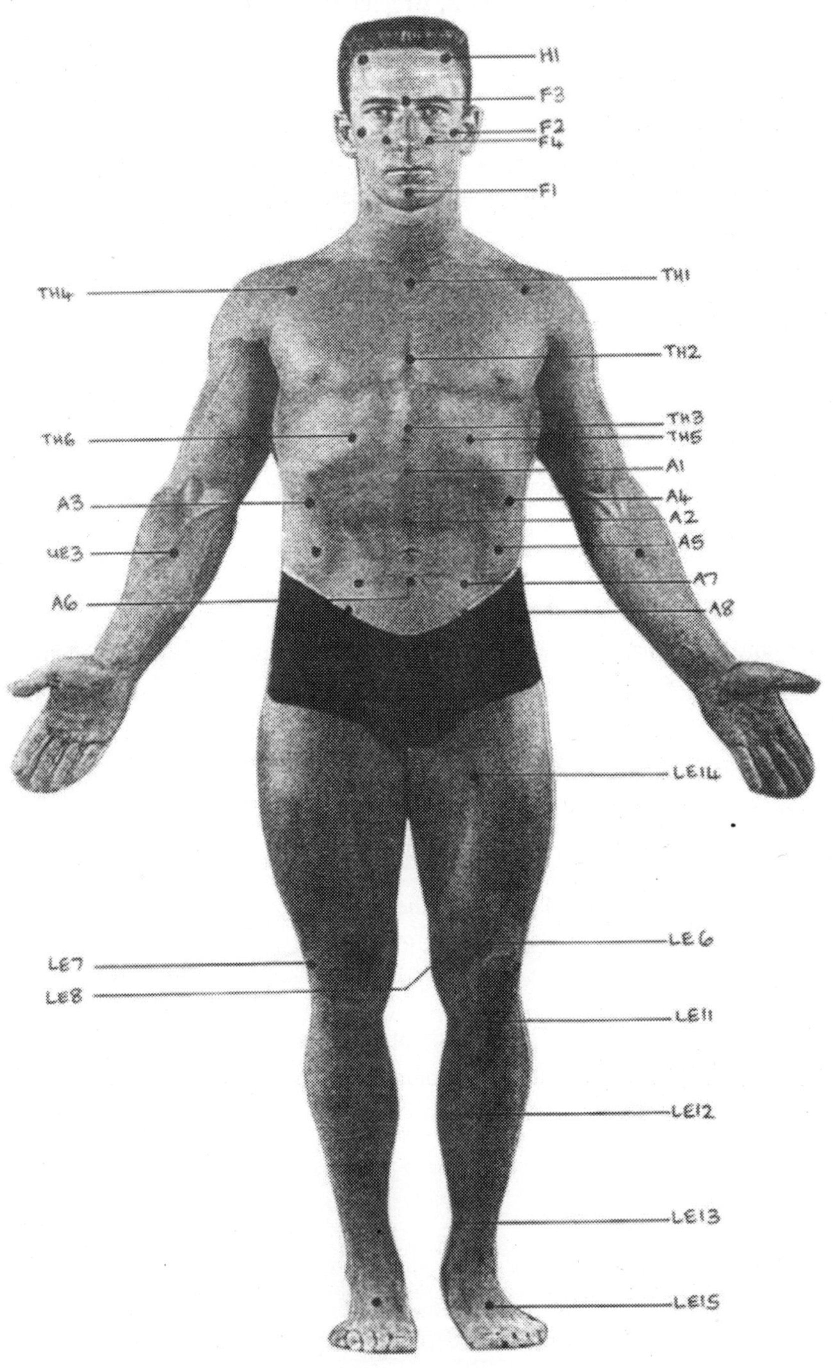

H1
F3
F2
F4
F1
TH4
TH1
TH2
TH6
TH3
TH5
A1
A3
A4
A2
UE3
A5
A6
A7
A8
LE14
LE6
LE7
LE8
LE11
LE12
LE13
LE15

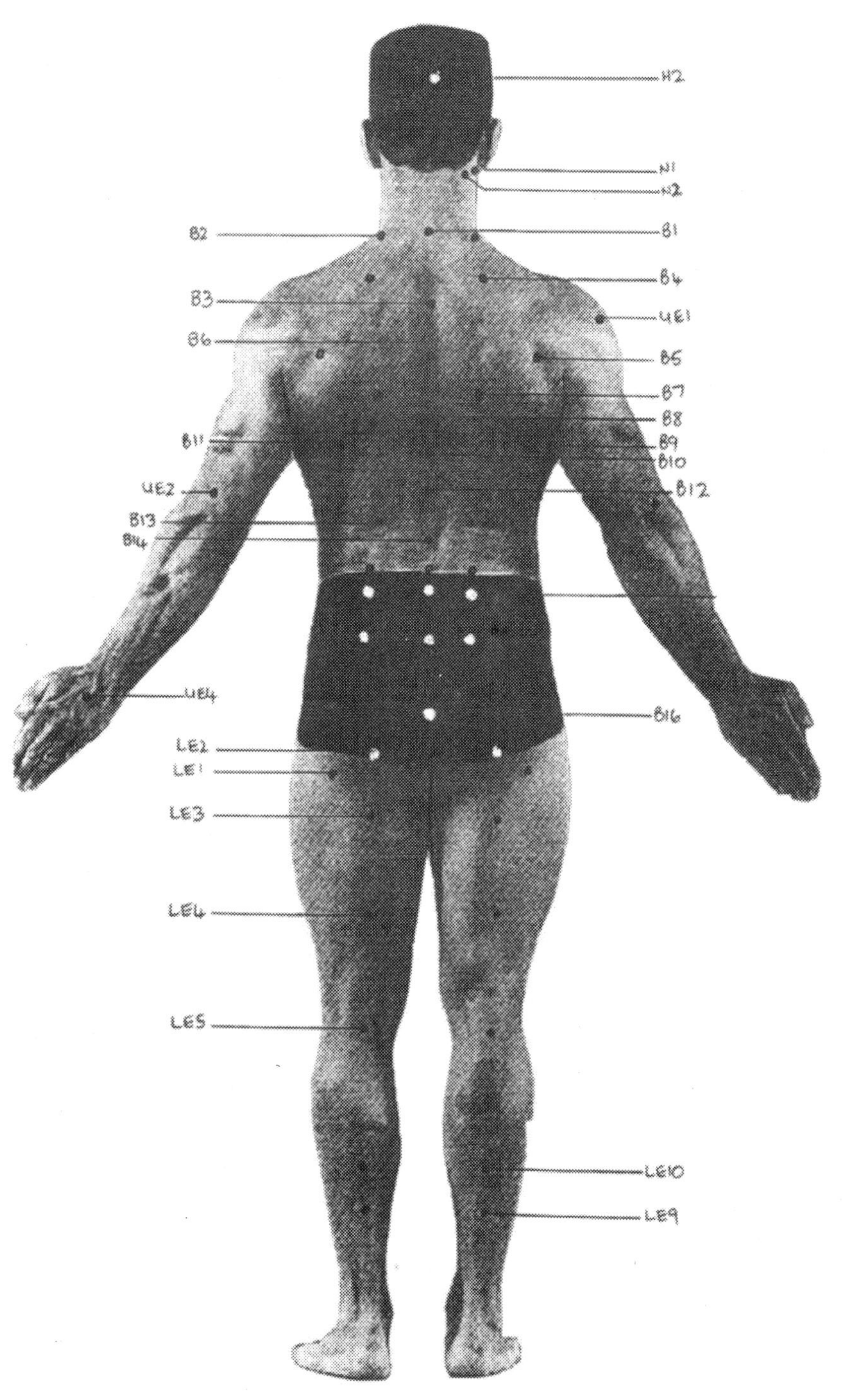
H2
N1
N2
B2
B1
B4
B3
UE1
B6
B5
B7
B8
B11
B9
B10
UE2
B12
B13
B14
UE4
B16
LE2
LE1
LE3
LE4
LE5
LE10
LE9

ABDOMINAL 2

- **Site:** Three fingers above the navel in the midline
- **Benefits:** Gastritis, peptic ulcer, stomach aches, bloated stomach, hiccups, vomiting, indigestion, dysentery and improves the function of the pancreas (diabetes)

ABDOMINAL 3

- **Site:** At the anterior end of the eleventh rib to the right
- **Benefits:** Enlarged liver and hepatitis

ABDOMINAL 4

- **Site:** Five fingers below the navel and 6 fingers lateral on both sides
- **Benefits:** Hernia, endometriosis, cystitis and irregular menstruation

BACK 1

- **Site:** Just below the spinal process of C7
- **Benefits:** Clears the brain, improve the memory, mental conditions, bronchitis, asthma, pulmonary TB, emphysema and neck pain

BACK 2

- **Site:** 2 fingers lateral to the spinal process of C7
- **Benefits:** Stiff neck, tension in the neck, cough, tuberculosis, pain in the back and shoulder, pinch nerve

BACK 3

- **Site:** Between the spinal 3rd and 4th thoracic vertebrae
- **Benefits:** Bronchitis, tuberculosis, asthma, pneumonia, chest and back pain and pleurisy

BACK 4

- **Site:** Between T2 and T3 3 fingers lateral to the midline
- **Benefits:** Bronchitis, tuberculosis, asthma, pneumonia, chest and back pain and pleurisy

BACK 5

- **Site:** In the centre of the shoulder blade
- **Benefits:** Shoulder pain and fluid accumulation in the lungs

BACK 6

- **Site:** Three fingers left lateral to the 5th thoracic vertebra
- **Benefits:** Heart illnesses, tachycardia, chest pain, angina

BACK 7

- **Site:** Three fingers left lateral to the 7th thoracic vertebra
- **Benefits:** Heart illnesses, tachycardia, chest pain, angina

BACK 8

- **Site:** Between T7 and T8

- **Benefits:** Hepatitis cholecystitis gallstones enlarged liver sluggish liver

BACK 9

- **Site:** Four fingers to the right from T9
- **Benefits:** Hepatitis, cholecysitis, gallstones, enlarged liver and sluggish liver

BACK 10

- **Site:** Between T9 and T10
- **Benefits:** gastritis, diabetes, spleenomegaly and sluggishness of pancreas

BACK 11

- **Site:** Four fingers to the left from T9
- **Benefits:** Gastritis, diabetes, splenomegaly and sluggishness of pancreas

BACK 12

- **Site:** Between T12 and L 1
- **Benefits:** Kidney pain, kidney stones, nephritis, kidney insufficiency, lower backache, impotence, irregular menstruation, strengthens the kidneys, endometriosis and spermatorrhoea

BACK 13

- **Site:** Fingers lateral to the L2 on the both sides
- **Benefits:** Kidney pain, kidney stones, nephritis, kidney insufficiency, lower backache, impotence, irregular menstruation, strengthens the kidneys, endometriosis and spermatorrhoea

BACK 14

- **Site:** Between L3 to L4
- **Benefits:** Kidney pain, kidney stones, nephritis, kidney insufficiency, lower backache, impotence, irregular menstruation, strengthens the kidneys, endometriosis and spermatorrhoea

BACK 15

- **Site:** Between L3 to L5 and on the sacrum and 3-4 fingers lateral
- **Benefits:** Leucorrhoea, spermatorrhoea, narrowing or vertebra and degeneration of inter-vertebral discs

Lower extremity 3

- **Site:** Three fingers below to the LE2 on the crease of the buttock
- **Benefits:** Lower backache, sciatica, paralysis of the low extremity, piles, retention of urine, herniated disc, leg pains and blood purification

Lower extremity 4

- **Site:** In the back mid of the thigh, 6 fingers below the LE3
- **Benefits:** Lower backache, sciatica, paralysis of the low extremity, piles, retention of urine, herniated disc, leg pains and blood purification

Lower extremity 5

- **Site:** In the fossa of the back of the knee joint
- **Benefits:** Lower backache, sciatica, paralysis of the low extremity, piles, retention in urine, herniated disc, leg pains, blood purification, knee pain, stiffness and swelling of the knee

Lower extremity 6

- **Site:** Three fingers above the knee cap (on the front)
- **Benefits:** Knee pain, knee swelling, knee cap problem, clicking of knee joint, stiffness and calcification of knee joint, hip and thigh pain

Lower extremity 7

- **Site:** On the lateral epicondyle of the femur
- **Benefits:** Knee pain, knee swelling, knee cap problem, clicking of knee joint, stiffness and calcification of knee joint, hip and thigh pain

Lower extremity 8

- **Site:** On the medial epicondyle of the femur
- **Benefits:** Knee pain, knee swelling, knee cap problem, clicking of knee joint, stiffness and calcification of knee joint, hip and thigh pain, irregular menstruation and abnormal uterine bleeding

Lower extremity 9

- **Site:** On the back of the leg, 12 fingers from the heel
- **Benefits:** Sciatica, piles, paralysis of the lower extremity, leg pain, twisted muscles of the calf, calf muscle cramps, growing pain, rheumatoid arthritis and blood purification

Lower extremity 10

- **Site:** On the back of the leg and 15 fingers from the heel
- **Benefits:** Sciatica, piles, paralysis of the lower extremity, leg pain, twisted muscles of the calf, calf muscle cramps, growing pain, rheumatoid arthritis and blood purification

Lower extremity 11

- **Site:** Four fingers below the LE8 (on the tibial side)
- **Benefits:** Irregular menstruation, abnormal uterine bleeding, dysmenorrhoea, nocturia and wet dreams

Lower extremity 12

- **Site:** Four fingers below the LE11 (on the tibial side)
- **Benefits:** Irregular menstruation, abnormal uterine bleeding, dysmenorrhoea, nocturia and wet dreams

Lower extremity 13

- **Site:** Four finger above the medial malleolus of the ankle
- **Benefits:** Strengthens the spleen, improves liver function, benefits of kidneys, amenorrhoea, irregular menstruation and incontinence of urine

Lower extremity 14

- **Site:** Make a line from anterior superior iliac spine along the Sartorius muscles to the medial epicondyle of femur, 15 fingers below the iliac spine
- **Benefits:** Toning up the thigh muscles, paralysis and the uterus.

Lower extremity 15

- **Site:** Upper side of the foot in the middle
- **Benefits:** Arthritis of lower extremities

Upper extremity 1

- **Site:** On the deltoid muscle
- **Benefits:** Arthritis of the shoulders and upper extremities

Upper extremity 2

- **Site:** Three fingers above the lateral epicondyle
- **Benefits:** Arthritis of the shoulders and upper extremities

Upper extremity 3

- **Site:** In between radius and ulna (frontal) 5 fingers from of humerus elbow joint
- **Benefits:** arthritis of the shoulders and upper extremities

Upper extremity 4

- **Site:** On dorsal side of the hand
- **Benefits:** Arthritis of the shoulders and upper extremities

Guidelines for Application

Cups applied should be of appropriate size. If the size is too small, the effect is not as desired, and on the other hand, if the cup is too big either the pressure cannot be maintained or it can affect areas that do not require cupping. Cupping should be done on an area where the skin is thick and free from hairs since hairs ruin the vacuum. There should be no bony protuberance in the cupping area as this can cause pain and again destroys the vacuum. It is advisable that the cupping area should be flat as far as possible. The person who is to undergo cupping must be informed about the expectations from the cupping. His anxieties and doubts should be laid to rest by explaining the procedure and the outcome of cupping. He should be comfortable and relaxed. Cupping must not be done in a warm and humid environment as it perpetuates early blister formation. When multiple cups are applied simultaneously, it is advisable that they should be separated by several centimeters. The patient must stay as still as possible. Movement of the area entrapped under the cup can cause muscular injury. Sudden jerky movements can result in tear in the musculature.

Precautions

There are certain precautions to be kept in mind while doing cupping. It involves blood and body sera so the chances of transmission of infection are there, hence, the use of surgical gloves is mandatory. The cups must be sterilized after every use to check infection. In case of wet cupping, the incision must be made by disposable needles or blade and the incisions must be superficial and not very deep. Since cupping causes a change in the haemodynamic status, hence, pulse and blood pressure must be noted before and after each cupping session. The patient should be enquired for any unusual sensation or feeling after cupping as patients rarely feel a little dizzy after the session. Cupping should never be done on full stomach. Meals should be taken at least one hour after cupping.

What to expect from Cupping

On first sitting or the initiation sitting, there is a sensation of pull coupled with pain. If the patient is suffering from intense pain then he experiences relaxation and heat under the cup. After initiation sitting, the skin is sensitized to this amount of pressure and he does not feel any discomfort. The patients experience such immense relief in pain during the session itself that some even doze off, and as a result of cupping, there is an increase in blood supply or hyperemia.

The principle behind cupping is that vacuum pulls the skin causing hyperemia at that site. The vascular changes brought about are similar to those seen in acute inflammation. There is dilatation of arterioles and subsequent opening of new capillary beds in that area causing a gush of blood leading to heat and redness. Subsequently, there is slowing of circulation due to increased permeability of the microvasculature. Consequently, there is stasis in capillaries with high concentration of red cells and increased viscosity within.

The corium layer of skin has the elastic fibers, which protect the epidermis in shearing stress. This layer is tough, flexible and highly elastic by nature. This layer is also home to subcutaneous capillary network, lymphatic vessels and nerves. During cupping, corium is stretched to its limits but the sustained negative pressure and subsequent congestion causes breach in the vascular epithelium or blood vessel wall leading to minute haemorrhage of capillaries known as bruise or ecchymosis.

Increased permeability leads to leaking of sera into the tissue, whereas, discrete capillary haemorrhage cause outflow of blood into the tissue causes edema at the site. Meanwhile, a host of chemical substances is released from the leucocytes or the white blood cells like histamine, bradykinine, cytokines and other chemical mediators. This cause further vasodilatation and hypotension among other changes attributed to these chemical mediators. These chemical mediators have the ability to change the cells that produced them (autocrine effect) and alter other cells close to them (paracrine effect). Besides these effects, they also produce certain chemicals that diffuse into the tissue and act specifically on cells of that area like release of adrenocorticotrophic hormone (ACTH), corticosteroids and slow wave sleep. A study done recently in Britain showed that the effect of cupping on inflammation causing chemicals was remarkable, proving that cupping does have the potential to relieve inflammation. This is called the Immune System Theory.

The skin has pressure sensitive buds called the baro receptors; these and the nerve endings of the skin transmit signals to the brain, which cause an increase in the secretion of enkephalin. Enkephalin is a protein polypeptide chain produced in the brain and works like analgesics like morphine. There is release of other chemicals like the serotonin, cortisol and adrenocorticotrophic hormones. Their release is triggered by the stimulation of nerves, hence, this type of release is known as neurotransmitter release.

Cupping is done at a site for approximately 15minutes. The site cupped bears a circular ecchymosed area corresponding to the circumference of the cup. The ecchymosis clears on its own without any treatment within a week. There is also edema in them. In certain cases, there may be pain for sometime.

After wet cupping, light dressing with antiseptic is sufficient and should not be burdened with heavy dressing. Cupping on the site of blisters or where there is wet cupping has been done is suspended for 3-4 days until the wound heals.

Certain sites like the shoulder girdle are susceptible to blister formation. The blisters thus formed are similar to burn or less commonly known pressure blister. They have straw-coloured fluid in them. The blisters formed are 4-5 in number, and sometimes, there are multiple small blisters. The total fluid in the all the blisters combined is 1-2 ml. These blisters resemble second grade pressure sores or bedsores. They are sub-corneal in nature and contain squamous epithelial cells.

Contraindications

Wet cupping is not advisable for anaemic and patients with bleeding disorders. Similarly, it is contraindicated for children below the age of 6 years and beyond the age of 65 years. However, dry cupping can be done to children aged 3 years and the elderly. It is improper to cup patients with high-grade fever and convulsions or those seriously ill with low blood pressure. It is against the rule to perform cupping over large blood vessels. Cupping is not recommended on the abdomen and sacral area of pregnant women. Similarly, wet cupping should not be done on female breasts.

Recent Studies

Studies on various aspects of cupping have been carried out in various parts of the world. A study was conducted by the St. Georges University of London by Kaleem Ullah and his group on *the effect of cupping therapy as a treatment for Anterior Knee Pain and its potential role in Health.* The result is as follows *there was statistically significance difference between the level of pain, well being and Range of Motion for patients with anterior knee pain pre- and post-Cupping (P <0.05)*, meaning that there were substantial improvement in terms of pain, mobility and general well-being of the patient.

Another study was carried out by Ilkay Chirali, Mark Bovey, Roz Gibbs *entitled Cupping for Patients with inflammatory complaints- Clinical and Biochemical outcomes* and the result is 95% patients reported an improvement in their symptoms with the therapy. Significant improvements were seen in the main and secondary complaints and in the activity levels of the patients. This indicates that a majority of the patients benefited by cupping sessions in the inflammatory symptoms, *i.e.*, pain, swelling, redness, heat and loss of function together with an increase in their activity levels. Erythrocyte Sedimentation Rate, the biochemical marker of inflammation also reduced with cupping. Other changes observed were in lymphocytes, platelets, fibrinogen and iron concentrations in the blood.

Yet, another study by Andreas Michalsen, Silke Bock, Rainer Ludtke, Thomas Rampp, Marcus Baecker of Immanuel Hospital Berlin, Germany on the *effects of Traditional Cupping Therapy in Patients With Carpal Tunnel Syndrome: A Randomized Controlled Trial.* The results of which are that cupping therapy was more beneficial than heat, according to the primary outcome measure, change in the total symptom score after day 7. This can be translated as the effect of cupping on wrist pain is far more better than heat fomentation and that all the symptoms showed change after day 7.

Majid Niasari, Farid Kosari and Ali Ahmadi studied *the Effect of Wet Cupping on Serum Lipid Concentrations of Clinically Healthy Young Men: A Randomized Controlled Trial* and reported their findings as the results of the study have shown that wet cupping is an effective method in reducing LDL cholesterol and the LDL/HDL ratio. Wet cupping did not show a significant influence on total cholesterol, triglyceride and HDL cholesterol, therefore, any changes in the LDL/HDL ratio seem to be related to the reduction in LDL cholesterol, which means that cupping showed a positive effect on the low-density lipoprotein (LDL), which is harmful for the body together with a decrease in low-density lipoprotein and high-density lipoprotein ratio. High-density lipoprotein is the good fat and helps in fighting cholesterol. However, there was no change in the general cholesterol, triglycerides and healthy cholesterol. The only change that occurs is on low-density cholesterol, which is bad for us.

Jong-In Kim, Myeong Soo Lee, Dong-Hyo Lee, Kate Boddy and Edzard Ernst from Korea Institute of Oriental Medicine, Korea and Universities of Exeter and Plymouth, United Kingdom surveyed the literature on the topic *Cupping for Treating Pain: A Systematic Review* and they concluded that cupping can give relief to pains arising from trigeminal neuralgia, Brachialgia paraesthetica nocturna (BPN), herpes zoster low back pain and cancer pain.

Observations on the Efficacy of Cupping for Treating 30 Patients with Cancer Pain

The study on the effects of dry cupping on cancer pain with conventional drug therapy was carried out by Huang *et al.* (2006) who reported favourable effects of cupping after 3 days intervention (RR 67% versus 43%, $P < 0.05$). Another comparative trial on the effect of moving cupping therapy on non-specific low back pain was carried out by Hong *et al.* (2006) who compared dry cupping with non-steroidal anti-inflammatory drugs in non-specific low back pain and suggested a significant difference in pain relief on VAS after treatment duration (MD, 22.8 of 100 mm VAS; 95% CI, 11.4–34.2, $P < 0.001$).

Observation on therapeutic effects of bloodletting puncture with cupping in acute trigeminal neuralgia recorded by Zhang (1997) who found that wet cupping reduced pain compared with analgesics in acute trigeminal neuralgia after the intervention period (RR, 93% versus 47%, $P < 0.01$). Trigeminal nerve is amongst the twelve nerves arising from the brain. Its compression causes pain. This pain is very severe and debilitating and is known as trigeminal neuralgia.

An experiment on the effectiveness of wet-cupping for nonspecific low back pain in Iran conducted by Farhadi *et al.* (2009) to prove wet cupping plus usual care for pain reduction compared with usual care in nonspecific low back pain suggested significant differences in pain relief (McGill Pain Questionnaire) at 3 months after three treatment sessions (MD, 2.2 of 6 points present pain intensity; 95% CI, 1.7–2.6, $P < 0.01$).

Branchialgia parasthesia is pain in the arm coupled with numbness. A study on this type of pain was carried out by Ludtke *et al.* (2006) who suggested that Brachialgia paraesthetica nocturna (BPN) could be relieved by wet cupping, which is the result of a randomized pilot study. One session of wet cupping plus usual care significantly reduced pain during a week compared with usual care alone in patients with Brachialgia Paraesthetica Nocturna (MD, 1.6 of 10 points score, 95% CI, 0.13-3.07, $P = 0.03$). Another study on the same topic entitled effectiveness of cupping therapy in brachialgia paraestetica nocturna was carried out by Michalsen *et al.* (2007). The results were comparable to the earlier results and had been summed up as the favourable effects of one session of wet cupping on pain reduction compared with a heat pad in patients with BPN at 7 days after treatment (MD, 22.9, 100 mm VAS; 95% CI, 10.5-35.3, $P < 0.001$).

Therapeutic effect of aciclovir combination with callateral-puncturing and cupping in the treatment of 40 cases of herpes zoster was observed. Wet cupping plus conventional drugs on pain reduction compared with conventional drugs alone in patients with herpes zoster failed to show favourable effects of wet cupping after interventions (RR, 100% versus 88%, P = 0.065). Herpes Zoster is a viral infection, which causes blisters with severe intense debilitating pain, which is treated with antiviral treatment of acyclovir.

These studies indicate that this old tradition still lives not only in the minds of the people as faint memories and mentioned in dusty memories of some ancient scholars but this science is still alive and kicking. It has demanded the attention of not just physicians but also scientists and general public to a large extend. It thrives in the hospitals, laboratories and the hearts of the masses. The expectations of all are being fulfilled by this ancient yet most scientific science.

■ ■ ■

26 Hereditary Foundations and Prenatal Care and Development

Shanti Balda[1] and Usha Turan[2]

[1]Human Development and Family Studies, I.C. College of Home Science,
CCS Haryana Agricultural University, Hisar, Haryana, India,
E-mail: <sbalda@rediffmail.com>
[2]CIS Kanya Mahavidhyalya, Fatehpur Pundri, Kaithal, Haryana, India,
E-mail: <ushaturan@gmail.com>

Imagine, how does the union of tiny male and female sex cells result in the development of a 6 to 7 pounds human being? Why does the sex of a baby male or female? Why do some children have curly hair rather than straight? Why does one child resemble a parent, while other does not? Why do some children born with inherited or chromosomal abnormalities? The reader will get answers of these questions in this chapter, which begins with hereditary foundations, and then, hereditary and chromosomal abnormalities, genetic counseling and methods of prenatal diagnosis are described. It follows with prenatal stages of development and hazards of prenatal period. Prenatal environmental influences on the child's health and survival are considered and reflected. This is followed by care of mother during pregnancy including common discomforts and warning signals of pregnancy. Then physical and psychological hazards of prenatal period are described in detail. Finally, childbirth and complications have been discussed.

1. Hereditary Foundations

1.1 Hereditary Transmission

Each human being is made up of cells. Within every cell, there is a nucleus, which is a controlling centre and contains rod like visible structures called *chromosomes*. Chromosomes store and pass on hereditary information from one generation to the next. Chromosomes come in matching pairs, except XY pair in male. Human beings have 23 pairs of chromosomes. One comes from the mother and the other from father.

Each chromosome is made up of *deoxyribonucleic acid* (DNA), which is a database of chemical information. Each chromosome contains thousands of segments of DNA called *genes* that are basic physical and functional units of heredity. The genetic make up of the individual is known as *genotype*.

1.2 The Sex Cells

The male sex cell is sperm and the female sex cell is ovum. In general, the sex cells are called gametes. Fertilization takes place when male and female sex cells, *i.e.*, sperm and ovum, combine. Before the life can begin, the male and female reproductive cells must go through several preparatory processes. Male cells go through two preliminary stages, *i.e.*, (i) maturation and (ii) fertilization. In case of female cell, there are three preliminary stages, *i.e.*, (i) maturation, (ii) ovulation and (iii) fertilization.

- **Maturation of Reproductive Cells:** The maturation of sex cells consists of chromosome reduction through cell division. In each sex cell, whether the male cell or the female cell, there are 23 pairs of chromosome. During maturational process, the chromosomes of each cell arrange themselves into pairs on opposite side of the cell's nucleus. Each pair contains one chromosome from the mother and one from the father when pair separates, one chromosome goes to one of the newly formed cells and its mate goes to the other. The two cells thus formed split again but this time lengthwise, so that, one-half of the original number of chromosomes is retained.

  After the maturation process is completed, there are 4 mature cells of each sex having 23 chromosomes. In case of sperm cells, each one of the 4 mature cells that come from the original cell is capable of fertilizing an ovum, but in the ovum, only one cell that contains most of yolk is capable of being fertilized, the other three cells, the polar bodies are absorbed and secreted.

- **Ovulation of Female Reproductive Cell:** The second preparatory process necessary for the beginning of life is ovulation. It is the process of escape of one mature ovum during the menstrual cycle. It is believed but not medically proved that the two ovaries are alternate in producing one ripe ovum during each menstrual cycle. Once released from one of the follicles of ovary, the mature ovum finds its way to the open end of the fallopian tube nearest the ovary. After it enters the tube, the ovum is propelled along the length of the tube by:

  - Cilia or hair like cells, which line the tube
  - Fluids composed of estrogen
  - Mucus from lining of the tube
  - Rhythmic contractions of walls of the tube

  If the length of menstrual cycle is normal, *i.e.*, approximately 28 days, ovulation occurs between fifth and twenty-third day of the cycle, with the average on the eleventh day. However, there are marked variations in ovulation among different women and in the same common from time to time.

- **Fertilization:** The third preparatory process to the beginning of a new life is fertilization of the ovum by a spermatozoon. In normal fertilization, the ovum is in one of the fallopian tubes, on its way from ovary to the uterus. As a result of coitus, a large number of spermatozoa enter the uterus. They are attracted to the ovum by a strong hormonal force. After one sperm cell enters the ovum, the surface of the ovum is so changed than no other sperm can enter it. After fertilization, the nuclei from the two cells approach each other and they merge after breaking down their membranes. The new cell formed has 23 pairs of chromosomes, one-half from the male cell and the other half from the female cell.

1.3 Importance of Fertilization

The time of conception can be regarded as one of the most important in person's life. The reason for this is that certain conditions are determined at that time.

- **Hereditary Endowment:** The newly created child's hereditary endowment is determined when the sperm unites with the ovum. It has been estimated that there are 40,000-60,000 genes in the chromosomes of mature male cell and a similar number in the mature ovum. The newly created baby's hereditary endowment contains 80,000-120,000 genes each of which carries potentials for physical and mental traits. It is entirely a matter of chance, how many chromosomes from different ancestors on the maternal or parental side will be passed down to a child. After conception, nothing can be done to add or subtract from the individuals heredity endowment.

- **Sex determination:** The second important condition determined at the time of conception is whether the child will be male or female. Fertilized ovum has 23 pairs of chromosomes and twenty-two of them are matched pairs, called *autosomes*. The twenty-third pair consists of *sex chromosomes*. A female has XX pair and male XY pair. One-half of all mature sperms carry the X chromosome and the other half the Y chromosome but all mature ova carry the X chromosome. Thus, in any fertilized ovum, there is always an X chromosome from the ovum and either X or Y chromosome from the sperm. If the sperm that unites with the ovum has Y chromosome, the result will be a male child, but on the other hand, if the sperm that unites with ovum has X chromosome, the result will be a female child.
- **Number of offspring:** The third important condition determined at the time of conception is whether the birth will be single or multiple. *Singletones* are children who are born alone. The term *multiple births* refer to the birth of two or more babies, *i.e.*, twins, triplet, or quadruplets, with in a few hours or days. Twins are of two types as described below:
- **Identical twins:** Identical or uniovular twins come from a single ovum fertilized by a single sperm. When ovum is fertilized by one spermatozoon, it occasionally happens that at the time of first division of cell, the new cells separate instead of remaining together. The twins formed from subdivision are called identical twins. They are always of same sex and carry the same genes. Because of their identical hereditary endowment, identical twins resemble each other very closely in all hereditary traits.
- ***Fraternal twins*:** Fraternal twins are known as non-identical twins. They are the product of two ova fertilized simultaneously by two spermatozoa. The individuals develop from these two ova are non-identical twins. There is lack of similarity in the physical and mental make up of these twins. They may be of same sex or of opposite sexes. Non-identical twins are not really twins rather they are the result of simultaneous pregnancies.

1.4 Multiple Offspring

When ovum is fertilized by one sperm, it occasionally happens that at the time of first division of cell, the new cells separate instead of remaining together. Twins formed from this division are called *identical* or *monozygotic twins*. They are always of the same sex and carry the same genes. *Fraternal* or *dizygotic twins* are created when two ova released at the same time and fertilized simultaneously by two sperms. The individuals develop from these two ova are non-identical twins. There is lack of similarity in the physical and mental make up of these twins. They may be of the same sex or of opposite sexes.

1.5 Hereditary Inheritance

Dominant-recessive and Codominance inheritance: A single pair of genes, one from each parent, contributes to a particular inherited characteristic. A child is said to be *homozygous* if he/she has inherited two genes of the same kind for a trait from the two parents. If the genes are different, the child is *heterozygous*. A homozygous child will display the particular characteristics associated with that gene. In case of a heterozygous child, one of the three outcomes is possible for a character:

- When only one gene affects the child's characteristics, it is called *dominant gene*. The second gene, which has no effect, is called a *recessive gene*. Colour of hair is an example of *dominant-*

recessive inheritance. Heterozygous individuals with one recessive gene can pass on that characteristic to their children and are called carrier of that characteristic.

- Child displays characteristics that are *intermediate* between the characteristics to be displayed by the two genes, *e.g.*, colour of the skin.
- Child displays characteristics that are affected by both genes, but instead of being intermediate, the characteristics will be entirely different from those linked to either contributing gene. This pattern of inheritance is called *codominance.*

For example, if the mother's blood type is A (with AO genotype) and the father's blood is B (with BO genotype) and the child inherits one type A gene from the mother and one type B gene from the father, it will exhibit a codominant outcome, *i.e.*, type AB blood, which is qualitatively different from type A and type B blood.

X-Linked Genetic Inheritance: Human characteristics affected by genes found on X-chromosomes are called *X-linked characteristics.* Males are more likely to get affected because their sex chromosomes do not match. Females receive two sets of corresponding X-chromosomes, one from mother and the other from father. Any recessive gene on one X-chromosome is being suppressed by a dominant gene on the other X-chromosome, thus, it does not cause a disease. A male child, who inherits a harmful recessive gene on his X-chromosome, lacks corresponding gene to override the recessive gene's harmful effects.

Red-green colour blindness is an example of X-linked recessive trait. A male child with this X-linked recessive gene will be unable to distinguish red from green because there is no corresponding gene on the Y chromosome. A female child will exhibit this trait only if she has father who is colour blind and mother who is either colour blind or carrier of that.

2. Hereditary and Chromosomal Abnormalities

Some children are born with defects and disorders that follow the rules of dominant-recessive inheritance. Many of the hereditary abnormalities tend to be the product of recessive genes. Besides inheriting recessive genes that are harmful, another major cause of developmental irregularities is abnormalities of chromosomes.

2.1 Genetic Disorders/Diseases

There are thousands of genetic disorders in humans. Some are common, whereas, quite a few are rare. Hereditary/genetic diseases are passed from parents to their children. Some of these diseases are discussed as under:

- **Achondroplasia:** It is a dominant gene disorder, which is the most common genetic cause of dwarfism. Height of individuals suffering from achondroplasia varies from 4 feet to 4 feet 4 inches. They have disproportionately short limbs. However, there is no intellectual disability. This disorder is usually the result of a mutation in sperm or egg of a healthy parent. Achodroplasia can be detected before birth with the help of prenatal ultrasound. However, there is no treatment for this disorder.
- **Albinism:** Albinism is a congenital disorder in which there is little or completely no production of melanin in hair, skin and iris of the eyes. Hence, albinos have light coloured skin, hair and eyes. It is caused due to the inheritance of recessive genes from parents. This disorder cannot be cured. In rare cases, albinism is caused due to the gene mutations. The presence of altered gene prevents the body from synthesizing the normal amount of melanin pigment. The growth and development of a

person with albinism is the same as that of a normal individual. They also have the ability to reproduce offspring.

- **Cystic fibrosis:** Cystic fibrosis is an inherited disease of the glands that secrete mucus and sweat. Cystic fibrosis causes the mucus to become thick and sticky that clogs various organs of the body and results in other complications. It mostly affects lungs, liver, pancreas, sinuses, intestines and sex organs. Cystic fibrosis also causes excess loss of salts through sweat and results in dehydration, tiredness, weakness and elevated heart rate. There is no cure for cystic fibrosis.
- **Galactosemia:** It is a metabolic genetic disorder, which impairs body's ability to break down galactose. Infants suffering from galactosemia show symptoms within a few days after birth. The symptoms include yellowing of skin, eyes, diarrhea, vomiting, refusal to drink milk, malnourishment and also mental retardation. If milk or milk products are given to infants suffering from galactosemia, the accumulation of galactose in their system damages brain, eyes, liver and kidneys. Eliminating lactose and galctose from diet of an individual is the only way to treat galactosemia.
- **Hemophilia:** Hemophilia is a disorder in which blood fails to clot and is an example of X-linked disorder. Mothers who carry the abnormal X-chromosome are more likely to have male children affected by the disorder. If mother is carrier and father is normal, these parents can have one carrier daughter, one affected son, one normal daughter and one normal son.
- **Muscular Dystrophy (MD):** It is a disorder of progressive degeneration of skeletal muscles. It causes defects in muscle proteins and death of muscle cells and tissues. Primarily skeletal muscles are affected. It may damage the functions of other body systems. In some cases, the symptoms appear in infancy of childhood, in certain instances muscular degeneration sets in during adulthood. There is no specific treatment to cure muscular dystrophy.
- **Neurofibromatosis:** Neurofibromatosis is a genetically inherited disorder in which nerve tissue grows tumors and may cause medical complications by compressing nerves and tissues around them. In this disorder, the bones, nervous system, the spine and the skin are affected. Tumors under the skin may appear as bumps and are associated with skin discoloration. Learning disabilities are also associated with this disorder.
- **Osteogenesis Imperfecta:** This disorder of the connective tissue is caused by dominant gene in which bones break easily. It is also known as *brittle bone syndrome*. Genetic mutation impairs synthesis of collagen, which is a protein that makes the bones strong. There is no cure for this disorder. Exercise, physical therapy, medicines and orthopedic devices are the only treatment available for people suffering from this disorder.
- **Phenylketonuria (PKU):** Phenylketonuria is a genetic disorder that interferes with a child's ability to metabolize an amino acid, *i.e.*, phenylalanine. Without this enzyme, phenylalanine quickly builds to toxic levels that damage the central nervous system. It is a treatable genetic disease, *e.g.*, if father and mother both are heterozygous carriers of PKU, these parents can have two carrier children, one normal child and one PKU affected child.
- **Sickle-cell anemia:** This genetic disease occurs in full form when child inherits two recessive genes. Sickle cell anemia affects the red blood cells. In sickle-cell anemia, the red blood cell is shaped like a sickle. The shape of the cell makes it difficult to transport oxygen. In addition, sickle cells are brittle and break easily. As a result, the body cannot manufacture new ones fast enough and it results in anemia.

- **Tay-Sachs disease:** It is a recessive genetic disorder, which progressively destroys nerve cells in brain and spinal cord. This leads to neurological and physical disabilities. Infants suffering from this disease lose motor control like ability to turn, sit, or crawl. As the disease progresses, there is mental disability, vision and hearing loss, occurrence of seizures and paralysis. Such infants usually die within the initial years of childhood. Tay-Sachs disease that may affect adolescents and adults are rare and less severe. There is absolutely no cure for the disease.

2.2 Chromosomal Abnormalities

Sometimes, the children are born with congenital birth defects, which are not hereditary though they are caused by the improper union of genes or error in development.

- **Down syndrome (trisomy 21):** One of the most common and widespread consequences of hereditary/genetic errors is Down syndrome. The primary cause of Down syndrome is the failure of 21 chromosomes to separate properly while undergoing cell division during the production of the egg and sperm. Down syndrome children have three of these chromosomes rather than two. Therefore, they have a total of 47 chromosomes rather than the normal 46. Down syndrome is frequently referred to as *trisomy 21*. Parents do not pass Down syndrome to their children as an inherited characteristic. The disease results from an error in cell division. The behavioural consequences of Down syndrome include mental retardation, speech difficulties, limited vocabulary and slow motor development.
- **Sex chromosome abnormalities:** These abnormalities are caused by an incomplete or incorrect division of sex chromosomes. The examples are:
  - Females born with XO combination instead of the typical XX combination are called to have *Turner's syndrome (TS)*. These single-X girls lack the second sex-determining chromosome, and therefore, have only 45 chromosomes. These individual are not likely to be mentally retarded but also they usually have specific cognitive deficiencies. Administration of female hormone (estrogen therapy) enables TS females to develop sexually in a normal way, although they remain sterile.
  - Males with *Klinefelter's syndrome* carry an extra X chromosome, resulting in an XXY combination rather than the normal XY male chromosome pattern. These males have problems with language skills as well as reading disabilities and neuro-motor problems. Males born with the *XYY syndrome* carry an extra Y chromosome. These boys typically are unusually tall and are mildly retarded.

3. Genetic Counselling and Prenatal Diagnosis

Earlier, people with history of genetic disorders in their family chose not to marry or bear a child at all. In recent years, genetic counselling and prenatal diagnosis helped the people in taking decisions whether or not to have children.

3.1 Genetic counselling

Genetic counselling is a process of providing information about genetic disorders in the family and their preventive options. The best time to seek genetic counselling is before conception so that the counsellor can assess the risk factors involved in transmission of a particular disorder. Genetic counselling can be of help to pregnant mothers as well.

One should consider genetic counselling only if there is either a family history of genetic disorders or physical defects at birth such as spina bifida, cleft lip or palate, clubfoot, congenital heart defects, *etc.* Genetic counselling is also required in cases where prenatal screening tests present abnormal results.

A couple needs genetic counseling if they have had repeated miscarriages, a still birth, have had difficulties in bearing children, have one child with a genetic disorder, one close relatives who has a genetic problem, are older and want to have children and the woman has been taking drugs or other dangerous substances during pregnancy.

The genetic counsellor helps the couple in understanding the problem by preparing a picture of the family tree in which the affected family members are identified. Parents are explained the likelihood of having an abnormal child, particularly if they had consanguineous marriages or belong to specific racial/ethnic groups or geographic areas with a certain genetic disorder most commonly found in their group or area. A couple then can consider appropriate options such as taking a chance and conceiving or adopting a child.

3.2 Alternative Forms of Conception

There have been a number of alternatives for a couple experiencing difficulty in conceiving. There are methods of overcoming problems in sperm production, egg production and fertilization. There are other alternative methods of conception such as donor insemination, *in vitro* fertilization and surrogate motherhood.

- **Donor insemination method (DIM):** It is used if a male has reproductive difficulty. In this method, injection of sperm from an anonymous man is given to a woman.
- **In vitro fertilization (IVF):** This is another reproductive choice, which is used to treat women whose fallopian tubes are permanently damaged. In this method, hormones are given to a woman to stimulate ripening of several ova. These ova are surgically removed and placed in a dish of nutrients. To this dish, sperms are added. Once an ovum is fertilized, it is then injected into the woman's uterus.
- **Gamete intra-fallopian transfer (GIFT):** In this method, a doctor inserts both the eggs and the sperm into a woman's fallopian tube for fertilization inside the woman's body. The resulting fertilized egg then travels to her uterus, as in the natural process of conception. The success rate for this method is about 25%.
- **Zygote intra-fallopian transfer (ZIFT):** This is the combination of IVF and GIFT. In this method, the eggs and sperms are fertilized in the laboratory, and then, the fertilized eggs are inserted in the fallopian tube so that they may travel to the uterus, as they would be in natural conception. The success rate is about 25%.
- **Intra-cytoplasmic sperm injection (ICSI):** In this method, a doctor injects a single sperm into a single egg for fertilization in the laboratory. The fertilized egg is then implanted in the woman's uterus. This procedure is useful for men with low sperm counts, low sperm motility, or sperm that cannot penetrate the egg on their own. The success rate with this method is about 25%.
- **Donor or frozen eggs and sperm:** Donor eggs, donor sperm, or both donor eggs and sperm may be used in fertilization techniques.
- **Surrogate motherhood:** A recent controversial form of alternate method of conception is surrogate motherhood. If wife is infertile, then sperms from her husband or donor are used to

inseminate a woman who agrees for her childbearing services for money. After delivery, she returns the baby to the natural father whose wife then adopts the child. In some cases, the surrogate provides the egg and undergoes one of the above fertilization techniques with the father's sperm. A surrogate may be a relative, friend, or hired woman. The legal involvement is complicated and the necessary arrangements require serious discussions.

3.3 Prenatal Diagnosis

Availability of prenatal diagnosis for a number of genetic disorders has made counselling and decision making process much easier. Prenatal diagnosis is the most useful application as it offers prospective parents the assurance of having an unaffected child in situations of high recurrence risks. Prenatal diagnostic methods, which help in detection of problems before birth, include amniocentesis, ultrasound, fetoscopy and maternal blood analysis. The prenatal diagnosis is recommended in the following cases:

- The pregnant woman is 35 years or older at the time of conception.
- Family history of child with a chromosomal abnormality.
- A history of repeated abortions.
- A history of parental consanguinity.
- The couple is carriers of a chromosomal abnormality.
- The pregnant woman is affected with type 1 diabetes mellitus or epilepsy.
- The pregnant woman is exposed to viral infections, such as rubella.
- The mother is exposed to excessive medication or to environmental hazards.
- A history of single gene disorder is present in her or her spouse's family.
- Her male relatives have muscular dystrophy or severe hemophilia.
- Pregnant mother is suspected of having some other harmful gene on her X chromosomes.
- The fetus is diagnosed to have some hereditary error of metabolism.
- The fetus is detected to be at increased risk for neural tube defects.

The benefits of prenatal diagnosis are:

- Prenatal diagnosis determines the outcome of pregnancy.
- It is helpful for couples to decide whether to continue the pregnancy or not.
- It indicates possible complications that can arise at birth.
- It prepares the couple for the birth of a child with an abnormality.
- Congenital abnormalities account for 20-25% of prenatal deaths. Many genetic and other disorders can be diagnosed early in pregnancy.

Prenatal diagnosis uses various techniques to determine the health of, the condition of, or any abnormality in an unborn fetus. Some of these techniques are outlined below:

- **Ultrasound:** It permits assessment of fetal age, detection of multiple pregnancies, fetal position, placental location, fetal growth and development, fetal movement and identification of physical defects. It is harmless to both the fetus and the mother. It also can assess amniotic fluid volume.
- ***Fetal echocardiography*:** It is recommended in cases where cardiac defects are suspected.

- **Screening for neural tube defects (NTD):** Screening for NTDs is recommended if the ultrasound findings indicate neural tube defects, a family history of NTDs exists, the mother has type 1 diabetes mellitus during pregnancy and if the mother is exposed to drugs.
- **Embryoscopy:** In this technique, a rigid endoscope is inserted *via* cervix in the space between amnion and chorion, under sterile conditions and ultrasound guidance to visualize the embryo for the diagnosis of structural malformations.
- **Fetoscopy:** This is usually performed between 15 to 18 weeks after conception to inspect the fetus for the defects of limbs and face. Fetal blood can also be obtained to diagnose disorders such as hemophilia, sickle cell anemia and other neural defects.
- **Amniocentesis:** In this method, a hollow needle is inserted through the abdominal wall to obtain a sample of fluid in the uterus to examine the cells for genetic defects. It can be done by 11 to 14 weeks after conception.
- **Chorionic villus sampling (CVS):** Chorionic villus sampling is a very accurate antenatal test that detects chromosomal abnormalities such as Down's syndrome. Some women choose CVS because they already have a child with a birth defect, or they have a family history of a genetic disease such as Tay-Sachs. Its main advantage over amniocentesis is that it can be performed very early in gestation between 9 and 12 weeks, ideally at 10 week's gestation.
- **Detection of Chromosomal Aberrations:** Chromosomal aberrations, such as deletions, duplications, translocations and inversions diagnosed in affected parents or siblings, can be detected in a fetus by chromosomal analysis.

Today, with the improvement in prenatal diagnosis, some medical problems are treated before birth. Genetic engineering and therapy also offer new hope for correcting genetic defects.

4. Prenatal Stages of Development

The process of human development begins at conception with the fusion of male and female gametes. Prenatal period is divided into three stages, *i.e.*, the period of zygote, the period of embryo and the period of fetus.

4.1 The period of zygote (conception to two weeks)

The germinal period or the period of zygote begins with conception, when the sperm and egg cell unite in one of the two fallopian tubes. This is the period of about 2 weeks, *i.e.*, from conception to implantation. Fertilized egg cell is known as *zygote*, the cell formed is floating free in the fallopian tube, and through ciliary action and muscle contractions, it moves down the fallopian tube. Cell division begins approximately 24 to 36 hours after conception, the period of rapid cell division is called *cleavage*, and by the fourth day, the cells form a *blastocyst*- a hollow ball of cells. The blastocyst consists of two layers of cells. The outer layer forms the placenta and the inner mass forms the embryo. The zygote proceeds down and it becomes implanted on the wall of the uterus. Implantation occurs when the outer layer forms finger like projections called *villi*, which attaches to the uterus. A membrane, called the *amnion*, covers the developing organism in *amniotic fluid*. Implantation is usually completed from 7 to 10 days after fertilization and it marks the beginning of pregnancy. The usual site of implantation is the upper, posterior part of the uterus. The placenta and umbilical cord begin to develop.

4.2 The period of embryo (two to eight weeks)

The second major phase of prenatal development lasts from implantation until about 8 weeks after conception. During this period, inner cells of the blastocyst *differentiate* into three layers, *i.e.*, the ectoderm, the mesoderm and the endoderm. The *ectoderm* develops skin, nails, hair, eyes, mouth and tooth enamel, and the *mesoderm* produces the circulatory system, muscles, bones, kidneys and the inner layer of skin, and the *endoderm* develops the internal organs, glands, lungs, digestive system, tongue, heart, eustachian tubes and liver.

In addition, in this period, three other important structures develop, *i.e.*, the amniotic fluid, the placenta and the umbilical cord.

- **Amniotic fluid:** The amnion and its tough membranous sac lining inner portion of the uterus contain amniotic fluid, which is clear and light yellow liquid. It protects the fetus from injury or physical shocks and maintains relatively constant fetal body temperature.
- **Placenta:** The placenta is the prenatal life support system. It separates the circulatory system of fetus and mother. Nutrients, oxygen and waste are exchanged across the placenta. The placenta serves all of the functions provided by the lungs, kidneys, stomach and intestines.
- **Umbilical cord:** The umbilical cord is the link between the fetus and the placenta through which it passes the fetal blood supply. The umbilical cord contains two umbilical arteries and one umbilical vein. The vein carries oxygenated blood to the fetus and arteries return deoxygenated blood to the placenta. At full term, the umbilical cord is about 2 cm in diameter and 5 cm long.

By the end of this period, the embryo has eyes, the beginnings of ears, a mouth that opens and closes, a nose, hands and feet, arms with elbows and legs with knees, a primitive circulatory system including a heart that beats, some kidney and liver function and spinal cord.

4.3 The period of fetus (third month until birth)

The final stage of prenatal development, the period of fetus, extends from beginning of the third month until birth. The transition from the embryo to the fetal stage is significant, because the fetus is now recognizable as a member of the human race. During this stage, the fetus grows and develops those structures formed during the embryonic stage. Body size and length increase rapidly and muscular development is rapid. The differentiation of external genital continues and the central nervous system develops rapidly. Between sixteen and twenty weeks, the mother often feels fetal movements, known as *quickening*. By the end of the 12th week, the fetus is covered with *vernix*, a greasy cheese like substance and by *lanugo*, a down like hair. The vernix protects the delicate fetal skin.

At about 20th week, the fetus reflexes such as sucking and swallowing usually appears. After 20th weeks, the fetus develops nails and sweat glands. By 22nd weeks, the eyes have developed and opening and closing of the eyes occur. Fetal movement can become quite strong and can be disruptive to the woman. This period of pregnancy can be uncomfortable, causing symptoms like weak bladder control and backache.

The age between 22 to 26 weeks sometimes referred to as the *age of viability*. At this time, the child has a probability of survival if gets proper medical care. In spite of this, premature birth remains a major threat to the fetus and may result in ill health in later life, even if the baby survives.

5. Prenatal Environmental Influences

Although prenatal environment is more constant than the environment outside the world but there are many factors, which can affect the developing embryo and fetus. Some of these factors are given as under:

5.1 Teratogens

Teratogen refers to any environment agent that causes damage during the prenatal period. Teratogens are in essence environmental factors. However, genetic factors within the mother or child affect the response of each to any teratogenic agent. The impact of teratogens depends on the dosage. Damage from teratogens is not always evident at birth but it appear later in life. The impact of teratogens often depends on the quality of prenatal environment.

Harm done by teratogens depends on many factors, *i.e.*, (i) the amount and length of experience makes a difference and larger doses over longer time usually have more adverse effects, (ii) the genetic make-up of mother and child play an important role as some individuals are better able to withstand harmful environments, and (iii) the effect of teratogens varies with age of the organism at the time of exposure, *e.g.*, damage is more harmful during *sensitive period.* Sensitive period is a limited time span in which a part of the body or behaviour undergoes rapid development.

During the period of zygote, teratogens rarely have any impact. If they do, the mass of cell dies. The period of embryo is a sensitive period when serious defects are most likely to occur since the foundations for all body parts are being laid down. During fetal period, the damage caused by teratogens is usually minor. Some organs, such as brain, eye and genitals, can still be strongly affected. Varieties of teratogens that can cause damage during the prenatal period are:

- **Prescription and non-prescription drugs:** Any drug a mother takes can enter the embryonic or fetal blood stream and can produce gross deformities in developing embryo. These children when grow older are more likely to score below average in intelligence. Some of the examples of drugs causing disastrous effects are thalidomide, antibiotics streptomycin and tetracycline, anticoagulants, anticonvulsants and artificial hormones.
- **Illegal Drugs:** Intake of illegal drugs is associated with pre-maturity, low birth weight, physical defects, breathing difficulties and death at the time of birth, and such infants become drug addicted. They are irritable and have sleeping problems. Behaviour problems are likely to persist when these infants grow older. These infants are less attentive and their motor development is slow. Some of the examples of illegal drugs are heroin, methadone, marijuana and cocaine.
- **Caffeine:** Intake of caffeine in large doses during pregnancy can result in spontaneous abortion and low birth weight.
- **Alcohol:** Children born to alcoholic mothers may suffer from *fetal alcohol syndrome* (FAS), which is a set of defects that results when pregnant women consume large amounts of alcohol during pregnancy. Fetal alcohol syndrome includes mental retardation, slow physical growth, poor attention and over activity. Children have distinct physical symptoms, *i.e.*, widely spaced eyes, short eyelid, a small upturned nose and a thin upper lip. Their small heads indicate that the brain has not developed fully.
- **Smoking:** The most well known consequences of smoking during pregnancy are low birth weight, pre-maturity, miscarriage, infant death and cancer later in childhood. Longitudinal studies show that these children have shorter attention spans and lower intelligence quotient (IQ).

- **Radiation:** Frequent exposure to X-rays can cause mutation and damage the DNA in ova and sperm. When mothers are exposed to radiation during pregnancy, additional harm can come to the embryo or fetus. Radiation is associated with miscarriage, slow physical growth, an under developed brain and malformations of eyes and skeleton.
- **Pollution:** Pregnant women can absorb lead from car exhaust, paint coming off the walls in houses and other materials used in industrial occupations. High levels of lead exposure are linked to pre-maturity, low birth weight, brain damage and physical defects.
- **Maternal Diseases:** The placenta cannot filter extremely small disease carriers, such as viruses. Children can be born with malaria, measles, chicken pox, mumps, syphilis, or other venereal diseases that have been transmitted from the mother. Rubella (German measles) is a well-known teratogen. Greatest damage occurs when rubella strikes during sensitive period, *i.e.*, the period of embryo. Damage includes mental retardation, heart defects, deafness, genital, urinary and intestinal abnormalities. Infection during fetal period is less harmful, *i.e.*, low birth weight, hearing loss and bone defects may occur.

The human immunodeficiency virus (HIV), which causes acquired immunodeficiency syndrome (AIDS), is affecting increasing number of children. Children born to AIDS affected mothers are likely to have AIDS.

- **Rh incompatibility:** The Rh-positive factor is an inherited genetically dominant trait in the blood that can result in a dangerous situation for the fetus. When an Rh-negative woman conceives a child with an Rh-positive man, the resulting child may have Rh-positive blood. Any small rupture in the capillaries of the placenta will release the Rh factor into the mother's bloodstream. The mother's immune system produces antibodies to fight it. The antibodies in mother's blood will then cross the placenta into the fetal bloodstream and attack its red blood cells. This condition is called fetal erythroblastosis. The result may be a miscarriage, brain defects and even death of the fetus or newborn child. In past, erythroblastosis was always fatal, but now, the medical techniques can minimize the harmful effects of Rh incompatibility.
- **Anoxia:** Every child undergoes some oxygen deprivation and retention of carbon dioxide during birth process. The conditions during labor and delivery can lead to severe anoxia, which results in brain damage, functional defects, or even death of the infant.

5.2 Other Maternal Factors

Maternal Nutrition: Development is more rapid during prenatal period than at any stage of development. During this period, they depend totally on the mother for nutrition. Mother who suffers from malnutrition during first trimester of pregnancy is more likely to have miscarriages or give birth to babies with physical defects. Prenatal malnutrition can cause serious damage to central nervous system. During last trimester, when brain is growing rapidly in size, malnutrition can cause loss in brain weight.

Maternal Age: Women who delay having children, particularly those who give birth to their children after the age of 35 years, face a greater risk of giving birth to babies with chromosomal abnormalities.

Teenage pregnancy is not directly associated with prenatal problems. Infants of teenagers are born with a higher rate of problems for quite different reasons. Many young mothers do not have access to medical facilities because most of these mothers come from low-income group. In lower income-groups stress, poor nutrition and health problems are common.

Women in their twenties seem to be at the best age for child bearing. Many younger and older women bear children of course but the chances of disorders for both mother and child are greater in very early and very late pregnancies. The probability of miscarriages, delivery difficulties fetal defects and pregnancy complications occur more often outside the age span of 18 to 38 years.

Psychological Stress: Intense maternal anxiety during pregnancy is associated with a higher rate of miscarriage, pre-maturity and low birth weight. It is also related to certain physical defects, such as cleft lip and palate. Prolonged psychological stress weakens the immune system and mother is more prone to infectious diseases. Women who experience long-term anxiety are more likely to smoke, drink and eat poorly that harm the embryo or fetus.

Number of Pregnancies: The women who had more than four pregnancies are at greater risk than the women who have had fewer pregnancies. Their babies are more likely to be still born and to be smaller at birth.

6. Care of Mother during Pregnancy

The first sign of pregnancy is a missed period. If the regular cycle is 10 days late, it is probable that conception has taken place but if the second period is missed, the pregnancy must be taken as confirmed. Another sign of pregnancy is morning sickness. It may occur at any time of the day. It usually begins about the 4th week but may occur earlier. Morning sickness is more common and more severe in first pregnancies. Early in pregnancy, there is increased desire to pass urine because of the enlarged uterus on the urinary bladder. The symptoms disappear in about 3 months when the uterus expands upward into the abdomen.

6.1 Important Points for Maternal Care

Expectant mother should keep in mind the following points for her care during pregnancy:

- **Diet:** A good diet during pregnancy is most important. Diet should be rich in proteins, vitamins and iron as there is excessive loss of blood during pregnancy.
- **Exercise:** Household work is good exercise during pregnancy. Activities, which require jumping or sharp jerky motions, should be avoided. It is not advisable to lift or push heavy objects. Avoid standing still for long periods.
- **Rest:** Plenty of rest is important. One should have at least 8 to 9 h sleep every night and a rest during the day. During early months whenever one feels drowsiness during the day, it is better to sleep or take rest. It is important not to be over-tired or over strained.
- **Clothing:** During pregnancy, one should wear loose clothes so that she may feel easy. If clothing is tight, it will create difficulty in breathing and blood circulation.
- **Medical check-ups:** Pregnant woman should go for regular medical check-ups. In first trimester, she should go once in a month, and after 4th or 5th months, she should go once in a fortnight or week as the doctor suggests. One should consult the doctor about taking iron, calcium, or vitamins tablets. During first trimester or pregnancy, no drug should be taken without the advice of doctor as these may have adverse effect on the developing fetus.
- **Personal hygiene:** A daily bath is desirable during pregnancy. However, it should be neither steaming nor so cold. There is tendency toward decay in the teeth of pregnant women it may be

because developing fetus is taking calcium from the mother's body. Extra quantity of milk may be included in the diet or one should take calcium tablets by consulting the doctor. Teeth should be brushed after every meal. Sunlight is necessary for the bones of developing fetus so one must get lots of fresh air and sunshine.

- **Psychological care:** A pregnant woman should live in a happy and cheerful mood. If she is worried or tense all the time, it will affect the developing fetus and health of the mother as well.

6.2 Common Discomforts of Pregnancy

Constipation: Most of mothers face the problem of constipation during pregnancy. One or two extra cups of tea at breakfast, fresh fruits, green leafy vegetables and a glass of warm or cold water in the morning will help in the prevention of constipation.

Digestive discomforts: Heartburns and indigestion are often caused by something in the diet. The first step towards relieving these symptoms is to discover what food might be causing it and to eliminate that food from the diet. Even then, if the heartburn persists, consult the doctor.

Varicose veins: Sometimes, veins in the legs become swollen and painful. This results from poor circulation and the effects of hormones on their walls. The pregnant woman should lift her legs up so that blood can return easily from the legs and should avoid standing for long periods and fastening of clothes around the waist.

Shortness of Breath: In last weeks of pregnancy, exertion may lead to considerable breathlessness since the enlarged womb hampers the working of one of the muscles of respiratory called the diaphragm, thus, a pregnant woman should avoid heavy exertion, do light household jobs only and take rest when feel tired.

Itching of the skin: Sometimes, a pregnant woman has complaint of itching. This may be helped by adding baking soda to the bath water but should be reported to the doctor since it may be a symptom of a food or drug allergy. If there is itching of vagina, the doctor should be consulted so that treatment measures can be taken.

Cramps: Cramps may occur in legs, especially in the back of thighs. This may be very painful and may disturb sleep. This may be either due to pressure of the enlarged womb on nerves of the leg or due to the deficiencies of calcium.

Vaginal discharge: In later parts of the pregnancy, one may have a thin pale yellow discharge from the vagina. This is normal and requires only the wearing of a sanitary pad and a thorough washing of the area once or twice daily. If the discharge is thick or accompanied by itching, the doctor should be consulted.

Backache: This is common especially in the later months of pregnancy since during pregnancy, the ligaments, which hold the joints together, become loose. The joints need special care. Walking on high heals alter the position of spine and muscles so that they can support efficiently. It is especially important to wear low healed shoes.

Frequency of passing urine: Frequent desire to pass urine is often present at two stages of pregnancy, *i.e.*, in first weeks and in last two or three weeks. Pain when passing urine, however, is not common and may be due to infection of the bladder or the kidneys, therefore, difficulty or inability to pass urine suggests that something is wrong and one should see the doctor immediately.

6.3 Warning Signals of Pregnancy

One must see the doctor at once if any one of the following signals is present:

- The slightest bleeding from the vagina (Haemorrhage)
- Abdominal pain before the expected date of delivery
- Swelling of feet and marked fall in the quantity of urine passed (Toxemia of pregnancy)
- Excessive or poor weight gain
- Pain or burning while passing urine
- Chills and fever
- Severe and persistent nausea and vomiting
- Continuing or sever headache
- Spots before eyes
- Sudden gush of water from the vagina before the expected date of delivery
- Movement of the baby diminishes.
- Anemia is one of the common danger signals in pregnancy.

7. Hazards of Prenatal Period

Hazards of prenatal period are most serious in nature than any other stage of life-span. These may be physical or psychological in nature.

7.1 Physical Hazards

Period of ovum

- There is possibility that the fertilized ovum will die of starvation before it becomes implanted if it takes long time to pass down the fallopian tube.
- The uterine wall is not prepared to receive the fertilized ovum due to hormonal imbalance.
- The fertilized ovum is attached to the wall of the fallopian tube- a *tubal pregnancy* resulting in lack of nourishment and space for growth.
- The fertilized ovum is attached to a small area of fibroid tissue in the uterine wall, which would prevent it from receiving nourishment.

Period of embryo

- Miscarriages or spontaneous abortions occur due to prenatal unfavorable conditions.
- Developmental irregularities are more likely to occur during this period.

Period of the Fetus

- There is possibility of miscarriage up to the 5th lunar month, especially at times when the woman's menstrual period would normally occur.
- Pre- and post-mature birth with complications may result in death of fetus.

- Complications of delivery may result in death of fetus.
- Developmental irregularities caused by conditions similar to those of embryo period retard the development of fetus.

7.2 Psychological Hazards

Psychological hazards during prenatal period can have long-term effects on child's development and can influence the treatment the child gets from parents and other significant people. Most important prenatal psychological hazards are:

- **Traditional Beliefs:** People, particularly in rural areas, still believe that it is within their power to control the sex of their child. They believe that they can do this by having sexual intercourse at certain periods during the menstrual cycle. They believe that by producing an acid environment in women's uterus girl child can be conceived and alkaline environment would result in a boy. The effects of these beliefs can be very hazardous when the sex of the child turns to be of the opposite sex. Parents are bitterly disappointed. Some people believe that women are responsible for the sex of a child. If she cannot produce the child of the desired sex, husband and in-laws may develop negative attitude towards her. There are also traditional beliefs about causes of developmental irregularities. Many people believe that these are caused by solar eclipse and evil spirits. Acceptance of these beliefs can lead to strong feelings of guilt on the part of mother. These mothers tend to overprotect the child to compensate the harm she believes she is responsible for that. Some people believe that multiple births are caused by evil spirits, and thus, are rejected by some social groups.
- **Preference for a child of a particular sex and dream-child concept:** The majority of parents want their first born to be a boy. Also, the family members usually have a dream child concept that a child should be physically, emotionally and mentally normal. They are disappointed when child born is of the opposite sex or born with developmental abnormalities.
- **Not wanting the child:** The expectant mother may not want the child because it is illegitimate or it will interfere with her educational or vocational career, or because she is already overburdened with family work. The father may not want the child because of financial burden. Siblings may not want the child because they do not want to share the love and affection and other possessions.
- **Not wanting the child presently:** The parents do not want child currently, as it may interfere with their educational and vocational plans or they feel that they are still too young to care for a child. Grand parents may not want the child since they believe that their son cannot afford the financial burden and may fear that it will be a financial burden on them.
- **Not wanting children of multiple births:** Many people believe that multiple births are caused by evil spirits. They also believe that children of multiple births may suffer from developmental irregularities. A favourable attitude is intensified if conditions after birth are similar to those they believed before birth.
- **Scorn for the child:** Friends, neighbours and relatives may have an unfavourable attitude toward the unborn child because it is illegitimate or because it is child of an inter-caste or inter-religious marriage.
- **Maternal Stress:** Maternal stress may be caused by many factors. Some of these factors include not wanting the child because of economic difficulties or marital conflicts, or they feel that child

will interfere with the educational and vocational career, or the mother feels that it would be a stage of discomfort for her. Maternal stress affects the child before and after birth. Severe maternal stress before birth may result in developmental irregularities and complications at the time of delivery or premature birth. Anxiety may lead to over eating and excessive gain in weight, which would result in complicated birth. After birth, children born to mothers who experienced stress during pregnancy are more likely to be hyperactive, irritable and show poor adjustment to postnatal life.

8. Child Birth and Complications

8.1 Child Birth Process

The birth process often referred to as labour is divided into three stages:

- **Dilation and Effacement of the Cervix:** The first stage of labour or birth process is the longest stage, lasting on an average 12 to 14 hours in a first birth and 4 to 6 hours in later births. Stage first covers the period during which two important processes occur, *i.e.*, *dilation* and *effacement*. The cervix (the opening at the bottom of uterus) must open up (dilation) and also flatten out (effacement). At the time of actual delivery of infant, the cervix must normally be dilated to about 10 cm.

  First stage includes two phases. In *Early Phase* (*Latent phase*), contractions are relatively far apart and are not typically too uncomfortable. In *Late Phase* [which begins when the cervix is about halfway dilated (5 cm)], the contractions are closer together and are more intense. This phase continues until the cervix is dilated to 8 cm. Finally, last 2 cm of dilation are achieved during the *transition phase*. This is the shortest phase in which the contraction are closely spaced and strong and it is the time when mother has natural urge to push the baby out.
- **Delivery of the baby:** Stage 2 lasts from half an hour to two hours. Strong contractions of the uterus continue but they do not do the entire job. The most important factor is the natural urge that mother feels to squeeze and push. Once the cervix is fully open, the baby is ready to be born. This is the actual delivery, when the baby's head moves and pass through the stretched cervix, into the birth canal, and finally, out of the mother's body.
- **Delivery of the Placenta:** The third stage of labour is the delivery of placenta (also called the after birth) and other material from the uterus.

There are different approaches to childbirth. In some families, home delivery is preferred, while other parents opt for hospital delivery. Mostly mothers prefer natural delivery until and unless there are some complications. A few mothers who do not want to undergo the process of labour pains prefer to deliver their babies by caesarian section.

8.2 Complications of birth

The ease or difficulty with which a baby is born and how quickly the newborn begins to breathe can affect the infant's well being. Some problems associated with birth are as under:

- **Anoxia:** Reduced oxygen supply is called anoxia. Anoxia may occur during the period immediately after birth because the umbilical circulation fails to continue the supply of blood oxygen until the baby breaths or because the umbilical cord has been squeezed in some way during delivery. Prolonged anoxia is often associated with such major consequences as cerebral palsy or mental retardation.

- **Hemorrhaging:** This is usually caused when very strong pressure on head of the fetus breaks blood vessels in the brain. It affects the supply of oxygen to nerve cells of the brain.
- **Low birth weight:** Another serious complication is to have an infant born weighing less than the normal. Those below 1500 g (about 3.3 pounds) are usually called as very low birth weight. There are different categories of low birth weights.

Some infants have low birth weights because they are born too soon. These are usually called *Preterm Infants* (any birth before 37 week's gestation is usually considered Preterm). Second important type of low birth-weight infant is the *Small for date* baby. These infants are usually light in weight, given the number of weeks of gestation they have completed. They may even have completed the full 38 weeks gestation but weigh under 2500 g.

Low birth baby have high infant mortality, lower level of responsiveness at birth, and in early months of life, high risk of respiratory problems and lower intelligence quotient.

■■■

27 Nutrition in Pregnancy and Lactation

Asha Kawatra and Salil Sehgal

Department of Foods and Nutrition, College of Home Science,
CCS Haryana Agricultural University, Hisar-125004, Haryana, India,
E-mail: <ashakawatra@gmail.com>

Pregnancy is a period of great psychological stress for the woman as she is nurturing a growing foetus in her body. For most women, pregnancy is a time of adjustment. A pregnant woman is undergoing a major transition and growing in many ways. Physically and emotionally, her needs are changing. If this is her first pregnancy, she senses that her lifestyle will have to change since she takes on the new responsibility of caring for a child. Ideally, she will be encouraged to develop this sense of responsibility by caring for herself during pregnancy. A large part of this care includes eating nutritious foods, staying physically active and getting plenty of rest. Foetal development is accompanied by many physiological, biochemical and hormonal changes occurring in the maternal body, which influence the needs for nutrients and the efficiency with which the body uses them. A completely new life begins at conception. Events follow quickly upon one another, and nutrition plays a supportive role in all of them. Nothing could be more extraordinary than the course of a pregnancy, yet this remarkable time is characterized as *normal.* Although this period is normal but it should be at proper period of life as World Health Organization (WHO) and United Nation International Children Fund have repeatedly emphasized that too early pregnancies, too late, too close and too many are not safe to the health of mother and her infant.

Growth and Development during Pregnancy

Physiological events during pregnancy bring major changes to the mother's body, creating an entire new organ within it (the placenta) and producing a new human being. Many dramatic changes are observed during pregnancy. Her uterus and its supporting muscles increase in size and strength, her breasts grow and prepare to produce milk, and her blood volume increases by half to carry the additional nutrients and other materials. The normal weight gain during pregnancy is about 12 to 15 kg. Increases in maternal tissues alone account for about two-thirds of this weight. Substantially, the basal metabolism rises during pregnancy since the tissues are actively synthesized for the foetus and it associated supportive structures. The maternal cardiovascular system, kidneys and lungs work hard to deliver needed substances and to clear wastes. Over the nine months of gestation, all this activity is estimated to expend about 55,000 kcal.

Placental Development and Function

In early days of pregnancy, the newly formed zygote is small and the nutrient-rich cells of the uterine lining provide adequate fuel and raw materials. To continue nourishing the developing embryo and foetus, a spongy structure placenta develops in the uterus. The placenta develops as an interweaving of foetal and maternal blood vessels embedded in the uterine wall. Two associated structures also form. One is the amniotic sac, a fluid-filled balloon like structure that houses the developing foetus. The other is the umbilical cord, a rope like structure containing foetal blood vessels that extends from the foetus's *belly button* (the umbilicus) to the placenta. These three structures serve crucial roles during pregnancy, and then, these are expelled from the uterus during childbirth.

The foetal portion of the placenta develops when the membranes that surround the embryo invade the uterine wall and form villi. These villi contain embryonic blood vessels. The uterus responds by allowing pools of maternal blood to form around these villi so that maternal and foetal blood will be in close proximity.

Maternal blood enters the placenta through small artery branches of the uterus, bathes the villi and leaves *via* the uterine veins. Foetal blood circulates from the umbilical artery into the villi, and back through the umbilical vein. An exchange of materials takes place within the sponge-like endometrium but no actual mingling of foetal and maternal blood occurs. Blood flow to the placenta increases dramatically during pregnancy, accounting for as much as 25% of the maternal cardiac output by term. The rates of maternal and foetal blood flow influence the exchange of substances across the placenta. The delivery of oxygen to the foetus improves as the flow of maternal blood increases.

The placenta transfers oxygen and nutrients to the foetus and returns carbon dioxide and other foetal waste products to the mother. Nutrients pass from maternal blood through the placenta into foetal blood. Waste products move from foetal blood through the placenta into maternal blood. By exchanging oxygen, nutrients and waste products, the placenta performs the respiratory, absorptive and excretory functions that the foetus's lungs, gastrointestinal tract and kidneys will provide after birth. Water, oxygen, carbon dioxide and electrolytes cross the placenta by passive diffusion. Glucose, calcium, phosphorus, magnesium and other nutrients cross the placenta *via* active transport. Foetal accumulation of these nutrients increases throughout pregnancy, becoming greatest during the third trimester.

A healthy placenta is essential for the developing foetus to attain its full genetic potential. Maternal nutrition, in turn, is crucial to placental development. Maternal malnutrition limits placental development, which reduces placental blood flow. A diminished blood flow curtails the transfer of energy and essential nutrients to the foetus. Thus, maternal malnutrition negatively affects the foetus first by impairing placental development and then by reducing the transfer of nutrients across the placenta.

Foetal Growth and Development

Foetal development begins with the fertilization of an ovum by a sperm. Three stages follow, *i.e.*, the zygote, the embryo and the foetus.

The Zygote: The newly fertilized ovum or zygote begins to divide within a day after fertilization. It floats freely within the uterus for several days, receiving nourishment from the intrauterine fluid. As cell division continues, the uterine wall or endometrium prepares to receive the zygote. Within two weeks, the zygote embeds itself in the endometrium- a process known as implantation. Cell division continues- each set of cells divides into many cells and these cells sort themselves into layers. The outermost layer forms the placenta and amniotic sac, and the innermost cells become the embryo.

The Embryo: The embryo accomplishes amazing development. The number of cells in the embryo doubles approximately every 24 hours at first. Later, the rate slows and only one doubling occurs during the final ten weeks of pregnancy. The embryo's size changes very little up to the eighth week of pregnancy but the events taking place during that time are momentous. As the cells divide, they sort themselves into three layers, *i.e.*, the outermost layer (ectoderm), the middle layer (mesoderm) and the innermost layer (endoderm). From the ectoderm, the nervous system and skin begin to develop, from the mesoderm, the muscles and internal organ systems, and from the endoderm, the glands and linings of the digestive, respiratory and excretory systems develop. At eight weeks, the 1¼ inch embryo has a complete central nervous system, a beating heart, a digestive system, well-defined fingers and toes and the beginnings of facial features. Already an embryonic tail has formed and almost completely disappeared again.

The Foetus: During the next seven months, each organ grows to maturity according to its own schedule, with greater intensity at sometimes than at others. Foetal growth is phenomenal and weight increases from less than an ounce to about 7½ pounds (3500 grams). The gestation period, which normally lasts from 38 to 42 weeks, ends with the birth of the infant. Malnutrition exerts its effects during critical periods of development during pregnancy. Maternal malnutrition can impair foetal development by impairing placental development and by reducing the nutrients available for transfer across the placenta.

Adverse Influences: Early malnutrition can have irreversible effects, although they may not become fully apparent until the person reaches maturity. A research study suggests that adverse influences at critical time set the stage for chronic diseases in adult life. Under-nutrition during *critical periods* may permanently alter body functions such as blood pressure, cholesterol metabolism and immune functions that influence disease development, *e.g.*, malnutrition during the critical period of pancreatic cell growth may be responsible for the development of diabetes mellitus in adulthood. The pancreatic cells responsible for producing insulin (the beta cells) normally increase more than 130 folds between 12 weeks gestation and 5 months postnatal. Nutrition is a primary determinant of beta cell growth, and infants who have suffered prenatal malnutrition have significantly fewer beta cells than the infants who are well nourished. One hypothesis suggests that diabetes may develop from the interaction of inadequate nutrition early in life with abundant nutrition later in life. The mass of beta cells developed in lean times may be insufficient during times of over nutrition when the body needs more insulin. Studies have found an association between low birth weight (an indicator of prenatal malnutrition) and glucose intolerance in adulthood. Furthermore, the degree of intolerance appears to be influenced by the adult's body mass index values.

Each organ and tissue is most vulnerable to nutrient deficiencies or toxins during its own critical period, *e.g.*, the critical period for neural tube development is from 17 to 30 days gestation. The neural tube is the beginnings of brain and spinal cord, which are the key structures in central nervous system. As in early weeks of embryo development, the nervous system and skin both derive from the ectoderm. Normally, by the fourth week of gestation, the ecotoderm that forms the neural tube of nervous system separates from the ectoderm that forms the skin. Neural tube development is most vulnerable to nutrient deficiencies or toxins at this critical time of differentiation. Any abnormal development of the neural tube or its failure to close completely can cause a major defect in the central nervous system.

In human beings, the total number of brain cells rises rapidly during prenatal life and continues to increase after birth up to the second year. Under-nutrition during this critical period may cause irreversible damage to the brain and adversely affect intelligence and behaviour in later years. A pregnant woman should continuously take good care of her health for achieving and maintaining a healthy body weight prior to conception and gaining sufficient weight during pregnancy to support a healthy infant.

Maternal Weight

Birth weight is the most reliable indicator of an infant's health. In general, higher birth weights present lower risks for infants. Two characteristics of the mother's weight, *i.e.*, her weight for height prior to conception and her weight gain during pregnancy, influence an infant's birth weight.

Weight before Conception

A women's weight prior to conception influences foetal growth. Even with the same weight gain during pregnancy, underweight women tend to have smaller babies than heavier women.

Underweight: An underweight woman has a high risk of having a low-birth weight infant, especially if she is unable to gain sufficient weight during pregnancy. In addition, the rates for preterm births and prenatal mortality are high in underweight women. An underweight woman maximizes her chances of having a healthy infant by gaining sufficient weight prior to conception or by gaining extra weight during pregnancy.

Weight gain prior to conception is best achieved by exercising to build muscle and eating energy-dense foodstuffs (food articles that provide many calories in a small volume). Following weight gain strategies can be followed by underweight women to gain weight.

- Select energy-dense foodstuffs regularly.
- Eat at least three hearty meals a day.
- Eat large portions of food articles and expect to feel full.
- Eat snacks between meals.
- Drink plenty of juice and milk.
- Exercise to build muscle.

Overweight: Overweight also creates problems related to pregnancy and childbirth. Overweight women especially have high risk of medical complications such as hypertension, gestational diabetes and post-partum infections. Compared with other women, overweight women are also more likely to have complications during labour and delivery.

Infants of overweight women are likely to be born post-term and to weigh more than 4 kg. These women are less likely to have premature infants, but if they do, the infants may be large for their gestational age. Recent researches suggest that obesity may double the risk for neural tube defects in the infant.

Weight loss dieting during pregnancy is never advisable. Overweight women should try to achieve a healthy body weight before becoming pregnant, hence, avoid excessive weight gain during pregnancy and postpone weight loss until after childbirth. Weight loss is best achieved by eating fewer amounts of nutrient-dense food items and exercising to lose body fat.

Weight Gain during Pregnancy

Women who plan to become pregnant, or who are already pregnant, often express concern about the weight gain that accompanies a healthy pregnancy. In a society such as ours, where body slimness is frequently equated with youth, vigor and good looks, many prospective mothers may view the weight gain of pregnancy in a negative light, which is unfortunate. Maternal weight gain during pregnancy supports foetal growth and correlates closely with infant birth weight, which is a strong predictor of health and subsequent development of the infant.

Current recommendations for weight gains during pregnancy are based on a woman's prepregnant weight and they offer a range that will support optimal foetal growth. Recommended weight gains for various prepregnant weights are shown in Table 1.

Table 1: Recommended weight gains based on prepregnant weight status

Prepregnant weight status	Recommended weight gain
Under weight[a]	12.5 to 18.0 kg
Normal weight[b]	11.5 to 16.0 kg
Over weight[c]	7.0 to 11.5 kg
Obese[d]	6.8 kg minimum (15 lb minimum)

[a]Under weight defined as BMI< 19.8; [b]normal weight defined as BMI 19.8 to 26.0; [c]over weight defined as BMI 26.0 to 29.0 and [d]obese defined as BMI > 29.0.

Source: Committee on Nutritional Status during Pregnancy and Lactation 1990, Food and Nutrition Board, Nutrition during Pregnancy (National academy Press, Washington, D.C., USA)

Components of weight gain: Weight gain during pregnancy supports both the growth of the foetus and associated tissues (placenta and amniotic fluid) and the expansion of maternal tissues. Foetal growth and maternal weight gain are shown in Table 2.

Risk of inadequate weight gain: Normal weight gains during pregnancy vary dramatically from individual to individual but certain maternal factors can predict a risk of an inadequate weight gain.

Risk of excessive weight gain: If a woman gains weight more than that is recommended early in pregnancy, she should not restrict her energy intake later in order to lose weight, however, a large weight gain over a short period indicates excessive fluid retention and may be the first sign of the serious medical complication preeclampsia.

Table 2: Foetal growth and maternal weight gains at different ages in gestation

Age (weeks)	Total foetal weight (g)	Maternal weight gain (g)
10	5	650
12	30	
20	300	4000
24	900	
28	1240	
30	1484	8500
32	1750	
34	2278	

Age (weeks)	Total foetal weight (g)	Maternal weight gain (g)
36	2750	
38	3052	
40	3230	12500
42	3310	

Component weight gain in pregnancy

Component	Period of Gestation			
	Up to 10 weeks (g)	Up to 20 weeks (g)	Up to 30 weeks (g)	Up to 40 weeks (g)
Foetus and placenta	55	720	2350	4750
Uterus and breast tissue	170	765	1170	1300
Extra cellular weight	600	1300	1350	1200
Fat	325	1915	3500	400
Total	650	4000	8500	12600

Weight Loss after Pregnancy

The pregnant woman loses some weight at delivery. In the following weeks, she loses more as her blood volume returns to normal and she sheds accumulated fluids, however, the typical woman does not return to her pre-pregnancy weight. In general, the more weight a woman gains beyond what she needs for pregnancy, the more she will retain. Even with an average weight gain, most women tend to retain a couple of kilogram with each pregnancy.

Importance of Nutrition during Pregnancy

From conception to birth, all parts of the infant, *i.e.*, bones, muscles, organs, blood cells, skin, and other tissues, are synthesized from nutrients in the food that the mother eats. For most women, the nutrients need during pregnancy and lactation is higher than at any other time. The better a woman takes care of herself nutritionally, the more successful her pregnancies are likely to be.

Maternal nutrition plays a role in certain types of foetal malformation or spontaneous abortion. Since these abnormalities occur, early in pregnancy modifications must be made before pregnancy begins. Due to undernourishment of the mother, the baby is at risk of being premature with low birth weight and developmental irregularities. Some of risk factors, which affect the pregnancy, are as follows:

Table 3: Nutritional risk factors in pregnancy

Risk factors present at onset of pregnancy	Risk factors occurring during pregnancy
Age less than 15 years or greater than 35 years	Low Hb- less than 12.0 per 100 ml blood
Frequent pregnancies- 3 or more during approximately 2 years	Inadequate weight gain

Risk factors present at onset of pregnancy	Risk factors occurring during pregnancy
Poor obstetric history or poor foetal performance	Any weight loss
Poverty	Weight gain less than 1 kg per month after 1st trimester
Bizarre food habits	Excessive weight gain- 1 kg per week after 1st trimester
Abuse of caffeine, nicotine and alcohol or drugs	
Therapeutic diet required for chronic disorders	
Inadequate weight : • Less than 85% of standard • Greater than 120% of standard	

Intrauterine nutrition is highly important for the growth of central nervous system and kidneys of the foetus, which mature during latter part of the pregnancy. Therefore, nutrition deficits before birth can never be wholly reversed after birth. The Figure shows the relationship between maternal and foetal nutrition.

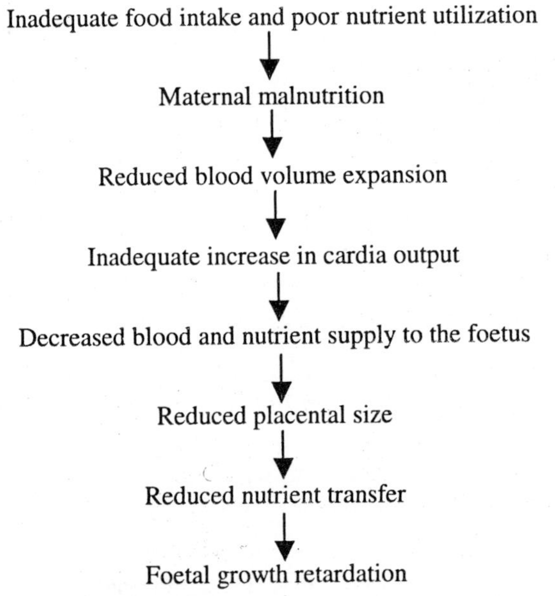

Figure 1: Relationship between maternal and foetal nutrition

This emphasizes the role of nutrition in successful pregnancy, which will only result in good pregnancy outcome, *i.e.*, a healthy baby and a well-nourished mother who enters the period of lactation with adequate nutritional stores.

Nutritional Requirements

Nutritional requirements increase during pregnancy. Increase in nutritional requirements depends on the nature of metabolic changes of pregnancy and nutrition reserves of the mother. The recommended dietary allowance (RDA) of the expectant mother for various nutrients is as follows:

Table 4: Recommended dietary allowance of an expectant mother

Nutrient	Normal adult woman	Pregnant woman
Energy (kcal)		
Sedentary	1875	+300
Moderate	2225	+300
Heavy	2925	+300
Protein (g)	50	+15
Fat (g)	20	30
Calcium (mg)	400	1000
Iron (mg)	30	38
Retinol (μg)	600	600
Beta-carotene (μg)	2400	2400
Thiamin (mg)		
Sedentary	0.9	+0.2
Moderate	1.1	+0.2
Heavy	1.2	+0.2
Riboflavin (mg)		
Sedentary	1.1	+0.2
Moderate	1.3	+0.2
Heavy	1.5	+0.2
Niacin (mg)		
Sedentary	12	+2
Moderate	14	+2
Heavy	16	+2
Pyridoxine (mg)	2.0	2.5
Ascorbic acid (mg)	40	40
Folic acid (mg)	100	400
Vitamin B_{12} (μg)	1	1

Energy

Energy needs during pregnancy increase because of the additional energy required for growth and physical activity of the foetus, growth of the placenta, normal increase in maternal body size, additional work involved in carrying weight of the foetus and extra maternal tissues along with slow but steady rise in basal metabolic rate during pregnancy. Any decrease in the activity of mother decreases the caloric requirement. For an Indian woman of 55 kg, the total energy cost of pregnancy is about 80,000 kcals. Of this, about 36,000 kcal is deposited as fat, which is utilized subsequently during location. The caloric requirement of pregnant woman is mostly increased in latter half of the pregnancy.

Protein

The normal protein requirement of an adult woman is 50 g/day. ICMR has recommended additional 15 g protein for pregnant women. Extra protein is required for rapid growth of the foetus, enlargement of the uterus, mammary glands and placenta, increase in maternal circulating blood volume and subsequent demand of increased plasma protein to maintain colloidal osmotic pressure and circulations of tissue fluids, formation of amniotic fluid and storage reserves for labour, delivery and lactation. The transfer of amino acids from mother to the foetus (up to 20 weeks all amino acids) must be provided to the foetus as it cannot oxidize amino acids as a source of energy.

Based on N accretion, protein requirement during pregnancy is computed, and depending on the body weight gain of 12 kg in a normal healthy well-nourished pregnant woman, the daily N deposited during the first, second and third trimesters of pregnancy is estimated to be 0.1, 0.5 and 0.9 g, respectively. After increasing by 50% to convert the factorial value into physiological value for N accretion and 25% for individual variation, the safe level of intake in terms of a high quality protein during the first, second and third trimesters will be 1.2, 6.0, 10.5 g, respectively. After adjusting the dietary protein quality of net protein utilization (NPU) as 65, the safe intake during the latter half of pregnancy recommended by the Nutrition Expert Committee is 15 g/day. If protein requirement is below RDA during pregnancy, the mother's growth will be affected and the foetus may grow at the expense of mother. As a result, maximum growth of the baby may not be obtained and brain growth will be less.

Linoleic acid

The requirement of linoleic acid during pregnancy is 4-4.5 en%. If invisible fat is 12.5 en% to meet EFA, the visible fat is to be 30 g/day.

Minerals

Calcium: As per ICMR recommendation, the calcium requirement of adult woman is 400 mg/day. Requirement increases during pregnancy to 1000 mg/day. Although the infant bones are poorly calcified at the time of birth, an appreciable amount of calcium is involved in foetal development. A full term foetal body contains 30 g of calcium. Increased intake of calcium by the mother is for the calcification of foetal bones and teeth and for building calcium stores of the mother to meet the high demands during lactation. Dairy products are a primary source of calcium. Green leafy vegetables also contain appreciable amount of calcium. To prevent *osteomalacia*, the mother's diet should contain less of phytic acid, adequate amounts of vitamin D and sufficient amount of calcium. Mothers should avoid repeated pregnancies.

Iron: The normal iron requirement of an adult woman is 30 mg/day. According to ICMR, the iron requirement during pregnancy is 38 mg/day. This increase in iron requirement is due to the following reasons:

- Infants are generally born with high Hb level of 18-22 g/100 ml of blood. Iron stores in liver of the infant lasts from 3 to 6 months. Iron is also required for growth of foetus and placenta. To achieve these levels, a mother needs to transfer 240 mg of iron to the foetus during gestation.
- It is also required for the formation of haemoglobin as there is 40-50% increased maternal blood volume, for which, 400 mg of iron is required.
- During entire pregnancy, the total loss of maternal iron through skin and sweat is about 170 mg.

The total iron requirement for the entire period of pregnancy is 810 mg. The actual increase in requirement of iron is only in second and third trimester but a single large dose of iron at the beginning of pregnancy is an effective way to build up iron stores and to protect against depletion of reserves. To avoid iron deficiency, a woman should enter pregnancy with a store of at least 300 mg of iron.

Liver, dried beans, dried fruits, green leafy vegetables, eggs, enriched cereals and iron-fortified salt can provide additional sources of iron.

Sodium: The normal adult woman's requirement of sodium should be maintained to prevent any defective disorders and deficiency. During pregnancy, there is an increase in extra cellular fluid, which calls for an 80% increase in body sodium. Restriction in the diet can cause a severe hormonal and biochemical changes. When blood sodium level drops, kidney produces the hormone rennin, as a result of which, the sodium, which is needed for use by the body, is retained.

Sodium deficiency in pregnant women can cause an increased risk of eclampsia, prematurity and low birth weight of infants. The use of diuretics should be discouraged during pregnancy as they can cause loss of sodium, which is restricted when there is oedema or hypertension.

Iodine: ICMR recommends additional requirement of 25 μg during pregnancy to the adult requirement of 100-200 μg. Iodine deficiency in mother can lead to abortion, stillbirths, congenital anomalies, increased prenatal mortality, cretinism and psychomotor defects. Mothers who are residing in goiter endemic areas should ensure that they get enough iodine through iodized salt and other means.

Zinc: Zinc deficiency during antenatal period leads to adverse effects on the newly born including foetal mortality, foetal malformations and reduced intra-uterine growth rate. Low zinc during pregnancy doubles the risk of low birth weight and trebles the risk of preterm delivery.

Fat-soluble vitamins

Vitamin A: A normal requirement of β-carotene for an adult woman is 2400 μg per day. Same level is suggested during pregnancy. Vitamin A requirement during pregnancy has been calculated based on the vitamin content of livers of the newly born. Additional intake of vitamin A required for this purpose is about 25 μg per day throughout the pregnancy. Since this constitutes a small fraction of the recommended allowance for normal women, no additional dietary allowance during pregnancy is suggested. Liver, egg yolk, butter, dark green and yellow vegetables and fruits are good source of vitamin A.

Vitamin D: Vitamin D is highly essential as it enhances the maternal calcium absorption. Its active forms, *i.e.*, calcidiol and calcitriol, cross the placenta with ease and play an important role in calcium metabolism of the foetus. Maternal deficiency of vitamin D results in neonatal hypocalcaemia and hypoplasia.

Excessive vitamin D can also result in complications such as atherosclerosis, hypercaleaemia, calcium deposits in various vital organs and mental retardation in the infants.

Vitamin E: In animals, vitamin E has an important role to play in the reproductive process and reduces the number of spontaneous abortions and stillbirths. In women, it has not been evidently. If vitamin E deficiency can cause pregnancy, failure although may prove beneficial to those who have had repeated spontaneous abortion or failure to conceive. Very little vitamin E crosses the placenta, thus, infant has low tissue concentrations that persist up to at least 6 years. Requirement of vitamin E increases with increased intake of polyunsaturated fatty acids (PUFA).

Vitamin K: Vitamin K is essential for the synthesis of prothrombin, which is necessary for normal coagulation of blood. It is highly essential for preventing neonatal haemorrahge.

Water-soluble Vitamins

Because little of water-soluble vitamins are stored, the pregnant woman must rely on her daily intakes that are high enough to meet the added requirements of pregnancy. Generally, maternal blood levels for water-soluble vitamins tend to fall and foetal blood levels tend to exceed those of mother by 50 to 100% during pregnancy.

Thiamin: According to ICMR recommendation, the thiamin requirement for a pregnant woman is increased by 0.2 mg/day. The relationship between thiamin and caloric intake remains the same during pregnancy, *i.e.*, for 1000 kcal, 0.5 mg of thiamin is required. The normal urinary excretion of thiamine drops indicating that more is being retained and used by tissues. In some cases, thiamin helps to relieve the nausea of pregnancy.

Riboflavin: According to ICMR recommendation, the riboflavin requirement for a pregnant woman is increased by 0.2 mg/day. This requirement is increased due to the increase in maternal body size and growth of foetus and accessory tissues. Riboflavin is present in higher amounts in the foetal blood than in maternal blood. Animal studies have shown that the lack of riboflavin interfere with cartilage formation resulting in skeletal malformations such as shortening of long bones and fusion of ribs.

Niacin: As per ICMR recommendation, the niacin requirement for a pregnant woman is increased by 2 mg per day, corresponding to the increase in calories. The conversion of amino acid into niacin is more efficient during pregnancy.

Vitamin B_6: ICMR suggested that the requirement of vitamin B_6 for pregnant woman is 2.5 mg per day. In hypertrophy, when the cells are growing in size, the vitamin B_6 and amino acids are required. During *toxaemia for pregnancy* (high blood pressure, oedema and proteinuria), the vitamin B_6 levels are lower.

Folic acid: ICMR recommendation for requirement of folic acid during pregnancy is 400 mcg per day. The recommended intake of folacin is based on its role in promoting normal foetal growth and preventing macrocytic anaemia of pregnancy. Folic acid is needed for the synthesis of essential components of DNA and RNA, which increase rapidly during growth, thereby, increasing the requirements. In addition, folacin is essential for the development of RBCs, which must increase as the mother's blood volume increases.

Deficiency of folic acid may be associated with spontaneous abortion and obstetric complications such as preterm labour and low birth weight. The greatest significance of folic acid and its potential influence on pregnancy outcome is its role in preventing neural tube defects, such as spina bifida. In this condition, a congenital defect in which the vertebral neural arches fail to close is developed, hence, exposing the contents of spinal canal posteriorly. Folic acid deficiency can lead to anencepthly (absence of brain). Folic acid deficiency has also been implicated in pregnancy-induced hypertension.

Women of childbearing age should be encouraged to include generous amounts of folic acid and sources of folic acid in their diets such as dark green leafy vegetables, legumes, orange juice, soybean, wheat germ, almonds and peanuts.

Vitamin B_{12}: The requirement of vitamin B_{12} remains the same even during pregnancy. The foetus has priority over the mother in vitamin B_{12} and foetal blood had twice the amount of vitamin B_{12} than the maternal blood, even when maternal levels are depleted. Low maternal levels are associated with prematurity. The capacity of a woman to absorb vitamin B_{12} is increased during pregnancy and a large amount is transferred to the foetus. Vegetarian mothers have more chances of getting vitamin B_{12} deficiency. Serum levels drop during pregnancy and return to normal without supplementation after delivery. Based on the vitamin B_{12} content of foetuses, it has been estimated that foetal demands may be of the order of 0.3 μg vitamin B_{12} per day.

Vitamin C: ICMR recommendations of vitamin C during pregnancy is 40 mg per day same as normal woman requirement. The foetal requirement is too small and sufficient safety margin is given for an adult woman. Extra allowance may not be necessary during pregnancy. In pregnancy, vitamin C can fully cross the placental barrier. The vitamin C content of foetal blood is thrice as much as maternal blood. There is some evidence to show that the placenta can synthesize vitamin C. This may account for higher levels in foetal tissue. Low maternal intake of vitamin C is associated with premature rupture of foetal membranes and increased neonatal death rates.

Diet and Feeding Pattern

The basic meal planning principles remain the same as for normal adults for planning meals for pregnant women. To meet the increased requirements during pregnancy, she needs to consume extra food and follow following dietary guidelines:

- Small and frequent feeding should be taken.
- Green leafy vegetables should be included in diet to ensure more calcium and iron.
- Diet rich in calcium should be taken to prevent osteomalacia.
- Iron rich food should be included in diet to prevent common problem of anaemia.
- Plenty of water should be taken to keep the bowels regular.
- Fluid should be taken between meals rather than along with meals.
- More fibre using whole cereals, pulses, green leafy vegetables and fruits need to be taken.
- Adequate amount of protein and energy by taking more amount of cereals, pulses, milk and milk products nuts and oilseeds, should be encouraged.
- Spicy and fried foodstuffs should be avoided.

A Sample Menu

Meal time	Foodstuffs
Breakfast	Missi roti
	Butter/curd/ butter milk/milk

Meal time	Foodstuffs
	Guava/apple
Mid morning	Sprouted chat
	Lemon water
Lunch	Salad
	Pea and cheese curry (any vegetable curry)
	Potato and fenugreek vegetable (any dry vegetable)
	Curd/raita
	Chapatti/rice
Evening tea	Banana shake
	Poha
Dinner	Salad
	Green gram (any pulse)
	Mix vegetable (any dry vegetable)
	Chapatti/rice
	Fruit custard

Alleviating Maternal Discomfort

Nausea, constipation and heartburn, which need to be alleviated, are very common during pregnancy.

Nausea: Not all women have uneasy stomachs in early months of pregnancy but many women have uneasiness. The nausea of *morning* (actually, any time) sickness ranges from mild uneasiness to debilitating nausea and vomiting. The hormonal changes early in pregnancy seem to be responsible for the sensitivities to the appearance, texture, or smell of food articles. The traditional strategies for alleviating nausea are listed in Table but some women benefit most from simply eating the foods they want when they feel like eating. They may also find comfort in a cleaner, quieter, and more temperate environment.

Constipation and hemorrhoids: As the hormones of pregnancy after muscle tone and the growing foetus crowds' intestinal organs, an expectant mother may experience constipation. She may also develop hemorrhoids (swollen veins of the rectum). These can be painful, and straining during bowel movements may cause bleeding.

Heartburn: Heartburn is another common complaint during pregnancy. As the growing foetus puts increasing pressure on the mother's stomach, acid may back up into the lower esophagus and create a burning sensation near the heart. Heartburn may be especially troublesome at night, when the woman is lying down; she may find relief by sleeping with her torso slightly raised.

A few simple strategies can help in averting the maternal discomfort.

To alleviate the nausea of pregnancy	To prevent or alleviate constipation	To prevent or relieve heartburn
On waking, arise slowly.	Eat foods high in fiber (fruits, vegetables and whole-grain cereals).	Relax and eat slowly and chew the food thoroughly.
Eat dry toast or crackers.	Exercise regularly.	Eat small and frequent meals.
Chew gum or suck hard candies.	Drink at least eight glasses of liquids a day.	Drink liquids between meals.
Eat small, frequent meals.	Respond promptly to the urge to defecate.	Avoid taking spicy or greasy foodstuffs.
Avoid foodstuffs with offensive odours.	Use laxatives only as prescribed by a physician.	Sit up while eating and elevate the head while sleeping.
When nauseated, drink no citrus juice, water, milk, coffee, or tea.	Do not use mineral oil since it interferes with absorption of fat-soluble vitamins.	Wait an hour after eating before lying down and wait two hours after eating before exercising.

Fitness during Pregnancy

Today, it is usual to see a pregnant women jogging, swimming, participating in aerobic dance classes, walking energetically, or doing other exercises with the same vigour as their non-pregnant peers. An active and physically fit woman experiencing a normal pregnancy can continue to exercise throughout pregnancy, adjusting the intensity and during as the pregnancy progresses.

Staying active throughout pregnancy can improve fitness, prevent gestational diabetes, facilitate labour and reduce stress. Women who exercise during pregnancy report fewer discomforts throughout their pregnancies. Regular exercise develops the strength and endurance a woman needs to carry the extra weight gain through pregnancy and to labour through an intense delivery. It also maintains the habit that helps a woman lose excess weight and get back into shape after the birth.

A pregnant woman should avoid sports in which she might fall or be hit be other people or objects. Swimming is ideal since it allows the body to remain cool and move freely with the water's support. Following guidelines for exercise can be followed during pregnancy:

- Do regular exercise (at least three times a week) and avoid spurts of heavy exercise followed by long periods of no activity.
- Avoid brisk exercise in hot and humid weather or when sick.
- Avoid jerky, bouncy, or high-impact motions. Activities, which call for jumping, jarring motions and quick changes in direction, may strain joints and cause pain. Low-impact exercise is best. A wooden floor or a tightly carpeted surface reduces impact and gives sure footing.
- Wear the proper shoes for the activity to be sure that your feet are well cushioned and to give the body good support. There are shoes designed just for walking, running, aerobics, or tennis.
- Avoid deep knee bends, full sit-ups, double leg raises and straight-leg toe touches.
- After 20 weeks of pregnancy, avoid exercises that require lying with the back on the floor for more than a few minutes.

- Always begin with 5 minutes of slow walking or stationary cycling with low resistance to warm up your muscles. Intense exercise should not last longer than 15 minutes.
- Follow intense exercise with 5 to 10 minutes of gradually slower activity that ends with gentle stretching in place. To reduce the risk of injuring the tissue connecting your joints, do not stretch as far as you possibly can.
- The extra weight you are carrying will make you work harder as you exercise at a slower pace. Measure your heart rate at peak times of activity.
- Get up from the floor slowly and gradually to avoid feeling dizzy or fainting and once standing, walk in place briefly.
- Drink water often before, during and after exercise to be sure that your body gets enough fluids. Take a break in your workout to drink more water, if necessary.
- If you did not exercise two to three times a week before getting pregnant, you should start with physical activity of very low intensity and move gradually to higher levels.
- Stop exercising and consult your doctor or nurse if you get any of these symptoms, including pain, vaginal bleeding, dizziness, feeling faint, shortness of breath, irregular or rapid heartbeat, difficulty in walking, pain in back, or pubic area.

All women should be advised about importance of eating a healthy diet during pregnancy. Some women may need intensive instruction and perhaps even supplements to correct for any inadequacies revealed through nutrition assessment. However, with small care and efforts, this normal physiological condition may remain normal.

Lactation

The period of lactation is a time of even greater nutritional stress for the mother than that of pregnancy. Adequate nutrition of the mother during lactation is also of vital importance since during first 6-9 months of life, the infant derives all his nutrition from the mother's milk generally. Any inadequacies in her diet influence both the quantity and quality of milk secreted, although the effect on quantity is more pronounced. A well nourishes mother on an average secretes about 850 ml of milk per day, whereas, it may go down to even 400 ml per day in severely malnourished mothers. As far as the quality is concerned, the mother has an excellent ability to breast feed the baby at the cost of her own health. Dietary deficiencies of water-soluble vitamins like ascorbic acid and B group vitamins lead to lower level of these vitamins in breast milk. The protein, carbohydrates and calcium content is not altered even when mother is malnourished. Therefore, it is important to provide adequate nutrition during lactation for proper quantity and quality of milk for infant and is able to maintain her own health and nutritional status.

Successful lactations influence by a number of factors such as neuroendocrine, psychological, social and nutritional factors. Suckling by the infant and the continuous emptying of breast provide the necessary neuroendocrine stimulation, which is important to maintain. The psychological factors, such as fear anxiety worry, grief, or anger, may retard milk secretion, while mother's willingness to feed the baby or even the cry or thought of the baby may lead to milk ejection. The effect of varied social factor in the modern times affects the lactation as many young mothers feel that breast-feeding may affect their fugues adversely, but in reality, a women feeds her baby properly help her to regain the normal posture easily if good food habits are followed. The overall importance of good nutrition in promoting lactation is must along with other factors.

Physiology of Lactation

The foundation for lactation is laid down with the enlargement during adolescence, adulthood and their further development during pregnancy, while actual milk secretion commences only after childbirth and it is controlled by several hormones.

The mammary glands stimulated by estrogen during puberty develop a system, *i.e.*, a ducts, lobes and alveoli. This system remains fairly inactive until pregnancy. During pregnancy, the hormones stimulate a proliferate stage of mammary gland activity. Estrogen promotes growth and branching of duct system, while progesterone stimulates the development of alveoli. During pregnancy, the elevated concentration of these hormone support mammary gland development while inhibiting the actual secretion of milk. These two hormones return to their basal concentration within few days of postpartum, thus, allowing the milk production to begin, and finally, prolactin and oxytocin coordinate lactation. Prolactin is responsible for milk production and inhibition, whereas, ovulation oxytocin stimulates milk ejection and promotes uterine contracting.

Nutrient Needs and the Recommended Dietary Allowances during Lactation

ICMR in 1990 has given the recommendations for a lactating mother for various nutrients based on the composition of breast milk and the fact that a mother generally secretes 600 to 850 ml of milk daily. However, milk secretion continues to increase in the early periods of lactation, reaching a maximum by about six months.

Energy

Lactating mothers need additional energy for the production of milk, which is computed from the volume of milk secreted, its energy content and the efficiency of conversion of food energy into milk energy. Based on those factors, the optimal milk output is generally 850 ml with conversion efficiency of 80%. Additional energy allowance recommended by ICMR during the first 6 months of lactation is 500 kcal per day. An extra allowance of 400 kcal per day is recommended for the period from 6 to 12 months of lactation as milk output is reduced after six months.

Protein

For the production of milk, protein needs also increase, as the milk contains 1.15 g of protein per 100 ml. The mother secretes on an average 850 ml of milk during the first 6 months and about 600 ml of milk during 6 to 12 months. This corresponds to a level of 9.4 g protein per day during 0 to 6 months and 6.6 g of protein per day in the next 6 months of lactation. Assuming the efficiency of conversion of dietary protein into milk protein to be 70% and taking an extra allowance of 25% for individual variation. The safe daily intake would be 16.8 and 12.0 g protein during 0 to 6 and 6 to 12 months period of lactation, respectively. In terms of dietary protein of NPU 65, the recommended additional protein during lactation is 25 g for 0 to 6 months and 18 g for 6 to 12 months. For proper milk production, adequate amounts of good quality protein or complete protein also need to be included in the diet.

Fat and Essential Fatty acids

The ICMR Expert Committee has suggested an intake of 45 g of visible fat per day during lactation.

Minerals

Calcium: Additional calcium is required for breast milk secretion, which in case of Indian mothers contains about 30 to 40 mg calcium per 100 ml milk. Based on an optimal output of 850 ml milk, the total calcium secreted through milk is about 300 mg per day. Therefore, in lactation, an additional intake of calcium is essential to enable the retention of 300 mg of extra calcium. A daily intake of 1 g calcium, which covers the total calcium needs of the lactating mother, *i.e.*, her own calcium needs as well as the extra calcium required for milk production, is recommended by the IMCR.

Iron: Iron requirement during lactation is the sum of the requirement of mother and that required for making up the iron secreted in breast milk. Since most of the mothers have lactational amenorrhoea, resulting in a saving of nearly 1 mg of iron per day or 16 μg iron per kg body weight per day, which would otherwise have been lost in the menstrual blood. This amount is more than sufficient to compensate the iron secreted in the mother's milk. Therefore, the total iron requirement is the same as for a non-pregnant non-lactating adult female, *i.e.*, 30 μg/kg (14 μg/kg basal iron need plus 16 μg iron per kg for secreting milk). Thus, for a woman, whose body weight is 50 kg, the total daily requirement of iron is 1.5 mg per day. Taking the dietary iron absorption during lactation to be 5% on a mixed cereal diet, the recommended dietary iron intake is 30 mg per day (ICMR, 1990).

Vitamins

Vitamin A: Since milk is a rich source of vitamin A (retinol), a lactating mother needs ample amount of this vitamin. The additional need during lactation is calculated based on vitamin A secreted in milk. The average vitamin A secreted in mother's milk is 350 μg of retinol per day. Considering this additional amount, a level of 950 μg of retinol or 3800 μg of β-carotene per day for a lactating mother up to 1 year of lactation has been recommended.

Thiamine: To compute the thiamine requirement for lactating mothers, studies have been conducted to assess the amount of thiamine secreted in mother's milk. The maximum concentration of thiamine, which can be achieved in the breast milk by providing supplements of the vitamin, has been reported to be 20 μg/100 ml. Thus, the total amount of thiamine secreted would be 0.14 mg/day (considering 700 ml of milk secretion daily). However, when the thiamine requirements are computed based on increased energy needs of lactation, *i.e.*, additional 550 and 400 kcal/day for one year of lactation, the calculated value is more than the amount of thiamine secreted in the mother's milk. The same level of 0.5 mg thiamine per 1000 kcal has been retained for lactating mothers also. The absolute amount of thiamine needs will increase with the increased demand of energy.

Riboflavin: The amount of riboflavin computed for the increased energy allowance would be more than sufficient to cover the amount of riboflavin secreted in the mother's milk.

Table 2: Recommended dietary allowances of various nutrients during lactation (0 to 12 months)

Nutrients	Lactation period 0-6 months			Lactation period 6-12 months		
	Sedentary worker	Moderate worker	Heavy worker	Sedentary worker	Moderate worker	Heavy worker
Energy (kcal)	2425	2775	3475	2275	2625	3325
Protein (g)	75	75	75	68	68	68

Nutrients	Lactation period 0-6 months			Lactation period 6-12 months		
	Sedentary worker	Moderate worker	Heavy worker	Sedentary worker	Moderate worker	Heavy worker
Fat (g)	45	45	45	45	45	45
Calcium (mg)	1000	1000	1000	1000	1000	1000
Iron (mg)	30	30	30	30	30	30
Vitamin A						
Retinol (μg)	950	950	950	950	950	950
β-carotene (μg)	3800	3800	3800	3800	3800	3800
Thiamine (mg)	1.2	1.4	1.5	1.1	1.3	1.4
Riboflavin (mg)	1.4	1.6	1.8	1.3	1.5	1.7
Niacin (mg)	16	18	20	15	17	19
Pyridoxine (mg)	2.5	2.5	2.5	2.5	2.5	2.5
Folic acid (μg)	150	150	150	150	150	150
Vitamin B_{12} (μg)	1.5	1.5	1.5	1.5	1.5	1.5
Ascorbic acid (mg)	80	80	80	80	80	80

Source: ICMR (1990)

Niacin: The amount of niacin secreted through mother's milk has been reported to be 100 to 150 μg/100 ml, *i.e.*, 0.9 to 1.2 mg niacin daily. The requirement of niacin also depends upon energy needs and for the increased energy needs. The amount computed is more than sufficient to cover the amount of vitamin secreted in milk. Therefore, ICMR (1990) has recommended 6.6 mg niacin equivalents per 1000 kcal during lactation.

Pyrodoxine: Pyrodoxine requirement is also increased during lactation, but in the absence of any specific studies, ICMR (1990) has recommended for a lactating mother the same level as that for a pregnant woman, *i.e.*, 2.5 mg of pyrodoxine per day.

Folic acid: Folic acid secreted in the breast milk by a mother is about 25 μg a day. Based on this, the ICMR (1990) has recommended an additional allowance of 50 μg of free folate per day during this period, *i.e.*, a total of 150 μg free folate per day.

Cynocobalamin (vitamin B_{12}): The amount of vitamin B_{12} (cynocobalamine) secreted in breast milk per day is between 0.25-0.3 μg. An additional intake of 0.5 μg of vitamin B_{12} per day, *i.e.*, 1.5 μg of the vitamin per day would thus meet the needs during lactation.

Ascorbic acid (vitamin C): As appreciable amount of ascorbic acid is secreted in the mother's milk, additional need during lactation is calculated based on vitamin C content in milk. Studies have shown that Indian women secrete about 20 mg vitamin C per day during lactation.

Taking into consideration the cooking losses on an average to be 50%, the ICMR has recommended an additional intake of 40 mg ascorbic acid per day during lactation, *i.e.*, a total of 80 mg of ascorbic acid per day.

Diet and Feeding Pattern

In order that a mother may breast-feed her baby without any undue strain on her own body reserves, she must continue to eat a well balanced diet to meet the enhanced nutritional requirement of lactation. The choice of food items is wide during lactation. No food is restricted except highly spiced and strongly flavoured. Special food items especially panjiri, ladoo prepared by using dry fruits, cumin seeds, ajwain, cereal flour and ghee are recommended for increasing the milk production. Additional amount of raw and cooked fruits, vegetables, whole grains and an adequate amount of water/fluid is also recommended in frequent and small feeds. A lactating mother should consume more milk and milk products, dry beans and groundnuts daily to meet her additional protein requirement. Liberal amount of yellow and green leafy vegetables and fruits will meet the requirement of vitamin A, C, iron and calcium.

Sample Menu

Meal time	Foodstuffs
Breakfast	Missi roti/Paushitik dalia
	Milk/curd
	Seasonal fruit (guava/apple)
Mid morning	Panjiri
	Tea
Lunch	Crushed/grated fresh vegetable salad
	Cowpea curry
	Seasonal vegetable (cabbage and peas)
	Raita (Mint)
	Chapatti/rice
	Kheer/custard
Evening tea	Ladoo
	Milk
Dinner	Tomato soup
	Green gram dhal
	Potato and brinjal/potato spinach vegetable
	Chapati/rice
Bed time	Milk

Things to be avoided for Breastfeeding Mother

Mothers who eat junk food during pregnancy and breastfeeding may be putting their children at risk of overeating and developing obesity.

The research suggests that pregnant and breastfeeding women should not indulge in fatty, sugary and salty foodstuffs under the misguided assumption that they are *eating for two.* Bakery products, confectionery and snacks are a major source of trans-fatty acids in maternal diet if consumed in excess. The levels of trans-fatty acids in human milk may reflect the current diet of mother as well as the diet consumed early in pregnancy.

Always consult with your doctor before taking any medications. Most medications pass into your milk in small amounts. If you take medication for a chronic condition such as high blood pressure diabetes or asthma, then also make the decision with the help of your doctor.

Beware of pollutants like nicotine and pesticide residues, which easily pass through mother's milk. If you are nursing, stay away from insecticides (especially in airborne forms such as aerosols or coils). Try to use natural insect repellents such as citronella. Eat primarily unsaturated fats. Sunflower, corn, rapeseed and olive oil provide fatty acids that are essential for building the baby's nervous system.

■ ■ ■

28 Development through Infancy and Childhood

Shanti Balda[1], Usha Turan[2] and Veena Shahi[3]

[1]Human Development and Family Studies, I.C. College of Home Science, CCS Haryana Agricultural University, Hisar-125004, Haryana, India,
E-mail: <sbalda@rediffmail.com>
[2]CiS Kanya Mahavidhyalya, Fatehpur Pundri, Kaithal, Haryana, India,
E-mail: <ushaturan@gmail.com>
[2]Krishi Vigyan Kendra, Vaishaili, Bihar, India
E-mail: <Veenashahi07@gmail.com>

This chapter begins with the principles of growth and development and factors affecting growth and development. Then development of children from infancy through late childhood will be discussed. Different developmental aspects discussed are physical, language related, social, emotional, cognitive and personality and moral development. It is followed by hazards in life span from infancy through childhood and roles and responsibilities of caregivers. Issues and challenges during infancy and childhood, and finally, discipline and child outcomes will be explained.

Principles of Growth and Development

Growth and development are complementary processes but these two terms cannot be used interchangeably. Growth indicates the quantitative changes in the body such as height and weight. In contrast, development refers to the qualitative changes, *e.g.*, acquisition of language. Growth and developmental changes follow certain principles. Some of these principles are as follows:

- Growth and development follow an orderly sequence and are continuous. Development is continuous from the moment of conception to death. Since development is continuous, what happens at one stage carries over into and influences the next stages. If one stage is delayed, it may delay the next stages.
- Development either physical or mental is not a uniform process. It is most rapid during the 9 months before birth (one cell to 9 pounds or 20 inches). Rapid growth continues until the age of 3 years, then slightly slower rate and again spurt during adolescence.
- Each child with its own essential characteristics normally passes through a number of stages. Different stages of human life span are prenatal period (conception to birth), infancy (birth to 2 years), early childhood (2 to 6 years), late childhood (6 to 10 years), puberty (10 to 12 years), adolescence (12 to 19 years), adulthood (19 to 40 years), middle age (40 to 60 years) and old age (60 years and above).

- There are individual differences in rate and pattern of development. Some develop in a smooth, gradual and systematic fashion, while others move in spurts. Individual differences are influenced by both internal and external conditions. Hereditary, food, general health, sunlight, fresh air, climate, emotions and stimulation, opportunities, facilities, encouragement, attitudes and social relationships all together create one child different from other.
- Although the human being develops as a unified whole, yet each part of the body develops at a different rate. Development follows two laws of the *directional sequence, i.e.*, the *cephalocaudal law* and the *proximodistal law*. According to first law, the development spreads over the body from head to foot. This means that improvements in structure and function come first in the head region. According to second law, development proceeds from near to far extremities. In fetus, the head and trunk are fairly well developed before the limb buds appear. Gradually, the arm buds lengthen and then develop into the hands and fingers. Functionally, the baby can use his arms before his hands and can use his hands as a unit before he can control the movements of his fingers.
- The patterns of physical and mental development show certain common and predictable characteristics. All children follow a similar developmental pattern, with one stage leading to the next. The baby stands before he/she walks and draws a circle before a square. In mental as well as motor responses, the general activity always precedes specific activity. Child moves whole body first then specific parts and uses hand first then fingers.
- Early development is more important than later development. Early development is like foundation house. Good physical and mental potentials can be seriously damaged by unfavourable environmental conditions during prenatal and early postnatal life. Later such conditions are less damaging. Many maladjusted people have history of maladjusted childhood.
- Development is essentially the result of an interaction between maturation and learning. Maturation is the unfolding of traits potentially present in the individual's genetic endowment. Learning refers to development that comes from exercise and effort. Development during the prenatal period comes mainly from maturation and is very little dependent upon activity. Postnatal maturation and learning are closely interrelated, and one influences the other. Development thus depends on the interaction of this hereditary endowment and the social and cultural forces of the environment. The biological determinants serve the physiological limits and the floor of potential in the organism. On the other hand, the socio-cultural factors are like soil in which the plant is nurtured. By enrichment or impoverishment, human behaviour potential can be made to blossom, or wither to achieve the limits or to fall short of them.

Factors Affecting Growth and Development

Heredity: Heredity means what we get from our ancestors. Heredity determines how tall or heavy one can be. Heredity determines our body-built and intellectual capacity, as well as many other physical, mental and psycho-social behaviour traits.

Prenatal environment: The environment during pregnancy is an important factor for prenatal and postnatal growth and development. If the mother is getting poor nutrition or emotionally upset or smokes, drinks, takes some medicine, or suffers from certain diseases, the growth of the child can be affected adversely.

Nutrition: Good nutrition is essential for proper development of a child. Malnutrition results in stunted growth.

Intelligence: Higher intelligence is associated with faster development, while lower intelligence is associated with retardation in different areas of development. Body and brain are closely associated as *healthy mind in a healthy body.*

Family environment: Parent-child relationship and parent-parent relationship has significant impact on family environment and development of children. A healthy parent-parent relationship is likely to result in healthy parent-child relationship. In a family where there is healthy or positive relation between both the parents and between parents and their children, the atmosphere in the family is happier than those families where such relationships do not exist. If there is a lot of conflict at home or the child is not given enough love and attention or if there is child abuse, then the child's development is affected adversely. The affectionate, tolerant, or respectful attitude towards each other in the family has a positive impact on children's development.

Health of the child: If the child is sick or suffers from some disorder, or is disabled or has disturbed physiological and endocrine functioning, his/her development is likely to suffer.

Stimulation and motivation: Stimulation means anything, which compels the person to act. Motivation is stimulation that comes from external and internal forces. External forces include parents, family, society and environment. Internal forces include urge of the individual to learn. Stimulation to the development of inborn abilities, which usually comes from the environment, is also very essential. The amount of stimulation an environment provides the opportunities for exploration of environment, opportunities of interaction with other people- all influence the rate of development. A child cannot learn until he/she is ready to learn, meaning that the necessary physical and mental foundations must be present before the new abilities can be built in them. If the child is not ready to learn, stimulation and teaching may be just wastage of time and effort. On the other hand, if the child is mature enough to learn but not permitted to, he/she may lose interest. Maturational readiness is the teachable moment when the child is physically and mentally mature, and society expects the child to learn certain developmental tasks, and the child him/herself is ready to achieve those tasks, then it is said that teachable moment has come. When a child is ready to learn, his/her interest will persist even when he/she meets obstacles and encounters setbacks. With practice, the child who is ready to learn will show improvement even if only slight and at a slow rate.

Socio-economic status: Socio-economic status of the family means social reputation and the financial conditions of the family. It determines the type of nutrition, stimulation, facilities and opportunities the child gets, and thus, affects the rate of his/her development.

Gender of the child: All children follow the same sequence of development. However, certain skills develop faster in girls than boys and *vice-versa*. Gender is also a factor that sometime decides the potential of the child in some aspects of development.

Development of Children from Infancy through Childhood

Development during Infancy: Infancy occupies first two years of life. Period of infancy is divided in two subdivisions, *i.e.*, period of *neonate* and period of *infancy* or *babyhood*. The period of neonate lasts from birth until approximately the end of 2nd week of life. Neonate period is *shortest* of all developmental periods. It is a time of adjustment to postnatal life and is a hazards period. The neonate must make quick adjustment to the postnatal life for survival. He/she should be able to suck, swallow, breathe, eliminate and adjust to temperature changes.

During first year of life, the infants are called as babies since they are quiet helpless and completely dependent on others. During second year of life, the infants are commonly called as *toddlers* as they have achieved enough body control. Babyhood is characterized by rapid growth and change, decreasing dependency and increased individuality. It is the age of socialization beginning. Infancy is considered as a hazardous age.

During infancy stage, the most infants learn to walk, talk, solve problems and relate to others. One major task for the infants during second year of life is to learn to be independent, that is why, they want to do things for themselves, have their own ideas about how things should happen and use *no* many times in a day.

During second year of children's life, the parental demands need to fit with the capacities of children, *e.g.*, they should not impose many *don'ts* on infants. Instead, they should put away breakable objects, place barriers across steep staircases and physically take away babies when they act in ways that is risky to them or disturb others. Different areas of development during infancy as described in the following section:

- ***Physical and motor development*:** During first six months after birth, the infants grow at a rapid rate and then at a slower rate. Pattern of growth is similar for boys and girls. On an average, a child begins walking between the age of 9 and 14 months. At two years, the weight is about 11-13 kg and height is about 32-35 inches. By the age of two and a half years, the child has full set of 20 temporary teeth, can run, climb upstairs, helps with dressing and undressing and tries to feed self with a spoon. Infants have big heads, protruding abdomens, small limbs and little hands and feet. Growth in height and weight results in changes in body proportions.

  Development of control over motor activities and muscles follow the *laws of developmental direction* as described earlier. According to these laws, control over body spreads from head to foot and from centre to periphery. Infants who sit early also walk earlier as compared to those who learn to sit later. Handedness also develops during infancy. During second year of life, the children show preference for one hand, preferable right hand, meaning they are dominantly right-handed. However, some children tend to be left-handed.

- ***Language development*:** The infant's first sentences consist of one word accompanied by gesture. Comprehension of words develops earlier than production. Infants can follow directions given by adults and show that they understand the meaning of words. An infant speaks first word around 12 months. The infant will then continue to add steadily to his or her vocabulary until around the age of 18 months when language increases rapidly. He or she may learn as many as 7-9 new words a day. Around this time, the toddlers generally know about 50 words. Infants can convey their ideas and emotional expressions through these one-word utterances called *holophrastic speech*, *e.g.*, pointing towards a ball, a child may say *ball.* The child may mean to say *I need ball.*

  By the age of 2 years, speaking vocabularies reach 50 words. Initially, the children put the two words together, *i.e.*, *see mummy*, *more water*, *etc.* This is called *telegraphic speech*, and like a telegram, the children speak only essential words. During infancy, young children are able to understand many more words than they can speak.

- ***Social development*:** Mothers fulfill demands and needs of young infants and they are centre of attention and affection of their mothers. Hence, the infants develop strong emotional ties with their mothers. The infants show a desire to become a part of group by protesting when left alone and by attracting attention of others. This attachment behaviour lays foundation for warm and lasting relationships with others. They imitate social behaviour such as hugging a toy and feeding a doll.

Early social experiences play an important role in determining infant's future relationships and patterns of behaviour toward others. Infant's life is centered around home, which is during this period that the foundations for later social behaviour are laid. At birth, the infants have no interest in people. Around the age of one and a half month, a social smile- a smile in response to a person- appears. This is the beginning of socialization. The social behaviour of infant with adults is different from that of social behaviour with other babies.

- ***Emotional development*:** Emotions are conscious processes that involve bodily changes and arise due to many causes. At birth, the emotions of infant appear in simple and undifferentiated form. With growing age, the emotional responses become more differentiated and can be aroused by a wide variety of stimuli. Characteristics of an infant's emotions are (i) they differ from those of older children and adults as their emotions are intense, (ii) the emotions are brief, transitory and appear frequently and (iii) during infancy, the emotions can be easily conditioned.

  The common emotional patterns of infancy are anger, fear, curiosity, joy and affection. Interference with infants' movements, not letting them to do what they want and not fulfilling their wishes are some of the stimuli that give rise to anger. To express anger, the infants scream, kick their legs, wave their arms and hit, and kick anything within the reach. During second year, the infants may also jump up and down, throw themselves on the floor and hold their breath, that is why, infancy is sometimes referred to as the *terrible twos* because of the temper outbursts for which they are famous. Toddlers tend to have temper tantrums because they have such strong emotions but do not know how to express themselves. Loud noise, strange persons, objects and situations, dark rooms, high places and animals are common causes to arouse fear in infants. Physical well being gives rise to joy. The infants react to being played with, being tickled and watching or listening to others. Joy is expressed by smiling, laughing, moving arms and legs. Bodily movements are intensified and infants coo and gurgle, and wave their arms. Caregivers and those who show them affection act as a stimulus for their affection. Toys may also become love objects for infants. They express their affection by hugging, patting and kissing. Anything new acts as a stimulus to make the child curious. In the beginning, newness may give rise to fear. Later, the fear is replaced by curiosity.

- ***Cognitive development*:** In beginning, the infants perceive things through sensory exploration. They look at, listen to, touch, smell and taste anything they can. As they grow older and are able to handle things, they acquire meanings through handling. This is *sensorimotor* stage in conceptual development (birth to 2 years). During this stage, babies think exclusively with their senses and motor skills and express this learning mainly through senses and motor skills. It is still difficult for them to distinguish between living and nonliving objects. They believe that all objects are living beings. Babies perceive small objects as light and a large object as heavy. Between the age of 6 and 24 months, infants begin to respond to different colours.

- ***Personality development*:** Personality includes those stable psychological characteristics that define each human being as unique. Both children and adults have personality traits such as *temperament*- a long-term characteristic and *states* such as moodiness a changeable characteristic. Personality traits remain relatively stable and therefore predictable over time.

  Infancy is a critical period in the development of personality, and the potentials of personality are present at birth. Personality is shaped by interplay of temperament and environment. During infancy, a child's environment is limited exclusively to home, and child's mother is the most continuous attendant, the kind of person she is and kind of relation the infant has with his/her mother will have profound influence on infant's personality development. Shyness tends to become stronger rather than weaker with age. Self-concept is the core of the personality and it remains the same. With the increase in age, it becomes less and less flexible. Thus, early experiences are very important in personality development.

- ***Moral development*:** Morality is the ability of a child to learn the difference between rights and wrong and to understand how to make the right choices. Infants have no conscience. Their behaviour is *non-moral,* as it is not guided by moral standards. Learning to behave in a morally approved behaviour is a long process. Infants learn right and wrong from the act in terms of happiness and pain it gives them.
- ***Development of play interests*:** Play has significant value in all round development of children. The most important value of play for infants is that it provides enjoyment to them. It provides stage for learning problem solving and creativity. Characteristics of infants' play are (i) there are no rules in their play and (ii) their play is more solitary than social. Different forms of play during this period are sensorimotor, exploratory, make-believe and imitative. Infants enjoy radio and television. They like games such as peekaboo, and hide and seek with their caregivers.

Development during early childhood: Early childhood is a time of remarkable physical, cognitive, social and emotional development. Early childhood period extends from 2 to 6 years. Parents label early childhood as the *problem age*, the *troublesome age*, or the *toy-age*, the educators label this stage as the *preschool age* and the psychologists as the *pre-gang age*, the *exploratory age*, or the *questioning age*.

Early Childhood is regarded as the *teachable moment* for acquiring skills because children enjoy the repetition, which is essential for learning new skills, they are adventuresome and enjoy trying new things and also they have few already-learned skills to make learning easy for acquiring new skills. During early childhood, children make developmental progress and express an interest in the world around them. They are enthusiastic to learn and to gain inner control. They learn by experiencing and by doing. They are more independent in comparison to toddlers. They can express their needs through meaningful language.

As the children enter early childhood, they explore, ask questions, start new tasks, and imagine. Parents of these children need to be open-minded and ready to grow to meet the challenging needs of children. They need to change their behaviour in ways that will be good for preschool children. Families with preschool children also experience frustration, anger and feelings of inadequacy while socializing these children.

It is not only that family and parents go through transition but children also go through transition from one stage to another. Children need to be prepared for a smooth separation or transition from family to preschool. The important developmental patterns during early childhood years are listed below:

- ***Physical development*:** During early childhood years, the physical development proceeds at a slow rate. At the age of 2-3 years, a child is 14-15 kg in weight and 37-38 inches in height, can jump of a step, ride a tricycle, use crayons, build an eight cubes tower, *etc.* At the age of 3-4 years, a child is 17-18 kg in weight, 40-41 inches in height and self-sufficient in many routines in house, *e.g.*, dressing and undressing. He/she can stand on one leg, jump up and down, draw a circle, a cross *etc.* At the age of 4-5 years, a child is 19-20 kg in weight, 43-44 inches in height, has mature motor control, skips, dresses on his/her self, do long jump, copy a square and a triangle. During these years, they can cut on a line with scissors and have more small muscle control. They can ties shoelaces.

- ***Language development*:** Language is the expression of ideas by means of speech-sounds combined into words. Words are combined into sentences to make ideas and thoughts meaningful. During early childhood, the children's ability to understand, process and produce language is increased in a significant manner. Young children experience a language explosion between the ages of 3 and 6. At the age of 3 years, their spoken vocabularies consist of roughly 900 words. By the age of 6 years, the spoken vocabularies expand dramatically to anywhere between 8,000 and 14,000 words. However, with this language explosion, their expressive (spoken language) abilities start to catch up with their receptive (ability to comprehend language) skills. They can tell long stories, read their own name and ask meaning of words. Once the children are able to speak with ease, they keep on talking continuously, which is why, the early childhood is also known as *chatterbox age*.

  During early childhood years, learning to speak is essential tool for socialization. Children who can communicate better are easily accepted in their peer group than those whose ability to communicate is limited. In addition, learning to speak is essential for achieving independence. Children who cannot make them understood or who cannot make others known their capabilities are treated as babies and they fail to achieve independence.

- ***Social development*:** Children during early childhood years get involved in parallel and cooperative play, enjoy being by others, take turns, know if he/she is a boy or girl, enjoy group activities and respond to verbal guidance. They enjoy a variety of activities, indoor and outdoor. Boys and girls have similar interests, so will often play together and share same toys such as dolls or trucks. They are highly social, talkative, and versatile and may play loosely organized group games. Children are learning to play games with simple rules. Also, they follow the rules and begin to learn that every body has equal right while playing games.

  During early childhood, the children begin to play more cooperatively. In cooperative plays, the young children engage in the same activity in a small group. Often, the first forms of cooperative play include pretend or symbolic play, *e.g.*, they may play *house* together and assign one child to be the mother and the other to be the baby. Pretend play is at peak for the majority of young children at the age of 4 and 5 years.

  As the young children continue to develop socially with peers, they often enter a stage of rough and tumble play, which includes running, racing, climbing, or competitive games. Often, this is the stage when social skills such as learning to take turns and follow simple group rules and norms are practiced. At this stage of social development, friendship usually means sharing toys and having fun playing together. Friendship at this age does not have the associated qualities of empathy and support.

- ***Emotional development*:** Emotions during early childhood years are usually extreme and short-lived. Heightened emotionality in early childhood is characterized by temper tantrums, intense fear and jealousy. Common emotional patterns of early childhood are anger, fear, jealousy, curiosity, joy, grief, envy and affection. Temper tantrums are at peak between the ages of two and four. Children become jealous when they feel that they are not given attention and it is diverted to someone else. They also become envious of the material possessions of other children. The most common causes of anger during early childhood years are conflict over play materials or when their wishes are not fulfilled. Physical aggression like hitting others decreases, however, verbal aggression, like name calling, may increase. They have imaginary fears of dark, monsters, injury, *etc.* They can differentiate facial expressions of anger, sorrow and joy, and have the sense of humour.

- They are learning to express their feelings with words. They can be aggressive but want friends and enjoy with other children. They appreciate praise for their achievements. They are developing ability to interpret the feelings of others. They are capable of some self-criticism.
- ***Cognitive development*:** Early childhood is the preoperational stage of thinking, which extends from two years of age until children are six years old. During early childhood, young children go through two sub-stages in cognitive development, *i.e.*, (i) they develop Symbolic Function between the ages of 2 and 4 years and (ii) the Intuitive thought sub-stage, which extends from the age of 4 to 7 years.

  During the Symbolic Function sub-stage, children master the ability to picture, remember, understand and replicate objects in their minds that are not immediately in front of them. In other words, children can create mental images of objects and store them in their minds for later use. During this stage, their thinking is still rigid, limited to one character of object's appearance.

  Preoperational children have not yet developed the mental abilities of conservation and transformation. *Conservation* is the ability to understand that some of the physical characteristics of objects remain the same, even if their outward appearance has changed. *Transformation* is the ability to understand that some of the physical characteristics change, while others remain the unchanged, *e.g.*, during preoperational stage, the children are unable to understand how things can change from one form to another. From two similar round balls of clay, if one is flattened, then children cannot understand that both the balls have same amount of clay. Another characteristic of preoperational stage is Egocentrism. *Egocentrism* is the inability to see the world someone else's point of view. Children with Egocentrism explain situations from their own perspective and understanding. Children in the preoperational stage also believe that things are alive or have human characteristics. This style of thinking called *Animism*, *e.g.*, young children may blame chairs or toys for causing them to fall.

  The second sub-stage in preoperational cognitive development stage is the Intuitive Thought sub-stage. Children at this sub-stage of development learn by asking questions such as, *why* and *how comes*? Children at this sub-stage tend to be confident of their knowledge and understanding but they are unaware that they have gained this knowledge by intuition. Another large gain during this sub-stage is the ability to comprehend dual relationships. Children now understand that something can be both an object itself as well as a symbol for something else, *e.g.*, a stuffed toy dog is a fun, furry toy as well as a representation of living dogs in general.
- ***Personality development*:** An individual's personality is firmly established by the end of early childhood. During early childhood, children gain some sense of being separate and independent from their parents. The task of preschoolers is to develop *autonomy* as well as *initiative*. Family members constitute the social world of young children, how family members feel about children and how they treat them are important in shaping self-concept of children, which is considered as core of the personality.

  By the end of early childhood, the children can be distinguished based on their personality traits. Physically and psychologically, the well-adjusted children are grouped under the category of *easy children*, and the children who are irregular in their physiological functions, intense in their reactions and slow to adapt to change are grouped under the category of *difficult children*. In third category are those children who have low activity level and do not adapt quickly. These children are called *slow to warm up*. These characteristics show how adjustable the children are in social relations.

- ***Moral development*:** Morality is the child's ability to learn the difference between rights or wrong and understand how to make the right choices. Children's experiences at home, the environment around them, and their physical, cognitive, emotional and social skills influence their developing sense of right and wrong. Between the age of 2 and 5, many children start to show moral-based behaviours and beliefs. Young children behave morally because they fear authority and try to avoid punishment. Children follow the rules because they do not want to get in trouble. Most young children can understand the difference between *good* and *bad* behaviour, and this understanding provides the basis for more complicated moral thinking in the future. During early childhood, the children have not yet developed conscience, so they do not feel guilty if caught doing something wrong. Children obey rules and conform to social expectations for gaining rewards.
- ***Development of play interests*:** During early childhood, toys are used in most of their plays. Hence, this stage is also called *toy age*. Availability of play equipments and motor skills of children influence play interests of children. Intelligent children prefer dramatic and creative play. In beginning of early childhood, the children prefer to play with toys. Throughout early childhood, children have interest in dramatic and construction plays. Children also like stories and books to read. Movies, radio and television are also enjoyed by children. During fourth year, the children begin to play games with their peers and their games have few rules.

Development during late childhood: Late childhood extends from 6 to 12 years. By the age of 12 years, the child becomes sexually mature. For most young children, there is a major change in the pattern of their lives. During late childhood, children enter into formal schooling. These children spend more time away from their families in schools and out of school with their peers. While adjusting to the new demands and expectations, most children are in a state of disequilibrium. They are emotionally disturbed. Many changes take place in attitudes, values and behaviour. During the last year or two, marked physical changes take place. The physical changes at the end of this stage create a disequilibrium resulting in the disturbance of accustomed life style.

Parents call late childhood a *troublesome age* by which the children are no longer willing to do what they are told to do. Older children, especially boys, are careless about their clothes and other material possessions. Parents also call this age as the *sloppy age*. In families where there are both, boy and girl as siblings, there are more fights and conflicts between siblings. Hence, some parents regard this age as a *quarrelsome age*.

Educators use different names for late childhood. During this age, the children are sent to schools for formal education, so educators call this age *elementary school age*. Psychologists use the term *gang age*, because during these years, the children are concerned about their acceptance by their age-mates and are willing to be the member of a gang. Psychologists also call late childhood as *play age*. During this age, there is overlapping of play activities of early childhood and those of adolescence.

The process of making children independent continues when the children go to school. As they learn new skills and rules, parents both teach and learn from them. Parents become members of Parents and teachers meetings (PTMs). Schoolteachers expect some commitment from parents, and parents have to adapt themselves for their new role. Children enlarge their parents' social contacts, bringing home friends and initiating contacts between families. At this stage of life, parents learn more about children and are able to see their children in relation to others. Children are more independent and may contribute some help with household chores. There are also possibilities for families and parents to feel like failures in launching their children into school.

For children going into school, there are many anxieties, nervousness and fears to deal with and overcome. During these years, the schoolwork is harder to do, and there are more classes and more teachers. Children have to interact with different personalities. If a parent has a child who is entering elementary or middle school for the first time, they should know that it is normal for their child to express concern and may even exhibit some acting out behaviour over school issues. The developmental characteristics of late childhood to help parents, teachers and caregivers to understand these children are discussed below:

- ***Physical development*:** Growth rate is slower during late childhood than during early childhood years. Children have better control of large muscles such as arms and legs than small muscles such as fingers. They enjoy testing muscle strength, skills and coordination. During this stage, hand skills and eye-hand coordination needed for activities such as writing and shoe lacing continue to develop. Children begin to lose baby teeth and acquire permanent teeth. Sexual organs grow at slower rate.

  During the elementary school years, the children grow an average of five or eight centimetres a year. Muscle mass and strength gradually increase. Legs lengthen and trunks slim down. Growth is slow and consistent. Motor development becomes smoother and more coordinated.

  Though some children will begin the changes of puberty before the age of 9 years, most will begin the process of rapid growth and physical change in the years between 9 and 12. On an average, the girls will begin puberty at the age of 10 or 11 years and boys will begin at the age of 11 or 12 years. There are a number of noticeable physical signs that a boy or girl has begun puberty. Boys begin their growth spurt normally a year later as compared the girls. Boys are usually better at gross motor skills and girls at fine motor skills.

  Body build affects height and weight in late childhood. Children with ectomorph body build have long and slender body. Their weight is lesser than mesomorph and endomorph. Mesomorph has heavier and rectangular body and endomorph has round and fatty body.

- ***Language development*:** During late childhood, vocabulary increases rapidly. Reading books, conversation, listening to radio and watching television helps in improving vocabulary of children. During these years, the children tend to talk as they learn and they learn best if active while learning. They are learning to read signs, make lists, count or write prices of objects they buy. They have greater ability to describe and generalize from their experiences and talk about thoughts and feelings.

  During late childhood, children discover that the simpler forms of communication, such as crying and gesturing, are socially unacceptable. They discover that comprehension of what others are saying to them, miscommunication occurs, as a result, they are not acceptable to the peer group. By the time, the children are in sixth grade, most children know approximately 50,000 words. Children not only learn many new words but they also learn new meanings for old words. Speech of children shifts from egocentric to socialized speech. During late childhood when children criticize others and make fun of them, they learn from experience that words can hurt others. There is progressively less and less talking as late childhood continues. With growing age, the children understand that endless talking annoys their peers. Girls talk more in comparison boys.

- ***Social development*:** Children during late childhood are still focused on self and able to have empathy. They have many best friends at the same time. Friends are likely to be of the same sex. They are becoming increasingly more aware of peer opinions and are concerned about being liked by their friends. Children tend to be competitive, bossy and unhappy if they lose in a competition. They like to win or be first in competition.

Middle childhood is often referred to as the *gang age* because it is characterized by interest in peer activities, an increasing strong desire to be an accepted member of a gang. They want to be with their peers and they are lonely and dissatisfied when they are not with them. Most children are aware of their socio-metric status. They play games and sports with their gang members. During middle childhood years, there is a tendency for children to interact mostly with same sex peers.

Boys tend to play in large groups that are hierarchically structured. Their groups have a leader who directs and tells the others what to do and how to do it and their games have winners and losers. On the other hand, girls prefer to play in small groups or pairs. Their relationships with friends and peers are intimate and persistent. As compared to boys, turn taking is a feature of girls' games. Girls prefer to sit and talk with each other and are more concerned about being liked by others.

At this stage of development, the friends serve six functions, companionship, stimulation, physical support, social comparison and intimacy or affection. Social knowledge plays important role in the children's ability to get along with their peers.

Peer status, which is unavoidable during childhood years, is an element of life. Five types of peer status, *i.e.*, popular children, average children, neglected children, rejected children and controversial children have been recognized. Popular children are good listeners and communicators, happy, have self-control, show concern for others and are self-confident, however, the rejected children are often shy and have low interactions with their peers.

- ***Emotional development*:** Emotional development develops significantly in middle and late childhood. Children begin to understand the significance of situations that they were not able to comprehend just a few years ago since their emotional stage becomes more complex by now. As these children grow, they become more exposed to stress and with experience, they learn to cope with stress. Learning to cope at a young age will help and provide the tools they will need as they grow into adulthood.

During middle childhood, children are egocentric but have the ability to feel for others. They are sensitive to criticism and do not know how to accept failure. The family is the main source of stability and security. They have strong desire for affection and attention of parents. They release tension through physical activity. New fears revolve around school, friendships and family. They act in order to avoid punishment and for them good and bad is what is approved by the family.

Common emotional patterns of middle childhood differ from those of early childhood. From experience, these children discover how others feel about various forms of emotional expression. In their desire to win social approval, they try to control the forms of emotional expressions, which are socially unacceptable such as crying. Temper tantrums become less frequent as these are considered babyish. Generally, middle childhood years are a period of relative emotional calm. There are times when children experience frequent and intense emotions. When children are ill or tired, they are likely to be irritable, fretful and generally difficult.

During late childhood, *emotional self-regulation* is developing. If the anxiety-producing event is controllable, *e.g.*, an examination, the child can work toward success. If the event is not controllable, they can distract themselves or redefine the situation. If they can manage this skill well, they develop a sense of *emotional self-efficacy*- feeling in control of their emotional experience. Children clear their system of pent-up emotional energy through *emotional catharsis*. It is a way to handle emotions in a positive manner to conform to social expectation. These children tend to be happier, more empathic, helpful and better liked by the peers. Less emotionally controlled children have peer problems and less empathy. Perspective taking increases during this period. Self-concept and self-esteem are important dimensions of the child's socio-emotional development during this period.

- ***Cognitive development***: Cognitive and intellectual development during late childhood progress and their reasoning become more logical. Their perspective taking skills become more mature and sophisticated. Elementary school children integrate knowledge from their interactions with teachers, peers and families. Asking questions helps the children to develop problem-solving skills. Children begin to form ideas mentally, and they can group things together that belong in one category, *i.e.*, they can handle only one operation at a time. The next level of mental development is sequencing and ordering that prepare them for math skills. They are very curious and trying to make sense out of the world. They are developing a sense of cause and effect.

  During middle childhood years, children enter the stage of *concrete operations* in thinking, a time when vague and unclear concepts of early childhood become specific and concrete. Concrete operations are mental actions that are reversible. Children's long-term memory, metacognitive knowledge and cognitive monitoring improve during childhood. It is important for children to engage in critical thinking. Children's creativity should be encouraged. During middle and late childhood, children become more analytical and logical in their approach to words and grammar. As children read books in school and consult encyclopedias and other sources of information, they learn new meanings for old concepts and correct faulty meanings associated with old concepts.

- ***Personality development***: Self-concept becomes more specific, including internal states and temperament. During late childhood, the children emphasize competencies and abilities. They describe aspects of personality, good and not so good. Self-esteem, which was very high in kindergarten, becomes more realistic as children compare themselves with other children. *Structure of the self* is affected by cognitive development and maturity. As perspective-taking skills improve, the children become able to read others' reactions and expectations. They develop an *ideal self* as their standard and assess their *real self* against it. High difference between the two leads to sadness, hopelessness and depression. Self-esteem is at peak between 4th and 6th grade, especially if they have good peer relations, or physical abilities. Those with high social esteem are better liked by their peers and teachers. Low self-esteem leads to anxiety, depression and antisocial behaviour. Poor health, overweight, peer rejection and failure lead to inferiority complex, which is dangerous for healthy personality development.

- ***Moral development***: During middle and late childhood, the development of children's moral development moves into a higher stage called *autonomous morality or the morality of* *cooperation*. Through peer interaction, the children learn that arguments and disagreements are part of life and that different people can have different perspectives. Children view that rules should be flexible and mutually agreed on and rules framed should suit the will of the majority. Children are learning about justice through sharing. *Distributive justice* is their beliefs, *i.e.*, material goods should be divided fairly. Sharing is a great challenge to younger children. Extra rewards should go to someone who worked hard or was exceptional in some way. As children grow, they believe that special consideration should be given to those with a disadvantage, *e.g.*, disabled children. Children recognize social conventions when they have a clear *purpose* and they regard deliberate violations of those as moral offences. During late childhood, children are at the level of *conventional morality*. This is the stage when social group makes rules that are appropriate for all group members and expect that all should conform to them to avoid social disapproval and criticism.

Development of conscience is one of the important developmental tasks of late childhood. It is an *internalized control*, which encourages children to act in a socially approved manner and to do what they believe, is right.

- ***Development of play interests***: During late childhood, children go for formal schooling. During this period, intelligent children prefer solitary play than social play. They prefer making construction and creative play. Older children also collect things that attract the attention. Children who belong to gangs and peer group play group games and sports and engage in exploratory activities. Favorite amusements of children during late childhood are reading, comics, movies, radio, television and day dreaming.

Hazards in Life Span from Infancy through Childhood and Roles and Responsibilities of Caregivers

Hazards in Infancy

Infancy is the foundation stage in human life span. During this stage, hazard can be physical and psychological and both are serious. The most important conditions influencing the infant's adjustment to postnatal life are prenatal environment, length of gestational period, kind of birth, postnatal care and parental attitude. An unfavourable prenatal environment is as harmful to normal development in postnatal life as is a complicated birth. Efforts should be made to minimize these hazards for appropriate development of infants.

Physical hazards: Physical hazards are serious for premature infants and infants who are born with developmental disabilities or those who are small for date. Small for date infants are those infants who are small or under weight with reference to time, and they have spent in their mothers' womb. During first year of life, death is usually caused by serious illness. During second year of life, deaths may occur due to accidents. Infants are protected from various diseases by immunization. Minor illnesses are common during infancy such as common cold and cough, and digestive discomforts. Parents need to consult doctors from time to time for vaccination and immunization of their infants.

Accidents are not very common during first year of life. During second year of life when infants can move around freely, they are more vulnerable to accidents. Malnutrition during infancy can result in stunted growth and development. The first two years of postnatal life are crucial for brain development. During this period, malnutrition can seriously damage the brain of infants.

Another physical hazard during infancy can be obesity. Infancy is one of the three critical periods of fat-cell development, and other two periods are last trimester of prenatal period and early part of adolescence. Obese infants tend to grow as obese adults.

Eating, sleeping and elimination are physiological habits established during infancy. Infants may develop unfavourable attitude if not properly cared and trained.

Psychological hazards: Psychological hazard during infancy stage is failure to master developmental tasks of infancy. These are explained taking into consideration different developmental domains during this period.

- ***Hazards in motor domain*:** All infants are expected to walk and gain control over their bodily movements. When their motor development is delayed and they lag behind their counterparts, they feel frustrated when they are unable to perform thing for themselves and fail. They develop negative attitude, which further delays learning and development.
- ***Hazards in language and speech domain*:** Due to various reasons language and speech, development is delayed during infancy. Some of these reasons are lack of motivation, lack of stimulation, poor intellectual skills and deprived conditions. To broaden social horizon, language is a key tool, which if delayed may result in social isolation as they find themselves outsiders if unable to communicate.
- ***Hazards in social domain*:** Infants are more egocentric. Due to their immature thinking, they are unable to understand perspectives of others. Lack of motivation and opportunities to learn to become social encourages the persistence of egocentrism. This leads to development of introversion and shyness. Shy infants tend to grow as shy adults. Long-term effects of shyness are loneliness, negative self-evaluation and self-consciousness.
- ***Hazards in emotional domain*:** During infancy emotional deprivation, stress, too much affection and dominance and emotions can be hazardous. Infants who are not given opportunities to express emotions such as affection, joy and curiosity may suffer from emotional deprivation. The emotionally deprived children lag behind in their development as compared to normal children of their age. Poor health, parental neglect and poor environmental conditions can cause stress in infants. Stressed children may exhibit eating and sleeping problems. If a shy and anxious infant is exposed to stressful situations and strangers too often then it is possible that shyness or anxiety will persist long after infancy.
- ***Hazards in personality domain*:** Foundations of personality development are laid during infancy. Self-concept during this stage is mirror image of what others have about them. If significant caregivers have negative attitude towards infants, it is more likely that they develop negative self-concept about themselves.
- ***Hazards in cognitive domain*:** During infancy, it is easy to replace wrong meaning associated with objects and people with correct meanings, *e.g.*, infants learn to associate bitter things with punishment and sweet with rewards. This may develop food likings and disliking, as they grow older.
- ***Hazards in morality domain*:** During infancy if children get more attention on their wrongdoing than when they behave in a socially approved behaviour, it could be hazardous for their moral development. It will lead to confusion for what is right and what is wrong and will lay foundations for behvioural problems.

- ***Family relationship hazards*:** Infants develop feelings of insecurity if they are separated from caregivers during infancy, particularly from their mothers. Feelings of insecurity results in failure to develop attachment behaviour with primary caregivers and in development of close relationships with their peers, as they grow older. During second year of life, infants become troublesome and are difficult to handle. This may trigger anger and resentment when parents are unhappy or when relationship between both the parents is strained. Sometimes, this may result in child abuse. Inconsistent discipline and child rearing practices may result in confusion in infants and they are unable to distinguish between right and wrong.

Roles and Responsibilities of Caregivers

Development of children is rapid during infancy. In addition, infancy is considered as the most hazardous postnatal stage. It is responsibility of caregivers to protect infants and do their maximum to minimize their severity. Infants should be immunized properly to protect them from various diseases. Prompt diagnosis and proper medical care are important to protect them from serious harm. During the second year of life when infants are able to move around in the house, they are prone to accidents. It is important that parents make the home accident proof. Parents should provide proper nutrition to their infant to make their body and mind healthy. Overfeeding is also dangerous as it could lead to obesity in children. From early years, it is important that caregivers develop good habits for eating, sleeping and elimination.

Parental pressure to achieve motor control and to learn motor skills before infants are physically and mentally ready to learn. Children may develop negative attitude and it may cause a delay in learning tasks. Therefore, parents should not pressurize infants for learning tasks until they are ready to learn. Parents should avoid pressuring children to be right or left-handed. A few 2 years olds will begin to show preference for one hand but many children will continue to use both hands for a few years. Parents should provide safe outlets for physical activity and space for exploration like small steps, boxes, barrels, tires, pulling and pushing toys, ride on and ride in toys.

Parents can play important role in acquisition of language skills of their infants. If infants are motivated and stimulated from early months to babble and to learn to say words, their language and speech development are accelerated. Parents should read aloud to children every day and encourage toddlers to look at books with large pictures and sturdy pages. Simple story lines are the best. Parents should encourage verbal skills by giving simple directions like *Close the door, please.*

Lack of parental motivation and opportunities to learn to become social encourages the perseverance of egocentrism. This leads to development of introversion and shyness. To overcome shyness in children, it is important that parents provide social opportunities to their infants. Parents should try to make family atmosphere cheerful and happy so that the children develop in a home base with healthy relationships. Parents should not expect infants to share or take turns. They are focused on learning how to handle themselves physically and on learning to talk. Learning to share will come later.

Hazards in Early Childhood

During early childhood, hazards could also be physical and psychological as discussed in the following section:

Physical Hazards: Preschool children are highly susceptible to all kinds of illness. Sickly children fall behind in their physical and motor skills. As a result, they find they are misfit in the playgroup. Children are also prone to accidents, awkwardness and obesity. Unattractive children have social disadvantage as they are less liked by peer group and other people. Left-handedness is also hazardous for children as they get embarrassed in many social situations and feel that they are different from other children.

Psychological Hazards: There are many psychological hazards, which could affect personal and social adjustment of children. Common psychological hazards of early childhood stage are the following:

- ***Hazards in language domain*:** Language is a tool for communication and is essential for social acceptance. Children, who cannot communicate with others, will be socially unaccepted, and this will lead to feelings of inadequacy and inferiority. Egocentric speech and bilingualism are hazardous during early childhood years and may become causes of peer rejection.
- ***Hazards in social domain*:** Children with poor language skills and poor emotional control are likely to be rejected by their peers in a social group. These children will not only feel lonely, even more important, they will be deprived of opportunities to learn to behave with peer-approved manner. As a result of the treatment they receive from peers, they may develop unhealthy social attitudes. Pets and imaginary companions are hazardous for social development because these leads to loneliness in children.
- ***Hazards in emotional domain*:** The major emotional hazard of early childhood years is the dominance of the unpleasant emotions, especially anger. If children experience too many of unpleasant emotions and very few of the pleasant ones, it will distort their outlook on life and encourage the development of an unpleasant disposition. This is likely to lead to negative parent-child relationship since parents may assess their child as with difficult temperament. If children are unable to develop emotional relationship with others than their family members and become too much dependent on one person could be hazardous for normal development of children. It will be difficult for children to develop friendly relations with their peers.
- ***Hazards in personality domain*:** One of the most serious personality hazards during early childhood is the development of an unfavourable self-concept. Unfavourable self-concepts once developed, they are hard to overcome. Many parents fail to recognize that young children are developing unfavourable self-concepts or they believe that children will outgrow these self-concepts, as they grow older and as their social horizons broaden.
- ***Hazards in cognitive domain*:** Concept development is a cognitive domain in development. Due to their limited vocabulary, children are unable to understand what is said to them. As in infancy, during early childhood also children associate wrong meanings to different objects. If understanding of different concepts is poor as compared to their peers, then it could be hazardous for children in their social and personal adjustment.
- ***Hazards in morality domain*:** During early childhood, inconsistent discipline interferes with the process moral development. In a family if father and mother have different rules, the children become confused in understanding what is considered as right or wrong. It is hazardous for understanding of social expectations and children develop negative attitude towards those who are over controlling.
- ***Hazards in family relationship domain*:** Parents are most significant persons in the life of children. Deterioration in family relationship is hazardous to good personal and social development. In some families, boys are preferred. Girls in these families will resent their parents and brothers. For boys, the major threat to family relationship in the lack of a father to identify with or the lack of emotional warmth between father and son.

Family relationships are also affected by stepparents if bonding between children and stepparents is poor. During these years, sibling quarrelling is caused by jealousy or by differences in interests among siblings. It can have negative impact on children's adjustment and development because it can deprive them of companionship at an age when their social world is limited mainly to family and when socially approved behaviour is learned. Children can carry their habit to peer group, which can endanger their chances for making friends.

Child-abuse in the family is also one of the serious family-relationship hazards. This may take forms varying from mild abuse in the form of slap to such serious abuse as to the permanent disability of child.

Roles and Responsibilities of Caregivers

During early childhood years, there is rapid development of children. It is the foundation stage for future development. The role of parents is to provide activities to promote all round development of children. During early childhood years, children spend most of the time in the family with their parents, siblings and grand parents. These children are sent to preschools, *balwadis*, or *anganwadis* for informal preschool education. This is the first time when they are away from their parents. Parents should remember that all children are different and reach the various stages at different times. Some children may achieve various developmental milestones earlier or later than the average. Parents are to teach children to carry out some responsibility and to get along with other children.

Parents need to immunize their children to protect them from illnesses and need to take care of health and diet so that adjustment is good in peer group, school and family. Parents can play important role in promoting language development of their children. If children are motivated and stimulated from early years to listen to others, to communicate with others and to express their thoughts, their language and speech development is accelerated. Once parents find that their child has poor language skills or that their child has speech disorder, they should consult the expert and provide stimulating environment to overcome problems associated with poor speech.

Parental guidance and a stimulating environment are very important to overcome some of behavioural problems and for better adjustment in family and preschool. Parents need to be very careful in their parenting practices to protect their child from the damage of negative emotions. Parents should try to make family atmosphere cheerful and happy so that children will learn to make their contributions to maintain this atmosphere.

In addition, parents need to be vigil in developmental aspects of their children and work best to eliminate habits and attitude, which predispose children to act in a socially unacceptable manner and protect them from developing distorted personality. Parents should make realistic aspirations in accordance with their capabilities. Success in these tasks will foster favourable self-concept in children.

During preschool years, parents need to teach children concept of right and wrong, how to respect the rights of others, which behaviours are acceptable and which are not, with a goal of helping to develop a child who feels secure and loved, is self-confident, self-disciplined and knows how to control impulses and who does not get overly frustrated with the normal stresses of everyday life.

Effective child-rearing practices should be adopted taking into consideration the developmental level of children. Authoritative parenting practices are most effective practices with reasonable expectations and willing to answer their children's questions. When children fail to meet the expectations, these parents are nurturing and forgiving rather than punishing. They want their children to be assertive, socially responsible, self-regulated, as well as cooperative.

Hazards in Late Childhood

Hazards of early childhood years may carry over during late childhood. Adjustment in different developmental domains includes physical, language, social, emotional, personality, cognitive and moral development. There are hazards in family relationship domain also. Physical and psychological hazards of late childhood are as follows:

Physical Hazards: Mortality during late childhood occurs less often as it was during early years. Children who are still prone to illness become irritable and more demanding. Sick children have to miss their school for number of days and as a result may become poor in their academics. During late childhood, children may also suffer from accidents. After effects of accidents could be permanent disability in the child. Unattractiveness, awkwardness and obesity could be hazardous during late childhood.

Parents may become intolerant in their attitudes towards the illness and complain about the extra work. Parents and family members need to be tolerant and supportive during their child's illness. Family also needs adjustment with obesity in children, which is not only a physical hazard to their health but they are also more prone to diabetes. Inappropriate body build and physical disabilities also have impact on family relationships. These may lead to personal and social maladjustment as these children are treated differently by their peers and others.

Psychological Hazards: There are many psychological hazards, which could affect personal and social adjustment of children. Common psychological hazards of late childhood are discussed in the following section:

- ***Hazards in language domain*:** Poor language skills handicap children in their schoolwork. Speech errors, such as mispronunciations and grammatical mistakes, stuttering, or lisping, may make children self-conscious. Their communication skills and performance are affected and they are assessed negatively by their teachers and peers leading to negative self-concept and poor adjustment.
- ***Hazards in social domain:*** During late childhood, children face adjustment problems when they are rejected or neglected by their peer group because they express unacceptable pattern of behaviour such as temper tantrums, anger and jealousy.

  There are a number of reasons that social development may be a challenge for a child, *e.g.*, the child may not be a *social thinker*, *i.e.*, the child may have difficulty in reading, understanding and responding to social cues, and therefore, does not think about social situations in the same way a more socially competent child would. Another explanation could be a learning disability or developmental disorder.

  Social relationships are important to the happiness, self-esteem and sense of belonging of schoolchildren and adolescents. Those who do not achieve social competence tend to exhibit more risky behaviourus and lower academic achievement as they enter into adolescence and young adulthood.
- ***Hazards in emotional domain*:** Dominant emotions of early childhood may carry over to late childhood and could be hazardous for children particularly if these emotions are unpleasant. If children express too much temper tantrums, anger and jealousy in their peer group, then these

children are considered as immature, and hence, rejected by the peer group. Shy children may express heightened social anxiety or may suffer from an anxiety disorder.

- ***Hazards in personality domain*:** One of the important serious hazards of personality development is unfavourable self-concept that leads to self-rejection. Another could be egocentrism, which is carried over from early years and which makes the children self-centered.
- ***Hazards in cognitive domain*:** Children who are dull in studies develop inferiority complex and feel rejected by their peers and teachers. They become shy and introverted. Children with high cognitive abilities may also develop poor self-concept because they are expected to perform very high.
- ***Hazards in morality domain*:** During late childhood, children may develop moral values and codes, which do not conform to family and social values. They may fail to develop super ego or conscience, which help them differentiate between right and wrong and control their behaviour. Inconsistent discipline in families and school make children confused about what is expected from them in similar situations. Acceptance of misbehaviour and misconduct for social approval in a peer group make the children habitual of wrongdoing.
- ***Hazards in family relationship domain*:** Family relationship domain is the most important for adjustment of family members with children during late childhood. Friction in family leads to weakness in family ties. It further leads to adjustment problem in family, which is carried outside the home.

Roles and Responsibilities of Caregivers

Role of families and parents is crucial for promoting development of children during late childhood. Caregivers, particularly parents, should give them quality time and take interest in their school and academic work. For parents, the monitoring of school experiences may involve more efforts as the number of teachers increases. Transition to middle childhood generates new tasks for children.

Social network of children expands significantly during middle and late childhood and children spend more time in the company of peers as compared to adults. These experiences outside the family necessitate them monitoring children's activities.

The problem of parenting during middle childhood becomes more difficult by increase in risks and stressors for children. Although these children are usually healthy but the chances of accidents increase. Tobacco, alcohol and other drugs use have become more common in middle childhood. Parents and family members play a significant role in grooming capacities of the children for functioning as responsible citizens. Parents can promote self-regulation by exposing children to standards of conduct and models of socially valued behaviours and by providing rewards and punishment in accord with those standards. Parents can encourage cognitive component of self-regulation through discussion and reasoning. As self-regulation increases, parents gradually allow children to assume more responsibilities.

Bullying tends to be at peak during late childhood years. The best way to protect the child is to sit down and discuss behaviours common during late childhood years such as relational aggression. Children who are being bullied may try to hide the fact from family members or teachers, so parents should be sure that they know the signs of bullying in order to take quick action.

In addition, the parents must model the appropriate behaviour for their children if they want their children to be successful in their behaviour. Developing clear expectation of what both parents want is the basis and the first steps to parenting during childhood. Once both parents have set appropriate expectations and rules for their child, the next step is to communicate those expectations clearly to their child in word and in deed. Children will not do what the parent says, and they will do what the parents have modeled. Children model the behaviours that the parent has presented to them repeatedly.

Good parent child relationship contributes successful adjustment in family and healthy family life. Children who are well adjusted, well liked by their peers and successful and happy in school are proof of good family adjustment and parental roles.

Issues and Challenges during Infancy and Childhood

Children of different ages present different sets of issues and challenges for their parents and families.

Issues and challenges during infancy: During first year excessive irritability, sleeplessness and feeding problems are likely to dominate parental concerns. During second year, when the child has become mobile and self aware, possibility of physical harm, destructiveness and aggression over power earlier worries. Issues and challenges during infancy are discussed below:

- ***Colic***: Some infants may cry continuously for often extending to more than 3 hours in a day. They may clench fists and flex legs or may make apparent sucking movements. This is called as colic. This colic may recur for several days, often more than 3 days a week. The problem usually starts within the first week after birth, reaches a peak by the age of 4-6 weeks and improves considerably by the age of 3-4 months. More commonly, it occurs in overactive infants, who are over stimulated by their parents. It could be a manifestation of hunger, improper feeding, swallowing of air during feeds or sucking, cow milk allergy, *etc.* The baby should be fed in the upright position and burped to promote egress of swallowed air. The infant should not receive undue attention during crying.
- ***Resistance to feeding***: The infant may exhibit resistance to feeding during weaning either due to dislike of the taste of the weaning food or as a reaction to the new situation. He/she may be physically uncomfortable because of painful ulcers in the mouth or blocked nose. The situation can often be remedied by adjustment in the feeding schedule.
- ***Stranger anxiety***: Infants relate to their mothers for getting their needs fulfilled, for comfort, pleasure and security. Infants show anxiety in the presence of strangers, especially in absence of their parents.
- ***Resistance to parental interference***: Infants like to explore their environment. In order to protect infants from injuries, parents make their home accident proof. They cover electric points with tapes or with covers, put some barrier in front of staircase or they lock the kitchen. These interferences by parents make the children frustrated because these are the obstacles in exploring their environment.
- ***Head banging***: Infants who are fatigued or under stress may bang their head against the bed or rock it in rhythmic movements. Apparently, this provides them with a pleasurable diversion. In these cases, bed should be padded to prevent injury to the child.
- ***Thumb sucking and nail biting***: Thumb sucking and nail biting is another pleasurable sensation that diverts the child from his or her environment. These behaviours are generally the signs of insecurity. They feel secure and happy when sucking their thumb. Thumb sucking can cause problems with alignment of the teeth. Parents should not to show excessive anxiety about thumb sucking during infancy.

- ***Masturbation***: A child may obtain pleasure by genital stimulation. This behaviour is usually harmless and it is not going to cause any sexual perversions or deviant behaviour later on.
- ***Breath holding***: Breath holding spasms usually occur between the ages of six months and five years and infants may turn blue or pale. Blue spasms are precipitated by anger and frustration, while pale spasms are associated with fear or minor injury. Child starts crying after he/she is hurt, admonished, frustrated, or frightened and after a bout of crying most of the air in the lungs is exhaled, breath is held in expiration for a few seconds, and the child becomes blue. There is rigidity followed by limpness. Thereafter, the child becomes less active for a short period. The child suddenly becomes unconscious.

Issues and challenges during early childhood: By the third year, parents are concerned about disobedience, resistance to routine, poor social skills and slow growth of verbal ability. During early childhood years, parents expect their children to develop cognitive skills, sociability, security, self-confidence, obedience and control of aggression during early years of preschool. A little later, parents hope to see gains in independence, autonomy, motivation for mastery and a willingness to compete with others.

Parents believe that the early development of these qualities will help their children to adapt to the social and cognitive challenges they will confront, as they grow old. If a preschool child is too timid with other children, the mother may initiate a playgroup. If the child is too dependent on the mother, she will encourage the child to play alone.

All parents sometimes have to deal with child behaviour problems. Young children can develop many behaviours that may trouble or annoy their parents. Some of these are explained in this section.

- ***Whining***: Nearly all children express whining at some time or another. The best way to handle whining is to ignore it as much as possible. Giving into whining can teach children that whining is an effective means of getting what they want. Consistency is extremely important if parent are to eliminate or reduce whining in their home.
- ***Poor eating habits***: Food battles are among the most common child behaviour problems. Eating habits begin forming during infancy and it is important for parents to offer healthy foods consistently and repeatedly. An eating schedule is also important.
- ***Hitting, biting and kicking***: Children sometimes express feelings of frustration or anger. Children who hit, bite and kick need to be taught gently that those behaviours are unacceptable. Parents may use time-outs to discourage aggressive behaviour but they should never hit them as punishment for this behaviour.
- ***Interrupting***: During early childhood years, interrupting adults when they are discussing with others, is also one of the behavioural problems. Children have a lot to say and they want to say it now even when their parents are busy. Children need to be encouraged to wait rather than interrupt but they can only wait for short periods.
- ***Pica***: Many children develop the habit of eating inedible things like sand and chalk. This is known as pica. There is an association between pica and anaemia. They are slow in motor and mental development and show neurologic defects and deviant behaviour. Children under 2 years of age usually put everything in their mouth. This is normal and the child tries to learn the taste but persistence of this habit beyond 2 years is worrisome. It can be a result of parental neglect, poor supervision, or lack of affection. It is common in children from lower socioeconomic strata. These children are prone to lead poisoning and often complain of chronic abdominal pain.

- ***Sleep disturbances*:** The child may suddenly awaken after a frightening nightmare. Reasons may include fear of dark, difficulty in falling asleep, night walking, sleep talking, or night terrors. If children frequently suffer these sleep disturbances, a child psychiatrist, or psychologist should be consulted.
- ***Unclear speech*:** Clarity of speech normally improves gradually during the first few years of life. It should become completely intelligible by 4 years of age. An unclear speech after this age often means that the child may be suffering from some major disorder in the language, cognitive development, or hearing. Such children may develop language based learning difficulties later in school years. These children should be carefully examined by a doctor and speech therapist together. A speech therapist will be able to pin point the reason for the unclear speech and provide a remedy for this problem.
- ***Bed-wetting*:** Some children empty the bladder involuntarily and wet the bed at night beyond an age by which bladder control is normally developed. An occasional lapse by a child should not cause undue parental concern. However, if bed-wetting occurs frequently then it is defined as enuresis. Enuresis or bed-wetting may represent a sub-conscious desire of the child to gain the care and attention of his/her parents as in earlier period of infancy or it may be a manifestation of sub-conscious resentment against parents. Thus, enuresis often has a psychological basis. This may be due to overzealous attempts at toilet training by the parents, emotional disturbances in the child or parent child mal-adjustment.
- ***Stuttering/stammering speech*:** Stuttering usually begins between the age of 2 and 5 years, a period in which the speech is not fluent. Stuttering is a defect in speech characterized by hesitation or stumbling and spasmodic repetition of some syllables with a pause. There is trouble pronouncing the initial consonants and is usually caused by spasm of palate and tongue muscles. Most children will show some degree of hesitation and repetition in their speech at some periods in early life, however, there are individual variations in the extant of such difficulties, whereas, some children can speak fluently, others are severely handicapped. It is possible that children who are unable to cope with the environmental and emotional stresses are more likely to develop stuttering. Many a times, the stress caused by the conflict between parental expectations and child's achievements may precipitate stuttering. Regular speech therapy if given early can help these children completely rid of stuttering. Children with stuttering should receive special attention from parents. They must be given plenty of emotional support and love.
- ***Sibling rivalry*:** It has been observed that there is intense rivalry amongst siblings. The child's emotional need for affection and security may appear to be threatened with the birth of another child. Parents have to give a part of their time and attention to the new arrival. The older sibling may feel deprived and this may initiate hostility in his/her behaviour. To avoid such situations, conscious effort should be made to involve the older child in the care of younger sibling under parental eye so that he/she can relate to the sibling with love, affection and a feeling of belonging.
- ***Autism*:** Children with autism have problems interacting and communicating with others. Autism appears before the third birthday, causing children to act inappropriately, often repeating behaviours over long periods, *e.g.*, some children bang their heads, rock, or spin objects. Symptoms of autism range from mild to severe. Children with autism may have a very limited awareness of others and are at increased risk for other mental disorders. It carries over to next stages of life span.

Issues and challenges during late childhood: Parents may face many challenging behaviours with children during middle and late childhood. It is important to keep in mind that these behaviours are common to this age group. Some of these challenging behaviours are explained in this section.

- ***Homework***: Some children have fear of doing homework. They may leave their homework or refuse to do it all together. Many parents handle this by trying to force their children to do their homework. However, getting into a power struggle with children is not effective. By having clear rules and expectations, parents can create a positive situation for both themselves and their child. Parents should avoid constantly nagging or lecturing their child to do homework.
- ***Television***: Television viewing can have both positive and negative effects. Children like to watch television because it is entertaining. However, watching television can become an addiction. Watching too much television can negatively affect children's health and behaviour. Parents should help their child in finding alternatives to watching television such as inviting a friend to come over to their house for play.
- ***Sibling rivalry***: Sibling rivalry is a normal part of growing up. Children could be resentful of the success or achievements of their siblings. Siblings argue about many things. In most cases, children can solve their disagreements with siblings if they are taught a few simple guidelines. Instead of rescuing children when they are arguing, parents should help them to get along better with each other so that they become more socially competent. Parents should not get involved in children's conflicts. When parents get involved in children's conflicts, they become dependent on parents to solve their arguments. Parents should get involved in children's conflicts if they notice sibling abuse. Sibling abuse includes behaviour such as hitting, biting, or kicking a sibling. Parents should not allow this to occur under any circumstances.
- ***Household chores***: It is important to assign children household chores. Household chores give children a sense of responsibility, pride and competence. As a family, decide how the chores will be assigned to different family members.
- ***Lying:*** Children lie for different reasons. Often, children prefer to lie than to face the consequences of making a mistake. Parents should model a positive attitude towards mistakes. If children learn that everyone makes mistakes, they will be less likely to lie when they make a mistake.
- ***Hyperactivity***: Hyperactive children find it very difficult to refrain from talking, moving and disturbing others around them. These children often blurt out comments, answers and questions at inappropriate times. Not only this is distracting for others in the classroom but also this behaviour can actually increase the child's own frustration because he/she knows his/her behaviour is unacceptable but he/she may not be able to stop what he/she is doing without help.
- ***Inattention***: The results of inattention can have a negative impact on academic performance of children. They are unable to pay attention to their teachers and may miss out an important instruction related to homework or class work assignments.
- ***Exposure to violence***: The ready availability of weapons to individuals of all ages increases the likelihood of being a victim or a perpetrator of violence during middle school years. The impact of violence surely concerns parents.
- ***Aggression***: Generally, physical aggression declines with age except for those children who are involved in more serious violence and criminal activity. Relational aggression, *i.e.*, verbal insults and social aggression, does not decline. In girls as well as boys, relational aggression becomes

more hurtful, more isolating and more important to deflect with humour, a shrug, or a counter-attack when possible.

- ***Bullying***: It is repeated systematic efforts to inflict harm through physical, verbal, or social attack. Boys who are bullies are often above average in size, whereas, girls who are bullies are often above average in verbal assertiveness. Bullying during middle childhood seems to be pervasive. Victims are generally rejected, withdrawn children, anxious and insecure, unable or unwilling to defend themselves, without friends who will take their side. Bullied children are anxious, depressed and underachieving during the months and years of their torment. Even years later, they still have damaged self-esteem as well as painful memories.

  Bullies become more hostile, challenging everyone who tries to stop them and get into trouble not only with peers but also with the police. Children who regularly bully other children often become violent criminals in future life.

- ***Learning difficulties***: During childhood years, learning difficulties are also challenging for parents and children. Learning difficulties could be mild, moderate and severe. Children with mild learning difficulties have some difficulties with normal school work but can cope with the normal curriculum. If the work is differentiated and support is available, their needs can be met. Children with moderate learning difficulties make very limited progress in basic academic skills such as literacy and numeracy, despite support being given. As a result, they need additional help, achieved by a statement of their special educational needs. Children with severe learning difficulties function at a low level in terms of basic skills. Communication skills and academic attainment are normally very limited.
- ***Severe depression***: Earlier people believed that severe depression did not occur in childhood. Today, experts agree that severe depression can occur at any age. Two of every one hundred children may have major depression. The disorder is marked by changes in emotions, *i.e.*, children often feel sad, cry, or feel worthless. They lose interest in play activities and their schoolwork declines. They may experience changes in appetite or sleeping patterns. They believe they are ugly, unable to do anything right, or that the world or life is hopeless. It is also important for parents and caregivers to be aware that some children with depression may not value their lives, which can put them at risk of suicide.
- ***Bipolar disorder***: Children who express exaggerated mood swings that range from extreme highs such as excitedness or manic phases to extreme lows, *i.e.*, depression may suffer from bipolar disorder. Periods of moderate mood occur in between the extreme highs and lows. During manic phases, children may talk nonstop, sleep very little and show unusually poor judgment. At the low end of the mood swing, children experience severe depression. Bipolar mood swings can recur throughout life.
- ***Schizophrenia***: Young people with schizophrenia have psychotic periods that may involve hallucinations, withdrawal from others and loss of contact with reality. Other symptoms include delusional or disordered thoughts and an inability to experience pleasure. Schizophrenia occurs in about five of every 1,000 children.

Discipline and Child Outcomes

Every child has certain basic emotional needs, which have to be satisfied to ensure his/her optimal development and to make him/her an emotionally mature individual who is able to relate with others in the family and society meaningfully. The child is dependent on parents and he/she should be able to trust them. Opportunities should be made available for play, activity and recreation. He/she should be able to take pride in his/her achievements.

Children should be disciplined. Discipline is not the same as punishment. Instead, discipline has to do more with teaching and involves teaching children right from wrong, how to respect the rights of others, which behaviour is acceptable and which is not, with a goal of helping to develop a child who feels secure and loved, is self-confident, self-disciplined and knows how to control his/her impulses and who does not get overly frustrated with the normal stresses of everyday life. Disciplinary techniques used by parents play important role in children's development. The parental discipline can generally be categorized into three as given below:

- **Power assertion:** Power assertion involves use of commands, physical power, or deprivation of privileges to control children's behaviour.
- **Love withdrawal:** In this type of discipline, parents use negative techniques such as ridicule, expression of dislike or disappointment and isolation of the child in order to threaten the child with loss of parental love.
- **Induction:** In this type of discipline, parents use reasons and explanations so that children can understand the rationale underlying the parents' reasons for engaging in or for prohibiting certain behavioural patterns.

Parents who use inductive disciplinary strategies their children display less disruptive behaviour than those who have power assertive parents. Children who experience coercive control at home are more likely to display aggressive behaviour in peer interactions. Children may model the aggressive behaviour of their parents in peer interaction and use unfriendly strategies that lead to rejection by peer group.

For effective parenting and positive child outcomes, discipline should be firm, fair and friendly. Firm means that consequences should be clearly stated and then adhered to when the inappropriate behaviour occurs. Fair means that the punishment should fit the misdeed or inappropriate behaviour. Also in case of recurring behaviour, consequences should be stated in advance so that the child may know what to expect. Harsh punishment is not necessary. Use a friendly but firm communication style when letting children know they have behaved inappropriately and let them know they will be responsible for the consequences of their behaviour. Encourage them to try to remember what they should do instead to avoid future consequences. Also praise them for appropriate behaviour.

In the following section different techniques of discipline, which could provide guidelines for parents and teachers working with infants and children, have been discussed below:

Disciplinary techniques during infancy (up to 2 years): Infants/toddlers are curious by nature. They try to experiment with everything they get. Therefore, it is wise to eliminate temptations, especially cleaning supplies and medications should be kept well out of their reach. They easily get irritated if they are disturbed in their activities. Followings are some of the techniques to discipline the toddlers:

- ***Change the activity*:** When children start doing something wrong, stop them. It is important to explain why they are stopped. Suggest something else they can do, *e.g.*, if they want to scribble on the wall, give them paper to scribble on.

- ***Ignore*:** Some misbehaviour is done to get parental attention. The best way to deal with this is to ignore it. Pay attention to good behaviour. Children need attention for good behaviour, not for misbehaviour.
- ***Be firm*:** Be clear and firm and tell the child what he/she needs to do.
- ***Use words*:** Talk calmly with the child. Talk about ways to work out the problem.
- ***Active listening*:** Since infants are not very verbally skilled, using active listening with them relies more on *hearing* what they are saying through their actions than does on listening to their words.
- ***Avoid physical punishment*:** Infants are unable to make connection between their behaviour and physical punishment. They will only feel the pain of punishment. Parents should not spank, hit, or slap children of any age. Children learn by watching their parents and parents are role model for their children.
- ***Set physical limits*:** It is difficult for toddles to understand only words, follow up verbal limits with physical limits. A physical limit is imposed when a child is stopped to do an activity, *e.g.*, from drawing on the wall. Physical limits like removing the pen from child's hand are different from physical disciplinary techniques like slapping the child.
- ***Provide choices*:** Choices are a part of daily life with toddlers and they are an essential part of teaching them discipline. Giving children choices, teach them that they are entitled to opinions and that they have right to say. It gives them respect, and it demonstrates parental trust. The more experience they have with making choices at a young age, the better they will do when they are teenagers faced with larger, more life-threatening choices.
- ***Natural consequences*:** Toddlers are rarely logical enough to understand logical consequences, thus, the best approach is to allow and point out natural consequences, *e.g.*, "You hit Swati, now Swati is sad and crying and she does not want to play with you anymore today."
- ***Time-outs*:** A time-out is a way of separating the child from the moment, person, or object, which is causing trouble. Time-outs can be effective discipline for toddlers. An infant, who has been hitting, biting, or throwing food, *e.g.*, should be told why the behaviour is unacceptable and taken to a designated time-out area, such as a chair for a minute or two to calm down. Longer time-outs are not effective for toddlers.

Disciplinary techniques during early childhood (2 to 6 years): As the child grows and begins to understand the connection between actions and consequences, parents should start communicating the rules of family to their preschool children.

- ***Set the rules and be consistent*:** The earlier that the parents set the rules and explain them to their children, the better for everyone. Consistency is the key to effective discipline, and it is important for parents to decide what the rules are and then uphold them.
- ***Reward good behaviour*:** Discipline is not just about punishment but also about recognizing good behaviour, *e.g.*, saying *I am proud of you for sharing your toys at playgroup* is usually more effective than punishing a child for the opposite behaviour, *i.e.*, not sharing.
- ***Natural consequences*:** When children see the natural consequences of their actions, they experience the direct results of their choices, but be sure that the consequences do not place them in any danger. Children will learn best when they learn for themselves and will not blame their parents for the consequences they receive. Natural consequences work best but they are not always appropriate. For preschool children, parents need to step in and create a consequence, which is

closely connected to their actions. When this method is used, it is important that parents mean what they say. They should let their child know that they are serious.

- ***Make a behaviour chart*:** If a child continues to misbehave or displays an unacceptable behaviour, make a chart with a box for each day of the week. Explain to the child how many times he/she can misbehave before getting punished or how long the proper behaviour must be displayed before it is rewarded. Keep a track of the good and unacceptable behaviour everyday. This will give the child and parent a concrete look at how it is going. Once this begins to work, praise the child for learning to control misbehaviour.
- ***Time-outs*:** Time-outs also work well for preschool children to teach them self-regulation. Establish a suitable timeout place, which is free of distractions and will force the child to think about how he/she has behaved. Consider the length of time that will best suit the child's age until child is calmed down.

Disciplinary techniques during late childhood (6 to 12 years): During late childhood, the children are mentally more mature and can understand the reasons behind parental discipline. They can understand the causes and consequences of their acts. It better that the parents explain the reasons for the behviour they expect from their children.

- ***Time-outs:*** Time-out is also one of the effective disciplinary strategies for this age group.
- ***Natural and logical consequences*:** Children in this age group can be disciplined with natural and logical consequences. As they mature and request more independence and responsibility, teaching them to deal with the consequences of their behaviour is an effective and appropriate method of discipline. Positive consequences are used to increase or encourage desirable behaviours. Positive consequences can range from short term rewards such as candy and extra play time to long term rewards such as trips and gifts. Doing extra chores in the house, taking away a positive and having no television for the night are examples of negative consequences. Children will soon realize that the behaviours that are causing these consequences are to be avoided.
- ***Reasoning*:** During middle and late childhood, children begin to think more logically. Due to children's growing ability to think logically, parents can use reasons and explanations more often when disciplining them. Children must be made to understand the long-term ramifications of their actions.
- ***Being firm*:** Being firm does not mean being harsh or rigid. It also does not mean yelling, threatening, or being unkind. Being firm is meaning what the parent says. This strategy is very effective if it is used consistently. Parents should avoid arguing with children. This can lead to a power struggle. Staying calm is an important part before applying any positive or negative consequences to your child's behaviour.
- ***Rewards*:** Rewards such as praise or special treat for handling difficult situation well, have a strong educational value, if they show children that they have behaved correctly and also motivate children to repeat the approved behaviour. Praise is a powerful reward. Praising child is one of the most important things one can do as a parent. Praise is nourishment. It helps the child grow emotionally, just as food helps the child to grow physically. The reason behind giving praise is to build up self-esteem, to give a belief of personal satisfaction, with the addition of feeling secure within ones self.

- ***Punishment*:** Too permissive or too rigid discipline is counterproductive. Like rewards, punishments must be developmentally appropriate and administered fairly; otherwise, it may result in resentment. Parents should never shout, shriek, or spank their children for small things. The best way to discipline a child is to take him in a separate room and explain firmly why his/her behaviour was wrong and make it clear that this misconduct will not be tolerated in future.
- ***Consistency*:** Consistency is the *backbone* to discipline. Good discipline is always consistent. A consistent discipline and parental attitude are conducive to a more balanced development. A wrong act should receive the same punishment every time it is repeated and the right acts the same reward. Consistency in discipline is essential and is followed through. Children have to believe that parents mean what they say. Also, parents should not make unrealistic threats of punishment in anger, since not following through could weaken all the threats. Huge punishments may take away the power of a parent. If parents cut down privileges of their child continuously, the child may not feel motivated to change behaviours because everything has already been taken away.

■ ■ ■

29 Diet and Health of Preschooler

Kanchan Sandhu

Krishi Vigyan Kendra, Panjab Agricultural University, Nurmahal, Jalandhar, Panjab
E-mail: < kanchansandhu1973@gmail.com>

Preschool is an age between 2 and 5 years. Preschool children want to touch, taste, and smell, hear and test things for themselves. They are eager to learn. They learn by experiencing and by doing. Preschoolers learn from their play. They are busy developing skills, using language and struggling to gain inner control. Preschoolers want to establish themselves as separate from their parents. They are more independent as compared to toddlers. Fears often develop during the preschool years. Common fears include new places and experiences and separation from parents and other important people.

Preschool children come under the group at greatest nutritional risks due to fast growth and development. Malnutrition is one of the major causes for the high death and disease rate among preschool children. Preschool children account for a large share of diseases and deaths in most of the developing countries. Malnutrition also poses a great threat to the physical and mental health of preschool children. A nutritional deficiency early in life causes loss of growth and limited capacity to overcome the resulting stunting. Physical, mental and emotional growth is at fastest pace in preschool age than any other age. They are energetic and need optimum nutrition for proper growth and development.

Preschool Children and Health Concerns in Today's Scenario

For children under the age of 5, under-nutrition is caused directly by the interaction between a lack of nutritious food, the prevalence of common childhood illnesses such as diarrhoea and respiratory infections and the impaired ability of caregivers to nurture their children because of resource or knowledge constraints. Eight of the 15 major risk factors of childhood diseases identified globally are nutrition related like underweight, obesity, iron deficiency, vitamin A deficiency, zinc deficiency, high blood pressure, high cholesterol and low fruit and vegetable intake.

The tragic consequences of malnutrition include death, disability and stunted mental and physical growth. About 60% of the 10.9 million deaths each year among children aged less than five years in the developing world are associated with malnutrition. Now a day, raised blood pressure, impaired glucose tolerance and dyslipidemia also tend to be clustered in children with unhealthy lifestyles and diets such as those with excessive intakes of saturated fats, cholesterol and salt and inadequate intake of fibre. Lack of exercise and increased television viewing add to the risk.

Obesity is increasingly prevalent among young children. The factors responsible for obesity are increasing industrialization, urbanization and mechanization making noticeable changes in diets, which have become richer in high-fat, high energy foods and lifestyles more sedentary. Children are busy in watching television or playing videogames at young age. Their diets are chips, aerated drinks and processed foods, fast food, energy-dense, micronutrient poor foods and beverages. Increase in obesity gives a dramatic rise in the prevalence of diabetes. For children, prevention of obesity implies the need to:

- Promote an active lifestyle
- Limit television viewing
- Promote the intake of fruits and vegetables
- Restrict the intake of sugars-sweetened soft drinks
- Restrict the intake of energy-dense and micronutrient-poor food articles, *e.g.*, packaged snacks

Feeding to Preschool Children to Affect Their Health

One of the life's most rewarding- and frustrating-challenges is feeding a healthy diet to a preschooler. After the explosive growth of infancy, the toddlers and preschoolers grow about 2 to 3 inches and gain about 4 to 5 pounds each year. Lifelong eating habits are being set, and lifelong good health may be at stake as bones, teeth, muscles, motor skills, vision and brain connections develop more. Good nutrition is critical.

Apart from lack of healthy food choices, wrong lifestyle affects child's health. Now a day, children are also being exposed to excessive amounts of toxins in the food they eat, the water they drink, the air they breathe and in the environment in which they live. Most of what we eat today is processed, sugar-laden, nutrient deficient, chemical loaded and damaging to our bodies. National dietary surveys showed that our children's diets are deficient in various nutrients. Children eat fruits or vegetables rarely. The calcium deficiency symptoms are commonly observed which has been exacerbated by the increased consumption of carbonated beverages, which interfere with calcium absorption. Studies show that one in every 13 children has asthma. Meats are injected with antibiotics and hormones, while our fruits and vegetables are loaded and grown with pesticides and herbicides. Over 60,000 food additives, *i.e.*, artificial colours, artificial flavours, preservatives, sweeteners, *etc.* are processed into the food mostly for cosmetic reasons and preservation, not nutrition.

Whatever our children eat, the prevailing lifestyles make it difficult to eat well. The food choices of children are influenced by television. As a parent, one must focus on meal timings as time of sharing, fellowship, enjoying each other and getting to know the family. This is a time to develop good eating habits for future optimal health and not watch television together at meal times.

Diet for Preschool Children

Diet plays an important role in preschool age. Children in this age should eat a healthy diet consisting of a variety of foods to avoid any nutritional deficiency. Food and nutrients help to form strong teeth and bones, muscles and a healthy body. A good diet can also help to protect your child against illness now and in the future. Young children's need for energy and nutrients is high but their appetites are small and they can be fussy too, and it can be a challenge to get your child's diet right. Pre-school children can normally eat the amounts they want, even if it seems they are not taking in very much. At this age, children are often good at regulating their appetite. If they are not hungry, insisting on larger amounts of food can create a battle, which you are likely to lose. Whatever we feed to the child should be healthy and balanced. Important nutrients and their sources are as follows:

- **Energy:** As the child becomes increasingly mobile, his/her energy requirement also increases. Dentition is fully developed by about 2 years of age. An overall good nutrition in the form of small energy dense meals should be planned for a growing preschooler. Insufficient intake of food will not only result in poor weight gain but will also hamper growth to a great extent. This may later affect development of cognitive skills and learning. Energy sources are cereals, sugars and fats. Cereals should be part of a pre-schooler's daily diet.

- **Cereals as source of energy:** A preschooler's diet must contain whole grains. Whole grains can be in the form of whole-wheat chapatti, brown rice, whole-wheat pasta, whole grain bread, cookies and biscuits made of whole wheat or *suji*, *etc.* Whole grains will provide vitamins as well as fibers. Many preschoolers do not eat enough fibers, and whole grain foods are an easy way to get fibers into your child's diet.
- **Sugars and fats as source of energy:** The fat intake should be around 25 g per day. 30-40% of total calories should be from fat sources in the diet of a preschooler since fats provide dense energy. Healthy fats at that age are essential for proper growth and development of a child. Monounsaturated fats, which are derived from plant sources like olive oil, nuts, seeds, avocado, *etc.*, are comparatively good. Sugars shall be consumed in moderation, as they are empty calories. Saturated fats shall be avoided.

  Fatty and sugary foods include spreading fats, cooking oils, sugar, biscuits, cakes, crisps, sweets, chocolate, cream, ice cream and sugary drinks. Do not give these often, and when you do, make sure they are in small amounts only. Many of these could affect your child's intake of more nutritious foods and lead to a less healthy diet. Sugary foods and drinks including fruit juice can also contribute to dental decay, especially when eaten or drunk between meals. Some sugar-free or diet drinks can also cause decay because of their acidity. Water is the best drink to have between meals.
- **Proteins:** Protein is a vital dietary component for preschoolers and toddlers, as it is needed for optimal growth. Bone growth and growth of muscle mass pave way for increased demands for protein. A preschooler requires adequate amount of protein as total body size increases during preschool years and haemoglobin formation also. We can say that protein is important for the building block of body especially in preschool age when a child is growing fast. Enough protein should be consumed every day to allow for proper development. A protein intake of 1.5 to 2 g/kg body weight is essential during this period. Deficiency of proteins can lead to stunted growth and development.
- **Calcium:** According to Mayo clinic, preschooler should consume 500 mg calcium every day. Inadequate calcium intake can adversely affect the bones of the growing children. Children need good amount of calcium for bone and teeth mineralization and maintenance. The amount of calcium a child needs is determined in part by the consumption of other nutrients, such as protein, phosphorus and vitamin D, as well as the child's rate of growth. During this period of development, children need calcium 2 to 4 times as much per kilogram of body weight as the adults need.
- **Milk and its products as a main source of protein and calcium:** Milk and dairy foods are an important source of calcium and proteins. Child should take 500 to 600 ml of milk per day. Since dairy foods are the primary source of calcium, children who do not consume enough dairy food or have an aversion to dairy products may be at risk for calcium deficiency. The diet of a preschooler should include around two glasses of whole milk every day. Milk also contains vitamin D that aids in calcium absorption. Milk and its products like curd, kheer, sevian, *etc.* are all good sources of calcium as well as proteins. Milk can be used on cereals or in drinks, puddings, sauces, cheese, yoghurt, *etc.* Grated cheese, cheese spread, or cheese portions can be used on sandwiches or toast.

Children suffering from lactose intolerance or milk allergy should meet their calcium requirements from other food sources. This includes tofu, yoghurt, ragi, fortified cereals, orange juice, broccoli and leafy greens.

- **Dals, beans and legumes as source of protein:** These are fair sources of proteins and other vitamins. Sprouted dals, legumes and beans are best to consume as a rich source of important nutrients. They should be a part of daily diet.
- **Meat, fish and poultry as a main source of protein:** A preschooler's diet must contain at least 20-30% of calories from the protein sources as growth rate is very fast and vital developments are taking place at this age. Egg is considered as a complete food. It is good source of protein. Fish is an excellent source of protein and omega-3 fatty acids. Boneless chicken, fish and mutton shall be introduced to preschoolers. Egg can be made and given in various forms like poached egg, boiled egg, half boiled egg, omelets, *etc.* Chicken can be boiled and chicken strips can be given. It can be filled in chapatti, omelets, *etc.*
- **Minerals and vitamins:** Minerals and vitamins are important as protective food. They are required for proper growth, development and for the absorption and utilization of other nutrients. Vegetables and fruits are important sources of minerals and vitamins along with fibres.
- **Vegetables:** Vegetables are coloured and children love coloured foods. Many of the essential minerals and vitamins, which are necessary for the development of preschooler, are present in different vegetables. Vegetables can be prepared in innovative ways and by right combination of colours and flavours to make them acceptable to the children, *e.g.*, carrots can be made with peas, beans with mushrooms, peas and mushrooms with cheese, lots of vegetables in a tikki form, all green leafy vegetables in the dough of *rotti*, *ghiya* in *dals*, *sambhar* with carrots, beans, *kaddu*, *etc.* Children love to consume vegetables in different shapes and in colourful plates. Carrot sticks, beans sticks and cucumber slices with curd dip are generally liked by the children. If your child does not like vegetables, try hiding them by pureeing into soups, sauces, casseroles and pizza toppings.
- **Fruits:** Fruits are good source of essential minerals, vitamins and fibres necessary for the child's growth and development. Fruits in raw form are best to consume. The next best form is fresh fruit juice. A colourful bowl of fruit is not only liked by the children but it provides lots of nourishment along with satisfying children's sweet tooth. Deficiency of any vitamin can lead to the occurrence of various deficiency diseases. Fruit does not contain the fibre that is in whole and cut-up fruits. Serve your preschooler no more than half cup to three-fourth cup (4 to 6 ounces) of juice a day. Choose 100% fruit juice. Aim for at least five servings a day, where a serving is about a handful in size. Use fruit in puddings and as snacks.

Table 1: Common nutritional deficiencies and their symptoms in preschool children

Nutrients deficiency	Role	Sources	Deficiency Symptoms
Vitamin A	Promotes healthy bone growth, vision and cell division and helps to regulate the immune system. It is needed for healthy skin,	Whole milk, liver, eggs, carrots, sweet potatoes, spinach, most dark and yellow orange colour fruits and vegetables	Bitot spots, night blindness, childhood blindness. Its deficiency increases the risk of severe illness even death from common childhood

Nutrients deficiency	Role	Sources	Deficiency Symptoms
	cell development and eye sight.		infections such as diarrheal diseases and measles
Vitamin B_1 (thiamine)	It is essential in converting foods into energy and helps in supporting the normal function of the nervous system, muscles and heart. It promotes normal growth and development and helps in regulating appetite.	It is present in moderate amounts in all food articles. The best sources are whole grains and fortified cereals.	Confusion, poor coordination, depressed hunger, irritable, sleep disorders, fatigue, depression, agitation, memory loss, headaches, emotional instability and sleeplessness
Vitamin B_2 (riboflavin)	It supports energy production necessary for red blood cell and antibody production, respiration and regulating human growth and reproduction. It is essential for healthy skin, nails and hair growth, thyroid activity, healing of wounds and general good health.	Liver, kidney, whole grains, green leafy vegetables, milk, yeast, cheese, oily fish, eggs, enriched cereals, almonds and mushrooms	Lack of energy, pale skin, brittle nails, hair fall and irritable
Vitamin B_3 (niacin)	It is involved in converting food into energy. It regulates circulation, hormone production, the digestive and nervous systems and promotes healthy skin.	Beef liver and kidney, pork, turkey, chicken, veal, oily fish, beets, peanuts and strawberries	Lack of energy, fatigue and irritable
Vitamin B_6 (pyridoxine)	It assists in the growth of new cells and the functioning of the immune system, It helps in converting food into energy, controlling moods and behavior. It is involved in cancer immunity, red blood cell production, preventing skin problems and in fighting certain heart difficulties.	Eggs, chicken, yeast, carrots, fish, liver, kidneys, peas and walnuts.	Irritable, lack of immunity, depression and mood disorders

Nutrients deficiency	Role	Sources	Deficiency Symptoms
Vitamin B_{12} (cobalamine)	It contains cobalt and it is known as cobalamin. The primary functions are to maintain a healthy nervous system and to produce red blood cells.	Meat, dairy products and eggs but no reliable plant sources	Unable to absorb the vitamin through normal food intake and will have to take ongoing regular injections to alleviate the problem.
Vitamin C (ascorbic acid)	The human body is unable to store vitamin C. It is important in forming collagen that gives structure to bones, cartilage, muscle and blood vessels, in maintaining bones and teeth and in the absorption of iron.	Most fruits and vegetables, amla, guava, tomatoes, oranges, strawberries and broccoli	Scurvy that can become fatal, depression, fatigue, bleeding gums, lethargy, weakness and aversion to work
Vitamin D	It promotes the calcium and phosphorous absorption that are vital in forming and maintaining strong bones. It is also involved in regulating cell growth and maintaining a healthy immune system.	Sunlight, oily fish, fortified cereals, beef liver, certain funguses (*i.e.*, mushrooms), milk and egg yolk.	It is associated with a large number of diseases, degenerative and genetic conditions, allergies, depression and fatigue
Vitamin E	A is a powerful source of antioxidants and as such acts as a barrier to poisons and diseases that can damage the body. Involved in immune system function, DNA repair, the protection of blood cells, the nervous system, muscles and the eye retinas from free radical damage.	Nuts, green vegetables (spinach, broccoli, sprouts) eggs, whole meal products, soybeans and vegetable oils and particularly olive oil.	Weak immune system
Iron	It is important constituent of haemoglobin. Iron deficiency is common in preschoolers. Food items rich in vitamin C, eaten at the same time, can help to maximize iron	Iron is found in meat, in dark green vegetables and whole grains. It is also found in some dried fruits.	Anorexia, fatigue, weakness, headaches, pallor, irritable and learning disabilities

Nutrients deficiency	Role	Sources	Deficiency Symptoms
	absorption from non-meat sources.		
Zinc	Zinc is essential for proper development. It is needed for wound healing, proper sense of taste, proper growth and normal appetite.	Preschoolers are at risk for marginal zinc deficiencies as the best sources are meats and seafood.	Poor appetite, irritability, failure to grow, sluggishness and emotional disorders
Calcium	It is essential for proper growth and development of bones. It maintains bones healthy and strong. It supports proper functioning of nerves and muscles and helps in avoiding blood clotting.	Excellent sources of calcium include spinach, turnip greens, mustard greens, and collard greens, yogurt, mozzarella cheese, cow's milk, and goat's milk, basil, thyme, dill seed, cinnamon, broccoli, sesame seeds, fennel, cabbage, green beans, garlic, tofu, oranges and mushrooms	Improper bone mineralization leads to rickets, a condition characterized by bone deformities and growth retardation. Low levels of calcium in the blood may cause a condition called tetany, in which nerve activity becomes excessive. Symptoms of tetany include muscle pain and spasms, as well as tingling and/or numbness in the hands and feet. Bed-wetting, fatigue, headaches, restlessness, hyperactivity, twitches, irritability teeth-grinding/night, poor sleeping habits, pins and needles sensation on skin
Magnesium	Magnesium relaxes nerves and muscles, build and strengthen bones and keep the blood circulating smoothly	Rich sources of magnesium are turnip greens, mustard greens, spinach, broccoli, green beans, cucumber and flax seeds.	Muscle weakness, tremor, muscle spasms, imbalanced blood sugar levels, headaches, elevated blood pressure, heart arrhythmia, irregular contraction or increased heart rate, softening and weakening of bone

Most Important Nutrient Element

Iron: Iron deficiency is common in preschoolers. Iron is a vital component of hemoglobin, which is the carrier of oxygen in the blood. As a young child grows, blood volume increases and so does the need for iron. Preschoolers typically eat less iron-rich foods than they did in infancy. In addition, the iron that children get is usually non-haem iron (from plant sources), which has a lower availability than haem iron (from animal sources). As a result, the children are at high risk for iron-deficiency anaemia. More than 30% of the world's children population is anaemic, may suffer from reduced learning ability and work capacity due to iron deficiency. Iron deficiency impairs the cognitive development of children. Iron deficiency damages immune mechanisms and is associated with increased morbidity rates. Iron deficiency impairs physical work capacity in children.

Sources: The sources of iron are meat, in dark green vegetables and whole grains. It is also in some dried fruits.

Factors influencing dietary Iron absorption: Amount of dietary haem iron, especially from meat or non-haem iron, content of calcium in meal (*e.g.*, from milk, cheese), food preparation (*i.e.*, time and temperature) all these factors influence the dietary iron absorption. The factors like ascorbic acid (vitamin C), fermented vegetables and legumes enhance iron absorption, whereas, phytate and other lower inositol phosphates (*e.g.*, bran products, bread made from high-extraction flour, nuts, soybeans and peas), iron-binding phenolic compounds (*e.g.*, tea, coffee and cocoa) and calcium (*e.g.*, from milk and cheese) inhibit its absorption.

Lunchbox Lunches

Lunch is really an important meal for your preschooler. It should provide more than one-fourth of the day's total calories and contain a mix of the healthy food groups. Whole grains, vegetables and fruits are the mainstays of a good lunch. Dairy products, which supply calcium, can enhance taste as well as variety. Eggs make a complete protein and vegetarians can use beans, nuts and legumes for protein. A lunch box with variety and colourful foods will be more interesting and relished by child. Wrap a sandwich in a lettuce leaf instead of bread. Include kid-sized cut up carrots, broccoli and snow peas with yogurt dip. Steamed foods like carrot added *idli*, *vada*, *upma*, *etc.* are healthy snacks for Tiffin.

A vegetable whole-wheat sandwich with added cheese and sprouts *chaat* with cucumber and tomato, cheese balls, sprout *parantha*, *missi rotti* with added greens, *etc.* add variety as well as nutrition to child's Tiffin.

Other Common Diseases in Preschool Age

Food Allergies: Food allergies are common among preschoolers. A child introduced to new food may develop some allergy. The common symptoms are as followings:

- Temper tantrums, whining and screaming
- Bags/wrinkles under eyes and dark eye circles
- Excessive fatigue and depression
- Abdominal pain
- Aggression glazed look
- Bloating and hyperactivity

- Leg aches and wiggly/restless legs
- Red earlobes and red cheeks
- Dry skin and eczema

Constipation: Children eat less of fibrous foods and fruits and veggies. They may also restrain from sipping water throughout the day. Due to avoiding nature's call for passing stools in hurry to go to school and faulty eating regime constipation is developed in preschoolers. The symptoms include hard stools, pain while passing stools, child avoid passing stools, lack of appetite and irritability. Constipation can be easily corrected by including fresh fruits, vegetables, whole-grains, *etc.* in the diet.

Dental problems: Dental caries are most commonly observed in children at around this age. This is specific because of high consumption of sticky, gummy candies and sweets. The longer the tooth is exposed to sugar in the sweets, the greater are the chances of tooth decay, thus, children should be taught habit of brushing their teeth twice daily. They should be encouraged to munch on raw crisp foods like chunks of apple/pear, cucumber, or carrots in the day.

Chicken pox: Chicken pox is an infection caused by varicella-zoster virus (VZV). It spreads from one person to other by air borne droplets. Majority of the kids suffer from this disease in preschool age. As they age, the antibodies that had developed in the course of prior infection help them avert a second episode of infection that may occur anytime in life. The immune system is not strongly developed in preschool age. As a result, the child gets more prone to infection.

Symptoms

- Pre-symptoms are high fever and abdominal pain, sore threat and headache.
- Chicken pox is characterized by the appearance of a rash generally on the abdomen.
- Rash is red, itchy and in the form of small bumps like an insect bite.
- Rashes start from abdomen, back, face and then spread all over the body.
- Irritation and itching
- In rare cases, complications like infection of lungs leading to pneumonia, swelling and infection of brain called as encephalitis is developed in children.
- In case of complications, the affected child may develop unconsciousness and convulsions that may prove fatal.

Treatment

- Soothing bath
- Give bland and simple food.
- Pat him dry and do not rub after bath.
- Do not give highly acidic, spicy and oily and very hot or cold foods.

Cough: Preschoolers are very sensitive to weather changes. They are susceptible to cold and cough very easily. When a child suffers from congestion due to infection of influenza virus or common cold, there is production of mucus (phlegm). Coughing action tries to get rid of excessive mucus formed in airways and helps the child breathe better. This type of cough is called productive cough. Whooping cough is caused due to a bacterial infection, which affects the lungs and airways. This is characterized by intense cough. A characteristic sound is there when a patient try to breathe after a bout of cough.

Symptoms

- Cold with cough and runny nose
- Cough for more than 2 weeks and it worsen with passing time making it difficult for child to breathe.
- Dry cough can be due to some allergy.

Asthma: Cough accompanied by wheezing is the symptom of asthma. Asthmatic child tend to cough more at night. In asthma, airways are constricted making breathing difficult. If cough persists for more than a week, one must consult a doctor.

Whooping cough: Whooping cough, a highly contagious respiratory infection, is considered a common cause of childhood illnesses. Whooping cough or pertusis is an infection of respiratory system caused by bacterium. It is characterized with severe coughing spells that may end up in a *whooping sound.*

Symptoms

- First symptoms can be of common cold for about 1 to 2 weeks, like sneezing, runny nose, mild cough, low-grade fever and diarrhoea.
- Episodes of severe coughing for about two to four weeks
- At times, these coughing spells may last for long durations.
- Prolonged spells of coughing followed by vomiting, turning red or blue, having difficulty in breathing, or showing signs of dehydration are the symptoms of emergency.

Upper respiratory tract infections: Upper respiratory tract infections include common cold, influenza, laryngitis, pharyngitis, sinusitis, tonsillitis and croups. The main causative organism for upper respiratory tract infection is generally virus and due to bacteria like sore throat. These infections are transmitted more readily as the nasal secretions from children with viral infections contain more viruses than those from infected adults. Generally, lesser attention is paid to the hygiene, which makes the spread of infection more vigorous. The possibility of infection spread is further enhanced when children interact with each other at crèches, play ways, childcare centres, parks, *etc.*

Symptoms

- Sneezing, coughing, runny nose, congestion, headache and sore throat
- Children suffering from flu develop the symptoms like chills, body aches, weakness, fatigue, high fever and headache.
- Sinusitis symptoms include cold, pain and pressure in the forehead, cheeks and nasal region along with the symptoms of cold.

Prevention of Upper Respiratory Tract Infections

- Take rest, increase fluid intake and use of petroleum jelly around the nose.
- Get children's hands washed several times a day.
- Teach them not to avoid touching their eyes, nose, mouth, *etc.* without washing their hands.
- Bronchitis and pneumonia are common lower respiratory infections in children.
- Commonly observed symptoms are fever with chills, muscle aches, cough, nasal congestion and sore throat.

Pneumonia: It is characterized by infection, inflammation and fluid in one or both lungs. The common symptoms are chills, cough, vomiting, rapid breathing, chest pain, abdominal pain and loss of appetite.

Prevention: One must give plenty of fluids to the child. Balanced diet and light foods are important in pneumonia.

Ear infection: The middle ear is connected to the throat by the Eustachian tube. The function of this tube is to drain fluids such as mucus from the middle ear and to allow air to reach the middle ear *via* throat. The presence of air at atmospheric pressure in the middle ear helps in equalizing pressure inside the ear to the outside. Children are more susceptible to develop middle ear infections because the Eustachian tube is not well developed in them. Inflammation of the middle ear is probably one of the most common childhood illnesses. The Eustachian tube can become blocked due to either congestion or improper drainage of the mucus during attack of upper respiratory infections such as common cold or due to allergies. Fluid may then build up in the tube and produce an environment conducive to the proliferation of bacteria, causing an infection. Sometimes, the infection may recur once again after disappearing on its own.

Symptoms

- If a child cries and complain earache and is fussy for no apparent reason, tugging and pulling at his ears, this can be a sign of ear infection.
- Ear infections generally occur right after a spell of common cold.
- Pressure builds up in the ear when there is an infection in the middle ear and is more painful when lying down. If child feel uncomfortable and cries when lying down then he might be suffering from ear infection.
- In severe cases, the pressure in the ear is so high that the eardrum (the membrane separating the outer ear from the middle ear) ruptures and releases the pressure in the middle ear. Some whitish or yellowish discharge from the ear is a sure indicator of an infection.
- Child is not able to hear soft sounds, turns up the volume of television, talks louder than usual, and then, he may be having mild hearing loss due to an ear infection.
- Other accompanying symptoms may be fever, dizziness, nausea and vomiting.

Prevention of ear infection

Few things that can prevent ear infection in children are:

- Maintain hygiene like washing hands frequently.
- Immunize your baby appropriately and routinely.
- Breastfed babies are less prone to ear infections that bottle fed babies.
- Ear infection in children is not contagious and kids often grow out of it as the years pass.

Diarrhoea: Every child has different bowel movements. If child visits the toilet more than usual and passes loose and watery stools, then he may be suffering from diarrhoea.

Causes: Diarrhoea in children may occur due to various reasons. It is an infection caused by bacteria, virus, or some parasite.

Diarrhoea due to bacterial infection: The symptoms are loose stools and cramps in the belly. There may be blood in the stool and the child may develop fever. Vomiting may or may not be present. *Escherichia coli*, *Staphylococcus*, *Shigella*, *Salmonella*, or *Campylobacter* are the main culprits in causing diarrhoea in children.

Diarrhoea due to viral infection: Viruses that affect the gut can cause diarrhoea. Rotavirus infection is very common in children. Others that may bring about diarrhoea are adenovirus, astrovirus and influenza virus. Symptoms due to viral infections are diarrhoea, vomiting, bellyache, fever, chills and general malaise. It is of utmost importance to maintain proper hygiene when dealing with a child who is suffering from viral diarrhoea.

Diarrhoea due to parasitic infection: Diarrhoea, accompanied by complains of gas, nausea and cramps, is an indicator of parasitic infection in his tummy. Another common symptom that is indicative of parasitic infection is that the child may pass stools that are greasy in nature.

Diarrhoea due to too much sugar: Drinking too much of sugary drinks like sweetened fruit juices or soda can lead to a child passing loose stool. There may not be any infection but the excess sugar upsets the balance of the gut and results in too much water being retained in the gut leading to loose stools.

Diarrhoea due to medications: Antibiotics may cause loose stools in children as they wash out the natural bacterial population in the gastrointestinal tract.

Diarrhoea due to food allergies/ intolerance: If the child is allergic to any food or is unable to digest some food like lactose intolerance, it may manifest in the form of diarrhoea.

Symptoms of dehydration

Sunken eyes, decreased frequency of urination (or dry diapers for babies), lack of tears when the child cries, dry or sticky mucous membranes (the lining of the mouth or tongue), lethargy (less than normal activity) and irritability (more crying and fussiness with inconsolability).

Treatment

As there is loss of water in the form of loose stools, the important is to give lots of clear fluids, ORS, squash, water, *etc.* Dehydration is a very serious problem in diarrhoea and if precautions are not taken, diarrhoea can become life threatening. Oral Rehydration Solutions (ORS) can be given immediately and carried on until the dehydration is corrected. The child should also be shifted to a cooler place (such as in an air conditioner room) so as to avoid loss of water due to sweating. Sugary drinks should be avoided and only clear fluids should be given. Eating bananas, rice and dry toast helps to tighten the stool.

Prevention

Maintaining hygienic habits ensures that child wash his hands frequently with soap and water before meals and after using the toilet. Cook food, especially meats and poultry thoroughly. Wash hands after handling raw meat and poultry. Clean cooking and eating utensils thoroughly after using them. Drinking water should be purified or filtered, especially in the monsoon season when the chances of water contamination are high.

Tonsils: Tonsils are glands made up of lymph tissue, which are located on the right and left sides at back of the throat. The lymph tissues in the tonsils contain infection-fighting cells and prevent the transmission of infection from areas like throat, mouth and sinuses to other parts of body. Tonsils help in the production of antibodies that help to fight the infections.

Tonsillitis: Inflammation of the tonsils is called tonsillitis. When the tonsils have to filter a large number of microorganisms, they increase in size but when they are unable to kill these germs, they are infected and inflamed by themselves. Tonsillitis is a very common complaint in preschool age children. Repeated occurrences of tonsillitis may result in the enlargement of tonsils, referred to as chronic tonsillitis.

Symptoms

- Enlarged tonsils and pain and difficulty in eating, drinking and swallowing
- Swelling of on either side of the neck or jaw
- Earache, stiff neck, bad breath, fever, chills and headache

Home Remedies for Tonsils

- Ensure child receives adequate amount of sleep to rest his/her voice.
- Give a solution made of one-fourth tablespoon of salt dissolved in four ounces of water to gargle. This will help in soothing his or her infected tonsils to some extent.
- During the day, give good amount of warm liquids like soups.
- Give soft and bland food so that it may not irritate the infected area. Give cold fresh fruit juices, or ice creams.

Bedwetting: Wetting the bed at night (nocturnal enuresis) is commonly observed in many three to eight years old children. With respect to the gender, 60% of all the bed-wetters are males and a major percentage of them wet the bed almost every night.

Causes: There can be many underlying reasons behind the problem of bed-wetting. Some of them are described as follows:

- Heredity can play an integral role in some cases. It has been observed that if one parent or both parents seem to have been faced with this problem in their childhood, then there are around 50% chances that their child will also be a bed wetter.
- Children with a small bladder than normal are more likely to suffer from this problem.
- Urinary tract infections or constipation may cause a full bowel to exert pressure on the bladder.
- It can be due to insufficient production of antidiuretic hormone (ADH) that has an important role in prevention of water loss from the body. Inadequate production of ADH also influences sleep, making it difficult for the person to wake up, thus, it has been observed that children who have this problem are usually the ones who also have deep sleep.
- Stressful events in a child's life such as moving to a new school, sudden shift in the residential place, quarrelling parents, arrival of a sibling, abuse of any kind, *etc.*

Steps to reduce the chances of child's bedwetting problem

- Refrain from rebuking your child for wetting the bed as it may affect his self-esteem to a great extent. A child already going through mental distress due to this problem should be rather comforted and reassured from your end that he is perfectly normal and that you both jointly will find a better way to deal with it.
- Limit your child's fluid intake two to three hours before bedtime.
- Follow a routine of visiting the washroom just before sleep.
- Make sure that the child is properly dressed and warm in winters.

■ ■ ■

30 Young Children and Healthy Eating

Shikha Trivedi

Lecturer Early Childhood Development, Department of Family and Consumer Sciences, University of Botswana; E-mail: <kirteshshikha@gmail.com>

Paying attention to nutrition is vital to overall childhood health. The ages between one and four are crucial period for learning good dietary habits. Good nutrition can hold off obesity, osteoporosis, diabetes and heart disease. These are now common diseases in obese children. Proper nutrition not only prevent many medical problems but also ensures that your child grow as a healthy child.

Healthy Eating

Healthy eating means eating a wide variety of nutritious food articles those help the body to grow and function well. Healthy food items contain all the carbohydrates, protein, minerals, vitamins and fiber needed by the body. It is very important to develop good habit of eating nutritious food from the childhood itself. Children need a regular intake of food to keep them active and help them in concentration while learning. Some children in this age group are very fussy about the food, so do not force the child to eat and offer a wide variety of foods.

Nutritional Requirements of Young Children

Calories

Schoolchildren have a high-energy requirement for growth and activity. The calorie intake has to be properly utilized by the body. If the child is taking plenty of calories and not involved in any physical activities or exercises, extra calories are going to build up body, resulting in overweight. If you realize that your child is putting on too much weight, follow the following points:

- Encourage him/her to do physical activity in any form such as sports, jogging, swimming, *etc.*
- Plan healthy meals based on the main food groups.
- Try to limit sugary and fatty food articles.

http://www.livestrong.com/lose-weight

- Encourage healthy eating to ensure children to maintain a healthy weight, and make sure that the whole family is eating healthy meals to provide good role models.

http://photobucket.com/images/healthy%20food/

Calcium

The calcium is important for healthy bone development. Dairy products such as milk, cheese, yoghurt and tofu are good source of calcium. It is important to give three servings of calcium rich food to the children, *e.g.*, 150 ml of milk, 100 ml of yoghurt and a small piece of cheese every day.

http://www.fotosearch.com/photos-images/calcium-rich.html

Iron

Iron helps to keep red blood cells healthy. If iron intake is insufficient, the individual can suffer from anaemia (lack of red blood cells in the blood). The good sources of iron are red meat, liver, beetroot, fortified breakfast cereals, beans and pulses. The food articles rich in vitamin C, eaten at the same time, can help to maximize iron absorption from non-meat sources.

Vitamins A, C and D

Vitamin A, C and D are important for the body in the following ways:

- Vitamin A is needed for healthy skin and cell development.
- Vitamin C is important for the immune system and growth.
- Vitamin D is essential for the body's calcium metabolism and can be produced by the body when skin is exposed to sunlight.

Young children can be given extra A, C and D vitamins in supplements as vitamin drops. Below are some common points that can help the parents to promote good eating habits in children:

Ways of Promoting Good Eating Habits in Children

You can encourage your child to eating healthy food by keeping following points in mind:

Parents being role model: Parents are the primary influence in the lives of young children. They can really inculcate good eating habits in their children by being the role model. Parents can promote good nutrition by setting healthy eating habits to be a regular part of the family's life. It is much easier if everyone in the house follows these guidelines, than if the child has to do it alone. Include variety of foods in your child's diet. Preschoolers need nutrients to provide energy, promote growth of body tissues, and regulate body functions.

Active involvement of children in various food related activities: Allow young children to help with menu planning, grocery shopping and meal preparation. Being actively involved teaches children the importance of a healthy diet. They will also appreciate that you value their opinions enough to let them assist when making healthy food choices.

Eating fresh fruits and vegetables as snacks: When a child feel hungry between meals, instead of giving them a packet of chips or biscuits have fresh fruits and vegetables readily available for them to eat. These will add nutrition to their diet.

Limit Fast Food Intake: Children like fast food, which is also easy to access and convenient for you but these foodstuffs have little or no nutritional value for a preschooler. To encourage healthy eating habits, always keep the children away from fast food.

Allow your child to eat according to their appetite: The child likes to eat as per his/her hunger. Allow the child to decide how much food he/she wants to eat. Never force the child to finish or clean the food since this will lead to disliking for the food. Parents can easily handle the situation by providing variety of healthy foodstuffs. Initially, the food should be offered in a small quantity first and later give more if the child asks for more food. Meal sizes will vary, as the amount of food a child needs depend on what else has been eaten during the day.

Caution: Never offer any kind of reward like chocolates for finishing the food.

Do not let children skip breakfast: It is very important to have breakfast in the morning to get all the energy to start a day. Children who skip breakfast generally have below average nutrition and they are unable to concentrate in school activities.

Skipping breakfast becomes more common as the children get older. Some schools have introduced breakfast programs because they were concerned about children who skip breakfast. Children did better in school once the program was introduced.

Quick and Easy breakfasts

Here are some easy to prepare healthy breakfast ideas:

- Fresh fruit with wholegrain breakfast cereal and reduced fat milk

- Toast with a thin spread of butter, half apple and milk
- Toast sandwich with cheese, tomato and milk
- Rolled oats with sultanas and milk
- Baked beans on toast and orange juice
- Semolina pudding, slice of apple and guava
- Boiled egg with buttered toast and mango shake

If you are not sure of the type and amount of nutrients needed by your child, you can follow the food guide pyramid for young children aging two to six.

The Traditional Healthy Mediterranean Diet Pyramid for Children

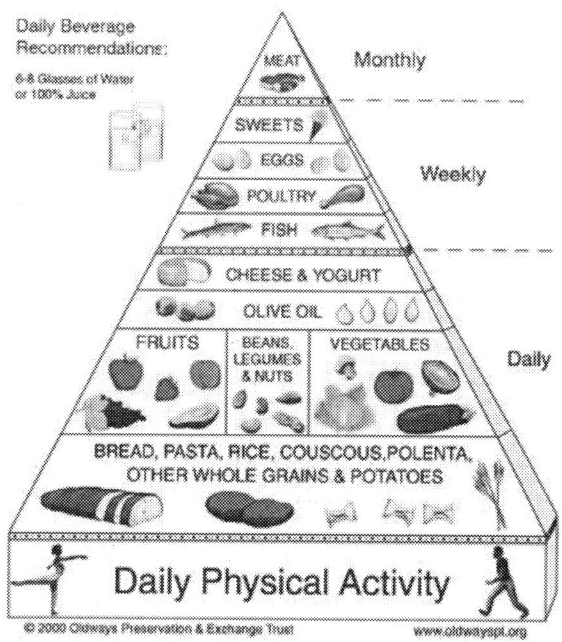

The United State Department of Agriculture recommends that healthy foodstuffs for children should include:

- **Grain group (6 servings):** These servings can include one slice of bread, 1/2 cup of cooked rice or pasta, 1/2 cup of cooked cereal and 1 cup of ready to eat cereal.
- **Vegetable group (3 servings):** These servings include 1/2 cup of chopped or raw vegetables, or 1 cup of raw leafy vegetables.
- **Fruit group (2 servings):** This includes 1 piece of fruit, 3/4 cup of 100% fruit juice, 1/2 cup of canned fruit, or 1/4 cup of dried fruit.
- **Milk group (2 servings):** It includes 1 cup of milk or yoghurt and 2 slices of cheese.
- **Meat group (2 servings):** includes 20 to 25 g of cooked lean meat, 1 egg, or 2 tablespoons of peanut butter, poultry, or fish, 1/2 cup of cooked dry beans.

By being a good role model, involving children in meal planning and preparation, having nutritious food items readily available, following the USDA guidelines and staying away from fast food, kids will quickly learn how to make healthier food choices for life.

Healthy Snacks for Preschoolers

Snacks

Snacks are an important part of a child's food intake for energy and nutrients. What children eat is more important than when they eat. Children who snack on lollypops and chips may not get all the nutrients needed for good health. It is important to note that sweets like lollypops, chocolates, *etc.* kill the appetite, and thus, affect the child's health. Suggestions for healthy snack include:

- Fresh and dried fruits, or fruit packed in natural juice
- Yoghurt or cheese
- Fruit bread, bun, or muffins
- Bread, rice cakes, or crackers with spread
- Vegetable pieces and dip

All children should be offered nutritious snacks in mid-morning and mid-afternoon. According to the National Network for Childcare, a preschooler's total caloric intake is about 1,000-1,200 calories and snacks should provide an estimated one forth of those calories. One can choose the snacks from the most important food groups, *i.e.*, fruits and vegetables, grains and milk.

- **Monday:** Whole wheat bread slice and peanut butter
- **Tuesday:** French toast and pear, plum, peach apple, or other in-season fruit
- **Wednesday:** Fruit slices and rusks
- **Thursday:** Boiled egg and toast
- **Friday:** Cereal and strawberry banana smoothie (1.5 cups low fat vanilla yogurt, two bananas, 2 cups of milk, 1.5 cups frozen strawberries blended until smooth)

Snacks to Avoid

Focus on variety so the children are exposed to a wide variety of food choices each day. Nutrition services recommend that we should shy away from the followings:

- Chocolates and candy
- Heavily sweetened desserts
- Fruit flavored drinks
- Soft drinks
- Cereals coated with sugar
- Potato chips

Beware of giving a child food he can easily choke on, such as:

- Peanuts
- Whole grapes
- Raw carrots
- Hard piece of meats

- Chewing gum
- Hard candies

Food tips for growing children

The general suggestions include the followings:

- Offer a variety of food articles everyday.
- Make mealtime interesting for children.
- Encourage healthy eating for everyone in the family.
- Let your child decide if they want to eat or not and how much he/she wants to eat.
- Involve children in meal preparation.
- Encourage water rather than fizzy drinks.
- Offer healthy snacks between meals.

Health tips: Treats are best kept for special occasions. Let the children enjoy the social aspects of eating. You can take your child occasionally to have meal at a fast food restaurant. They can enjoy parties as long as good nutrition is continued on most days. Food is an important part of special occasions for everyone and it should be enjoyed.

Healthy drinks

It is very important to encourage children to drink plenty of fluids. It is recommended to drink three glasses of milk a day to get enough calcium for bone development. Sweet drinks such as juice, soft drink, *etc.* are not very healthy for child's health in a child's diet.

Kindergarten meals

Children continue to learn new skills and ideas about food every day, especially once they start going to kindergarten. They can be involved in preparing their lunch box and this helps the parents in making healthy lunch boxes as per children's choices. Preparing meals together is a great opportunity to give children positive messages about nutrition, such as 'milk makes your bones strong' or 'rice gives you energy to play'. Few suggestions for lunches include:

- Sandwich, fresh fruit and a tub of yoghurt
- Carrot and raisin salad, slice of bread, fruit and a bottle of plain milk
- Boiled egg, buttered toast and fruit juice

Remember: Drinks can be frozen in summer for school.

Health tips: Encourage your child to drink plenty of water and try to eat fresh fruits than canned fruit juice.

Way forward: Prevention of feeding problems

- The best way is to teach your child to feed himself as early as possible, provide them with healthy choices and allow experimentation.
- Mealtimes should be enjoyable and pleasant and not a event full of struggle.

- Do not force your child to eat. Find out the reason of not eating and act accordingly.
- Give only recommended amount of milk and juice so that child has space for eating nutritious solid food.
- Avoid giving large amounts of sweet desserts, sugarcoated cereals, chips, or candy, as they have little nutritional value and kill the appetite.
- Though your child will probably be hesitant to try new food items, you should still offer small amounts of them once or twice a week. Most children will try a new food after being offered it 10-15 times.
- It is important to keep in mind that most children will only eat one or two full meals each day while you should provide three well-balanced meals each day. If your child has had a good breakfast and lunch, it is okay that he does not want to eat much at supper.
- Do not use food as a bribe or reward for desired behaviors.
- Avoid discussing or commenting on your child's poor eating habits while at the table.
- Measuring servings is important so kids do not get too much of one food and not enough of another as well as to control their food intake.

■■■

31 Adolescence and Health Issues

Shanti Balda and Sheela Sangwan

Human Development and Family Studies, I.C. College of Home Science,
CCS Haryana Agricultural University, Hisar-125004, Haryana, India,
E-mail: <sbalda@rediffmail.com>, <sangwan.sheela@yahoo.com>

In the human life span, adolescence refers to the second decade of the life span, roughly ranging from 10 to 20 years. The word adolescence is derived from the Latin word *adolescere,* which means *to grow into adulthood.* In all societies, adolescence is a time of growing up from childhood, an immature stage, to the adulthood, a mature stage. Adolescence is considered as a transition stage between childhood and adulthood. Some people use the terms *teenage years* and *puberty* for adolescence, whereas, puberty refers to the hormonal changes that occur in early youth, and the period of adolescence extends beyond the teenage years. Adolescence is marked by the major physical changes of puberty and important cognitive and social development. There is no scientific definition of adolescence or set age boundary. Psychologically, adolescence is the age when the individual becomes integrated into the society of adults. The individual feels that he or she is equal to his or her elders in legal rights. Many researchers and developmental specialists use the age span 10-24 years as a working definition of adolescence. This age span can be further divided into three sub-stages, *i.e.*, (i) early (11-13 years of age), (ii) middle (14-18 years of age) and (iii) late adolescence (18-24 years of age).

Characteristics of Adolescence

Adolescence has certain characteristics that differentiate it from other stages of life span. These characteristics are being described below:

Important stage: The transition between childhood and adulthood leads to rapidly changing behaviour, identity disturbances and strong emotions. These characteristics are developmentally normal and a natural part of an adolescent's growth. Adolescence is a stage of rapid mental development. Therefore, there is need for mental adjustment and to establish new attitudes, values and interests.

Transitional stage: Adolescence is considered as a transitional stage between childhood and adulthood. During transitional stage, individual's status is vague. Adolescents at this stage are confused about their role since they are considered neither children nor adults. When they behave like children, they are reminded to behave like adults and when they behave like adults, they are said that they are too young.

Stage of personal identity: During adolescence, the individuals begin to explore their personal identities. Adolescents engage in a process of searching for where they fit in with peers and others in the society. They are not clear about their identity, and they try out new personal labels and associate themselves with various peer groups.

Peer relationships: During adolescence, adolescents give more importance to peer relationships than those with the family. Though family interactions and relationships are very important and essential for adolescents' development, they often place a stronger emphasis on their friends' perceptions and values. During adolescence, teens are strongly influenced by their peers' beliefs and behaviour and are often vulnerable to negative peer pressure. They are very concerned about peer acceptance. They behave in ways associated with their gender as the gender role identification strengthens.

Stage of self-centered attitudes: Adolescents lack empathy skills. They find it difficult to look at situations from other person's perspectives. Thus, adolescents are more likely self-centered and focused on their own needs without considering needs of others. This lack of empathy is normal and resolves itself by the end of adolescence. They tend to be self-conscious, lacking in self-esteem and highly sensitive to criticism.

Stage of independence and testing boundaries: Adolescents often test rules and boundaries set by their parents and teachers. Although this behaviour is considered rebellious by parents and adults, but in most cases, this behaviour is driven by the adolescent's need to develop autonomy, experience new activities and to become independent.

Stage of storm and stress: Adolescents often experience mood swings, vacillating between happiness, distress, self-confidence and worries. Some of these mood changes result from physical and glandular changes that are taking place during adolescence. Conflicts with peers, friends and parents, school pressures and experimentation with opposite sex relationships can exacerbate the emotional state of adolescents. They believe that their personal problems, feelings and experiences are unique to them.

Adolescence Transition

No single boundary line can denote the end of childhood or the beginning of adolescence. The passage from childhood to adolescence is composed of a set of transitions that unfold gradually. These transitions are biological, cognitive, social and emotional described as under:

Biological transition: Puberty is the biological transition of adolescence and is the most important indication of beginning of adolescence. The term puberty refers to the period during which an individual becomes capable of sexual reproduction. Puberty is also used as a collective term to refer to all the physical changes that occur in individuals from childhood to adulthood.

The timing of physical maturation varies widely across cultures. In India, menarche, the first menstrual period, occurs around the age 12, although some girls start puberty when they are 8-9 years old, while some show signs of puberty in their teens. The duration of puberty varies greatly for girls and boys. In girls, it varies from fifteen months to six years and in boys from two to five years. In girls and boys, an increased growth hormone results in adolescent growth spurt, *i.e.*, sudden increase in height and weight.

During puberty, there is development of primary and secondary sex characteristics. Adolescents become capable of sexual reproduction through the development of primary sex characteristics. Through secondary sex characteristics, girls and boys begin to look like mature women and men. Adolescents experience rapid and irregular growth.

In boys, primary sex characteristics include growth of the testes and scrotum, appearance of pubic, facial and body hair. Secondary sex characteristics include producing and ejaculating sperms.

In girls, the first sign of puberty is a slight elevation of breasts, which is preceded by the appearance of pubic hair. Breast development continues during adolescence. Internal sexual changes include maturation of the uterus, vagina and other parts of the reproductive system. Menarche, the first menstrual period, does not start at the beginning of puberty though it starts relatively late. Regular ovulation and the ability to reproduce full term baby usually follow menarche by several years.

Cognitive transition: In addition to all physical and emotional changes, the adolescents also start to think differently. During adolescence, individuals go through cognitive transition. Cognitive development refers to development of the ability to think and reason. Adolescence marks the beginning of development of more complex thinking processes called formal logical operations. Adolescents' thinking include abstract thinking (thinking about possibilities), the ability to reason, the ability to consider many points of view and the ability to think about the process of thinking. Their abstract thinking permits them to apply reasoning and logical thinking in abstract concepts such as faith, honesty, justice, fairness and other social and ideological matters. They can look at problems from multiple perspectives. They are able to understand that social situations can have different interpretations, depending on one's point of view. Adolescents acquire the ability to think systematically about all logical relationships within a problem. The transition from concrete thinking to formal logical operations occurs over time. Adolescent progress at varying rates in developing their ability to think in more complex ways.

On a practical level, adolescents will start to challenge adults more as they begin to learn to reason, argue and respond. They also become better able to forecast and consider the consequences of actions before those actions occur. Their abilities to question will also lead them to start to question their own identity. They will begin to understand that they play different roles to different people, such as child, sibling, student, athlete, *etc.*

These cognitive changes do not happen over night. In early adolescence, the adolescents begin to question authority and express personal opinions about their own life. Once in the middle of adolescence, they begin to think about their future, start-making goals and become more involved in search of their identity. In late adolescence, they start making firm opinions about external issues, often quite idealistically. They will also begin to explore what their identity will be once they reach adulthood. Adolescents also begin to think about themselves and may get involved in increased introspection and self-consciousness. They may develop a sort of egocentrism and they may believe that others are watching and evaluating them.

Emotional transition: Adolescence is also a period of emotional transition. As the adolescents mature cognitively and intellectually, they become more interested in understanding their own personality and behaviour. During adolescence, some adolescents have low self-esteem and they are more insecure and self-critical. Self-esteem increases over time during middle and late adolescence. Adolescents evaluate themselves along several different dimensions. An adolescent may have high self-esteem when he or she performs well in academics, low self-esteem when he or she performs poor in games and moderate self-esteem when he or she is satisfied with his or her physical appearance.

Developing a sense of autonomy is an important part of the emotional transition during adolescence. It is as important as is establishing a sense of identity. During adolescence, there is shift from the dependency typical of childhood to the autonomy typical of adulthood. This can be seen in several ways. Adolescents do not generally seek parental support whenever they are upset or worried. They do not see their parents since they feel that they all know or are quite powerful. Adolescents often have a great deal of emotional energy in relationships outside the family. Girls may feel more attached with their boyfriends and boys with their girlfriends. They give more importance to their friends than to their parents. Adolescents are able to interact with their parents as people, not just as their parents. Similarly, parents can discuss with their adolescent children, something that was not possible when their children were younger.

During adolescence, the development of independence is understood in terms of the adolescent's developing sense of individuation. The process of individuation begins during infancy and continues until late adolescence. It is gradual and progressive sharpening of sense of self as autonomous, competent, separate and independent from parents. Individuation is very important in the development of a sense of identity, *i.e.*, how the adolescents see and feel about ourselves. Adolescents who have been successful in establishing a sense of individuation can accept responsibility for their choices and actions instead of looking to their parents to do it for them. The process of individuation may not cause stress and internal turmoil in these adolescents.

It is believed that adolescents are susceptible to peers' pressure. In some situations, peers' opinions are more influential, while in others, parents' opinions are more influential. Adolescents are more likely to conform to their peers' opinions when it is concerned about dress code, tastes in music and choices among leisure activities. When it comes to educational or occupational plans or values and ethical issues, adolescents are influenced by their parents. During middle adolescence, the behavioural independence emerges and conformity to parents as well as peers declines.

Social transition: Adolescence is the period of developmental transition between childhood and adulthood. It involves changes in personality as well as in physical, intellectual and social development. There are important changes in the adolescent's social relationships or the social transition during adolescence. As the individual moves through the adolescent years, changes take place with friends and with family members. During adolescence, there is a sharp increase in the amount of time the individuals spend with their peers, and there is more contact with opposite-sex friends. As the children grow, they begin to spend more time with their friends and less time with their parents. As a result, friends can influence adolescents thinking and behaviour.

During adolescence, identity formation is a crucial part of social development. Choices about relationships, appearance, school and career aspirations are all important features of identity formation. One aspect of relationships is friendships, group affiliation and roles in the family.

The importance of peers during adolescence is due to individuals' needs for intimacy. During adolescent years, the search for intimacy increases, and self-disclosure between best friends becomes an important pastime. The understanding that they think and feel the same as someone else becomes important basis of their friendship. One of the most important social transitions that take place during adolescence is the emergence of sexual and romantic relationships. In metropolitan society, adolescents begin dating during adolescence. Dating during adolescence can mean group activities that bring boys and girls together or to group dates, in which, a group of boys and girls go out jointly.

It is incorrect to characterize adolescence as a time when the family ceases to be important. Family relationships change most during puberty, with increasing conflict between adolescents and their parents, especially between adolescents and their mothers. Changes in the ways the adolescents view family rules and regulations may contribute to increased disagreement between them and their parents. Family conflict during this stage is more likely to take the form of bickering over day-to-day issues.

Middle and late adolescence is a period of significant change and reorganization in family relationships. In most families, there are changes in parent-child relationship during these years. Their interactions are bi-directional, horizontal and symmetrical, in which, the parents and their adolescent children are on a more equal footing. Family relationships may become less conflicted and more intimate during late adolescence.

Majority of adolescents are able to adjust with biological, cognitive, emotional and social transitions of adolescence successfully. Although the mass media bombard us with images of troubled youth, however, majority of individuals move from childhood through adolescence without serious difficulty.

Adolescence and Reproductive Health

Reproductive health is a state of physical, mental and social well-being, not merely the absence of diseases related to reproductive system, therefore, reproductive health means that people are able to enjoy safe sex life and to reproduce if they desire so. Men and women have right to have access to safe, effective, affordable and acceptable methods of family planning of their choice. They have right to choose methods of their choice for regulation of fertility. They also have the right of access to appropriate health-care services.

According to the definition of World Health Organization (WHO), youth is the period between the age of 10 and 24 years. Adolescence is the period between the age of 10 and 19 years, which is generally come when puberty occurs. It is estimated that the adolescent group constitutes about one fifth of India's population. There are almost 200 million adolescents in India, and it is estimated that this age group will grow to over 214 million by 2020. The Government of India Reproductive and Child Health programme declares 10-19 years olds as adolescents. The Constitution and Labour Laws of the country consider all people below the age of 14 to be children.

In India, about 35% of the population is in the age of 10-24 years. The health needs of adolescents have not been addressed adequately, particularly their reproductive health needs. Reproductive health needs of adolescents are often misunderstood and under recognized. Adolescents in developing countries are particularly vulnerable to sexual and reproductive health risks due to their young age, lack of awareness related to sexuality and reproductive health. Adolescent girls are a particularly disadvantaged group in relation to sexual and reproductive health due to early marriage and early childbirth. In this section, attempt has been made to discuss reproductive health problems of adolescents.

Dating: It is very common during adolescence. Though it is normal in metropolitan cities, dating relationships can lead adolescents to indulge in sexual relationship at an early age. Those who have sex at an early age are more likely to engage in risky sexual behaviour. Research shows that adolescents are indulging in premarital sex more frequently at an early age, the incidence of pregnancies among them is rising day by day, and most of them face the risk of induced abortions under unsafe conditions and contracting sexually transmitted infections including human immunodeficiency virus (HIV).

The likelihood of negative consequences for adolescent development and functioning may be greatest if adolescents begin to date very early or are over involved in dating. Young people who began dating early before the age of 14 or who dated more frequently are more likely to use alcohol and drugs and to participate in delinquent behaviour.

Menstruation problems in girls: Many of the adolescent girls suffer from menstrual dysfunctions. The age of the normal menarche may vary from 12 to 16 years, and the average length of menstrual cycle is 21 to 28 days. Some adolescent girls suffer from amenorrhoea or absence of menstruation. If a girl does not menstruate by the age of 16 years that means she is suffering from amenorrhoea. Normally, very obese or very slender adolescent may have late onset of menarche. The causes of amenorrhoea may be malnutrition, diabetes, some diseases of ovaries and ingestion of some drugs, especially hormonal preparations. All these causes need therapeutic intervention.

Abnormal vaginal bleeding, also called dysfunctional uterine bleeding, is common in adolescents in the first year of menarche. Painful menstrual cramps are experienced by nearly two third of the adolescents. During second half of the menstrual cycle, many adolescents experience breast pain, fatigue, headache, irritability, mood swings, depression, inability to concentrate, tearfulness and violent tendencies. Most of the times, it requires simple pain relievers, diuretics and reassurance from an expert.

Sexual abuse and rape: Sexual abuse and rape are the most painful psychosocial reproductive health problems of adolescent girls. This may be acquaintance rape, a most common form of rape in adolescent, where offender is close and known to the family and victim. The sexual abuse done by a close relative- uncle, cousin, or real brother- is called incest. Date rape, by the person dating with the victim is generally drug facilitated. Gang rape is an example of increasing sexual frustration among youth and disrespect of law and order in the society. In some backward societies, it is an illegal form of punishment and revenge. Statutory rape refers to sexual activity of an adult and an adolescent under the age of legal consent. The criminal offense of statutory rape is committed when an adult sexually penetrates a person who, under the law, is incapable of consenting to sex. These persons are considered deserving of special protection because they are especially vulnerable. These rapes may be prevented by avoiding high-risk situations, illicit drug use in heterosexual groups, overconfidence on acquaintances, sex education and training for self defense. Rape victims need psychosocial help from social workers and medical experts.

Teenage pregnancy and childbearing: Parenting at any age can be challenging but it can be particularly difficult for adolescent parents. Childbearing during adolescence negatively affects the parents, their children and society. Compared with their peers who delay childbearing, teen girls who have babies are less likely to finish high school, more likely to be poor as adult and more likely to have children who have poorer educational, behavioural and health outcomes. Teenage pregnancies are associated with increased rates of alcohol abuse and substance abuse, lower educational level and reduced earning potential in teen fathers. Babies born as a result of a repeat teen pregnancy are even more likely to be born premature and with a low birth weight.

Teenage pregnancy is a serious issue that may seriously affect the future of a young woman. Apart from the consequences, teens generally lack the basic requirements for motherhood, *i.e.*, patience, maturity and ability to handle stress during pregnancy and after the delivery of baby. Teenage pregnancy is certainly risky due to varied factors, and the most important is biological factor. At this age, the body is not ready though still it is a growing body. Socio-economical aspects are also extremely important. Lack of access to maternal and child health services can further lead to problems. Unwanted pregnancy in unmarried girls may stigmatize them, leading to poor self-esteem. Unplanned pregnancy in both married and unmarried girls may prompt them to go for illegal and unsafe abortions, which is more common in unmarried girls. Children that are born from a teenager mother have higher risk to die than newborns that are born from older mothers.

Sexually transmitted diseases: Sexually transmitted diseases (STDs) are serious health problems for adolescents, occurring in an estimated one-quarter of sexually active teenagers. A sexually transmitted disease (STD) is an infection caused by bacteria or a virus that can be transferred from one person to another through any type of sexual contact. Sometimes, STDs are referred to as sexually transmitted infections (STIs) since they involve the transmission of a disease-causing organism from one person to another during sexual activity. Sexually transmitted diseases spread through oral, genital, or anal sex. Human immunodeficiency virus and viral hepatitis are the most common sexually transmitted infections. Other examples of STDs include Chlamydia, syphilis and gonorrhea. Girls have higher rates of getting infection with Chlamydia, HIV and herpes.

The sexually transmitted infections (STIs) affect men and women of every age, race, religion, rich, or poor and irrespective of education level. Majority of STIs can be treated and cured but some of them live in the body forever. Untreated STIs can cause serious health issues affecting fertility and can cause health problems during and after pregnancy for both mother and their newborn baby. Pregnant women should be tested and treated before giving birth to decrease the risk of problems during pregnancy and delivery and complications for the baby. Numerous STDs have been identified but only the common types of STD are discussed below:

***Human immunodeficiency virus (HIV) and AIDS*:** HIV virus causes acquired immunodeficiency syndrome (AIDS) that destroys the body's ability to fight off infection. Human immunodeficiency virus is transmitted through exchange of certain bodily fluids such as blood, semen, vaginal secretions and breast milk. To produce an infection, the virus must pass through the skin or mucous membranes into the body.

Exposure to HIV can be a consequence of many of the risk-taking behaviours that occur among adolescents. Today, adolescents of both sexes face a serious risk of HIV infection, which is the cause of AIDS. Human immunodeficiency virus infection and AIDS can be prevented by avoiding risk behaviours. Knowledge about HIV is an important aspect of preventing HIV. Parents should educate their children to ensure that children and teens receive sex education and preventive drug abuse courses, which include material on HIV. Acquired immune deficiency syndrome is a chronic illness and often a fatal disease.

***Human papillomavirus (HPV)*:** HPV is a common sexually transmitted disease that can cause genital warts, which can occur on inside or outside areas of the genitals and may spread to the surrounding skin or to a sexual partner. Human papillomavirus is so common that nearly all sexually active men and women get it at some point in their lives even though they have sex with one person in their lifetime. Majority of people with HPV never develop symptoms or health problems. Most HPV infections go away by themselves within two years, but sometimes, HPV infections will persist and can cause a variety of serious health problems. Women with an HPV infection have an increased risk of developing cervical cancer. In rare circumstances, a pregnant woman with genital HPV can pass the HPV to her newborn baby during delivery.

Human papillomavirus is a different virus than HIV and causes different health problems. This virus does not live in the blood cells though lives on the skin. The vaccines of HPV can prevent many diseases and cancers caused by HPV. In addition to vaccination, there are other ways to lower the risk of health problems caused by HPV. Condoms may lower the risk of HPV and HPV-related diseases, such as genital warts and cervical cancer. Regular Pap smear tests can detect HPV infection as well as abnormal cervical cells.

***Chlamydia*:** The most common of all STDs is Chlamydial infection, which is caused by the bacteria *Chlamydia trachomatis*. When transmitted through sexual contact, the bacteria can infect the urinary and reproductive organs. In many cases, Chlamydia causes only mild symptoms or no symptom at all. Therefore, an infection can last for weeks or months before it is discovered.

It can affect both men and women. The infection causes an abnormal genital discharge and burning with urination. In women, untreated Chlamydial infection may lead to pelvic inflammatory disease (PID). It causes infection of the uterus, fallopian tubes and other reproductive organs. The symptoms include lower abdominal pain. Chlamydial infections can be treated with antibiotics. If untreated, Chlamydia can lead to serious complications even years after the person is infected. The most common and serious complications occurred in women are pelvic inflammatory disease, ectopic (tubal) pregnancy and infertility. Men may have urinary symptoms or no symptoms at all.

If anybody is diagnosed with Chlamydia, his or her sexual partner needs to be treated with antibiotics, even if they do not have any symptoms so that they may not develop any long-term complications or spread infection to others. Since Chlamydia is spread through sexual contact, the best way to prevent it is to abstain from having sex. Once sexual activity is part of a life, it is important to be screened yearly for Chlamydia by a medical practitioner. In addition, the use of condoms decreases the risk of STDs, including Chlamydia.

***Gonorrhea*:** It is very common infectious disease caused by a bacterium *Neisseria gonorrhoeae*. Gonorrhea can grow easily in warm moist areas of the reproductive tract, including the cervix, uterus and fallopian tubes in women and in the urethra in men and women. The bacterium can also grow in the mouth, throat, eyes and anus. Like Chlamydia, gonorrhea is also a *silent* disease. Many women who are infected have no symptoms of the disease. On the other hand, men are more likely to show symptoms. If symptoms of gonorrhea develop, they usually appear within two to 10 days after sexual contact with an infected partner although a small percentage of people may be infected for several months without showing symptoms.

Gonorrhea causes a discharge from the vagina or penis. The infected person suffers from painful urination. Symptoms of rectal infection include discharge, anal itching, and sometimes, painful bowel movements. The most common and serious complications occurred in women are pelvic inflammatory disease, ectopic (tubal) pregnancy and infertility. Gonorrhea infections can be treated with antibiotics. Gonorrhea cannot be diagnosed through a physical examination. Many doctors prefer to use more than one test for accurate diagnosis. A pregnant woman infected with gonorrhea is likely to give infection to her newborn as the baby passes through the birth canal during delivery. There is an increased risk of premature labour, stillbirth and postpartum uterine infection. Hence, it is important that the pregnant women should be tested for gonorrhea. Treating gonorrhea as soon as it is detected in pregnant women is important to protect her baby.

***Genital Herpes*:** Genital herpes is an infection of genitals, buttocks, or anal area caused by herpes simplex virus (HSV). The virus is very contagious and spreads from one person to another through skin-to-skin contact, such as during vaginal, anal, or oral sex. Genital herpes is a chronic, lifelong viral infection. One can catch genital herpes if he or she has oral sex with someone who has a cold sore. Cold sore is a blister-like lesion around the mouth caused by HSV. People with genital herpes have at least twice the risk of becoming infected with HIV if exposed to it than those without it.

Symptoms may include painful blisters or open sores in the genital area, which may be preceded by burning sensation in that area. The herpes sores usually disappear within few weeks but the virus remains in the body and the lesions may reoccur from time to time. Pregnant women are treated for herpes during the last month of pregnancy to reduce the chance of outbreak at the time of delivery. If there is an outbreak around the time of delivery, a cesarean section is suggested to decrease the chance of infecting the newborn. Fortunately, infection of a baby from a woman with herpes infection is rare.

Genital herpes can cause recurrent painful genital sores in many adults, and herpes infection can be severe in people with suppressed immune system. Regardless of severity of symptoms, genital herpes frequently causes psychological distress in people who know that they are infected. Using condom is the best method to have protection from the herpes virus if it covers the infected area. No treatment can cure herpes but antiviral medications can shorten an outbreak and reduce symptoms. Once a person is infected with HSV, it can be reactivated so often to cause a new episode of genital herpes, which is known as recurrence of HSV. Genital herpes is very easy to notice when an infected person has blisters or sores.

***Syphilis*:** Syphilis is a bacterial infection caused by a microscopic organism known as *spirochete*. Syphilis is a highly contagious disease spread primarily by sexual activity, including oral and anal sex. The disease starts as a painless sore, typically on person's genitals, rectum, or mouth. Syphilis spreads from person to person *via* skin or mucous membrane contact with these sores. The sore appears at the location where syphilis entered the body. Sore is usually firm, round and painless. Since sore is painless, it can easily be unnoticed. The infected person is often unaware of the disease and unknowingly passes it on to his or her sexual partner. If the infected person does not receive adequate treatment, the infection progresses to the secondary stage.

Women who are infected during pregnancy can pass on infection to the fetus through the placenta. This disease, called congenital syphilis, can cause abnormalities or even death to the child. Early syphilis can be cured without treatment. Untreated syphilis may go on to more advanced stages, including a transient rash, and eventually, a serious involvement of heart and central nervous system and can be life threatening. Use of condom greatly reduces but cannot eliminate the risk of catching or spreading syphilis completely. Special blood tests can also be used to diagnose syphilis. Depending on the stage of disease and the clinical manifestations, the treatment options for syphilis vary. Syphilis infections can be treated with antibiotics.

***Pelvic Inflammatory Disease*:** Pelvic inflammatory disease, commonly called PID, is one of the most serious complications of a sexually transmitted disease in women. Pelvic inflammatory disease is the inflammation or infection of the fallopian tubes or ovaries. In many cases, PID develops when the Chlamydial infection spreads into the uterine cavity and extends into the tubes. A woman who has PID experiences low abdominal pain, vaginal discharge and fever. This disease can be diagnosed by a pelvic examination and can be treated with antibiotics but hospitalization may be required in severe cases.

Pelvic inflammatory disease can cause wound in the fallopian tubes, which can block the tubes and prevent fertilization from taking place. In other cases, wound may interfere with the passage of fertilized egg down into the uterus. When this happens, the egg may implant in the fallopian tube, which is called ectopic or tubal pregnancy. This is life threatening for the mother and results in loss of the fetus. In severe cases, PID can cause permanent tubal damage and infertility.

Normally, the cervix prevents bacteria that go through the vagina from spreading to the internal reproductive organs. If the cervix is exposed to sexually transmitted diseases, such as gonorrhea or Chlamydia, the cervix is infected, and then, infection spreads to the internal organs. Pelvic inflammatory disease occurs when the disease-causing organisms travel from cervix to the upper genital tract. Untreated gonorrhea and Chlamydia cause PID in majority of cases.

The symptoms of PID can vary but may include pain in the lower abdominal area and abnormal vaginal discharge, which is yellow or green in color or that has an unusual odour, painful urination, nausea and vomiting, Chills or high fever and pain during sex.

***Hepatitis*:** Hepatitis B and C are viral infections that infect and can damage the liver. These are passed from one person to another when the blood or sexual fluids of an infected person gets into another persons blood stream through openings in the skin (punctures, cuts and sores) or through the mucosal membranes (nose and genitals). Hepatitis virus can live outside of the body for several days. Hepatitis B can be prevented by vaccination and is recommended for all age groups. There is no Hepatitis C vaccine. Screening is recommended for all pregnant mothers. Infected mother may pass on the virus to her newborn baby and child is born with congenital hepatitis.

For acute hepatitis B, no medication is available, but for acute hepatitis C, antiviral treatment is recommended. For chronic hepatitis B and C, a regular monitoring is recommended for the signs of liver disease progression. Some patients are treated with antiviral drugs.

Non-gonococcal urethritis: Non-gonococcal urethritis is an STI that causes swelling of the urethra- the tube carrying urine. The infection is spread by having unprotected anal, vaginal, or oral sex. Both men and women can get non-gonococcal urethritis but it is more common in men because a woman's urethra is less easily infected during sex. Most people do not have symptoms and are unaware of their infection. Even though no symptoms are seen, the infection can still be passed to other people. Pain or itching during or after urination is one of the symptoms in women. Symptoms in men include itching and irritation of penis, pain when urinating, discharge and drying of opening of penis.

Testing for non-gonococcal urethritis is done by a urine sample or testing the discharge. Antibiotics can cure non-gonococcal urethritis. Women who do not receive treatment can develop pelvic inflammatory disease (PID), which can result in pain, infection of fallopian tubes, or infertility. Pregnant mothers have a risk of having a miscarriage or a preterm delivery. The infection can be passed from a mother to baby during delivery. Men who do not receive treatment may develop inflammation of the testicles, which could lead to infertility or develop a prostrate infection.

***Trichomoniasis*:** Trichomoniasis, often called *trich*, is a sexually transmitted infection (STI) caused by a parasite. Trichomoniasis infection can cause harmful health concerns for both men and women, including increased HIV transmission and infertility. Even if no symptoms are present, the infection can still be passed to other people. In women, symptoms can range from no symptoms to severe inflammation and irritation. Other symptoms may include foul-smelling vaginal discharge, lower abdominal pain and painful or difficult urination. Majority of the men experience no symptoms although some may have inflammation of urethra or prostrate. Other symptoms may include painful urination, mild discharge and irritation or redness on top of penis.

Testing for trichomoniasis requires a physical examination or lab test to detect the parasite that causes the infection. Usually, it is treated by antibiotics. If left untreated, fertility of both men and women may be affected.

In untreated women, the parasite moves freely in the vagina and may disrupt normal sperm movement. This can make it difficult for sperm to fertilize the egg causing infertility. In untreated men, the sperm that travels through an infected urethra gets affected in movement and strength. It decreases the ability to reach and fertilize the egg and causes infertility. Infection during pregnancy can result in preterm delivery or premature rupture. Children born to infected mothers may have low birth weight, urinary tract infections, inflammation of the vagina and respiratory problems.

Although many STDs are treatable but effective cure is not there for HIV, HPV, and hepatitis B and C. Even gonorrhea, which could be easily cured, has become resistant to many antibiotics. Many STDs can be spread by people who do not have any symptoms of the disease and who have not yet been diagnosed with an STD. Therefore, it is essential to create public awareness and education about these infections and the methods of preventing them. There is no *safe sex*. The only and most effective way to prevent STDs is abstinence. Sex in a monogamous relationship where none of the partner is infected with STD is considered *safe*. Many people think that kissing is a safe activity. Unfortunately, syphilis, herpes and other infections can be contracted through kissing. Condoms are helpful in decreasing the spread of infections such as Chlamydia and gonorrhea. They do not completely protect from genital herpes, genital warts, syphilis, and AIDS. Prevention of the spread of STDs depends upon the counseling of at-risk individuals, particularly adolescents, and the early diagnosis and treatment of infections.

The health seeking behaviour also depends upon the marital status of the adolescent. Reproductive health services are more oriented towards adult married women. Unmarried adolescents hesitate to seek health services due to the fear that these services are not confidential, parental approval and negative attitude of health care providers. Even married adolescent girls hesitate to see doctors due to embarrassment and the taboo associated with reproductive health problems. Another reason could be limited mobility and lack of decision-making power in adolescent girls and young married women.

Adolescence and Other Health Problems

Beside the problems related to reproductive health, other health problems and issues also need attention during adolescence. Most of the adolescents present problems related to normal physical or psychological changes of this period. Environmental factors have significant impact on adolescents' health and well-being. Adolescents are extremely sensitive to environmental influences due to their limited cognitive abilities for evaluating information. The behaviour and health patterns of adolescents always reflect the status of their social environment.

A significant problem for adolescents is that they are bombarded from an early age with various social messages. Social marketing targeted at this age group combined with risk-taking behaviour common during adolescence they struggle towards independence. Risk taking behaviour includes problems such as accidents, eating disorders, sexually transmitted diseases, unwanted pregnancies, substance abuse and mental health issues. These conditions seriously affect physical and mental adolescent health and result in decreased future productivity and employment opportunities. Some of the health problems during adolescence are discussed below:

Chronic illness: Chronic illness or disability occurs in approximately 10% of individuals under the age of 17 years. Some of these adolescents have to cope with additional physical and mental disabilities as a result of their primary illness. All adolescents have to deal with psychological consequences of their illness. Illness, injury, medical treatments, hospitalization and surgery can all intensify concerns about physical appearance, interfere with efforts to gain independence and disrupt changing relationships with parents and friends. Physical health problems occurring during adolescence can often complicate their development. Adolescent developmental issues may complicate a teenager's ability to cope with illness and response to treatment.

Out of ten, one adolescent has a chronic condition that somehow limits his or her daily activities. Chronic conditions can affect adolescent development and that adolescent development can have an effect on chronic condition. Delayed growth and puberty is common in most chronic illnesses. Biological, psychological and social development are highly inter-related and the timing of one may affect development in other domains. The psychological and social effects of delayed puberty should not be ignored. The wellbeing of chronically ill adolescents is determined by severity of the disease and the amount of treatment required. Identity, self-image and ego-development are affected by chronic illnesses. This is particularly true when illness is more severe. Recurrent illness and the demand of treatment have significant impact on school attendance and educational achievement, which in turn may result in vocational impairments and loss of financial independence in adult life. Teenagers with chronic conditions are more likely in comparison to their healthy peers to miss school due to their condition or treatment they need. School attendance may be affected by severity of the illness, treatment adherence, and psychological problems.

The presence of an adolescent with a chronic condition imposes an increased burden on parents and other family members. The demands of managing a chronic illness increase dependence on family members. Due to chronic illness, adolescents are excluded from peer group activities.

Overweight and obesity: All individuals have fat tissue in the body since too much body fat results in overweight and obesity. Generally, a person is not considered obese until his or her weight is at least 10% higher than his or her average height and body type. Obesity usually begins during childhood or adolescence. A child who is obese between the age of 10 and 13 has 80% chance of becoming an obese adult. Obesity occurs when a person eats more calories than the body burns up. If one parent is obese, there is 50% chance that their children will also be obese, and when both parents are obese, their children have 80% chance of being obese. Overweight or obesity is identified using the Body Mass Index (BMI), which is calculated by measuring the proportion of weight to height. It is calculated the same way for adults and children. For adults, if a BMI score is between 25 and 29.9, a person is considered overweight and if the score is 30 or above, the person is said to be obese.

Obesity has become a serious public health concern affecting a significant portion of the population all over the world. Obesity affects all socio-economic groups, irrespective of age, sex, or ethnicity. Many factors can increase the risk of obesity in adolescents. Although some medical disorders can cause obesity but it is rare for overweight and obesity to be caused by hormonal or genetic defects. Overweight and obesity in childhood and adolescence can be related to poor eating habits, overeating, lack of exercise, family history of obesity, endocrine and neurological problems, stressful life events, family and peer problems, low self-esteem, emotional problems and use of medications such as steroids and psychiatric medications. Convenience food items that are high in fat and calories, such as candy, chips and sugary drinks, are produced and consumed more frequently. Overweight and obese adolescents eat more calories than their body need. Children and adolescents lead more inactive life styles than they lead in the past. Rather than playing outside, they watch television, play video games, sit at computer and exercise less. As the number of dual-earner families is increasing, parents are spending less time with their children, cooking less nutritious food and letting their children spend more time in front of computer or television.

The consequences of childhood and adolescent overweight and obesity are serious. The effects could be immediate and long-term on health and well-being. Obese youth are more likely to have risk factors for cardiovascular disease, such as high cholesterol and high blood pressure. Obese adolescents are more likely to have diabetes and are at greater risk for bone and joint problems, sleep apnea and social and psychological problems such as stigmatization and poor self-esteem. In long-term, children and adolescents who are obese are likely to be obese as adults and are therefore more at risk for adult health problems such as heart disease, high blood pressure, type 2 diabetes, stroke, several types of cancer and osteoarthritis.

Healthy lifestyle, including healthy eating and physical activity, can lower the risk of becoming overweight and obese. Eating meals together as a family promotes healthful eating habits among children and adolescents. This practice increases children's consumption of whole grains, fruits and vegetables, and decreases their consumption of soft drinks and fats. Schools also play an important role by establishing a safe and supportive environment and practices that support healthy behaviour. Schools have more continuous and intensive contact with children and adolescents during their first two decades of life than any other institution. Schools can make a positive impact on children's health by promoting physical activity, providing healthy food and educating children and parents on nutrition. The local community also plays significant role as it is the first social group that the adolescents encounter beyond their families and it plays significant role in their development. Many communities have developed parks and green spaces to encourage walking. Obesity frequently becomes a lifelong issue. One of the reasons is that most obese adolescents gain back their lost weight because they tend to go back to their old eating habits and poor exercising. Overweight and obese adolescents must learn to eat and enjoy healthy food and to exercise regularly to maintain their weight.

Treatment for overweight and obesity is more likely to be successful in childhood than in adolescence or adulthood. Intervention programs targeting youngsters are more effective and have more long-term success than adult programs. The treatment of over weight and obesity in children and adolescents requires a multidisciplinary approach. The team should include a physician, nurse practitioner, dietician, physical instructor, behavioural therapist and a social worker in addition to parents, caretakers, teachers and policymakers. Parents play a significant role in making sure that adolescents exercise and eat well, and parental support is essential if adolescents are to make changes in their weight and live a healthy lifestyle.

Alcohol and other drug use: Experimenting is a natural part of adolescence. Experimentation with alcohol, drugs and smoking can be part of their need for learning and establishing independence. At times, it can be a means of escape from social, educational and emotional challenges that the adolescents are facing. It may be an attempt to gain acceptance from peers and to fit in with a social group. Alcohol use and the abuse of prescription drugs are common among teens. The most common drugs used by the adolescents are alcohol, tobacco, caffeine, marijuana and pills that are not prescribed. The other less used drugs include opiates, cocaine, amphetamines, hallucinogens, depressants and inhalants.

Alcohol is a psychoactive drug that depresses the central nervous system. Initially, it makes the individuals feel relaxed and sociable, but with prolonged use, these feelings will be replaced by feelings of depression, anger, loss of control and drowsiness. Alcohol affects both the mind and the body. Alcohol lowers the ability of brain to control behaviour and impairs the ability to perform motor skills and speech. Alcohol can also cause memory loss, addiction, brain damage, *etc.* It can result in liver, kidney, stomach, and intestine damage. It can also cause death from breathing or heart failure and interactions with other drugs. It takes most people about one hour to burn up the alcohol in an average serving of beer, wine, or liquor. For adolescents, alcohol and drugs are especially dangerous. When they drink large quantities of alcohol, they are at a higher risk of problems like alcohol poisoning, car accidents and violence.

Use of drugs causes dependence, damage to lungs, block short-term memory, restlessness and mood swings, eating and sleeping disorders, personality changes and increase heart rate and blood pressure. Drugs also depress body systems, cause muscle rigidity, damage to nerve, kidney and liver. These may also increase injury proneness because of inaccurate perception of time and space, and trigger pain in chest, muscles and joints. Long-term usage may result in comas, seizures and brain damage.

The factors that protect adolescents from alcohol and drug abuse include involving parents, peers, community and school to work together for their welfare. Parents who model positive behaviour, have good communication skills, set limits and supervise their adolescent children are less likely to have children to indulge in alcohol and drug use. Having friends in the community who do not abuse alcohol and drugs helps protect adolescent. Schools help by providing after school activities, sports, teachers and coaches who are good role models. Alcohol and drug intervention programs are useful once an adolescent has been found to be using or abusing alcohol and drugs. The earlier an adolescent begins alcohol and drug use, they are at higher risk for future problems.

Adolescents who have been assessed and found to be chemically addicted or dependent on alcohol and drugs are most likely to be referred for inpatient treatment. Inpatient treatment begins with a medical detoxification process for those who are at risk of medical complications. Some adolescents enter inpatient treatment for individual and group therapy sessions. All of their medical, emotional and mental health issues are discussed in inpatient treatment. Many kinds of therapy are used with adolescents to help them learn how to cope without alcohol and drugs. Family therapy is also helpful for many adolescents who are trying to live without alcohol and drugs.

Smoking and tobacco use: Although smoking and tobacco use by adolescents have declined significantly in cities but smoking rates are still higher in rural areas. Tobacco products used by adolescents include cigarettes, cigars, pipes, hookahs, smokeless tobacco and oral products such as pouches, strips and sticks. All of these products have toxic effects of tobacco. Among adolescents as compared to cigarette smoking, the use of smokeless tobacco products such as snuff and chewing tobacco is less. Hookahs (water pipes) are also popular among some adolescents and are used in groups. Hookahs are as addictive and toxic as other forms of tobacco smoking. Oral tobacco products including pouches, strips and sticks are designed to dissolve in the user's mouth. These are offered in candy-like flavours and all contain dangerous toxins that can lead to cancer and other serious diseases.

Tobacco use in adolescence is associated with a range of health problems-compromising behaviours including being involved in fights, carrying weapons, engaging in high-risk sexual behaviour and using alcohol and other drugs. Cigarette smoking causes lung cancer deaths and is responsible for most cancers of the larynx, oral cavity, esophagus and bladder. The risk of developing smoking-related cancers as well as noncancerous diseases increases with total lifetime exposure to cigarette smoke. Tobacco is a type of central nervous system stimulant that makes a first-time user feels dizzy, light headed, excited, or nauseous. Over time, regular users develop a craving for nicotine, followed by relief and relaxation. Tobacco use affects both the mind and body. It increases feelings of alertness and relaxation and lowers stress and irritability. Tobacco also raises blood pressure, breathing and heart rate, increases muscle relaxation, reduces pain and decreases appetite. In addition, tobacco use can lead to chronic illnesses such as heart disease, cancer of throat, mouth and lungs, bronchitis and resistance to cold, flu and other infections.

An infant's risk of dying from Sudden Infant Death Syndrome (SIDS) increases five times if he or she is exposed to second hand smoke in the room. Second hand smoke also increases the risk of lung cancer and asthma. Smoking not only causes lung cancer, but also other respiratory problems like coughing, wheezing and shortness of breath. Children and adolescents who use tobacco have smaller lungs as compared to their peers, and therefore, tend to be less physically fit in comparison to their peers. Adolescents' use of smokeless tobacco increases their risk of oral cancer and nicotine addiction.

There are many factors associated with youth smoking and tobacco use in adolescents. Tobacco advertisements and movies promote smoking in adolescents. Adolescents are more likely to use tobacco if their peers use tobacco and if they perceive tobacco use it is acceptable or normative among their peers. Parental smoking may promote smoking among adolescents. Expectations of positive outcomes from smoking, such as coping with stress and controlling weight, are related to adolescents' tobacco use. It is more difficult to quit smoking once the adolescents have started it.

Parents play an important role in preventing adolescents from smoking and using tobacco or helping them quit smoking. Parents can set an example by choosing not to smoke themselves. Adolescents whose parents strongly disapprove smoking are less likely to smoke. Other significant adults, such as teachers, grandparents, aunts and uncles, influence whether adolescents start using tobacco or they stop. These adults play significant role in communicating a no-smoking message to adolescents.

Unintentional, intentional injuries and abuse: Unintentional injuries are those that occur as a result of accidents where there is no plan to harm and hurt self or anybody. Accidental injuries include automobile crash or fire and careless play near water. Majority of deaths in adolescents are the result of preventable accidents. Educational programmes warn individuals about the dangers of risky behaviours, such as careless driving, careless use of fire, or careless water play and can significantly decrease the risk of unintentional accidents in adolescents.

Intentional injuries result from interpersonal and self-directed violence that involves two or more individuals and also includes domestic violence, bullying, fighting, and homicide. Interpersonal violence means threatening or actually harming another person. The violence can be subtle as in bullying or as ultimately destructive as murder. Adolescents are threatened psychologically, emotionally, or physically because of the increasing occurrence of interpersonal confrontations in schools and homes.

Self-directed violence is the result of self-inflicted injury, including suicide. Suicide is one of the most important causes of death among adolescents, the reason being lack of awareness about the warning signs. It is very important for family members and friends to know the warning signs of suicide in order to protect the life of an individual. The warning signs include being preoccupied with death, talking about suicide and depression.

Abuse is very common and it causes long-lasting emotional damage. Therefore, it is important that it should be recognized as early as possible so that steps can be taken to prevent and handle it. Any act that harms the physical or emotional well-being of any child or adolescent is the abuse. This includes striking, hitting, slapping, pushing, shoving, throwing, whipping, kicking, shaking, choking, tying up or locking inside a small space, burning with a matchbox or cigarette, beating, pulling hair, biting, starving, or any sexual activity between a child and an adult whether consensual or not. Psychological abuse includes screaming, yelling, shouting, calling a child negative or offensive names and telling a child that he or she is bad, stupid, worthless, or shaming. The harm that results from the emotional damage may not be evident until later in the child's life during adolescence or even adulthood. The damage may manifest itself in mood disorders, alcohol or drug abuse, physical illness, behavioural disorders, academic problems, difficulty in relationships either with chosen partners or even work related relationships. Violence of any type, interpersonal or self-directed and abuse has overwhelming consequences on the individual and the family members.

Eating disorders: There are different types of eating disorder among adolescents. The most common eating disorders are anorexia nervosa, bulimia nervosa and binge eating. Eating disorders involve an unhealthy relationship with food and eating, and often the adolescents have intense fear of being overweight.

The causes of eating disorders include low self-esteem, problems with friends or family relationships and the death of near ones. Problems at school, college, university, or work, being bullied or abused, may also result in eating problems. Lack of confidence, a divorce in the family, or sexual or emotional abuse could be other causes. Many adolescents may eat extra food to help them cope with painful situations or feelings.

Anyone can develop an eating disorder, regardless of age, sex, cultural, or racial background. The people most affected tend to be adolescent girls, particularly between the age of 15 and 25. Not eating enough food can lead to a wide range of complications, some of which can be fatal, such as acute kidney failure, liver damage and heart failure.

***Anorexia nervosa*:** Anorexia nervosa is an eating disorder and mental health condition that can be life threatening. Adolescents with anorexia try to keep their weight as low as possible by restricting the amount of food they eat. They have a distorted self-image of themselves and believe that they are obese, while actually they are not. Adolescents with anorexia exercise excessively, and some eat a lot of food in a short space of time and then purge and vomit. The attempt to get rid of food taken is called purging. Anorexics often hide their behaviour from family and friends by lying about eating, or by pretending to have eaten earlier.

One of the biggest challenges in treating anorexia is helping adolescents to accept that their behaviour is not normal. The first step towards getting better is to recognize the problem and to have a genuine desire to get well. Once a person seeks help, it usually takes five to six years of treatment to recover fully and the chances of relapses are common. Approximately half of the adolescents with anorexia will continue to have eating problems. Anorexia is one of the leading causes of mental health-related deaths. Anorexia treatment usually involves nutritional feeding, medical monitoring and psychological treatment. Cognitive behavioural therapy helps by identifying and replacing inaccurate thoughts to help change behaviour and emotional state.

***Bulimia nervosa*:** Bulimia nervosa is commonly called bulimia, which is a serious and life-threatening eating disorder among adolescents. Adolescents with unrealistic expectations about weight may end up suffering from bulimia. A person with bulimia eats a lot of food in a short amount of time called binging. He or she then tries to prevent weight gain by getting rid of food by purging. The attempt to get rid of extra calories is usually referred to as purging. A person with bulimia feels that he or she cannot control the amount of food eaten. In addition, bulimics might exercise a lot, eat very little or not at all, or take pills to pass urine often to prevent weight gain. Many people think that eating disorders affect only adolescent girls and women from high socio-economic status. It is true that most bulimics are girls or women, approximately 85-90% but bulimia affects people from all income groups, including males and even older women.

Bulimia is more than just a problem with food. A binge can be triggered by dieting, stress, or uncomfortable emotions, such as anger or sadness. Purging and other actions to prevent weight gain are ways for people with bulimia to feel more in control of their lives and ease stress and anxiety. There is no single known cause of bulimia but some factors, *viz.* genes, hormones and chemicals in the brain, may play a part in developing bulimia. A person with bulimia may not like him or herself, hate the way he or she looks, or feel hopeless. He or she may be very moody, have problems expressing anger, or have a hard time controlling impulsive behaviours. Traumatic events like rape and stressful things like transfer can lead to bulimia. Parents who think that looks are important, diet themselves or criticize their adolescent children's bodies are more likely to have adolescents with bulimia.

Signs of bulimia nervosa include an obsessive attitude towards food and eating, an over critical attitude to their weight and shape and frequent visits to bathroom after eating. Bulimia nervosa can be prevented by awareness, early detection and intervention. This can reduce the severity of symptoms, enhance adolescents' normal growth and development and improve their quality of life. Encouraging healthy eating habits and realistic attitudes toward weight and diet are also helpful. A combination of treatments, including cognitive behavioural therapy and antidepressant medication, can be used to help adolescents to overcome bulimia.

***Binge eating*:** Binge eating is an eating disorder where a person feels compelled to overeat on a regular basis. Adolescents who are binge eat consume very large quantity of food over a short period. They often eat even when they are not hungry. Binges are often planned and sometimes adolescents buy special binge foods. Binge eating usually takes place in private, with the person feeling that they have no control over their eating. Unlike those with bulimia, people who are binge eat do not purge themselves to control their weight. Adolescents are more likely to limit their weight gain by severely cutting down the amount of food they eat, which can make the problem worse.

Anyone can have binge-eating problem, and binge eating affects men and women equally. The condition tends to be more common in older adults than in younger people. Binge eating is often associated with obesity, where someone is very overweight with a body mass index (BMI) of 30 or over. The causes of binge eating include low self-esteem, lack of confidence, anxiety and depression. These feelings worsen over time while the adolescent is still binge eating. Binge eating is a treatable condition, and the treatments include psychological therapy and antidepressant prescription in combination with therapy.

***Depression*:** Depression is a medical condition that can cause a wide variety of psychological and physical symptoms. It is estimated that between 4 and 8% of the adolescents are depressed. However, depression in adolescents is probably under diagnosed because the symptoms can be difficult to recognize. Many adolescents are unsure that why they are depressed. Depression sometimes occurs in response to a stressful event, such as a recent death or break-up, or it can occur for no apparent reason. Although anyone can develop depression, certain factors increase an adolescent's risk for becoming depressed, including a history of depression in a parent or sibling, prior history of depression, history of anxiety disorder, attention deficit hyperactivity disorder and learning disability, problems or conflicts with parents, difficulties with friends or peers, difficulties with school work, negative outlook or poor coping skills and chronic illness.

Many adolescents who suffer from depression are not aware that depression is causing changes in their behaviour, emotions, ability to interact with others and school performance. These changes can have serious consequences. There is increased risk for future depressive episodes, especially if the depression is not recognized or treated properly.

It is sometimes difficult to differentiate the signs and symptoms of depression from the occasional bad mood and negative attitude that the most adolescents experience at some point in their life. A depressed or irritable mood is the most common symptom of depression, although depression can cause other psychological and physical symptoms. Many people are not aware that depression can cause physical symptoms such as aches and pains or appetite and sleep disorders.

In adolescents, irritability may be a better indicator of depression. Signs of irritability include feeling angry by everything and everyone. The depressed adolescent may be moody, negative and quarrelsome, picking fights as a means to communicate his or her emotional distress. He or she often is unable to tolerate disappointment and frustration and responds with angry outbursts. Appetite changes and weight loss are common in people with depression. Sleep disturbance is common in depressed adolescents. A depressed adolescent may report feeling tired all the time. Parents can sometimes misinterpret this behaviour as laziness, a bad attitude, or avoiding responsibilities. During depression thinking and information, processing is slowed down. Adolescents are indecisive, which results in inability to take action. They take longer time to complete homework and class work.

Different methods are used to treat depression, including medications and psychotherapy. Family therapy is also helpful if family conflict is the cause of adolescents' depression. The adolescent also needs support from family and teachers if he or she has school or peer problems. If an adolescent suffers from severe depression, hospitalization in a psychiatric unit may be required. A combination of medication and cognitive behavioural therapy is most effective in treating depression among adolescents.

***Conduct disorder*:** Conduct disorder is a serious behavioural and emotional disorder that can occur in children and adolescents. A child with this disorder may display disruptive and violent behaviour and have problems following rules. The behaviour is considered to be a conduct disorder when it is long-lasting and when it violates the rights of others, goes against accepted norms of behaviour and disturbs the adolescent's or family's everyday life. The symptoms of conduct disorder vary depending on age of the child and intensity of the disorder whether it is mild, moderate, or severe. Conduct disorder falls into four general categories, *i.e.*, aggressive, destructive and deceitful and violation of rules.

Behaviour that threatens or causes physical harm including fighting, bullying, being cruel to others or animals, using weapons and forcing another into sexual activity is considered aggressive behaviour. Intentional destruction of property, such as deliberate fire setting and harming another person's property, is described as destructive behaviour. Repeated lying, shoplifting, or breaking into homes or cars in order to steal is deceitful behaviour. Violation of rules involves going against accepted rules of society or engaging in behaviour, which is not appropriate for the person's age. These behaviours may include running away, skipping school, playing pranks, or being sexually active at a very young age.

The exact cause of conduct disorder is not known but it is believed that a combination of biological, genetic, environmental, psychological and social factors play an important role. Treatment for conduct disorder is based on the child's age, the severity of symptoms as well as the child's ability to participate in and tolerate specific therapies such as psychotherapy, cognitive behaviour therapy and family therapy. Psychotherapy is aimed at helping the child learn to express and control anger in appropriate ways. Cognitive behavioural therapy aims to reshape the child's thinking to improve problem solving skills, anger management, moral reasoning skills and impulse control. Family therapy is used to help improve family interactions and communication among family members.

Gender identity disorder (GID): This disorder is sometimes called transsexualism and is a mental disorder. Gender identity disorder is a conflict between a person's actual physical gender and that to which they actually identify. Children with this disorder express an intense desire to be the opposite gender- may even insist that they are the opposite gender. They may prefer activities appropriate for the opposite gender and even prefer to wear clothing of the opposite gender. In adolescents and adults, the symptoms of GID include desire to be the other sex, desire to live or be treated as the other sex, or the conviction that he or she has the typical feelings and reactions of the other sex. Majority of GID cases turn out to be homosexual adults.

For young children, GID may be normal gender role experimentation such as a tomboy phase for a girl but if this behaviour persists for more than six months, treatment should be sought. If untreated, the disorder can develop into serious problems such as transsexualism. The child feels trapped in the wrong body, disgusted by their own genitals and wishes they could be altered to resemble the genitals of the opposite gender.

Gender identity disorder is more common in young children than in adolescents. Since 80-98% of the cases resolve once, the children experience the effects of their innate sex hormones. The psychological challenges are many in these patients. Gender identity disorder in children is a highly treatable condition. Treatment generally includes several types of intervention including individual counseling, family therapy and behaviour-modeling to teach appropriate gender role behaviours.

Oppositional defiant disorder: When an adolescent child regularly engages in patterns of defiance and hostility, and that behaviour begins to affect the child's social relationships and family life, may be diagnosed as oppositional defiant disorder. It differs from attention-deficit or hyperactivity disorder and conduct disorder in many ways. Children with attention-deficit or hyperactivity disorder are impulsive, while those with oppositional defiant disorder are mainly aggressive. Children with conduct disorder are unsafe when not supervised, while the children with oppositional defiant disorder do not have safety issues.

Symptoms of oppositional defiant disorder include losing one's temper, arguing with parents, teachers and authority figures, acting up, irritable and say hurtful things when angry, refuse to follow rules and blame others for mistakes.

A careful diagnosis is important because children with oppositional defiant disorder may also have other psychiatric disorders, such as attention-deficit or hyperactivity disorder, which must be treated as well. A therapist will conduct an assessment to determine the nature of disorder and plan a treatment, which includes play therapy, individual counseling and family counseling to help others in the family assist in the counseling process.

Phobias: Phobia is an identifiable and persistent fear, which is excessive and is triggered by the presence or anticipation of a specific object or situation. Phobia is an irrational fear of a specific situation or thing. Children and adolescents with one or more phobias consistently experience anxiety or even panic when exposed to specific object or situation. Sometimes, fear is so great that it interferes with school and social relationships. Common phobias include fear of animals, blood, heights, closed spaces, or flying. In children and adolescents, if the identified fear lasts at least six months, it is considered a phobia.

Different types of phobia seen in children and adolescents include specific phobia, panic attack and social phobia. Specific phobia is associated with a specific object or situation. An unpredictable, unexpected period of intense fear or discomfort compounded by shortness of breath, dizziness, lightheadedness, shaking, fear of losing control and an increased heartbeat is called a panic attack.

Usually, phobias are caused by some early childhood trauma. Genetic and environmental factors contribute to the onset of phobias. Specific phobias have been associated with a fearful first encounter with the phobic object or situation. Many a times, phobias, which begin in childhood, will fade away by adulthood. Children and adolescents usually begin to feel better when they receive proper treatment but since each phobia is triggered by a unique set of factors, it can be frustrating when treatment takes longer time than expected. Some clients feel better after a few sessions. Others may take longer.

Mental health professionals use a variety of treatments including cognitive therapy, behaviour therapy, hypnotherapy and relaxation training. The speed of recovery is linked with the willingness of client to try new behaviour and to practice new skills and techniques for controlling his or her feelings.

Sleep problems: Sleeping well is important for the adolescent's physical and emotional well-being. Sleep problems include a wide variety of experiences. Sleep terrors, sleepwalking and sleep talking are sleep disorders, called parasomnias. The child with sleep terrors screams uncontrollably and appears to be awake but is confused and cannot communicate. The child usually has no memory of sleep terror in the morning. Children who sleepwalk appear to be awake as they move around but are actually asleep and in danger of hurting themselves. Both sleep terrors and sleepwalking run in families and affect boys more often than girls. Most often, children with these sleep disorders have single or occasional episodes of the disorder. When the episodes take place several times a night or nightly for weeks at a time or interfere with the child's daytime behaviour, treatment by a child and adolescent psychiatrist may be necessary. Sleep wake reversal may occur in some adolescents and cause problems with daily life. Sleep can also be disturbed by mood disorders, substance abuse, attention deficit and hyperactivity disorder and anxiety.

A range of treatments is available for sleep disorders. A psychotherapist can help to rule out physical or emotional disorders that may interfere with an adolescent's sleep and can suggest things that the adolescents and family members can do to alleviate sleep problems.

Conclusion

In conclusion, adolescence stage is marked by the major physical changes of puberty. Psychologically, adolescence is the age when an individual becomes integrated into the society of adults. Adolescence has certain characteristics that differentiate it from other stages of life span. Adolescence is an important and transitional stage in which the adolescents begin to explore their personal identities and give more importance to peer relationships. Adolescents are more likely to be self-centered and often test rules and boundaries set by their parents and teachers. Adolescence is sometimes considered as a stage of storm and stress. Majority of adolescents are able to adjust with biological, cognitive, emotional and social transitions of adolescence successfully. Adolescents in developing countries are particularly vulnerable to sexual and reproductive health risks due to their young age, lack of awareness related to sexuality and reproductive health. Other health problems during adolescence include chronic illness, overweight and obesity, smoking, alcohol and drug use, unintentional and intentional injuries, eating disorders, depression, conduct disorder gender identity disorder, oppositional defiant disorder, phobias and sleep problems.

■■■

32 Microwave Ovens and Food Safety

Ram Singh[1] and Geetanjali[2]

[1]Department of Applied Chemistry, Delhi Technological University, Bawana Road, Delhi, India; E-mail: <singh_dr_ram@yahoo.com>

[2]Department of Chemistry, Kirori Mal College, University of Delhi, Delhi, India
E-mail: <geetanjalichem@yahoo.co.in>

Public health is directly related to the safety of food. Food safety takes into the account of all those factors that make the food injurious to health. This is a complete scientific discipline in itself and covers a wide area of understanding. This includes the description about handling, preparation and storage of food in different ways, which prevent food borne illness. These descriptions are based on a number of factors, which should be followed to avoid potentially severe health hazards. These factors usually arise from improper agricultural practices, poor hygiene at all stages of the food chain, lack of preventive controls in food processing operations, overuse of chemicals, contaminated inputs, inappropriate storage and handling or improper cooking.

Food can transmit disease from one person to other and can also serve as a growth medium for bacteria, which can cause food poisoning. In developed countries, there are difficult standards for food preparation, whereas, in underdeveloped countries, the main issue is simply the availability of adequate safe water and food. As per World Health Organization (WHO) report 2003, about 40% of the reported food poisoning outbreaks in the WHO European Region take place in private homes. According to WHO and CDC, in the USA alone, annually, there are 76 million cases of food borne illness leading to more than 325,000 hospitalizations and 5,000 deaths. Hence, proper food cooking is an important area that can be checked properly to avoid maximum cases of food poisoning. Food is cooked by the application of heat, which may be either moist or dry. No doubt, the art of cooking also includes preparation and serving of raw food but the part of cooking, which requires the most skill and experience, is that in which heat is involved.

The proper food cooking depends on the kind of stove or heating material and the cooking utensils applied. The heat not always helps in cooking rather sometime obstacles cooking also. Thus, a dish may be spoiled by subjecting it to heat, which is too intense by cooking it too long, or by not cooking it rapidly enough. All these points must be learned, and the best way to master them is to put into constant practice and principles, which are involved in cookery. Hence, this is always recommended to cook the food with full knowledge of cooking. The first step in gaining knowledge of cooking methods is to become familiar with different methods and processes, the ways in which they are applied and the reasons for applying them. There are numerous ways of cooking food but the principal processes include boiling, stewing, steaming, dry steaming, roasting, baking, grilling, or pan-frying. The choice of the method always depends upon the food that is to be cooked and the result desired. If a wrong method is employed, there will be a waste of food material or the food will be rendered less desirable in flavour or tenderness.

Cooking is therefore always involved in generating substantial heat in a way and place, which makes it easy to transfer most of that heat to a conveniently placed cooking vessel. Beginning with the open fire method, a man has evolved many ways to generate such heat. The two basic methods in modern times have been the chemical and the electrical method. In the former, one either burns some combustible substance like wood, coal, or gas, while in later one runs an electrical current through a resistance element (that, for instance, is how toasters work), whether in a *coil* or more recently, inside a halogen-filled bulb. The third important method introduced so far is induction. This is completely different from all other cooking technologies. This method does not involve generating heat, which is then transferred to the cooking vessel rather it makes the cooking vessel itself the original generator of the cooking heat.

Microwaving, an oven only technology, is a fourth method, wherein the heat is generated directly in the food itself. After the discovery of this cooking method, people knew that the microwave ovens are going to revolutionize the kitchen and cooking techniques. The present chapter is going to cover the basics of microwave oven, their characteristic features and potential applications.

Basic of Microwave (MW) Oven

Microwaves are a form of electromagnetic energy, like light or radio waves. The first formal use of the term *microwave* occurred in 1931, but the first time in astronomical context, this word appeared in 1946 in an article *Microwave radiation from the sun and moon* by R. Dicke and R. Beringer. This same article also made a showing in the New York Times issued in 1951. Normally, microwaves have wavelength between 1 nm and 1m (frequencies of 300 MHz to 300 GHz). These are similar to frequencies of radar and telecommunications. Hence, in our modern technological age, microwaves are used to relay long distance telephone calls, television programs and computer information across the globe or to an orbiting satellite in space. Microwaves are good for transmitting information from one place to another because microwave energy can penetrate haze, light rain and snow, clouds and smoke. In recent times, the microwave is most familiar to us as an energy source for cooking food in microwave (MW) ovens. The microwave oven is one of the great inventions of the 20th century. The first personal microwave oven was introduced in 1967 by the Amana Corporation. The year 2007 marks the 40th anniversary of the domestic compact microwave oven.

Microwaves from the sun are based on pulsed direct current (DC), which does not create frictional heat, while microwave ovens use alternating current (AC) creating frictional heat. Energy from the sun operates in a wide frequency spectrum. A microwave oven produces a narrow wavelength of energy with all of the power going into only one narrow frequency of the energy spectrum. Every microwave oven contains a device called a magnetron, a tube in which electrons are affected by magnetic and electric fields in such a way as to produce micro wavelength radiation at about 2450 Mega Hertz (MHz) or 2.45 Giga Hertz (GHz). This microwave radiation interacts with the molecules in the food. All wave energy changes polarity from positive to negative with each cycle of the wave. In microwaves, these polarity changes happen billions of times every second. Food molecules, especially water molecules, have a positive and negative end, in the same way a magnet has both a north and a south polarity. In microwave ovens, as these microwaves generated by the magnetron bombard the food, they cause the polar molecules to rotate at the same frequency, millions of times per second. The molecules within the food, especially the polar water molecules but also amino acids, lipids and proteins, are forced to align themselves with the rapidly changing alternating electrical field. They oscillate around their axis in response to a reversal of the electric field, which occurs billions of times per second. This oscillation creates considerable intermolecular friction that further results in the generation of heat. All these agitation creates molecular friction, which heats up the food. Therefore, in general, the microwaves are a form of energy, which is absorbed by the food placed inside the MW oven cavity and follow the simple flow chart given below:

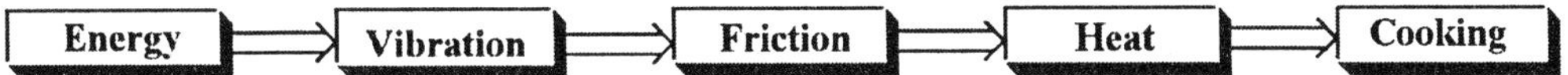

Thus, increasing the amount energy increases the heat inside the food. To produce more energy one either increases the power level, if the machine has variable power settings, or has to increase the cooking time. Cooking in a microwave oven is different in comparison to any other type of cooking. This is because, in a microwave oven, the heat is produced inside the food. In all other cooking methods, heat is applied from outside. The energy present in microwave oven penetrates the food to a depth of about 25 mm. Microwaves are similar to light and they are emitted downwards from top of the microwave oven cavity. The waves bounce off the metal sides of the cavity, hitting the food from different angles. These microwaves do not hit the food evenly. To minimize this effect, the oven has either a rotating turntable or a paddle wheel in the roof to stir up the microwaves but there are still problems with uneven heating.

A general public misconception is that microwave ovens cook food *from the inside out*, meaning the heating from center of the entire mass of food outwards. In reality, microwaves are absorbed in outer layers of the food. This is similar to the heating from other methods. The misconception arises because microwaves penetrate dry non-conductive substances at surfaces of the many common food articles, and thus, often induce initial heat more deeply than other methods. The heating also depends on the water content in food. Depending on the water content, the depth of initial heat deposition may be several centimeters or more with microwave ovens. Penetration depth of microwaves is dependent on food composition and the frequency, with lower microwave frequencies (longer wavelengths) penetrating further. Microwaves cook from the inside out only in the sense that each molecule is generating heat from *inside* and radiating it *outward*.

Based on the nature of applications, microwaves ovens have been classified into three types, *i.e.*, (i) Simple or Standard Microwave oven, (ii) Microwave oven with Grill and (iii) Convection Microwave oven. In standard microwave oven, the heat is produced only with microwaves and can be used for all sorts of everyday tasks such as defrosting, reheating and general cooking. In second type, microwave exists with a grilling option, which allows to brown and crisp food after it has been cooked. The convection ovens provide radiant heat as well as microwaves. These ovens can be used as conventional oven with any type of cookware. In combination (radiant and microwave) or microwave mode, only special microwave proof cookware should be used. The combination mode provides a better cooking time. The radiant mode in reality may take more time to cook than the conventional ovens. Convection ovens are good for items such as Pizza. The commercial model is used for fast food service, for dealing with large numbers of people and supporting business practices. It is often a higher wattage than domestic models and is built with the appropriate components to meet these requirements.

Research is always going on to develop microwave oven with high efficiency and less health hazards. An inverter technology is the recent addition. The conventional microwave ovens use power transformers to increase the household line voltage with 120 volt alternate current (VAC) at 60 Hz to a level high enough to operate the magnetron. This technique has its drawbacks. Operating at a low frequency of 60 Hz, the transformer is relatively inefficient. Power is lost through heat dissipation in converting the line voltage to the higher magnetron level, and the transformer operates at a constant power that can only be changed by switching the power on or off repeatedly, whereas, in inverter-equipped microwave ovens, the power transformer is replaced by a circuit board, which converts the 60 Hz incoming line frequency to a variable rate of 20 to 45 KHz. A relatively small transformer is then required to increase the voltage to the level required by a magnetron. Conventional technology uses just a single power level, which is regulated by switching pulses. In contrast, inverter technology directly controls the power output. This constant soft penetration of microwave energy prevents the common problems of shrinkage, overcooking and loss of nutrients. The result is even food temperature and textures throughout.

Like any other devices, microwave oven has its own advantages and disadvantages too. The brief description about the pros and cons of using microwave oven has been discussed in the following sections:

Features of Microwave Oven

So many microwave ovens are available in the market with varied design and make but the basic features present in them are almost common. In general, a microwave oven consists of the following basic components (Figure 1).

- **Power supply:** This supplies power to the device, which further fed to the magnetron and helps in cooking the food.
- **Magnetron:** This consists of a vacuum tube in which electrical energy is converted to an oscillating electromagnetic field. In general, frequency of 2450 MHz has been set aside for microwave oven for the domestic use.
- **Stirrer:** It is commonly used to distribute the microwaves from waveguide and allow more uniform heating of food.
- **Waveguide:** This is a rectangular metal tube, which directs the microwaves generated from magnetron to the cooking cavity. This helps in preventing direct exposure of the magnetron to any spattered food, which would interfere with function of the magnetron. This gives added safety to the magnetron.
- **Turntable:** In microwave oven, equal amount of heat would not be dissipated all over the food. Some part may get more and other part may get little. The turntable is a revolving glass plate to ensure even cooking. Most ovens have round removable plate that can be cleaned. Some have rectangular plate that slides back and forth. This rotates the food products through the fixed hot and cold spots inside the cooking cavity and allows the food products to be evenly exposed to microwaves.
- **Cooking cavity:** This is a space inside, where the food is placed and heated when exposed to microwaves. This is the place where cooking takes place.
- **Door:** This allows the access of food to the cooking cavity. The door is specially engineered that it may prevent the microwaves from leaking through the gap between the door and the cooking cavity.
- **Interlock:** If one opens the door when oven is running, there is a chance for burning. To avoid that oven should automatically stop when door is opened. If there were interlock, the oven would not run if door were opened. Most ovens have two interlocks but some have three interlocks, depending upon the requirements.
- **Automatic Sensors:** This part is the recent addition to the microwave ovens. In modern ovens, there is no need to set timer for cooking. Cooking would stop automatically by measuring steam in the oven. In a microwave oven attached with automatic sensor, 100 or more preprogrammed menus are available. If one cooks these menus, the oven automatically sets time for cooking. There is no need to set the time manually.

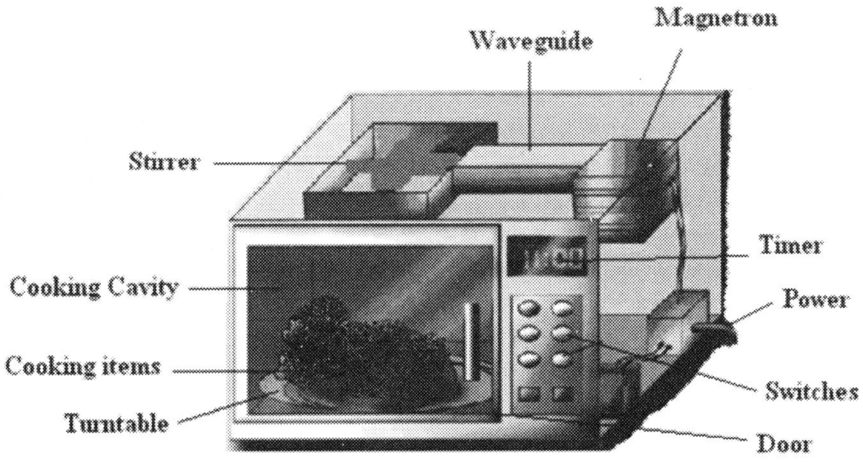

Figure 1: Basic features of microwave oven

Advantage of Microwave Oven in Cooking

This is true that a microwave oven has many positive features and this positivism more depends on the usage and need of the individual to go for type of oven, along with necessary features and options. Over 90% of homes in the United States have microwave ovens used for either food cooking or re-heating cooked food. Since this method of cooking is convenient and energy-efficient compared to conventional cooking, very few homes and restaurants are without them. In general, people believe that whatever a microwave oven does to the food items during cooking, it does not have any negative effect either on the food or on them. These believe is true also. Because the microwave ovens we use today are safer, less powerful than those were used in the 1970s but still not full proof. All microwave ovens manufactured after October 1971 are covered by a radiation safety standard enforced by the Food and Drug Administration (FDA). The standard limits the amount of microwaves that can leak from a microwave oven throughout its lifetime. The limit is 5 mW (milliwatts) of microwave radiation per square centimeter at approximately 2 inches from the oven surface. As one moves away from a microwave oven, the level of any leaking microwave radiation, which might be reaching, decreases exponentially (Figure 2). The microwave method of cooking made the life easier. Cooking healthy and appetizing meals very quickly changes the attitude in life, especially in recent times when life is becoming faster. Earlier, this was considered as a luxurious item, but now, millions of homes around the world have one oven. This indicates the microwave oven to be a necessary cooking apparatus. Indians are slowly joining the bandwagon.

The basic advantages of using microwave oven can be summarized in the following points:

- The microwave oven cooks food evenly, quickly and efficiently.
- The nutrients are preserved and actual taste of food is retained to a higher degree while cooking.
- The amount of oil is required for cooking is very less, and hence, good from the health point of view.
- The microwave oven involves shorter and controlled cooking time.
- The food may be cooked and served in the same utensil.
- The food is cooked without any smoke, grease and heat, and hence, it keeps the kitchen clean.

- Microwave oven has diverse uses like de-frosting, re-heating, *etc.*
- The radiation from microwave oven is useful in reducing some disease causing microorganisms also. This point can be elaborated with an example of meat cooking.

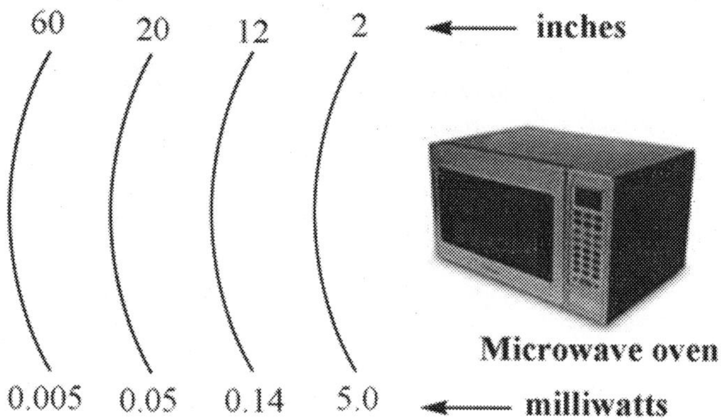

Figure 2: Radiation release

The International Agency for Research on Cancer (IARC) of the World Health Organization has classified one of the heterocyclic amines (HCAs), 2-amino-3-methylimidzo [4,5-*f*] quinoline, as probable human carcinogen and comprehensively genotoxic, whereas, few other HCAs as possible human carcinogens. As per one study, a group of compound belonging to HCAs is present in cooked muscle meat after high-temperature cooking such as grilling/barbecuing, broiling, or pan-frying. The cooking methods like microwaving can result in food with lesser amounts of HCAs leading to protection from cancer. A study found that the formation of certain types of HCAs in chicken legs could be reduced by microwave cooking when compared with frying. Also, precooked meat in microwave oven before barbecuing would significantly reduce the formation of HCAs. Another study also found that compared with non-microwaved beef patties fried under identical conditions, the amount of selected HCAs decreased 3 to 9 folds after microwave pretreatment.

Disadvantage of Microwave Oven in Cooking

As per the law of nature, every positive has more or less negative aspects and it depends on how an individual takes part in that. Similar is associated with microwave ovens also. Though it is pre-mature to say, in conclusion that microwave oven is disadvantageous to the human health. Even the FDA admits that much is unknown and much research is under way on microwaves and their effect on human body. However, this is also a known fact that microwave radiation can heat body tissue the same way as it heats food. Exposure to high levels of microwaves can cause painful burns. Again, less is known about what happens to people when they are exposed to low levels of microwaves.

We have read in the earlier section that the microwave oven does unusual type of heating. This unusual type of heating sometimes causes substantial damage to the surrounding molecules, often forcefully deforming them or tearing them apart. It is the friction and heat, which can destroy the fragile structure of vitamins and enzymes in the food. Hence, microwave heating appears to produce the greatest losses. Microwaves may also cause pathological changes in our bodies. Once the structure of a food is altered, it is unable to perform the desired function in our bodies. Clinical studies have shown that microwave heating of milk or cooking of vegetables is associated with a decline in hemoglobin levels.

A group of scientists at the Stanford University School of Medicine in California discovered that microwaving breast milk at high temperature (72 to 98°C) caused a marked decrease in activity of all the tested anti-infective factors. *Escherichia coli* growth at >98°C was 18 times that of control human milk. Even at 20 to 25°C, the *E. coli* growth was 5 times that of control human milk. Due to this significant loss of the immunological properties of milk, the authors of the study concluded that microwaving is definitely not a suitable heat treatment modality for breast milk. Their study on the effects of microwaved food on human beings in comparison to conventionally prepared food showed that food, which had been cooked in a microwave oven, caused significant changes in the blood immediately after ingestion by the test individuals.

Microwaving may even result in the development of new or unknown substances. There is a possibility that the produced chemical substances are toxic and lead to adverse effect on the body. The microwaves-induced reversal of the polarity causes the cells in the nutrients to become destructively polarized, possibly allowing for the creation of free radicals. All free radicals have a strong tendency to cause reactions. They can interact with enzymes thus causing a disruption of biological processes. In addition, through induction the food itself becomes a carrier and secondary source of technically generated radiation. In 1973, two American scientists, P. Czerski and W.M. Leach also proved that microwaves cause cancer in animals. The January/February 1990 issue of *Nutrition Action Newsletter* reported the leakage of numerous toxic chemicals from the packaging of common microwavable foods, including pizzas, chips and popcorn. The chemicals involved in microwave cooking include polyethylene terpthalate (PET), benzene, toluene and xylene. Microwaving fatty foods in plastic containers leads to the release of dioxins (known carcinogens) and other toxins into the food. In another study, it was found that microwaving food products damages the blood components also. Blood is normally warmed before being transfused into a person. In fact, one woman died after receiving a transfusion of microwaved blood in 1991, which resulted in a well-publicized lawsuit.

Another important study was done by the Russian investigators published by the Atlantis Raising Educational Center in Portland Oregon. According to them, carcinogens were formed in virtually all foodstuffs tested. All the food articles tested were subjected to microwaving for only specified time necessary for their cooking. The result showed that the microwave oven prepared meats contain a well-known carcinogen. The microwaving milk and cereal grains converted certain of their amino acids into carcinogens. The thawing frozen fruits converted their glucoside- and galactoside- containing fractions into carcinogenic substances. In addition, very short exposure of raw, cooked, or frozen vegetables converted their plant alkaloids into carcinogens. The carcinogenic free radicals were formed in micro waved plants, especially root vegetables. The studies also showed that certain trans-amino acids are converted into their synthetic cis-isomers, which are not biologically active. Further, one of the amino acids, *L*-proline, was converted to its *d*-isomer, which is known to be neurotoxic (nervous system) and nephrotoxic (kidneys).

The liquids in microwave ovens need special care. They can superheat when heated in a microwave oven in a container with a smooth surface. In other words, the liquid reaches a temperature slightly above its normal boiling point without bubbles of vapour forming inside the liquid. The boiling process can start explosively when the liquid is disturbed, such as when the user takes hold of the container to remove it from the oven or while adding solid ingredients such as powdered creamer or sugar. This can result in spontaneous boiling, which may be violent enough to eject the boiling liquid from the container and produce severe scalding. This behaviour can be shown by any liquid. There have been many reports of injury to people using microwave ovens to heat water to make hot drinks. Water heated in a microwave oven may be superheated and when objects (*e.g.*, a spoon) or granulated materials (*e.g.*, instant coffee) are put into it, the water may boil very vigorously or even appear to explode out of the container. The vigorously ejected boiling water can cause serious burns. Sometimes, even the act of taking the container out of the oven and/or putting it on the bench can cause the boiling.

A few reports have emerged of jawbreakers, a type of hard confectionery, exploding after being heated. A 2004 episode of the Discovery Channel television program MythBusters demonstrated that heating a jawbreaker in a microwave oven could cause the different layers inside to heat at different rates, yielding an explosive spray of very hot candy when compressed. MythBusters crewmembers Adam Savage and Christine Chamberlain received light burns after a jawbreaker exploded. A year before also, there was a report of burning of a 9 years old girl in Starke, Florida due to this jawbreaker heating in microwave oven.

Safety Tips for Microwave Oven

Modern microwave ovens must comply with the standards of safety and reliability as established by various standard institutions in different countries. Using the microwave oven in careful manner is very important as this is going to reduce maximum disadvantages. As the maintenance of any device is necessary for its proper functioning, similar is true for the microwave ovens also. In general, the following safety measures must be taken care of for minimizing the adverse effects.

- The microwave oven door should not be subject to any strain. It should be closed and open as softly as possible. This will avoid any misalignment leading to the leakage of the microwave radiation.
- The quantity of food to be cooked is also very essential. Small quantities of food with low moisture content can burn, spark, or catch fire, if re-heated for long time. In addition, oven should not be operated without food.
- The proper placement of microwave oven in the house is equally important. Usually, a microwave oven should be kept at least 5-6 cm away from the wall. This will help the proper exhaust of air. The oven must not be placed near gas burners or near radios, televisions, or any electronic items.
- The type of frying is also important. Food should not be deep-fried in the oven as it is not possible to control the temperature of the oil and it may result in catching fire easily. If somehow food catches fire, the oven must turn off as soon as possible and the plug should be removed.
- The interior of the oven must be clean as small pieces of food particles inside can reduce its efficiency. It is advised not to cook eggs with the shells as they explode. In addition, the liquid food in bottles with lids closed must not be used in the microwave oven.
- Milk or any food for kids must be heated on simmer mode only.
- The utility should be cleaned immediately after the use. The filters provided are easily removable to facilitate periodical cleaning. The turntable should be taken out and cleaned. The user should carefully read all the instructions in the manual issued by the manufacturers along with the product.

- The use of utensils for cooking is also very important. The cookware to be used must be those meant for microwave cooking only. One should not use metal or plastic cookware to cook but metal can be used for the convection mode. Metallic dishes should not be used in the combination mode. Pottery and plastic ware declared microwave-safe could be used in the common microwave oven. Dishes must be ½ or ¾ filled depending on the liquid content of the food. If the dish is too small, the food will boil over and if the dish is too large, the thick curry will spread out and overcook. Shallow vessels make cooking faster. Cakes, chicken, meat and rice dishes cook well in large bottomed vessels, while round dishes are ideal for cooking vegetables and curry dishes. The ideal container for heating food in the microwave oven is one that transmits the microwaves but does not absorb, reflect, or refract them. Many current plastic containers and food wraps are specially designed to withstand microwave radiations. Some products may use the term *microwave safe* and may carry a microwave symbol, *i.e.*, three lines of waves, one above the other or simply provide instructions for proper microwave use. Any of these is an indication that a product is suitable for microwaving when used in accordance with the directions provided.
- The user must not forget to wear gloves while taking out the food from the oven since, sometime, the hand may get the burn. One can find these gloves easily in the market.
- If the microwave oven is not working, then the user must not try to repair it by oneself, rather the service provider must be contacted.
- The position of food inside the cooking cavity is another important aspect. Food must be placed off-center as the microwave energy disperses more to the corners and sidewalls. Items like idlis should be arranged in a circle with space among them. Chops should be arranged with the thinner areas pointing to the middle of the utensils.

Conclusion and Future Prospects

In India, it is a misnomer that delicious dishes cannot be cooked in the microwave oven unlike the conventional methods, but slowly and slowly, this line of thought is vanishing. Gone are the days when women had to toil in the kitchen from sunrise to sunset churning out dishes. Now, the time has changed. The use of microwave ovens has led to substantial savings in time for many types of food preparation. The cooking of food takes one-fourth the time in the microwave than that it takes in the conventional cooking. Cooking in the microwave oven is very clean and efficient. The number of utensils involved is also very less, thus, making the life easy for the womenfolk. Hence, women, especially those working, find it suitable for their family. With the prices of microwave ovens dropping very low with the increase in technology, everyone can buy one these days without the need of any finance to own something high priced.

At the commercial microwave oven market, one can identify different types of feeding habits. In broad, food market can be divided into following types where microwave ovens are playing major role:

- Fast food outlets
- Full menu restaurants
- Hospital feeding systems
- Food vending
- Convenience store, super markets

Many studies in the area of food nutrition seem to conflict each other. Some studies indicate that the microwave oven cooking completely destroys the nutritional value of foods while others claim no change to the nutritional value. Many more studies are required to assess the safety as well as possible nutrient loss associated with microwave cooking. However, as per one report of Department of Scientific and Industrial Research, Government of India, Research and Development in the field of microwave ovens is considered as low priority area by national laboratories and the industry is not equipped to invest in in-house Research and Development, operating in a low-volume market. The best available evidence supports that the use of microwave cooking resulted in foodstuffs with safety and nutrients quality similar to those cooked by conventional methods, if the consumers followed the given instructions.

One cannot foresee a sudden death of microwave ovens due to some adverse effects, rather support them to improve and increase in the features they possess. For today, a cooperative effort between the food industry and microwave oven producers is highly required for mutual benefit and rapid market expansion. However, one day, a new technological breakthrough may occur that may make microwaves obsolete, but until then, we must continue to enjoy the food cooked with microwave oven with some added precautions.

■■■

33 Cooking for Good Health

Rajni Modgil[1], Asha Kawatra[2], S. Sehgal[2] and Veenu Sangwan[2]

[1]Food Science and Nutrition, College of Home Science, CSK Himachal Pradesh Krishi Vishvavidhyalya, Palampur, Himachal Pradesh, India

[2]Department of Foods and Nutrition, CCS Haryana Agricultural University, Hisar, Haryana, India, E-mail: <rajni_modgil1@yahoo.com>

Cooking food is an art and important step in meeting the nutritional, health, social and psychological needs of the family. Good nutrition is not only a matter of selecting right food to eat but it is also important to prepare the food items in ways that they may retain maximum nutrients and promote health. Careful decisions about cooking techniques may have a profound effect on nutrients content of the prepared food. The cooking strategies, ingredient substitution and innovating new ideas for cooking can help to retain, and in some cases, improve the nutritional value of dishes.

Cooking is the process of preparing raw food for consumption so that maximum health benefits can be obtained from it. Cooking of raw food articles makes them tasty, of acceptable flavor, texture, easy to swallow and digest. Cooking process involves selecting and combining right ingredients in right proportion using a wide range of tools and methods so that a healthy and nutritious dish may be developed. During the process of cooking, the flavour, texture, appearance and chemical properties of the ingredients are changed, which make the food more palatable. Cooking techniques and ingredients used during cooking vary widely across the world, depending upon the environmental and economic conditions along with the cultural traditions of people living in that particular region.

Prerequisite for healthy cooking is good quality raw material and choice of appropriate cooking method so that a good quality and health-promoting product may be developed. The raw materials particularly fruits and vegetables should be washed thoroughly so that all dust and pesticide residues could be removed. If possible, fruits and vegetables should be dipped in hot water and washed with hot water only. Personnel cooking and handling food should also maintain hygiene and be free from disease causing organisms.

When heat is used in the preparation of food, it kills or inactivates the potentially harmful organisms including bacteria and viruses. The temperature ranging from 5 to 57°C is the food danger zone because at this temperature range, the bacteria can grow rapidly, *e.g.*, *Escherichia coli* under optimal conditions can double in number every twenty minutes. The food may not appear spoiled but can be harmful to anyone who eats it.

Cooking brings about chemical changes in raw materials under the influence of heat. Cookery method employed also has direct bearing on nutritional quality of food, which in long term affects human health. Food, which is being consumed over a long time and dietary habits of a person play an important role in preventing the life style diseases like cardiovascular diseases, diabetes, obesity, hypertension, cancer, *etc.* It is a proven fact that a healthy diet can help in preventing the risk of numerous diseases like high cholesterol, obesity, arteriosclerosis, heart attack and diabetes. However, many people do not know that eating healthy foodstuffs is not enough but the way the food is cooked, *i.e.*, the cooking methods employed have an important effect on health of person. No cooking method is necessarily wrong but some are better for health than the other methods of cooking. There are pros and cons of every cooking method, thus, selecting appropriate cooking method is an important step towards healthy life style. Different cooking methods along with their advantages and disadvantages for health are discussed below:

Boiling: Boiling refers to cooking food in liquids like water or milk by heating at boiling temperature. If food is to be cooked by boiling method, it should be covered with adequate quantity of water and heated to boiling. Intensity of heat should be decreased once the boiling starts and the food should be cooked until it is done. This is known as simmering. If food is cooked on high heat during boiling, the water will evaporate quickly, thereby, causing nutritional loss and food will start burning. Water left after boiling, the food should not be discarded otherwise most of the water-soluble minerals and vitamins are lost. Boiling does remain preferable for food items with more dietary fibres and which have strong flavours, as these may be imparted to other food articles during steaming, *e.g.*, mustard greens, green leafy vegetables, *etc.*

Advantages

- This is an economical method.
- Minimum or no use of fat is used.
- All nutrients except vitamin C are maintained in food provided excess boiling water is not discarded.
- Boiled foodstuffs are suitable for patients suffering from gastrointestinal tract diseases, fever, heart diseases and post-operative stages.

Disadvantages

- Water left after boiling, if discarded, could lead to loss of water-soluble minerals and vitamins.
- More time is required for cooking.

Steaming: It is a moist heat (steam) cooking method with or without pressure using the natural convection of heat, which is traveling in air, steam, or liquid. Steaming is done in special pots meant for the purpose or any normal vessel filled with water on which strainer is placed above the water. This gives tender results because food is not exposed to intense dry heat as with other cooking methods. Steaming protects the nutritional value of foods, which are placed on strainer or within sealed containers, as food never comes in contact of heat source or steam-producing liquid. The container must be at least 2.5 cm above the liquid so that it may not touch the food and overcook it. Steam produced from the heat of liquid builds up inside the steamer, and food is cooked in a very moist atmosphere.

Flavorings may be added to food items in a variety of ways during steaming. These can be added in the cooking liquid, which may be water flavoured with herbs and spices, fresh stock, wine, or by marinating foods before steaming. As the flavour is kept within the cooking vessel, the resulting dishes are quite intense and enjoyable. Vegetables like roots and tubers can be steamed.

Advantages

- Steamed food is easy to digest.
- Steaming is suitable for most of food items and a surprising variety of recipes can be made by this method.

- Like boiling, steaming has great advantages over other cooking methods. Higher levels of nutrients, minerals and vitamins can be retained by this method of cooking. During steam cooking, only 40% of vitamin C is lost as compared to boiling where 70% of it is lost.
- Steaming does not immerse the food in water into which nutrients can leach out.
- Foodstuffs are more nutritious and healthy when steamed since no additional fat is required for cooking. Even in recipes where higher fat content ingredients are called for, lower fat alternatives can generally be substituted, such as low-fat milk, cream and cheese and poultry is always skinned to reduce fat content.
- Overcooking of proteins is avoided.
- Steamed food items are suitable for patients suffering from gastrointestinal tract diseases, fever, heart diseases and post-operative stages.

Baking: It is a method of cooking inside the oven using dry heat. This is done to make the breads, cookies, cakes, desserts and to cook seafood, poultry, meats and vegetables. For baking, the leavening agents like baking powder, soda, eggs, or yeast are required for the release of gases and formation of desired texture and product. While baking, plain fat-free, low-fat yogurt, fat-free, or low-fat sour cream should be used.

Advantages

- This method does not necessitate the addition of excess fat or oil, thus, excess cholesterol and calories are not added to the food.
- When baking bread, pastries, *etc.*, they should be taken out of the oven as soon as the top or crust is light brown. The less they are exposed to heat, better will be the retention of vitamins.
- For fishes, baking is a good way to go. Nutrients and flavour are better retained in fishes when baked.

Disadvantages

- If batter were not prepared well, it would not be possible to correct it and get the product of desired quality.

Poaching: Poaching is cooking food in liquids like water or milk by heating below the boiling temperature..

Advantages

- Poaching is a wonderful cooking technique for fish, chicken, eggs and other foodstuffs that becomes too hard if boiled. It maintains tenderness of fish, chicken and eggs.
- No extra/unwanted fat or oil is required for cooking.

Disadvantages

- The only drawback is that since the temperature of cooking is below boiling point, it requires more time for cooking, and thus, may have some negative effect on nutritive value of food.

Stir-frying: Food is chopped into small pieces and cooked in a caldron over medium-high heat with continuous stirring of the ingredients, keeping them nice and crisp without burning. It is best to cook over medium heat to prevent exposure to oxidized or rancid fats. To avoid oils all together, a small amount of broth (chicken or vegetable) can be used to replace the oil.

Advantages

- Very little oil is used.
- Cooking temperature is less, thus, nutrient retention is more.

Sautéing: Sautéing is similar to stir-frying. It is used to cook food quickly in a pan or skillet over high heat with a small amount of fat (oil or butter). However, the problem of oxidizing oils arises. To minimize this risk, the oil can be replaced with a broth. Alternatively, oils with high heat thresholds can be used. Healthy oils such as extra-virgin olive oil or expeller-pressed canola oil can be used at lower heat points.

Advantages

- Cooking time is less than that of stir-frying.
- Food develops characteristic aroma and crust.
- This is the best method for cooking vegetables.
- Very little oil is used.
- This method makes very colourful and attractive dish if different coloured vegetables are used.

Microwaving: This method makes use of electromagnetic waves developed in microwave oven on application of electricity. It is used for cooking as well as re-heating the cooked food in specially designed bowls and also for defrosting frozen food.

The electromagnetic waves act on the water or food particles, and thus, activate them, which produce heat for cooking or re-heating food. Microwave cooking of food articles creates molecular friction, which heats up the food. Microwave cooking of food creates molecular friction, which heats up the food. The friction also causes substantial damage to the surrounding molecules, often tearing them apart or forcefully deforming them. The scientific name for this deformation is structural isomerism.

Advantages

- Food is cooked in less time.
- Food particularly re-heated food is safe as all food is heated at the same time and temperature.
- Use of fat for cooking is the minimum.
- Loss of nutrients is the minimum.

Disadvantages

- Research has demonstrated that the food items cooked in microwave oven can decrease nutrients bioavailability.
- The structure of proteins can be altered, *e.g.*, in formula milk for infants and cause damage to ingested food molecules.
- Special utensils are needed.
- Not all kind of food articles can be cooked, *e.g.*, frying cannot be done and eggs in shell cannot be boiled.
- Browning of surface of cake is not achieved which is a desirable organoleptic attribute.
- It is not suitable for simmering.

Grilling: Grilling is also a dry heat method of cooking and is best suitable for meat. The food comes in direct contact with heat of the grill. Barbecue is one of the variant of grilling in which food does not come in direct contact of heat and smoke of grills.

Advantages

- While grilling, excess fat is dripped away, thus, making the food healthier.

- Cooking time is less, thus, retention of nutrients and flavours is maximum.

Disadvantages

- Grilling foodstuffs that contain fat is less damaging than frying but browning or burning of food during grilling can creates free radicals. The free radical is an atom or molecule that has one or more unpaired electron(s) in its outer orbit. These free radicals can cause damage to cells and tissues. Free radicals are to some extent responsible for ageing and can be cancer causing and inflammatory, thus, one should try to avoid barbecued food, or at least ensure that what so ever is eaten is not charred. Eating excessive amounts of grilled meat or chicken can increase the risk of developing cancer too.

Deep-frying: In this method, food is cooked in deep vessel fully immersed in oil at high temperature.

Advantages

- Deep fried food items are tasty and crispy.
- Cooking is fast.

Disadvantages

- When such food items are consumed on regular basis, they have detrimental effect on health. Excessive use of fat leads to elevation of low-density lipoprotein (LDL-C), bad cholesterol and lowers high-density lipoprotein (LDL-C), good cholesterol leading to hardening of arteries and development of cardiovascular diseases (CVDs). Moreover, when food is fried repeatedly in same oil without changing the oil, such food is carcinogenic. Deep-frying of potatoes specially leads to the formation of acrylamides having carcinogenic effects.
- It is one of the worst ways to cook. In addition to exposing to inflammatory oxidized fats, frying foodstuffs adds calories, usually destroys the fat-soluble vitamins and produces cancer-causing chemicals.
- This type of cooking should be avoided since it can cause inflammation, contribute to elevation of cholesterol and clog the arteries resulting in heart problems.
- Excessive use of vanaspati for frying, which contains trans-fatty acids, is also detrimental for healthy heart and could also lead to diabetes, insterility, hypertension and related health problems.

Cooking for health

Cardiovascular disease is the world's leading killer. With modernization, a large proportion of Asians are trading healthy traditional diets for fatty foods, physical jobs for deskbound sloth and the relative calm of the countryside for the stressful city. Nearly 95% of the people who developed a fatal cardiovascular disease had at least one of these major risk factors, *i.e.*, high blood cholesterol, high blood pressure, smoking, diabetes besides a poor diet and overweight. However, diet plays an important role in heart problems. Much can be done about healthy cooking for preventing many diseases.

Oil and Health

Oil is an indispensable part of cooking but excessive consumption of fats and oils can clog up the arteries, which increases the chances of developing health problems. Low fat or selective fat if used properly in selective cooking method is useful for the management of many of the diseases like obesity, atherosclerosis, diabetes, *etc.*

The fat intake should not be more than 25 to 30 gram per day. Hidden fats taken through processed foods should be avoided. Dietary fats are the best when they come from the unrefined natural fats found in nuts, seeds, fish, soy, olives and avocado since these fats are accompanied by other good nutrients.

While purchasing food items, look for low fat foodstuffs like skimmed milk, low fat cheese, yoghurt, salad dressings and choose leaner meat cuts. The fats containing all three types of fatty acids, *i.e.*, polyunsaturated fatty acids, monounsaturated fatty acids and saturated fatty acids should be used, meaning one should use one part refined oil, about two parts mustard oil and one part ghee or butter for daily cooking.

Fats should not be avoided completely because they help the body in absorbing fat-soluble vitamins and antioxidants.

Low fat cooking can be achieved in following ways:

- A very little oil should be used during cooking.
- It is better to cook vegetables in liquids (such as water or lemon juice) instead of oil.
- When cream has to be used as a thickener, low fat yoghurt, low fat soymilk, or evaporated skim milk can be used in place of oil.
- When browning vegetables, they should be put in a hot pan, and then, oil can be sprayed on these instead of adding the oil in the beginning of cooking. This will reduces the amount of oil, which is absorbed by the vegetables during cooking.
- When serving meat and fish, pesto, salsas, *etc.*, use chutneys and vinegars in place of sour creams, butter and creamy sauces.
- Use oil or fats as lubricants should be avoided.
- It is better to use non-stick cookware.
- Remove chicken skin before eating since it contains very high amount of fat. However, to retain moisture in chicken meat, the skin should be removed at the end of cooking.

Salt and Health

For controlling hypertension, decrease the amount of salt in the food since high salt content can cause hypertension. Reducing salt will not hinder the taste, if a part of the salt is replaced with any herb, spice, citrus juice, or peel and flavored vinegar. If possible, salt with low sodium content should be used. For any meat, soup and sauces, garlic or onion powder should be incorporated. Even a mixture of garlic, paprika, onion and coriander leaves can be used in food. Eating preserved, canned, or processed vegetables should be avoided since these have very high sodium content. Salt is a traditional flavour enhancer but research suggests that a high salt diet could contribute to a range of disorders including high blood pressure.

Some tips to reduce salt content in the diet are given below:

- Additional salt should not be added to the food without tasting it first.
- A splash of olive oil or lemon juice should be incorporated close to the end of cooking time or to cooked vegetables since these can enhance flavour in the same way as the salt can.
- Fresh or frozen vegetables should be preferred over canned and pickled vegetables since processed vegetables contain higher salt content.
- The consumption of salty processed meats, such as salami, ham, corned beef, bacon, smoked salmon, frankfurters and chicken loaf should be done less frequently.

- The bread and breakfast cereals containing low salt should be used since breads and cereals are a major source of salt in the diet.
- Iodized salt should be used always as a major dietary source of iodine in the food.
- Salt-laden processed foods, such as flavoured instant pasta, canned or dehydrated soup mixes, chips and salted nuts should be avoided in diet menu.
- Margarine and butter contain a lot of salt so should be avoided to include in diet.
- Most cheeses are very high in salt so their intake should be restricted or preference should be given to lower salt varieties.
- The use of soy sauce, tomato sauce and processed sauces and condiments (for example mayonnaise and salad dressings) should be limited since they contain high levels of salt.

Sugar and Health

Diabetes is a disease that affects the body's ability to utilize glucose (sugar). Normally, a diabetic must be careful about the foodstuffs that they eat and the method they follow to prepare them.

For the diabetic with a sweet tooth, artificial sweeteners can be used. Sugar should be used judiciously.

Tips for Retaining Nutrients

Water-soluble vitamins are delicate and easily destroyed during preparation and cooking so to retain these nutrients, the following tips should be adopted:

- The vegetables should be cleaned properly rather than peeling them, as many nutrients are found close to the skin.
- Very little water should be used for boiling vegetables, and over boiling of vegetables should be avoided.
- After boiling vegetables, pulses, rice, *etc.*, the left over water should not be thrown away since this water contains important water-soluble vitamins. This water may be used for coking legumes, making soups, or kneading the flour, *etc.*

Use of Herbs

Culinary herbs are leafy plants that add flavour and colour to all types of meal. In many cases, they can replace the flavour of salt and oil. While cooking herbs, the following points should be kept in mind:

- Herbs are delicately flavoured, thus, they should be added in last few minutes of cooking.
- Dried herbs are more strongly flavoured than fresh ones. As a rule, one teaspoon of dried herbs equals to four teaspoons of fresh ones. Hence, these should be used in fewer amounts.
- Apart from boosting meat dishes, the herbs can be added to soups, breads, mustards, salad dressings, vinegars, desserts and drinks.
- Herbs such as coriander, ginger, garlic, chilli and lemongrass are especially complimentary in vegetable-based stir-fried recipes. These have the medicinal value also.

Tips for Healthy Cooking

Healthy cooking starts with healthy ingredients. Cooking is an important step for satisfying hunger and meeting the nutritional requirements of family. Cooking renders food palatable, easily digestible, free from microorganisms, safe, lowers antinutritional factors and provides variety. Indian cooking is an intimidating and time-consuming affair. It does utilize many fresh ingredients, sauces and spices. By introducing few tips and techniques into the process, Indian cooking can be a fun endeavor. Preparing healthy meals is not difficult if the basics of nutritious cooking are known. Some of the healthy cooking tips to be kept in mind for cooking are as follows:

- The menu should be planned, and the food, which is freshly cooked, should be served immediately to prevent reheating since reheating destroys the nutritive value of food articles and tampers with original taste.
- While buying pulses and legumes, it should be kept in mind that they should be unpolished and small sized since some vital nutrients are lost during polishing and processing.
- Most of the vegetables are covered with fertilizers and pesticides so to remove them, the vegetables should be washed in a solution of potassium permanganate, and then, rinsed with plain water. Thereafter, these should be cut and cooked.
- Vegetables should not be over-cooked. The more they are cooked, the less nutritious they will be. Fresh and tender vegetables cook quickly and do not need a lot of seasoning and spices.
- The nutrients and colours in vegetables should be preserved. They should be cooked quickly by steaming or stir-frying.
- Green leafy vegetables should not be cooked using lime or tamarind.
- Use non-sticking utensils and thick bottom utensils since they get uniformly heated. For electric stoves, use flat bottom utensils.
- To ensure that food is not overcooked in a pressure cooker, first boil the water without putting the lid on the cooker, and when it starts boiling, the raw food is added and pressure-cooked. The cooking should be done on low flame for some time until the food is cooked instead of pressure-cooking on high flame with many whistles of the cooker.
- Milk purchased from a dairy, which is not pasteurized, should be boiled for at least 10-15 minutes at boiling temperature to eliminate the effect of any bacteria or other germs.
- To ensure that ladyfingers may not turn black or stick to the vessel while cooking, add a spoonful of curd to them.
- To boil potatoes faster, add a little turmeric and oil to the water.
- Add a cup full of grated beetroot or carrot to the coconut while making *burfy*. You will have a natural colouring and added nutritional value too.
- For soft cakes, add a teaspoon of glycerin to the batter.
- Avoid the use of ghee. If it is necessary, substitute it with canola oil. Even for making *halwa*, you can partly substitute it with oil.
- Place a betel (*paan*) leaf over the leftover *idli* and *dosa* batter to prevent them sour. Do not beat *idli* batter too much, the air, which has been incorporated during fermentation, will escape. If you add half a teaspoon of fenugreek seeds to the lentil and rice mixture while soaking, *dosas* will be crisper.

- Avoid deep-frying. Substitute deep-frying with stir-frying or oven bake. Do not pour the oil but make a habit of spraying the oil in the utensil for cooking. Heat the utensil first, and then, add oil. This way oil spreads well. You will use less oil this way.
- Bake *papad* in microwave oven. Wrap the *papads* in polythene sheet and place with *dal* or rice will prevent them from drying and breaking.
- Do not throw away the rice water after cooking. Use it to make soup or add it in making *dal* (lentils).
- Chop the vegetables only when you are ready to use them. Do not cut them in too advance. It would spoil their food value.
- Make a small hole in the egg by piercing a pin before boiling it. You will be able to remove its skin very easily.
- Garlic skin comes off easily if the garlic cloves are slightly warmed before peeling.
- If the lemon or lime is hard, put it in warm water for 5-10 minutes to make it easier to squeeze.
- Moisten the base of the vessel with water to reduce the chances of milk to stick at the bottom. Keep a spoon in the vessel while boiling milk at medium heat. It will avoid sticking the milk at the bottom of vessel. Adding half a teaspoon of sodium bicarbonate in the milk while boiling will not spoil the milk even if not kept in the refrigerator.
- When the noodles are boiled, drain out all the hot water and add cold water. This way all the noodles will be separated.
- Do not fry the cheese; immerse it in boiling water to make it soft and spongy.
- Use sprouted mustard seeds (*rayee*) and fenugreek (*methi*) seeds for seasoning. Both of them when sprouted have more nutritional values. In addition, this adds flavour to the dish and can be more beneficial, besides giving decorative look to the dish.
- The food before storing in the refrigerator should be covered well to prevent mixing of flavour.
- To keep cheese fresh for several days, wrap it in a blotting paper while storing in the refrigerator.
- To keep celery fresh for long time, wrap it in aluminum foil and place in the refrigerator.
- To store chilly powder longer, keep a piece of asafetida in the container.
- Add 5 g of dried powdered mint leaves to 1 kg of rice. It will keep insects at bay. Put a small paper packet of boric powder in the container of rice to keep insects at bay. Put a few leaves of mint in the container of rice to keep insects at bay.
- To remove the skin of tomatoes, place them in warm water for 5-10 minutes. The skin can then be easily peeled off.
- When tomatoes are not available or too costly, may be substituted with tomato puree or tomato ketchup/sauce.
- Place overripe tomatoes in cold water and add some salt. Overnight, they will become firm and fresh.
- Do not discard the water in which the vegetables are soaked or cooked. Use it in making soup or gravy.

- Place a raw peeled potato in the bowl since it will absorb the extra salt. Set yogurt in winter and place the container in a warm place like oven or over the voltage stabilizer.
- If the yogurt has become sour, put it in a muslin cloth and drain out all the water. Then add milk to make it as good as fresh in taste. Use the drained water in making tasty gravy for vegetables or for *besan* curry.
- To keep the yogurt fresh for many days, fill the vessel-containing yogurt with water to the brim, refrigerate and change the water daily.
- To ripen fruits, wrap them in newspaper and put in a warm place for 2-3 days. The ethylene gas they emit will make them ripe.
- To avoid crying, cut the onions into two parts and place them in water for 15 minutes before chopping them.
- Onions are a rich source of quercetin, which is a very important phytochemical and antioxidant. Onion should be cut and kept for at least 10 minutes before serving since oxidation helps the quercetin to be activated.
- Immerse the egg in a pan of cool salted water. If it sinks, it is fresh, while if it rises to the surface, it is certainly quite old.
- Use herbs, vinegar, tomatoes, onions and/or fat-free or low-fat sauces or salad dressings for better health.

Painted and decorated pots and pans should be avoided for cooking as some are made of lead and silver toxin material, which can lead to severe digestive disorders. Using aluminum vessels for cooking can cause high levels of aluminum in the brain tissue because it is difficult to remove aluminum from the body. People suffering from Alzheimer's disease have high aluminum in brain tissue. Selection of proper cookware is also as important as the cooking method used. The metal of which cooking ware is made generally gets leached into the food being cooked in it. The amount of metal leached depends on the duration of cooking, pH of food, *etc.* Pyrex and stainless steel is the safest cookware.

■■■

34 Sucrose Surpassed- Alternatives Abound

Sheeba Menon[1], S. Bhanumati[2] and S. Narayan[3]

[1]Department of Biology, K.K. Shah Jarodwala Maninagar Science College, Rambaug, Maninagar, Ahmedabad, Gujarat, India E-mail: <ksm51171@yahoo.com>
[2]Chemistry Department, Gargi College (Delhi University), Siri Fort Road, New Delhi, India, E-mail: <sbhanumati@gmail.com>
[3]Agricultural Biochemist, E-mail: <naren16in@yahoo.co.in>

Food is the source of energy for all bodily functions. About 10% of all the energy supplied is expended in the digestion and assimilation of the very food we eat. Different types of food require different amounts of energy for digestion and assimilation. The energy spent is in decreasing order starting from proteins to fats, oils, carbohydrates and then from other nutrients such as vitamins, minerals, salts, *etc.* Our working ability is hampered by overeating or gormandizing occasionally or habitually. Experimentally, it is established that a meal of rice and milk is digested in a few hours, while a flesh based meal takes twice of this time for digestion and assimilation. In other words, a large quantity of energy is wasted in unnecessary digestion and assimilation work. The surplus food consumed is less utilized in a constructive way. Protein rich food consumed in excess in fact becomes a toxic burden on the digestive system and is soon excreted as urea, uric acid and other catabolic products. Often these are putrefied in the intestines producing harmful chemicals such as indole, skatol and pyrrole like toxins. These chemicals enter into the circulatory system and produce undesirable side effects, which affect the physical fitness of an individual.

Fitness is not only essential for normal metabolism but also essential to burn extra calories acquired inadvertently. Proteins, fats, carbohydrates minerals and vitamins make a wholesome diet. Cereals and veggies like potatoes are very rich in carbohydrates, which are first broken down into glucose by saliva present in the mouth. This is followed by digestion in stomach and small intestine. The energy packed glucose reaches all body parts through the circulatory system. Glucose on decomposition liberates carbon dioxide and water along with several calories of energy per gram molecule of glucose. The remaining undecomposed glucose is stored in the liver as glycogen, which serves as reserved food and releases energy in times of emergency.

The glucose in adipose tissues as triglycerides is stored in the form of fat. This explains why excess carbohydrates when consumed lead to obesity. Maintenance of blood glucose level is therefore extremely important for strong body, functional brain and a normal healthy heart.

Pancreas, Sugar Metabolism and Diabetes

A special group of cells called the Islets of Langerhans (IL) present in pancreas has a significant role in sugar metabolism in the body. The pancreas and its secretions (hormones) form a part of the *Endocrine system*. There are four types of cells present in Islets of Langerhans. The *A* type cells secrete glucagon, *B* type insulin, *D* type stomatostanin and *F* type pancreatic polypeptide. A, B and D type cells play an important role in balancing blood sugar (glucose) level. Diabetes mellitus is caused when any one or more of these hormones is either insufficient or in excess. Diabetes is a complex disease, and if not treated well in time and managed subsequently, can even be fatal. Excess of insulin causes hypoglycaemia, a condition in which blood sugar level is present below the normal level, *i.e.*, below 80 mg per 100 ml of blood. This condition may lead to convulsions and coma but is relieved by administering glucose.

Insulin is a polypeptide containing two amino acid chains, which are linked by disulphide bonds as bridges. The endoplasmic reticulum of *B* type cells of IL secretes insulin. The insulin gene is located on chromosome 11 in human beings. Insulin is a protein hormone, which regulates the blood glucose level. In diabetics, due to insulin deficiency, the glucose transport is affected leading to its accumulation in blood. Hence, insulin is administered as injection to the patients to keep them healthy. There are different types of diabetes. The maturity onset diabetes is non-insulin dependent diabetes mellitus (NIDDM), which is usually found after 40 years age due to over eating. Obesity is a result commonly noticed in these patients. The second type is the juvenile onset diabetes, which is insulin dependent (IDDM). It is believed that coxackie viruses destroy the *B* type cells of IL, and hence, this condition results. Prolonged stress sometimes may trigger diabetes. Excessive secretion of the stress hormone adrenaline and cortisone in a stressed individual induces diabetic syndromes.

The Food market, Demand, Supply and Its Impacts

The natural food market is an ever-growing industry in present times. With the great reservoir of scientific back up with positive and negative worth of each and every food item whether natural or synthetically designed, a parallel stream of innovations in alternative products is emerging with a fast pace. A large segment of consumers is the victim of affluence-generated life style, which in turn is inviting health problems due to temptations coupled with non-discretion. One such temptation is the sweet to taste food. People who suffer from diabetes have therefore no option but to curtail on sugary eatables and reduce their caloric intake. This can be managed by substituting the sugars with non-caloric sweeteners.

Traditional Sweeteners and Sugar Substitutes

Sucrose is the common table sugar, which is a high-energy sweetener (100 gram= 394 kcal energy). Chemically, it is made up of one molecule of glucose and one molecule of fructose. Sucrose is obtained from sugar cane and from sweet potatoes, sugar beet, *etc.* The cane juice is concentrated and the crystals obtained are refined by bleaching to remove the colour. The various forms of white sugar thus obtained are chemically identical. Table sugar is granulated. Castor sugar has finer and smaller crystals than table sugar. Icing sugar is powdered and ground sugar with some calcium phosphate to prevent the lump formation. Cube sugar is the compressed form of table sugar. There is also a quick dissolve variety, which is less compressed than the cubes. Brown sugar is less refined and strongly flavoured. In its raw form, cane sugar is also popular as Barbados, mclasses, Muscovado, *etc.* Beet molasses is very bitter, hence, less suited for human consumption. Cane molasses is rich in metals and minerals like iron, calcium and their salts. Refined sugar combined with molasses is often called *syrup*. Golden syrup contains some hydrolyzed sucrose. Hence, the two constituents, glucose and fructose are freely present in it. Glucose is grape sugar also called dextrose and is present in honey and some fruits. It is often administered to patients intravenously for providing energy instantaneously. Fructose, also called laevulose, is a constituent of some fruits and honey. Interestingly, glucose is just 3/4th times as sweet as sucrose, while fructose is twice as sweet as sucrose. The mammalian milk contains lactose, which is only 1/6th as sweet as sucrose. Beer and malt also contain a sugar called maltose, which is only 1/3rd times as sweet as sucrose. Developing sweet tooth is endangering one's own health.

Sugar consumption in India is constantly on the increase. Over consumption of sugar based items, such as biscuits, cakes, pie, chocolates, toffees, jams, ice cream, sweet meats, condensed milk and a good lot of drinks bottled or brewed, is leading to alarming number of diabetics in India. Besides, obesity, shorter life span, heart problems, blood pressure and eye related problems are the interlinked maladies, which need to be correlated and addressed.

There are some polyhydric alcohols known as *Polyols* or sugar alcohols, which have less sweet taste than sucrose but have the same bulk properties. Glycerol (glycerine) is one such polyol.

Honey, which is rightly called the ancient and mysterious sweetener, is a wonderful sweetener. It is said that no one has discovered a purer and sweeter food than honey. Because of the presence of laevulose or fruit sugar, honey is almost twice as sweet as cane sugar.

Table sugar like table salt has but one taste, while every type of honey has its own. The taste and flavour of a specific brand of honey is largely governed by the variety of flowers from which the bees collect the nectar.

Properly ripened honey is the purest sweet. Under the high sugar concentrations and the high osmotic pressure, the bacteria fail to survive. When diluted with water, honey proceeds with fermentation. Honey fermentation is undesirable, and thus, fermented honey should be discarded. Honey is a special ingredient in religious rituals and in dietician's recommendations.

Each drop of nectar collected from flowers is an ocean of minerals, nutrients and enzymes. It is also rich in traces of iron, copper, potassium, sodium, phosphorous, proteins and various vitamins. Amazingly, many diabetics can tolerate honey contrary to the table sugar. Calcium retention in babies is facilitated by honey feed.

The conversion of nectar into honey is probably still a subject of research, but the very enzyme, which brings about this awesome change, is again supplied by the tiny skilled engineers- the honeybees. The news would surprise a common man that the tobacco companies buy millions worth of low grade honey annually for preserving, flavouring, moistening and mellowing tobacco. Besides being a food delicacy, honey is also an ingredient in cosmetics, cough medicines, *etc.*

Chemically, in addition to other minor constituents, honey comprises of glucose and fructose not hooked together as in the case of sucrose- the common sugar. The relative ratios of these in honey are fructose 38.2: glucose 31.7 and sucrose 0.7.

Sugar Substitutes

The sugar substitutes are known by various names, *i.e.*, sugar free, artificial sweeteners, non-sugar sweeteners, tabletop sweeteners, *etc.*

An important class of sugar substitutes is known as high intensity sweetener. Artificial sweeteners are compounds with sweetness, which is many times that of sucrose. The sensation of sweetness caused by these compounds is sometimes notably different from sucrose, thus, they are often used in complex mixtures that achieve the most natural sweet sensation. If the replaced sucrose has contributed to the texture of product, a bulking agent is often needed. This may be seen in soft drinks labeled as *diet* or *light*, which contain artificial sweeteners, and often, have notably different mouth feel, or in table sugar replacements, which mix maltodextrins with an intense sweetener to achieve satisfactory texture sensation. In a complex food, sweetener is one element. It is also important to note that the sugar substitutes lack the browning, tendering and moisture retaining properties. Hence, the limitations of one type of sweetener can be overcome by using blends.

Definition of Sugar Substitutes

Sugar substitute is a food additive that duplicates the effect of sugar in taste, usually with less food energy. Some sugar substitutes are natural and some are synthetic. Those that are not natural are called artificial sweeteners.

Brief Overview of Usage of Artificial Sweeteners in USA

In United States, six intensely sweet sugar substitutes, *i.e.*, saccharin, aspartame, sucralose, neotame, acesulfame potassium and stevia, have been approved for use. There is some ongoing controversy over whether the usage of artificial sweetener poses health risks. The majority of sugar substitutes approved for food use are artificially synthesized compounds. However, some bulk natural sugar substitutes are known including sorbitol and xylitol, which are found in berries, fruits, vegetables and mushrooms. It is not commercially viable to extract these products from fruits and vegetables. Hence, they are produced by catalytic hydrogenation of the appropriate reducing sugar, *e.g.*, xylose is converted to xylitol, lactose to lacitol and glucose to sorbitol. Still, other natural substitutes are known but they await the official approval for food use.

Some non-sugar sweeteners are polyols, also known as *sugar alcohols*. These are less sweet than sucrose but have similar bulk properties and can be used in a wide range of food products. Sometimes, the sweetness profile is *fine-tuned* by mixing with high intensity sweeteners. As with all food products, the development of a formulation to replace sucrose is a complex proprietary process.

Food and Drug Administration

Artificial sweeteners as food additive are regulated by the United States Food and Drug Administration (FDA), thus, all the food additives must be approved by the FDA, which publishes a Generally Recognized as Safe (GRAS) list of additives. The safe conclusions are based on a detailed review of a large body of information, including hundreds of toxicological and clinical studies. Acceptable daily intake (ADI) is the amount prescribed per kilo of body weight and is a conservative level, which reflects an amount 10 times less than the maximum level at which no observed adverse effect occurs in animals. There is also an herbal supplement, *i.e.*, stevia, which is widely used. Due to a lack of research on stevia's safety, there was a battle over its approval as a sugar substitute.

Food Industry and Artificial Sweeteners

The Food and Beverage Industry is increasingly replacing sugar or corn syrup with artificial sweeteners in a range of products traditionally containing sugar. Artificial sweeteners cost the food industry only a fraction of the cost of natural sweeteners, in spite of the extremely high profit margin for manufacturers of artificial sweeteners. Thus, it is not surprising that the food industry is promoting its *diet* or *light* products heavily, thus, moving the customers over to its even more profitable artificially sweetened products.

Market Value of Artificial Sweeteners

According to market analysis Mintel, total 3,920 products containing artificial sweeteners were launched in the United States between the year 2000 and 2005. In 2004 alone, 1,649 artificially sweetened products were launched. According to market analysis Fredonia, the United States artificial sweetener market is set to grow at around 8.3% per year to US $ 189 million in 2008.

Aspartame is currently the most popular artificial sweetener in the United States Food Industry, as the price has dropped significantly since the Monsanto Company patent expired in 1992.

Reasons for Use

The sugar substitutes are used for a number of reasons given as under:

- **Assist in weight loss:** Some people choose to limit their food energy intake by replacing high-energy sugar or corn syrup with other sweeteners having little or no food energy. This allows them to eat the same foodstuffs they normally would like to eat, while allowing them to lose weight and avoid other problems associated with excessive caloric intake. Obesity management through artificial sweeteners is a successful mantra for practical and aesthetic reasons.
- **Dental care:** Sugar substitutes are teeth friendly, as they are not fermented by the micro-flora of dental plaque. Dental decay is also largely minimized by avoiding the traditional sugars and by using the sugar substitutes.
- **Diabetes mellitus:** People with diabetes have difficulty regulating their blood sugar levels. By limiting sugar intake with artificial sweeteners, they can enjoy a varied diet while closely controlling their sugar intake. In addition, some sugar substitutes do release energy but they are metabolized more slowly, allowing blood sugar levels to remain more stable over time. Maximizing insulin conservation and minimizing insulin dependence are the double benefits derived from artificial sweeteners.
- **Reactive hypoglycemia:** Individuals with reactive hypoglycemia will produce an excess of insulin after quickly absorbing glucose into blood stream. This causes their blood glucose levels to fall below the amount needed for proper body and brain function. As a result, like diabetes, they must avoid intake of high glycemic foodstuffs like white bread, and often, choose artificial sweeteners as an alternative.
- **Avoiding Processed foods:** Individuals may opt to substitute refined white sugar with less processed sugars, such as fruit juice or maple syrup.
- **Cost:** Many sugar substitutes are cheaper than table sugar that is why people prefer to use the artificial sweeteners.

Health Issues

Controversy over health: A population-based study by the University of Texas and Health Science Center at San Antonio in 2005 showed that increased weight gain and obesity was associated with increased use of diet soda. Animal studies have indicated that artificial sweeteners can cause body weight gain. A sweet taste induces insulin response, which caused blood sugar to be stored in tissues (including fat) but since blood sugar does not increase with artificial sugars, there is hypoglycemia and increased food intake the next time there is a meal. After a while, rats given sweeteners have steadily increased calorie intake and increased adiposity (fatness).

Artificial Sweeteners and Cancer: As mentioned earlier, artificial sweeteners, the sugar substitutes, are substances that are used instead of sucrose (table sugar) to sweeten food articles and beverages. The Food Additives Amendment to the Food, Drug and Cosmetic Act, which was passed by United States Congress in 1958, requires Food and Drug Administration to approve food additives, including artificial sweeteners, before they can be made available for sale in the United States. However, this legislation does not apply to products that are generally recognized as safe. Such products do not require approval from the Food and Drug Administration before being marketed.

Questions about artificial sweeteners and cancer arose when early studies showed that cyclamate in combination with saccharin caused bladder cancer in laboratory animals. However, results from subsequent carcinogenicity studies of these sweeteners have not provided clear evidence of an association with cancer in humans.

Some very popular sugar substitutes are described below with their significant features and their side effects, if any:

Cyclamate: Wayback in 1937 while working on antibacterials, Michael Sveda at the University of Illinois accidentally discovered this sweetener. The compound is used as its sodium and calcium salts and also as cyclamic acid. In the United States, the Food and Drug Administration banned the sale of cyclamate in 1970 after laboratory tests in ratsinvolving a 10:1 mixture of cyclamate and saccharin indicated that large amounts of cyclamate causes bladder cancer, a disease to which the rats are particularly susceptible.

Saccharin: Aside from sugar of lead, saccharin was the first artificial sweetener and was originally synthesized in 1879 by Remsen and Fahlberg. Its sweet taste was discovered by accident. It had been created in an experiment with toluene derivatives. A process for the creation from phthalic anhydride was developed in 1950, and currently, saccharin is produced by this process as well as the original process by which it was discovered. It is 300 to 500 times as sweet as sugar (sucrose) and is often used to improve the taste of toothpastes, dietary foods and dietary beverages. The bitter after-taste of saccharin is often minimized by blending it with other sweeteners.

Fear about saccharin increased when a study during 1980 showed that high levels of saccharin might cause bladder cancer in laboratory rats. Subsequently, it was discovered that saccharin causes cancer in male rats by a mechanism not found in humans. At high doses, saccharin causes a precipitate to form in rat urine. This precipitate damages the cells lining the bladder (urinary bladder Urothelial Cytotoxity) and a tumour forms when the cells regenerate (regenerative hyperplasia). According to International Agency on Cancer (part of the WHO), saccharin and its salts were downgraded from Group-2B, possibly carcinogenic to humans to groups not classifiable as to carcinogenicity to humans, despite sufficient evidence of carcinogenicity to animals since it is carcinogenic by a non-DNA reaction mechanism, which is not relevant to humans because of critical interspecies differences in urine composition. Studies in laboratory rats during early 1970s linked saccharin with the development of bladder cancer. Subsequent studies in rats showed an increased incidence of urinary bladder cancer at high doses of saccharin, especially in male rats. However, epidemiology studies have shown no consistent evidence that saccharin is associated with bladder cancer incidence.

As the bladder tumour seen in rats are due to a mechanism not relevant to humans and because there is no clear evidence that saccharin causes cancer in humans, saccharin was delisted from the United States Toxicology Program's Report on Carcinogens in year 2000. Saccharin is an *o*-toluene sulfonamide derivative and causes certain dermatological reactions. In earlier days, saccharin was thought to cause cancer. However, that idea has changed since studies have proved that it is not connected to cancer in humans. Therefore, saccharin is now present in substantial amounts in certain drugs. The safe limits of saccharin consumption have been advocated as 2.5 milligrams for every kilo of body weight.

Side Effects

- Since there has been cross-sensitivity with sulfonamides, children who suffer from sulfa allergies should avoid saccharine.
- Some side effects in both adults and children are hypersensitivity, photosensitivity and eczema.
- Some other reactions to saccharine include nausea, tongue blisters, diarrhea, wheezing, headache and sensory neuropathy.
- Saccharine-infused milk causes irritability and insomnia.

Acesulfame Potassium: Acesulfame potassium is the newest alternative sweetener approved in the United States and being marketed under the names acesulfame K, Ace-K, or Sunett. Food and Drug Administration approved acesulfame-K in 1988. When compared with sugar, it is 200 times sweeter than sucrose and is often used to either enhance or preserve sweetness of foodstuffs. The blends of acesulfame-K with other nutritive and non-nutritive sweeteners can synergize the sweetness potential and the bitter taste. It shelf life is 3 to 4 years since it is heat stable, hence, can be used in baked foods. It does not provide any energy since it is not metabolized by the body though excreted unchanged by the body through urine. It is used as a sweetener in soft drinks, chewing gums, coffee, tea, candies, puddings, yogurt, gelatins, non-dairy creams, dairy items and salted foods.

Side Effects

- Due to the presence of methylene chloride, it causes cancer.
- Apart from cancers, long-term usage of methylene chloride can cause nausea, headaches, depression, kidney and liver problems, mental confusion and visual disturbances.

Neotame: Neotame is much like aspartame with regard to its chemical combination but does not have the phenylalanine dangers that the aspartame has. It is between 7,000 and 13,000 times sweeter than table sugar.

Side Effects

Not too many side effects are known since the studies conducted on neotame are not available to the masses. However, since neotame is a type of aspartame, it is recommended that caution should be exercised when using it, since side effects are assumed similar.

Aspartame: Aspartame was discovered in 1965 by James M. Schlatter at the G.D. Searle Company (later purchased by Monsanto) while he was working on an anti-ulcer drug and spilled some aspartame on his hand by accident. When he licked his finger, he noticed that it had a sweet taste. Aspartame is an odourless white crystalline powder, which is derived from two amino acids, *i.e.*, aspartic acid and phenylalanine. It is about 200 times sweeter than the sugar and can be used in frozen desserts, gelatins, beverages and chewing gums. When cooked or stored at high temperature, aspartame breaks down into its constituent amino acid. It is more stable in acidic condition, such as in soft drinks. It does not have a bitter after taste like saccharin and it may not taste exactly like sugar. When consumed, aspartame is metabolized into amino acid. It releases some energy similar to protein but it is so intensely sweet that only a very little is needed to sweeten a food product. Thus, it is useful for reducing number of calories in a product.

The safety of aspartame has been studied extensively since its discovery with research that includes animal studies, clinical and epidemiological research and post- marketing surveillance. Aspartame has been the subject to multiple claims against its safety, including supposed links to cancer as well as complaints of neurological or psychiatric side effects.

Aspartame distributed under several trade names, *e.g.*, nutrasweet and equal, was approved in the year 1981 by the Food and Drug Administration after numerous tests showed that it did not cause cancer or other adverse effects in laboratory animals. Questions regarding safety of aspartame were renewed in 1996 report, suggesting that an increase in a number of people with brain tumours between the year 1975 and 1992. Moreover, the increase in overall brain cancer incidence occurred primarily in people with the age of 70 years and older, a group that was not exposed to the higher doses of aspartame since its introduction. These data do not establish a clear link between the consumption of aspartame and the development of brain tumors. This is a dipeptide of a methyl ester of phenylalanine and aspartic acid. It is a predominant ingredient of sugar-free formulae and chewable disorders tablets.

Side Effects

- Headaches, especially migraines have been triggered off by aspartame.
- It links to various neuro-psychiatric disorders such as mood swings, panic attacks, visual hallucinations, isolated dizziness and manic episodes.
- Those who have valve prolepses are placed at a greater risk for neuro-psychiatric.
- Prolonged aspartame consumption leads to seizures, especially in children.

Sucralose: Sucralose is a chlorinated sugar, which is about 600 times sweeter than sugar. It is produced from sucrose, when three chlorine atoms replace three hydroxyl groups. It is used in beverages, frozen desserts, chewing gum, baked goods and other foodstuffs. Unlike other artificial sweeteners, it is stable when heated and can therefore be used in baked and fried goods. About 15% of sucralose is absorbed by the body and most of it passes out of body unchanged.

Sucralose is a chlorinated sugar prepared from sucrose or raffinose. Safety concerns pertaining to sucralose revolve around the fact that it belongs to a class of chemicals called organochlorides, some types of which are toxic or carcinogenic. However, the presence of chlorine in an organic compound does not ensure toxicity in any way. The way sucralose is metabolized suggests a reduced risk of toxicity. Sucralose is insoluble in fat and does not break down into smaller molecules. Sucralose is also known as splenda

Side Effects

- Sucralose causes shrinking of thymus glands up to 40% and enlargement of liver and kidneys.
- It is assumed to cause renal mineralization and caecal enlargement.
- Contrary to the claims, sucralose is known to break down into small amounts of 1,6-dichlorofructose, a substance that has not been completely tested on humans.
- The presence of chlorine in sucralose might lead to serious chronic neurological or immunological or problems.

Lead Acetate: Lead acetate, sometimes called sugar of lead, is an artificial sugar made from lead, which is of historical interest. It was widely used by ancient Romans. The use of lead acetate as a sweetener eventually produced lead poisoning in any individual ingesting it habitually. Lead acetate was abandoned as a food additive throughout most of the world after high toxicity of lead compounds became apparent.

Natural Sugar Substitutes

Brazeein	Lacitol	Pentadin
Curculin	Mabinlin	Sorbitol
Erythritol	Maltitol	Stevia
Glycyrrhizin	Malto	Tagatose
Glycerol	Mannitol	Thaumatin
Hydrogenated starch hydrolysates	Miraculin	Xylitol
Inulin	Menatin	
Isomalt	Monellin	

Artificial Sugar Substitutes

Ace sulfame potassium	Dulcin	Saccharin
Alitame	Glucin	Sucralose
Aspartame	Neohesperidin dihydrochalcone NEO-DHC	
Cylamate	Neotame	

Other Artificial Sweeteners

Fructo-oligosaccharide: Fructo-oligosaccharide (FOS) refers to short chain oligosaccharide, which comprises of D-fructose and D-glucose, containing 3 to 5 monosaccharide units. Similar molecules are obtained by partial enzymatic hydrolysis on inulins. These are called oligosaccharides. Fructo-oligosaccharide is also called neo-sugar or short chain FOS and is produced on a commercial scale from sucrose using fructosyl-transferase enzyme. It is 0.8 times sweeter than sugar and contributes less than 1 calorie per gram, *i.e.*, 0.25 times of sugar.

Fructo-oligosaccharide looks like honey and tastes like honey and are indigestible, and hence, contribute very low calorie to the body. Fructo-oligosaccharide is non-digestible oligosaccharide. They are fermented by a limited number of colonic bacteria, which are good to health.

The excellent stability and quality taste of FOS means it can be used in a wide range of foodstuffs and beverages. It is unaffected by ethanol, thus, it is suitable for use in alcoholic drinks.

Fermented milk products containing FOS lowers LDL cholesterol levels in human beings. Fructo-oligosaccharide was found to reduce incidence of colon tumors.

Uses

- **FOS acts as dietary fiber:** Fructo-oligosaccharide improves the microbial ecology of the intestine and protects against some harmful bacteria particularly in large intestine.
- **FOS improves absorption:** In body, the fructo-oligosaccharide increases calcium and magnesium absorption and eliminates toxic compounds from the body. Fructo-oligosaccharide overall plays vital role in rejuvenation, improving strength, stamina and contributes to fit and healthy health.

- **FOS as prebiotic:** Fructo-oligosaccharide promotes the growth of beneficial bacteria in the colon, which helps in increasing immunity level.

Polyols: Polyols are alcohol derivatives of sugar, which provide fewer calories than sucrose. It is synthesized from dextrose and remains stable up to 160°C temperature. They are absorbed by passive diffusion and are metabolized slowly.

Polyols produce low glycerine response and they are popularly used as bulking agent in food industry to improve taste and texture of food.

Side Effects

Polyols in large amounts exerts laxative effect causing abdominal cramps and diarrhoea.

Glycyrrhizin: Glycyrrhizin is a root extract of an Asian and European shrub and is triterpene glycoside by nature. It is widely used in India under the name Atimadhuram. Glycyrrhizin is mainly used as coating for candles and chewing tobacco.

Side Effects

- Glycyrrhizin in large amount can cause headache, lethargy and sodium and water retention.
- Glycyrrhizin intake in large amount can even lead to high blood pressure, and sometimes, heart failure.

Chemistry of Saccharin

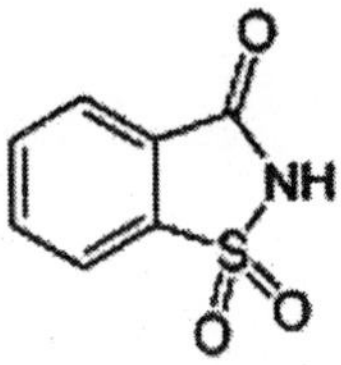

NH

N⁻ Na⁺

(i) 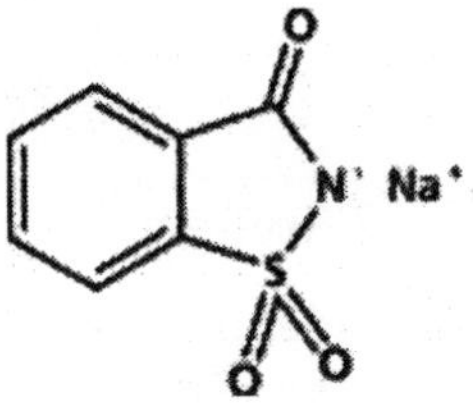 (ii)

The chemical formula of saccharin is $C_7H_5NO_3S$ and it can be produced by different methods. The original route starts with Toluene but the yield in this synthesis is low. An improved synthesis was developed in 1950, where anthranilic acid reacts with nitrous acid, sulphur dioxide, chlorine and ammonia to yield saccharin.

Chemistry of Aspartame

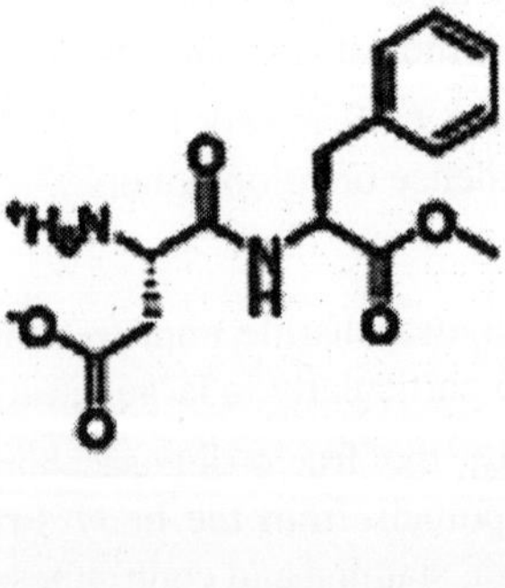

Aspartame is 200 times sweeter than sugar. It is marketed under the brand names *Nutrasweet* and *Equal*. Aspartame is the methyl ester of dipeptide of the amino acids, *i.e.*, aspartic acid and phenylalanine. Aspartame loses its sweetness, when heated and is usually not suitable for baking.

Side Effects

People with rare disorder phenylketonuria (PKU) cannot metabolize phenylalanine. Thus, they should avoid aspartame.

Chemistry of Neotame

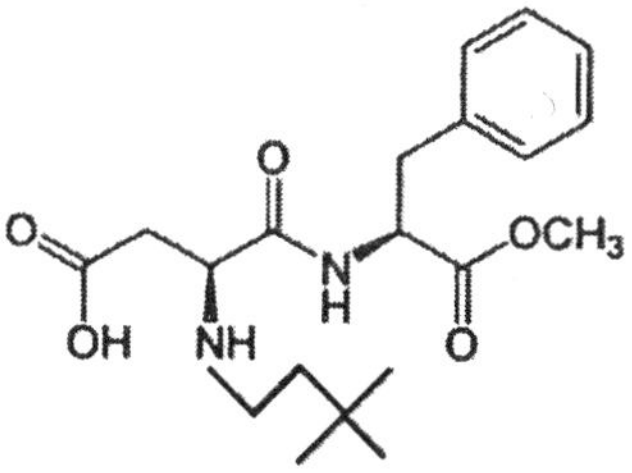

Neotame is 8000-13,000 times sweeter than sugar. Neotame is similar to aspartame but sweeter and more stable. Neotame is hydrolyzed to produce methanol. Since neotame is used only in small amounts, the amount of methanol produced is less than what is found in fruit juices.

Chemistry of Acesulfame Potassium

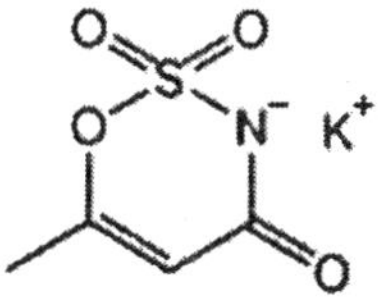

Acesulfame potassium (Acesulfame K or Ace K) is non-nutritive sweetener and is 200 times sweeter than sugar. It is marketed under brand name *Sunett*. At high concentration, it has a bitter after taste like saccharin. The chemical structure of acesulfame potassium is the potassium salt of 6-methyl-1,2,3-oxathiazine-4 (3H)- one 2,2-dioxide.

Side Effects

Some results have oblique that acesulfame potassium may increase the occurrence of breast tumors in laboratory animals.

Chemistry of Sucralose

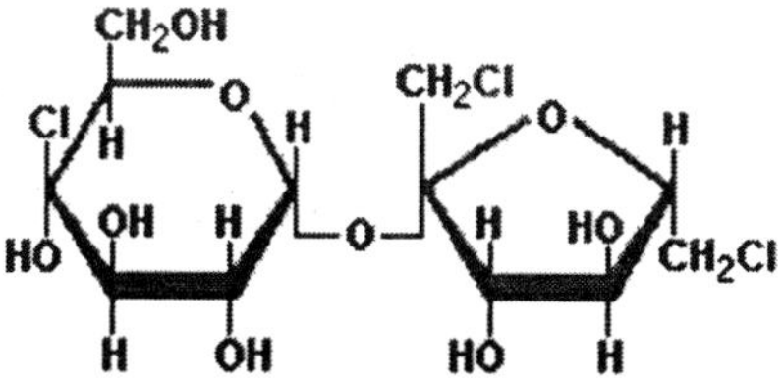

Sucralose is marketed as *Splenda*. It is available as a tabletop sweetener and as an ingredient in food processing. Sucralose is about 600 times sweetener than sugar. Sucralose is stable at hot and cold temperature and can be used in cold and hot drinks, as well as in baked goods.

Chemistry of Cyclamate

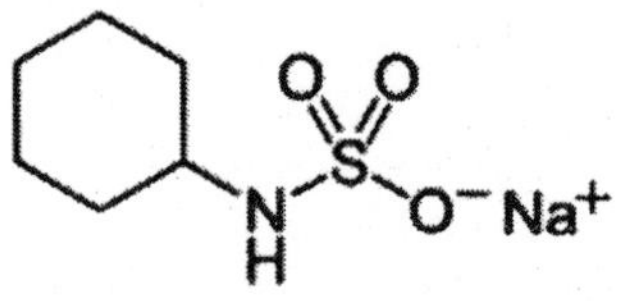

Cyclamate is 30 to 50 times sweetener than sugar and is sold under the trade names *Sucrayl* and *Sugar Twin*. Cyclamate is the sodium or calcium salt of cyclamic acid (cyclohexanesulfamic acid).

Side Effects

Cyclamate was banned in United States of America in 1970, as large doses caused bladder cancer in rats.

Chemistry of sorbitol, xylitol and erythritol

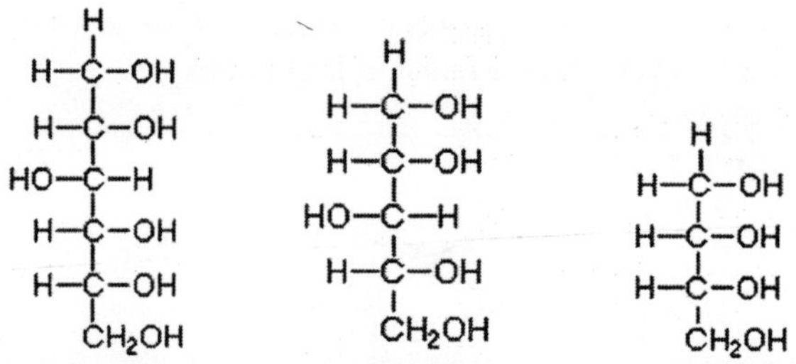

Sorbitol **xylitol** **erythritol**

Sorbitol, xylitol and erythritol are natural sugar alcohols and are found in fruits and vegetables. They can be produced commercially by catalytic hydrogenation from the sugars. Xylitol is produced from xylose. Sorbitol, also known as glucitol, is produced from glucose. Erythritol is produced commercially by fermentation of glucose with yeast *Moniliella pollnis.*

Sorbitol and xylitol are common ingredients in *sugar free* candies and chewing gums. Other sugar alcohols include mannitol, maltitol, lactitol and erythritol. Erythritol is 60 to 70% as sweet as the sugar.

Side Effects

- Sugar alcohols are not well absorbed in the intestinal tract and they are fermented by micro-flora that may produce gas and diarrhea.
- Sorbitol about 10 g may cause gastrointestinal distress.
- Xylitol can cause seizures, liver failure and death in dogs in relatively small doses.

Chemistry of Brazzein

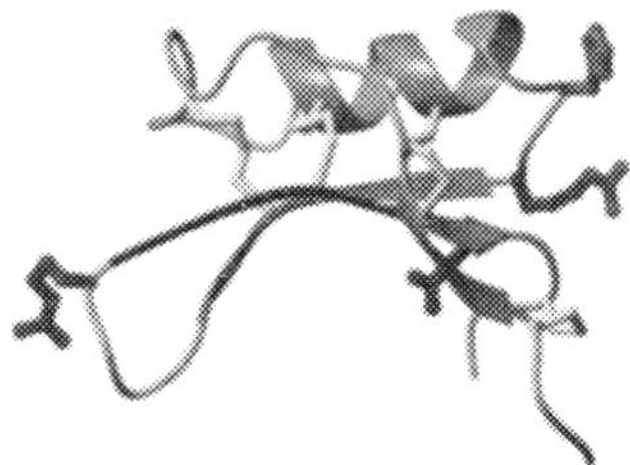

Brazzein is a sweet-tasting protein extract from the fruit of West African vine *Pentadiplandra brazzeana*. The chemical structure of brazzein consists of 54 amino acid residues arranged in one alpha-helix and three strands of anti-parallel beta sheets. Brazzein is about 1000 times sweeter than sugar. Brazzein is stable over a broad pH range 2.5 to 8 and is heat stable at 98°C for two hours.

Chemistry of Steviol

Stevia is a non-caloric green sweetener. Whatever is closer to nature is more evolutionary for us. The sweet herb *Stevia rebaudiana* is a native of Paraguay and the leaves of this plant are endowed with sweet taste and are therefore rightly called *sweet leaf plant* and *honey leaf plant*. Stevia belongs to the family compositae and is a relative of ester, chamomile and chrysanthemum. It is 300 times sweeter than sucrose with 1/300th caloric value of sucrose. Stevia no doubt commands a special place amongst the naturally occurring alternative to sucrose.

Stevia is obtained from plant, whose leaves are used to produce extracts having 70 times sweeter taste than sucrose. Robinson in 1930 showed that a glycoside *steviol* is responsible for the sweet taste of stevia. Stevioside and Rebaudioside are two of the sweet steviol glycosides in stevia leaves. Stevioside has one beta-D-glucose molecule replacing the bottom hydrogen atom of steviol. Stevia products are marketed under the brand names *Truva* and *PureVia*.

In Japan artificial sweeteners are banned, hence, stevia crops are cultivated. This priceless plant can readily be propagated from cuttings or seeds. It is valued as a perennial transplanted crop. An aqueous extract of stevia is made from whole leaves. The extract is greenish blue and is exceedingly sweet to taste. After concentration, the white powder obtained is 200 to 300 times sweeter than sugar. It is heat stable up to 200°C temperature. The stevia glycosides were found to be stable even in carbonated drinks. stevia is endowed with triple qualities, *i.e.*, as a flavouring agent, a sweetening agent and as a medicinal herb and is abundantly used in bland foods, drinks and low calorie preparations. It is also a flavour enhancer when mixed with lemon.

Side Effects

Experiments on rats have shown that high doses of Stevioside decrease the production of sperm in males, and the females had fewer and smaller offspring.

Monellin

The serendipity berries (a South African plant, *Dioscoreophyllum cumminsii*) contain a sweet sap, which comprises of a protein. There are two polypeptide chains having 42 and 50 amino acid units, which render an astounding 2000 times sweteer taste than the table sugar, *i.e.*, sucrose. The sweetness lies in the undissociated protein, while none of the two protomers has any sweet effect.

A similar South African plant Katemfe is also intensely sweet due to the two proteins present in it, which together are 1600 times sweeter than sucrose. Research studies on their side effects are nonconclusive.

Alitame

This is a peptide containing d-aspartic acid, d-alanine and a novel C-terminal amide moiety. It is devoid of any metallic taste and is supposed to be 2000 times sweeter than sucrose. The side effects caused by this substitute are still under scan.

Neohesperidine dihydrochalcone

This is a bitter component of the grape fruit skin, which gives sweet menthol like after taste. It is used in juices, beverages and alcoholic drinks in small amounts as it is 1500-1800 times sweeter than sucrose. There are little known side effects.

D-Tagalose

It is present in some mushrooms, shrimps and lobsters. It is a disaccharide having two glucose units combined, which is different from that pressent in maltose.

Mannitol

Mannitol, which is widely used to mask the bitter taste of vitamin supplements, has a pleasant taste. Due to its great mouth feel, it is used widely in chewing gums. Similarly, erythritol is a colorless, odourless, crystalline, non-hygroscopic thermo-stable solid, which is only 70% that of sucrose in sweetness.

Lactitol

Lactitol, which is derived from milk, has only 0.4 time sweeter that of sucrose, without any after taste. Because of low calorie sweetener, it is needed in some applications to provide desired sweetness.

The age old saying of wisdom that $^1/_3$rd of what you eat makes you live and $^2/_3$rd of what you eat makes the doctor live goes a long way. Let not your waistline shorten your lifeline so count your calories judiciously.

■■■

35 Food Additives and Health

Alok Kumar Srivastava[1], Manisha Guha[1] and Veenu Sangwan[2]

[1]Central Food Technological Research Institute, Mysore, Karnataka
(Council of Scientific and Industrial Research, New Delhi, India)
[2]Department of Foods and Nutrition, CCS Haryana Agricultural University,
Hisar, Haryana, India; E-mail: <aloksriva@yahoo.com>, <guha84@yahoo.com>

Food additives in food processing industry and at household level are used since time immemorial. They play different roles in food processing and preservation, give body, flavour, colour, *etc.* to food and they are added for their technological necessity. Salting, smoking and use of colors and flavours by Egyptians, saltpeter by Romans, turmeric by Indians and baking powder as raising agents, thickeners for sauces and gravies are few such examples of ancient usage of additives. Since prehistoric times, honey, wine, salt and spices have been added to foodstuffs to perform specific functions. Lemons and vinegars are classified additives as citric and acetic acids, respectively.

Food additives are intentionally added to natural foods for increasing their shelf life, preserving them against microbes, oxidation and other chemical changes and for making the food's appearance, taste, or mouth feel better and appealing. Food additives play a key role in maintaining the food quality and characteristics keeping the foods safe, wholesome and appealing from farm to fork.

Though food additives play important role in maintaining the quality and safety of food, consumers need to guard themselves against the misuse of food additives, which may cause adverse health effects. Chemicals used in food additives may cause headaches or alter the energy levels. They may also affect mental concentration and immune response. Few harmful additives with short- and long-term effects could increase risk of cancer, cardiovascular disorders and degenerative diseases. Instances of food allergies and food intolerances are very common and affect many people today, causing wide range of adverse health reactions to sensitive individuals, which could be toxic in nature. Food and Agricultural Organization and World Health Organization Food Standards Programme, Codex, which is responsible for laying down food safety standards, defines food additives as any substance, not normally consumed as a food in itself and not normally used as a characteristics or typical ingredient of food, whether or not it has nutritive value, the intentional addition of which to a food for technological purpose in the manufacture, processing, preparation, treatment, packaging, transport, or storage of such food results, or may be reasonably expected to result, in it or its by-products becoming a component of or otherwise affecting the characteristics of such foods (Codex STAN 192-1995, 89/107/EEC).

Food Additives and Food Safety for human health

Food additives used in food processing industry and at household level shall be of food grade and conform to standards of identity, safety and purity of Codex. According to Codex General Standard, the use of food additives is justified under following conditions:

- Fulfill the technical function and necessity and useful for product development.
- Safe and do not cause hazard to the human health as evidenced by scientific data.
- Shall be used within prescribed Maximum Residue Levels (MRLS) and as per Good Manufacturing Practices (GMP).
- They shall be used only in such products where they are approved for use.
- They should be used under proper labeling declaration.
- They shall not deceive the quality of product and mislead the consumers.

Thus, the food additives permitted by Food Safety and Standards Authority of India (FSSAI) shall be used under the conditions of Good Manufacturing Practices (GMP):

- The quantity of additives added to food articles shall be limited to the lowest possible level, necessary to accomplish the desired technological effect.
- The quantity of additive that becomes the component of food, as a result of its use in manufacture, processing, or packaging of a food and which is not intended to accomplish any physical or other technical effect in food itself is reduced to the extent reasonably possible.

As the food additives include preservatives, antioxidants, acidulants, neutralizers, buffers, colouring agents, flavouring agents, sweeteners, *etc.*, they serve following basic objectives and functions:

- To preserve the nutritional quality of food
- To enhance the shelf life of food
- To improve the organoleptic properties of food
- To provide necessary ingredients or constituents for food, manufactured for group of consumers, having special dietary needs
- To provide aids in the manufacturing, processing, preparation treatment, packaging, transport, or storage of food, provided that the additive is not used to disguise the effects of the use of faulty raw material or of undesired practices during the course of any of above activities.

International Numbering System (INS)/E Numbers of Food Additives

Based on their function, Codex Alimentarius Commission (CAC) classifies food additives into 23 functional classes. These functional classes are based on the Codex class number and International Numbering System (INS) for food additives (CAC/GL 36-089). The Joint FAO and WHO Expert Committee on Food Additives (JECFA) is the main International body that has addressed the issue of safety of food additives. Further, based on the safety evaluation of European Commission Scientific Committee on Food (SCF)/European Food Safety Authority (EFSA), the European Union (EU) has approved food additives, assigning E Number. In most cases, INS and E numbers are similar.

In United States, the Food and Drug Administration (FDA) is responsible for all food products including food additives. Provisions of Act are enforced through detailed regulations laid down in Title 21 of the Code of Federal regulations (21 CFR). The United States categorize the ingredients intended in the manufacturing of food items into three categories, *i.e.*, food additives, prior sanctioned food additives and Generally Recognized as Safe (GRAS). GRAS status is based on experiences derived from common use, substances used in foods prior to January 1958, or on scientific procedures. In India FSSAI lays down the standards for use of food additives and follows Codex after assessing full scientific justification and impact on human heath. It is well documented that the use of food additives within prescribed limits in given food items is safe for human health.

Different Classes of Food Additives and their safety aspects

The common classes of food additives are as follows:

- **Acids:** Acidulants include organic acids and their safety, *e.g.*, phosphoric acid, citric acids, acetic acid, malic acid, tartaric acid, ascorbic acid, lactic acid, *etc.* They not only increase the acidity, impart sour taste and flavour to foods like soft drink, jam, cheese and beer but also act as sequestrants, buffering agents, preservatives, viscosity modifiers, curing agents, *etc.* They play important role to control microbial action and thus enhance shelf life of products, making foods safe and available for longer periods.
- **Acidity regulator/Modifier:** Buffers, bases, acids, alkali and other pH adjusting agents alter or control the acidity or alkalinity of food articles. The additives like potassium, sodium, or calcium acetate, calcium or potassium citrate, calcium oxide and calcium carbonate act as acidity regulators and buffering agents. They modify the taste (sometimes as flavourings) and soften the sharpness of acids, making acidified foods more palatable.
- **Anti-caking agents:** Aluminium, magnesium, calcium silicate, carbonates or phosphates of calcium, magnesium and sodium, potassium or calcium salts of fatty acids (palmitate or stearate) are used as anti-caking agents. They improve the free flow properties of products like dry soup, fruit and vegetable powders, sugar, salt, *etc.*
- **Anti-foaming agent:** Additives like propylene glycol, triethyl citrate (in eggs) and dimethyl polysiloxane (silicon) are used as antifoaming, anti-sticking, or anti-frothing agents in foods like jam and soft drinks.
- **Antioxidants:** FSSAI defines antioxidants as substances which when added to food retard or prevent oxidative deterioration of foods and does not include sugar, cereal, oils, flours, herbs and spices. No antioxidant other than lecithin, ascorbic acid and tocopherol shall be added to any food unless other wise provided in the Food Safety and Standards (FSS) Rules, Government of India. Fats, flavouring substances, vitamins and colours all can be oxidized, when exposed to air, leading to rancidity, loss of flavour, color and nutritive value of food. Oxidative initiators like air oxygen, metals (iron, copper, *etc.*), enzymes (lipase and lipoxygenase), heat and light are the causative agents for oxidation process in foods. Free fatty acids (hydrolytic rancidity) and peroxides, aldehydes and ketones (oxidative rancidity) are the reaction products of such oxidation. Antioxidants retard the oxidation process and improve the quality and shelf life of susceptible foods. They cannot restore the oxidized foods. Ascorbic acid actually removes oxygen by self-oxidation, whereas, butylated hydroxyanisole (BHA), butylated hydroxytoluene (BHT), tertiary-butyl-hydro-quinone (TBHQ), or tocopherols and galic acid esters interfere in the oxidation mechanism. Vitamin C and tocopherols (natural anti-oxidants) are less restrictive in use but less effective too. Use of sequestering agents like citric acid along with antioxidants has shown synergistic potential by reducing the availability of catalytic metallic ions.

- **Bulking agents:** Additives like polydextrose and maltodextrin add to the bulk of foods, without significantly adding to calories. In baked foods, chocolates, candies, dairy deserts, fruit spreads, diabetic foods and beverages, polydextrose is used as replacement of sugar or fat. They are partially metabolized by gut flora and ascribed the caloric value of 1 kcal/g.
- **Colours:** Colour is important in consumer perception of food, and it enhances the visual appeal of food. Like other additives, use of colours is restrictive and permitted in selective foods to restore colour lost in processing, to ensure consistency of colours in inter-batch and raw material variations. Food colours are mostly used in confectionery, convenience foods, beverages and snacks. Care should be taken that colours are not used to deceive the nature and quality of food, which could adversely affect wholesomeness of food.
- Food colours are of two types, *i.e.*, natural and synthetic colours.

Natural colours are extracted from the seeds, flowers of plants, vegetables, *e.g.*, bright yellow from turmeric or safflower, orange from carrot or paprika, green from spinach (rich in chlorophylls), brown from caramelized sugar and red, blue or purple from beetroot, berries or fruits (rich in anthocyanins) and also some insects. Compared to synthetic colours natural colours suffer from fastness deficiency, involve high cost, restrictive shade range, presence of distinct aroma and variation from batch to batch. Synthetic colours are derived from synthetic dyes and highly purified petro-chemicals. They are superior to natural colours in terms of colour intensity, varieties of shades available and stability but need to be used with caution for protecting food safety and human health. Common synthetic food colours belong to five chemical classes namely azo (ponceau 4R, carmoisine, sunset yellow FCF), xanthene (erythrosine), pyrazolone (tartrazine), indigo (indigo carmine) and tri-aryl methane (brilliant blue FCF and fast green, FCF).

Misuse of non-permitted colours and usage of excess quantity (more than the prescribed limits) may lead to several adverse effects on human health. Studies have suggested that artificial colours have detrimental effect on children behaviour including hyperactivity, hence, use of synthetic colours in children foods are discouraged. Artificial food dyes have shown the potential to cause behavioural disorders in susceptible individuals, *e.g.*, tartrazine leads to food intolerance cases. It is well documented that use of unpermitted colour dyes or excess use of permitted ones leads to allergies, palpitation, blurred vision, heat sensation, pruritus, suffocation feel, weakness, nausea, exfoliative dermatitis, bronchospasm, migraine, irritable bowel syndrome, asthma, rhinitis and urticaria. Caramel, commonly used as colouring agent in host of formulations including alcoholic and non-alcoholic beverages has led to convulsions in animal models, mainly due to the presence of an impurity, *i.e.*, 4-methylimidazole. Hence, they should be used judiciously and as per law of the land.

Emulsifiers and Stabilizers: Emulsifiers and stabilizers are substances, which when added to food are capable of facilitating a uniform dispersal of oils and fats in aqueous media or *vice versa* and/or stabilizing such emulsions. They are also referred to as surface-active agents or surfactants. Based on the Hydrophillic-Lipophillic Balance (HLB) value, specific emulsifiers aid in formation and stability of water in oil or oil in water emulsions. These emulsified mediums of water and fat are then maintained by stabilizers. These additives are virtually essential for the production of margarine, mayonnaise, fat spreads, ice creams, cakes and chocolates. Emulsifiers have contributed in the development of reduced- and low-fat food products.

Agar, alginates, pectin, modified starches, lecithin, monoglycerides, or diglycerides of fatty acids, polyoxyethylene sorbitan or polyglycerol esters of fatty acids, acetic acid, lactic acid, citric acid or tartaric acid esters of mono- and di-glycerides of fatty acids, sodium or calcium stearoyl lactylate and glyceroyl mono stearate are the examples of emulsifiers.

Generally, emulsifiers with their permitted dosage are safe to use. However, Brominated Vegetable Oil (BVO), with bromate as its main constituent used as an emulsifier in soft drinks was found to have several adverse health effects including loss of muscle coordination, fatigue, memory loss and drooping eyelids.

Sodium, potassium, ammonium, calcium alginates, or common stabilizers are polysaccharides extracted from seaweeds. They provide high water binding and viscosity to liquid media and are useful in solid foods in preventing water loss, syneresis and phase separation. Calcium sulphate, phosphate, or citrate may function as sequestrants in setting process.

Guar gum, acasia gum, xanthan gum, karaya gum and gellan gum are thickeners and stabilizers used in food industry. Pectins are commonly used as gelling and thickening agents in range of acidic foods like jam, jellies, fillings or toppings and yoghurts. High methoxyl pectins stabilize proteins in yoghurts and soya products. Celluloses (microcrystalline and powdered) bind water to reduce stickiness, freeze-thaw damage (frozen foods) and improve extrusion properties (pasta and puffed snacks). Its ability to bind water and oil allows their use as flavour carrier and free flow aids in instant foods. Carboxymethyl cellulose (CMC) with its solubility in hot as well as cold water has great stabilizing property in frozen products (ice creams) to inhibit growth of ice crystals and improve smooth texture. Thickening function of CMC is used in drinks, toppings and sauces to stabilize fruit pulp.

Flavours: Flavouring agents include flavour substances, flavour extracts, or flavour preparations of defined chemical substances, capable of imparting flavouring properties, namely taste, odour, or both to food.

There are three distinct classes of flavour given as under:

- Natural (obtained exclusively by physical, enzymatic, microbiological process from plants, or animal sources), *e.g.*, citra and peppermint
- Nature identical (chemically isolated from aromatic plants or obtained and are chemically identical to natural products), *e.g.*, vanillin
- Artificial flavouring substances, which have not been identified in natural products, *e.g.*, ethyl vanillin

Lactic acid is widely used flavourings in dairy and meat flavours. Flavourings may contain permitted antioxidant, emulsifiers and preservatives. In recent years, much flavourings are sold as complex mixture of individual flavouring preparations and substances. Product labels need to indicate *added flavours*. However, *Natural* label must meet the legal definition.

Flavour enhancers: Class of additives that may not have their own pronounced flavour or taste but bring out and improve the flavour of foods, to which they are added. Flavour enhancers have attracted quite adverse attention, particularly Mono Sodium Glutamate (MSG), or salt of amino acid, *i.e.*, glutamic acid. Their use in specific food like infant food is not allowed. MSG has been linked to damaged nerve cells, headaches, burning sensation, nausea, central nervous and vascular system problems and brain edema.

Humectants: Additives like glycerol, polydextrose and propylene glycol keep the foods moist without increasing their water activity. Humectants reduce the chances of mould or bacterial growth, retard staling or drying-out, inhibit sugar crystallization and maintain viscosity and texture of products.

Sweeteners: Sweeteners include the followins:

- Nutritive sweeteners, which are obtained from natural sources, are sorbitol, mannitol, isomalt, xylitol, *etc.* They are comparatively less sweet and provide body and mouth feel to finished product. They have about half sweetness of sucrose and low glycemic index, and these sweeteners can be used for the manufacturing of food for people having diabetes, sugar free candies, chewing gums, soft drink, desserts, *etc.*

- Non-nutritive sweeteners, which are mostly synthetic compounds and have very high sweet intensity compared to sucrose, are saccharine, acesulfame K, aspartame, sucralose, cyclamate and rebiana. Aspartame is not stable at high temperature and not suitable for the manufacturing of bakery products, while sucralose is extremely water soluble and has ability to withstand high temperature processes. Intense sweeteners are used in beverages, ice cream, confectionery, chewing gum, jam, *etc.* The artificial sweeteners can also be used in combination where permitted.

The artificial sweeteners when used in excess can cause behavioural problems, hyperactivity in children, allergies and possible carcinogenesis. Children and pregnant and lactating women are cautioned against their use. No artificial sweeteners shall be added to any article of food unless otherwise permitted in FSS Rules and that too under label declaration. Artificial sweeteners like aspartame and those containing aspartame are not suitable for people suffering from phenylketonuria as it could lead to hyperactive and aggressive behaviour.

Preservatives: They prolong the shelf life of food items by protecting against deterioration caused by microorganisms, arresting the processes of their fermentation, acidification, or other decomposition. Preservatives may include anti-microbial, anti-mycotic and chemosterilant, and thus, protect against toxins, which may have food poisoning potential. FSS Rules of India classify the preservatives into following categories:

- Class I preservatives, which include salt, sugar, dextrose, spices, acetic acid (vinegar), honey, *etc.*, are generally not restricted in food applications.
- Class II preservatives include benzoic acid and their salts, sulphurous acid and their salts (sulphites or bisulphites and SO_2), nitrates or nitrites of sodium or potassium (meat products), sorbic acids and their salts, propionic acid and their salts, lactic acid and their salts, nicin, sodium diacetate, *etc.* They have restrictive use in food industry, *i.e.*, they can be used only in foods in which they are allowed to be used and that too within prescribed limits or as per GMP under label declarations.

Sulphur dioxide and its salts find their major use as anti-microbial agent in fruit products and beverages. They are effective against acid producing bacteria and spoilage yeast and also control non-enzymatic browning. Benzoic acid is mainly directed against yeast and molds. Benzoic acid acts synergistically with sorbic acid and effective at lower pH. Sorbic acid is effective even at high pH, and hence, preferred in baked and confectionery products.

Curing agents: Sodium nitrite is used to preserve/cure meat giving them organoleptically acceptable colour and flavour. Nitrate is non-toxic but its metabolites may produce adverse effect on human health. It is reported that 5-10% of nitrate is converted into more toxic nitrite in human by gastrointestinal reduction. Nitrite is reported to reduce blood pressure and can be carcinogenic. Hence, high nitrate intake for prolonged period should be avoided.

Firming agents: Calcium chloride and aluminium sulphate are employed in manufacture of canned fruits and pickles. They give firm texture to finished products, and thus, improve aesthetic value of food products.

Clarifying agents: Like gelatin, the clarifying agents are used to clarify beverages like juices, alcohol, wine, beer, *etc.* and shall be used judiciously.

Specific additives like azo colors, antioxidants, intense sweeteners and glutamate have been reported to cause allergic intolerant reactions in sensitive individuals, especially hyperactivity in children. Sulphur dioxide and sulphites are reported to cause allergies in asthma patients. Some scientific studies have linked nitrates and nitrites to carcinogenesis. Appropriate labeling on foods to declare potential allergens is vital to avoid unnecessary sufferings. Recent scientific studies have reported varying levels of adverse effects of using excessive or non-permitted food additives, which may include histopathological lesions, carcinogenicity, impaired fertility, behavioral changes and lowered immunity. There are additives, generally of natural origin, like carotene (color), tocopherols and vitamin C (antioxidants), lecithin (emulsifiers), pectin (stabilizers), *etc.*, which are safe and one needs not have any apprehensions in their usage. In recent years, health conscious consumers have craved for replacement of artificial food additives with natural ingredients, which perform similar functions. Lecithin and soya proteins (emulsifiers), grape seeds and olive leaf extract (antioxidant), lycopene, anthocyanins and chlorophyll (colors), lysozyme, cinnamic acid and chitosans (preservatives) fall into natural and clean label ingredients category. Specific protective bacterial cultures produce anti-microbial metabolites like organic acids (lactic or acetic acid), which have the ability to inhibit growth of pathogenic bacteria and micotoxin producing moulds. Food waxes, protective and preservative coatings of fruits and vegetables may trigger allergies and other health complications if used indiscriminately.

Use of Food Additives in Indian and Global Context

India has adopted CODEX INS system of food additives, which is available on the website of Ministry of Health and Family Welfare, Government of India. Recently, the Prevention of Food Adulteration (PFA) Act is integrated into Food Safety and Standards (FSS) Act with nodal agency as FSSAI, an autonomous statutory authority under Ministry of Health and Family Welfare, Government of India.

The use of new food additives in India is governed by FSSAI, Ministry of Health and Family Welfare. Before allowing the use of any new food additive or new ingredient FSSAI lays emphasis on risk assessment, risk management and risk communication aspects to achieve Appropriate Level of Protection (ALOP) to human health. Use of food additives and their permissible doses are categorized in following mechanism, *i.e.*, list of food to which they are permitted, list of foods to which the use of additives is not permitted or restricted and maximum usage level of additives in various food groups to ensure that its cumulative uses level do not exceed its Acceptable Daily Intake (ADI).

Globally, Joint Expert Committee of FAO/WHO Food Standards Programme on Food Additives and Contaminants (JECFA) of Codex Alimentarius Commission (CAC) evaluates the safety of food additives based on available scientific data on chemical, biochemical and toxicological studies and sets International Standards. World Trade Organization (WTO) specifies that Codex standards would apply in case of any international trade dispute.

At European level, the additives approved for their use are evaluated by the scientific committee of European Food Safety Authority (EFSA). There may be examples, wherein an additive is permitted in a country, whereas, it may be prohibited in other countries but this has to be supported by scientific justification. Stevioside, extracted and refined from *Stevia rebaudiana Bertoni* leaves, is permitted as sweetener in countries like Japan, however, the same is not permitted in EC, United States and India due to insufficient scientific data on safety.

Regulatory authorities across the globe are increasingly relying to use quantitative risk assessment to define food safety related to usage of additives and contaminants in order to devise intervention strategies. Safety evaluations are based on extensive toxicological tests, identification of any critical toxic effects, their dose response and setting of Acceptable Daily Intake (ADI). ADI (mg/kg body weight) specifies the amount of food additives on body weight basis, which can be ingested daily over lifetime without appreciable health risk. Safety of food additives is determined by Risk Analysis. Scientific Risk Assessments have become foundation for food safety, worldwide. General acute toxicity studies are mandatory for safety assessment of food additives. In setting, ADI of food additives often long-term, chronic toxicity, or carcinogenicity studies in rat and mouse species or multi-generation reproductive and developmental toxicity studies are utilized. The ADI is derived by applying a safety factor of 100 to the lowest No Observed Adverse Effect Level (NOAEL) in the toxicity study. Safety factor of 100 comprises a factor of 10 to take account of inter-species differences, extrapolating animal studies to human consumption and further factor of 10 to account for inter-individual differences among human, hence, the Acceptable Daily Intake is derived by dividing NOAEL by a safety factor of 100.

After accounting for our dietary pattern, India generally follows Codex guidelines and JECFA monographs for toxicological evaluation. Food additives are very carefully regulated for their functionality and safety. Usage of food additives shall follow Good Manufacturing Practices, wherein quantity of additives shall be limited to lowest possible level to accomplish desired effect and additive shall be of food grade quality prepared and handled in the way as food ingredient (Codex STAN 192-1995).

It can be concluded that although food additives are an important invention for sustainability of food industry, yet even minute dosage of unpermitted harmful additives, when taken continuously may have irreversible toxic effects but justified usage of permitted additives are not harmful for human health. Thus, the consumers and Food Safety regulators need to guard against the unscrupulous usage and ever-increasing misuse of additives by unscrupulous vendors for economic gains. With increasing awareness, the market and consumers have become more sophisticated and knowledgeable about quality and safety of foods. Scientific and technological advancements have helped in development of novel additives and usefulness of such additives cannot be denied. Scientific safety evaluation and strict regulation or enforcement pertaining to use of food additives would be necessary in ensuring the variety and convenience of safe food supply.

■■■

36 Food Adulteration: The Demonic Onslaught on Health and Wellness

Vinita Katiyar[1], S. Bhanumati[2] and M.K. Rana[3]

[1]Guest Faculty (Biosciences) Delhi University, Counselor (Food Microbiology), IGNOU, New Delhi, India; E-mail: <vinita.katiyar@gmail.com>

[2]Chemistry Department, Gargi College (Delhi University), Siri Fort Road, New Delhi, India; E-mail: <sbhanumati@gmail.com>

[2]Department of Vegetable Science, CCS Haryana Agricultural University, Hisar, Haryana, India; E-mail: <mkrlotus@gmail.com>

Food provides various nutrients to regulate body processes, protects from diseases and satisfies hunger, which give us a sense of psychological satisfaction. According to great *Shreemad Bhagwat Geeta*, the "*Anant Bhawati Bhootanee*" means that the human being is made from food. This statement is true because everything in our body was once the food we ate. Pure food gives us health, happiness, efficiency and longevity. Pure food is what it is represented to be, it should be wholesome and do all the good it should do and be free from anything that does any harm or might impair its goodness. The nature, quantity and quality of food play a vital role in the maintenance of health of individuals. Preparation and processing of food products may modify their nutritional value, and the use of food additives may introduce direct or indirect toxicity but the real hazards to health are caused by food adulteration.

Food, which is essential for life, should be pure, nutritious and free from any type of adulteration for proper maintenance of human health. Scientific revelations have made quick impact the world over on the concept of a healthy diet. There is also an emphasis on *optimal nutrition* rather than *adequate nutrition*. The very food, which once was being developed for taste, hunger and commerce, has now made a transition to health, nutrition and specific performance with an added value of being *Functional Foods*. In short, the nutritionists are serious about inventing *foodstuffs that counter diseases*. The genuine efforts made by the health constructors are being counteracted by the health destructors in the form of food adulteration. Despite the improvements in production, processing and packaging, more poisons seem to be entering our food chain every moment. Every consumer wants to get maximum quantity of a commodity in as low a price as possible. This attitude of the consumer is being coupled with the intention of the traders to increase the margin of profit, where quality of the commodity is compromised with the addition of a baser substance and/or even removal of vital elements. This practice is commonly known as food adulteration, *e.g.*, Indian spices or condiments add taste and flavor to the food and help in digestion too. Some spices like turmeric have an antiseptic effect on the body but what is most important is the quality of these ingredients. When a chemical color is added to make the turmeric look more attractive, this deed is adulteration and the chemical added is an adulterant.

Concern of Food Quality

The handling, preparation and storage of food in ways that prevent food borne illness lead to food safety issues. Awareness of food safety issues is necessary among producers, retailers and consumers. Food safety issues should be considered to avoid potentially severe health hazards. The ongoing debates on genetic food safety are significant issues, which need to be addressed seriously from the food safety angle. Food safety can be defined in both absolute and relative terms. Absolute food safety is the assurance that damage or injury from the use of a substance is impossible. Relative food safety is the assurance that damage or injury will not result from a food or from the ingredients, which are used in a reasonable quantity and also in a customary manner.

An understanding of food safety is further improved by defining two other basic concepts, *i.e.*, *hazard* and *toxicity*. *Hazard* is the relative probability that harm or injury will result when the substance is used in a proposed manner and quantity, while *toxicity* is the capacity of a substance to produce harm or injury of any kind under any condition. Food *quality* refers to the attributes that influence a products value to a consumer. This includes both negative attributes such as spoilage, contamination, adulteration and food safety hazards, while the positive attributes include qualities, such as the origin, color, flavor, texture, *etc.* Food *hygiene* is also a term, which is used while talking about food safety. Food *hygiene* refers to all conditions and measures necessary to ensure the safety and suitability of a food at all stages of the food chain.

Food *contamination* is caused by any substance not intentionally added to food, which is present in food as a result of production (including operations carried out in crop husbandry, animal husbandry and veterinary medicine), manufacture, processing, preparation, treatment, packing, packaging, transport, or holding of such food as a result of environmental factors of contamination. Contamination in foodstuffs can be physical, chemical, or biological.

The fact that a food contaminated with pathogens (harmful microorganisms such as bacteria, viruses, or protozoa) may or may not render it adulterated. Generally, for ready-to-eat food items, the presence of pathogens will render the food adulterated, *e.g.*, the presence of *Salmonella* on fresh fruits or vegetables or in ready-to-eat meat or poultry products (such as luncheon meats) will render those products adulterated.

Food Adulterants

Adulteration is defined as a process by which the quality or the nature of a given substance is reduced through (i) the addition of a foreign or an inferior substance (addition of water to the milk) and (ii) the removal of vital vitamins from a food articles (removal of fat from the milk). Figure 1 illustrates the criteria of any foreign to be considered as adulterants.

Under the Prevention of Food Adulterant Act (1954), an adulterant is any material, which is employed for the purposes of adulteration. Any article of food is adulterated if:

- Any inferior or cheaper substance has been substituted wholly or in part.
- Any constituent of the article has been abstracted wholly or partially.
- The article has been prepared, packed, or kept under unsanitary conditions.
- The article consists of part filthy, rotten, decomposed, diseased animal or vegetable and it is infested with insects.

- The article is obtained from diseased animal.
- The article contains any poisonous ingredient.
- The article has un prescribed coloring substance or the coloring substance is in excess of the prescribed limits.
- The article contains any prohibited or excessive preservative.
- The quality or purity of the article falls below prescribed standard limits.

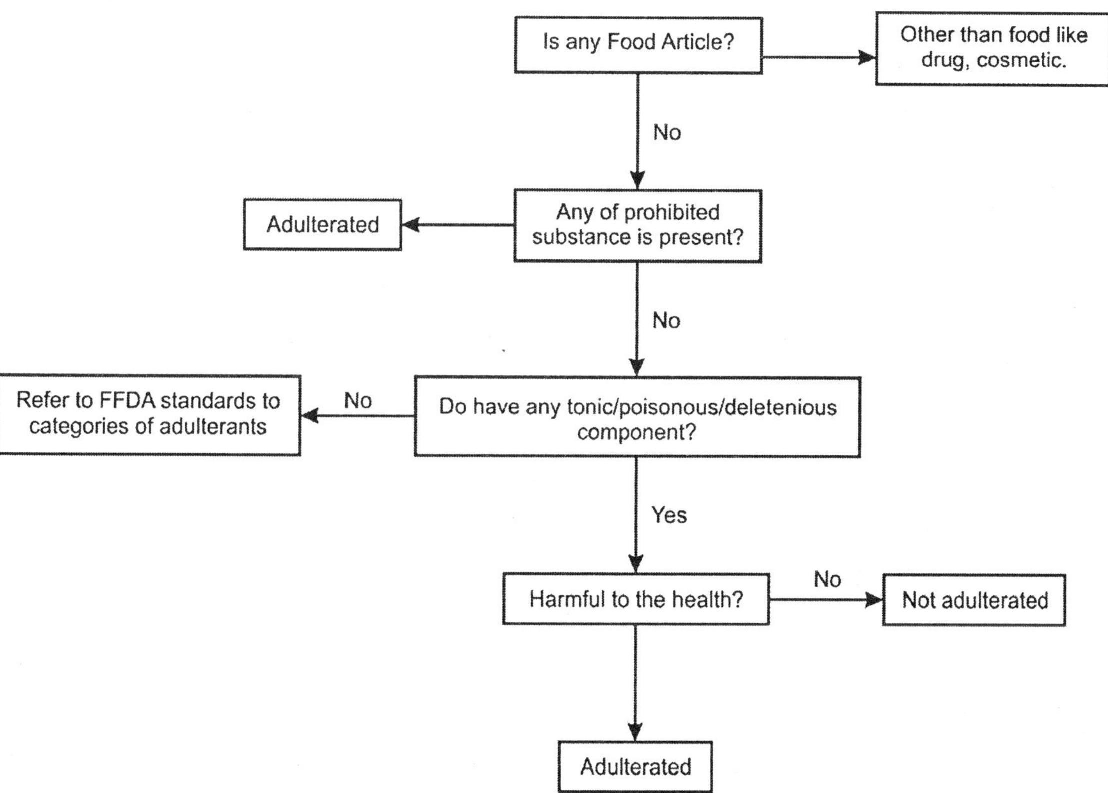

Figure 1: A flow chart of section 7 of the Pure Food and Drug Act of 1906

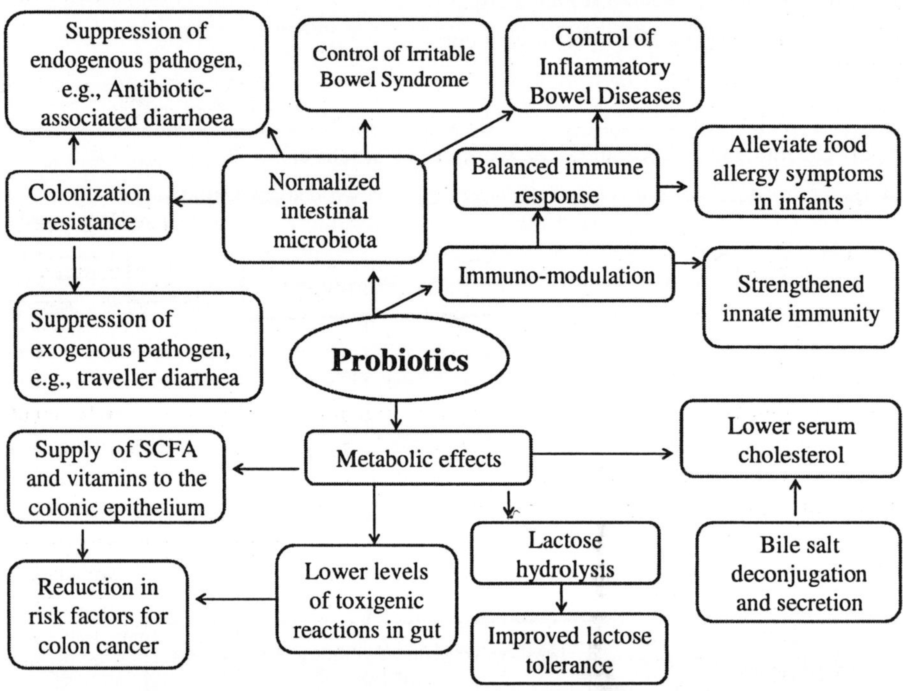

Adulterated Food

According to the American Federal Food, Drug and Cosmetic Act (1938), the food is *adulterated* if it meets any one of the following criteria:

- It bears or contains any *poisonous* or *deleterious substance*, which may render it injurious to health.
- It bears or contains any *added* poisonous or deleterious substance (other than a pesticide residue, food additive, color additive, or new animal drug, which are covered by separate provisions), which is unsafe.
- Its container is composed, in whole or in part, of any poisonous or deleterious substance, which may render the contents injurious to health. It bears or contains a pesticide chemical residue, which is unsafe. *Note*: The Environmental Protection Agency [EPA] establishes tolerances for pesticide residues in foods, which are enforced by the FDA.

Food also meets the definition of adulteration if:

- It is, or it bears or contains, an unsafe food additive.
- It is, or it bears or contains, an unsafe new animal drug and it is, or it bears or contains, an unsafe color additive.
- It consists, in whole or in part of *any filthy*, *putrid*, or *decomposed substance* or is otherwise unfit for food.

- It has been prepared, packed, or held under unsanitary conditions (insect, rodent, or bird infestation) whereby it may have become contaminated with filth or rendered injurious to health.

Further, food is considered adulterated if:

- It has been irradiated and the irradiation processing was not done in conformity with a regulation permitting irradiation of the food in question. *Note*: FDA has approved irradiation of a number of foodstuffs, including refrigerated or frozen uncooked meat, fresh or frozen uncooked poultry and seeds for sprouting [21 C.F.R. Part 179].
- It contains a dietary ingredient that presents a significant or unreasonable risk of illness or injury under the conditions of use recommended in labeling, *e.g.*, foods or dietary supplements containing aristolochic acids, which have been linked to kidney failure, have been banned.)
- A valuable constituent has been omitted in whole, or in part or replaced with another substance. Damage or inferiority has been concealed in any manner or a substance has been added to increase the product's bulk or weight, reduce its quality or strength, or make it appear of greater value than it is (this is *economic adulteration*).
- It is offered for import into the United States and is a food that has previously been refused admission unless the person reoffering the food establishes that it is in compliance with U.S. law [21 U.S.C. § 342 (Figure 2)].

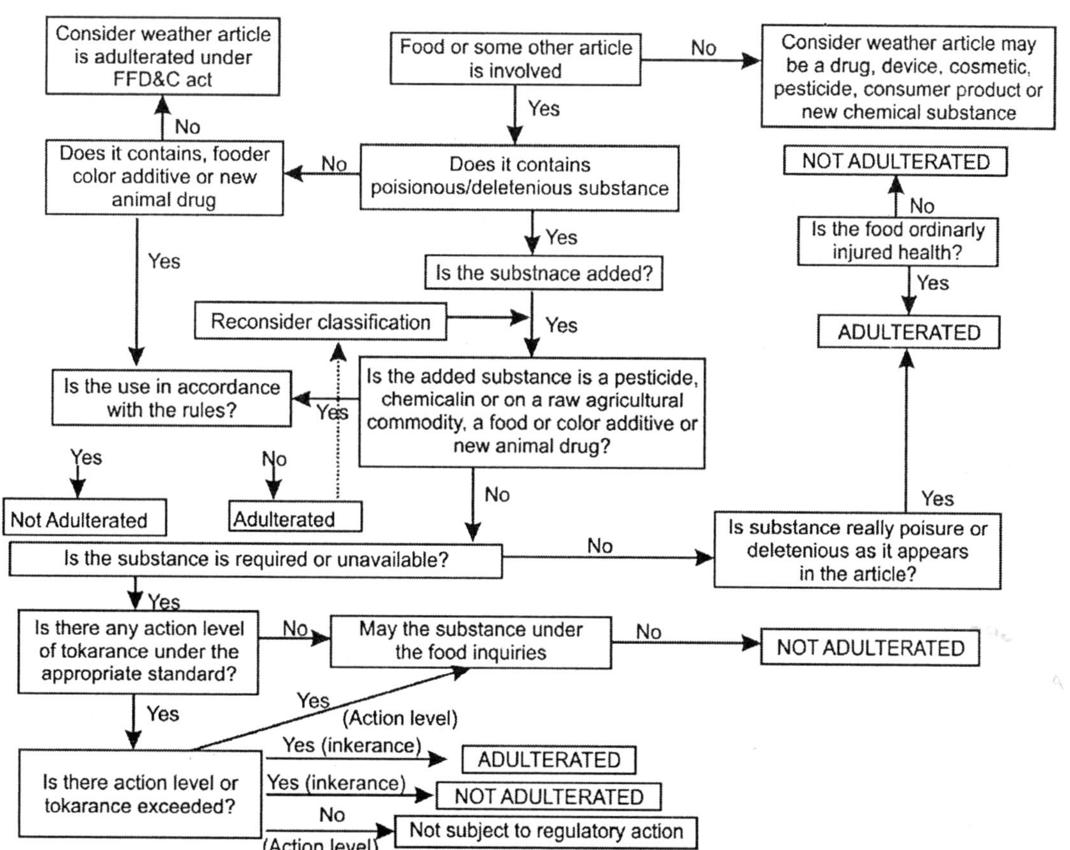

Figure 2: A flow chart of section 402(a)(1). (2) and 402 of FDCA

Age of Adulteration

The adulteration of a food can take place at any of the stages in supply chain of the foodstuff. The food chain could be as follows:

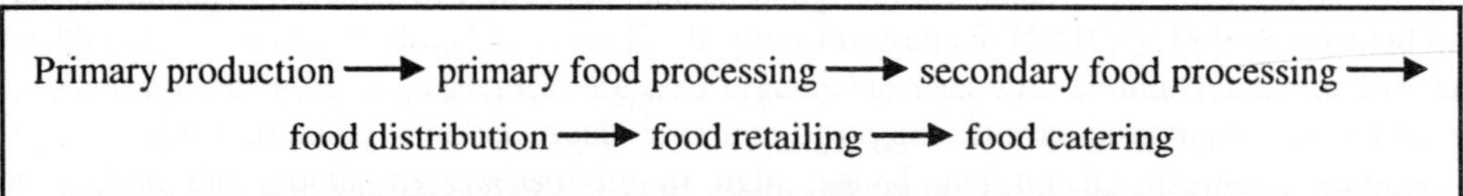

Agricultural inputs during primary food production such as pesticides and veterinary drug residues, whose residue could be found in foods, are also considered as adulterants. During both primary and secondary food processing, adulterants may enter the food chain. Besides, adulterants may also find their way during food distribution, food retailing and catering.

Typical Supply Chain for a Food Community

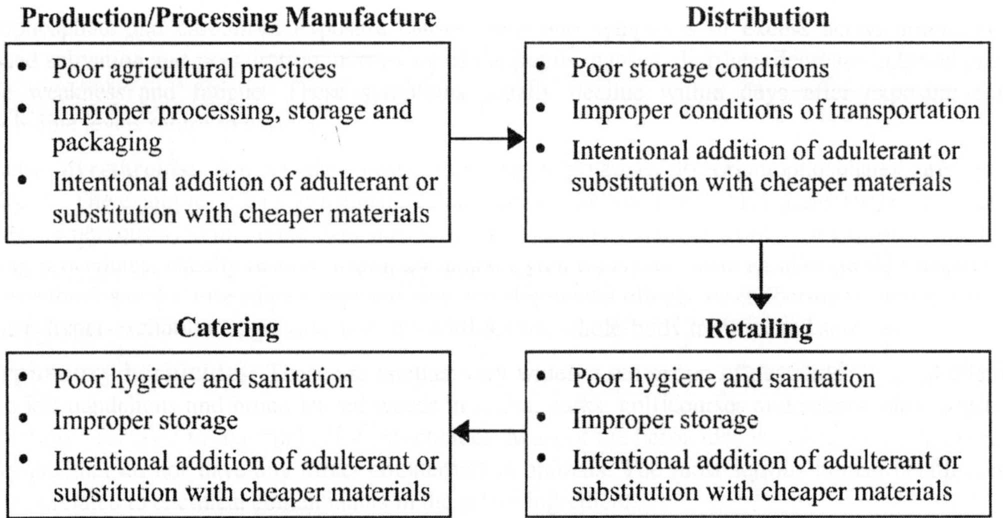

Source: Food Microbiology and Safety (MFN 003) IGNOU

Causes of Adulterations

Adulterants may enter in the food supply chain from any source and from direction. In most cases, however, the distributor or the retailer introduces the adulterant into the food. Inappropriate storage and transportation conditions may also contaminate the food at this stage. In fact, any commodity that commands a premium in the market and is either expensive or has a high volume sales, is a target for adulteration. Adulteration of milk and milk products, edible oils and fats and spices is most common due to the total sales value of these food items is high. Food articles, which are in a powder, minced, or paste form, are more susceptible to adulteration. This is because it is more difficult for the naked eye to detect adulteration in these foodstuffs. Adulteration of foods sold loose by the retailer is also more common as compared to the packaged food items. Sometimes, packaged food may also be adulterated if package is easy to tamper with. The causes of adulteration may be:

- Availability of too many products in the market
- Poor buying practices of consumers
- Consumer mentality of bargaining
- Consumer psyche
- Availability of adulterants

Food adulteration is the act of purposely debasing the quality of food offered for sale by either the admixture or substitution of inferior substances or by the removal of some valuable ingredient. Food is declared adulterated if:

- A substance is added which depreciates or injuriously affects it.
- Cheaper or inferior substances are substituted wholly or partially.
- Any valuable or necessary constituent has been abstracted wholly or partially.
- It is an imitation.
- It is colored or otherwise treated to improve its appearance or if it contains any added substance injurious to health.

Category of Adulteration

Prevention of Food Adulteration act clearly distinguishes the types of adulterants- *intentional* or *unintentional*. The former is a willful act on the part of adulterator who intended to increase the margin of profit. On the other hand, adulteration may be incidental contamination, which is usually due to ignorance, negligence, or lack of proper facilities.

Intentional added adulterants (IAAs): Intentional adulterants are sand, marble chips, stones, mud, chalk powder, water, mineral oil and coal tar dyes. This type of adulteration causes harmful effects on the body.

Accidental adulterants (AAs): Incidental adulterants are present as contaminants due to improper storage, handling, transportation, or processing of foodstuff, *e.g.*, pesticide residues.

Category of adulterants can also be defined as based on the source of adulterants it includes both *intentional* and *accidental*.

- Antibiotic residues, *e.g.*, milk and milk products, meat and meat products
- Pesticide residues, *e.g.*, milk, grains, oil, bottled water, vegetables and fruits
- Metallic contaminants, *e.g.*, nickel due to the improper processing of vanaspati, iron fillings in tea and semolina (suji) from rollers used in processing
- Microbial contaminants, *e.g.*, fungi and fungal toxins, bacteria and their toxins due to improper storage or processing of the food products

Adulterants in Different Food Articles

Vegetables and Spices

Many of the spices, ready to eat condiments and commonly used products are found contaminated or adulterated. The adulteration problem in India has attained massive dimensions.

Parsnip or turnip adulteration has been detected in prepared horseradish. Red chili powder is found adulterated with cheap dye *Rod amine B* culture and bricks powder. Likewise, turmeric powder is adulterated with a yellow color dye *metanil* yellow to provide the brightness in appearance. Same dye is also used to polish the arhar and moong dal and washed channa. On the other hand, green leafy food articles are found adulterated with basic dye *malachite green.*

Cheaper agricultural products like wheat starch, sorghum, rice, corn and arrowroot starch are used in a number of expensive food articles like ground spices (red chilly powder, turmeric powder, coriander powder, condiments, *etc.*). The starch, which is white in color, is usually dyed to the color of the spice to which it is being added. Sometimes, essential oil, which is derived from expensive spices like cloves, is extracted, and the exhausted spice is sold as such. *Cinnamon bark* may be mixed with the bark of another similar looking tree and asafetida (*Heeng*) may be mixed with a foreign resin, soapstone, or earthy matter. Seeds of black pepper may be mixed with *papaya seeds*, and *mustered seeds* may be mixed with *Argemone* seeds, which look similar. Saffron is often mixed with colored dried tendrils of maize cob, *eucalyptus*, or with dried flowers of harsinghar, which are also orange in color like saffron.

Fruits and Fruit Juices

Often, *metal salts* are used as sprays for better appearance or for increasing the shelf life of fruits and vegetables. The invisible contaminants are slow poisons and may cost substantially as health erasers on long time consumption. Fruits such as apples sprayed over with *lead arsenate*, rat poisons (*Barium carbonate*). Fruit juices, soft drinks, *etc.* when come in contact with cadmium plated vessels or equipment may be adulterated with *cadmium (Cd)*. Cadmium also gets entry as adulterants in fruit juice components with contaminated water, *e.g.*, shellfish.

Sometimes, chemicals used for the ripening of fruits deteriorate the fruit quality. Increase in the concentration of calcium salts (calcium chloride and calcium sulphate) resulted in delayed ripening though improves aroma and color of fruits but has negative effect on fruit quality by increasing skin shriveling and lowering flavor and taste of the fruits indicating poor eating quality. These chemical treatments might be left over as residue on the surface of the foodstuffs and may be consumed along with fruits.

Although calcium carbide (CaC_2) has been frequently used since long times to enhance ripening process of mango fruits, however, some other calcium salts especially calcium chloride and calcium nitrate have been reported in literature to delay the ripening and senescence in fruits by lowering the respiration rate. Considering its hazardous aspects, the use of calcium carbide must be strictly monitored and controlled. Food adulterations have been isolated from fruits juice by Dionex Corporation. The analysis of fruit juice shows that they are adulterated with medium invert sugar extracted from beets.

Oil and ghee

The most common adulterant in edible oils is cheaper oil, which may or may not be edible. The cheaper oils generally used to adulterate expensive cooking oils are *caster oil, mineral oil, argemone oil, palmolein, cottonseed oil* and *rapeseed oil.*

Ghee is usually found adulterated with *vanaspati* or *hydrogenated oils*. *Lard*, a cheaply available animal body fat, may also be added. Contaminants, which have been detected in oils, are the pesticides residues and *aflatoxins* especially in unrefined groundnut oil. The process of refining the oil destroys the aflatoxins and makes it safer to consume.

Adulteration has been a perennial problem in the oil and fat trade for a long time. It is sometimes deliberate as well as occasionally accidental. Indeed accidental contamination is hard to avoid in bulk handling installations, where oils of different qualities must be pumped through common valves and pipelines. However, generally expensive oil is intensively adulterated with the cheaper one. A number of adulterants in edible oils have been detected in the range of 0.1 to 2%.

Milk

In milk, the most widely used adulterant is water. Addition of preservatives like formalin, boric acid, hydrogen peroxide and neutralizers like sodium bicarbonate and caustic soda is also prevalent to increase the shelf life of milk. Contaminants usually found in milk are pesticide residues (from pesticides sprayed in cattle sheds), antibiotic residues (from medications given to the cattle) and aflatoxin (from contaminated feed given to the cattle).

Miscellaneous: Various other types of food menace are given below in Table 1:

Table 1: Miscellaneous adulterants

Adulterants	Food articles
Colors	Non-alcoholic beverages, confectionary and sweets
Non-permitted synthetic colors: orange II, Sudan dyes and auramine	Spices, fruits and vegetables
Sand, husk, saw dust, wood pieces, stones, straw, *etc.*	Cereals and pulses
Metal adulterants: iron fillings and nickel	Suji and vanaspati ghee
Artificial sweeteners: saccharine	Non-alcoholic beverages and ice creams

Adulteration and Health

Food adulterants are intentional addition, substitution, or abstraction of substances, which adversely affect the nature of substance and quality of food. Food adulteration is global concern about food safety and unfair food trade practices in quality and quantity. The impact of chemical contaminants on consumer health and well-being is often apparent only after prolonged exposure at low levels, *e.g.*, the problem of adulteration of mustard oil, which has been adulterated by *argemone* seeds, exists as evidenced by dropsy at Delhi in 1998. The toxicity of *argemone* oil has been attributed to its physiologically active alkaloids, *i.e.*, sanguinarine and dihydrosanguinarine. There are wide ranges of potentially toxic food adulterants, which might be present in food products. Health hazards due to the consumption of adulterated food are given below in Table 2.

Table 2: Health hazards of adulterants

Sr.No.	Adulterant	Food articles	Effects on health
1.	*Argemone* oil	Oils and fats	Epidemic dropsy, glaucoma, blindness and cardiac arrest
2.	Pesticide residue	All types of food	Acute or chronic poisoning with damage to nerves and vital organs
3.	Mineral oil (used motor oil)	Oils and black pepper	Diarrhea, vomiting and cancer
4.	Methyl alcohol	Alcohol liquors	Blurred vision, blindness and death
5.	Lead chromate	Turmeric powder and mixed spices	Anemia and brain damage
6.	Metanil yellow	Turmeric, mixed spices, saffron, dehusked pulses, rice and golden beverages	Tumours, cancer and testicular degeneration in males
7.	Lead	Tap water and some processing foodstuffs	Lead poisoning, causing foot drop, anemia and brain damages
8.	*Kesari dal*	Pulses and *besan*	Paralysis of legs
9.	Dung	Coriander powder	Tetanus
10.	Iron filling	*Suji* and tea leaves	Possibility of tetanus

Source: YOJANA, April 16, 1980, page 16, Government of India

Adulteration of food causes several health problems in humans. Some of the health hazards include stomachache, body ache, anemia, abortion, paralysis, increased incidence of tumors, pathological lesions in vital organs and abnormalities of skin and eyes. Hence, the food adulteration should be given great importance due to its effect in the health significance of the public.

Table 3: Adulterants or contaminants in foods and their health effects

S.No.	Adulterant	Foods commonly involved	Diseases or health effects
Adulterants in food			
1.	*Argemone* seeds and oil	Mustard seeds and edible oils and fats	Epidemic dropsy, glaucoma and cardiac arrest
2.	Artificially colored foreign seeds	As a substitute for cumin seed, poppy seed and black pepper	Injurious to health
3.	Foreign leaves or exhausted tea leaves and saw dust artificially colored	Tea	Injurious to health and cancer

S.No.	Adulterant	Foods commonly involved	Diseases or health effects
4.	Rancid oil	Oils	Destroys vitamin A and E
5.	Sand, marble chips, stones and filth	Food grains, pulses, *etc.*	Damage digestive tract
6.	*Lathyrus sativus*	*Khesari dal* alone or mixed in other pulses	Lathyrism (crippling spastic paraplegia)
Chemical contamination			
7.	Mineral oil (white oil and petroleum fractions)	Edible oils and fats and black pepper	Cancer
8.	Lead chromate	Turmeric whole and powdered and mixed spices	Anemia, abortion, paralysis and brain damage
9.	Methanol	Alcoholic liquors	Blurred vision, blindness and death
10.	Arsenic	Fruits such as apples sprayed over with lead arsenate	Dizziness, chills, cramps, paralysis and death
11.	Barium	Foods contaminated by rat poisons (barium carbonate)	Violent peristalsis, arterial hypertension, muscular twitching, convulsions and cardiac disturbances
12.	Cadmium	Fruit juices, soft drinks, *etc.* in contact with cadmium plated vessels or equipment. Cadmium contaminated water and shell-fish	*Itai-itai* (ouch-ouch) disease, Increased salivation, acute gastritis, liver and kidney damage and prostrate cancer
13.	Cobalt	Water and liquors	Cardiac insufficiency and mycocardial failure
14.	Lead	Water, natural and processed food	Lead poisoning (foot-drop, insomnia, anemia, constipation, mental retardation and brain damage)
15.	Copper	Food	Vomiting, diarrhoea
16.	Tin	Food	Colic and vomiting
17.	Zinc	Food	Colic and vomiting
18.	Mercury	Mercury fungicide treated seed grains or mercury contaminated fish	Brain damage, paralysis and death
19.	Food flavors beyond safe limit	Flavored food	Chances of liver cancer

S.No.	Adulterant	Foods commonly involved	Diseases or health effects
20.	Brominated vegetable oils	Cold drinks	Anemia and enlargement of heart
21.	Sulphur dioxide and sulphite beyond safe limit	In variety of food as preservative	Acute irritation of the gastro-intestinal tracts, *etc.*
22.	Artificial sweeteners beyond safe limit	Sweet foods	Chances of cancer
Bacterial contamination			
23.	*Bacillus cereus*	Cereal products, custards, puddings and sauces	Food infection (nausea, vomiting, abdominal pain and diarrhea)
24.	*Salmonella* spp.	Meat and meat products, raw vegetables, salads, shell-fish, eggs and egg products and warmed-up leftovers	Salmonellosis (food infection usually with fever and chills)
25.	*Shigella sonnei*	Milk, potato, beans, poultry, tuna, shrimp and moist mixed foods	Shigellosis (bacillary dysentery)
26.	*Staphylococcus aureus* Entero-toxins-A, B, C, D, or E	Dairy products, baked foods especially custard or cream-filled foods, meat and meat products, low-acid frozen foods, salads, cream sauces, *etc.*	Increased salivation, vomiting, abdominal cramp, diarrhea, severe thirst, cold sweats, prostration
27.	*Clostridium botulinus* toxins A ,B, E, or F	Defectively canned low or medium acid foodstuffs, *i.e.*, meats, sausages, smoked vacuum-packed fish, fermented food, *etc.*	Botulism (double vision, muscular paralysis, death due to respiratory failure)
28.	*Clostridium perfringens* (Welchii) type A	Milk improperly processed or canned meats, fish and gravy stocks	Nausea, abdominal pains, diarrhoea and gas formation
Fungal contamination			
29.	Aflatoxins	Aspergillus flavus-contaminated foods such as groundnuts, cottonseed, *etc.*	Liver damage and cancer
30.	Ergot alkaloids from *Claviceps purpurea* Toxic alkaloids, ergotamine, ergotoxin and ergometrine groups	Ergot-infested pearl millet, rye meal, or bread	Ergotism (St. Anthony's fire-burning sensation in extremities, itching of skin and peripheral gangrene)

S.No.	Adulterant	Foods commonly involved	Diseases or health effects
31.	Toxins from *Fusarium sporotrichioides*	Grains (millet, wheat, oats, rye, *etc.*)	Alimentary toxic aleukia (ATA)- an epidemic panmyelotoxicosis
32.	Toxins from *Fusarium sporotrichiella*	Moist grains	Urov disease (Kaschin-Beck disease)
33.	Toxins from *Penicillium inslandicum*, *Penicillium atricum*, *Penicillium citreovirede*, *Fusarium*, *Rhizopus* and *Aspergillus*	Yellow rice	Toxic mouldy rice disease
34.	Sterigmatocystin from *Aspergillus versicolour*, *Aspergillus nidulans* and *bipolaris*	Food grains	Hepatitis
35.	*Ascaris lumbricoides*	Any raw food or water contaminated by human faces containing eggs of the parasite	Ascariasis
36.	*Entamoeba histolytica* Viral	Raw vegetables and fruits	Amoebic dysentery
37.	Virus of infectious Hepatitis (virus A)	Shell-fish, milk, unheated foods contaminated with faeces, urine and blood of infected human	Infectious hepatitis
38.	Machupo virus	Foods contaminated with rodents urine, such as cereals	Bolivian haemorrhagic fever
Natural Contamination			
39.	Fluoride	Drinking water, sea foods, tea, *etc.*	Excess fluoride causes fluorosis (mottling of teeth, skeletal and neurological disorders)
40.	Oxalic acid	Spinach, amaranth, *etc.*	Renal calculi, cramps and failure of blood to clot
41.	Gossypol	Cottonseed flour and cake	Cancer
42.	Cyanogenetic compounds	Bitter almonds, apple seeds, cassava, some beans, *etc.*	Gastro-intestinal disturbances
43.	Polycyclic Aromatic Hydrocarbons (PAH)	Smoked fish, meat, mineral oil-contaminated water, oils, fats and fish, especially shell-fish	Cancer
44.	Phalloidine (Alkaloid)	Toxic mushrooms	Mushroom poisoning (Hypoglycemia, convulsions,

S.No.	Adulterant	Foods commonly involved	Diseases or health effects
			profuse watery stools, severe necrosis of liver leading to hepatic failure and death)
45.	Solanine	Potatoes	Solanine poisoning (vomiting, abdominal pain and diarrhea)
46.	Nitrates and nitrites	Drinking water, spinach rhubarb, asparagus, *etc.* and meat products	Methaemoglobinaemia especially in infants, cancer and tumors in the liver, kidney, trachea oesophagus and lungs. The liver is the initial site but afterwards tumors appear in other organs.
47.	Asbestos (may be present in talc, kaolin, *etc.* and in processed foods)	Polished rice, pulses, processed foods containing anti-caking agents, *etc.*	Absorption in particulate form by the body may produce cancer.
48.	Pesticide residues (beyond safe limit)	All types of food	Acute or chronic poisoning with damage to nerves and vital organs like liver, kidney, *etc.*
49.	Antibiotics (beyond safe limit)	Meats from antibiotic-fed animals	Multiple drug resistance hardening of arteries and heart disease

Source: http://agmarknet.nic.in/adulterants.htm

Detection of food adulterants

Adulteration in food is normally present in its most crude form. Prohibited substances are either added or substituted partly or wholly. In India, normally, the contamination or adulteration in food is done either for financial gain or due to carelessness and lack in proper hygienic condition of processing, storing, transportation and marketing. This ultimately results that the consumer is either cheated or often become victim of diseases. Such types of adulteration are quite common in developing countries or backward countries. However, adequate precautions taken by the consumer at the time of purchase of such produce can make him alert to avoid procurement of such food. It is equally important for the consumer to know the common adulterants and their effect on health. Detection of adulteration in common food is given in Table 4.

Table 4: Food adulterants and methods of their detection

Food article	Adulteration	Test
Vegetable oil	Castor oil	Take 1 ml of oil in a clean dry test tube, add 10 ml of acidified petroleum ether, shake vigorously for 2 minutes and add 1 drop of ammonium molybdate reagent. The formation of turbidity indicates the presence of castor oil in the sample.

Food article	Adulteration	Test
	Argemone oil	Add 5 ml concentrated nitric acid (HNO_3) to 5 ml sample, shake carefully and allow to separate yellow, orange yellow, crimson color in the lower acid layer. It indicates adulteration.
Ghee	Mashed potato, sweet potato, *etc.*	Boil 5 ml of the sample in a test tub, cool and a drop of iodine solution. The occurrence of blue color indicates the presence of starch. Colour disappears on boiling and reappears on cooling.
	Vanaspati	Take 5 ml of the sample in a test tube, add 5 ml of hydrochloric acid (HCl) and 0.4 ml of 2% furfural solution or sugar crystals, insert the glass stopper and shake for 2 minutes. Development of a pink or red color indicates presence of vanaspati in Ghee.
	Rancid stuff (old ghee)	Take one teaspoon of melted sample and 5 ml of HCl in a stoppered glass tube, shake vigorously for 30 seconds and add 5 ml of 0.1% of ether solution of Phloroglucinol. Restopper, shake for 30 seconds and allow to stand for 10 minutes. A pink or red color in the lower (acid layer) indicates rancidity.
	Synthetic coloring Matter	Pour 2 g of filtered fat dissolved in ether, divide into 2 portions, add 1 ml of HCl to one tube and 1 ml of 10% NaOH to the other tube. Shake well and allow it to stand. The presence of pink color in acidic solution or yellow color in alkaline solution indicates added coloring matter.
Honey	Invert sugar/jiggery	Fiehe's Test: Add 5 ml of solvent ether to 5 ml of honey and shake it well. Decant the ether layer in a Petri dish and evaporate completely by blowing the ether layer. Add 2 to 3 ml of resorcinol (1 g of resorcinol resublimed in 5 ml of concentrate HCl). Appearance of cherry red color indicates the presence of sugar/jiggery.
		Aniline Chloride Test: Take 5 ml of honey in a porcelain dish, add aniline chloride solution (3 ml of aniline and 7 ml of 1:3 HCl) and stir well. Orange red color indicates the presence of sugar.
Pulses/Besan	Kesari dal (*Lathyrus sativus*)	Add 50 ml of dilute HCl to a small quantity of dal and keep on simmering water for about 15 minutes. The pink color, if developed indicates the presence of *Kesari dal.*
Pulses	Metanil Yellow (dye)	Add concentrated HCl to a small quantity of *dal* in a little amount of water. Immediate development of pink color indicates the presence of metanil yellow and similar color dyes.
	Lead Chromate	Shake 5 g of pulse with 5 ml of water and add a few drops of HCl. Occurrence of pink color indicates the presence of lead chromate.
Bajra	Ergot infested pearl millet	Swollen and black Ergot infested grains will turn light in weight and will float in water.

Food article	Adulteration	Test
Wheat flour	Excessive sand and dirt	Shake a little quantity of sample with about 10 ml of carbon tetra chloride and allow it to stand. Grit and sandy matter will collect at the bottom.
	Excessive bran	Sprinkle on water surface. Bran will float on the surface.
	Chalk powder	Shake sample with dilute HCl. Effervescence indicates chalk.
Common spices like turmeric, chilly, curry powder, *etc.*	Colour	Extract the sample with petroleum ether and add 13N H_2SO_4 to the extract. Appearance of red color (which persists even upon adding little distilled water) indicates the presence of added colors. However, if the color disappears upon adding distilled water, the sample is not adulterated.
Black pepper	Papaya seeds/ light berries, *etc.*	Pour the seeds in a beaker containing carbon tetra-chloride. Black papaya seeds float on the top while the pure black pepper seeds settle down.
Spices (ground)	Powdered bran and saw dust	Sprinkle on water surface. Powdered bran and sawdust float on the surface.
Coriander powder	Dung powder	Soak in water. Dung, which floats, can be easily detected by its foul smell.
	Common salt	To 5 ml of sample and add a few drops of silver nitrate. White precipitate indicates adulteration.
Chillies	Brick powder grit, sand, dirt, filth, *etc.*	Pour the sample in a beaker containing a mixture of chloroform and carbon tetrachloride. Brick powder and grit will settle at the bottom.
Large cardamom seeds	Small cardamom seeds	Separate out the seeds by physical examination. The seeds of large cardamom have nearly plain surface without wrinkles or streaks, while the seeds of small cardamom have pitted or wrinkled ends.
Turmeric powder	Starch of maize, wheat, tapioca and rice	A microscopic study reveals that only pure turmeric is yellow colored, big in size and has an angular structure, while foreign/added starches are colorless and small as compared to pure turmeric starch.
Turmeric	Lead chromate	Ash the sample and dissolve it in 1:7 sulphuric acid (H_2SO_4). Filter it and add 1 or 2 drops of 0.1% dipenylcarbazide. A pink color indicates the presence of lead chromate.
	Metanil yellow	Add few drops of concentrated hydrochloric acid (HCl) to sample. Instant appearance of violet color, which disappears on dilution with water, indicates pure turmeric. If color persists, metanil yellow is present.

Food article	Adulteration	Test
Cumin seeds (black jeera)	Grass seeds colored with charcoal dust	Rub the cumin seeds on palms. If palms turn black, adulteration is indicated.
Asafetida (Heeng)	Soap stone, and other earthy matter	Shake a little quantity of powdered sample with water. Soap stone or other earthy matter will settle at the bottom.
	Chalk	Shake the sample with carbon tetrachloride (CCl_4). Asafetida will settle down. Decant the top layer and add dilute HCl to the residue. Effervescence shows the presence of chalk.
Food grains	Hidden insect infestation	Take a filter paper impregnated with Ninhydrin (1% in alcohol.), put some grains on it and then fold the filter paper and crush the grains with hammer. Spots of bluish purple color indicate presence of hidden insects infestation

Source: http://www.go2inbox.com/info/information.php?readinfo=2013

Table 5: Food adulterants and methods of their detection

Food article	Adulterant	Simple method for detecting the adulterant
Milk	Water	Put a drop of milk on polished vertical surface. The drop of pure milk either stops or flows slowly leaving a white trail behind it, whereas, milk adulterated with water will flow immediately without leaving a mark.
	Urea	Take 5 ml of milk in a test tube and add 2 drops of bromothymol blue solution. Development of blue color after 10 minutes indicates the presence of urea.
Mustard seeds	*Argemone* seeds	*Argemone* seeds have rough surface and mustard seeds on pressing is yellow inside, while *argemone* seed is white.
Ice cream	Washing powder	Put some lemon juice, bubbles are observed on the presence of washing powder.
Sugar	Chalk	Dissolve sugar in a glass of water, chalk will settle down at the bottom and similarly for salt.
Silver foil	Aluminum foil	On ignition genuine silver foil, burns away completely leaving glistening white spherical ball of the same mass, while aluminum foil is reduced to ashes of black grey color.
Honey	Water	A cotton wick dipped in pure honey burns when ignited with a matchstick. If adulterated, presence of water will not allow the honey to burn. If it does will, it produces a cracking sound.
Coffee	Chicory	Gently sprinkle the coffee powder on surface of water in a glass. The coffee floats over the water but chicory begins to sink down within few seconds. The falling chicory powder particles leave behind them a trail of color, due to large amount of caramel they contain.

Food article	Adulterant	Simple method for detecting the adulterant
Tea	Colored leaves	Rub leaves on white paper, artificial color comes out on paper.
	Used tea	Tealeaves sprinkled on wet filter paper, pink or red spots on paper show color.
	Iron fillings	Move a magnet through the sample, the iron will stick to the magnet.
Red Chilly Powder	Rodamine culture	Take 2 g sample in a test tube and add 5 ml of acetone. Immediate appearance of red color indicates the presence of Rodamine.
	Brick powder	Brick powder settles fast, while chilly powder settles slowly when put in water.
Turmeric Powder	Metanil yellow	Add a few drops of HCl to the extract of turmeric from water. Instant appearance of violet, when the color persists when diluted with water indicates the presence of metanil yellow.
Dal arhar, moong and washed channa	Metanil yellow	Extract the color with luke warm water from the sample of pulses and add drops of HCl. A pink color indicates the presence of metanil yellow.
Green vegetables like chilly	Malachite green	Take a small part of the sample and place it over a moistened white blotting paper. The impression of the color on paper indicates the presence of malachite green.
Pure ghee or butter	Vanaspathi	Take one teaspoonful of melted ghee or butter with equal quantity of concentrated hydrochloric acid in a test tube and add to it a pinch of cane sugar. Shake well for one minute and let it stand for five minutes. Appearance of crimson color in lower layer shows the presence of vanaspathi.
Black Pepper	Papaya seeds	Float the sample in alcohol. The mature black pepper berries sink, while the papaya seeds and light black pepper float.
Hing	Soap stone or earthy matter	Shake a little portion of sample with water and allow it to settle. Soap stone or earthy matter will settle down at the bottom.
Saffron	Colored dried tendrils of maize cob	Pure saffron will not break easily like artificial. Pure saffron when allowed to dissolve in water will continue to give its color so long as it lasts.
Common Salt	White powdered stone	Stir a spoonful of sample in salt water. Chalk will make the solution white and other insoluble impurities will settle down.

Source: http://www.cleanindia.org/resoucewatch/foodaduteration.htm

The Food Safety and Standards Authority of India (FSSAI)

The food safety and standards authority of India has been established under Food Safety and Standards Act 2006 that consolidates various acts and orders that have hitherto handled food related issues in various Ministries and Departments. FSSAI has been created for laying down science-based standards for articles of food and to regulate their manufacture, storage, distribution, sale and import to ensure availability of safe and wholesome food for human consumption.

An Act to consolidate the laws related to food and to establish the Food Safety and Standards Authority of India for laying down science-based standards for articles of food and to regulate their manufacture, storage, distribution, sale and import and to ensure availability of safe and wholesome food for human consumption and for matters connected therewith or incidental thereto.

Various central Acts like Prevention of Food Adulteration Act 1954, Fruit Products Order 1955, Meat Food Products Order 1973, Vegetable Oil Products (Control) Order 1947, Edible Oils Packaging (Regulation) Order 1988, Solvent Extracted Oil, De-Oiled Meal and Edible Flour (Control) Order 1967, Milk and Milk Products Order 1992, *etc.* have to be repealed after commencement of FSS Act 2006.

The Act also aims to establish a single reference point for all matters relating to food safety and standards by moving from multi-level, multi-departmental control to a single line of command. To this effect, the Act establishes an independent statutory Authority, *i.e.*, the Food Safety and Standards Authority of India with head office at Delhi. Food Safety and Standards Authority of India (FSSAI) and the State Food Safety Authorities shall enforce various provisions of the Act.

Consumer Protection Act India 1986

An Act is to provide better protection of consumers' interests and to make provision for the establishment of consumer councils and other authorities for the settlement of consumer disputes and for matters connected therewith.

Conclusion

It is true that adulteration primarily thrives in a period of shortages. The real income of consumer is falling due to rising prices of even essential commodities. Psychologically, consumers pay less attention to the quality of products during this period. He/she faces disadvantages in the form of adulteration, which is always a health menace. Thus, food adulteration takes many forms, *i.e.*, mixing, substitution and abstraction, concealing the quality sale of decomposed foods and using false labels. The pity is that the so-called modernization has brought with the evils of adulteration. Somehow, the Indian consumer has become accustomed to live with adulteration. Even educated consumers do not pay attention to the menace of adulteration.

Consumer is the largest economic group and central point of all marketing activities. With the rise in income of people, the quality, quantity and sophistication of consumer goods have also increased. The market is literally overflowing with the new products based on intricate technology. It is very difficult for the consumer to select one food item because of misleading advertisements, improper media emphasis and food adulteration. As a result of these malpractices, the ultimate victim is a consumer, who innocently takes adulterated food articles and suffers. Therefore, a consumer's best defense is knowledge of his/her rights as a consumer and of the remedies, which exist to resolve these problems when they occur. Knowledge and awareness about adulterated foodstuffs, laws and rights related to adulterated food are crucial in a society where technology heightens opportunities for perpetrators of fraud deception and misrepresentation.

Consumer awareness is a campaign, which needs to be promoted. The culprits promoting this malpractice should be booked and punished. This mission has to be propagated as a mass movement. Even a check on the imported foodies has to be judiciously clamped for building a healthy nation and a force of physically, mentally and spiritually vibrant citizens.

■■■

37 Pesticides vs. Human Health

Beena Kumari

Department of Entomology, CCS Haryana Agricultural University, Hisar-125 004, Haryana, India
E-mail: <beena@hau.ernet.in>

Since the dawn of time, humanity has had two primary goals, *i.e.*, obtaining enough food to survive and improving the quality of life. The single most important task facing the society is the production of food to feed its population before it can devote resources to education, arts, technology, or recreation. A basic fac of green revolution in India, which converted a food deficient state to food surplus state was the ability to control pests, weeds, insects and diseases by providing effective plant protection umbrella coupled with soil-water management and high yielding varieties. Today, farmers regard pesticides as an essential tool to ensure production of crops of quality and quantity to satisfy an increasing human population. Agricultural experts believe that these food and fiber needs can be met, but to do so, it will require the increased use of pesticides. Thus, these are considered as an indispensable component in pest management strategies for food production and public health in today's context.

Pesticides are designed to kill and because their mode of action is not specific to one species, they often kill or harm organisms other than pests, including humans. The World Health Organization (WHO) estimates that there are 3 million cases of pesticide poisoning each year and up to 220,000 deaths, primarily in developing countries. The application of pesticides is often not very precise, and unintended exposures occur to other organisms in the general area where pesticides are applied. Regardless of the method of application, pesticides ultimately reach the soil, which serves as a reservoir for these chemicals and in turn contaminate the environment.

Pesticides

According to Food and Agricultural Organization (FAO) 1986, the term pesticide has been defined as any substance or mixture of substances intended for preventing, or controlling any pest or any substance or a mixture of substances intended for use as plant growth regulator (PGR), defoliant, or desiccant. The term excludes fertilizers, antibiotics, or other chemicals administered to animals to stimulate their growth or to modify their reproductive behaviour.

Classification of Pesticides

Classification Based on Their Use

Depending on the purpose for which they are used, the pesticides are categorized into the following broad groups:

- **Algicides:** These chemicals control algae in lakes, canals, swimming pools, water tanks and other sites.
- **Antifouling agents:** These chemicals kill or repel organisms that remain attached to underwater surfaces, such as bottoms of the boat.
- **Antimicrobials:** These chemicals kill the microorganisms such as bacteria and viruses.
- **Attractants:** These chemicals attract pests, *e.g.*, luring an insect or rodent to a trap. However, food is not considered a pesticide when used as an attractant.
- **Biopesticides:** These chemicals are certain types of pesticide derived from such natural materials as animals, plants, bacteria and certain minerals.
- **Biocides:** These biological agents kill the microorganisms.
- **Disinfectants and sanitizers:** These chemicals kill or inactivate the disease-producing microorganisms on inanimate objects.
- **Fungicides:** These chemicals kill fungi including blights, mildews, molds and rusts.
- **Fumigants:** These chemicals produce gas or vapour intended to destroy pests in buildings or soil.
- **Herbicides:** These chemicals kill weeds and other plants that grow where they are not wanted.
- **Insecticides:** These chemicals kill insects and other arthropods.
- **Miticides (acaricides):** These chemicals kill mites that feed on plants and animals.
- **Microbial pesticides:** These microbial pesticides kill, inhibit, or out compete with pests, including insects or other microorganisms.
- **Molluscicides:** These chemicals kill snails and slugs.
- **Nematicides:** These chemicals kill nematodes (microscopic worm-like organisms that feed on plant roots).
- **Ovicides:** These chemicals kill eggs of insects and mites.
- **Pheromones:** These biochemicals are used to disrupt the mating behaviour of insects.
- **Repellents:** These chemicals repel pests, including insects (such as mosquitoes) and birds.
- **Rodenticides:** These chemicals control mice and other rodents.

Classification Based on Mode of Action

Based on mode of action, the insecticides are divided into following groups:

- Physical poison
- Protoplasmic poison
- Metabolic inhibitors
- Neuroactive agents
- Hormone mimics
- Stomach poisons

Classification Based on Chemical Groups

Based on chemical groups, the pesticides are broadly divided into four major groups, *i.e.*, organochlorines, organophosphates, carbamates and synthetic pyrethroids, which are being discussed below in details:

Organochlorines: These pesticides are chlorine-containing compounds including dichlorodiphenyltrichloroethane (DDT), aldrin, dieldrin and lindane. They operate by disrupting the sodium to potassium balance of the nerve fiber, forcing the nerve to transmit continuously. Their toxicity varies greatly but they have been phased out because of their persistence and potential to bioaccumulation. The pesticides of this group are mostly insecticides and are termed as hard pesticides because they do not degrade in the environment easily. They are lipophilic in nature. In India, DDT and hexachlorocyclohexane (HCH) except lindane (γ-isomer of HCH) are banned in agriculture since 1993 but DDT is still in use for public health purpose under National Malaria Eradication Programme (NMEP). Polychlorinated biphenyls (PCB's), which are not used as pesticides, are also organochlorines with similar human action, and thus have the potential for an additive effect. Acute ingestion of organochlorine insecticides can cause a loss of sensation around the mouth, hypersensitivity to light, sound and touch, dizziness, tremors, nausea, vomiting, nervousness and confusion.

Organophosphates and carbamates: Examples of these groups are chlorpyriphos, malathion, quinalphos, carbofuran, carbaryl, aldicarb, *etc.* Both organophosphates and carbamates bind cholinesterases and block their action in the hydrolysis of acetylcholine neurotransmitters, thus, acting principally in the parasympathetic and central nervous system. These have now become the most widely used agricultural pesticides. Research studies have shown persistent deficits in cognitive function in workers chronically exposed to pesticides. Newer evidence suggests that these pesticides may cause developmental neurotoxicity at much lower doses and without depression of plasma cholinesterase levels. Acute organophosphate and carbamate exposure causes signs and symptoms of excess acetylcholine, such as increased salivation and perspiration, narrowing of the pupils, nausea, diarrhea, decrease in blood pressure, muscle weakness and fatigue. These symptoms usually decline within days after exposure ends as acetylcholine levels return to normal.

Synthetic pyrethroids: Among the most promising alternatives to organophosphates are synthetic pyrethroids. The examples are cypermethrin, deltamethrin, allethrin, permethrin, fenvalerate, *etc.* These are toxic to the nervous system. Acute exposure in humans usually resulting from skin exposure due to poor handling procedures, usually resolve within 24 hours. Pyrethroids can cause an allergic skin response, and some pyrethroids may cause cancer, reproductive, developmental effects, or endocrine system effects. They can cause hyper-excitation, aggressiveness, uncoordination, whole-body tremors and seizures.

Chlorphenoxy herbicides: These are another very widely used group of pesticides in most commonly used to kill dandelions and broad-leaved weeds in lawns, parks, golf courses and school yards. Herbicides are generally less toxic to mammals than insecticides. Most of the herbicides interfere with plant hormones or enzymes that do not have any direct counterpart in animals. The most serious human health concerns have been related to chemical contaminants in the active ingredient.

Classification Based on Pesticide Toxicity

The World Health Organization (WHO) classifies pesticides according to acute toxicity, using the LD_{50} (Lethal Dose 50%) benchmark. LD_{50} denotes the amount of a chemical required to kill 50% of an exposed population of laboratory rats. There are two measures for each product, oral LD_{50} (the product is given orally) and dermal LD_{50} (the product is given through the skin).

WHO Category	Oral LD_{50}		Dermal LD_{50}	
	mg per kg body weight required to kill 50% of rat population			
	Solid	Liquid	Solid	Liquid
Ia. Extremely hazardous	5 or below	20 or below	10 or below	40 or below
Ib. Highly hazardous	5-50	20-200	10-100	40-400
II. Moderately hazardous	50-500	200-2000	100-1000	400-4000
III. Slightly hazardous	Over 500	Over 2000	Over 1000	Over 4000

Pesticides are toxic by design. The Bhopal disaster occurred when a pesticide plant released 40 tonnes of methyl isocyanate (MIC) gas, a chemical intermediate in the synthesis of some carbamate pesticides. The disaster immediately killed nearly 3,000 people, and ultimately, caused at least 15,000 deaths. Some of the most potent pesticides are insecticides, those targeting the nervous system of insects. Unfortunately, the basic neural mechanism is similar in insects and mammals, making humans susceptible to these potentially lethal chemicals as well. In both groups of organisms, messages are transmitted along nerve cells using electrical impulses. When these reach the end of a nerve, a chemical *neurotransmitter* activates the next cell in the chain. Each new release of neurotransmitter can be detected by the recipient cell as enzymes exist which break down and remove any neurotransmitter left from previous signals.

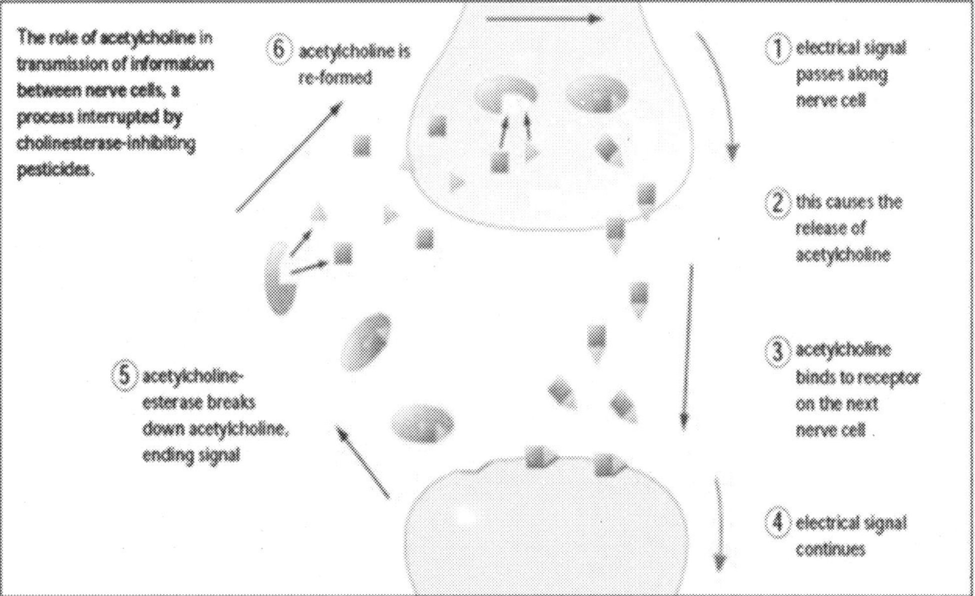

One important neurotransmitter, *i.e.*, acetylcholine, is broken down by the enzyme acetylcholinesterase. Two major classes of pesticides, *i.e.*, the organophosphates and the carbamates, inhibit acetylcholinesterase and cholinesterase. Acetylcholine accumulates in the synapse and there is a *jamming* of information preventing messages from being passed properly between nerve cells. Correct working of nerves is prevented by such inhibitors. Effects are normally dose dependent, which may be minor or fatal. This can affect the nerves in the brain, responsible for release of hormones or controlling hormones actions. Since hormones are especially important in early stages of human development and in reproduction, such endocrine disruption can be particularly damaging to human embryos or children. Organochlorines and pyrethroids (two other major classes of pesticide) also attack the nervous system but these chemicals are not cholinesterase-inhibitors. Their main effect is on individual nerve cells, interfering with the transmission of messages along their length.

Effects of Pesticides on Human Health

Pesticides can affect the human health mainly by two ways given as under:

(A) By Direct Contact to the Pesticides

The extent of the problem

An estimated three million reported cases of pesticide associated acute poisoning occur annually, resulting in 220,000 deaths among which 99% occur in the developing world, in spite of these countries accounting for only 20% of the global pesticide use. The true number of cases is likely to be considerably higher. In 1990, as per the WHO report, an estimated 25 million agricultural workers are poisoned by pesticides every year. Up to date global estimates are lacking, but today, there are 1.3 billion agricultural workers and it is likely that millions of pesticide poisoning cases still occur every year.

The potential adverse impact on human health from exposure to pesticides is likely to be higher in countries like India due to easy availability of highly hazardous products and low risk awareness, especially among children and women. The general conditions of use in very hot climates where personal protection gear is not always used, increase the risks to health. Overexposure to pesticides can occur before spraying because of easy access for children, lack of adequate labeling and during mixing during spraying and after spraying operations. Spray operators and bystanders can be affected. Particularly at risk are women at home who are generally in charge of washing the used clothes of spray operator, and usually, do not know about risks from pesticide-contaminated clothes. Pesticides can cause depression, insomnia, blindness, cancer, hyperflexia and other gastrointestinal problems. The chemicals can also mix with the ground water through seepage into the soil that may be consumed ultimately as drinking water. According to researchers from the National Institute of Health (NIH), the licensed pesticide applicators that used chlorinated pesticides on more than 100 days in their lifetime were at greater risk of diabetes. One study found that associations between specific pesticides and incident diabetes ranged from a 20 to 200% increase in risk. New cases of diabetes were reported by 3.4% of those in the lowest pesticide use category compared with 4.6% of those in the highest category. Risks were greater when users of specific pesticides were compared with applicators who never applied that chemical.

Due to highly persistent nature of organochlorine pesticides, the organophosphate pesticides have increased in use. These are associated with acute health problems, such as abdominal pain, dizziness, headaches, nausea, vomiting, as well as skin and eye problems in workers that handle the chemicals.

Farmers and Workers

As pesticides are inherently toxic to living organisms, these are likely to affect the health of farmers and workers. Besides users, they can be dangerous to consumers, workers and close bystanders during manufacture, transport, or during and after use. The use of chemicals in farming and agriculture has resulted in an increased intake of chemicals by humans. This increased intake of chemicals has in turn resulted in many deleterious effects to the human body. The effects of chemicals on human health are very damaging. It can lead to a host of health problems that can even be fatal. Insecticides and pesticides used for agriculture can be eaten if the food especially vegetables and fruits are not thoroughly washed. Chemicals can also enter the human body through the skin, nose, eyes and mouth.

Pesticide-related health problems result from exposures, which occur chiefly *via* one or more of the following routes:

- Oral ingestion
- Inhalation
- Dermal (through the skin)

Within the body, pesticides is metabolized or stored in the fat or secreted unchanged, sometimes, metabolism increases toxicity, *e.g.*, the hydrolysis of carbosulfan and furathiocarb produces more toxic water soluble compound, *i.e.*, carbofuran. Some fat-soluble substances, *i.e.*, DDT and HCH, are not metabolized readily but stored in fatty tissues. In time of poor nutrition, these are released into blood stream with the possibility of toxic effects if concentration reaches a high level. The recognized effects of pesticides on human health include the followings:

- Cataract formation is caused by exposure to diquat herbicide.
- Cellular proliferation in the lungs is caused by paraquat.
- Effect on immune system is caused by dicofol, organotin compounds (fungicides) and trichlofon (insecticide).
- Uncoupling of oxidative phosphorylation is caused by dinitrophenols and dinitrocresol herbicides.

Besides the above-recognized effects, some other ill prominent effects of pesticides are given below:

Birth defects

Many birth defects have been seen in babies born in places near agricultural areas. The babies can have circulatory and respiratory problems. There can also be babies born with malformations in their extremities. The commonly used pesticide, chlorpyriphos (brand name Dursban), causes severe birth defects in children exposed in uterus. Chlorpyriphos is widely used as an agricultural chemical but is also the most common pesticide used indoors to kill termites, fleas, cockroaches and in pest control strips. Hydrocephaly cleft pallets are some of the common birth defects due to exposure of the pregnant woman to chemicals. It can also result in low birth weight of babies. It can also invade the milk produced by the lactating mothers. This leads to a high-risk nutrition to the infants leading to more health problems. The potential effects of chemicals on human health have far-reaching effects. This makes it imperative that we cut down on the chemicals used in food.

Neurological effects

Pesticides can be potent neurotoxins. When people are exposed to neurotoxins, they may feel dizzy, lightheaded, confused and may have reduced coordination and ability to think. These are the short-term effects, while the long-term exposure can result in reduced intelligent quotient (IQ) and learning disability, associated with permanent brain damage. The children born in these areas where there is an increased use of chemicals are prone to have decreased intelligence due to brain damage. The chemicals mainly affect the nervous system leading to these problems.

Hormone disruption

Some substances cause physical birth defects, others can cause subtle hormonal effects on the developing fetus or affect a child's functional capacities. Hormone disruptors have been linked to many health problems including reproductive cancers. The drug diethylstilbestrol (DES), which was given to pregnant women to prevent miscarriage between 1941 and 1971, worked as an endocrine disrupting chemical on the developing fetus. Decades later, many of these DES exposed daughters developed cervical cancer. Twenty-four pesticides still in the market, including 2,4-D, lindane and atrazine, are known endocrine-disrupters.

Steroid hormones, such as oestrogens, androgens (*e.g.*, testosterone) and progesterone, are crucial for primary sex determination, foetal development and acquisition, and maintenance of secondary sexual characteristics in adults. Chemicals, including many pesticides, with similar structures to these hormones can interfere with their function and lead to a variety of developmental and reproductive anomalies.

Mechanisms of Action of Hormone Disruption

- Binding to and activating hormone receptors. For example, the oestrogenic chemicals like DDT can create the same response in a cell as the natural oestrogen. Though they are often weaker than natural equivalents, the effects of different chemicals can be cumulative.
- Binding to and deactivating hormone receptors, and thereby, preventing natural hormones from binding.
- Modifying rates at which hormones are metabolized, and thereby, affecting their abundance. Organochlorines, such as DDT and endosulfan, have been found to affect paths by which the female sex hormone oestradiol is broken down, creating higher levels of a metabolite linked to breast cancer.
- Modifying the hormone production rate.
- Affecting cells number of hormone receptors. Exposure to endocrine disruptors is especially harmful to foetuses and young children, as the body's systems are particularly sensitive to hormones when they are developing. Effects produced in animals include abnormal sexual behaviour, structural deformities of the reproductive tract, including cross-gender type conditions and undescended testes, deficits in sperm counts and distorted sex ratios. These effects have been reported both in laboratory animals and in wildlife and humans heavily exposed to endocrine disrupting pesticides. Some examples of these effects from studies carried out in 1996-97 in United States, England and Canada are given below:
  - **Endometriosis:** A disease in which the uterine tissue grows outside the uterus and a common cause of infertility is virtually unheard of 20 years ago. It now affects 5.5 million women in the United States and Canada, about 10-20% of women of childbearing age.
  - **Hypospadias:** A condition in which the urethra is near the base of the penis, not the end as it should be, has doubled in the last 10 years.
  - **Undescended testicles:** It is linked with later risk of testicular cancer, and it is increasing day by day. Researchers reported a doubling in cases between 1962 and 1982 in England and Wales.
  - **Precocious puberty in girls:** It is now common in girls. A study of girls in the United States found that the onsét of puberty for white girls was 6-12 months earlier than expected and African-American girls experienced puberty 12-28 months earlier than the onset of puberty in whites.
  - **Reduced sperm counts:** Low sperm count is a marker for testicular cancer and it is documented now a day.

- **Fetal death:** Exposure of pregnant women to pesticides can increase the risk of fetal death. There can be stillbirths occurring as a consequence of these chemicals.
- **Miscarriage:** There is an increased risk of miscarriage if the pregnant woman is exposed to chemicals in the third to eighth week of pregnancy.
- **Brain tumor:** Tumor or cancer of the brain is more common in people who have been exposed to various chemicals.

Effects of Some Selected Pesticides

Lindane

Acute Effects: Acute inhalation exposure to lindane in humans results in irritation of the nose and throat and effects on blood (anemia) and skin (elevated itchy patches of skin). The major effects noted from oral exposure to lindane in humans are effects on the nervous system, such as seizures and convulsions. Vomiting, nausea and effects on the cardiovascular and musculoskeletal systems have also been reported.

Chronic Effects (Non-cancer): Chronic exposure to lindane by inhalation in humans has been associated with effects on the liver, blood and nervous, cardiovascular and immune systems.

Chlorpyriphos

This pesticide inhibits the activity of acetyl cholinesterase, the enzyme responsible for inactivating acetylcholine, the compound ultimately responsible for most of the toxic symptoms seen with chlorpyriphos. After chlorpyrifos enters the body, it is changed by the liver into other forms of compound that may or may not be less toxic than the original material. The major nontoxic chlorpyriphos metabolic product formed by the liver is 3,5,6-trichloro-2-pyridinol (TCP), which is primarily eliminated from the body in the urine.

Monocrotophos

Monocrotophos can be absorbed following ingestion, inhalation and skin contact. When inhaled, it affects the respiratory system and may trigger bloody or runny nose, coughing, chest discomfort, difficulty in breathing or shortness of breath and wheezing due to constriction or excess fluid in the bronchial tubes. Skin contact with organophosphates may cause localized sweating and involuntary muscle contractions. Eye contact causes pain, tears, pupil constriction and blurred vision. Repeated daily high-level exposure may gradually lead to poisoning. Several studies on occupationally exposed workers have been conducted in countries with a hot climate like India where the workers usually did not wear protective clothing. In most cases, plasma cholinesterase was inhibited. It has been reported that absorption of 20 mg of monocrotophos causes inhibition of plasma acetyl cholinesterase.

Malathion

Malathion, which has been shown to have serious health effects upon humans, is a pesticide used for the control of medfly and mosquito. The observed effects include weakening of immune system (thereby resulting in increased colds, flu and infections), birth defects, genetic damage, accelerated aging of certain body organs, increased neurological damage to the elderly and serious harm to wildlife.

(B) Through Pesticide Residues in Environment and Food Commodities

Before explaining the effects of residues, the definition of pesticide residues is given below:

Pesticide Residues: Substances, which remain in or on a feed or food commodity, soil, air or water following use of a pesticide. For regulatory purposes, it includes the parent compound and any specified derivatives such as degradation and conversion products, metabolites and impurities considered to be of toxicological significance (FAO, 1986).

The toxic residues of the chemicals can affect the central nervous system, liver and respiratory organs. Indiscriminate and improportionate use of pesticides is comparatively more in certain crops, while in some, it is negligible. The farmers use pesticides more frequently and in increased doses than the recommended doses or procedures. It leads to the presence of high amount of residues in food commodities. The pesticides are used improportionately in India in relation to places, and the amount of pesticides residue varies from one place to another. Tamil Nadu consumes 1.2-2.0 kg/ha of land followed by Andhra Pradesh and Punjab where 0.8-1.2 kg is the rate of consumption. Pesticide residues in the feed and fodder are solely responsible for their accumulation in animal and poultry. The states like Tamil Nadu, Andhra Pradesh, Punjab, Haryana and Karnataka have highest use of pesticides in order to get more production, while on the other hand, the states like Bihar, West Bengal and North Eastern states have lowest use of pesticides. It is because of illiteracy of farmers, poor economic conditions, or due to lack of awareness. Hence, the food commodities in high using states have more residues of pesticides.

Repercussion of Long-Term Exposure to Pesticide Residues through Contaminated Food

Contrary to the common belief, the chances of cancer at confronted levels of residues are very remote because of safety factor employed in calculation of maximum residue limit (MRL) values and uneven distribution of residues in the food commodities. However, long-term exposure to toxic residues may adversely affect the immune system, which in turn makes human being more susceptible to bacterial, viral and parasitic infections. Additionally, the long-term exposure may lead to promotion in tumor growth in many animal species and exert adverse effects on reproduction, brain, liver and kidney.

The greatest concern with the organochlorines is the long-term effects. The United States, Environmental Protection Agency (EPA) has concluded that dichloro diphenyl trichloro ethane (DDT), dichloro diphenyl dichloro ethylene (DDE) and dichloro diphenyl dichloro ethane (DDD) are probable human carcinogens. On this basis, both Canada and the United States banned the organochlorines. However, they continue to be very prevalent posing long-term health risks.

The organochlorines are still widely used in developing countries including Central and South America, India, China and many other countries. Products imported from these countries are obvious sources of DDT and other organochlorines. They are also transported in air, oceans and bioaccumulate in organisms. Pesticide residues from sprayed crops are consumed by animals, being lipophilic nature of organochlorine pesticides, are secreted into milk which in turn consumed by human beings including mothers which ultimately reach in mother's milk and in newly born babies. This cycle goes on and in turn bioaccumulation also takes place. Quite high level of organochlorine pesticide residues like DDT, HCH and endosulfan have been reported in mother's milk in various parts of India like Haryana, Uttar Pradesh and Himachal Pradesh. This is because of formation of food chain.

Food is the primary route of exposure. The foodstuffs containing DDT include meat, fish, poultry, dairy products, and root and leafy vegetables. Fish from the Great Lakes Basin and Inland water is a large food source of organochlorine exposure.

The provincial and federal health departments report that there are instances where maximum allowable levels of DDT intake may be exceeded in breast-fed infants. Similarly, they report that people living near hazardous waste sites have been found to have an increased exposure to organochlorines in large part because of leaching into the soil.

In fish and wildlife, there is evidence of reproductive and developmental effects as a consequence of chronic exposure, and there is increasing concern that exposure of humans to these chemicals may be causing adverse effects on reproductive function. A number of chemicals in the environment possess estrogenic activity and these compounds include pesticides as well as plasticizers, estrogenic agents administered to livestock and a variety of other chemicals.

The use of organochlorine pesticides (OCPs) like dichlorodiphenyl trichloroethane (DDT) and hexachlorocyclohexanae (HCH) has either been banned or restricted in agriculture in India but are still being used in public health programmes for the control of vector borne diseases such as malaria, filarial, kala azar, *etc.* In view of the classical stability and persistency of organochlorine pesticides in environment and biological media, their residues through food chain enter the human body. The principal route of exposure is the consumption of food particularly leafy and root vegetables, fatty meat, fish and poultry. Monitoring studies carried out in 2005 in India have revealed wide spread contamination of fruits, vegetables, milk, meat and their products with pesticide residues. DDT was found in all the 55 samples with a frequency occurrence of 100% and a mean concentration of 6.86 mg kg^{-1}, while HCH was detected in 46 samples with a frequency occurrence of 83.6% and a mean concentration of 5.72 mg kg^{-1}. Distribution of DDT and HCH residues based on sex, food habits, occupation, obesity and urban/rural population did not clearly indicate any pattern. An interesting finding of the present study was the presence of residues in a 7 months old female child with the level of DDT and HCH as 1.92 and 0.59 mg kg^{-1}, respectively. Substantially, higher level of DDT (15.2 mg kg^{-1}) has been reported from adipose tissue of a day old child from Mexico. High level of pesticide residues at this age might be due to the exposure to organochlorine compounds both in uterus and through breast-feeding. Infants who were breast-fed had higher concentration of organochlorine pesticides (OCPs) than those who were formula fed.

The primary concern of the chronic low dose toxicity in man and animals is related to the carcinogenic, teratogenic, mutagenic, immunotoxic, immunopathological and/or neuropathic effects of pesticides. Various pathological effects of low doses of pesticides in animals and man are as under:

- **Immunopathological effects:** The iImmunopathological effects of pesticides in animals and man are classified under acquired immunodeficiency or immunosuppression, autoimmunity and hypersensitivity. They are considered the causes of vaccinal failures or occurrence of disease epidemics in animals and man due to lowered immunocompetence. It has also been reported that a state of immunosuppression for a longer period may also lead to the development of neoplasms as the immune surveillance mechanism becomes defective. Such animals also exhibit recurrent bacterial infections due to defective phagocytic machinery of the body. Organochlorine group of pesticides binds with certain proteins of the body to become antigen leading to initiation of an autoimmune response in body. Autoimmune glomerulonephritis, autoimmune hemolytic anemia, or autoimmune rheumatoid arthritis is such manifestations in animals and man. Eczema in man was found due to maneb, 2,4-dichlorophenoxy acetic acid (2,4-D) and 2,4,5-trichlorophenoxyacetic acid (2,4,5-T). Dichlorodiphenyl trichloroethane (DDT) has also been known to cause type I hypersensitivity reaction. Dust of pesticides is the cause of allergic respiratory disorders like asthma. Cutaneous allergy has been known to occur due to contact of pesticide-contaminated food items.
- **Carcinogenic effects:** Most of organochlorine pesticides like dieldrin, gamma isomer of benzene hexachloride/hexachlorocyclohexane BHC/HCH, dichlorodiphenyl trichloroethane (DDT) and

polychlorinated biphenyls (PCB) may cause cancer in liver and lung. Indirectly, a state of immunosuppression for a longer period is helpful in increasing the susceptibility of an animal for malignancy. Since many pesticides are known to cause mutation in chromosomes of man and animals, it is considered that they may also lead to carcinogenicity.

- **Mutagenicity:** Pesticides may cause alterations in structure or number of chromosomes resulting in translocations, mutations and chromosomal breakage. The altered chromosomal number may become lethal during fetal stage. Several pesticides like DDT, Endrin, PCB and HCB are known to cause chromosomal aberrations. The mutagenic effect of pesticide poses a more serious threat to the future of human race.
- **Teratogenicity:** The accumulation of pesticides in body tissue and congenital birth defects in children have not been well correlated so far. However, there are certain pesticides, which cause teratogenic defects in animals. Carbaryl, thiram, propoxur, parathion, leptaphos, 2,4-D, lindane and diazinon are having teratogenic defects in animals. In mice, cypermethrin, alphemethrin and malathion are found to exert birth defects in baby mice.
- **Neuropathy:** Most of the organophosphates, organochlorines carbamates may cause neutrotoxic effects in man and animals including increased irritation, loss of memory, in coordination of movement, ataxia, delayed response, convulsions, spasms and paralysis. Such changes appear due to demyelination of nerves in central and peripheral nervous system. Pesticide residues are also responsible for marked behavioural changes in man and animals.
- **Nephropathy:** The pesticide residues present in foodstuff may act as hapten and when they bind with certain body proteins, they may become antigenic. This antigenicity is responsible for initiation of immune response in body and a continuous presence of antigen and antibodies in body may lead to the formation of immune complexes. The immune complexes when produced in excess are deposited in glomerular basement membrane leading to glomerulonephritis, commonly known as renal failure for which patient needs dialysis after a regular interval to survive.
- **Hepatotoxicity:** The pesticide residues in food may harm liver tissue since they are metabolized here. There are instances of chronic liver disorders leading to cirrhosis. Certain pesticides are not so dangerous but their metabolites cause severe damage to hepatic parenchyma. The cirrhosis once starts it never stops even after withdrawal of the primary cause.
- **Reproductive Disorders:** It has been observed that the pesticides are lethal to dividing cells of genitalia. They may cause abnormalities in sperms leading to decrease their ability for fertilization. On the other hand, the ova become defective and not able to implant on the uterine surface leading to early abortion or miscarriage. DDT has been found to cause weak eggshell in birds leading to their decreased population. The pesticide residues in food, thus, may ultimately lead to sterility, early abortion, stillbirths, or repeat breeding.

Pesticides *vis-a-vis* Human Cancer

The International Agency for Research on Cancer (IARC) has evaluated the potential carcinogenicity of a number of pesticides. IARC has classified mineral oil (used as pesticide) as carcinogenic to human beings and two pesticides ethylene dibromide and ethylene oxide as probably carcinogenic to human beings. There is strong evidence that arsenical pesticides are associated with respiratory cancer in human subjects. Fourteen pesticides were classified as possibly carcinogenic to human beings. These were amitrole, aramite, hexachlorobenzene, hexachlorocyclohexane, mirex, nitrofen, sodium orthophenyl phenate, sulfallate and toxaphene, possibly carcinogenic to human beings. None of the synthetic pesticides was classified by IARC as carcinogenic to human beings. The general lack of data from human studies is striking. Evidence of carcinogenicity of organochlorine pesticide for human beings is not strong. However, the persistence of these agents in the environment, the continued use of DDT in developing countries and animal evidence consistent with carcinogenicity make further study imperative.

In some occupational groups exposed to certain pesticides, epidemiological studies have demonstrated an increased risk of lung cancer, lymphopoietic cancers and possibly other forms of cancer. In addition, there is a potential cancer risk from ingesting certain pesticides.

Cancer risk assessment suffers from many sources of uncertainty, the knowledge of how chemicals cause cancer is far from complete. Uncertainty arises from the need to extrapolate from the high doses necessary to induce tumors in animals to the lower doses to which humans are likely to be exposed. Data from cancer studies of many pesticides are limited, and data on both pesticide use and residues on raw crops and in processed foods are insufficient to conduct aggregate risk analysis. Up to date information on dietary patterns of various population groups is also lacking.

Carcinogenesis (cancer formation) involves irreversible alteration of a stem cell, its uncontrolled proliferation, and finally, invasion of other tissues. In this sequence, the pesticides by various mechanisms may contribute to cancer development. The most obvious mechanism is genotoxicity, direct alteration of DNA turning harmless cells into cancer cells. Even levels of exposure to organophosphates too low to decrease cholinesterase levels significantly increased the chromosomal aberrations found in blood samples from farmers. Pesticides thought to cause cancer in this way include the fumigants ethylene oxide and ethylene dibromide.

Hormonal effects of some pesticides, may also promote tumours in certain tissues by stimulating hormonally sensitive cells to carcinogenesis, *e.g.*, the herbicide atrazine is generally non-genotoxic (does not directly affect DNA) but it is associated with increased incidence of cancers in sites such as uterus lining and mammary tissue in animal experiments. Recent studies have shown that the incidence of hormone-related organ cancers, or hormonal cancers, is elevated among farmers, with exposure to endocrine-disrupting pesticides, particularly to DDT and phenoxy herbicides, suspected to the involvement in some of these cancers.

Another way in which pesticides may promote cancer is through immunotoxic effects, disturbing the body's normal cancer surveillance mechanisms, *e.g.*, organophosphates can inhibit serine esterases, critical enzymes in the immune system's T-lymphocytes and natural killer cells.

Lung cancer

Farmers tend to have lower risks of contracting lung cancer than the general population, as they are less likely to smoke, *e.g.*, the studies on workers in other occupations exposed to pesticides, pest control workers and pesticide-production workers have shown increased risks. Also, where studies on farmers have controlled for the effect of smoking, lung cancer excesses have been found. A study showed that exposure to DDT is to be a factor increasing lung cancer risk. A study of pest control workers in Florida (USA) found lung cancer to be associated with carbamates, organophosphates and phenoxyacetic acids and more specifically with DDT, diazinon, carbaryl and propoxur.

Breast cancer

Breast cancer has been estimated to have increased in incidence by 1% annually since the 1940s in the United States of America and increased by 50% in Denmark from 1945 to 1980. Excess breast cancer risks have been noted in a number of occupations and industries, notably those that entail exposure to solvents and pesticides, particularly crop farmers and the fruit/vegetable industry. An American study showed that among women farmers, breast cancer risks were elevated for women present in fields during or shortly after pesticide application but not among those who reported using protective clothing.

Pancreatic cancer

Associations between pesticide exposure and pancreatic cancer have been revealed in studies from Spain, Italy and the United States of America. Increased prostate cancer risks related to pesticide exposure have been suggested in a number of other epidemiologic studies, including one which found excess risks for occupational exposure to fungicides or herbicides rather than insecticides. Among chemical manufacturing workers, a significantly increased risk of prostate cancer was found with DDT exposure.

Bladder cancer

Fruit farming was found to be associated with an increased risk of colon and bladder cancer, which may be due to the use of mildly refined mineral oils as insecticides. Bladder cancer is generally found slightly reduced among farmers but some studies have shown that there may be a slight excess if the effect of smoking is removed. A significant link between bladder cancer and exposure to pesticides was revealed in a study of French vineyard farmers.

Environmental effects

Pesticide use raises a number of environmental concerns. Over 98% of sprayed insecticides and 95% of herbicides reach a destination other than their target species, including non-target species, air, water and soil which in turn human health. Pesticide drift occurs when pesticides suspended in the air as particles are carried by wind to other areas, potentially contaminating them. Pesticides are one of the causes of water pollution, and some pesticides are persistent organic pollutants and contribute to soil contamination.

Conculsion

Every year, pesticides are estimated to cause tens of millions of cases of accidental poisoning, largely in the developing world. Farmers of developing nations are at special risk of harmful exposures due to the nature of the pesticides they use and the widespread lack of awareness of the risks they pose. Adequate safety measures are often not employed and pesticide poisoning is a frequent consequence. Very high doses can result in unconsciousness, convulsions, or death. Children and unborn babies are particularly susceptible to pesticide poisoning. Their widespread exposure to pesticides in developing nations is of serious concern. Exposure to pesticide leads to respiratory and skin diseases, cancers, birth defects and reproductive and neurological disorders.

■■■

38 Sanitation in Food Establishments

M. Pal[1] and Vijay J. Jadhav[2]

[1]Department of Microbiology, Immunology Epidemiology and Public Health, Faculty of Veterinary Medicine, Addis Ababa University, Post Box No. 34, Debre Zeit, Ethiopia E-mail: <palmahendra1947@yahoo.com>

[2]Department of Veterinary Public Health and Epidemiology, College of Veterinary Sciences, Lala Lajpat Rai Veterinary and Animal Sciences University, Hisar, Haryana, India; E-mail: <vetviju@gmail.com>

Some anthropologists believe that humans have the natural tendencies to live in a clean and orderly environment. This species-specific behaviour is noticed particularly with the food and eating habits of human beings. The hygienic aspect of food has been identified by prehistoric human beings. This must be originated due to intellectual capability of human beings to identify link between unhygienic food and foodborne diseases. With the advancement of science, the people are taking food hygiene aspect more seriously. Market research revealed that consumers avoid unclean food environment. Food hygiene that was earlier limited to kitchen hygiene at family level is now expanded at industry level and already has been a subject of academic interest covered under the discipline of food hygiene and food sanitation.

The *sanitation* is derived from the Latin word *sanitas*, meaning *health*. Concerning food industry, food sanitation refers to the hygienic measures for ensuring food safety. More specifically, it is an attempt to utilize scientific knowledge to provide wholesome food processed, prepared, merchandised and sold in a clean environment by healthy workers, with an aim to prevent contamination that can cause biological, physical, or chemical hazard to the consumers. The major sources of contamination in any food plant can be personnel, building, equipments, water, air, packaging material and animal (in case of abattoir). Sanitation as an applied science incorporates the principles of design, development, implementation, maintenance and improvement of hygienic practices to exert control over these sources of contamination.

Sources of Food Contamination

The food, which is meant for man, contains many of the same nutrients, which are also essential for survival and proliferation of microorganism. Therefore, once get entered the food, microorganisms are easily established inside the food resulting into its spoilage, and in worst case, they may cause food borne infection or intoxication to the consumer. In order to prevent the contamination of food due to these microorganisms, it is very necessary to identify the sources of food contamination. It is known fact that microorganisms are everywhere, *i.e.*, in soil, feces, air and water. Therefore, food can be contaminated any time if it comes in contact of such sources during its production, processing and distribution.

In order to exert control over contamination of food, it is very much important to know the sources of contamination in the food establishment as well as extent to which contamination can occur due to each of the identified sources. The most rational approach for carrying out this exercise, which is the one proposed by Kraft Foods, is now well adopted by many other food establishments. The said approach is known as *Zonal Approach*, which involves identification of all the source of contamination in the food establishment and then classifying them in four zones based on their potential to cause contamination of food. These zones are depicted as four concentric circles as in *bull's eye* with zone I at the innermost place and other zones encircling it in ascending order. Zone 1 represents most critical areas for cleaning and sanitizing primarily direct food-contact surfaces, *e.g.*, production equipment, utensils and containers contacting directing with foods. Zone II target includes the areas of concern for cleaning and sanitizing of indirect food-contact surfaces, *e.g.*, drains, utility pipes, heating ventilation, air conditioning system equipment, other utensils used in food preparation *etc.* Zone 3 includes floors, walls and other items in contact with them. Finally, zone IV includes maintenance equipment and areas further away from production such as hallways, entrances and recreational facilities. Further examples of sources of contamination in food establishment are given below:

Raw material: In case of dairy industry, raw milk acts as one of the principle sources of contamination. Contamination of raw milk occurs mostly at the farm level itself but when care is not taken, transportation of milk from the farm to the dairy plant can increase the contamination. Though subsequent pasteurization is designed to kill all spoilage and pathogenic bacteria, it cannot assure the safety of milk obtained from unsanitary raw supply.

Production of meat involves slaughter of live animals, and therefore, meat is always prone to acquire bacteria that live on the external surface of the animal or within gut. Carcass of animal is contaminated by contact with the hide, feet, manure, dirt and visceral contents from punctured digestive organisms.

Poultry products are prone to contamination during processing. The processing of poultry involves step of defeathering and evisceration that provide an easy opportunity for microorganisms to contaminate the carcasses.

In case of preparation of ready to eat or ready to cook products, ingredients are potential vehicles of harmful or potentially harmful microorganisms and toxins. The amounts and types of these agents vary with place and method of harvesting, type of food ingredient, processing technique and handling.

Equipments: Microorganisms in the equipments and utensils are originated from the air, as well as from employees and materials. They remain adhere to the metallic surface of equipments and utensils. Certain bacteria even have tendency to form biofilm, which offers them resistance to disinfectants and sanitizers. Thus, equipments and utensils if not cleaned properly, act as constant source of food products contamination.

Employees: Employees especially food handlers are the most important source of food contamination. The hands, hair, nose and mouth harbour microorganisms commensally, which are transferred to the food during its handling in the food establishment by touching, breathing, coughing, or sneezing. Thus, employees can transfer spoilage and pathogenic microorganisms in the food. Employees suffering with communicable disease provide serious threat of spreading disease amongst sizable population of consumers.

Water: Water is very cheap and efficient cleansing medium, however, it is prone to acquire contamination from animal, human beings and environmental sources. If contaminated water is used for cleansing operation in food establishments, it is bound to contaminate premises and equipments. Therefore, it becomes necessary to monitor the quality of water before its use.

Air: Unclean air from the surrounding enters the food processing area containing dust particles and aerosols that carry microorganisms. Thus, the contamination can result from airborne microorganisms in food processing, packaging, storage and preparation areas.

Insects and Rodents: Flies, cockroaches and rodents are mechanical vectors of many different microorganisms. These are responsible for the transfer of microorganisms from their source of contamination to the food articles.

Objectives of Food Plant Sanitation

As detailed above, various sources of contamination are liable for the contamination of food, which is being prepared in the food establishments. In order to prevent reduced shelf life of the product or outbreak of foodborne diseases caused due to contaminated food, it becomes necessary to exert stringent control measures to check all the sources of contamination. This is achieved by sanitation, which is of pivotal significance in food safety program. The major objectives of food plant sanitation are as follows:

- To control food poisoning
- To increase shelf life of the products
- To avoid foreign body contamination
- To check pest infestation
- To project a hygiene visual image
- To ensure safety of operatives and maintenance staff
- To meet customers quality expectations
- To allow maximum plant productivity
- To reduce the risk of litigation
- To meet national and international legislation

General Sanitation Procedure for Food Establishments

In order to achieve the above-mentioned objectives of food plant sanitation, there is a scientifically designed sanitation programme, which includes cleaning procedure and application of cleaning agents and sanitizers in a pre-determined manner based on scientific evidence and experience. Some basic supporting services and infrastructural facilities required for the establishment of food plant are as follows:

- A continuous electric power source to provide heat, light, refrigeration and to operate equipments in the plant is necessary.
- An abundant supply of hot, cold, potable and running water is imperative. Washing in clean water containing a detergent is basic step in sanitizing procedure without which the food industries cannot be operated.
- A plumbing system that transports water to the points of use such as toilet and sinks and also removes sewage, installation of traps on drains and construction that prevent cross connection and back flow from sewerage to water line and fixing of values for control of temperature and flow rate.
- Facilities should be sufficient for sanitary collection and disposal of solid wastes. Ventilation system that distributes clean air and removes aerosols within the building should also be there.

- Physical separation of raw material from cooked food is also essential to prevent contamination.

Cleaning Procedure

Food plant-sanitation procedure followed for daily use is a sequential process, meaning every step should be followed after successful completion of previous one, and if these steps are not taken in sequence, sanitation will be ineffective. The details of each of these steps are given below:

Step I: Gross clean (preparation): This step involves removal of food residue, which is left on the equipment, surface and floor. In addition, it includes sweeping floors, removing of unused materials and tools and removal of bulk soils and debris from the area to be cleaned. During this step, it is also important to cover the equipments if necessary; otherwise, it may be disassembled to such an extent to facilitate thorough cleaning and sanitization in successive steps.

It is important to note that if the food residue is not removed before further cleaning, it will protect the surface as well as bacteria from the attack of detergent. It also reacts with and consumes the detergent thereby weakened its function and wastage of valuable chemical. Further, it will eventually be washed into the drainage system, causing blockage in traps and thereby create interference in the cleaning operation. Therefore, in a properly managed cleaning programme, all the pieces of food products, which are larger than a fingernail, are removed before application of detergent. It can be carried out by handpicking, scraping and shoveling. The collected rubbish should be put into bins, which are subsequently removed from the area.

Step 2: Pre-Rinse: The purpose of this step is to remove stubborn deposits, which cannot easily be removed by picking or scraping. In this step, the area and equipment surfaces are rinsed with water under proper pressure and temperature until they are visually free of soils. The pressure should not be very high so as to avoid the risk of cross-contamination associated with aerosol migration and overspray. In addition, the temperature should not be more than 50°C to avoid coagulation of the food residue or any other type of organic dirt.

Step 3: Detergent application: The purpose of this step is to remove the thin layer of proteins and fats. These cannot be identified easily by visual observation, and thus, they may remain present on the surface that may already look clean. This technique involves use of appropriate detergent at right concentration, use of mechanical action, appropriate water temperature and adequate contact time. After application of detergent, foaming should be carried out in such a way so that all surfaces, *i.e.*, top and bottom, are covered with foam. During this operation, once again, the sequence is important which should be followed in the order of walls and floors at first followed by the equipment to reduce the potential for cross-contamination and to prevent detergent from drying on equipment surfaces.

Step 4: Post-rinsing: It is the step that involves application of water to remove dirt material, which remains suspended in detergent solution once it has been separated from the surface due to action of detergent. During this operation, care should be taken to minimize the amount of splash and aerosol formed, which may re-contaminate previously cleaned surfaces. The operation is carried out in the same sequence followed, while detergent application, *i.e.*, walls and floor, at first followed by equipments to avoid recontamination.

Step 5: Remove and assemble: In this step, the instruments previously dissembled for cleaning are assembled to their ready to work condition, but before doing this, it is important to ensure that equipment is sufficiently dry. It is also important to carry out prior sanitization of those equipments or components of equipments that remain inaccessible once reassembled.

Step 6: Disinfection: Disinfection should only be carried out on a visually clean and well-rinsed surface. The concentration as well as reaction time of disinfectant is equally important for effective sanitation.

Step 7: Terminal rinsing: Most but not all disinfectants are safe to leave on surfaces without final rising. In fact, disinfectants remain active for considerable time at its residual concentration and thereby maintain low microbial count. However, if disinfectant is not allowed to remain on food contact surfaces according to food regulations then post-rinsing should be carried out. This step should be carried out in a timely manner that will permit drying of the food contact surfaces but does not allow microorganism to recover and multiply before the commencement of production work. Criteria for the selection of appropriate chemical sanitizers include their spectrum of antimicrobial activity (it should have broad-spectrum activity), environmental resistance (effective in the presence of organic matter, detergent and soap residues and in water with varying degree of hardness and pH), nontoxic and nonirritating properties, good water solubility, devoid of odour, stability in concentrated and diluted form, ease of use, availability, *etc.*

Cleansing and Sanitizing Agents

Cleansing agents are designed for specific tasks like washing floors and walls, suitable for use in high-pressure washer and cleaning in place system, whereas, sanitizing agents are designed to exert lethal effect on microbes and at the same time non-toxic to the person using it. Desirable characteristics of cleaning agents are nontoxic nature, economical for regular use, non-corrosive, non-dusting and availability of test or method to measure lowest possible concentration in food or on the inanimate objects. For sanitization of food establishment, the suitable sanitizer should have broad-spectrum activity against microorganisms and ability to demonstrate potency in adverse situations such as presence of organic matter, detergent and soap residues and hard water with varying pH. In addition, it should be nontoxic and nonirritating for the personnel, stable in concentrated and diluted form, easily available and should not impart its odour or colour to the food. A brief overview of such cleansing agents and sanitizers is provided in Table 1 and 2, respectively.

Table 1: Cleaning compounds used for sanitation in the food establishments

Group of cleansing compound	Examples	Use
Strongly alkaline cleansers	Sodium hydroxide (caustic soda)	It can remove heavy soils, such as those from food processing equipments like ovens and smoking cabinets.
Heavy-duty alkaline cleansers	Sodium metasilicate, sodium hexametaphosphate, sodium pyrophosphate, sodium carbonate and trisodium phosphate	It leads to emulsification of soil especially fats which are then removed easily. It is compatible for use with high pressure or other mechanized systems.
Mild alkaline cleansers	Sodium bicarbonate, sodium sesquicarbonate, tetrasodium pyrophosphate, phosphate water conditioners (sequesters) and alkyl aryl sulfonates (surfactants).	It acts as water softening agent and therefore increase the efficiency of detergents used.

Group of cleansing compound	Examples	Use
Chlorinated alkaline cleansers	Alkaline cleaners fortified with hypochlorites	These are designed to breakdown proteinaceous deposits which helps in easy removal. These are used in cleaning of pipelines, tanks and vats
Acid cleansing compounds	Phosphoric acid, nitric acid, sulfuric acid and sulfamic acid	These help in removal of encrusted surface and mineral deposits.
	Organic acids such as citric, tartaric, sulfamic and gluconic acid	These acts as water softeners and these are not corrosive or irritating to the skin.
Strongly acid cleansers	Hydrochloric acid, hydrofluoric acid, sulfamic acid, sulfuric acid and phosphoric acid	These are used to remove the encrusted surface especially in the boilers and some processing equipment.
Mildly acid cleansers	Levulinic, hydroxyacetic, acetic and gluconic acids	Due to their mildness, these are used as manual cleaning products. They also function as water softeners.
Synthetic detergents	*Cationic*: Quaternary ammonium salts such as cetyl trimethylammonium chloride *Anionic*: Alkyl benzene sulphonate, alpha olefine sulphates and sulphonates, *i.e.*, sulphated monoglycerides *Non ionic*: Pentaerythrityl palmitate	Synthetic detergents lower the surface tension of the solution, promotes wetting of particles, and deflocculates and suspends soil particles.
Enzyme based cleaners	Lypolase and protease P	They break down soil particles into smaller pieces and help in its removal by attacking site of attachment.
Solvent cleaners	Mineral spirits and 91% isopropyl alcohol	These are used to remove oil and greases in the area where mechanical equipments are in operation.

Source: Marriott, N.G. and Gravani, R.B., 2006. *Principles of Food Sanitation*, 5th Edition. Springer, U.S.A.

Table 2: Sanitizers used in the food establishments

Group of chemical sanitizers	Examples
Chlorine compounds	Liquid chlorine, hypochlorites, inorganic and organic chloramines and chlorine dioxide
Iodine compounds	Iodophors, alcohol-iodine solutions and aqueous iodine solutions
Bromine compounds	Bromine has been used alone or in combination with other compounds
Quaternary ammonium compounds	Alkyl dimethyl benzyl ammonium chloride and alkyl dimethyl ethyl benzyl ammonium chloride
Acid sanitizers	Organic acids such as acetic, peroxyacetic, lactic, propionic and formic acid

Source: Marriott, N.G. and Gravani, R.B., 2006. *Principles of Food Sanitation*, 5th Edition. Springer, U.S.A.

Equipments Used in Food Sanitation

Cleansing activity in food establishments is a routine operation and is an integral part of production. It is labour intensive and involves use of chemicals in the form of cleansing and sanitizing agents. However, its efficiency is increased by the use of equipments appropriate for the task. The range of equipments include as simple as brush to elaborate sophisticated installed instrumentation. In general, following equipments are used in well-established and quality cautious food processing organizations.

Mechanical abrasives: These include scrubbing devices such as steel wool and copper chore balls to remove dirt adhered to walls, floors and surface of the equipments. These are not intended to use on the surface that has direct contact with food since the metal particles or pieces released from such device can contaminate food resulting into physical hazard to the consumers.

Water hoses: These are long pipes usually fitted with nozzles to spray water on the area to be cleaned. It can be used for combined operation of washing and brushing when fitted with a spray head brush.

High-pressure water pumps: It is a stationary or portable device accompanied with a storage tank for cleaning solution. It serves the purpose of cleaning and sanitation by the use of relatively less volume of water.

Steam guns: This device applies mixture of steam and water or steam and cleaning compound. Although this is very efficient in cleaning but it is considered as energy consuming method of sanitation.

Cleaning-in-place system: Manual cleaning is usually practiced for easily accessible utensils and equipments. Cleaning-in-place system (CIP) is practiced in big food establishment where it is very difficult to dismantle and reassemble the equipments. It saves time and labour, minimizes damage to the equipment and also appears to be cost effective. In this method, detergents and sanitizers are circulated for a specific period at a specified speed and in a specific sequence.

Evaluation of Food Plant Sanitation

It is well known fact that quality of food products manufactured in the food establishments depends upon sanitary conditions prevailing in the plant besides the quality of raw material. Therefore, it is very much important to ensure that cleaning and sanitization procedures carried out really helped to serve the purpose. This is achieved by microbiological analysis of samples collected from various locations in the food establishments. These samples include food samples collected at various processing points as well as samples collected from surface of equipments, walls, floors and hands of personnel. Further, water and air samples are also subjected for microbiological analysis.

Microbiological Examination of Sanitized Surfaces

Swab/swab-rinse methods: This is quite old method for microbiological examination of the surfaces. It uses either cotton or calcium alginate swabs applied on definite area of surface, *e.g.*, 1 cm^2. If possible, the sterile template is placed over the surface, and the area within is rubbed with a moistened swab. The used swab is replaced into its container that contains sterile diluent. The swab containers are kept in refrigerator until subjected for microbiological examination.

Contact plate: These are nothing but commonly used microbiological media plates, which are excessively filled with medium to form a raised surface. While taking the sample, the plate is inverted in such a way so that agar medium makes direct contact with the surface. Then lid of plate is replaced and plates are subjected to incubation at required temperature.

Microbiological Examination of Containers

In order to evaluate sanitization of containers used for carrying the food items, the samples are taken by rinse method. For this, a measured quantity (20-50 ml) of normal saline solution is poured into the container. The container is shaken vertically, horizontally and in circular direction so that rinsing solution may come in contact of entire surface. This solution is collected in sterile vials for subsequent microbiological examination.

Indicator Organisms

Since it is very difficult to look for each bacterial pathogen that might present in the food and food plant environment, indicator organisms may be employed to reflect the microbiological quality of foods relative to product shelf life or their safety from foodborne pathogens. A brief overview of these indicators is given below:

Coliforms: These are gram negative, facultative anaerobic and rod-shaped bacteria, which belong to the family Enterobacteriaceae. This includes *Escherichia coli*, and *Enterobacter aerogenes* besides *Citrobacter* and *Klebsiella* spp. The presence of some coliforms in food is associated with animal or human feces. Therefore, coliform count can indicate fecal contamination or contamination from equipment that has not been properly cleansed or sanitized. The coliform count is the quantitative method that identifies and enumerates the number of distinctive colonies on violet red bile agar grown at 37°C during 24 hours.

Enterococci: These are Gram-positive organisms present in the faeces of man and animals. These were earlier known as faecal *Streptococci*. Like coliforms, their presence in food products shows the contamination with faecal matter directly or through contaminated water so they are indicator of hygienic quality. These include *Enterococcus faecium, E. faecalis, E. asini, E. avium, etc.* Due to their resistance to heat and freezing, their presence indicates poor hygienic practices in frozen foods and inadequate thermal processing of some canned products.

Bifidobacteria: Genus *Bifidobacterium* consists of at least 25 species that are Gram-positive, anaerobic and catalase negative non-motile rods. They are most commonly found in the stools, and therefore, serve as indicator of fecal pollution especially of waters. The *Bifidobacteria* have been found in human feces at much higher levels than *Escherichia coli.* By using *Bifidobacteria,* it is possible to determine their origin among one of the three sources such as human feces, animal feces, or environmental conditions. The difficulty in following the laboratory method is their strict anaerobic nature due to which they tend to grow slowly and require many days for results.

Besides these indicator organisms, certain group of microorganism based on their temperature requirement also serves as indicators of sanitary quality and shelf life of the products. These include psychrotropic, psychrophiles and thermophiles bacteria.

- **Psychrotrophic and psychrophilic bacteria:** The presence of these bacteria is of much importance for the products, which are stored at refrigeration temperature as they reduce the shelf life of products.
- **Thermoduric bacteria:** Bacteria of this group have the ability to survive at higher temperature. Since most of the bacteria of this group are originated from the environment, their presence in the final product indicates unsanitary conditions during harvesting of the raw material. These are also responsible for the reduction of shelf life.
- **Fungi:** These are eukaryotoic, heterotropic, Gram positive, non-acid fast and saprobic living organisms. The estimation of fungi (yeast and mould) in food can be done to assess the potential of fungal spoilage.

Significance of Personal Hygiene in Sanitation of Food Establishments

Personal hygiene is an important component of public health programme. It includes all the protective measures, which influence the health and well-being of individual. The persons engaged in food processing plants should be educated about the possible health hazards caused by contaminated foods particularly of animal origin. They should be made aware about the need of hygiene practice and routine medical check up to ensure that sick personnel are not allowed to work. They should wear clean protective clothing including headgear. Employees with purulent skin lesions or sores should not be allowed to handle the food items at any stage of manufacture. The wound should be properly covered with antiseptic dressing. Food handlers suffering from some infectious diseases such as scarlet lever, septic sore throat, typhoid, diphtheria, coughs and colds contribute to the transmission of these diseases to susceptible individuals by contaminating the food articles. The persons who are carrier of some infectious disease-producing agent should not be permitted to handle the foodstuffs. All the persons should wash their hands with chlorine solution frequently to check contamination. It is very important to mention that staff of the food establishments must adopt the high standards of personal hygiene and follow the clean habits such as proper washing of hands with antiseptic solution or detergent soap after toilet and before handling the foods, regular trimming of nails, avoid inserting fingers in the nose, avoid spilling on the floor, *etc.* The workers must wash hands thoroughly with soap and warm water before starting the work and after handling the contaminated material such as raw meat and poultry. After eating, drinking, smoking, nose blowing, toilet using, the employees must properly wash their hands before returning to work.

Dairy Plant Sanitation

The prerequisite for production of high quality milk and milk products is the cleaning and sanitation of contact surfaces that contribute to 60% of the total contamination in a dairy plant. Not only the dairy equipments should be cleaned but also the dairy plant atmosphere be kept free from microbes. It is also important to ensure that clean air and water are available and that surfaces in contact with dairy food items do not react with the products. Dairy plant has characteristic soils, which include deposits of minerals, proteins, lipids, carbohydrates, dust, lubricants and microorganisms. Effective sanitation practices can reduce soil deposition microorganisms by the use of appropriate cleaning and sanitizing agents. Modern dairy plants are invariably provided with advanced cleaning in place system, which take care of the cleaning and sanitation of pipelines and instruments.

Sanitation of abattoir

The high standards of sanitation in the abattoir help to improve the quality and shelf life of meat. The major sources of contamination in abattoir are animals, building, equipments, tools, water, soil and personnel. The dirt and skin of animal contribute about 33% contamination of meat in the abattoir. An average weight of manure, soil and other dirt adhering to cattle hides is 4.0 kg (average 0.9-15.8 kg). The scrapings from the skin contain $3x10^9$ bacteria per gram of scrapings. After skinning, as many as 10,000 to 100,000 bacteria per square centimeter can be found on the tissue. The contamination is found maximum at the point of incision. Blades of knives can carry 80,000 to 40 million bacteria per blade. Visual cleanliness and absence of deposits in the plant are important for the control of actual microbiological risks.

Factors contributing to sanitation of the abattoir include the selection of site, design of the abattoir, quality of material used for construction, type of equipment and tools, health of the animal, effluent treatment system, disposal of waste and awareness of the workers about hygiene and sanitation. The abattoir should be fly proof and vermin proof. A thorough cleaning of all surfaces is necessary before the application of sanitizers. Floors, walls, equipments and tools must be thoroughly cleaned at the end of a day's operation. The potable water should be used for cleaning, hand washing and carcass spraying. The water flow rate and water pressure should be high to dissolve dirt from surfaces. Sufficient contact time should be given for the sanitizers and detergents for achieving optimum cleaning effect. For effective cleaning of areas where grease or flat is adhered, hot water (82°C) should be used. In addition to daily cleaning, a complete cleaning operation must be undertaken every week. Microbiological assessment by using surface swab technique should be done regularly to monitor the abattoir sanitation.

Fruit and Vegetable Processing Plant Sanitation

In fruit and vegetable processing plant, raw materials are principle source of sanitation. The dominant microflora belongs to heat-resistant bacteria, which are present in the soil. Microbial population of fruits and vegetables also depends upon prevailing conditions like humidity, sunlight, temperature, as well as contamination from domestic and wild animals, irrigation water, bird droppings, harvesting equipment and workers. However, once brought to the processing plant, washing and peeling operations helps to minimize microbiological load on fruits and vegetable, provided, the operations are carried out in hygienic manner. Chlorinated water is the most frequently used sanitizer for the washing of fresh produce. In order to achieve high degree of cleanliness in the fruit and vegetable plant, it is essential to provide suitably designed facilities and equipments, training of sanitation personnel, use of appropriate cleaning compounds and sanitizers, adoption of effective cleaning procedures and effective administration of the sanitation program.

The effectiveness of a sanitation program can be evaluated through the establishment of standards as guidelines, visual inspection and laboratory tests.

Food Sanitation at Home

The attempts to produce high quality food items in food establishments may become futile, if consumer handles it in unhygienic manner. In case of foodstuffs procured from the market in raw condition such as milk, fruits and vegetables, meat, fish, *etc.*, the risk is more if proper care is not taken during their handling. It is therefore very important to follow sanitation and hygienic practices during handling of food at domestic level. Following guidelines are useful for consumers to avoid food borne health hazards due to mishandling of food in the kitchen:

Maintain cleanliness: Consumers should wash their hands with soap and water before preparing food and after using the bathroom, changing diapers and handling pets. Hand wash is most effective when performed by applying soap and rubbing hands for 20 seconds before rinsing thoroughly in warm water. After washing the hands, they should be dried with a disposable tissue paper rather than repeatedly using a towel.

All raw food materials should be washed thoroughly in clean water before use or subsequent storage in the refrigerator. While preparing food, it is very important to wash cutting boards, knives and utensils in hot soapy water after preparing each food item and before going on to the next one.

After completion of cooking operation, the kitchen surfaces should be cleaned and disinfected. If cloth towels are used during this operation, consumers should wash them often.

Avoid cross-contamination: Cross-contamination leads to transfer of bacteria from one food product to another. This is especially true for raw meat, poultry and seafood. Cross contamination can be avoided by separating raw meat, poultry and seafood from other ready to eat food items, fruits and vegetables. If possible, use one cutting board for raw meat products and another for salads and other foods that are ready to be eaten. The plates used for handling raw foods should not be used for covering or serving cooked material.

Cooking food at proper temperature: Heating of food for a long time at sufficiently high temperature can kill not all but most of the harmful bacteria that cause foodborne illness. This is especially important for raw milk, as it requires boiling for 30 minutes to make it safe. Cook chicken and mutton to 82°C and eggs until the yolk and white become firm, not runny for safety.

Store the cooked food in refrigerator: The food products prepared in the kitchen awaiting consumption or food products remaining and kept for further consumption should be kept in the refrigerator. If this is not possible then avoid consumption of such products after long holding period and without reheating. Refrigerate or freeze perishables, prepared food and leftovers within two hours.

■■■

39 Personal Hygiene

Shikha Trivedi[1], Kanchan Sandhu[2] and Kalpana Sharma[3]

[1]Lecturer, University of Botswana, Gaborone, E-mail: <kirteshshikha@gmail.com>
[2]Kishi Vigyan Kendra, Nurmahal, Punjab Agricultural University, Ludhiana, Punjab, India, E-mail: <kanchansandhu1973@gmail.com>
[3]Ginni Devi P.G. College, Modi Nagar, Uttar Pradesh, India
E-mail: E-mail: yogesh_krsharma@yahoo.com

Personal cleanliness and good hygiene practices are essential for good health and social reasons. People who are perceived as unclean are often ostracized by others. Keeping oneself clean and well groomed is not only a key to success but also a way to excel in life. Good personal hygiene practices boom to self-esteem confidence. Most of the time, *hygiene* and *cleaning* are interchangeably used but there is a slight difference between these two terms. Hygiene is a practice of keeping the body clean and cleaning is a process to clean the body parts. Personal hygiene practices include seeing a doctor, a dentist, regular bathing, washing and healthy eating. Personal hygiene is achieved by using various products like soap, shampoo, toothbrush, cotton swabs, antiperspirant, facial tissues, mouthwash, skin cleansers, toilet paper and other such products.

What is personal hygiene?

The term hygiene refers to the set of practices perceived by a community to be associated with the preservation of health and healthy living. Hygiene mostly means practices that prevent spread of disease causing organisms, whereas, cleanliness is a way to achieve hygiene. Cleanliness involves cleaning processes, like hand washing, feet washing, *etc.* that remove infectious microbes as well as dirt and soil particles.

Personal hygiene is explained as the proper way of maintaining cleanliness of the external body or personal cleanliness. Personal hygiene is the first step to good health. Every external part of the body needs a basic amount of attention regularly. If we do not take care of personal hygiene, we may develop physical, emotional and social problems, *e.g.*, bad body odour due to sweat can lower down the self-esteem of a person.

Importance of Personal Hygiene

Failure to keep up a standard of hygiene can have many implications. There is not only an increased risk of getting infection or illness but number of social and psychological aspects can be affected.

- Personal hygiene helps to keep the one self clean, healthy, wanted and happy.
- It gives the sense of well being.
- It helps in reduction of personal illness and healing from personal illness.

- It helps in preventing the spread of disease and pandemic outbreaks, such as coughs and colds that can easily be passed from person to person due to unhygienic practices.
- Personal hygiene enhances the self-confidence and improves chances of success in many areas of life. In the absence of it, the social aspects of life can be affected. On the other hand, absence of personal hygiene like bad body odour leads to lack of self-confidence and affects the social life of a person.
- Poor personal hygiene can have significant implications on the chance of promotion. If an individual is unable to look after him or herself, it is unlikely for that person to be promoted because no company, institution, or organization wants to be represented by such person.
- Lack of personal hygiene can cause many health problems. People who do not bathe regularly are more susceptible to fungus infections.

Grooming

Personal grooming extends personal hygiene as it pertains to the maintenance of a good personal and public appearance, which needs not to necessarily be hygienic. Here are important practices to maintain proper hygiene for various body parts:

Skin

A healthy and glowing skin is a sign of good personal hygiene. A proper blood circulation is important for skin. When the blood flow is hindered to the skin for any reason, the skin hardens, thickens and loses its normal appearance. The blood circulation to the skin decreases with the progression in age thereby losing its youthful appearance. The speed of aging starts at 25, it slows between the age of 30 and 35 and it is very fast between the age of 35 and 50. The regular attention and care of skin is essential.

Deficiency of vital nutrients like proteins, vitamin B and vitamin C cause lack of proper blood circulation. The inadequate supply of protein and amino acids in diet makes the tissues to sag causing wrinkles. Vitamins are important for rebuilding the tissues and muscles and for the prevention of wrinkles. Facial muscles are living elastic tissues, which can be nourished and built up from inside only with a good and healthy diet.

Skin is a vital barrier against abrasions, chemicals and pathogens. Proper skin hygiene is the best way to keep the skin healthy. Good hygiene of the skin can be achieved by removing dead cells, dirt and microbes on the skin surface. At the same time, it becomes essential to maintain natural fats and oils in the outer skin layer. Skin needs to be clean and clear because of the following reasons:

- Skin needs a healthy mix of resident microbes to protect it from attacking by disease-causing pathogens.
- Overly harsh and frequent skin cleansing can strip the resident microbes and allow the disease causing pathogens to multiply.
- Skins need both hydration and lipids to maintain the upper layer and its protective barriers.
- It is important to keep the skin clean and maintain its health since stress level, imbalanced diet and environmental factors all damage the skin.

Ways to Keep the Skin Healthy

- The simplest way to take care of skin is to keep the skin always clean by removing daily dirt and grime.
- Take proper bath once or twice in a day as it gives glow to the skin and relaxation to mind by improving blood circulation.
- Limit your bath time. Do not use very hot or very cold water since hot water or long shower removes oils from the skin. Using hot water can over-dry and damage skin. Use only lukewarm water for washing hands and face.
- Ensure proper rinsing of skin after applying soap or washing.
- Do not use harsh soaps and detergents as they can remove skin oils and fats and strip protective layers of the skin. Use scrub and gentle soaps to clean the body and its parts. Avoid harsh scrubbing actions.
- Dry your hands and skin thoroughly.
- Sports person should take bath with antiseptic soap after playing activities.
- Give proper attention to the armpits, back of the ear, in between toes and private parts.
- Dry the private parts thoroughly with clean cotton towel and avoid sharing soaps and towels.
- Change undergarments everyday and wear washed clothes.
- Protect the skin from sunburns as exposure damages the skin. Using a good sunscreen prevents the skin from sunburns and wrinkles.
- Choose skin cleansers that contain moisturizers and are not overly acidic or alkaline.
- A moisturising oil or cream can be used if the skin is dry. It is better to use this at night, because if you go out in the sun or go on dusty roads if the skin is wet, dust sticks to it and oils may give a tan to the skin.
- Excessive application of soaps, creams and ointments can also adversely affect certain natural processes of the skin, *e.g.*, soaps and ointments can deplete the skin of natural protective oils and fat-soluble contents such as vitamin D_3 and external substances can be absorbed to disturb natural hormonal balance.
- Scratches or cuts on the skin shall be taken care of. Be sure to wash the injured area with lukewarm water and mild soap and apply some antiseptic cream with physician's advice.
- Do not scratch bug bites, bee stings, *etc.* as scratching tears the skin and the germs may get in there causing possible infection..
- Hay fever, asthma and other allergies can cause eczema, *i.e.*, an itchy problem so moisturize your skin properly.
- Dandruff is another skin problem. The skin on scalp gets flaky and fall off so it becomes important to keep the skin properly clean and moisturized.
- Excessive body hygiene of the skin can result in skin irritation. The skin has a natural layer of oil, which promotes elasticity and protects the skin from drying so use aqueous creams to protect this layer.
- In winters as the temperature drops, skin problems soar. A dry itchy skin gives unhygienic feeling.
- Dryness and dehydration are biggest threats to skin in winters. Lipids in the intercellular cement begin to disperse. They are responsible for cellular cohesion.
- Use a cream low in water and high in oil content. It deposits hydrating agents on the skin's surface leaving a protective layer.

- When purchasing skin acne products, make sure that they are of standard company.

Teeth

Regular brushing and cleaning between teeth after eating is a good oral hygiene. It helps to prevent dental problems, tooth decay and gum diseases. For maintaining healthy teeth, the following points should be considered:

- Brush your teeth at least twice a day, especially before going to bed and preferably after every meal to keep them healthy and shiny. If you do not brush the teeth, bacteria will keep developing in teeth and cause gingivitis.
- Use proper toothbrush with soft bristles. Hard bristle brushes are not advised to use since they can damage the healthy gums.
- While brushing follow the up and down movement of brush to get rid of the food materials in between the teeth.
- Use toothpaste with green tea extracts as its properties help in preventing bacterial enzyme from turning sugar into a sticky substance that allows plaque to stick to teeth.
- Clean the tongue properly since bacteria can accumulate on the tongue and can contribute to the foul small when you open your mouth or breathe.
- Rinse your mouth thoroughly with fresh water. If you wish, you can also gargle with any mouth fresheners.
- Floss to remove food that is stuck in between your teeth and decay since not everything from teeth can be removed with brush.
- Chew sugarless gum or suck sugarless mint to stimulate saliva production from a dry mouth.
- Avoid the use of tobacco, beer and cold drinks as they are harmful for the teeth.
- Avoid eating sweets at night and brush your teeth if you eat sweets at night.
- Visit dentist regularly.

Hair

Hair is considered to enhance the beauty of an individual. Each strand of hair has its own sebaceous (oil) gland, which keeps the hair shiny and waterproof, but during puberty, when the sebaceous glands produce extra oil, it can make the hair look too shiny, oily and greasy. It is must that you know your type of hair and take care of them accordingly. Here are some common tips to look after your hair:

- Keep them at length at which you can maintain them.
- Oil your scalp with oil preferably an hour before washing.
- Wash your hair at least once a week using soap or shampoo as per your hair texture. If you are not sure of your hair texture, go to a saloon or dermatologist, he or she can assist you in identifying the hair texture.
- Rinse thoroughly and remember that any left soap or shampoo in your hair will damage them.
- Dry your hair properly with clean towel.
- Brush your hair three to four times a day with soft brush or comb with smooth bristles and do not forget to wash your hairbrush or comb regularly.

- If you do not wash your hair, you could develop dandruff problems and lice. Getting rid of lice can be difficult.

Hands

The most common way by which the germs spread is the people's hands. Germs can cause illnesses such as colds, flu and tummy bugs. Hand hygiene is the single most important way of reducing cross-infection. Frequent hand washing reduces the spread of germs, bacteria and food borne diseases. Hand hygiene is central to prevent the spread of infectious diseases in home and everyday settings. According to the United States Centres for Disease Control and Prevention, one should wash hands in the following way:

- Wet your hands with lukewarm clean water, apply soap and leather for at least 20 seconds, making sure to clean fingers, fingernails, cuticles, palms and wrists rinse thoroughly and dry with a fresh paper towel or cotton towel.
- One must wash hands before handling or eating food, after trips to rest rooms, after handling animals, before and after changing a dipper or handling a small child, after handling garbage or other untidy objects, before and after contact with a sick person and after blowing your nose, sneezing, or coughing.

It is very important to take care of hygiene of the hands using the following tips:

- Use an antibacterial soap when washing hands to ensure killing the germs that could be present on your hands and forearms.
- Rub wet soapy hands together for at least 20 seconds.
- Rinse thoroughly and then dry with a clean towel.
- Use moisturizing lotion after every wash to keep the hands from drying out.

Nails

Nails are as important as beautiful hair and skin. Nails should also be kept well and healthy to show confidence of a person. Long nails are good for fashion deeds but to keep long nails you have to take extra good care of them. Keeping short nails are advised because they do not need as much care as long nails need. Here are some points to follow to keep the nails clean and healthy:

- Get any hard dirt out from under the nails with one of those sharp nail cleaners that sometimes come with nail clippers, or something similar.
- Soak your nails in a bowl of warm soapy water for about three minutes.
- Wash your hands again very thoroughly with soap and warm water.
- Pat them dry, do not rub them and rub on some rich hand cream that is really good for your skin.
- Cut your nails if you keep them short, or file them if you are growing them.
- Make sure that the nails are all completely even.

Feet

Feet are most important part of the body, which are commonly ignored though it is a matter of care and hygiene. They are as important as hands, skin and hair. Feet have as many as glands as your hands. Generally, our feet are boxed from morning to evening in shoes made of synthetic materials, creating damp and low oxygen climate, which is ideal for the breeding of fungus and skin diseases. Foot care is not only important but also essential. Followings are the few tips to keep the feet clean and healthy:

Feet wash: Wash your feet daily twice, especially once before going to bed. A pumice stone can be used to clean the soles and a soft brush to clean the space between the toes. Pat dry feet and space between toes with cotton towel. Followings are the tips to wash feet:

- Take footbath as many times as possible as it enhances blood circulation.
- Water used to wash feet should be lukewarm.
- Use alkali free soaps to wash feet, as it does not destroy the acid layer of the skin.
- Dry your feet completely especially between the toes.
- Do not cut toenails round. Always leave them straight to stop the both unattractive and painful risk of ingrown toenails.
- Apply moisturizer if you have dry skin since the excessive drying and dehydration cause the heels to crack.
- Look for foot problems, like cuts, corns, blisters ingrown toenails, swelling, *etc.* and report any concerns to the physician.
- Whenever possible, walk barefoot on grass and natural ground.

Choose appropriate footwear: While buying footwear, it is very important to think about your feet apart from looks and brands.

- Wear comfortable shoes, which are not too tight.
- Prefer to buy lightweight shoes.
- Shoes should allow the feet to breathe and remain dry.
- Prefer leather shoes instead of rubber or plastic.
- Change your shoes occasionally

Choose right socks: It is always better to use cotton socks since they absorb sweat. The socks should be well washed and daily changed.

- Avoid using high heels footwear.
- Take 3-4 bags of tannins tea in one litre water, boil this for 10 minutes, add enough cold water and use it for feet bath. Thirty minutes twice a day of feet relaxation in this tea bath will help your feet in remaining dry during the long summer days.

Watch your diet: Watch what you eat since food is responsible for lot of smell the body spreads. Avoid eating onion, garlic and other spices as their essential oils are excreted by the sweat glands of the feet leading to an unpleasant smell.

Ways to prevent foot odour: Foot odour is caused when the sweat does not evaporate and bacteria feed on it. Prevent foot odour by following ways:

- Wash the sole of your feet and in between toes with water and antibacterial soap regularly.
- Soak your feet in kosher solution.
- Use antiperspirant or foot deodorant.
- Use clean and comfortable socks and change them daily.

- Take care of your feet and shoes. Choose right shoes, wash and air them. Sprinkle your shoes to absorb moisture.

Hygiene of Private Parts

It is very important to take care of private parts by an individual since private parts are the channels to get rid of the body waste. Various types of infection can occur if proper hygiene is not maintained. It is worse in case of females because of their biological structure. Sometimes, these infections can even lead to infertility in human beings. Here are some general points to maintain hygiene of personal parts:

For man

- The pubic region should be showered, cleansed and rinsed thoroughly to avoid infection and bad smell.
- A fresh pair of underwear should be changed everyday after taking bath.
- When showering, be sure to assess all the regions of which your genital area is composed of penis, testicles, pubic mound and inner thighs.

For woman

- The pubic region should be showered, cleansed and rinsed thoroughly to avoid infection and bad smell.
- A fresh pair of underwear should be changed everyday after taking bath.
- When bathing, be sure to wash all the regions of your genital area, *i.e.*, inner vagina wall, vulva, pubic mound and inner thighs.
- During menstruating period, the subject should change pads or tampons frequently since blood decomposes very fast and gives bad odour. The vagina should thoroughly be washed after removal of used pads and before replacing the new one to avoid any kind of infection and bad smell. It is advisable to change pads or tampons before going to bed to have peaceful sleep.

Urinary Tract Infection

Urinary tract infections develop when bacteria enter the urinary tract. Several factors contribute to urinary tract infections, and the causes vary for males and females. However, one of the important causes is poor hygiene. Moisture in the genital area can promote the development and spread of bacteria. The most common cause of urinary tract infections is the bacteria from the bowel, which live on the skin near the rectum or in the vagina, which can spread and enter the urinary tract through the urethra. Once these bacteria enter the urethra, they travel upward, causing infection in the bladder, and sometimes, in other parts of the urinary tract. For women, wiping from front to back after a bowel movement prevents bacteria in the anus from entering the urinary tract.

Some Common Problems due to Improper Hygiene

Head lice

Head lice are insects that live on the scalp of your head. They suck blood to nourish themselves. They stay mostly on unclean hair. You can never find them on clean and shiny hair. Followings are the ways to prevent lice:

- Encourage your kids to avoid sharing things including hairbrushes, combs, hats, scarves, towels, helmets, pillows, sleeping bags, *etc.* that have been on or near another child's head.
- Regularly clean things, such as car seats, pillows, head phones, *etc.* that your child's head has direct contact with.
- Wash your hair with anti-lice shampoo 2-3 times a week until your hairs are clean of lice.

Dandruff

This is dried dead skin on the scalp of your head, which comes off as tiny flakes. This is associated with some disturbance in tiny glands of the skin called the sebaceous gland. They excrete oil or sebum. When there is too little oil, the skin becomes flaky and dry. When there is too much oil, the dandruff is possible. Followings are some tips to prevent dandruff:

- **Shampoo often:** The foremost consideration in the treatment of this disorder is to keep the hair and scalp always clean to minimize the accumulation of dead cells. If you have an oily scalp, daily shampooing to remove the excess oil may help in preventing dandruff.
- **Cut back on styling products:** Hair sprays, styling gels, mousses and hair waxes can all build up on hair and scalp, making them oilier, while hair dryers, artificial hair colouring agents, *etc.*, too pose a potential threat by drying the scalp. Natural alternatives like henna may be of help for hair colouring.
- **Eat a healthy and well-balanced diet:** A diet that provides enough zinc, B-complex vitamins and essential fatty acids may help in preventing dandruff. Adopt a well-balanced diet, with emphasis on raw foodstuffs, especially fresh fruits and vegetables. Short periods of all fruits diet for three days or so may be helpful at monthly intervals until the skin condition improves. Some may find that reducing the intake of food articles containing sugar and yeast can help in controlling dandruff.
- **Get a little sun:** Sunlight may be good for dandruff. Do not expose the skin too much to sun because its rays could harm the skin and the hair texture as well. The exposure to ultraviolet light for a long period can increase the risk of skin cancer, hence, do not take sunbathe. Instead, just spend a little time outdoors, may be in the morning or evening, a couple of days a week, about 5-10 minutes.

Bad breath

Poor oral hygiene and infection of gum can result to bad odour emanating from the mouth. This is called *Halitosis*. Proper brushing of teeth and oral care can get rid of bad breath. Below are the some important causes of bad breadth:

- Food trapped in cavities after eating.
- Consumption of food like onion, garlic, tobacco and beer.
- Plague deposits.
- Gum diseases.

Ways to detect bad breath

- Breathe into a handkerchief.
- Ask your parents or family who are close to you and can tell you the truth.

Means to Maintain Fresh Breath All the Day

- Maintain a good oral hygiene.
- Brush your teeth at least two times a day. It is important to brush after eating and before sleeping.
- Prefer to use toothpaste with green tea extracts as its properties help to prevent bacterial enzyme from turning sugar into a sticky substance that allows plague to stick to teeth.
- Floss to remove food that is stuck between teeth, as it will cause bad breath otherwise.
- Clean your tongue with tongue cleanser everyday. Generally, the bacteria accumulate on the tongue and contribute to the foul smell.
- Gargle or rinse with water when you cannot brush or use natural mouthwash but avoid commercial mouthwashes.
- Avoid consuming food articles containing strong odour like onion, garlic, tobacco and beer.
- Drink plenty of water and chew sugarless gum to stimulate saliva production from dry mouth.
- Eat crisp food items and vegetable salads.
- Find the exact cause of bad breath and cure it accordingly.
- Visit dentist regularly.

Some Common Dental Problems Related to Poor Oral Hygiene

Caries (tooth decay)

The caries develop when holes form in parts of the enamel of a tooth. Mainly, the caries develop due to a build-up of plaque. Dental plaque is a soft whitish deposit that forms on the surface of teeth. It forms when bacteria (germs) combine with food and saliva. In plaque, the bacteria react with sugars and starches in food to form acids, which are kept next to the teeth by the sticky plaque and they dissolve the tooth enamel. If there is tooth decay, it needs fillings, crowns, or inlays.

Gum disease (periodontal disease)

Gum disease means infection or inflammation of the tissues that surround the teeth. Most of gum diseases are related plaque, which contains many different types of bacteria, and a build-up of some types of bacteria is associated with developing gum disease. Gum disease is the most common cause of tooth loss in adults. It is also a main cause of bad breath (halitosis).

Tooth (dental) Erosion

Tooth erosion is a common problem. It is the gradual erosion of tooth enamel by the action of acid on the teeth. Dentine is sensitive so erosion can lead to your teeth being more sensitive to hot, cold or sweet foodstuffs and drinks.

Note: It is advised to have a regular dental check up at intervals (at least once in a year). A dentist can detect any tooth related problem such as plaque, calculus, gingivitis at an early stage, thus, preventing severe damage of the teeth.

Earwax

Ceriman or earwax accumulates in the ear canal that leads from outer ear to the eardrum. As the secretion comes out of the ear, it collects dust particles which might have got in from outside. Putting in hairpins, safety pins or blunt edged things for cleaning purposes might harm the ear.

Excessive hygiene of the ear can result in infection or irritation. The ear canals need less hygiene as compared to other parts of the body. Most of the times, the ear canals are self-cleaning. Old earwax is constantly transported from deeper areas of the ear canal out to the opening where it generally dries up, flakes and falls out. Repeated attempts to clean the ear canals through removal of earwax in fact reduce the cleanliness by pushing debris and foreign material into the ear. If you feel that wax has accumulated and is plugging your ears and interfering with hearing, consult your doctor.

The ways to prevent earwax are as follows:

- Keep the body clean.
- Soften the earwax so that it may not become impact by putting bio-oil into the ear.
- Use the earwax removal kit to remove the wax.
- Ear bud can be used to remove any remaining wax from the ear.
- Clean the ear regularly. This involves the wetting of one side of the cloth with warm water.

Body Odour

The body has nearly two million sweat glands. These glands produce three quarts to one pint of sweat in a day. In tropical countries, naturally, more sweat is produced. The perspiration level increases with an increase in physical exertion or nervous tension. The body odour is usually caused by sweat, which is produced under the armpits and the groins. The bacteria that feed upon it cause the stinking odour. One of the major reasons for having body odour is the lack of attention to body hygiene. Excessive perspiration can lead to the scaling of skin or inflammation. Usually, this is no cause of worry. Some people sweat more and some less due to hereditary and body composition.. Body odour makes you feel unsafe and insecure in the company of others. Person's social, personal and professional life will suffer, if actions are not taken to eliminate it.

- Perspiration, or sweat, comes from sweat glands that you have always had in your body, but thanks to puberty, these glands not only become more active than before, they also begin to secrete different chemicals into the sweat that has a stronger smelling odor. You might notice this odor under your arms in your armpits. Your feet and genitals might also have new smells.
- The best way to keep clean is to bathe or shower every day using a mild soap and warm water. This will help wash away any bacteria that contribute to the smells. Wearing clean clothes, socks, and underwear each day can also help you to feel clean. If you sweat a lot, you might find that shirts, T-shirts, socks, and underwear made from cotton or other natural materials will help absorb sweat more effectively.
- If you are concerned about the way your underarms smell, you can try using a deodorant or deodorant with antiperspirant. Deodorants get rid of the odor of sweat by covering it up, and antiperspirants actually stop or dry up perspiration. They come in sticks, roll-ons, gels, sprays, and creams and are available at any drugstore or supermarket. All brands are similar (and ones that say they're made for a man or for a woman are similar, too, except for some perfumes that are added).

- If you choose to use deodorant or antiperspirant, be sure to read the directions. Some work better if you use them at night, whereas, others recommend that you put them on in the morning, but keep in mind that some teens don't need deodorants or antiperspirants. So why use them if you do not have to? Deodorant and antiperspirant commercials may try to convince you that you'll have no friends or dates if you don't use their product, but if you don't think you smell and you take daily baths or showers and wear clean clothes, you may be fine without them.

Ways to prevent body odour

- Bathe daily.
- Clean underarm and groin area with water and bacterial soap.
- Use deodorant or antiperspirant.
- Wash the clothes after using once.
- Avoid the consumption of pungent food items that are carried in your sweat from your diet, like onion, garlic and beer.

Athlete's foot

This is very common problem due to improper hygiene. The skin of the feet becomes scaly. Sometimes, sores or blisters are developed between toes, and often it spreads to the soles. This infection is caused by a fungus. This is a minor irritation and often disappears by itself, but sometimes, these cracks and sores become the site for other infections. Proper foot care can alleviate the condition responsible for the development of athlete's foot.

- Rub off peelings gently.
- Wash feet well and apply powder.
- A mild fungicidal ointment at bedtime will help you to heal fast.
- Keep the feet exposed. If you have to wear shoes, wear cotton socks.
- Sometimes, the blisters begin to ooze. Soak your feet in a warm bath for 10 minutes and then apply calamine lotion.

■ ■ ■

40 Yoga, Meditation and Spirituality

Purvi Dalal

43, Mistry Park Annexe, 77 Bhulabhai Desai Road, Mumbai-400 036, Maharashtra, India,
E-mail: <dalal_purvi21@yahoo.co.in>

The word Yoga has been derived from the Sanskrit word *yuj* meaning to bind, join, attach, yoke, or to direct and concentrate one's attention on. It also means union, thus, it truly means *to be one with or in union with the Divine.* The yoking of all powers of the body, mind and soul to God by making the intellect, the mind, the emotions and the will power disciplined. The poise of the soul enables one to look at life with detachment and an even equipoise.

With the understanding that there is something beyond this mind and body, the ancient yogis established means by which the soul (*Jivatma*) could be united with the God (*Paramatma*), and thereby, they attained salvation (*Moksha*), thus, Yoga became a path by which one could be liberated from all suffering and pain. To attain this, an 8 folds path (*Ashtanga Yoga*) consisting of certain codes of conduct was established by the renowned Sage of Patanjali Yogpeeth.

Ashtanga Yoga

These eight limbs consist of certain codes of conduct and practices, which when diligently followed, could release a person from bondage in the ego-consciousness to liberation or self-realization. They are as follows:

***Yamas*:** These are the universal moral commandments, which when practiced diligently raise the practitioner's consciousness from pain and ignorance to peace and happiness. The five *yamas* have been described as under:

- ***Ahinsa*:** Non-violence or non-killing is to be practiced in a broad and spiritual sense of oneness and loving kindness in thought, word, deed and action. Living with the feeling of oneness towards all, the practitioner is freed of hostility, aggression and violence and is steeped in peace and love.
- ***Satya*:** It is truth in one's conduct, speech and thought. Untruthfulness in any form brings about disharmony in the practitioner. Thus purified by the practice of *satya*, the mind of the person becomes like a mirror, which reflects the divine mind. He becomes a mirror of truth.
- ***Asteya*:** Non-stealing or non-covetousness reduces the yogi's craving for the material comforts like excess food, wealth, power and fame, and endows tranquility on the practitioner and enables one to ward off all temptations and lead a simple life.
- ***Brahmacharya*:** According to Sankaracharya- a *brahmachari*, is one who constantly moves in Brahman and knows that all exists in Brahman. He is a man engrossed in the study of sacred texts. In other words, one who sees Divinity in every creature of God is a *brahmachari*. By not

dissipating energies and consciously laying stress on continence of the body, speech and mind and by remembering the Lord's name constantly, one develops vitality, strength and vigour in the body.

- ***Aparigraha*:** Non-hoarding or non-collection of things that one does not require immediately makes life simple and trains the mind not to feel the lack or want of anything in one's life.

Niyamas: These are the individual codes of conduct to be adopted by the practitioner. The five niyamas are as under:

- ***Saucha*:** It is the purity and well-being of the body and mind.
- ***Santosha*:** It is total contentment, *i.e.*, enjoying or feeling of complete satisfaction with all that one has.
- ***Tapas*:** Tapa means to burn or consume by heat, therefore, *Tapa* refers to the burning effort to achieve a particular goal through purification, self-discipline and penance.
- ***Svadhyaya*:** It is self-study. With the study of self, one eliminates ignorance and acquires true knowledge.
- ***Isvara Pranidhana*:** It is the complete surrender of one's thoughts, actions and will to the Divine power.

Asana: Asana or posture endows the practitioner with good health, mental equilibrium and a balanced personality. By reducing fatigue, soothing the nerves and disciplining the mind, a yogi can conquer the short-comings of the body and make it a fit vehicle for the spirit.

Pranayama: It is the science of breath control and rhythmic deep breathing. Pranayama strengthens the respiratory system, soothes the nervous system and reduces cravings, thereby, setting the mind free to be a fit vehicle for concentration.

Pratyahara: It is the withdrawing of mind from external objects. With rhythmic control of the breath, the senses instead of dwelling on external objects are turned inwards. This helps to eliminate desires and fears, making the mind pure and tranquil.

Dharana: It is the practice of concentration. Here the practitioner concentrates wholly on a single point or task in order to achieve the state of complete absorption.

Dhyana (meditation): When the flow of concentration is uninterrupted, it leads to dhyana (meditation).

Samadhi: It is the way of self-liberation and the way to achieve salvation (*moksha*). At the peak of meditation, the practitioner looses awareness of the body and mind and just becomes aware of the One All Pervading Consciousness or Light. All dualities between the knower and the known cease, and one achieves Union with the Divine.

Thus, Yoga, with all its therapeutic benefits also enables one to improve quality of life and help one to evolve as a better human being.

Physical and Therapeutic Benefits of Yoga

Acknowledging the necessity to have a healthy body, the yogis formulated a practice that involved doing a set of *asanas* and *pranayama* everyday. These *asanas* are to be practiced in a particular sequence and for a prescribed period to enable the practitioner to get the full therapeutic benefits. It is desirable to begin the practice daily at a scheduled hour and at a designated place. The last meal should have been consumed at least 3 hours before the practice is commenced. With this regularity of routine, the mind and the body are prepared for the practice and fall into a rhythm quite effortlessly. The most recommended time for the practice is the *Brahma Muhurat*, *i.e.*, the early hours of dawn between 4 and 6 a.m.

At the chosen place and at the scheduled hour, one sits on the floor or a mat with legs loosely crossed, eyes closed and simply observing the breath. After chanting the word *OM* three times and seeking the guru's blessings, one should start with the practice. While sitting with legs loosely crossed, back of the hands resting on the knees, spine erect and chin dropped to the chest, the practitioner becomes aware of the breath. With each inhalation and exhalation, one feels certain vibrations or little impulses or currents running from brain to the neck, shoulders and arms to the centre of palms and fingertips. The currents flow upwards with the inhalation and downwards with the exhalation. It is as if the palms are laden with energy and when one raises the palms to face each other, there seems to be a force, which acts like a magnetic pull between the palms. It feels like a cushion or sponge between the palms, which does not allow them to meet, similar to like poles repelling. As the palms are drawn apart from each other, the force of energy seems to diminish. This energy or *Prana Shakti* is then felt flowing through the entire body and one experiences a great sense of joy and peace while sitting thus. It is fitting that this pose is called *Sukhasana*. It is not by rote but by actually experiencing this joy and happiness (*Sukham*) that one realizes the difference between a posture and an *asana*. *Sthiram Sukham Asanam* or in other words, *any posture taken up by the body for a long period keeping the mind completely calm, without any reaction and experiencing joy in the pose is considered as asana.*

A sequence of certain *asanas* can be formulated and arranged in such a way, that by simply practicing these *asanas*, the practitioner could be cured of a number of ailments like back problems, weak digestion, cervical spondylitis, hormonal imbalance, arthritic tendencies in the joints, hypertension, lapses in memory and a host of other ills. Though some of these *asanas* are expounded upon here, it would still be advisable, if they were first learnt under the guidance of a yoga teacher one trusts.

Asanas

The foremost among the list of *asanas* is the practice of traditional Sun Salutations (*Surya Namaskar*). The *Surya Namaskars* comprise of a series of 12 dynamic *asanas* performed in a single graceful flow. Each movement is coordinated with the breath and its respective mantra. Often used as a warm up, the *Surya Namaskar* builds strength, increases flexibility and promotes healthy circulation. Each step of the *Surya Namaskar* is illustrated below with its respective mantra.

- **Namaskarasana:** Stand erect with feet together, hands folded in a prayer position (*namaste*) in front of the chest, at your heart centre and *exhale* in this position. The mantra to be recited is *Om Mitraya Namaha*, meaning salutation to the friend of all. Allow the feeling of oneness, love and compassion to flow within while chanting this mantra.
- **Ardhachandrasana:** Inhaling and keeping the legs straight, stretch up the arms and arch back from the waist. The mantra to be recited is *Om Ravaye Namaha*, meaning salutations to the shining one who spreads brightness and joy through the consciousness.
- **Uttanasana:** Exhaling, bend forward, placing the hands by the side of feet, and keeping the knees straight, try to touch the forehead to the knees. The mantra to be recited is *Om Suryaya Namaha*, meaning salutations to the one who induces activity. In addition, I salute the Supreme Being who dispels the darkness within.

- **Vanarasana:** Inhaling, take the right leg back, placing the knee on floor and toes pointing back and look up by raising the chin. The mantra to be recited is *Om Bhanve Namaha*, meaning salutations to the one who illumines the consciousness, while expressing gratitude for the visible world.
- **Phalakasana:** Exhaling, take the left leg back to the plank position. The mantra to be recited is *Om Khagaya Namaha*, meaning salutations to one who flies through the sky, giving us a sense of direction.
- **Ashtanga Namaskara:** Inhaling, lower the knees, chest and forehead to the floor, and keeping the hips high, point the toes back. Retain this position through the exhalation. The mantra to be recited is *Om Pushne Namaha*, meaning salutations to the giver of strength, nourishment and peace.
- **Sarpasana:** Inhaling and keeping the feet together and toes pointing back lower the hips to the floor. Bending the arms at the elbows, raise the head and shoulders like a snake. The mantra to be recited is *Om Hiranyagarbhaya Namaha*, meaning salutations to the golden protective womb.
- **Adhomukhaswanasana:** Exhaling, curl your toes under and raise the hips to resemble a downward looking dog. Try to keep the heels to the floor, lowering the head between the arms and

keeping the arms straight, try to touch top of the head to the floor. The mantra to be recited is *Om Marichaye Namaha*, meaning salutations to the Lord of Dawn who with its single ray illumines the world.

- **Vanarasana:** Inhaling, bring the right leg forward letting the left leg stay as described in position 4. The mantra to be recited is *Om Adityaya Namaha*, meaning salutations to the first born of Aditi (the Divine Cosmic Mother).
- **Uttanasana:** Exhaling, bring the left foot forward as in position 3. The mantra to be recited is *Om Savitre Namaha*, meaning salutations to the Lord of Creation and Generator of life and energy.
- **Ardhachandrasana:** Inhaling, stretch your arms forward, and then, up and back over the head as in position 2. The mantra to be recited is *Om Arkaya Namaha*, meaning salutations to the Praiseworthy One.
- **Namaskarasana:** Exhaling, lower the arms back to a *namaste* as in position 1. The mantra to be recited is *Om Bhaskaraya Namaha*, meaning salutations to the one who leads us to enlightenment.

These sets of 12 *asanas* comprise one round or one complete *namaskar*. The ideal number to be practiced on a regular basis is twelve *namaskars*. Just the mindful practice of the *Surya Namaskar*, in the above manner, which includes *asana*, *pranayama* and *mantra*, becomes a complete spiritual practice in itself.

The *Surya Namaskars* along with the mantra should be performed with the feeling of gratitude and humility towards the sun. The entire body should be open and offered to the Sun, so that it can be bathed by the divine energy. Even while chanting the mantras, it is extremely beneficial to chant each mantra tuning into its energy and understanding the essence of each mantra for maximum benefit. A sustained practice of the namaskars creates a more balanced and dynamic life.

Surya Namaskars engender the practitioner with suppleness, strength and mental peace. The heart and lungs are energized due to the controlled breathing. It is an effective way of loosening, stretching, massaging and toning of muscles and joints, thereby, improving flexibility. Blood circulation improves to all parts of the body. The endocrine system is harmonized. *Surya Namaskars* stimulate internal organs of the body making them function more effectively. The practice of the *Surya Namaskar* increases the practitioner's powers of concentration. The entire physiology is invigorated and one gains in confidence. The rhythmic sequence along with mantras generates transforming alpha waves in the brain, which dissipate tension and stress in the body.

Other Asanas

Vrikshasana or the tree poses

Stand with the feet together, the heels and the big toes touching each other and the weight of the body distributed evenly on both the feet. Let the hands at first rest in a *Namaste* at centre of the chest. Bend the right leg at the knee and place the right heel at root of the inner left thigh close to the pubis. Rest the foot on left thigh toes pointing downwards. Balance on the left leg and raise the arms straight over the head in a *Namaste*. Stay for 6 to 12 breaths and lower the right leg and arms. Repeat the same, now raising the left leg and staying for the same length of time.

This pose tones the leg muscles and strengthens them, gives a sense of balance and poise and improves concentration.

Vajrasana or the thunderbolt pose: The *vajra* or the thunderbolt is said to be the weapon of Lord Indra.

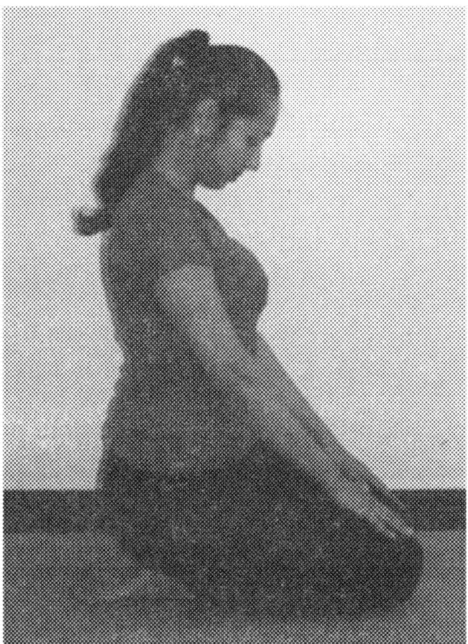

Kneel on the floor. Keeping the knees together and keeping the big toes touching each other, lower the buttocks on the heels. Place the hands on the knees and the palms down. Retain this position for about 12 breaths, keeping the spine erect and the chin dropped. With practice, the time can be extended to about 5 to 15 minutes, especially after meals.

If there is pain in the thighs, knees, or ankles initially, sit on a mat or place a folded blanket or small cushion between the buttocks and heels until one can comfortably sit without the assistance of any props.

Vajrasana alters the flow of blood and nervous impulses in the pelvic region, thereby strengthening the pelvic muscles. It increases the efficiency of entire digestive system, relieving stomach ailments such as hyperacidity, flatulence, indigestion, *etc.* It stimulates and activates the *prana* to rise in the spine, making it a very important asana for meditation.

Padmasana or the lotus pose: *Padma* means lotus. This is an important pose for meditation.

Sit on the floor with the legs stretched out in front of the body. Bend the right leg at the knee and place the right foot at root of the left thigh such that the right heel may be close to the navel. Now, bend the left leg and place it over the right foot such that the left heel may be close to the navel. The soles of the feet should be turned upwards. From base to the neck, the spine should remain erect. Allow the arms to be stretched out such that the wrists may rest on the knees. After holding this position for 6 to 12 breaths, change the leg position by placing the left foot on the right thigh first and then placing the right on the left. Hold again for 6 to 12 breaths, so that flexibility in both legs may develop evenly.

At the start of practice, there might be excruciating pain in the knees and ankles. However, with a regular practice, the pain will gradually subside.

Padmasana is very beneficial for increasing flexibility in the knees and ankles. The spine and abdominal organs are toned. The erect spine keeps the mind attentive and alert, and thus, it is often used during pranayama and meditation.

Sarvangasana:*Sarva* means entire and *anga* means body or limbs. Hence, *sarvangasana* means the entire body or all the limbs. Since the entire body benefits from this pose, it is rightly called *sarvangasana.*

Lie flat on the back keeping the legs stretched out on the floor. Bend the knees and move the feet towards the buttocks. Raise the hips from the floor and rest the hands on back by bending the arms at the elbows. Raise the trunk perpendicularly supported by the hands, legs straight, toes pointed upwards and the chest touching the chin. Only the back of head, neck, shoulders and backs of arms up to the elbows should rest on the floor. If one cannot do this asana without support, it can be done by propping the legs up on the wall and then raising hips off the floor, with the help of a stool or any other prop as recommended by your teacher. Stay in this position for 12 to 24 breaths before gently coming down. This asana can also be safely practiced for about 5 to 10 minutes.

This asana should be practiced only under the guidance of a professional teacher if one is suffering from hypertension, cervical spondylitis, slipped disc, or any heart problems.

Due to the chin lock, this asana stimulates the thyroid and parathyroid gland, thus, balancing the circulatory, digestive and reproductive, nervous and endocrine systems. Due to the enriched blood flow to the brain, it also tranquilizes the mind, relieves mental and emotional stress, fears and headaches and helps clear psychological disturbances. It boosts the immune system and ensures normal development and regeneration of bones, preventing premature calcification. It helps to clear phlegm from the respiratory passages and lungs.

Setubandhasana: *Setu* means bridge and *bandha* means closely held. In this pose, the whole body arches to resemble a bridge, and the feet are closely held with the hands, hence, the pose is called *setubandhasana.*

Lie down on the back, keeping the head flat on the floor, and bend the knees, keeping the feet shoulder width apart. The feet should be parallel to each other. Hold the ankles with your hands, raising the hips and arching the back such that the chest rises and comes close to the chin. Let the belly be completely relaxed. Hold this position for 12 to 24 breaths.

This asana strengthens and tones the dorsal, lumbar and sacral regions of the spine. It helps in relaxing the brain and the nervous system.

Pavanmuktasana: *Pavan* means wind (*prana*) and *mukta* means release, thus, *pavanmuktasana* refers to a group of *asanas*, which remove blockages preventing the free flow of energy in the body and mind. However, here, we will only explore the basic pose and one dynamic variation of the same.

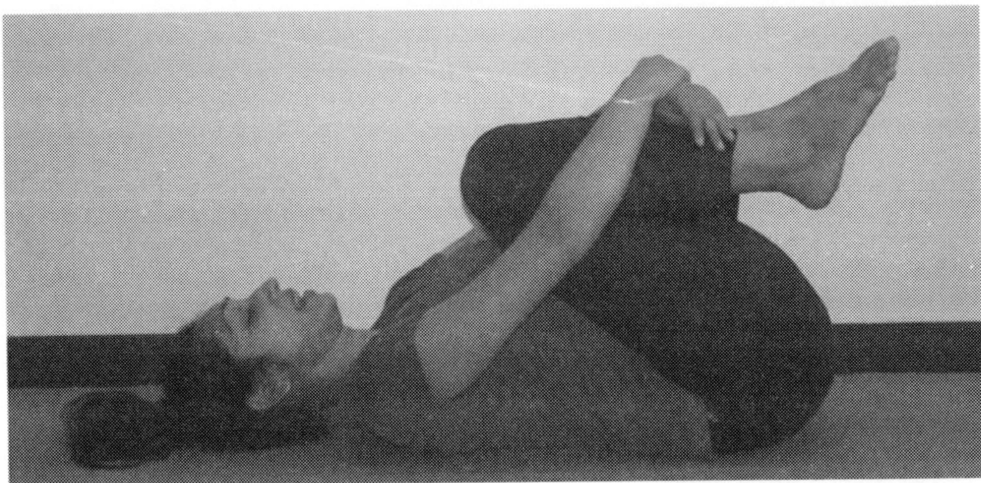

Lie down on the floor and bending the knees, bring them close to the chest. Keeping the tailbone and the neck on the floor, wrap the arms around the knees. This forms the base position and it should be practiced for 6 to 12 breaths. This time can be increased, as one gets more relaxed and comfortable in the pose.

Dyanamic Variation

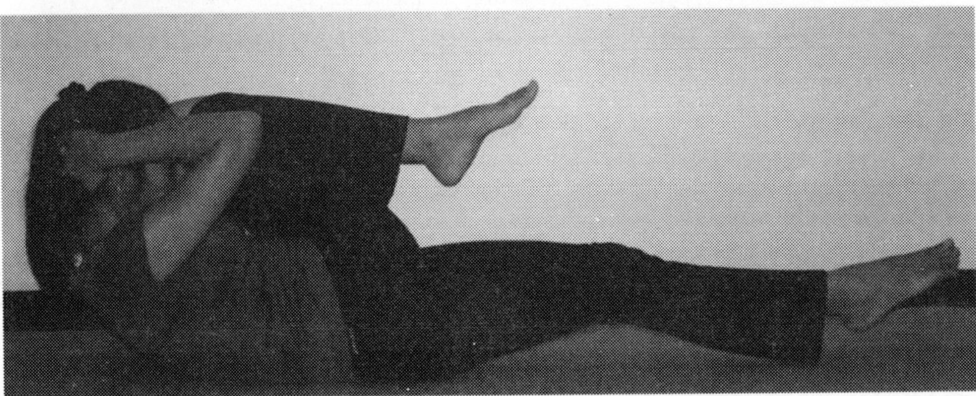

Lie down on the floor and bending the knees bring them close to the chest and try to keep the tailbone on the floor. Interlock the hands behind the head and raise the head, neck and shoulders off the floor, try to touch the knees to the elbows. Exhale and stretch out the right leg keeping the left knee close to the left elbow. Inhale, stay and hold this position. Exhale and change sides. Stretch out the left leg bringing the right knee close to the right elbow. Repeat this action 12 to 24 times on each side.

It strengthens the lower back muscles and loosens the spinal vertebrae. It massages the abdomen and digestive organs and is therefore effective in removing wind and constipation.

Savasana or corpse pose

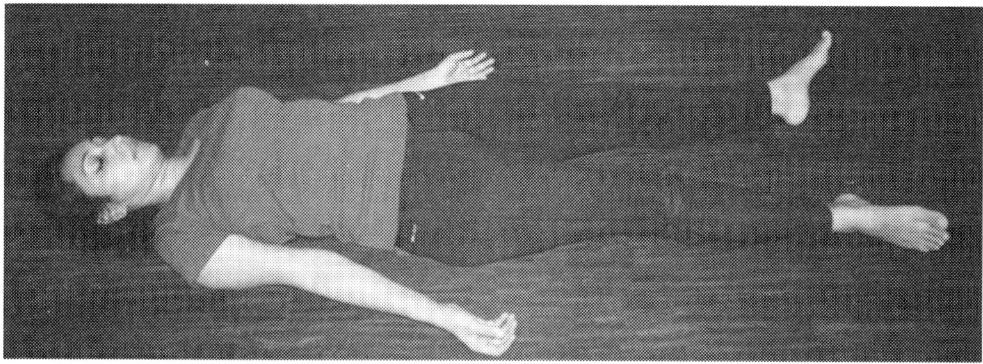

Lie flat on the back with the arms dropped outwards (about 15 to 20 cm) away from the body, palms facing upwards. Allow the feet to drop outwards keeping the head and spine in a straight line. Close the eyes, allowing the eyeballs to look deep within the heart and relax the whole body. Let the tongue drop softly to the lower palette and relax the jaws completely. Become aware of the natural breath and focus on the exhalation. Keeping it long and smooth, breathe out slowly. On inhalation, feel the breath rising from toes to the head. On exhalation, feel the breath flowing down from head to the feet. Stay in this pose for 12-24 breaths or longer if deeper relaxation is desired.

This asana relaxes the complete psycho-physiological system, and thereby, reduces fatigue and induces calmness of mind. For stresses of modern day civilization and the resultant strain on nerves, *savasana* with its fine, smooth and steady breathing without any jerky movements, soothes the nerves, calms the mind and it is therefore the best antidote for stress.

A regular practice of the *asanas* will strengthen the immune system, digestive system and nervous system, making the body healthy and strong. It will relieve the practitioner from lower back problems, arthritic joints, will reduce fatigue and boost the energy levels. However, it is always best and most advisable to learn Yoga first from a qualified Yoga teacher, who would be able to guide you according to the needs of your body.

Mudras

As in the universe, our bodies are also made up of five basic elements. The five elements are Fire (*Agni*), Air (*Vayu*), Space (*Aakash*), Earth (*Prithvi*) and Water (*Jal*). These five elements are represented in our body through our five fingers.

Name of the fingers		**Elements they represent**
Thumb	:	Fire [*Agni*- Sun]
Index	:	Air [*Vayu*- Wind]
Middle	:	Space [*Aakash*- Ether]
Ring	:	Earth [*Prithvi*]
Little	:	Water [*Jal*]

Any imbalance created by an increase or decrease in any of the elements can thus be brought to balance by joining tips of the fingers to the thumb or placing them as illustrated below:

Gyan Mudra: Touch tip of the thumb to tip of the index finger, with the other three fingers stretched out.

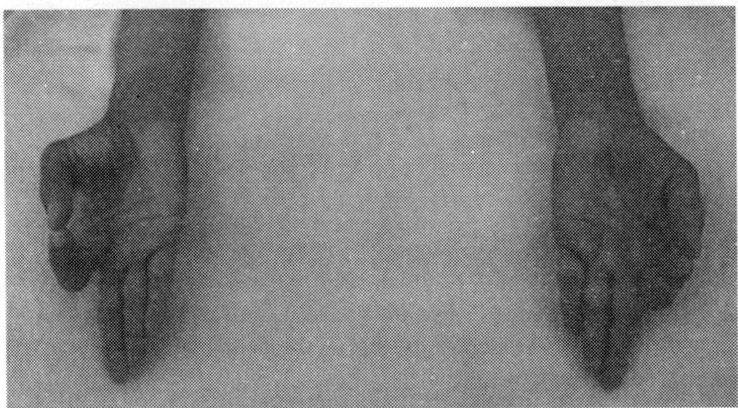

This *mudra* enhances the knowledge and activates the pituitary and pineal glands. It increases memory power and concentration. This *mudra* can be practiced at any time of the day and for any time duration.

Vayu Mudra: Touch tip of the index finger to base of the thumb and press down the index finger with the thumb, keeping the other three fingers straight.

It prevents all diseases that occur due to the imbalance of air, the feeling of restlessness or gas. Other *vayu* diseases such as Parkinsons disease, sciatica, rheumatism, arthritis, cervical spondylitis, *etc.* can also be healed with this mudra.

The practice of this mudra for 45 minutes reduces severity of the disease. It can be done 2 to 3 times a day for 15 minutes but should be discontinued after the symptoms subside.

Prana Mudra: Bend the ring finger and the little finger to touch tip of the thumb, keeping the other two fingers straight. This *mudra* helps in the progress of vital energy and improves immunity. It also helps in cleansing the blood and removing impurities present in the blood vessels. It improves eyesight and reduces any kind of vitamin deficiency.

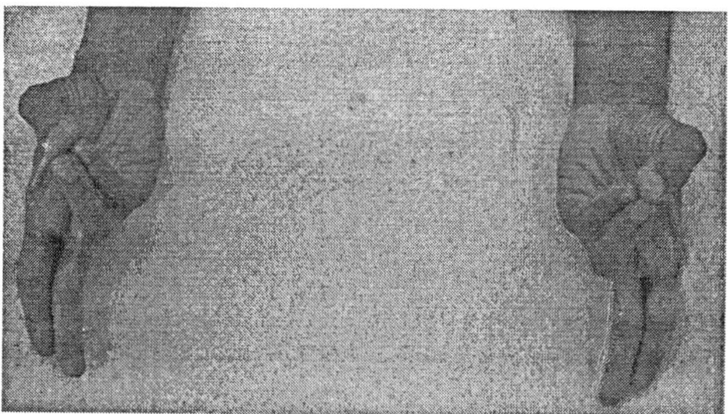

There is no specific duration to practice this mudra and it can be practiced at any time of the day.

Apaanvayu Mudra: Touch tips of the middle and ring finger to tip of the thumb, while the index finger touches the base of the thumb keeping the little finger stretched out.

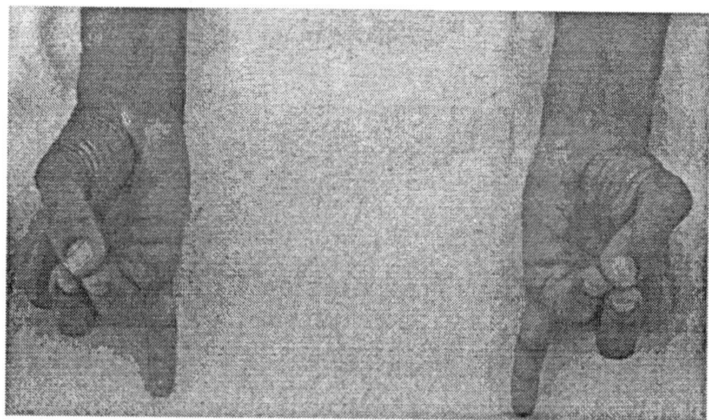

This *mudra* benefits and strengthens the heart, regulates the excretory system and reduces gastric trouble. For best results, this mudra should be practiced for 15 minutes twice daily.

Pranayama

While practicing any asana, the practitioner should be extremely aware and conscious of the breath. With this constant awareness of breath, one realizes that after one inhales, there is a tiny pause before one exhales completely and there is a tiny pause at the end of exhalation. Thus, one becomes aware that the breath is not just inhalation and exhalation but comprises of four parts as given below:

- **Puraka:** Inhalation
- **Antar Kumbhaka:** Pause at the end of the inhalation
- **Rechaka:** Exhalation
- **Bahya Kumbhaka:** Pause at the end of the exhalation

Simply by becoming aware of these tiny pauses, one finds that one does not just mechanically inhale and exhale any more. By just relaxing and allowing the breath to pause effortlessly after the inhalation, and then, pause effortlessly after the exhalation, one finds that the breaths automatically become longer. Thus, it is through experience that one understands the meaning of *Pranayama*.

According to the *Upanisads*, *Prana* is the principle of all life and consciousness and is equated to the vital life force, which sustains all life on the universe. Realizing the importance of this life force, the yogis utilized and tapped this energy to provide specific benefits. Thus, *Prana* is also used as breath, respiration, life, vitality, energy, or strength. *Ayama* means stretch, expand, regulate, restrain, or control, thus, *Pranayama means, the prolongation, stretching and restraint of the breath.*

Pranayama can also be therefore termed as the controlled intake and outflow of the breath and its retention. With the practice of this kind of breathing, *i.e.*, breathing with awareness, one experiences a rising, expanding and stimulating energy with every inhalation (*puraka*), a calm peaceful energy with every exhalation (*rechaka*) and an even distribution of energy (*pranashakti*) with every pause (*kumbhaka*). Thus, it is found that while *asanas* remove the obstructions, which impede the flow of *Prana*, Pranayama regulates the flow of *Prana* throughout the body. Simply by even becoming aware of the pauses and effortlessly allowing these to become longer, one experiences a great sense of peace and quiet.

With this kind of abundant intake of oxygen and the controlled and disciplined method of breathing, chemical changes take place in the practitioner's body. The practice of Pranayama helps to cleanse the *nadis or the tubular channels through which energy flows.* It helps the mind to concentrate, reduces the number of thoughts, regulates the quality of thoughts and endows the practitioner with equanimity, poise and a peaceful will power and confidence.

All Pranayama practices should be undertaken after having completed the asana practice. Relax the body completely by lying down in *Savasana* for about 5 to 7 minutes, and then, sit up in a comfortable position of your choice like *sukhasana, padmasana,* or *vajrasana*. Keeping the mind completely relaxed, one should start observing one's breath. Many types of Pranayama have been devised to meet the physical, mental and spiritual requirements of the practitioner under fluctuating conditions and as per each individual's requirement. As the breaths become longer, one should undertake Pranayama as required or suitable to one's needs.

Important Pranayamas

Ujjayi Pranayama: The prefix *uj* means upwards or expanding and *jaya* means conquest or triumph. *Ujjayi* is the process in which the lungs are fully expanded and the chest is puffed out completely.

Sit in a comfortable position (choose any one from those mentioned above) keeping the spine rising and erect, the head lowered to the trunk and the chin dropped between the collarbones. Stretch out the arms allowing the back of wrists to rest on the knees, palms facing the sky. Close the eyes and exhale completely. Take a deep breath, feeling the breath rising up from roof of the palette making a *sa* sound. This sound should be audible. However, do not force the sound. At first, this may be a little difficult but with practice, it becomes easier. Exhale slowly and steadily, relaxing the diaphragm completely. Again, feel the air on the roof of palette and the brushing of air on the palette should make an aspirate sound of *hum*. Repeat this cycle of breathing for about 12 to 24 breaths to begin with. This is the only Pranayama, which can be done at any time of the day and night.

This Pranayama aerates the lungs, removes phlegm, gives endurance and increases strength. It soothes the nerves and calms the brain thereby toning the entire system.

Nadi Sodhana Pranayama: It is a process of alternate nostril breathing. *Nadis* are tubular channels for the passage of energy (*prana*). *Sodhana* means purifying or cleansing, thus, *nadi sodhana* is the purification of nerves and channels throughout the body.

Sit in a comfortable position, keeping the spine rising and erect, head dropped and chin resting on the collarbone. Stretch the left arm and allow it to rest on the left knee with palm facing the sky. Bend the right arm at the elbow and place the index and middle finger on top of the palm. Place the right thumb on right side of the nose just below the nasal bone, and then, place the ring and little fingers on left side of the nose also just below the nasal bone. Press the ring and little finger to block left side of the nose completely. Keeping the left nostril closed, inhale through the right nostril. After a complete inhalation, close the right nostril completely, release pressure from the left nostril and exhale completely through the left nostril. After a complete exhalation, retain opening of the left nostril and inhale through the left nostril. Close the left nostril and exhale out through the right nostril. This completes one cycle of *nadi sodhana*. Thus, one cycle of *nadi sodhana* comprises of an inhalation through the right nostril, exhalation through the left nostril, inhalation through the left nostril and exhalation through right nostril. One can practice from about 6 to 24 cycles as required.

Due to the penetration of *prana*, the blood receives more oxygen. The nerves are calmed and purified. The mind becomes still clear and tranquil and brings serenity and balance in the consciousness.

Bhamari Pranayama: *Bhamara* means a large black bumblebee. This *Pranayama* is so called because during exhalation, a soft humming sound like that of a bumblebee is made. This *Pranayama* is most effective if practiced with the *Sanmukhi Mudra*.

Sanmukhi Mudra: Raise the hands to the face and the elbows to the level of shoulders. Place the thumb tips in the ear holes to shut out all external sounds. Close the eyes allowing the eyeballs to drop down. Draw the upper lids down with the pads of middle fingers and cover the remaining space above it with the pads of forefinger to shut out light. Press the nostrils gently with the tips of ring fingers. Place little fingers on the upper lip. The practitioner now can listen only to the inner sounds and is aware of only the inner world.

Technique of *Bhamari pranayama*: Sit in a comfortable position and place the hands in the *Sanmukhi Mudra*. Inhale with the *ujjayi* sound and exhale with a humming or murmuring sound. Repeat the practice 12 to 24 times.

This Pranayama clears the brain of unwanted thoughts, as one just listens to the internal sounds and brings a sense of calm and peace. Openness is experienced in the heart. With a regular practice of *Pranayama*, one finds that one's breath becomes automatically longer. With this lengthening of breath, the entire nervous system becomes calmer, clarity is established in the consciousness, the cravings and desires for the worldly pleasures diminish and the mind becomes a fit instrument for concentration.

The regular practice of *asanas* and *pranayama* makes the practitioner aware of every nuance of the body. With the inner awareness increasing, one finds that any emotional turbulence due to fear or joy could affect the rate of breathing and with deliberate regulation of the breath and emotional excitement could be curtailed. Now, the practice advanced from its physical aspect to its subtler and greater truth, *Yoga is the union of mind with the Divine*.

Yoga from Physical to the Subtle

With the awareness of breath and a settling sense of equanimity in the consciousness, one experienced another truth *Yogas citta vrtti nirodhah*. *Citta* or *chitta* refers to mind, reason and ego, *vrtti* refers to oscillation and *nirodhah* refers to comfortable control. In other words, *Yoga is the complete and comfortable control over oscillations of the mind.* The mind (*citta*) with its ego sense is like a chariot.

When this chariot of mind is yoked to the horse of breath, one can control the mind by freeing it from external distractions and stilling its constant wayward movements.

Mind is an instrument that classifies, judges and coordinates impressions from the outside world in relation to its ego ('I') sense and the thoughts that arise within it. These thoughts are subtle, feeble and difficult to restrain. With this feeling of 'I', each time that one comes into contact of anything said or done that differs from the personal viewpoint of 'I', there is an almost instant reaction, which is sometimes followed by repentance. Thus, initially the consciousness is always Stimulus → Reaction → Repentance.

However, with the increased awareness of each movement and each thought, and its effects on the consciousness as also due to the greater sense of balance and equipoise, the reactivity in the mind is replaced by a calm and collected stillness, and the *pranasakti* can be felt pervading the entire consciousness. As the breath flows up and down through the spinal column, one becomes aware of certain energy centers or *chakras* situated in the spine. A blockage in any of these chakras has an adverse effect on the physical, physiological and psychological aspects of a person.

Chakras or Psychic Energy Centers

The chakras are the centers of activity of subtle and vital forces and are inter-related with the parasympathetic, sympathetic and autonomous nervous system. Each chakra is characterized as a lotus-each with a differing number of petals and each with its own seed sound. There are totally eight major chakras given as under:

Muladhara Chakra (root chakra): *Mula* means root and *adhara* means support. This chakra helps in grounding the consciousness, providing it with a sense of stability, security and balance. Blockages in this chakra could cause problems in bones and knees.

- **Position:** Base of the spine, close to the perineum
- **Element:** Earth
- **Colour:** Red
- **Bij Mantra:** Lam

Svadisthana Chakra (Sacral chakra): *Sva* means self and *sthana* means seat, abode. This chakra brings us fluidity, grace and the important ability to accept change. It is also connected with fear, anxiety and sexuality. Thus, blockages could result in uterine, bladder, or kidney problems.

- **Position:** Lower abdomen, between pubis and navel
- **Element:** Water
- **Colour:** Orange
- **Bij Mantra:** Vam

Manipura Chakra (Solar plexus chakra): *Mani* means jewel and *pura* means town. This chakra rules our personal power, will power and metabolism and brings us energy, confidence and non-dominating power. Thus, blockages result in anger, depression, skin eruptions and poor digestion.

- **Position:** Navel
- **Element:** Fire
- **Colour:** Yellow
- **Bij Mantra:** Ram

Surya Chakra: This chakra endows brightness and clarity of thought on the practitioner.

- **Position:** Five finger widths above the navel.
- **Element:** Light
- **Colour:** Yellowish green
- **Bij Mantra:** Kham

Anahata Chakra (heart chakra): *Anahata* means soundless. Centeredness, oneness, love and compassion flow freely from this center. Heart and lung problems and a weak immune system could result from a weak heart chakra.

- **Position:** Centre of the chest
- **Element:** Air
- **Colour:** Green
- **Bij Mantra:** Yam

Vishuddhi Chakra (throat chakra): *Vishuddhi* means purity. Being the centre of sound and creativity, it aids in communication, self-expression, rhythm and creativity. Difficulty in self-expression, sore throats and thyroid problems, and allergies reflect a weak throat chakra.

- **Position:** Throat
- **Element:** Ether
- **Colour:** Blue
- **Bij Mantra:** Ham

Ajna Chakra (third eye): *Ajna* means knowledge. Being the centre of wisdom and inner vision, it endows one with clarity of thought, wisdom and enhanced intuition. Headaches, confusion, mental stress, migraines and rigidity of thought are caused due to blockages in this chakra.

- **Position:** Between the eyebrows and centre of forehead
- **Element:** Light
- **Colour:** Indigo
- **Bij Mantra:** Om

Sahasrara Chakra (crown chakra): *Sahasrara* means thousand-petalled. Complete knowledge, all wisdom and cosmic consciousness connecting the body to the transcendent self and a supreme oneness reflects a healthy and open crown chakra.

- **Position:** Top of the head
- **Element:** Light
- **Colour:** Violet
- **Bij Mantra:** Ah (sound of visarga)

By simply chanting the *bij* mantras, one can experience a rising flow of energy and bring about balance by removing the blockages created by a weak chakra. These mantras can be chanted 7 to 21 times as required by the practitioner.

Having a deeper and clearer understanding of the body, a pure, calm mind, unhindered by any external stimulus, a presence is felt at ones' own heart centre. Whenever one connects with this presence, one experiences a deep sense of peace and calm. Whenever the connection is present and when one moves or acts with the awareness of this presence, the ego-consciousness diminishes. There are also occasions when one loses all senses of the body and is just conscious of this *All Pervading Consciousness* or *Light.* The small egocentric consciousness now widens to a God-centric consciousness. Thus, *Yoga*, as *Sri Aurobindo* rightly said, *is the exchange of egoistic for a Universal consciousness or with some deeper consciousness within which one is aware of one's own soul, one's own inner being and of the truth of existence.*

Yoga now becomes a process by which one can shed the present mode of being, our habitual unconsciousness and rise to our ever-present higher Self. Initially, it is expected that to practice yoga, one has to lead life of an ascetic and had to renounce the path of a householder. What then are the choices if people like us wanted to pursue the path of yoga and seek the truth?

Karma Yoga

By stating that *'All life is Yoga' Sri Aurobindo* in his *Integral Yoga* declared that to be free of all desires, attachments and egoism and still live in this world and carry on the work of the Divine is the finer, though often more challenging yoga. Offering all our actions to the Divine, offering the fruits of our actions to the Divine and even offering desire for the fruits of our actions to the Divine, Sri Aurobindo tells us to enjoy this world. In this desireless state, free from all bondage, one can perform all actions as a service to the Divine. By keeping a completely calm mind and equanimity in success or failure, one can overcome any obstacle created by desire, attachment and aversions. This is known as Karma Yoga. By not renouncing the world but living in this world and acting with the consciousness that the Divine is working through the spirit, one can be free of Karmic consequences.

Bhakti Yoga

Did detachment indicate that one also has to overcome the emotions of love and companionship? With a growing awareness of an all-pervading consciousness and the feeling of oneness (*maitri bhav*), all the hearts' impulses are channelized to a higher level, *i.e.*, love for the Divine. By a complete surrender to the Divine, of all that one is, all that one does, thinks and feels, it becomes easier for the devotee (*Bhakta*) to purify and transform his nature. By a constant recitation of the Lord's name (*Japa*), feeling His presence in the heart, a total consecration of all action to His Will and an ever-widening faith in the Divine, the devotee opens his consciousness to the Divine.

Thus, in Bhakti Yoga, all love is raised to its highest level, *i.e.*, complete faith and self-surrender to the Divine. Realizing God and feeling His presence in the innermost core of our beings, all vital desires may soon be transformed to realize and live the widest vision of ourselves. Everything that the *Bhaktas* experience is considered a gift from the Lord. They experience the Divine presence in all things and in everybody. By and by, their sense of self completely merges with the object of worship, ending all duality and experiencing just oneness and union.

Gyana Yoga

Through discrimination and a constant self-study (*Svadhyaya*) of all one's reactions, thoughts, emotions, feelings and through this self-study rejecting all falsehood and egoism, contemplating on the truth or the one Universal Divine being, one could be liberated from the bonds of egoism and ignorance and could acquire the knowledge of Self. This could be achieved by contemplation or concentrating the mind on a single idea.

Dhyana (Meditation)

Meditation (Dhyana) means the concentration of mind on a single thought, vision, idea, or an unbroken flow of thought toward the object of concentration for a prolonged period.

Meditation and Some of its Techniques

Concentration in every lifestyle (meditative or mindful living)

In its simplest and most practical form, meditation also means mindfulness or living with mindfulness. Having observed the fluctuations of mind, this seems the easier option to begin with.

Mindful living involves doing every chore with the mind focused on the task in hand. Say for example, while eating food, we find that rather than savouring the taste of food with each morsel, noticing and enjoying the different textures of food, the mind is distracted with conversations at the table or is contemplating some tasks to be done after the meal, then, this is not mindful eating or meditative eating. If the mind could be so trained as to remain focused on the task in hand, *i.e.*, in complete awareness of all the different aspects of the task, then it results in mindful living and helps one to *overcome one of the biggest obstacles to meditation, i.e., a wandering mind.*

If one practices mindfulness in every little task one undertakes, it helps the mind get accustomed to focusing single-mindedly on the task. Mindfulness prevents the dissipation of one's mental energies. This not only helps one to accomplish the task faster than usual but also increases the efficiency with which the task is done. One also experiences a sense of peace and calm in whatever one undertakes to accomplish.

Meditative Self Observation

This technique was practiced by *Swami Vivekananda* in his *Raja Yoga*. Taking up the stand of an observer, this method involves stepping back from one's thoughts by simply observing them as they occur in the mind. Often one thought leads to another with such rapid progression that the original thought or parent thought is forgotten.

Let as suppose that the original thought was an invitation to a party. Now, the mind starts contemplating on the dress to be worn, the accessories to be carried on, who would comprise the guest list, same past disagreement with one of the potential guests or any other related thought until one has forgotten what the parent thought was.

In this method, thoughts are just allowed to pass rather than reacting to them. This observation of thoughts reveals workings of the human mind and the alterations to be done in one's consciousness. Slowly one finds that the number of thoughts decrease until they come to a complete standstill and one develops a one-pointed mind.

Meditation by Visualization

Visualize the Deity of your choice with your eyes closed. Try to see this image in all its details. Let us say that one is meditating on the *form of Lord Krishna*. Visualize his crown, the peacock feather, his eyes, lips, ears, earrings, *etc.* until you travel all the way down to the yellow pitamber and the details of the toes. Visualization can also be only on the face or just eyes, or whatever it is that one chooses and connects with the object. The visualization should be made by trying to observe the minutest details of each part so keenly that after a while, all senses of self dissolve and there is just consciousness of the selected form.

Visualization can also be on an *Effulgent Light* at the heart centre, which has *no fixed form.* Then, one may concentrate all one's energies on the image, vision, or Light at the heart centre. With the offering of oneself to the Lord or Light within and with an aspiration for an inward opening and oneness with that presence, all thoughts may be stilled completely.

Meditation on One's Breath

When one focuses one's mind on one's breath, listening sound of the *ujjayi* breath it allows the mind to flow with the breath as it moves through the body. Dwelling in the silence during the pause (*kumbhaka*) between the breaths, helps to internalize and calms the mind. This enables ones' surface mind to quieten itself and throws light on the inner space. When you enter this silence and tune into the inner energies, one temporarily looses sense of the environment, all external sounds, and even, the passing of time. This is an extremely powerful meditation in itself.

Meditation on a Mantra

Just simply by reciting your chosen mantra silently in your mind constantly also helps in stilling the mind. Meditation on the most sacred mantra *OM* with the understanding of its omnipotence, omnipresence and omniscience shuts out all mundane thoughts. As the practitioner meditates on the mantra (*Nam*), bondage to the world, fear and injustice is overcome. The mantra has the power of freeing one from self-centeredness to liberation. The mantra can be recited while walking, driving, or performing household chores, and thus, it is a powerful meditative practice.

Some Experiences of Meditation

In initial stages of meditation, one finds it very difficult to sit for a prolonged period in the chosen meditative posture. Sometimes, one also experiences profuse sweating. However, with a prolonged practice, these initial problems settle and one experiences a one-pointed mind, and a balanced and peaceful consciousness.

Sometimes, one can smell sweet fragrance of the flowers or fragrance of the incense sticks arise in the consciousness. Occasionally, mystic sounds of conch shells or bells pealing, and sounds of lightening and thunder can also be heard. Lights of the most dazzling colours are seen flashing before the eyes or at the third eye centre. Images of Gods and Goddesses also appear during deep meditation.

In deep meditation, while concentrating at third eye centre or at heart centre, a lightness is experienced in the body. One feels that one has moved or separated beyond the physical body into the subtle or astral body. One experiences an ecstasy and an immense joy very difficult to explain. This is accompanied with an awareness of an All-pervading light or a presence, *i.e.*, a Light, a Force by whose Will and Grace the universe functions.

Being very deeply connected with the Deity chosen for meditation, one experiences their presence in ones' heart. One also becomes aware of an inner guidance. Whenever one listens to this inner voice and follows its dictates, there is perfection and harmony in every action. Many a times, one is filled with a light, a deep sense of calm and a widening of consciousness but what is most intriguing is the feeling *that when one is in complete surrender to this Light,* there is a distinct feeling of being carried, lifted, or moved by an external *Force.* It is as if the *Force* is raising ones' arms, speaking through one and performing every task undertaken. If it is the *Force,* who is the doer, then, *who am I*?

Spirituality

If this consciousness Force, Light, or Presence is the truth, then who am I? Why am I born? Why does one die? Where does one go after death? Faced with these questions, one suddenly feels inadequate, and feels that life no longer holds any real purpose.

What is here today may no longer be there tomorrow or might undergo a change in shape, size, place, or appearance, making one conscious of the transient nature of things. To all appearances, it seems that change is the only thing constant in life.

Looking at larger picture of the universe, what is it that holds the Sun in its place, ever spewing forth millions of tonnes of energy and yet appearing unmoved and unchanged? What is it that makes the planets revolve around the Sun with such accurate precision or what is the force that keeps the earth rotating on its axis day after day, year after year and so on?

To the questioning mind, it seems that beyond the logic and understanding of our diminutive human intelligence there is a Force that works with an exactness and precision way beyond the realms of man's rational thought and reason.

It is this need to know and understand a thirst to seek for something, which has a meaning or a purpose and an indication that beyond our mundane lives in this habitual unconsciousness, life serves a greater purpose. With this seeking, the awareness grows from ego-consciousness and the transient nature of things to God- consciousness and the larger truth of self.

The truth takes us beyond the realms of petty egoism, beyond the mind and body, to our inner selves, to our highest reality and to truth of the spirit. It carries us to a reality beyond ourselves and pervading the entire universe, which also inhabits our own being as that one all-pervading consciousness, *i.e.*, the *Spirit. Spirituality, in its essence, can thus be defined as an awakening to the inner reality of our being.*

Spirituality is an awareness of a spirit, soul, or self, which is greater than our mind, body and life. Conscious that there is something greater, something larger and more significant and beyond what till now seemed the only truth, there arises in us an inner aspiration to know, to feel to *BE* that vastness and to enter into contact with this greater reality. By a constant seeking through knowledge, by contemplating on this Force, which can be called God or Divinity; through devotion or by simply experiencing the truth of the Force performing its entire works through ones personality, one may arrive at a union with the Force. Shedding our smallness, liberated from the bonds of egoism and ignorance and opening up to the truth of our being, we attain self-realization or oneness with the Divine.

All life is a process of evolution from the protozoa to the animal, animal to the man and in man to evolve from our lower nature to our highest potential- from man to Super manhood. Hence, all life consciously or sub-consciously is Yoga. As *Swami Vivekananda* tells us, *Yoga may be regarded as a means of compressing one's evolution into a single life, a few years, or even a few months of bodily existence.*

It is through this constant seeking that the deeper meaning of existence reveals itself. *Dharma* or following one's religion, *Artha* or earning ones' livelihood, *Kama*, or leading a life of a householder and *Moksha* or self-realization is the purpose of all existence. As *Sri Aurobindo* says, *"a Will for growth-growth into the true Divine Individual with disappearance of the ego, a growth of the spirit in nature, affirming and developing itself in mind, life and body, a true spiritual essential change of consciousness-this is life's true objective."*

The path is long but *self-surrender* makes it easier. To strike a balance between family life and spirituality, is sometimes like walking on the razor's edge, but with the Guru's grace and Divine Blessings, self-realization is not meant for only an exalted few. It is the Destiny of all mankind and it is our collective Destiny.

Today, the world is heading for a change- A world based on oneness, love, light and peace. Will you be a part of that change? If you are one of those seeking a change, are looking for a deeper meaning to your existence, then come, let us walk this *Sunlit Path* together. Let us make our mundane existence into *a life divine*.

■■■

41 Anti-nutritional Factors and Human Health

Anita Bhatnagar

Department of Zoology, Kurukshetra University, Kurukshetra, Haryana, India
E-mail: <anitabhatnagar@gmail.com>

Every person and every other living organism needs certain substances called nutrients to build and maintain the body's structure and to supply the energy that powers all vital activities. Food is made up of different nutrients needed for growth and health, which include carbohydrates, proteins, fats, minerals, vitamins, fibers and water. All nutrients needed by the body are available through food and many kinds and combination of food items can lead to a well balanced diet. Each nutrient has specific uses in the body. It is the relationship of food to health of the body, determining nutritional status. Proper nutrition means that all the nutrients are supplied and utilized in adequate amount to maintain optimal health and well-being. At some level, all food articles have something, which for the variety of reasons act anti to the nutrients are called as *Anti-nutritional Factors* (ANFs). These anti-nutritional factors have great concern with human health. Nutritionists must be aware not only about the dietary requirements but also about the anti-nutritional factors associated with foodstuffs.

What actually the anti-nutritional factors are?

Anti-nutritional factors are the compounds, which reduce the nutrients utilization and/or food intake of plants or plant products used as human food, or animal feed and they play a vital role in determining the use of plants for humans and animals. There is a wide distribution of biologically active constituents throughout the plant kingdom, particularly in plants used as animal feeding stuff and in human nutrition. The knowledge that these compounds elicit both toxic and advantageous biological responses has given rise to several investigations in recent times as to their possible physiological implications in various biological systems.

Anti-nutritional factors (ANF), also referred to as biologically active substances, are essentially evolutionary adaptations by plants to enable some level of protection against being eaten. The variety of anti-nutritional factors found in different plant species varies quite widely, both in diversity and relative concentration.

Characteristics of Anti-nutritional Factors

- An anti-nutritional factor when present in human or animal foods reduces growth.
- They are natural or synthetic compounds that interfere with the absorption of nutrients.
- Anti-nutritional factors decrease digestibility.

- Decrease food intake.
- Cause low mineral availability.
- Anti-nutritional factors also act as hepatotoxin and decrease plasma levels.

How anti-nutritional factors are generated?

The anti-nutritional factors are generated in natural feedstuffs by the normal metabolism of species and by different mechanisms such as inactivation of some nutrients, diminution of digestive process and diminution of metabolic utilization of feed, which affect contrary to optimum nutrition. Being an anti-nutritional factor is not an intrinsic characteristic of a compound but depends upon the digestive process of ingesting animal, *e.g.*, trypsin inhibitors are anti-nutritional factors for monogastric animals, which do not exert adverse effects in ruminants since they are degraded in the rumen.

Classification of Anti-nutritional Factors

The anti-nutritional factors in plants may be classified based on their chemical structure, the specific actions they bring about, or their biosynthetic origin (Figure 1).

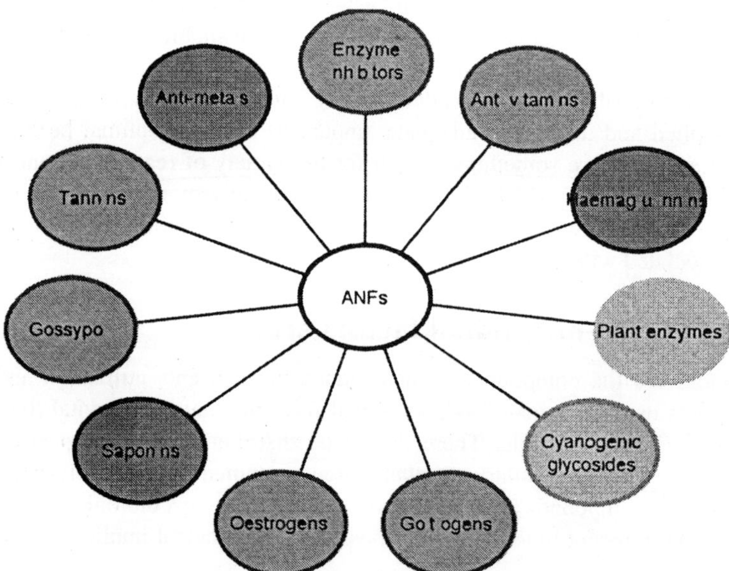

Figure 1: Classification of anti-nutritional factors

Based on temperature sensitivity, the anti-nutritional factors may be divided into two major categories. Though this classification does not encompass all the known groups of anti-nutritional factors, it does present the list of those found frequently in human foodstuffs and animal feedstuffs.

They are the proteins such as lectins and protease inhibitors, which are sensitive to normal processing temperature. The other substances, which are stable or resistant to temperature, include polyphenolic compounds (mainly condensed tannins), non-protein amino acids and galactomannangums. More often, a single plant may contain two or more toxic compounds, generally drawn from the two categories, which add to the difficulties of detoxification. According to Aletor (1993), the anti-nutritional factors that are very significant in plants and used for human foodstuffs and animal feedstuffs are given as under:

- **Enzyme inhibitors:** Trypsin and chymotrypsin inhibitors, plasmin inhibitors and elastase inhibitors
- **Haemaglutinnins:** Concanavalin A and ricin
- **Plant enzymes:** Urease and lipoxygenase
- **Cyanogenic glycosides:** Phaseolunatin, dhurrin, linamarin and lutaustralin
- **Goitrogens**: Pro-goitrins and glucosinolates
- **Oestrogens:** Flavones and genistein
- **Saponins:** Sapogenin and gossypol from *Gossypium* species (cotton)
- **Tannins:** Condensed and hydrolysable tannins
- **Amino acid analogues:** BOAA, DAP, mimosine and N-methyl-1-alanine
- **Alkaloids:** Solanine, chaconine, solasodine, fenugreekine, trigonelline, diosgenin, cucurbitacine, *etc.*
- **Anti-metals:** Phytates and oxalates
- **Anti-vitamins:** Anti-vitamins A, D, E and B_{12}
- **Favism factors:** In broad bean

Liner (1980) has classified endogenous anti-nutritional factors within plant feedstuffs under five main groups:

- **Protein:** Protease inhibitors, haemaggutinin, toxic amino acids and food allergens
- **Glycosides:** Goitrogens, cyanogens, saponins and estrogens
- **Phenols :** Gossypols and tannins
- **Miscellaneous:** Phytic acid, anti-vitamins, anti enzymes and toxic fatty acids

Biochemical Effects of the Anti-nutritional Factors

Anti-nutritional factors diminish animal productivity but may also cause toxicity during the periods of scarcity or confinement when the feed rich in these substances is consumed by animals in large quantity. Some of the specific effects are as under:

- **Cyanogenic glucoside:** Cyanogenic glycosides are present in a number of food plants and seeds. Hydrogen cyanide is released from the cyanogenic glycosides when fresh plant material is macerated as in chewing, which allows the enzymes and cyanogenic glycosides to come together, releasing hydrogen cyanide. In other words, on hydrolysis, it yields toxic hydrocyanic acid (HCN). The cyanide ions inhibit several enzyme systems and depress growth through interference with certain essential amino acids and utilization of associated nutrients. They also cause acute toxicity, neuropathy and death. At small doses, it causes headache, tightness in throat and chest and muscle weakness. The effects of chronic (long-term) exposure to cyanide are less well known. Food sources of cyanogenic glycosides include bitter almonds (250 mg/100 g), cassava root (53 mg/100 g), whole sorghum (250 mg/100 g) and lima bean (10-312 mg/100 g).
- **Alkaloids:** Organic compounds normally containing at least one nitrogen atom in a heterocyclic ring are alkaloids. Some alkaloids act as anti-nutritional factor and they cause gastrointestinal and

neurological disorders. The glycoalkaloids, solanine and chaconine present in potato and *Solanum* spp. are haemolytically active and toxic to fungi and humans. Potato glycoalkaloids also cause gastrointestinal upsets and neurological disorders, especially when they are consumed in excess.

- **Coumarins:** They are constituents of forage and have been associated with so-called bleeding disease in cattle consuming spoiled or putrid sweet clover. It is believed that cinnamic acid or its derivatives are the precursors of coumarin and that when plant tissues containing derivatives are disrupted by mastication, freezing, drying, or microbial spoilage, coumarin is transformed to haemorrhagic factor, *i.e.*, dicoumarol. Coumarin has appetite-suppressing properties and is moderately toxic to the liver and kidneys.
- **Dicoumarol:** Dicoumarol is a natural chemical substance of combined plant and fungal origin. It is a derivative of coumarin, a bitter substance made by plants that does not itself affect coagulation but which is transformed in mouldy feeds or silages by a number of species of fungi into active dicoumarol. It depresses blood prothrombin concentration, and by implication, it reduces clotting time. It is a competitive inhibitor of vitamin K in the formation of prothrombin. Administration of vitamin K is therefore the antidote for dicoumarol toxicity. The toxicity and the antidote effectiveness are measuring with the prothrombin time (PT) blood test.
- **Tannins:** Tannin is a type of an astringent and bitter plant polyphenolic compound that binds to and precipitates proteins and various other organic compounds including amino acids. They cause decreased feed consumption in animals and bind dietary protein and digestive enzymes to form complexes that are not readily digestible. They also cause decreased palatability and reduced growth rate. In sensitive individuals, a large intake of tannins may cause bowel, kidney and stomach irritation, liver damage and gastrointestinal pain. With the exception of tea, long-term and/or excessive use of herbs containing high concentrations of tannins is not recommended. If ingested in excessive quantities, tannins inhibit the absorption of minerals such as iron, which may, if prolonged, lead to anaemia. Tannins interfere with iron absorption through a complex formation with iron when it is in the gastrointestinal lumen, which decreases the bioavailability of iron. There is an important difference in the way in which the phenolic compounds interact with different hydroxylation patterns and the effect on iron absorption. The content of iron-binding groups may be the major determinant of the inhibitory effect of phenolic compounds. However, condensed tannins do not interfere with iron absorption.
- **Trypsin (protese) inhibitor:** Trypsin inhibitors are the group of peptides that reduce the availability of trypsin (a proteolytic enzyme formed in the intestine by the cleavage of trypsinogen and is essential to nutrition of many animals, including humans. They are present in varied sources as soybeans, egg white and human colostrums and inhibit the active site of trypsin molecule. They are also called kunitz inhibitor. They cause pancreatic enlargement and growth depression. These anti-nutritional factors are not completely deactivated during ordinary cooking, and thus, they can produce serious gastric distress, reduce protein digestion and chronic deficiencies in amino acid uptake. In test animals, diets high in trypsin inhibitors cause enlargement and pathological conditions of the pancreas, including cancer.
- **Protease inhibitors:** Protease inhibitors block the enzymes that digest protein (proteases). They can affect the stomach protease enzyme *pepsin* and the small intestine protease enzymes *trypsin* and *chymotrypsin*. When cooked, most of the enzyme blockers are destroyed but some are not. In animal experiments, high levels of protease inhibitors lead to increased secretion of digestive enzymes by the pancreas. This is because the body can sense that the enzymes have been knocked

out and orders to pancreas to make more. They inhibit the activity of trypsin and other proteases. They are found in legumes such as beans and peas but also in cereals, potatoes, and other products. Their presence results in impaired growth and poor food utilization. They are considered as effective poisons. Experiments have shown that rats cannot gain weight if they have substantial amounts of enzyme blockers like protease inhibitors in their diets.

- **Amylase inhibitors:** Some enzyme blockers affect the enzymes, which digest starch (amylase), and they are called amylase inhibitors. They have activity against amylases, which are important in breaking down the structure of carbohydrates. They hydrolyze sugar and starches, ultimately improving digestion of carbohydrates. These anti-nutrients can rapidly inactivate amylase in human intestinal lumen in a dose-dependent manner and reduce postprandial rises in glucose and insulin. The aqueous/ saline protein extract of wheat seed is called the albumin fraction. Within the albumin fraction, a very large number of protein components from insect, mammalian, avian and marine species are capable of inhibiting alpha-amylases. Alpha-amylase inhibitors make up as much as 80% of the total albumin fraction and may represent 1% of wheat flour. Because of their thermostability, alpha-amylase inhibitors persist through bread baking and are found in large amounts in bread, breakfast cereals, pasta and other wheat products. Alpha-amylase inhibitors are ubiquitous in the cereals of gramineae family plants, and in addition to their presence in wheat, they have been found in rye, barley, oats, rice and sorghum.
- **Haemagluttinins:** They are proteins known for agglutinating red blood cells. They depress animal growth by interfering with the digestion and absorption of nutrients in the gastrointestinal tract. Haemagluttinins are found in plants as Phytohaemagglutinin (PHA), which are lectins especially available in legumes. Actually, PHA consists of two closely related proteins, called Phytohaemagglutinin L, which agglutinates leukccytes (PHA-L) and PHA-E, which agglutinates erythrocytes. They are found in the highest concentration in uncooked red and white kidney beans and in lower quantities in many other types of green beans and other common beans. In high doses, it is a toxin. As a toxin, it can cause poisoning in human beings. Symptoms occur within three hours, beginning with nausea and then vomiting followed by diarrhea. Recovery occurs within four or five hours of onset, usually without the need for any medical intervention.
- **Saponins:** Saponins are glucosides with foaming characteristics. They are found in many plants but get their name from the soapwort plant (*Saponaria*), the root of which was used historically as a soap (Latin *sapo* > soap). Saponins consist of a polycyclic aglycones attached to one or more sugar side chains. Saponins are phytochemicals, which can be found in most vegetables, beans and herbs. The best-known sources of saponins are peas, soybeans and some herbs with names indicating foaming properties such as soapwort, saoproot, soapbark and soapberry. They cause hypocholesterolaemia by binding cholesterol, making it unavailable for absorption. They also cause haemolysis of red blood cells and are toxic to rats. Saponins from *Bulbostermma paniculatum* and *Pentapamax leschenaultii* species have also been demonstrated to have anti-spermal effects on human spermatozoa. They significantly inhibited acrosine activity of human sperms, and the spermicidal effect was attributed to strong damage of the spermal plasma membrane.
- **Phytates:** Phytates are the salts of phytic acid (also named as inositol hexakisphosphate). The acids (or salts) are phosphorus-based compounds found in many plants specifically whole grains, legumes, nuts and seeds. They are found in plants, which are high in fiber, with concentrations generally being highest in foods containing large amount of insoluble fiber. Phytate is not

digestible to humans or non-ruminant animals so it is not a source of either inositol or phosphate if eaten directly. Moreover, it binds to certain dietary minerals including iron, zinc and manganese and to a lesser extent to calcium and magnesium and slows their absorption. This process can therefore contribute to mineral deficiencies in people whose diets rely on these foods for their mineral intake. It also acts as an acid, chelating the vitamin niacin the deficiency of which is known as Pellagra. Probiotic *Lactobacilli* and other species of the endogenous digestive microflora as well are an important source of the enzyme phytase, which catalyses the release of phosphate from phytate and hydrolyses the complexes formed by phytate and metal ions or other cations, rendering them more soluble, ultimately improving and facilitating their intestinal absorption.

- **Glucosinolates:** The glucosinolates are a group of organic compounds that contain sulfur and nitrogen and are derived from glucose and an amino acid and found in several vegetables particularly broccoli, Brussels sprouts and cabbage. They are secondary metabolites and also called as plant thioglucosides. They are well known for their toxic effects (mainly as goitrogens) in both man and animals at high doses. The ingestion of large amount of glucosinolates containing feeds may reduce feed intake, cause thyroid gland hypertrophy and reduce levels of circulating thyroid hormones, mainly by inhibiting the iodine uptake by the gland. The glucosinolate-containing crops are primary food source for animals. Such effects may adversely affect the productive performance of livestock. Poultry and pigs are more susceptible than cattle, sheep and goat because the latter group of animals can detoxify glucosinolates hydrolysis products to some extent.
- **Oxalates:** Oxalate salts are poorly soluble at intestinal pH and oxalic acid is known to decrease calcium absorption. Calcium in spinach with high oxalate content is very poorly absorbed, while kale, a low-oxalate vegetable, exhibits excellent calcium absorbability. They bind minerals like calcium and magnesium and interfere with their metabolism. They also cause muscular weakness and paralysis. Oxalates also cause gastrointestinal tract irritation, blockage of the renal tubules by calcium oxalate crystals, development of urinary calculi and hypocalcaemia. Oxalates cause nephrotic lesions in the kidney. Excess of soluble oxalate in the body prevents the absorption of soluble calcium ions as oxalate binds the calcium ions to form insoluble calciumoxalate complexes. As a result of this, people with the tendency to form kidney stones are advised to avoid oxalate-rich food items.
- **Plant oestrogens: Plant oestrogens or phytoestrogens** are plant-derived xenioestrogens functioning as the primary female sex hormone (estrogen) not generated within the endocrine system but consumed by eating phytoestrogonic plants. Also called "dietary estrogens", they are a diverse group of naturally occurring nonsteroidal plant compounds that, because of their structural similarity with estradiol (17-β-estradiol), have the ability to cause estrogenic or/and antiestrogenic effects. Phytoestrogen content are found in nuts and oilseeds, followed by soy products, cereals and breads, legumes, meat products, and other processed foods that may contain soy, vegetables, fruits, alcoholic, and nonalcoholic beverages. Flax seed and other oilseeds contained the highest total phytoestrogen content, followed by soybeans. In human beings, phytoestrogens are readily absorbed, circulate in plasma and are excreted in the urine. They also cause toxicity in animals, *e.g.*, it has long been recognized that sheep grazing subterranean clover are prone to poor reproductive performance. The causal chemical agent has since been identified as the isoflavone and genistein. Chronic exposure to these oestrogenic principles in plants may lead to

biochemically and physiologically active levels. Uncontrolled use of plant oestrogens could produce various types of tumour in animals.

- **Gossypol:** Gossypol is a natural phenol derived from the cotton plant (genus *Gossypium*). Cotton plants produce gossypol in order to slow down the reproduction of insects that eat cotton bolls and seeds. Chemically, it is a phenolic aldehyde, which permeates cells and acts as an inhibitor for several dehydrogenase enzymes. It is also anti-nutritional toxic compound, which causes low potassium levels (hypokalemia), and thus, it causes temporary paralysis. It also has effect on digestive system. Cottonseed oil and cottonseed meal (a by-product of oil extraction industry used for live stock feeding) are the main sources of gossypol toxicity. In animals, gossypol primarily affects the heart and liver. The reproductive tract, abomasums and kidney are also affected.
- **Cyanogens** Cyanogenic glycosides or cyanoglycosides account for a wider group of plant toxins known as cyanogens. The key characteristic of these toxins is cyanogenesis, the formation of free hydrogen cyanide, and is associated with cyanohydrins, which have been stabilized by glycosylation (attachment of sugars) to form cyanogenic glycosides. The major edible plants in which cyanogenic glycosides occur are almonds, sorghum, cassava, lima beans, stone fruits and bamboo shoots. Potential toxicity of cyanoglycosides arises from enzymatic degradation to produce hydrogen cyanide, resulting in acute cyanide poisoning. Cyanide can be lethal to humans and the acute dose is in the region of 1 mg/kg body weight. Cases of acute poisoning have been associated with misuse, particularly of preparations from apricot pits, bitter almonds and cyanide rich apple seeds. Symptoms of acute cyanide poisoning include rapid respiration, drop in blood pressure, rapid pulse, headache, dizziness, vomiting, diarrhoea, mental confusion, stupor, blue discolouration of skin due to the lack of oxygen (cyanosis), twitching and convulsions.
- **Mycotoxins:** Mycotoxins are toxic secondary metabolites of fungi belonging essentially to *Aspergillus, Penicillium* and *Fusarium* genera. One mold species may produce many different mycotoxins and/or the same mycotoxin as another species. These naturally occurring chemical compounds can be produced on a wide range of agricultural commodities and under a diverse range of situations worldwide. The accumulation of mycotoxins in foods and feeds represents a major threat to human and animal health as they are responsible for many different toxicities including the induction of cancer, mutagenicity and estrogenic gastrointestinal, urogenital, vascular, kidney and nervous disorders. Some mycotoxins are also immuno-compromising and can thus reduce resistance to infectious disease. Mycotoxicoses is the term used for poisoning associated with exposures to mycotoxins. The symptoms of a mycotoxicosis depend on the type of mycotoxin, the concentration and length of exposure as well as age, health and sex of the exposed individual. There have been outbreaks of pet food containing mycotoxins in North America.
- **Anti-vitamin factors:** Raw kidney beans are believed to contain an antagonist to vitamin E as evidenced by liver necrosis in rats and muscular dystrophy and low concentration of plasma tocopherol in chicks. Anti-vitamin E has also been noted in isolated soy protein, which is suspected to be α-tocopherol oxidase.
- **Antigenic compounds:** A few protein components of some legume seeds and cereals elicit antigenic effects in animals. These compounds are capable of inducing intestinal mucosal lesions, abnormalities in the villi, specific and non-specific immune responses and abnormal movement of digesta through the gut. Soybean protein contains compounds such as glycinin (G) and β-conglycinin (βC), which act as allergens to several animals.

- **Exorphins:** Exorphins are food chemicals, which have morphine-like activity. They are found in dairy products and wheat. Our body has its own natural morphine like substances called endorphins. Endorphins work by stimulating a type of nerve cell surface receptor called endorphin receptors. Endorphins are very important in controlling pain and addictive behaviour. Exorphins also act on endorphin receptors and may stimulate them or block them. It is logical that exorphins may therefore affect chronic pain and also affect addictive behaviour.

Mode of Action of Anti-nutritional Factors

These compounds are known to interfere with carbohydrates, protein and mineral metabolism. On reaching the intestine, these anti-nutritional compounds undergo extensive microbial and enzymatic biotransformations. The possible route of action is shown in Figure 2. However, different anti-nutritional factors behave in a different way:

- Oxalic acid and its content have deleterious effects on human nutrition and health, mainly by decreasing calcium absorption and aiding the formation of kidney stones.
- Phytic acid is an organic acid found in plant materials. Phytate binds minerals in the gastrointestinal tract, making the dietary minerals unavailable for absorption and utilization by the body. It decreases calcium bioavailability and form complexes of calcium phytate that inhibit the absorption of elements like iron and zinc. Phytates inhibit several proteolytic enzymes and amylases.
- Tannins form complexes with proteins and reduce their digestibility and palability damaging the intestinal tract. Furthermore, tannins could impair iron availability.
- Saponins are naturally occurring oily glycosides occurring in wide variety of plants when eaten. They are dangerous when injected into the blood stream and quickly haemolyse red blood corpuscles.
- Tripsin inhibitor when ingested by man in large quantity disrupts the digestive process and leads to undesirable physiological implications.
- Dietary polyphenols have been reported to be capable of modulating oxidative damage in the gastrointestinal tract.
- Gossypol form indigestible gossypol-protein complexes, which may produce deficiencies of some amino acids, such as methionine, which are essential for the normal fat metabolism.

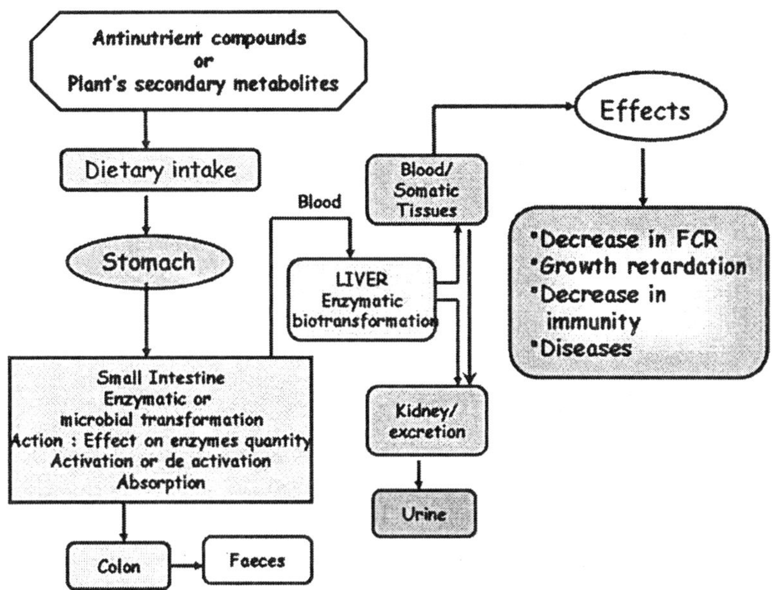

Figure 2: Possible routes of intake of anti-nutritional compounds and their biotransformation and absorption

Processing to reduce Anti-nutritional Factors

Substances containing anti-nutritional factors are processed to reduce anti-nutritional factors, thus, increasing the digestibility and nutritive value of the food item. Nutritive value is the ability of food to provide a usable form of nutrients, *i.e.*, carbohydrates, protein, minerals and vitamins.

Information on the effect of different processing techniques on the anti-nutritional factors in human foodstuffs and animal feeds is essential for human and animal nutritionists, dieticians, veterinarians, public health and food regulatory bodies/authorities on the exploitation of these techniques so that the nutritive values of food items and feeds could be efficiently maximized. Some of the common techniques are:

- **Effect of Gamma irradiation on anti nutritional factors:** Irradiation treatment has a significant effect on glucosinolate content of the canola meal. Gamma irradiation treatment with its radiolytic effects can destroy glucosinolate molecules. Some investigators reported that anti-nutritional factors, such as protease inhibitors (α-amylase inhibitors, phytohamagglutinins and oligosaccharids) and tannin can also be inactivated by gamma irradiation.
- **Soaking:** Soaking the seeds of beans like cowpea, moong bean, chickpea, *etc.* in plain water and mineral salt solution for 12 h decreased phytic acid to the maximum (46-50%), whereas, sprouting for 60 h had the most pronounced saponin lowering effect (46%). The other methods of processing were less effective in reducing the levels of these anti-nutritional factors.
- **Germinating and Sprouting:** When food like legumes are properly soaked and germinated, their nutritive value increases greatly, usually to the levels equal to or exceeding those of the cooked bean. The increase in nutritive value in the raw sprouted seed is due to an explosion of enzyme activity, which breaks down the storage-protein and starch in the seeds into amino acids, peptides

and simpler carbohydrates needed for the seed to grow. The seed is literally digesting its own protein and starch and creating amino acids in the process. Because of this process, sprouted seeds are essentially a predigested food. At the same time, the anti-nutritional factors such as enzyme inhibitors and other anti-nutritional are greatly decreased to insignificant levels or to nothing. Germinating or sprouting the seeds of chickpeas can also eliminate the anti-nutritional factors like trypsin inhibitors.

- **Soaking:** Seeds soaking alone causes a significant decrease in anti-nutritional factors as these factors are leached into the soak water and as the germination proceeds, the anti-nutritional factors are degraded further to lower levels. Soaking for 12 hours and 3-4 days of germination completely removes all hemagglutinating activity in mung bean and lentil. Soaking for 10 hours and germination for 3 days eliminate the amylase inhibitor completely in lentils.
- **Heat treatment:** Wet heat treatment like pressure-cooking, boiling, microwaving and autoclaving can reduce or even eliminate the trypsin inhibitor activity. Some of the heat labile anti-nutritional factors, which can be inactivated by thermal processing, are depicted in Table 1. In foodstuffs given below, the nutritional value is greatly enhanced by thermal processing.

Table 1: Some heat-labile anti-nutritional factors inactivated by thermal processing

Anti-nutritional factor	Common food sources	Effects of anti-nutritional factor
Avidin	Egg whites	Binds biotin, making it biologically unavailable
Hemagglutinins	Red kidney beans and yellow wax beans	Induces red blood cell clumping
Lathyrogens	Chickpeas	Disrupts collagen structure
Goitrogens	Sweet potatoes, beans cabbage and turnips	Causes goiters by limiting iodine absorption
Alpha-amylase inhibitors	Cereal grains, peas and beans	Slows starch digestion
Trypsin inhibitor	Legumes, egg whites and potatoes	Inhibits activity of trypsin
Thiaminases	Fish, shellfish, Brussels sprouts and red cabbage	Destroys thiamin

Source: Table 111.2 adopted from the book Modern Nutrition in Health and Disease 9th Edn. Williams and Wilkins, Baltimore (1999)

- **Blanching:** Levels of both tannic acid and phytic acid can be significantly reduced by conventional and microwave-blanching methods, while oxalic acid levels cannot be reduced significantly by blanching methods. In general, blanching is recommended as an effective method for reducing the anti-nutritional factor in green vegetables.
- **Extrusion:** Extrusion is supposed to be the best method to abolish chymotrypsin, trypsin, α-amylase inhibitors and haemagglutinating activity without modifying protein content. Furthermore, this thermal treatment is most effective in improving protein and starch digestibility when compared with dehulling, soaking and germination.

- **Fermentation:** Fermentation is an anaerobic (without oxygen) cellular process in which organic foodstuffs are converted into simpler compounds. Fermentation makes the foods easier to digest and the nutrients easier to assimilate and also it retains enzymes, vitamins and other nutrients that are usually destroyed by food processing. Fermentation alleviates the effect of anti-nutritional factor and accordingly mineral extractability, protein digestibility and amino acid contents are improved.
- **Dehydration:** Dehydrated food articles are considered texture-modified food, which may be rehydrated with the appearance and texture of canned or conventionally prepared food items. Hence, dehydration of foodstuffs is a controlled effect to preserve the structure or create a new one that serves for functional purposes. In addition, these food items present the advantages such as maximum microbiology security (low food manipulation), soft and homogenous texture, prolonged preservation time, lesser production of food by-products, easy use and storage food.
- **Roasting:** Roasting is the cooking in dry heat in an oven or on a spit with the aid of fat or oil. Roasting is similar to cooking/boiling but involves higher temperature and reduced time. Most of the anti-nutritional factors become ineffective as a result of roasting. Roasting reduced the phytate content by up to 62.35% in cowpea.
- **Enzyme treatment:** Enzymes provide additional powerful tools that can inactivate anti-nutritional factors and enhance the nutritional value of plant-based protein in feeds. They provide a natural way to transform complex feed components into absorbable nutrients. Endogenous enzymes found in the fishes digestive system help to break down large organic molecules like starch, cellulose and protein into simpler substances. The addition of enzymes in feed can improve nutrient utilization reducing feed cost and the excretion of nutrients into the environment.

  Phytic acid is one of the most powerful anti-nutritional factors in plant ingredients. The anti-nutritional activity of phytic acid can be eliminated by the addition of relevant enzymes, *e.g.*, phytase. The phytic acid or phytate found in cereals, legume grains and oil seeds is bound with phosphorus and also with calcium and magnesium, trace elements like iron and zinc, protein and amino acids. Some animals like fishes do not possess their own enzymes to break down the phytate and release the nutrients so they pass through the animal undigested, this is why, higher proportions of valuable nutrients from vegetable sources are not utilized by the animals and are wasted as excreta. The feed enzyme phytase not only releases phosphorus from the phytate but also releases minerals and amino acids that are also bound, paving the way for maximum utilization of nutrients.
- **Genetic manipulation:** The evolution of new genes to synthesize novel secondary products in plants is an ongoing process that might account for most differences in gene function among the plant genomes. Among the thousands of metabolites, only a few are part of the "primary" metabolic pathways and the rest are termed as secondary metabolites as they have no specific function in the plants. Levels of secondary metabolites are both environmentally induced as well as genetically controlled. The ability to synthesize toxic compounds has evolved to

A Combination of Anti-nutritional Factors Reducing Methods

- Soaking of beans overnight reduced the trypsin inhibitor activities (TIA) by 6.3% and cooking of the soaked beans caused further reduction in the trypsin inhibitor activities content by 66.7%.
- Cooking treatments and germination separately decreased the concentrations of lysine, tryptophan and total aromatic and sulfur-containing amino acids. However, cooked and germinated chickpeas were still higher in lysine, isoleucine and total aromatic amino acid contents.

- The losses of minerals and B group vitamins in chickpeas cooked by microwaving were smaller than in those cooked by boiling and autoclaving. Germination resulted in greater retention of all minerals and B group vitamins compared to cooking treatments. *In vitro* protein digestibility, protein efficiency ratio and essential amino acid index were improved by all treatments. The chemical score and limiting amino acid of chickpeas subjected to the various treatments varied considerably, depending on the type of treatment. Based on these results, microwave cooking appears to be the best alternative for legume preparation in households and restaurants.

General Effects of Processing for Reducing Anti-nutritional Factors

- Eases absorption of nutrients.
- Helps in synthesis of B-complex vitamins, biotin, thiamine, riboflavin, nicotinic acid vitamin B_{12} and vitamin K.
- Prevents auto-intoxication and digestion upsets.
- Relieves weakness and stress.
- Reduces diseases.
- Improves gut conditions and restores the health of bowel.
- Enhances Food Conversion Ratio (FCR).
- Improves feed consumption and thereby growth.
- Prevents adverse effects from anti-nutritional factors.

Therapeutic Effects of Anti-nutritional Factors

Small amount of anti-nutritional factors like protease inhibitors, tannins, and phytates is beneficial and can be considered a normal part of our nutritional ecology.

- Some of the substances commonly referred to as anti-nutritional factors are actually powerful cancer-protecting phytochemicals. These include protease inhibitors and tannins. The problem in most diets is that we do not get enough of these substances. Substantial research shows that protease inhibitors are one of the most powerful anti-carcinogens we have in our arsenal. They have proven to be particularly protective against cancer of the colon, breast and prostate.
- Tannins have also been shown to give substantial protection against cancer specially cancer of the stomach and lungs, when ingested orally. Tannins and other polyphenols also play a role in fighting tooth decay. Evidence shows that some of the tannins inhibit the growth of bacteria that cause tooth decay.
- Phytates, like tannins, may also interact with digestive processes in a beneficial way. Small amount in food slows down the absorption of sugars and regulates insulin levels. This is beneficial in the prevention and treatment of diabetes and hyperlipidemia (high blood fats).
- Saponins are a compound found in legumes and legume sprouts. They are toxic to red blood cells only *in vitro* (outside of the body in a test tube) but harmless when ingested. In fact, saponins appear to be beneficial, being responsible for a major part of the cholesterol-lowering effect of legumes. Saponins also seem to be anticarcinogens. In one study, it has been shown that they inhibited colon cancer.

- Concomitant action of both phytic acid and catechin can act as hepato and down regulate total sialic acid. Catechin also protects the liver against increased lipoperoxidation induced by anticancer or antibiotic treatment.
- Polyphenols have the properties including antioxidant, antimutagenic, anticarcinogenic, antiinflammatory and have the antimicrobial effects that might potentially be beneficial in preventing diseases and protecting stability of the genome.
- Gastrointestinal parasites, especially the nematodes, pose a severe threat to animals, especially those having a poor nutritional status. Consumption of dietary catechins may affect gastrointestinal nematode number.
- Tannic acid enriched aqueous extract of certain medicinal plants such as bearberry and cowberry was found to be bacteriostatic against *Helicobacter pylori*, the bacteria responsible for most ulcers and many cases of stomach inflammation and chronic gastritis.
- Phenols, tannins and flavonoids present as major active constituents of some medicinal and edible plants have been reported to be antagonistic toward multidrug-resistant human pathogenic bacteria and yeast.
- Phytoestrogens having a potential benefit in the prevention of atherosclerosis and arterial degeneration, as anticancer agents, or reduced risk of hormone dependent cancers such as prostate and breast cancers. Eating food items containing phytoestrogens like soybean products, plant oils and a range of different fruits and vegetables appears to reduce the risk of developing lung cancer even in smokers.
- Another important factor to be considered is the interactions between various anti-nutritional factors in a particular substance as these interactions in some instances lead to a decrease in toxic effect of interacting anti-nutritional factors, *e.g.*, saponin-tannin tannin-lectin and tannin-cyanogen interactions have all been shown to result in a reduction in the individual toxicity of the anti-nutritional factors.

Conclusion

Anti-nutritional factors exhibit diverse pharmacological and biochemical actions when ingested by animals and men along with food. Most of the toxic and anti-nutritional factors effects of these compounds in feed or feed ingredients could be removed by several processing methods such as soaking, germination, boiling, autoclaving, fermentation, genetic manipulation and other processing methods. Food processing methods such as boiling reduce the amount of these phytochemicals. However, small amounts of protease inhibitors, tannins and phytates are beneficial and can be considered a normal part of our nutritional ecology.

■ ■ ■

42 Vegetables from Natural Surroundings and Human Health

Kamla Kulshreshtha[1] and M.K. Rana[2]

[1]National Botanical Research Institute, Rana Pratap Marg, Lucknow, Uttar Pradesh, India; E-mail: <dr_kamla@rediffmail.com>

[2]Department of Vegetable Science, CCS Haryana Agricultural University, Hisar, Haryana, India; E-mail: <mkrlotus@gmail.com>

A number of plants, which are fundamental to life and life processes, are eaten as food. With the help of chlorophyll, they convert solar energy into chemical energy in the form of carbohydrates or sugar. Plants provide nutrition and possess health promoting and curative powers. Most of the plants today eaten as food are in the form of vegetables. All plant parts such as fruits (tomatoes, pumpkin, gourds, melons, *etc.*), leaves (amaranth, beet leaf, fenugreek, cabbage, *etc.*), roots (carrot, radish, turnip, *etc.*), stalks (onion, celery, *etc.*) and flowers (cauliflower, broccoli, *etc.*) are used as vegetables. A balanced diet contains various groups of foodstuffs, such as energy yielding food articles, bodybuilding foodstuffs and protective food items in adequate and appropriate amounts so that an individual's minimum requirement of all the nutrients may be met. The components of a balanced diet depend on age, sex, physical activity, economic status and the physiological state of an individual.

Vegetables constitute the health promoting and protecting components of food. Location-specific vegetables are considered the best since they contain specific active principles that help in adjusting and adapting with the local environment/season. Our ancestors, by trial and/or experimentation, selected specific vegetables that were naturally available within the region in a particular season. Thus, the vegetables that are naturally available in the summer have properties of being diuretic, antioxidant and protective against solar radiation, and therefore, best suited for that season. Similarly, the vegetables available in winter have contained health protective properties that enable the organism to adjust and to adapt itself to the cold climate.

During last few decades, the time-tested dietary practices, including the use of seasonal vegetables and fruits, are fast disappearing due to modern means of preservation and transportation technology. As a result, this has caused many health problems. It is, therefore, very important to have a closer look into the traditional practices for ensuring best health to humankind. With increasing population, the availability of vegetable is decreasing. In general, people are spending relatively little on this component of diet, whereas, the ground reality is that an increase in the consumption of green leaves and other vegetables could play a major role in reducing deficiency diseases like anemia, protein-energy malnutrition, vitamin deficiency, night blindness, ulcer, wounds, *etc.* Nonetheless, vegetables contribute to the requirement of calcium, iron, vitamins and essential minerals. They also provide various antioxidant components like polyphenols and flavanoids, which are beneficial for protection against radiation. Vegetables are good source of fibers, which act as good source of diuretics that help flushing of the kidneys, purifying blood and expelling many toxic and undesirable substances from the body.

Vegetables remain an important source of nutrients in many parts of the world and offer advantages over dietary supplements because of low cost and easy availability. Vegetables occupy an important place in vegetarian diets of Indians. If consumed in appropriate proportion, the percentage of vitamin C (91%), vitamin A (48%), vitamin B_6 (27%), thiamine (17%), niacin (15%), magnesium (16%), protein (5%), and minerals and fibers in appreciable range may be assured. Other contents include riboflavin, zinc, calcium, potassium and phosphorus. Green leafy vegetables are rich in carotenoides, calcium, ascorbic acid, riboflavin, folic acid and appreciable amounts of other minerals. It has been noted that of the total beta-carotene (a pigment, which is converted into vitamin A by the human body) available from all sources. In the country, 95% beta-carotene is derived from fruits and vegetables, and out of this, green leafy vegetables contribute 90% of β-carotene. A wide variation in ecological habitats and varying climate and season found in Indian perspectives, the growth of a number of species of edible leafy vegetables, such as amaranth, beet leaf, Bengal gram leaves, cauliflower leaves, mint and coriander, appeared. Carrot is also considered a good, cheaper and easily available source of β-carotene and can be as useful as green leafy vegetables in supplying vitamins in the diet. These plants are also a good source of iron, which is needed to build blood and provide energy. Too little iron in the diet can cause anemia. Iron is highest in beet leaf, peas, beans, beetroot, whole potatoes, *etc.*

It has been reported that climatic conditions have strong effect on nutritional quality of the vegetables. Low temperature favours synthesis of sugars and ascorbic acid, and at the same time, it decreases the rate of ascorbic acid oxidation. Food articles rich in beta-carotene can help in preventing cancer. This group of vegetables includes carrot, winter squash, spinach, broccoli, sweet potatoes, *etc.* Aonla and lemons are also rich in vitamin C and can help in preventing cancer of throat and stomach. Other vegetables rich in vitamin C include peppers, cabbage, cauliflower, coriander, drumstick leaves, tomatoes, green vegetables and water chestnuts, and boost up the body immune system. The maximum β-carotene (vitamin A) content occurs at a range of 15-21°C but β-carotene content is reduced if temperature is higher or lower than this range, principally due to the temperature sensitivity of lycopene, the precursor of β-carotene and a yellow or orange pigment lutein found in some fruits and vegetables. This component is known to concentrate in retinas of eyes and is necessary for good vision. A diet rich in lutein may lower the risk of developing cataracts and macular degeneration. Lutein may also help in preventing or slowing down arteriosclerosis and the thickening of arteries, which is a major risk for cardiovascular disease. Carrot and other fruits and vegetables are sources of lutein. Turnip greens and beet leaf contain large amounts of vitamin A and lutein.

Though we all realize the numerous health benefits of vegetables, very few of us actually follow a diet that includes it in appropriate quantity. Fresh vegetables can help in getting rid of numerous diseases, such as heart disease and stroke, blood pressure, cancer, painful intestinal ailment called diverticulitis, cataract and macular degeneration and vision loss. Apart from these, vegetables have a low fat content and low calorie value, whereby, they minimize the problem of excessive weight or obesity. They even have anti-ageing properties also. Vegetables have a rich content of nutrients, right from proteins, vitamins, potassium, phosphorus, magnesium, calcium to selenium, iron, manganese, copper and zinc. A high vegetable diet definitely assures a relief from all the major and minor health problems. Dark green vegetables, such as spinach, mustard greens and cabbage are high in calcium, which is needed for healthy bones and teeth.

Fiber is the part of the plant that cannot be digested or absorbed by the body. A high fiber diet may protect the body against colon cancer. All vegetables are sources of fiber. It is best to get fiber from food rather than from pills or other supplements. According to experts, people consuming greater amounts of vegetables are high on energy and feel less lethargic or stressed. The nutrition thus provided helps the body to perform all the activities by providing the body cells and organisms all the necessary requisites for supporting life. Some vegetables are strong antioxidants and function to modify the metabolic activation and detoxification/deposition of carcinogens or even influence processes that alter the course of tumor cell. Although antioxidant capacity varies greatly among fruits and vegetables but it is better to consume a variety of commodities rather than limiting consumption to a few with the highest antioxidant capacity. The flavanoids are phenolic compounds, which occur in all fruits and vegetables, are known for their antioxidant activities. The major sources of flavanoids are capsicum, pumpkin, brinjal, red and orange colour carrots, tomatoes, sweet potato, black carrots, cherries, greens of winter season, mint leaves, aonla, round melon, green beans, red and yellow onions, broccoli, *etc.* The anthocyanins found in black carrot, drumsticks, banana flower, *etc.* might help in reducing the risk of heart disease by inhibiting cholesterol formation. Dietary intake of potassium has an important role in energy metabolism since its deficiency results in muscular weakness and mental confusion, and it is reflected in heart malfunction.

Nutritive constituents of vegetables that have positive impact on human health and their resources are energy value, protein, vitamins, minerals and dietary fibers.

Energy Value

Food supplies energy to our body. Carbohydrates provide about 70 to 80% of the energy required by the body. Carbohydrates are synthesized by plants and occur in several forms, out of which, *structural carbohydrates* of vegetables and fruits are celluloses and other polysaccharides, which are not digested but aid in the movement of food through digestive tract by maintenance of muscle tone and their physical capacity to bind water. A minimum of 100 g carbohydrates are needed in the diet to ensure efficient oxidation of fats. The plants rich in energy are soybean (*Glycine max*), cowpea (*Vigna unguiculata*), French bean (*Phaseolus vulgaris*), pepper (*Capsicum annuum*), onion (*Allium sativum*), banana (*Musa paradisiaca*), *etc.*

Protein

Protein is the most abundant component for being vital part of nucleus and protoplasm. Amino acids, the simpler component of protein, are essential for the maintenance of human body but cannot be synthesized into it. Therefore, these have to be supplied through food. However, the concentration of proteins in most vegetables is small. Even then, the plants are considered the primary source of protein. Though the quality supplied may be small, these proteins do enhance the *quality of protein* in the diet. Deficiency of proteins affects the growth and repair of the body tissues adversely. Soybean (*Glycine max*), cowpea (*Vigna unguiculata*), French bean (*Phaseolus vulgaris*) and pepper (*Capsicum annuum*) are some of the vegetables known for their rich protein content.

Vitamins

Vitamins are important for their regulatory and protective functions. Unlike most other nutrients, these are required in very small amounts but it is necessary to provide it in the diet since the body cannot synthesize many of these. The lack of vitamins results in definite deficiency disorders. The main vitamins are:

Vitamin C (ascorbic acid): It is an essential nutrient and a strong reducing agent. It is involved in collagen synthesis, bone and teeth calcification and many other reactions in the body. Vitamin C deficiency causes scurvy characterized by weakness, bleeding of gums and defective bone growth. It aids wound healing, provides healthy immune system and prevents cardiovascular diseases. Ascorbic acid occurs widely in plant food, particularly in fresh vegetables especially in green varieties. Because of its reducing action when vegetables become dry and stale, or cut and exposed to air, most of the vitamin C originally present is destroyed. When dry pulses and beans are allowed to germinate, vitamin C is formed. The fruits like aonla and guava are the richest source of vitamin C. It is found in cabbage, citrus fruits, pepper, potato, tomato and all the leafy green vegetables.

Vitamin A (carotenoids): Plants provide a source of vitamin A for animals in the form of orange-yellow pigment called carotenoides. The human body converts beta-carotene to vitamin A in the intestinal mucosa during absorption. Night blindness occurs due to the severe deficiency of vitamin A. Beta-carotene is present in only food of vegetable origin. It is found in dark green vegetables like spinach, amaranth, coriander, drumstick leaves and turnip and also in orange vegetables like carrot, pumpkin, tomato, *etc.* It prevents night blindness, chronic fatigue, psoriasis, heart diseases, stroke and cataract.

Vitamin E (tocopherols): It is found in corn, dry beans, lentils, chickpea and dark green leafy vegetables and acts as very good remedy for heart disease, diabetes and also for cancer.

Vitamin K: This vitamin is found in green onions, cabbage, leafy greens and sprouts and reported very effective in the problems of osteoporosis.

Niacin: It includes both nicotinic acid and nicotinmide. Niacin works in close association with co-enzymes that contain riboflavin and is necessary for growth. Vegetables contain very small amount of niacin.

Minerals

Above twenty years of age, micronutrients are essential for human health and for proper growth and development of human beings. About 4 to 6 per cent of body weight is made up of mineral elements. The largest concentration of minerals is found in bones and teeth. These do not act singly in their regulation of body process. Therefore, a certain concentration of each element must be present for efficient functioning of the body. Micronutrients deficiency is prevalent as public health problem in developing countries. Over two hundred million people experience micronutrients deficiency in the world. Besides calcium and iron, phosphorus, potassium, sodium, chlorine, sulphur, copper, magnesium, manganese, iodine, fluorine, zinc, cobalt and selenium are important for human health. Surprisingly the micronutrients wave has not yet crested and special emphasis is to be given for alleviation of malnutrition. Some of the important micronutrients are:

Calcium: Green leafy vegetables (amaranth, beet leaf, spinach, cabbage, *etc.*), other vegetables (green beans, onion, *etc.*) are good source of calcium though only a part of the calcium from these is available in the body. Some spices such as cumin and coriander seeds contain enough calcium to contribute if these are used frequently in sufficient quantity in the diet. It is found in all cooked vegetables, beans, tomatoes, peas, papaya, pumpkin and cauliflower. It is a proven remedy of osteoporosis, teeth problems and blood pressure.

Folate (folic acid): Though the amount of iron in the body is only about 2-3 g, its presence is essential. Iron combines with proteins for the development of haemoglobin, the red colouring matter of the blood. Among the plant foods, dark green leafy vegetables such as amaranth, beet leaf, spinach, fenugreek leaves, *etc.* are good source of iron. Presence of ascorbic acid helps in iron absorption. If the proper dose of iron is not available to the body, it results in causing anaemia due to insufficiency of haemoglobin. Persons lacking iron get tired easily or feel faint due to inability of the body to carry sufficient oxygen to the cells for respiration. It is found in dark leafy green vegetables, legumes, chickpea and green peas. It is also an important remedy for cancer, heart diseases and problems of nervous system.

Phosphorus: It is necessary for the body, as an essential part of all living cells. Phosphorus compounds are important co-factors in various metabolic processes like energy metabolism, *etc.* Perhaps no mineral element has as many widely differing functions as phosphorus. It is usually considered that one gram of phosphorus should be consumed in the diet. Phosphorus is widely available in plants. A diet, rich in protein and calcium, will provide sufficient phosphorus. Although milk, meat and cereals are the main source of phosphorous in our diet but carrot, beans (legumes like soybeans), oil seeds, *etc.* contain fairly good amount of phosphorus.

Magnesium: The amount of magnesium, present in the body, is much smaller than that of calcium and phosphorus. In plants, magnesium is present in chlorophyll in a chemical structure similar to the iron in haemoglobin. Vegetables provide 19.8% magnesium in the diet. Green leafy vegetables, *e.g.*, *Amaranthus gangeticus*, spinach, *etc.* and dry beans, peas, nuts and soybeans are especially good sources of magnesium. It is also reported from spinach, potato, banana and corn. It is useful in the problems of osteoporosis and it also improves the immune system.

Iodine: About one third of the iodine in the adult body is found in the thyroid gland and is stored as thyroglobulin. In food, it is ingested as inorganic iodides and as organic compounds. The water of the oceans contains iodine. The amount of iodine in plants is dependent on the presence of minerals in the soil. Okra fruits contain iodine in considerable amount. The ash of dry sea plants, which contains iodine, has been successfully used in the treatment of goiter.

Potassium: This element is found in potato, banana, beans and cooked greens. It has been reported as an effective remedy of the problems of hypertension, stroke and arterial-sclerosis.

Dietary Fibers

Besides providing macro- and micronutrients, vegetables contain digestible residue, which adds bulk fiber to the diet. Dietary fibers have the ability to absorb water, thus, serve useful purpose in the elimination of intestinal wastes. Fiber stimulates peristaltic (rhythmic) movements of the gastrointestinal tracts by adding bulk to the intestinal contents. Fiber is known to be associated with reduced incidence of coronary heart diseases, reduces the risk of colon cancer and is reported to be a proper cure of diabetes. It is suggested that an intake of 40 g dietary fibers per day, which can be obtained from most fresh vegetables, beans and peas, is desirable.

Medicinal uses of some vegetables grown in North India

(i) Amaranth

Hindi name : Chaulai

Scientific name : *Amaranthus blitum* Linn.

Family : Amaranthaceae

Edible part : Leaves and stem

Medicinal uses : Its leaves are found rich in potassium nitrate, calcium and iron, hence, makes the bones and teeth strong and cures osteoporosis. The grains of amaranth are a rich source of protein and essential amino acids like lysine, leucine, isoleucine, *etc.*, which are essentially required for growth of children. In Kumoan hills of India, local people consider that the grains help in curing measles in children by bringing down the body temperature when the children sleep over the grains of amaranth. In Tamil Nadu hills, the tribes use the juice of leaves and stems of *Amaranthus spinosus* in the treatment of kidney stones. *Amaranthus viridis* is used in treating snakebites.

(ii) Arum

Hindi name : Arvi

Scientific name : *Colocasia esculenta* (L.) Schott

Family : Araceae

Edible part : Leaves and corms

Medicinal uses : *Poi* (fermented taro paste) is reported to be an excellent food against food allergies, and consumption of cooked corm helps in relieving pain from morphia. Taro is reported to reduce ulcers due to tuberculosis, and it expedites healing of fungal abscess in the animals. Corm juice is used as poultice, and the petiole juice is used as stypic. It is very useful for those who are suffering from hammertoes.

(iii) Beat leaf

Hindi name : Palak

Scientific name : *Beta vulgaris* Linn. Var. orientalis

Family : Chenopodiaceae

Edible part : Leaves

Medicinal uses : It is diuretic and a very good source of chlorophyll. Lipids in the leaves possess antibacterial action. Its leaves are good source of iron and calcium, hence, good for anemic and subjects suffering from osteoporosis. Its seeds are demulcent, diuretic and employed in fever. Night blindness especially in children, which is mainly due to the deficiency of vitamin A, can be corrected by the consumption of beet leaf. The plant is reputed in Indian medicine as a tonic and used to treat liver and spleen ailments. The leaves of beet leaf because of having very high fiber content are laxative, thus, their inclusion in diet prevents constipation and colorectal cancer. They provide roughage, which helps in the movements of bowel. Therefore, they have a significant place in a balanced diet.

(iv) Bitter gourd

Hindi name : Karela

Scientific name : *Momordica charantia* Linn.

Family : Cucurbitaceae

Edible part : Fruits and whole plant

Medicinal uses : It has anti-carcinogenic and anti-leukemia properties. The seeds of ripe fruits are used as condiment. Its leaves are rich source of calcium, carotene, riboflavin and ascorbic acid. Its fruits are easily digestible and have laxative property. From the *Ayurvedic* perspective, bitter gourd is an excellent medicine for balancing cough. It helps in purify blood tissue, enhances digestion and stimulates the liver.

Bitter gourd is highly beneficial in the treatment of blood disorders, like blood boils, scabies, itching, psoriasis, ringworm and other fungal diseases. Since ancient times, bitter gourd roots are used as folk medicine for respiratory disorders. Root paste mixed with equal amount of honey or *Tulsi* leaf juice acts as an excellent medicine for asthma, bronchitis, pharyngitis, colds and rhinitis. Fresh juice of its leaves is an effective medicine against cholera at an early stage and other types of diarrhea, particularly during summer. Juice of the fresh leaves is used for the cure of piles.

(v) Cabbage

Hindi name : Band gobhi

Scientific name : *Brassica oleracea* var. capitata Alef.

Family : Brassicaceae

Edible part : Leaves

Medicinal uses : Cabbage has some medicinal properties and is useful in curing digestion problems, thus, the health researchers have become increasingly concerned about glucosinolates in cruciferous crops and their effects on human body. It is used against ailments like gout, diarrhea, stomach and celiac troubles. It is included in the diet of patients who are suffering from liver problems. Raw cabbage has vermifungal properties. Leaves are applied to blisters and in gout and rheumatism. The consumption of cabbage and other cruciferous vegetables is associated with reduced incidence of some forms of cancer due to the glucosinolates-derived bioactive substances but there is no evidence that the glucosinolates present in brassica vegetables are detrimental to human health. Cabbage juice containing glucosinolates (vitamin U) is very useful in curing the peptic ulcer caused by *Helicobacter pylori* bacteria. Sauerkraut prepared by fermenting the shredded cabbage leaves under pressure has a curative effect in scurvy diseases.

(vi) Carrot

Hindi name : Gajar

Scientific name : *Daucus carota* Linn.

Family : Umbeliferae

Edible part : Roots and seeds

Medicinal uses : Its roots are rich source of carotene, hence, good for vision and in night blindness. Carrot infusion is being used a folk medicine for jaundice and thread worms since ancient time. Its juice is diuretic, thus, found to be very effective to eliminate the uric acid through urine. The poultice is used to cure ulcers, burns and scalds. The roots and seeds both are used as an aphrodisiac and nerve tonic. The essential oils extracted from roots of carrot possess antibacterial property. Its seeds are ingested for the abortion of fetus. Carrot seeds are aromatic stimulant, diuretic and carminative and have been found useful in kidney and dropsy diseases. Ingestion of carrot roots in large quantity helps in maintaining nitrogen balance in the body. Its juice can cure dermatitis, which is intensified with the exposure to ultraviolet light and results in skin discolouration. Being antioxidant, a combination of carrot and orange juice also reduces the oxidation of low-density lipo-proteins in habitual cigarette smokers. The β-carotene has anticarcinogenic, photo-protective, chemo-preventive, anti-mutagenic and immuno-enhancing properties in particular. It is very useful in reducing the risk of cancer of lungs, cervix, esophagus and stomach, and also the risk of cataract formation. The extract provides protection against oxidative stress in body tissues by inhibiting lipid oxidation and making the membrane stabilize. Carrot juice application gives relief to the patients suffering from sun burning effects on skin.

(vii) Cauliflower

Hindi name : Phool gobhi

Scientific name : *Brassica oleracea* var. botrytis (Linn.) Alef.

Family : Brassicaceae

Edible part : Immature flowers (curd)

Medicinal uses : Immature flowers known as curd are used as vegetable. The inflorescence extract of cauliflower is used as a blood purifier for the treatment of scurvy and as an antacid. Extract of cauliflower has been reported to be effective in the inhibition of initiation and promotion of carcinogenesis in *in-vitro* assays. A chemical compound sulforaphane produced by glucosinolate through the action of myrosinase enzyme has anticarcinogenic effect. It has the ability to detoxify a number of carcinogens, and thus, might have the ability to protect against a variety of cancers. It is reported that three days old sprouts of cauliflower contains 10 to 100 times higher levels of glucoraphanin, *i.e.*, glucosinolate of sulforaphane than the mature cauliflower sprouts. It is thought that they play an important role in the prevention of various diseases, most importantly aging diseases like cancer and coronary heart diseases.

(viii) Cluster bean

Hindi name : Guar

Scientific name : *Cyamopsis tetragonoloba* (L.) Taub.

Family : Leguminosae

Edible part : pods and seeds

Medicinal uses : Boiled seeds are used for the treatment of plague, enlarged livers, and swellings of head and broken bones. Importance of cluster bean as potential hypoglycaemic agents for the treatment of diabetes has recently been emphasized. In food, guar gum acts as bulk laxative and cholesterol controlling agent. Guar gum is one of the most widely stipulated food fibers that make the food more viscous in human digestive tract and slow down the absorption of glucose in intestine. It controls sugar metabolism in diabetes patients and in those having high serum lipids and high blood pressure.

(ix) Coriander

Hindi name : Dhania

Scientific name : *Coriandrum sativum* Linn.

Family : Umbelliferae

Edible part : Leaves and seeds

Medicinal uses : Its fruits are diuretic, laxative, carminative, tonic, stomachic, refrigerant and aphrodisiac. Seeds are chewed to correct breath. Its leaves are best source of vitamin C, hence, good for healthy gums and heeling wounds. Coriander seeds possess immense medicinal properties such as carminative, diuretic, tonic, anti-stomachache, aphrodisiac, laxative, antibillious and refrigerant, thus, being used in *Ayurvedic* and *Unani* system of medicine. An infusion of coriander seeds is taken thrice a day to lower down the blood sugar. Its infusion in combination with cardamom and caraway seeds is useful medicine against flatulence, vomiting, indigestion and other intestinal disorders and decoction in milk after adding sugar is beneficial in bleeding piles, mucous diarrhea, rheumatism, neuralgia and cephalagia and locally in eye infection. Coriander water is useful in indigestion and other bowl complaints. The seeds are chewed to correct the foul breath and along with jaggery to remove constipation. Its seeds are reported to lessen the intoxicating effects of spirituous liquors. The paste of coriander, nutmeg, clove, honey, lemon peel and rose water lightens and tones up the skin. Smearing of coriander leaves paste with juice of apple, milk and *Multani mitti* (a special type of loam) tones up the skin during summer, and for similar action, a paste of its leaves with apricot is also used. Coriander and tender coconut mixed together are popularly used as a cosmeceutical application in India to cure pimples.

(x) Cowpea

Hindi name : Lobia

Scientific name : *Vigna unguiculata* Walp.

Family : Papilionaceae

Edible part : Pods and seeds

Medicinal uses : The green pods are rich in protein, vitamin B and minerals. Cowpea is a rich source of lysine and tryphtophan. The proteins of cowpea are of high biological value. It can fortify the carbohydrates rich diet appreciably to improve the nutritive value of food. Chemically, the mature cowpea seeds contain an average of 23% protein, 60% starch and 2% oil. The high quality protein of cowpea is also a natural supplement to that of staple grain crops. Its ripe seeds are boiled and eaten, considered antibilious and prescribed in hepatic troubles and jaundice.

(xi) Drumsticks

Hindi name : Sahjan

Scientific name : *Moringa oleifera* Lam.

Family : Moringaceae

Edible part : Pods, leaves, flowers and seeds

Medicinal uses : Its pods are rich in leucine. Whole plant is used in treatment of ascites, venomous bites, rheumatism, snakebites and as a cardiac and circulatory stimulant. *Moringa* has many medicinal properties such as abortifacient, antidote, bactericide, diuretic, ecbolic, estrogenic, expectorant, purgative, rubefacient, stimulant, tonic, vermifuge and vesicant. Aqueous extract of roots and leaves are known to have anti-malarial and anti-impanation properties. Drumstick leaf extract is also known to have anti-carcinogenic property as its leaf extract has the ability to inhibit mutagenicity. Its leaves are rich in vitamin A and C, hence, useful in catarrhal infections and scurvy. Its flowers are used as tonic, diuretic, cholagogue and seeds antipyretic. The drumstick seed oil is also known to enhance the absorption of nutrients and beta-carotene.

(xii) Eggplant

Hindi name : Baingan

Scientific name : *Solanum melongena* L.

Family : Solanaceae

Edible part : Fruits

Medicinal uses : Its fruits are excellent remedy for those suffering from liver troubles and high blood cholesterol. The de-cholesterolizing action is attributed to the presence of poly-unsaturated fatty acids (linoleic and lenolenic) in fruits. Piercing brinjal fruits with a needle and fried in sesame oil is a remedy for toothache. Aqueous extracts of fruit juice inhibits choline esterase activity of human plasma and dry fruit is reported to contain goitrogenic principles. White eggplants are said to be a remedy for diabetic patients.

(xiii) Fenugreek

Hindi name : Methi

Scientific name : *Trigonella foenum-graecum* Linn.

Family : Papilionaceae

Edible part : Leaves and seeds

Medicinal uses : Its seeds are used as a condiment and also used raw or cooked. Seeds are aromatic, carminative, tonic and galactogogue and useful as poultice for boils. The seeds have the good therapeutic properties against digestive disorders, diabetes and have the anti-fertility and diuretic actions too. It prevents constipation, removes indigestion, stimulates spleen and liver, and is appetizing. Usually, the seeds are also used in *Ayurvedic* system as carminative, antipyretic, anthelmintic, tonic and aphrodisiac and customarily used for the treatment of colic, flatulence, dysentery, diarrhea, dyspepsia with loss of appetite, chronic cough, dropsy, rickets, gout and are good for the elimination of bad breath and body odour. It is believed that regular intake of fenugreek seed reduces the level of glucose, total cholesterol and triglycerides in blood.

(xiv) French bean

Hindi name : Rajma

Scientific name : *Phaseolus vulgaris* Linn.

Family : Leguminosae

Edible part : Green pods and seeds

Medicinal uses : The dried leaves are nutritious, comparing closely with meat as a source of protein. Its green pod shells are used as diuretic especially in kidney and heart troubles. French bean is said to be anti-diabetic and kidney resolvent and good for bladder burns and rheumatism.

(xv) Garlic

Hindi name : Lehsun

Scientific name : *Allium sativum* Linn.

Family : Liliaceae

Edible part : Bulb and leaves

Medicinal uses : Its bulbs are, juice used in skin troubles and even used in the treatment of leprosy. As per the *Ayurvedic* system of medicine usually practiced in India, garlic is anthelmintic, antiseptic, carminative, strong natural antibiotic, diaphoretic, expectorant, diuretic, anthelmintic, anti-malarial, antidiabetic, hypocholesterolaemic, fibrinolytic and various other biological actions. It is gastric stimulant, and thus, helps in the digestion and absorption of food. It is given in breathing problems, cancer, flatulence, worms-infection, leucoderma and skin diseases. Garlic is considered good to cure epilepsy and convulsive *vata* diseases and especially good to contract the uterus after delivery, thus, the Indian women take garlic preparations after delivery. Garlic also has a blood-sugar lowering capacity due to allicin and other related disulfide containing compounds, however, its excess consumption may lead to anemia, hence, its per day consumption should not be more than 10 g. Garlic juice is used in various ailments of stomach, as a rubefacient in skin diseases and as eardrop in earache. The juice diluted with water can be used against peptic ulcers. Use of garlic in diet eliminates obnoxious waste matter from the body. The digestive disorders can be cured if crushed garlic cloves are taken with milk. Use of garlic this way kills the harmful bacteria in the intestines by leaving the beneficial ones untouched. A regular use of garlic capsules can eliminate the internal infection and purify the blood. Because of its antiseptic properties, garlic juice mixed with boiled water in the ratio of 1: 3 is an effective lotion for cleansing infected wounds. A formulation containing 1% garlic extract provides protection to persons against mosquitoes and black fly for about 8 h. Extract of garlic along with chilli and ginger has been found beneficial against soil nematodes.

(xvi) Mustard

Hindi name : Sarson greens

Scientific name : *Brassica campestris* Hook. f. and Thoms.

Family : Cruciferae

Edible part : Leaves, shoots and seeds

Medicinal uses : Mustard greens provide good to excellent amount of protein, dietary fibers, ten minerals, nine vitamins and health promoting phytochemicals, *i.e.*, glucosinolates. In addition, mustard greens are an excellent source of three notable antioxidants, *viz.* vitamin A (through their concentration of β-carotene), vitamin C and vitamin E. Mustard greens contain numerous nutrients that can contribute to a healthy cardiovascular system, including the antioxidants, vitamin E, β-carotene and vitamin C. The vitamin E supplied by mustard greens is instrumental to a host of different mechanisms that reduce the development of atherosclerosis including protecting low-density lipoprotein (LDL) particles from oxidation and decreasing platelet clumping. mustard greens also have concentrated amounts of other nutrients like vitamin B_6, folic acid and magnesium that can keep the heart healthy. Mustard greens are also a very good source of magnesium, a mineral that helps smooth muscle cells, like those lining the bronchial tubes and lungs, to stay relaxed rather than constricting themselves and the airways of which they are a significant part. Seeds yield edible oil, which is used in muscular rheumatism, stiff neck, *etc.* Mustard oil is good for those who have the weak heart as it contains the least amount of saturated fatty acids, making it safe for heart patients. Mustard oil provides both linolenic and linleic acid to the consumer, while many other edible oils lack one of the two essential fatty acids.

(xvii) Okra

Hindi name : Bhindi

Scientific name : *Abelmoschus esculentus* (L.) Moench.

Family : Malvaceae

Edible part : Fruits

Medicinal uses : The immature tender fruits, which have been found to be very useful for increasing virility and vogour in humans, act as a tonic. Its raw fruits consumed on an empty stomach every morning nourish the body, cure peptic ulcer and increase the semen content. As medicinal uses, its roasted seeds are handy for curing chronic dysentery, spermatorrhoea, genitourinary disorder, renal colic problems and fever based weakness. The ground powder of its roots is given with sugar for providing relief in leucorrhoea backache. *Ayurveda* system of medicine considers it a good vegetable for people suffering from stone in kidney. Okra fruits are very much useful in prevention of goiter due to high contents of iodine. The leaves are potential source for the preparation of medicine to reduce inflammation in human and animal body.

(xviii) Onion

Hindi name : Pyaz

Scientific name : *Allium cepa* Linn.

Family : Liliaceae

Edible part : Bulbs, leaves and seeds

Medicinal uses : Its fresh juice is bactericidal. Roasted bulbs are used to cure boils and abscesses. It possesses innumerable medicinal properties like expectorant and stimulant, tonic, stomachic, diuretic, stimulant, expectorant, aphrodisiac, antimalarial, antirhumatic and contains volatile oil, allyl-propyl disulfide, which rejuvenates the body, removes toxins, revitalizes the blood and stimulates blood circulation in the body. It is regarded in antiquity as a diuretic of the first order. Freshly extracted juice has antiseptic and moderate bactericidal properties due to the presence of several sulfur compounds and is used as inhaler in faintness on hysterical fits and used to relieve sore throat too. Onion is said to be very effective in the prevention/ treatment of atherosclerosis and coronary heart disease. Its extract can inhibit the aggregation of human blood platelets for arterial blocking, and is beneficial in the treatment of cancer, tuberculosis and asthma. It is also claimed to be of value as a digestive stimulant, as an anti-fermentative and as an antidiabetic. Inclusion of onion in diet results in fewer tendencies to form blood clots and lower levels of cholesterol and lipoproteins in blood serum generally seems to be associated with heart disease.

(xix) Peppers

Hindi name : Mirch

Scientific name : *Capsicum annuum* Linn.

Family : Solanaceae

Edible part : Fruits

Medicinal uses : It is a good source of vitamin C and used as prophylactic, therapeutic, stimulant and carminative. It is an excellent remedy for flatulence, dyspepsia and dipsomania. Eating hot pepper is good, especially for people living in hot climate since capsaicin in hot pepper produces stabilizing effect on red blood corpuscles, resulting in more stress resistance. Green hot peppers eaten with foodstuffs increase the flow of saliva by stimulating the taste buds. The vitamin P in hot pepper provides protection against damage due to secondary irradiation caused by atomic explosion. The green fruit decoction with opium and fried asafoetida is also equally good in cholera, and infusion with cinnamon and sugar is a useful drink for patients suffering from delirium. In diphtheria, its application is said to hasten the separation of false membranes but its large doses produce gastric enteritis and irritation in intestine. The whole plant soaked in milk is applied to reduce swellings and hardened tumors. The capsaicin/capsicutin is used in many pharmaceutical preparations like tincture, pain balms, vaporous, lineaments, plaster, medicated wool and ointment for cold, chest congestion and sore throat. It is also used in cosmetics like prickly heat powders and skin ointments. The tincture prepared from it is used to stimulate appetite, digestion and mitigate flatulence. Medicinally, hot peppers are used internally as a cure for several maladies including colic, diarrhea, dropsy, black vomit, gouts, toothaches and as counter irritants in neuralgia, lumbago and rheumatic disorder and are especially useful in dyspepsia. It has a tonic and carminative action when taken with food and being having a powerful action on mucus membrane it may be added to gargles for pharyngitis and sore and putrid throat. Its paste is used externally as rubefacient and as a local stimulant for the tonsils in tonsillitis.

(xx) Potato

Hindi name : Aloo

Scientific Name : *Solanum tuberosum* L.

Family : Solanaceae

Edible part : Tubers

Medicinal uses : Potato is considered a protective food since it supplies a substantial amount of vitamin C and B group vitamins, which are essential for nervous system, normal digestion and skin health. Its consumption plays an important role in preventing teeth decay because of its lack of stickiness and cleansing effects. A cup of its fresh juice taken three times a day cures the peptic ulcer caused by *Helicobacter pylori*. Potato is a boon to those who wish to reduce blood pressure by limiting the salt intake. The minerals in potato affect the actions of nerves and muscles, maintaining osmotic pressure and keeping the body neither too acidic nor alkaline. Being good source of iron, its regular consumption prevents anemia.

(xxi) Pumpkin

Hindi name : Petha

Scientific Name : *Cucurbita moschata* Duch. Ex. Poir.

Family : Cucurbitaceae

Edible part : Fruits

Medicinal uses : Its fruit is used to cure stomach catarrh, hyperacidity and stomach ulcers, especially peptic ulcers. Raw pulp of pumpkin has purgative effect and it is recommended in case of obesity and deterioration in the function of gall bladder. Pumpkin juice as well as the raw and boiled fruits has diuretic and cholagogue properties. Its juice or decoction with honey is also advisable for people suffering from insomnia or from uneasy and troubled sleep.

(xxii) Radish

Hindi name : Mooli

Scientific Name : *Raphanus sativus* var. sativus L. (Asiatic radish)

Family : Cruciferae

Edible part : Roots

Medicinal uses : Its leaves are a source of protein extraction on commercial scale. Eating radish increases appetite, produces cooling effect and prevents constipation. Its roots are considered good for patients suffering from liver trouble, gall bladder ailments, hemorrhoids, jaundice and enlargement of spleen. It is a very good natural alkalinizer. In homeopathy, it is used for neurological headaches, sleeplessness and chronic diarrhea. Its roots are said to be useful in urinary complaints and gastric troubles. The juice of its fresh leaves is consumed as diuretic and laxative and the seeds are said to be carminative, expectorant and digestive. Radish root contains ascorbic acid, which acts as chromosomal stabilizer among others stated under different items discussed. Its roots juice is virucidal. It contains caffeic acid, which functions in chromosome stabilization, antitumor promotion, antioxidant and antiviral to HIV and to other viruses. Caffeic acid also exerts actions such as antiulcer and antihypercholesterolaemia and serves as chemical trap for highly electrophilic agents such as DMBA and other similar compounds. Radish contains maleic acid, which is antibacterial and keeps the urinary tract disinfected from the bacteria usually attack the same and *p*-coumaric acid, which exerts in human body an antipyretic effect. It contains crocetin, which increases oxygen diffusion in blood plasma by 80%, binds serum albumin and lowers the serum cholesterol. Ferulic acid present in radish has various bioactions such as anti-platelet forming action, stimulation on broad cardiovascular system, anti-allergic and restriction of damaging action of lipid peroxidase. Cyanin present in radish is antiviral against poliovirus.

(xxiii) Spinach

Hindi name : Vilayati palak

Scientific name : *Spinacia oleracea* Linn.

Family : Chenopodiaceae

Edible part : Tender leaves and petioles

Medicinal uses : Its dark green leaves are innately rich in beta-carotene (pro-vitamin A) is strongly recommended by the nutritionists and dieticians to cure the vitamin A deficiencies. Being rich in fibers, spinach supplies roughage essentially required in the daily diet for normal movement of bowel. Its consumption prevents constipation, which causes colourectal cancer. A daily dose of spinach extract prevents some loss of long-term memory and learning ability and spinach is the most potent in protecting different types of nerve cells in various parts of brain against the effect of aging. Spinach because of having specific phytonutrients is considered more effective than the strawberries.

(xxiv) Tomato

Hindi name : Tamatar

Scientific name : *Lycopersicon esculentum* Mill.

Family : Solanaceae

Edible part : Fruits

Medicinal uses : Tomato is a powerhouse protective food. It has protective effects against cancers of digestive tract and prostrate. The lycopene present in tomato fruits is effective in inhibiting low-density lipoprotein oxidation, and thus, has been associated with low risk of cardiovascular diseases. The pulp and juice both are easily digestible, mild-appetizer, promoter of gastric secretion and blood purifier and considered as an intestinal antiseptic. It removes constipation, stimulates torpid liver and is good in chronic dyspepsia. It is one of the best vegetables, which keep our stomach and intestine in good condition. Tomato is said to be useful in curing cankers and sores in mouth. However, the people suffering from hyperacidity and stones in gall bladder should avoid the excessive use of tomato in daily diet.

(xxv) Mushroom

Hindi name : Khumbi

Scientific name : *Agaricus bisporus*

Edible part : Tender shoots, leaves and seeds

Medicinal uses : Mushrooms have long been considered to have medicinal value. Due to unique chemical composition, mushrooms are suited to specific groups suffering with some ailments/disorders. As a low calorie high protein diet with almost no starch and sugars, mushrooms are the 'delight of the diabetic'. Cholesterol content of mushroom is low. Certain varieties of mushroom can inhibit growth of cancerous tumors. Due to high potassium to sodium ratio, few calories and low fat (rich in linoleic acid and lacking cholesterol), mushrooms are suited to persons suffering from obesity, hypertension and arthrosclerosis. Alkaline ash and high fiber contents make them suitable for those suffering from hyperacidity and constipation. It is anti-hypertensive, and remedy for hepatitis, bronchitis, piles, asthma and diabetes and improves immunity. The Concord International Trading Pvt. Ltd., Australia is preparing Concord Sunchih from fresh cultivated *Gandoderm* fruit-bodies. Its prescribed dosages are 1 to 2 caps twice a day for healthy and three caps thrice a day for individuals suffering from afore said ailments.

(xxvi) Banana

Hindi name : Kela

Scientific name : *Musa paradisiaca* Linn.

Family : Musaceae

Edible part : Fruits and flowers

Medicinal uses : It is rich source of carbohydrate, minerals and vitamins (B group). Inflorescence before opening is used as a vegetable. Its fruit is laxative and used in intestinal disorders, uremia, nephritis and hypertension.

(xxvii) Chenopod

Hindi name : Bathua saag

Scientific name : *Chenopodium album* Linn.

Family : Chenopodiaceae

Edible part : Tender shoots, leaves and seeds

Medicinal uses : It is used as laxative, anthelmintic and antiscorbutic. It is the richest source of iron, hence, consumed to cure anemia.

(xxviii) *Ipomoea* leaf

Hindi name : Naari saag or Kalmi saag

Scientific name : *Ipomoea reptans* Poir

Family : Convolvulaceae

Edible part : Shoots and leaves

Medicinal uses : Its leaves are good source of minerals and vitamins, *i.e.*, carotene. Its plant juice is used as emetic in case of opium and arsenic poisoning, in nervous and general debility, in piles and its buds are used in treatment of ringworm.

(xxix) Mint

Hindi name : Pudina

Scientific name : *Mentha arvensis* Linn.

Family : Labiatae

Edible part : Leaves

Medicinal uses : Its leaves are used as carminative, antiseptic and stimulant. Its decoction is used in fever and heat apoplexy.

(xxx) Pigeon pea (Bengal gram)

Hindi name : Chana saag

Scientific name : *Cicer arietinum* Linn.

Family : Leguminosae

Edible part : Leaves, Edible part tender shoots, leaves and seeds

Medicinal uses : Germinated gram is used as prophylactic against deficiency diseases, particularly in scurvy.

(xxxi) Soybean

Hindi name : Soya bean

Scientific name : *Glycine max* Merrill.

Family : Papilionaceae

Edible part : Pods and seeds

Medicinal uses : Its seeds are rich in protein, specially used in malnutrition.

Vegetables are recognized as essential to equilibrate our diet preventing numerous diseases. Fruits and vegetables are a food group containing a diversity of micronutrients (vitamins and minerals) and other bioactive substances with polyphenols, glucosinolates, carotenoids, *etc.* It is always good to include at least 50-70 g of these vegetables in the diet. However, scientific more evidences are required for understanding the optimal levels of their intake and their functionality.

Clearly, the ultimate goal in attainment of nutritional food security is to ensure that people consume a diversity of vegetables and fruits that provide them with the required quantities of all essential macro- and micro-nutrients on a continuing basis. Improving the quantity of food consumed and taking steps to ensure that diets are nutritionally well balanced will certainly help in removing malnutrition. Food-based approaches are emphasized as the only sustainable way to improve the nutritional status of the individual, family, society and the nation.

■■■

43 Vegetables as Source of Phytochemicals

A S. Shekhawat[1], M.K. Rana[2] and Manoj Kumar Nalla[2]

[1]Nuziveedu Seeds Private Limited, Sonipat, Haryana, India
E-mail: <shekhawatdr@gmail.com>
[2]Department of Vegetable Science, CCS Haryana Agricultural University, Hisar, Haryana, India, E-mail: <mkrlotus@gmail.com>

Under the circumstances of hunger and undernourishment, particularly in the developing countries, vegetable crops can play a pivotal role in alleviating the malady. Life cannot be sustained without adequate nourishment. Food is eaten for the sustenance of body. Among the five major nutrients, namely carbohydrates, proteins, fats, minerals and vitamins, the former three are required in larger quantities, whereas, minerals and vitamins are required in small amounts but are essential for physiological processes and metabolic activities of our body. Vegetables and fruits are the only natural sources of protective food supplying all the nutrients especially minerals, vitamins and crude fibers. However, in general, vegetables are not considered the primary source of carbohydrates, protein and fat.

Hunger and malnutrition are the most threatening global problems today, and the worst situation exists in the developing countries. According to Food Organizations of Unite Nation, about 25 million people die from hunger and exhaustion every year. In India, the ill effects of chronic malnutrition are of greater magnitude. Nearly half of all preschool children in the developing countries are deprived of normal nourishment, which causes various physiological disorders even physical and mental retardation. According to World Health Organization, the xerophthalmia (dryness of eyes due to the deficiency of vitamin A) afflicts nearly 13 million infants and preschool children every year.

Indian Council of Medical Research (ICMR) has given a recommendation that an average man needs daily a diet, which can provide 2800 calories, 55 g protein, 450 mg calcium, 50 mg vitamin C, 20 mg iron, 1.4 mg riboflavin, 3000 μg β-carotene, 100 μg folic acid, 19 μg niacin, 5 μg vitamin D and 1.0 μg vitamin B_{12}. According to the recommendation given by ICMR, an average man with vegetarian or non-vegetarian food habit should consume 125 g of green leafy vegetables, 100 g of roots and tubers and 75 g other vegetables. An increase of nutritional importance and availability of vegetables have increased their consumption per capita per day from 145 g in 1991 to 200 g in 2011-12.

Table 1: Vegetable sources of carbohydrates availability (mg/100 g of edible part)

Sources	Contents	Sources	Contents
Tapioca	38.1	Pea	15.9
Sweet potato	28.2	Cluster bean	10.8

Sources	Contents	Sources	Contents
Potato	22.6	Carrot	10.8
Arum (arvi)	21.1	Beetroot (sugar beet)	8.8

Table 2: Vegetable sources of fibers availability (mg/100 g of edible part)

Sources	Contents	Sources	Contents
Chilli	6.8	Cauliflower	1.2
Green pea	4.0	Okra	1.2
Cluster bean	3.2	Celery	1.2
Pointed gourd	3.0	Fenugreek	1.1
Cowpea	2.0	Amaranth	1.0
Knol-kohl	1.5	Beet leaf (palak)	0.7

Table 3: Vegetable sources of protein availability (mg/100 g of edible part)

Sources	Contents	Sources	Contents
Soybean	43.2	Cowpea	4.3
Lima bean	8.3	Amaranth	4.0
Pea	7.2	Indian bean	3.8
Broad bean	4.5	Beet leaf (palak)	3.4
Fenugreek greens	4.4	Cluster bean	3.2

Table 4: Vegetable sources of vitamin A availability (μg/100 g of edible part)

Sources	Contents	Sources	Contents
Ragi leaves	14190	Radish leaves	5295
Arum black leaves	12000	Celery	3990
Arum green leaves	10278	Parsley	1920
Coriander leaves	6918	Carrot (European)	1890
Beet leaf (palak)	5862	Carrot (Asiatic)	11571
Spinach	5580	Lettuce	990
Amaranth	5520	Pumpkin	1998

Table 5: Vegetable sources of vitamin C availability (mg/100 g of edible part)

Sources	Contents	Sources	Contents
Parsley	281	Hot pepper	111
Agathi	169	Bitter gourd	96
Sweet pepper	137	Cauliflower	88
Coriander leaves	135	Beet	56
Cabbage	124	Tomato	31
Brussels sprout	124		

Antioxidants are substances (proteins/enzymes) or nutrients (vitamins/minerals) in food articles, which can prevent or slow down the oxidative damage to the body. Antioxidants act as *free radical scavengers*, and hence, prevent and repair damage done by these free radicals. Health problems such as heart disease, macular degeneration, diabetes, cancer, *etc.* are all contributed by oxidative damage. Indeed, a recent study conducted by research scientists from London indicated that five servings of fruits and vegetables reduce the risk of heart stroke by 25%. Antioxidants may also enhance immune defense, and therefore, lower the risk of cancer and infection. Perhaps, the most well known phytochemicals are the antioxidant group, which protect cells from damage caused by the byproducts (free radicals) of metabolism, as well as toxic substances in the environment and foods. At high levels, reactive species, such as reactive oxygen and nitrogen species, can be damaging to cells and may contribute to cellular dysfunction and disease. Antioxidants significantly decrease the adverse effects of reactive species by eliminating free radicals as they circulate throughout the body. Beta-carotene and the other carotenoids are not antioxidants but influence the biochemical reactions involved in the oxidative process. Other phytochemicals include sulfides (allium), indoles, phytosterols, protease inhibitors, phenols, tannins and terpenes.

Phytochemicals are part of a plant's natural systems of defense and repair. The body utilizes a small fraction of these components by incorporating them into various cellular and metabolic processes that enhance cellular growth, regeneration and repair. Some phytochemicals act as antioxidants, some protect and regenerate essential nutrients, and still others work to deactivate cancer-causing substances. Phytochemicals are also important to a plant's reproductive system, in which, some of these components are responsible for the bright beautiful colors that make the plants appearance more attractive to insects and animals, encouraging cross-pollination. The specific phytochemical content of fruits and vegetables tends to vary by the fruit or vegetable's color, with each component possessing unique functions. There are thousands of phytochemicals in plants but not all have specific benefits to human health.

Fruits and vegetables play an important role in human diet. According to many research reports, the increased amount of vegetables, especially cruciferous vegetables, in diet has a strong correlation with cancer prevention and inhibition. The detailed mechanism has not been fully realized. However, the high amount of nutritional and non-nutritional antioxidants in vegetables reducing the possibility of tumor formation has been confirmed. Studies of cruciferous vegetables only forced on some of the edible botanical portions, like heads of broccoli and cauliflower.

In last few decades, many studies have studied relationship between fruits, vegetables and health. Early studies on diet and cancer focused on dietary components that increase disease risk bur the recent studies have focused on the protective benefits of fruits and vegetables, *e.g.*, eating a diet rich in fruits and vegetables has been associated with the prevention and treatment of the ghastly leading causes of death, *i.e.*, cancer, heart disease, diabetes and high blood pressure. Phytochemicals, unique components found only in plants, are an important source protective effect. Fruits and vegetables are excellent source of carbohydrates, protein, minerals, vitamins and dietary fiber. Hence, nutritionists and health professionals now stress the need for change in dietary pattern, recommending a shift to one with less saturated fat and refined carbohydrates, particularly obtained from plant food items. With the progression of science, researchers have identified several chemical compounds in vegetables, which have therapeutic significance against several ailments. These compounds are known as phytochemicals, which impart health benefits to humans in addition to those provided by minerals and vitamins alone. Phytochemicals differ from minerals and vitamins but they are not considered essential nutrients. With the exception of carotenoids that are similar to vitamins are not stored in the body for later use, hence, they must be consumed consistently over time.

About 34% reduction in mortality is due to cardiovascular disease among those who consume vegetables rich in vitamins E and C. A 30-40% reduction in risk of colon cancer in populations with higher vegetable consumption, especially garlic and dietary fibers has also been reported. A diet high in vegetables reduces rheumatoid arthritis, anaemia, diabetes, muscular degeneration and gastric ulcer. The phytochemicals in vegetables are responsible for mitigating some other diseases also. Constant supply of phytochemicals containing fruits and vegetables with desirable health benefits, beyond basic nutrition, is essential to activate the defensive mechanism to reduce the risk of chronic diseases in humans. The benefit cannot be attributed to one particular component or even to one vegetable but rather due to a synergism among phytochemicals. The availability of a large number of vegetables round the year has enabled consumers to include a variety of health-promoting phytochemicals in the diet. Therefore, research on health benefits of vegetables needs to focus on two main key areas in near future. The first area is to identify the genetic mechanisms that regulate the synthesis of key phytochemicals, such as carotenoids, flavonoids, glucosinolates and thiosulfides. The second key area for future research is related to identify optimum conditions for maintaining the optimum levels of these compounds after harvesting and processing. The special role of particular phytochemical in human body has been given as under in Table 6.

Table 6: Antioxidants in vegetables their source and possible actions

Substance	Food source	Possible action (s)
Alpha-carotene (Carotenoids)	Pumpkin, carrot, kale and water melon	• It slows the growth of cancer cells. • It reduces the risk of lung cancer. • It boosts the immune response. • An antioxidant inhibits cell proliferation.
Beta-carotene (Carotenoids)	Sweet potato, carrot, corn, pumpkin, green and yellow pepper, spinach and other green leafy vegetables, squash, mango and apricots	• Protects HDL lipids from free radical damage. • Reduces the risk of heart diseases, bladder, colon and skin cancers. • Reduces risk of cardiovascular disease especially in smokers. • Stimulates immune response. • An antioxidant, precursor to vitamin A, helps in differentiation of normal epithelial cells and inhibits cell proliferation.

Substance	Food source	Possible action (s)
Lycopene (Carotenoids)	Asiatic carrot, tomato and water melon	• Protects the protein, fat and DNA from free radical damage. • Reduces the risk of cardiovascular diseases, colon, pancreatic, prostrate and other cancers.
Lutein and Zeaxanthin (Carotenoids)	Kale, spinach, beet, mustard greens, sweet pepper, carrot, tomato, paprika and orange pepper	• Form a pigment in the eye, *i.e.*, out damaging forms of light and functions as an antioxidant. • Reduce the risk of macular degeneration, the most common cause of blindness in the elderly. • Reduce the risk of lung, colon and prostrate cancers and improve the immune response. • An antioxidant protects against cataract and macular degeneration.
Anthocyanins	All purple colour berries	• Offer protection against heart disease and act as antioxidants.
Capsaicin	Chilli or pepper	• It impedes carcinogens seen as nitrates and cigarette smoke from attacking to cellular and preventing formations of cancer cells. • It may also kill bacteria that can cause ulcer. • It increases circulation of blood and reduce the risk of lung and other cancers. • An antioxidant prevents carcinogens from binding to DNA.
Dithiolthiones, glucosinolates and indoles	Cruciferous vegetables like Chinese cabbage, broccoli, brussels sprouts, cabbage, cauliflower, collards, kale, kohlrabi, mustard green, radish, rutabaga and turnip	• Increases activity of enzymes involved in detoxication of carcinogens and other foreign compounds. • Stimulate enzymes that diminish the effect of hormone estrogen and improve immune response. • Reduce the risk of breast and ovarian cancers. • Stimulate production of enzymes that break down cancer causing agents. • Protects against estrogen-promoted cancers and induces protective enzymes.
Isothiocyanates/sulphorophane	Cruciferous vegetables	• Increase production of enzymes that block carcinogens from damaging cells. • Slows tumor growth and reduce lung cancer. • Neutralization of free radicals that cause cell damage and protection against some cancers. • Exceptionally potent inducer of detoxification enzyme.

Substance	Food source	Possible action (s)
Vitamin C	Spinach, green pepper, Brussels sprout, broccoli, tomato, leafy vegetables, potato and cabbage	• Inhibits nitrosamine formation, a potentially dangerous carcinogen
Anthocyanins	Red/purple pepper, eggplant red cabbage and carrot	• Antioxidant cell protection may help prevent binding of carcinogens to DNA. • Delay several diseases associated with ageing. • Improvement of vision, inhibition of nitric oxide production, induction of apoptosis, decreased platelet aggregation and neuroprotective effects.
Allicin (allyl sulfides)	Garlic, onion, shallot, leek, chives and scallions	• Have anticancer effects, reduce formation of blood clots and lower cholesterol levels. • Stimulates anticancer enzymes, detoxifies carcinogens and acts as antibacterial. • Activities may inhibit conversion of nitrate to nitrite, thereby reducing formation of nitrosamines which are thought to be carcinogenic
Isoflavones (genistein and daidzein)	Soybean, soy milk, textured vegetables	• Reduce risk of breast, colon, ovarian, prostate and other oestrogen-sensitive cancer. • Help in bone maintenance and in reducing risk of osteoporosis. • A reduction in blood pressure and increased vessel dilation • Inhibit cell replication in gastrointestinal tract. • An antioxidant inhibits growth of cancer cells and lowers blood.
Resveratrol	Red wine, peanuts, gapes, *etc.*	• Offsets artery-damaging effects of high fatty diet. • An antioxidant protects against heart disease.
Catechins	Green and black tea	• Reduces risk for cancer and heart disease
Kaempferol	Green and black tea	• An antioxidant may reduce cell proliferation. • Extends action of vitamin C. • Inhibits blood clot formation. • It has antinflammatory action.
Tannins	Black-eye pears, grapes, lentil, tea and white and red wine	• Inhibit carcinogen activation, cancer promotion and act as antioxidant. • Prevents carcinogens from binding to target sites.

Substance	Food source	Possible action (s)
Saponins	Alfalfa and other sprouts, green vegetables, potato and tomatoes	• Interfere with DNA multiplication, prevent cancer cell multiplication and stimulate immune response.
Monoterpenes	Parsley, carrot, broccoli, cabbage, cucumber, squash, eggplant and peppers	• Aid protective enzyme activity.
Flavonoides	Onion, parsley, carrot, broccoli, cabbage, cucumber, squash, eggplant, peppers, tomato and tea	• Block receptor sites for hormones that promote cancer. • Anti-inflammatory, anti-microbial properties, cardioprotective, neuroprotective, anti-carcinogenic and other activity
Phenolic acids	Parsley, carrot, broccoli, cabbage, tomato, eggplant and pepper.	• They act as antioxidant, inhibit nitrosamine formation and help from protective enzymes.
Vitamin E	Dark leafy vegetables and sweet potato	• It prevents oxidation of beta-carotene and vitamin A in the intestine. • It prevents lipid peroxidation of polyunsaturated fatty acids in cells.
Selenium	Onion and garlic	• It plays various anticarcinogen functions in the body.
Cryptoxanthins	Red capsicum and pumpkin	• Antioxidant properties
Glutathione	Green leafy vegetables	• Antioxidant properties
Gingeberin	Ginger	• Carminative and stimulant of gastrointestinal tract
Curcumin and tannin	Turmeric	• Iron absorption

The significance of some important colourful fruits and vegetables containing health-promoting phytochemicals has been given below:

Orange Colour Food

Sweet potatoes, carrots, pumpkins, cantaloupe, mangoes and peaches all contain beta-carotene, a common phytochemical within a group of carotenoids, which are converted into vitamin A by the human body, which is essential for the growth and repair of body tissues, formation of bones and teeth, resistance of the body to infection and development of healthy eye tissues. Citrus fruits, such as oranges, grapefruits and tangerines contain the antioxidant vitamin C and bioflavonoids, a group of phytochemicals that work together with vitamin C to help in reducing the risk of cancer, strengthening bones and teeth, healing wounds, keeping the skin healthy and lowering the risk of heart attack.

Green Colour Food

Green leafy vegetables, such as beat leaf, coriander, fenugreek, spinach, lettuce, collard greens, kale and broccoli, green peas, honeydew melon, kiwifruit and avocado are the powerful source of lutein, a powerful antioxidant that helps in reducing the risk of cataracts and macular degeneration. Cruciferous vegetables, such as broccoli, cauliflower, cabbage and Brussels sprouts contain indoles, which help in protecting against breast cancer in female and prostate cancer in male up to 42%.

Red/Pink Colour Food

Water melon, grapefruits and tomatoes are good source of lycopene, which is one of the carotenoids. The tomato-based processed products, such as ketchup, sauce, soup and juice have the most concentrated source of lycopene. Cooked tomato sauces are associated with greater health benefits, compared to uncooked since heating process increases the bioavailability of lycopene to the body. Lycopene is fat-soluble, thus, it should be eaten with at least a small quantity of fat.

Blue/Purple Colour Food

Blueberries, blackberries, black grapes, plums, raisins and eggplant contain disease-fighting phytochemicals, *i.e.*, anthocyanine and polyphenols. These powerful antioxidants help in reducing the risk of several diseases including cancer, heart disease and Alzheimer's and may even slow down the ageing process. Research findings suggest that eating blueberries, in particular, may prevent some of the effects of ageing by improving cell communication in the central nervous system.

White Colour Food

Garlic, onion, leek, shellac and tree onion are white in colour but they contain powerful phytochemicals, *i.e.*, allicin, the most common phytochemical in plants of this group. Research shows that eating garlic and onions every day lowers cholesterol and blood pressure, increases the body's ability to fight infections, speeds up recovery from colds, reduces the risk of heart attack and even stops the spread of certain cancers, particularly stomach and colon cancer.

APPROXIMATE NUTRIENT COMPOSITION OF VARIOUS VEGETABLES

	% Water	Calories	CHO	Pro	Fat	Fib	Ash	A	C	B_1	B_2	Niacin	Ca	P	K	Na	Mg	Fe
FUNGI																		
Agaricus bisporus Button mushroom	91.5	23	4.5	2.4	0.20	0.80	1.00	–	4	0.10	0.50	4.40	8	112	392	10	11	1.1
Auricularia polytricha Jew's ear mushroom	92.6	25	6.8	0.5	0.40	2.10	0.20	–	1	0.80	0.20	0.07	16	14	43	9	25	0.6
Cantharellus cibarius Chanterelle mushroom	91.5	25	3.0	1.5	0.50	–	0.80	–	6	0.02	0.23	6.50	8	44	507	3	14	6.5
Lentinula edodes Shiitake mushroom (dried)	9.5	296	75.4	9.6	1	11.5	4.6	–	4	0.30	1.30	14.10	11	294	1534	13	132	1.7
Morchella spp. Morel mushroom	90.0	31	5.4	1.7	0.30	0.80	1.00	–	5	0.13	0.06	–	11	162	390	2	11	1.2
Tuber melanosporum Truffle mushroom	75.5	56	7.4	5.5	0.50	6.40	1.90	–	–	–	–	–	24	62	526	77	24	3.5
MONOCOTS																		
ALLIACEAE																		
Allium ampeloprasum Great headed garlic (bulb)	86.3	45	10.3	2.2	0.30	–	0.90	–	–	–	–	–	52	50	–	–	–	1.1
Allium ampeloprasum Leek (pseudo-stem & leaves)	83.0	56	12.6	1.9	0.30	1.40	1.10	75	15	0.08	0.05	0.45	55	43	230	16	25	1.9
Allium cepa Shallot (bulb)	79.8	72	16.8	2.4	0.10	0.70	0.80	–	8	0.06	0.02	0.20	37	60	334	12	–	1.2
Allium cepa Shallot (leaves)	91.0	30	5.0	1.8	0.90	1.60	–	–	19	0.07	0.12	0.40	86	25	–	–	–	3.7
Allium cepa Onion (bulb)	90.8	36	8.0	1.3	0.18	0.52	0.40	40	9	0.05	0.03	0.15	26	33	156	6	11	0.4
Allium cepa Onion (leaves "scallions")	91.9	25	5.6	1.7	0.14	0.90	0.70	5000	45	0.07	0.14	0.20	60	33	257	14	20	1.1

	% Water	Calories	CHO	Pro	Fat	Fib	Ash	A	C	B_1	B_2	Niacin	Ca	P	K	Na	Mg	Fe
Allium Chinese																		
Ricky (bulb)	86.0	50	12.0	2.2	0.30	0.50	0.50	–	10	0.05	0.03	1.0	22	66	–	–	–	0.5
Allium fistulosum																		
Japanese bunching onion	90.5	34	6.5	1.9	0.30	1.00	0.70	380	27	0.05	0.09	0.40	55	49	200	10	–	1
Allium sativum																		
Garlic (bulb)	64.0	131	27.9	6.20	0.35	1.00	1.30	0	11	0.20	0.08	0.50	32	187	465	17	25	1.3
Allium sativum																		
Garlic (leaves)	86.4	12	9.5	2.60	0.50	1.50	1.00	680	38	0.08	0.16	0.70	82	66	326	4	–	0.5
ALLIACEAE																		
Allium schoenoprasum																		
Chives (leaves)	92.0	26	3.9	2.80	0.60	0.90	0.80	6400	70	0.10	0.12	0.60	82	46	250	6	55	1.2
Allium tuberosum																		
Chinese chives	93.1	23	2.8	2.10	0.10	0.90	1.00	1800	25	0.06	0.19	0.6	50	32	–	–	–	0.6
Allium wakegi																		
Wakegi onion	91.2	29	5	1.90	0.30	1.00	0.60	500	30	0.05	0.15	0.4	38	35	–	20	–	1.2
ARACEAE																		
Alocasia macrorrhiza																		
Giant taro, "Alocasia" (corm)	84	71	17	0.60	0.20	1.50	1.10	–	5	0.08	0.02	0.4	50	50	–	–	–	1
Amorphophallus camponulatus																		
Elephant foot yam (corm)	78	81	18	1.60	0.20	0.70	1.00	434	6	0.07	0.06	1	50	36	–	–	–	1.1
Colocasia esculenta																		
Taro (corm)	72.9	99	23.4	2.00	0.20	0.80	1.20	20	5	0.11	0.04	0.9	30	63	512	10	33	0.9
Colocasia esculenta																		
Taro (leaves)	86	42	7.4	4.40	0.90	1.60	1.50	8100	45	0.18	0.4	1.4	156	80	648	3	45	1.7
Colocasia esculenta																		
Taro (shoots)	94	18	15	0.60	0.40	0.50	–	107	10	0.04	0.05	0.5	50	25	332	1	8	1.1
Cyrtosperma chamissonis																		
Giant swamp taro (corm)	65	116	32	0.80	0.30	1.30	1.00	–	1	0.04	0.1	0.9	370	55	–	–	–	1.2

	% Water	Calories	CHO	Pro	Fat	Fib	Ash	A	C	B_1	B_2	Niacin	Ca	P	K	Na	Mg	Fe
Cyrtosperma chamissonis Giant swamp taro (leaves)	86	45	4	5.00	0.70	1.50	2.00	–	–	–	–	–	–	–	–	–	–	–
Xanthosoma sagittifolium Tannia (cormels)	65	134	32	2.10	0.30	0.90	1.20	–	10	0.10	0.90	0.5	14	48	–	–	–	0.8
Xanthosoma sagittifolium Tannia (leaves)	89.9	34	5.3	2.50	1.00	2.10	1.30	3300	37	–	–	–	95	388	–	–	–	2
Xanthosoma sagittifolium Tannia (shoots)	89.5	33	5.7	3.10	0.60	3.20	1.10	–	82	–	–	–	49	80	–	–	–	0.3
ARECACEAE (PALMAE)																		
Sabal palmetto Swamp cabbage	85	21	3.9	2.20	0.20	0.09	–	5200	16	0.05	0.08	0.5	55	32	88	–	–	1.5
CYPERACEAE																		
Eleocharis dulcis Chinese water chestnut	76.4	86	19.6	1.45	0.18	0.85	1.20	–	10	0.17	0.2	1	11	65	542	17	17	2.1
DIOSCOREACEAE																		
Dioscorea spp. Tubers/bulbils (various spp.)	70	108	25.1	2.05	0.20	0.95	1.00	–	12	0.11	0.04	0.5	20	62	600	9	26	0.7
LILIACEAE																		
Asparagus officinalis Asparagus (green spear)	92.2	22	3.8	2.60	0.21	0.77	0.79	950	33	0.20	0.14	2	22	67	271	2	18	0.8
Asparagus officinalis Asparagus (white spear)	93.6	20	2.9	1.90	0.14	0.84	0.62	50	21	0.11	0.12	1	21	46	207	4	20	1
MARANTACEAE																		
Maranta arundinacae West Indian arrowroot (rhizome)	64.9	142	29.7	1.90	0.10	1.70	1.30	–	9	0.08	0.03	0.7	20	40	–	–	–	3.2
MUSACEAE																		
Musa spp. Plantain and cooking banana	64.7	127	31.4	1.10	0.40	0.40	0.90	990	18	0.07	0.04	0.6	7	35	420	1	33	0.7

	% Water	Calories	CHO	Pro	Fat	Fib	Ash	A	C	B_1	B_2	Niacin	Ca	P	K	Na	Mg	Fe
POACEAE																		
Bambusa and other spp. Bamboo shoots	91	27	5.2	2.60	0.30	0.70	0.90	20	4	0.15	0.07	0.6	13	59	532	4	3	0.5
Zea mays Sweet corn	75.3	95	20.3	3.60	1.20	0.75	0.71	480	2	0.17	0.12	1.7	35	109	273	8	45	1.1
Zizania latifolia Water bamboo	92.6	26	5.5	1.20	0.20	1.00	0.50	–	2	0.09	0.04	0.02	5	36	–	–	–	0.06
ZINGIBERACEAE																		
Zingiber officinale Ginger (fresh)	86.2	50	9.8	1.45	1.00	1.15	1.10	10	4	0.02	0.04	0.75	22	37	264	6	–	1.1
DICOTS																		
AMARANTHACEAE																		
Amaranthus spp. Amaranths (leaves)	87	36	5.6	3.80	2.90	1.40	2.20	4320	56	0.05	0.3	1	305	77	527	20	55	5.5
APIACEAE [UMBELLIFERAE]																		
Anthriscus cerefolium Chervil (leaves)	–	57	11.5	3.40	–	–	–	–	9	–	–	–	–	–	–	102	–	0.5
Apiumgraveolens var. *dulce* Celery, green (leaves & petioles)	94.7	17	3.8	0.80	0.11	0.65	0.89	184	8	0.03	0.03	0.6	52	32	313	107	17	0.4
Apium graveolens var. *dulce* Celery, blanched (leaves & petioles)	93	18	3.8	0.85	0.15	0.70	1.10	90	8	0.04	0.05	0.6	35	34	308	116	14	0.5
Apium graveolens var. rapaceum Celeriac, celery root (root & leaves)	88.3	40	8.4	1.70	0.32	1.28	1.00	16	8	0.04	0.07	0.8	50	107	305	94	15	0.6
Apium graveolens var. secalinum Smallage (leaves)	90.9	27	4.6	2.20	0.60	1.40	1.70	2685	49	0.08	0.12	0.6	326	51	318	151	–	1.5

	% Water	Calories	CHO	Pro	Fat	Fib	Ash	A	C	B_1	B_2	Niacin	Ca	P	K	Na	Mg	Fe
APIACEAE [UMBELLIFERAE]																		
Coriandrum sativum Cilantro, Chinese parsley	90	28	5	2.40	0.50	1.00	1.60	2800	–	0.1	0.1	1.1	134	48	542	28	37	5.5
Daucus carota Carrot	87.8	42	9.9	1.10	0.20	1.00	0.90	1100	9	0.35	0.06	0.8	32	40	332	41	19	0.6
Foeniculum vulgare var. dulce Florence fennel	88	33	6.1	1.90	0.23	0.50	1.70	100	20	0.3	0.1	0.2	73	51	445	88	36	1.5
Pastinaca sativa Parsnip (root)	80	74	17	1.47	0.44	2.00	1.10	30	17	0.09	0.09	0.5	47	75	355	10	29	0.7
Petroselinum crispum Parsley (leaves)	85	46	8.5	2.80	0.46	1.40	2.00	6050	123	0.12	0.23	1.1	160	57	508	90	61	4.3
Petroselinum crispum Parsley (root)	88	26	2.3	2.88	0.60	–	1.62	trace	41	0.1	0.09	2	–	57	–	–	–	–
ASTERACEAE [COMPOSITAE]																		
Cichorium endive Endive, escarole (leaves)	93.8	19	3.8	1.50	0.15	0.90	1.40	2140	8	0.08	0.11	0.5	66	41	304	50	13	13
Cichorium intybus Chicory (leaves)	92	22	4.3	1.80	0.30	0.80	1.30	4000	24	0.06	0.01	0.5	93	43	420	45	22	0.9
Cichorium intybus Chicory (wit-loof)	95.1	16	3.2	0.90	0.10	–	0.60	–	–	–	–	–	16	20	177	8	13	0.5
Cichorium intybus Chicory (root)	80	73	17.5	1.40	0.20	1.95	0.89	6	8	0.05	0.03	0.4	41	61	290	50	22	08
Cynara cardunculus Cardoon (petioles)	94	20	4.9	0.70	0.10	–	0.31	120	2	0.02	0.03	0.3	70	23	400	170	42	0.7
Cynara scolymus Globe artichoke	85.7	45	10.2	2.80	0.19	2.00	1.03	250	10	0.08	0.05	0.9	50	88	388	61	41	1.3

	% Water	Calories	CHO	Pro	Fat	Fib	Ash	A	C	B_1	B_2	Niacin	Ca	P	K	Na	Mg	Fe
Helianthus tuberosus																		
Jerusalem artichoke	79	72	16.7	2.20	1.60	0.85	1.60	20	7	0.19	0.08	1.2	18	80	478	–	15	3
Lactuca sativa																		
Lettuce (head or iceberg)	95	14.5	2.4	1.05	0.16	0.51	0.70	470	7	0.06	0.07	0.3	22	26	166	7	11	15
Lactuca sativa																		
Lettuce (butter head)	95.5	13.8	2.4	1.25	0.21	0.50	0.90	1065	8	0.06	0.06	0.3	35	26	260	7	11	1.8
Lactuca sativa																		
Lettuce (cos)	94.9	17.4	2.9	1.45	0.25	0.70	0.90	1925	22	0.08	0.09	0.5	44	35	277	9	9	1.3
Lactuca sativa																		
Lettuce (leaf)	94	18.3	3.5	1.30	0.30	0.70	0.90	1900	18	0.05	0.08	0.4	68	25	264	9	11	1.4
ASTERACEAE [COMPOSITAE]																		
Lactuca sativa (stems)																		
Celtuce, asparagus lettuce	94.5	22	3.7	0.85	0.30	0.40	0.70	70	6	0.06	0.07	0.6	39	39	330	11	28	0.6
Taraxacum officinale																		
Dandelion	85.7	45	8.9	2.70	0.70	1.60	1.90	13900	34	0.19	0.22	0.8	179	67	409	76	36	3.1
Tragopogon porrifolius Salisfy,	77	34	5.2	3.30	0.20	–	–	–	8	0.08	0.22	0.5	60	75	380	30	23	0.7
BASELLACEAE																		
Basella alba, B. rubra																		
Ceylon, Malabar spinach	93.1	20	3.6	1.80	0.30	0.70	1.40	7000	95	0.05	0.12	0.5	130	52	505	21	–	1.5
BRASSICACEAE [CRUCIFERAE]																		
Armoracia rusticana																		
Horseradish	71	100	21	3.30	0.23	2.50	2.20	trace	107	0.17	0.1	0.6	111	53	442	11	57	2.2
Brassica juncea																		
Indian mustard	91.6	27	4.9	2.69	0.42	0.99	1.26	6000	93	0.1	0.19	0.8	181	46	374	33	29	2
Brassica napus																		
Rutabaga	89.3	41	9.6	1.10	0.12	1.20	0.79	455	38	0.07	0.07	1	59	41	257	8	16	0.4

	% Water	Calories	CHO	Pro	Fat	Fib	Ash	A	C	B_1	B_2	Niacin	Ca	P	K	Na	Mg	Fe
Brassica napus Siberian kale	87	42	8.	2.80	0.60	1.23	1.28	3100	130	0.07	0.06	1.3	205	62	450	70	88	3
Brassica oleracea var. acephala Collard	89.7	36	6.6	3.40	0.60	0.95	1.20	5350	86	0.15	0.24	1.2	202	55	420	43	45	1.2
Brassica oleracea var. acephala Kale	85	41	6.8	4.00	0.80	1.40	–	7150	119	0.11	0.21	1.7	177	71	383	55	35	2
Brassica oleracea var. botrytis Cauliflower	91.1	27	5.2	2.60	0.22	0.88	0.80	75	75	0.1	0.1	0.7	25	57	328	16	20	1
Brassica oleracea var. capitata Cabbage, white (**f.** alba)	92.5	24	5.1	1.40	0.22	0.09	0.66	150	50	0.06	0.05	0.3	49	29	272	17	16	0.5
Brassica oleracea var. capitata Cabbage, red (**f.** rubra)	91.7	29	6.1	1.60	0.22	1.00	0.68	40	58	0.08	0.05	1.4	43	36	257	17	17	0.7
Brassica oleracea var. sabauda Cabbage, Savoy	90.5	27	4.9	2.50	0.24	0.60	0.90	600	47	0.06	0.06	0.3	54	51	262	20	20	0.8
Brassica oleracea var. gemmifera Brussels sprouts	84.8	46	8.6	4.20	0.40	1.60	1.36	520	98	0.11	0.15	0.9	61	80	404	22	26	1.5
Brassica oleracea var. gongylodes Kohlrabi	91.0	29	6.2	2.00	0.10	1.10	–	25	64	56	0.38	0.5	42	50	369	17	27	0.05
Brassica oleracea var. italica Broccoli	89.2	32	5.7	3.60	0.35	1.40	1.10	3150	109	0.1	0.2	1	101	77	389	23	23	1.2
Brassica chinensis Chinese cabbage (Pak choi)	93.8	17	3.1	1.70	0.23	0.67	0.83	3300	57	0.06	0.1	0.8	118	39	234	71	19	0.8
BRASSICACEAE [CRUCIFERAE]																		
Brassica chinensis Choy sum, mock (Pak choi)	95	–	1.2	1.20	0.20	–	–	5800	53	0.04	0.07	0.5	102	37	181	100	27	2.0
Brassica peknensis Chinese cabbage (Pet sai)	95	14	2.7	1.20	0.15	0.57	0.69	625	25	0.05	0.04	0.6	71	35	235	44	13	0.6

	% Water	Calories	CHO	Pro	Fat	Fib	Ash	A	C	B_1	B_2	Niacin	Ca	P	K	Na	Mg	Fe
Brassica rapa Turnip (root)	91.5	30	6.6	1.00	0.19	0.90	0.74	trace	30	0.05	0.06	6	39	32	207	50	15	0.5
Brassica rapa Turnip greens	90.5	28	5.2	2.10	0.35	0.80	1.60	5740	93	0.14	0.33	0.8	224	66	309	32	45	1.5
Crambe maritime Sea kale, blanched	94	13	1	2.00	0.30	–	–	100	26	0.04	0.04	0.3	35	34	–	–	–	0.5
Crambe maritime Sea kale, green leaves	90	22	1.3	3.50	0.30	–	–	4600	87	0.16	0.1	0.5	110	63	360	30	64	0.9
Lepidium sativum Garden cress	88.2	40	5	3.75	1.10	1.10	1.83	9300	73	0.11	0.16	1.2	179	48	592	12	–	2.5
Nasturitium officinale Watercress	93.5	19	3	2.15	0.30	0.85	1.10	4800	67	0.07	0.15	0.6	143	57	293	39	24	1.9
Raphanus sativus Radish (root)	94.4	18	3.7	0.90	0.18	0.66	0.72	9	26	0.03	0.03	0.3	29	28	237	25	11	1
Raphanus sativus Radish (pods)	90.5	–	4.7	2.30	0.30	1.40	–	50	69	0.07	0.05	0.2	80	100	–	–	–	2.8
Raphanus sativus Radish (sprouted seed)	90	41	3.1	3.80	2.50	–	0.53	391	29	0.1	0.1	2.9	51	113	86	6	44	0.09
Raphanus sativus Black radish	93.5	21	3.9	1.05	0.15	0.70	0.75	trace	30	0.30	0.03	0.4	33	29	322	18	15	0.8
R. sativus cv. Longipinnatus Daikon	94.6	19	4.2	0.75	0.10	0.67	0.58	10	27	0.03	0.02	0.3	31	24	203	21	16	0.5
CACTACEAE																		
Opunita ficus-indica Cactus (cladode)	91.9	37	6.9	1.30	0.40	2.10	0.90	–	21	0.27	0.5	0.3	110	20	–	–	–	1.4
Opunita ficus-indica Cactus (tuna)	85	–	–	0.50	0.10	1.80	1.60	50	30	–	–	–	60	–	–	–	–	–

	% Water	Calories	CHO	Pro	Fat	Fib	Ash	A	C	B_1	B_2	Niacin	Ca	P	K	Na	Mg	Fe
CHENOPODIACEAE																		
Beta vulgaris var. crassa Table beet	87.7	43	9.6	1.67	0.11	0.90	0.97	20	12	0.03	0.05	0.36	20	40	310	105	23	0.8
Beta vulgaris Table beet greens	92.3	2.3	4.5	2.00	0.35	1.70	1.90	6250	31	0.1	0.21	0.4	121	40	565	148	89	3.3
CHENOPODIACEAE																		
Beta vulgaris var. orientalis Palak, spinach beet	86.4	–	6.5	3.40	0.80	0.70	–	9770	70	0.26	0.56	3.3	380	30	–	–	–	16.2
Beta vulgaris var. cicla Swiss chard	92.3	24	4.2	2.20	0.28	0.77	1.64	5000	35	0.07	0.17	0.5	93	39	380	205	73	2.3
Chenopodium album, C. quinoa Lambsquarter, quinoa	84	44	7.3	4.30	0.80	2.10	3.30	11300	90	0.15	0.4	1.3	280	81	–	–	–	2.1
Spinacia oleracea (Spinach)	91.5	25	4	3.20	0.33	0.65	1.70	7045	50	0.4	0.2	0.6	107	57	605	110	92	2.7
CONVOLVULACEAE																		
Ipomoea batatas Sweet potato, orange root	69.4	118	27	1.63	0.40	0.83	0.85	14000	23	0.90	0.06	0.6	35	60	343	60	30	0.8
Ipomoea batatas Sweet potato (leaves & shoot tips)	87	48	8	3.50	0.60	1.70	1.40	3300	23	0.12	0.28	1.2	147	78	540	9	35	2.6
CUCURBITACEAE																		
Benincasa hispida Mokwa, immature fruit	96	13	2.9	0.71	1.17	0.70	0.30	210	39	0.05	0.07	0.4	22	20	111	6	–	0.5
Benincasa hispida Chinese wax/winter gourd, mature fruit	96	9	2.3	0.20	0.10	–	–	trace	14	0.02	0.03	0.5	14	7	200	2	16	0.4
Citrullus colocynthis Colocynth (dried seed)	6.7	556	19.5	23.60	47.20	1.50	3.00	–	–	0.11	–	–	46	580	–	–	–	–
Citrullus lanatus Watermelon (fruit)	92.8	24	6.4	0.50	0.16	0.50	0.30	235	6	0.04	0.05	0.2	8	11	106	2	10	0.2

	% Water	Calories	CHO	Pro	Fat	Fib	Ash	A	C	B_1	B_2	Niacin	Ca	P	K	Na	Mg	Fe
Citrullus lanatus																		
Water melon (dried seed)	5.7	567	15.1	25.80	49.70	4.00	3.70	–	–	0.1	0.12	1.4	53	–	–	–	–	7.4
Coccinia grandis																		
Ivy gourd (fruit)	93.5	18	3.1	1.20	0.10	1.60	–	–	15	0.07	0.08	0.7	40	30	–	–	–	1.4
Cucumis anguria																		
West India gherkin	93	17	2	1.40	0.30	0.60	0.50	270	51	0.1	0.04	0.4	26	38	290	6	32	0.6
Cucumis melo																		
Melon, Casaba	90	26	7.3	1.00	0.10	0.50	0.50	trace	19	0.06	0.02	0.4	15	14	210	10	10	0.5
Cucumis melo																		
Melon, Honeydew	87	41	10.3	0.90	0.10	0.50	0.50	500	32	1.06	1.02	0.6	14	14	330	17	10	0.3
Cucumis melo																		
Melon, cantaloupe (orange flesh)	93.7	20	4.5	0.50	0.18	0.43	0.55	2245	31	0.06	0.04	0.6	20	206	6	12	–	0.7
Cucumis sativus																		
Cucumber (fruit)	96.3	14	2.4	0.75	0.17	0.60	0.50	250	11	0.30	0.40	0.2	19	25	166	6	11	0.9
CUCURBITACEAE																		
Cucumis sativus																		
Cucumber, pickling fruit	96	12	2.4	0.70	0.10	0.60	0.50	270	19	0.04	0.40	0.4	13	24	190	6	14	0.6
Cucumis sativus																		
Cucumber (leaves)	90.4	26	3.4	4.20	0.40	1.50	1.60	–	58	0.17	0.17	1.8	127	96	–	–	–	5.8
Cucumis sativus																		
Cucumber (flowers)	94.8	16	2.7	1.40	0.30	0.60	0.80	–	18	0.02	0.11	0.6	47	86	–	–	–	1
Cucurbita spp.																		
Winter squash (fruit)	88.8	43	10.5	1.40	0.17	1.32	0.80	1980	10	0.06	θ.1	0.6	25	36	300	2	22	0.7
Cucurbita spp.																		
Winter squash (leaves)	88	34	4.9	3.90	0.50	1.80	1.70	1942	45	0.07	0.22	0.9	303	117	436	11	38	1.5
Cucurbita spp.																		
Winter squash (flowers)	92.2	27	4.5	1.60	0.08	0.65	0.48	1947	28	0.04	0.08	1.7	80	55	173	5	24	0.7

	% Water	Calories	CHO	Pro	Fat	Fib	Ash	A	C	B_1	B_2	Niacin	Ca	P	K	Na	Mg	Fe
Cucurbita spp. Winter squash (dried seeds)	4.3	548	19.7	27.10	44.70	2.00	4.20	–	2	0.2	0.13	1.7	45	1095	–	–	–	2.8
Cucurbita ficifolia Malabar gourd	93.6	21	5.1	0.80	0.10	0.40	0.40	4000	11	0.40	0.03	0.3	15	1	–	–	–	0.4
Cucurbita pepo Pumpkin (fruit)	91.4	27	6	1.00	0.10	1.20	0.80	1600	9	0.16	0.25	0.6	22	44	360	1	10	0.8
Cucurbita pepo Summer squash	93.5	20	4.3	1.10	0.15	0.65	0.65	319	19	0.05	0.09	0.8	22	27	188	2	19	0.5
Lagenaria siceraria Bottle gourd (immature fruit)	94.7	15.5	3.6	0.46	0.10	0.50	0.40	10	12	0.04	0.02	0.4	23	18	143	3	11	0.5
Lagenaria siceraria Bottle gourd (dried seed)	3.2	574	15	28.20	49.80	2.00	4.20	–	–	0.4	0.26	4.6	75	1100	–	–	–	5.3
Lagenaria siceraria Kanpyo (dried fruit wall strips)	19.97	258	65	8.58	0.56	9.13	5.86	–	1	–	0.04	2.9	280	188	1582	15	125	5.1
Luffa acutangula Angled loofah (immature fruit)	94.6	17	3.7	0.90	0.15	0.50	–	315	12	0.04	0.04	0.4	18	26	–	–	–	0.5
Luffa aegyptiaca Smooth loofah, sponge gourd (fruit)	93.7	20	4.1	1.00	0.20	1.00	0.40	430	10	0.03	0.05	0.4	24	26	118	3	19	0.7
Momordica charantia Balsam pear, bitter melon (fruit)	93.3	21	4.7	1.00	0.16	1.20	0.80	380	137	0.05	0.06	0.4	29	32	296	5	17	0.7
Momordica charantia Balsam pear (leaves)	80.1	60	12.3	5.10	0.40	0.50	2.10	–	247	0.05	0.47	–	264	666	–	–	–	7.1
Momordica charantia Balsam pear (shoot tips)	89.2	30	3.29	5.30	0.69	2.28	1.47	1735	88	0.18	0.36	1.1	84	99	608	11	85	2
Momordica cochinchinensis Sweet gourd (fruit)	90.2	29	6.4	0.60	0.10	–	–	–	–	–	–	–	27	38	–	–	–	–

	% Water	Calories	CHO	Pro	Fat	Fib	Ash	A	C	B_1	B_2	Niacin	Ca	P	K	Na	Mg	Fe
CUCURBITACEAE																		
Momordica dioica																		
Balsam apple (fruit)	84.1	52	7.7	3.10	1.00	3.00	–	–	–	0.05	0.18	0.6	33	42	–	–	–	4.6
Sechium edule																		
Chayote (fruit)	91.7	26	6.3	0.71	0.15	0.70	0.40	43	20	0.60	0.1	0.5	19	20	107	4	14	0.5
Sechium edule																		
Chayote (leaves)	89.1	60	4.7	3.70	0.40	3.00	1.50	4560	16	0.08	0.18	1.1	58	108	–	–	–	2.5
Telfairia occidentalis																		
Fluted gourd (leaves)	86.4	47	7	2.90	1.80	1.70	1.90	–	–	–	–	–	–	–	–	–	–	–
Trichosanthes cucumerina																		
Snake gourd (immature fruit)	94.3	18	3.6	0.70	0.30	0.80	–	235	12	0.04	0.05	0.5	26	34	–	–	–	0.8
Trichosanthes dioica																		
Pointed gourd (immature fruit)	92	20	2.2	2.00	0.30	3.00	–	–	29	0.05	0.06	0.05	30	40	–	–	–	1.7
EUPHORBIACEAE																		
Manihot esculenta																		
Cassava (root)	60	148	35	1.20	0.30	1.40	1.56	trace	36	0.05	0.03	0.7	35	75	–	–	–	0.7
Manihot esculenta (leaves)	84.2	91	18.3	7.00	1.00	3.20	2.30	12450	316	0.26	0.5	3	297	116	–	–	–	7.8
FABACEAE [LEGUMINOSEAE]																		
Canavalia gladiata																		
Sword bean (dry seeds)	10.7	347	59	24.50	2.60	7.40	3.20	–	1	–	0.15	1.8	158	298	–	–	–	–
Canavalia gladiata																		
Sword bean (fresh pods)	83.6	59	10.7	4.60	0.40	2.60	0.70	40	32	0.22	0.1	2	33	66	–	–	–	1.2
Cyamopsis tetragonolobus																		
Cluster bean, guar (fresh pods)	81	–	10.8	3.20	0.40	–	–	316	47	0.09	0.09	–	–	–	–	–	–	–
Lablab purpureus																		
Hyacinth bean (fresh pods)	87	46	8	2.90	0.45	1.50	0.64	210	11	0.9	0.08	0.6	130	59	163	2	37	1.2

	% Water	Calories	CHO	Pro	Fat	Fib	Ash	A	C	B_1	B_2	Niacin	Ca	P	K	Na	Mg	Fe
Lablab purpureus Hyacinth bean (dry seeds)	10.9	340	62.1	22.80	1.00	8.60	3.20	–	–	0.54	0.14	2.3	90	328	–	–	–	9
Phaseolus acutifolus Tepary bean (dry seeds)	9	353	63.6	20.70	1.30	4.10	3.60	–	–	0.33	0.12	2.8	112	310	–	–	–	–
Phaseolus coccineus Scarlet runner bean (fresh seeds)	34.2	250	46.3	16.40	0.30	12.20	2.80	–	–	0.54	0.14	2.3	61	227	–	–	–	4.1
Phaseolus lunatus Lima beans (fresh seeds)	68.3	105	22.2	7.90	1.20	1.60	1.75	250	29	0.22	0.11	1.4	102	165	460	3	51	2.8
Phaseolus lunatus Lima beans (dry seeds)	12.7	334	61.3	21.00	1.40	4.30	–	–	2	0.42	0.17	2	68	381	–	–	–	7.5
Phaseolus vulgaris Snap beans (fresh green pods)	90.9	33	6.6	2.00	0.19	1.20	0.76	550	19	0.11	0.11	0.7	50	41	220	8	34	0.9
Phaseolus vulgaris Snap beans (dried seeds)	12.4	340	57	21.00	1.50	4.00	4.20	–	24	0.5	0.27	2.8	150	425	1540	4	132	6.5
Pisum sativum Peas (immature seeds)	74.6	87	15.4	6.50	0.55	2.20	0.86	780	33	0.33	0.15	2.6	30	118	285	4	35	1.8
Pisum sativum Peas (dried seeds)	11.3	343	60.1	23.80	1.35	4.75	2.80	120	2	0.75	0.29	3	60	362	798	25	128	3.1
Pisum sativum Edible podded peas	87.6	38	7.9	2.60	0.13	1.80	0.58	405	52	0.13	0.09	0.7	50	50	85	4	24	1.3
Psophocarpus tetragonolobus Winged bean (fresh pods)	90	27	4.9	2.60	0.50	1.90	0.67	332	15	0.21	0.1	0.8	64	37	214	4	34	0.8
Psophocarpus tetragonolobus Winged bean (dry seeds)	11	1697	32	33.00	16.00	5.00	3.00	–	–	–	–	–	–	–	–	–	–	–
Psophocarpus tetragonolobus Winged bean (fresh leaves)	76.9	74	14.1	5.85	1.10	2.50	2.10	–	–	–	–	–	224	63	176	–	8	4
Psophocarpus tetragonolobus Winged bean (root)	66.4	125	24.2	7.20	0.75	2.40	1.50	–	–	–	–	–	30	45	–	–	–	0.2

	% Water	Calories	CHO	Pro	Fat	Fib	Ash	A	C	B_1	B_2	Niacin	Ca	P	K	Na	Mg	Fe
Trigonella foenum-graecum Fenugreek (leaves)	86.9	35	6.1	4.50	0.55	1.20	1.40	–	54	0.50	–	1.1	255	50	51	8	67	17.2
Vicia faba Broad bean (fresh seeds)	75.4	88	15.8	7.30	0.47	3.10	1.00	350	33	0.18	0.17	1.6	30	121	200	50	38	2.1
Vigna unguiculata var. sesquipedalis Yard long bean (immature pods)	88.6	40	6.8	3.00	0.33	1.30	0.63	750	26	0.09	0.11	0.8	56	60	240	4	44	1.1
Vigna unguiculata var. unguiculata Cowpea (immature pods)	85.3	39	7.9	3.00	0.28	1.80	1.25	1250	20	0.1	0.1	0.9	64	60	205	3	54	1.7
Vigna unguiculata var. unguiculata Cowpea (fresh seeds)	66.8	127	22	9.10	0.75	1.80	1.60	365	29	0.34	0.13	1.1	29	145	518	3	51	2.1
Vigna unguiculata var. unguiculata Cowpea (mature seeds)	10.8	343	61.6	22.90	1.45	4.30	3.30	30	2	0.99	0.19	2.2	75	438	987	37	230	6.2
MALVACEAE																		
Abelmoschus esculentus Okra (immature pods)	88.9	36	7.8	2.20	0.24	1.00	0.80	610	30	0.18	0.15	0.9	89	57	234	7	46	0.9
Abelmoschus manihot Sunset hibiscus (leaves)	90	150	4	4.10	0.40	1.00	–	900	118	–	–	–	580	–	–	–	–	3
MORINGACEAE																		
Moringa pterygosperma Horseradish tree (leaves)	75	75	11.7	6.70	1.10	0.90	2.38	10615	177	0.06	0.08	3.2	472	120	337	20	147	6.7
Moringa pterygosperma Horseradish tree (fruit)	87.6	37	6.1	2.30	0.15	3.00	1.50	129	120	0.05	0.07	0.4	30	80	461	30	45	4.5
POLYGONACEAE																		
Rheum rhaborbarum Rhubarb (petioles)	93.3	18	3.8	0.74	0.13	0.75	1.10	100	10	0.03	0.04	0.3	130	21	360	6	21	0.9

	% Water	Calories	CHO	Pro	Fat	Fib	Ash	A	C	B_1	B_2	Niacin	Ca	P	K	Na	Mg	Fe
Rumex acetosa, R. patientia Sorrel and dock (leaves)	92.8	25	4.63	1.93	0.40	0.82	–	8000	62	0.07	0.16	0.5	63	50	364	5	103	2.8
PORTULACACEAE																		
Portulaca oleracea Purslane	93.9	19	3.6	1.50	0.25	0.85	1.25	1320	25	0.04	0.1	0.5	84	41	494	45	68	2.7
SOLANACEAE																		
Capsicum annuum Pepper (mature pungent red fruit)	86	65	15.8	2.30	0.40	2.30	–	11000	260	0.1	0.2	2.9	9	49	420	17	27	1.1
Capsicum annuum Pepper (mature green fruit)	92.8	24	5.06	1.05	0.33	1.30	0.62	500	144	0.08	0.07	0.5	8	22	204	5	16	0.8
Capsicum frutescens Tabasco pepper	82.6	61	12.9	2.60	1.00	1.70	0.90	770	163	0.14	0.13	1.5	33	60	340	7	25	2
Cyphomandra betacea Tree tomato (fruit)	85.5	53	12.9	1.50	0.30	2.60	1.00	150	22	0.04	0.04	1	33	22	–	–	–	0.7
Lycopersicon lycopersicum Tomato	93.5	22	4.75	1.05	0.20	0.55	0.50	900	25	0.06	0.04	0.7	12	26	244	3	14	0.5
Lycopersicon pimpinellifolium Cherry tomato	93.2	22	4.9	1.00	0.20	0.40	0.70	2000	50	0.05	0.04	–	29	62	–	–	–	1.7
Solanum macrocarpon African eggplant (fruit)	89	40	8	1.40	1.00	1.50	0.60	–	–	–	–	–	13	–	–	–	–	–
Solanum macrocarpon African eggplant (leaves)	86	42	6.2	4.60	1.00	1.60	2.40	–	–	–	–	–	391	49	–	–	–	–
Solanum melongena Eggplant, aubergine (fruit)	91.8	26	6.1	1.10	0.25	1.05	0.06	30	6	0.05	0.05	0.7	18	28	214	2	16	0.8
Solanum muricatum (Pepino fruit)	92	–	7	–	–	1.00	0.30	–	47	0.06	0.04	0.4	4	166	–	–	–	0.6
Solanum torvum (Turkey berry)	89	–	7.9	2.00	0.10	–	–	750	80	0.08	–	–	50	30	–	–	–	2.0
Solanum tuberosum (Irish potato)	78.4	82	17.9	2.10	0.10	0.50	0.90	20	18	0.1	0.04	1.4	12	51	420	10	27	0.7